WITHDRAWN BY THE
UNIVERSITY OF MICHIGAN

Computational Methods in Physics and Engineering

Samuel S. M. Wong

University of Toronto

Prentice Hall

Englewood Cliffs, New Jersey 07632

Library of Congress Cataloging-in-Publication Data

Wong, S. S. M. (Samuel Shaw Ming),
    Computational methods in physics and engineering / Samuel S.M.
Wong.
      p.   cm.
    Includes bibliographical references and index.
    ISBN 0-13-155953-2
    1. Physics--Data processing. 2. Mathematical physics.
3. Engineering--Data processing. 4. Engineering mathematics.
I. Title.
QC52.W66 1992
530'.0285--dc20   92-7345
                                                                 CIP

Editorial/production supervision: MARY P. ROTTINO
Cover design: BRUCE KENSELAAR
Pre-press buyer: MARY E. MCCARTNEY
Manufacturing buyer: SUSAN BRUNKE
Acquisitions editor: BETTY SUN

 © 1992 by Prentice-Hall, Inc.
A Simon & Schuster Company
Englewood Cliffs, New Jersey 07632

LIMITS OF LIABILITY AND DISCLAIMER OF WARRANTY:
The author and publisher of this book have used their efforts in preparing this book and software. These efforts include the development, research, and testing of the theories and programs to determine their effectiveness. The author and publisher make no warranty of any kind, expressed or implied, with regard to these programs or the documentation contained in this book. The author and publisher shall not be liable in any event for incidental or consequential damages in connection with, or arising out of, the furnishing, performance, or use of these programs.

The publisher offers discounts on this book when ordered in bulk quantities. For more information, write: Special Sales/Professional Marketing, Prentice-Hall, Inc., Professional & Technical Reference Division, Englewood Cliffs, New Jersey 07632

All rights reserved. No part of this book may be
reproduced, in any form or by any means,
without permission in writing from the publisher.

Printed in the United States of America
10 9 8 7 6 5 4 3 2 1

ISBN 0-13-155953-2

Prentice-Hall International (UK) Limited, *London*
Prentice-Hall of Australia Pty. Limited, *Sydney*
Prentice-Hall Canada Inc., *Toronto*
Prentice-Hall Hispanoamericana, S.A., *Mexico*
Prentice-Hall of India Private Limited, *New Delhi*
Prentice-Hall of Japan, Inc., *Tokyo*
Simon & Schuster Asia Pte. Ltd., *Singapore*
Editora Prentice-Hall do Brasil, Ltda., *Rio de Janeiro*

Contents

Chapter 1 Introduction
1-1	Programming computers	1
1-2	Integers and floating numbers	8
1-3	Examples of integer calculations	10
1-4	Algorithm for computation	19
1-5	A calculation for the value of π	24
	Problems	26

Chapter 2 Integration and Differentiation
2-1	Numerical integration	29
2-2	Rectangular and trapezoidal rules	32
2-3	Simpson's rule	37
2-4	Gaussian quadrature	42
2-5	Monte Carlo integration	49
2-6	Multidimensional integrals and improper integrals	53
2-7	Numerical differentiation	62
	Problems	66

Chapter 3 Interpolation and Extrapolation
3-1	Polynomial interpolation	69
3-2	Interpolation using rational functions	77
3-3	Continued fraction	88
3-4	Fourier transform	96
3-5	Extrapolation	115
3-6	Inverse interpolation	122
3-7	Cubic spline	135
	Problems	142

iii

Chapter 4 Special Functions
4-1 Hermite polynomials and quantum mechanical harmonic oscillator   145
4-2 Legendre polynomials and spherical harmonics   155
4-3 Spherical Bessel functions   171
4-4 Laguerre polynomials   179
4-5 Error integrals and gamma functions   195
4-6 Clebsch-Gordan coefficients   204
    Problems   209

Chapter 5 Matrices
5-1 System of linear equations   211
5-2 Matrix inversion and LU-decomposition   224
5-3 Matrix method to solve the eigenvalue problem   236
5-4 Tridiagonalization method of Givens and Householder   251
5-5 Eigenvalues and eigenvectors of a tridiagonal matrix   265
5-6 Lanczos method of constructing matrices   283
5-7 Nonsymmetric matrices and random phase approximation   296
5-8 Complex matrix, band matrix, and the generalized eigenvalue problem   310
    Problems   316

Chapter 6 Methods of Least Squares
6-1 Statistical description of data   319
6-2 Uncertainties and their propagation   330
6-3 The method of maximum likelihood   337
6-4 The method of least squares   342
6-5 Statistical tests of the results   348
6-6 Linear least-squares fit   358
6-7 Nonlinear least-squares fit to data   368
    Problems   380

Chapter 7 Monte Carlo Calculations
7-1 Generation of random numbers   383
7-2 Molecular diffusion and Brownian motion   407
7-3 Data simulation and hypothesis testing   412
7-4 Percolation and critical phenomena   421
7-5 The Ising model   434
7-6 Random matrix ensembles   444
7-7 Path integrals in quantum mechanics   455
7-8 Fractals   471
    Problems   483

Chapter 8 Ordinary and Partial Differential Equations
8-1	Types of differential equations	487
8-2	Runge-Kutta methods	499
8-3	Predictor-corrector methods	510
8-4	Solution of initial value problems by extrapolation	517
8-5	Solution of boundary value problems by shooting methods	524
8-6	Relaxation methods	536
8-7	Boundary value problems in partial differential equations	549
8-8	Parabolic partial differential equations	554
8-9	Hyperbolic partial differential equations	565
8-10	Nonlinear differential equations	576
8-11	Stiff problemss	584
8-12	Finite element methods	588
	Problems	604

Chapter 9 Graphical Representation of Results
9-1	Basic ingredients of computer graphics	607
9-2	Functions with one independent variable	617
9-3	Functions with two independent variables	622
	Problems	632

Chapter 10 Computer Algebra
	633
Problems	644

Appendix A Synopsis of the FORTRAN Language	645

Appendix B List of programs	653

References	655

Index	661

Preface

Computational methods form an increasingly important part of the undergraduate curriculum in physics and engineering these days. This book is mainly concerned with the ways that computers may be used to advance a student's understanding of physics. A large part of the material is common to engineering as well.

The subject matter covered in this volume may be classified also under the title of "computational physics." There are several ways to organize the material that should be included. The choice made here is to follow the traditional approach of mathematical physics. That is, the chapters and sections are grouped around methods, with physical problems used as the motivation and examples. One attractive alternative is to group around physical phenomena. The difficulty of following this way of organization is the heavy reliance on the physics background of the readers, thus making it harder to follow for students at early stages of their education. For this reason, such an approach is rejected.

The intimate relation between physics and mathematics may be seen by the way that physics is usually taught in the undergraduate curriculum. With a knowledge of calculus, for example, the subject of mechanics is discussed in a more rigorous manner. By the time the student is introduced to differential equations, topics such as harmonic oscillators and alternating current circuits, are brought in. Long experience in the community has shown this way of teaching to be very successful. The major problem here is the delay in introducing certain other basic concepts, because of the need to acquire first a certain maturity in mathematics. The fault is not with the mathematics required but the way it is used. For example, discussions on a pendulum are usually limited to small amplitudes at the early years. For finite amplitudes, the differential equation is nonlinear and the necessary skill to solve such equations by analytical methods comes only in later years. On the other hand, it is possible to use the same numerical methods to solve both types of differential equations for the pendulum problem and they are no more difficult than analytical methods. In this way, the discussion on pendulum does not have to be limited to small amplitude oscillations.

Until quite recently, the mathematics required in undergraduate physics and engineering, to a large extent, consists of analytical techniques to manipulate algebraic equations, to carry out integrals, and to solve simple differential equations. In addition to these "algebraic" methods, there is also a large class of numerical approaches that can be used to solve physical problems. While it is true that computers have made numerical calculations popular, many of the methods have their origins with the same group of mathematicians as the algebraic methods, such as Gauss and Newton. In the intervening years since the introduction of these mathematical techniques, numerical calculations have lost out to "algebraic" ones, perhaps because of the tedium of carrying out numerical calculations by hand. This reason is certainly no longer true, as attested by the explosion of numerical solutions in research papers. In spite of its popularity in research, the introduction of computational physics to the undergraduate syllabus is only starting.

To carry out numerical calculations on a computer, the usual practice is to write one or more programs in one of the high-level languages, such as FORTRAN or C. To simplify the process, one may make use of standard subroutine libraries to take over specific tasks, such as inverting a determinant or diagonalizing a matrix. In contrast, the general approach to algebraic computation is to make use of one of the symbolic manipulation packages. Most of these packages are very powerful and are able to carry out a large variety of complicated calculations. Although a substantial amount of "programming" can be done in most cases, the symbolic manipulation instructions are not programming languages on the same level as, for example, FORTRAN or C. Furthermore, there has been little attempt to standardize the instructions between different packages. As a result, any discussions of algebraic calculations in a volume on computational methods must either be very abstract or very specific in terms of one of these packages. The choice made here is the former, as it is not clear what is available to the average reader.

The results obtained from computer calculations often appear in the form of a large table of numbers. For human beings to comprehend such huge quantities of information, graphical presentations are essential. For this reason, graphical techniques are essential parts of computational methods. On the other hand, it is not possible to cover all three aspects, numerical, symbolic, and graphical, in a single volume. The choice made here is to select a few of the standard topics in physics covered in the undergraduate curriculum and present them in ways that computers may be useful for their solutions. Even this is too big a task. The compromise is to put the emphasis on numerical techniques. For reasons mentioned in the previous paragraph, only an introduction is made to symbolic manipulation techniques. Computer graphics is perhaps one of the fastest growing areas in computing. For this reason also, it is best to leave everything beyond an introduction to volumes specializing on the subject.

There is quite a bit of interest in the physics community in developing courses in computational physics, as has already been done in many places. Such courses should be regarded as in parallel with the more traditional ones in mathematical physics and experimental physics. It is one of the aims of this volume to serve as a textbook or major reference for such a course. At the same time, many graduate students and senior undergraduates may not have benefited from such a course. It is also the intention here to serve this group of physicists. Engineers and other professional people who make use of computational techniques in physics may also find it useful to examine some of the background involved in solving some of the physical problems on computers.

Although a set of computer program is present here, it is not the primary intention of the author to provide a library of subroutines for common problems in physics. The computer programs used as examples and included in the accompanying diskette are intended as illustrations for some of the materials discussed. They can be modified for other applications. However, before one does that, as with any computer program, the programs should be thoroughly tested first for the intended purpose.

S. S. M. Wong

Chapter 1
Introduction

Computers are used in a variety of ways in physics. They help us to take the large quantity of data in the laboratories, they carry out many of the calculations that are impossible otherwise, and they assist us in presenting the results by plotting the graphs and typesetting the reports. Computational physics covers only a part of this wide range of activities. Traditionally, computers have been known primarily as the tools for making calculations. The high speeds achieved by some of the modern machines are such that we now can solve problems that were unthinkable even a few years ago. This capability gives us not only the opportunity to examine some of the complicated theories and enormous quantity of data but also allows us to explore a large number of different ideas in a relatively painless way. However, this is not the complete story.

In many instances, the computer serves also as the place to explore new physics. The most obvious example in this category is the study of chaos. Here, experimentation using a computer supplies us with "data," and we make use of them in ways that are not too different from those for which we normally depend on experimental results. For this reason, computational physics is now recognized as a new branch of physics that is different from both theoretical and experimental physics. This is a good idea since there is also a set of specialized knowledge associated with computational physics that cuts across many of the traditional divisions, such as atmospheric physics, biophysics, and condensed matter physics. In this chapter, we shall only be concerned with a few of the general problems in using computers to do physics. Later, we shall deal with topics that are more specific to different aspects of computational physics.

§1-1 Programming computers

For our purpose, a computer may be regarded as a digital electronic device capable of carrying out mathematic operations such as addition, subtraction, multiplication, and division; logical operations such as if, and, or; and input-output operations such as reading from the keyboard, storing a number in a particular memory location, and sending the result to a printer. With these basic operations, a machine can take over from the user a large fraction of the computational labor, both numerical and algebraic.

In principle, all the instructions to a computer may be entered through the key board or some other input devices. However, at the machine level almost all the computers are constructed to understand only machine language instructions. These are usually expressed in a binary representation, that is, a string of zeros and ones. This is done for the convenience of designing and building the machine, but is hardly the preferred way for human beings to communicate with the computer. For this reason, each instruction in machine language is usually given a symbolic name. For example, we may use the symbol STR to represent the machine code for storing a number and the symbol ADD to add two numbers. This is the start of an *assembly language,* and it is very easy to image *programming,* meaning giving a set of instructions to a machine to translate an assembly language program into the corresponding machine codes.

To make a machine that is fast in its operations and easy to construct, the number of basic instructions must be limited. In fact, the trend in the development of fast scientific computers is to reduce the number of basic instructions to a minimum, and this results in the new generation of computers called reduced instruction set computers (RISC). On the other hand, as machines become faster, we wish to perform more complicated operations with them. This calls for simplifications in programming a computer. In other words, we want our computers to be able to carry out very complicated tasks with a minimum amount of coding. To meet these two diametrically opposite goals of reducing the number of basic machine instructions and increasing the ability to carry out complicated computations, a large software industry has developed. At the heart of this industry is the programming language that translates instructions, which are easy for the researcher to formulate for the problem of interest, to machine codes, with which a computer can operate.

High-level computer languages. Even with one of the assembly languages, it is still a nightmare to write a computer program for anything other than simple calculations. As a result, high-level computer languages are constructed that are closer to the working languages of various applications (see, for example, "Programming Languages," by J.A. Feldman in the December 1979 issue of *Scientific American*). For scientific works, both FORTRAN and Lisp belong to this category and they have been in use for more than 30 years. More recently, the C language is also widely used for many of the programs that traditionally would have been written in FORTRAN or Lisp.

The conversion of a computer program written in one of the high-level languages to a set of machine instructions is the function of a *compiler*. The output of a compiler is called the *object code* of the program so as to distinguish it from the *source code,* which is written in a high-level language. The compiler itself is also a computer program. Often, the power of a scientific computer depends to a significant extent on the strength of the compilers that accompany it. We shall return later with a few comments on the choice of computer language for scientific applications.

Even with high-level languages, it may take many instructions to express a problem of interest to us in a form understandable to the compiler. For example, it is not uncommon to find that certain calculations take several thousands of lines of coding. This is not a trivial task. Before we embark on such a project, a few basic considerations must be examined. The first point is the question of how to find errors, commonly referred to as "bugs," in the program. Many beginners have the false sense of security that one can write an error-free program from the start. This is seldom true. Since it is almost impossible to avoid errors in writing a large program, one of the necessary requirement of a good program is that it is easy to *debug*, that is, to find the errors and to correct them.

A second point is that the program should be readable. A set of thousands of instructions to a computer is as hard for a human reader to understand as the instructions to assemble a complicated toy in an unfamiliar foreign language. It is not just difficult for someone else to read the program; it may also become a problem for the original programmer, especially after some time has elapsed since the program was written.

Finally, a program should be efficient. The question of efficiency, like any judgment of economy, depends to a large extent on the environment. If computer use is charged in terms of real money, it is relatively easy to calculate whether it is worthwhile to spend a day in improving the program. However, for most computer uses these days, charging is no longer the central consideration. Nevertheless, it is usually desirable to have the results we are interested in as soon as possible. It is therefore always worthwhile, for example, to reduce a calculation from taking an hour on a given computer to a minute. On the other hand, it may not pay to speed up a 5-second calculation unless one wants to run the same program a large number of times. In general, it is hardly necessary to emphasize efficient codes, since it often involves the pride of a physicist.

Structured programs. A good practice in programming is to group the instructions into convenient segments. A long program should be divided into a number of sections or subprograms, called subroutines or functions in FORTRAN and procedures in several other languages. Each subprogram carries out a specific task in the calculation. For example, in Monte Carlo calculations, random numbers are needed. Since the generation of random numbers (see §7-1) is a task that can be isolated, it is usually written in the form of a subroutine or a function. There are three good reasons for doing this. The first is that it makes the program more readable, as it can be made quite clear through comments in the program that this is an independent unit of the program concerned with only random number generation. The second is that we can debug the random number generator part of the program independently of the rest of the calculation. In this way, we can make sure that it is functioning properly before incorporating it as a part of the complete Monte Carlo calculation. The third reason is even more important. A common task, such as generating random numbers, is used in many other calculations as well. As a result, it may not be necessary to write it at all. Such a subprogram may exist already as a part of some other calculations or as a part of standard libraries that are commonly available. In general, if a subprogram can be found in a standard library, it is advisable to make use of it rather than write one's own.

Within a subprogram, the instructions should be divided into groups, similar to paragraphs in an essay. Nowadays, it is a common practice to indent the lines of a program in such a way that each group of instructions is clearly identified even in a causal glance at the program listing. Furthermore, the logical flow in the program should be as straightforward as possible. For example, a program with a large number of branching instructions, such as GO TO statements in FORTRAN, is difficult to follow and should be avoided whenever possible. It is often claimed that most programs can be written without such branching statements, and this is often quite correct if a little time is spent in planning the calculations.

Another important part of a program is the comments. These are statements that the compiler normally ignores. They serve as explanations of what the programmer has in mind at particular points of the program. Until a truly conversational programming language is developed, it is usually difficult to translate the programming instructions back to the formulas and algorithms we are trying to get the computer to carry out for us. For this reason, the addition of helpful comments is essential for a program to be readable.

Structured programming is an important part of writing computer codes, and all high-level languages, including FORTRAN, are evolving in directions that encourage this good practice. It is again a mistake of the novice to produce a long code without considering whether a human reader will understand the program.

Libraries of subprograms and linkage editor. In most computer calculations, certain common tasks form more or less independent units, such as taking the square root of a number, expressing the output in some specific format, or diagonalizing a matrix. It is useful to collect such subprograms together and put them into a *program library*. A program library differs from ordinary libraries in that, instead of books, we have programs stored in forms accessible to the computer. Program libraries may be organized into four different categories. At the level that is closest to the machine, there is a set of functions that interacts with the operating system of the computer, such as reading the clock inside the computer or making requests for more memory. These are the *system subroutines*. Usually, these subprograms are not visible to the average user since they

Table 1-1 Intrinsic FORTRAN functions.

Name	Function	Name	Function
Type conversion		**Trigonometric functions**	
INT	To integer	COS	Cosine
REAL	To real	SIN	Sine
DBLE	To double precision	TAN	Tangent
CMPLX	To complex	ACOS	Arc cosine
AINT	Truncation to nearest integer	ASIN	Arc sine
ANINT	Nearest whole number	ATAN	Arc tangent (1 argument)
NINT	Nearest integer	ATAN2	Arc tangent (2 arguments)
Numeric functions		**Hyperbolic functions**	
ABS	Absolute value	COSH	Hyperbolic cosine
MOD	Modulo	SINH	Hyperbolic sine
SIGN	Transfer of sign	TANH	Hyperbolic tangent
DIM	Positive difference		
MAX	Maximum	**Character functions**	
MIN	Minimum	LGE	Lexical $\geq$
DPROD	Double precision product	LGT	Lexical $>$
AIMAG	Imaginary part	LLE	Lexical $\leq$
CONJG	Conjugate	LLT	Lexical $<$
Transcendental functions		CHAR	Integer to character
SQRT	Square root	ICHAR	Character to integer
EXP	Exponential	INDEX	Location of string
LOG	Natural logarithm	LEN	Length of string
LOG10	Logarithm base 10		

are usually invoked by the compiler and other related services of the high-level languages.

As a part of a compiler for any one of the high-level languages, there is usually a library of *intrinsic functions*. For languages designed with numerical applications in mind, such a library will include, for instance, trigonometric functions and other transcendental functions. As an example, the basic set of intrinsic functions for FORTRAN compilers is given in Table 1-1. In addition to the intrinsic functions, there are also subroutines that are specific to a particular type of computation, for example, eigenvalue problems. Several such *subroutine libraries* are more or less standardized for scientific applications, and they are valuable in computational physics. Finally, as the fourth category of library, each user may have his or her own set of routines that form the *private library* of the individual user.

There are two ways to incorporate into a program this vast resource of subprograms residing in various libraries. The most elementary way is to include them explicitly as a part of the program. That is, a copy of the source code of the required subprograms is put into the programs we are writing. To do this, it is necessary that the subprograms be written in the same high-level language as the rest of the program we wish to develop. The use of library subroutines in the form of source codes is often inconvenient and sometimes impossible. The alternative is to incorporate them in the form of object codes. In this case, the library consists of the outputs of the compiler for the subprograms. The advantage to the user here is that this is a faster way to bring in the required subprogram, and the source code is kept to a minimum size. Furthermore, some of the subroutines may be written in a different high-level language so as to take advantage of the convenience offered by that language. In general, it is relatively easy to mix object codes compiled from source codes that are written in a different high-level language. However, since the compilers are different, it is not possible to mix source codes written in different languages.

Once the compiler translates our source code into an object code, it is the role of a linkage editor, or linker for short, to group the various subprograms into a coherent single unit. This includes the intrinsic library, the subroutine libraries, and the private library. The output of the linker is the executable form of the program, a set of machine language instructions that the computer can follow.

In summary, there are three distinct steps in constructing a computer program in general. The first step is to use a text editor to input those parts of the program that must be written. At this stage we may also wish to incorporate those parts of the program that exist in source code libraries. The second step is to compile the source code into an object code. This is followed by the third step of linking the object code, together with the intrinsic and other object-code libraries, to form an executable unit. It is this executable unit that the computer follows to carry out the calculations we wish it to do for us.

In some cases it is also possible to use an *interpreter* instead of the compiler-linker combination to transform a source code into an executable form. This is a much simpler way. However, interpreters are usually limited in the programs they can handle, and we shall not be concerned with this alternative here.

Choosing a computer language. We have a choice of several high-level computer languages for scientific applications. Among these, the C language is perhaps the most powerful. It is modern in its design and has, for example, the capability of handling characters as well as numerical variables. The FORTRAN language, on the other hand, is designed primarily for numerical computations. The name comes from **for**mula **tran**slation, and the language has been developed over the years into a form that is easy to learn and powerful enough for most needs in physics. The Lisp language is aimed at symbolic manipulation and is therefore the language of choice for algebraic calculations. Other languages, such as ALGOL, APL, BASIC, and Pascal, are also used in physics but they are not as popular as the three mentioned previously.

Most of the material discussed in this book is independent of the computer language used. However, all the programming examples, including one in algebraic computing, are given in FORTRAN. The reasons for making this choice are quite strong. In the first place, it is an easy language for people in the physical sciences to learn. With a convenient computer set up, for example a personal computer with a good editor and a reasonable compiler, it is not unusual for a student in physics to learn enough FORTRAN in an afternoon to be able to write programs for simple calculations. A second reason is that there are far more physicists who are familiar with the FORTRAN language than any of the others. The third reason is perhaps even more important. The library of FORTRAN programs for numerical calculations in physics is extensive, and it would be a shame not to make good use of this vast resource.

Programs one must write. In computational physics, our interest is primarily in the physics that computing can help us to learn. For this reason, the writing of a computer program itself is not the main interest. On many occasions, packages are available to carry out the specific calculations we wish to do. Each of these packages consists of one or more computer programs that are put together as a unit to perform a set of specific computations. For example, many of our plotting and graphics functions may be met by such packages. For most routine computational needs, it is often possible to make use of one of the available packages instead of developing one's own programs. On the other hand, the variety of computations we wish to carry out is usually too broad for standard packages to cover completely. It is almost inevitable that we must write some computer programs for our interests in physics. When such needs arise, it is worthwhile, as far as possible, to make use of standard subroutine libraries. In addition to savings in program development time, the professionally written subroutine libraries are usually well thought out and very efficient. In fact, the range of subroutines available is so broad these days that it is difficult to imagine a large calculation that cannot make some use of these libraries. We shall come back to this in §1-4 for a few brief comments on some of the more widely used subroutine libraries.

If most standard calculations are available to us in the form of library subroutines, what is the purpose of a book on computational methods? Certainly, the aim should not be to try to produce a library of subroutines for calculations in physics and engineering. A number of examples are given later. They are used mainly for purposes of illustration and to provide some practical exercises on the various topics. The primary purpose here is to supply the background that goes into the more common types of calculations in physics and to show the powers and limitations of various methods. Beyond computer programs, we hope that the reader will develop a feeling for making intelligent decisions on what to compute, the method to use for a calculation, and, of equal importance, when not to use a computer.

§1-2 Integers and floating numbers

Nearly all computers make a clear distinction between integers and floating numbers. An integer is a number, such as 5, −213, and 0, that does not have a part that must be represented by a decimal point. A floating or *real* number is any other type of number, such 3.1415926..., 3.0×10^{23}, and −9.9, that requires a decimal point for its specification. The reason for differentiating between these two types of numbers comes from the structure of the computer memory.

The basic unit for storing a number in a computer is a *bit*, the state of an electronic component that is either *on* or *off*. The two possible states of a bit may be used to represent two numerical values 0 (off) and 1 (on). Since a single bit is too small for most interests, 8 bits are grouped together into a unit called a *byte*. The status of a byte may be represented by an eight-digit binary number $b\sqcup\sqcup\sqcup\sqcup\sqcup\sqcup\sqcup\sqcup$, where we have added the prefix b in front of the number to indicate that it is in the binary representation. For example, the integer 5 is shown as b00000101 or simply as b101. The largest integer that can be represented in one byte of storage is then b11111111 $= 2^8 - 1 = 255$.

Before we leave the subject of internal representation of integers in the computer memory, we shall define two other representations. The binary representation used above is inconvenient to use in many cases because of the large number of digits required to express most of the numbers of interest to us. The hexadecimal representation is based on powers of 16 (in a similar way as the decimal system is based on powers of 10). Each hexadecimal digit can take on values 0 to 15, and they are usually written as z0, z1, z2, z3, z4, z5, z6, z7, z8, z9, zA, zB, zC, zD, zE, and zF. To indicate that the number is in the hexadecimal basis, the usual procedure is to add a prefix z, as we have done here. In terms of computer memory, each hexadecimal digit represents one of the possible values stored in four binary digits ($2^4 = 16$). The value that can be stored in a byte is then represented by two hexadecimal digits. Examples of numbers in the hexadecimal representation and their corresponding values in decimal and binary representations are given in Table 1-2.

Another way of displaying binary-based numbers is the octal representation. To distinguish it from other representations, we shall prefix a number in the octal representation with the letter o. (The lowercase o is used here instead of the uppercase so as to make it easy to differentiate it from the number zero.) Each octal digit represents the value given by 3 bits. This is convenient for some models of computers whose memories are made of multiples of 3 bits, such as 36 and 60. In addition, octal representation is also a convenient way for carrying out many types of manipulations.

In most calculations in physics, a byte, which can only store integers up to 255, is still too small. For this reason, each integer is often assigned either two or four bytes of memory. Such a grouping of bytes is sometimes called a

Table 1-2
Decimal, hexadecimal, octal, and binary representations of numbers.

Decimal	Hexa-decimal	Octal	Binary	Decimal	Hexa-decimal	Octal	Binary
0	$z0$	$o0$	$b0$	10	zA	$o12$	$b1010$
1	$z1$	$o1$	$b1$	11	zB	$o13$	$b1011$
2	$z2$	$o2$	$b10$	12	zC	$o14$	$b1100$
3	$z3$	$o3$	$b11$	13	zD	$o15$	$b1101$
4	$z4$	$o4$	$b100$	14	zE	$o16$	$b1110$
5	$z5$	$o5$	$b101$	15	zF	$o17$	$b1111$
6	$z6$	$o6$	$b110$	16	$z10$	$o20$	$b1\,0000$
7	$z7$	$o7$	$b111$	17	$z11$	$o21$	$b1\,0001$
8	$z8$	$o10$	$b1000$	18	$z12$	$o22$	$b1\,0010$
9	$z9$	$o11$	$b1001$	19	$z13$	$o23$	$b1\,0011$
255	zFF	$o377$	$b1111\,1111$				
256	$z100$	$o400$	$b1\,0000\,0000$				
257	$z101$	$o401$	$b1\,0000\,0001$				

computer word, or just *word* for short. Before we go into the question of the range of integer values a computer word can store, we must recall that most numbers we are interested in have a ± sign associated with them. It is common practice to designate the first bit of a two-byte or four-byte integer word as the sign bit. Thus, for a two-byte integer I_2, only 15 bits are available to store the magnitude of the number. The possible value that can be represented by such a word is then

$$-32,768 \leq I_2 \leq +32,767$$

That is, -2^{15} to $(2^{15} - 1)$. The reason that the maximum positive integer value is one less than 2^{15} comes from the fact that $+0$ must also be considered as one of the integers. (Why, then, is the maximum absolute value of a negative integer one larger?) For a four-byte word, the possible value of an integer is then

$$-2,147,483,648 \leq I_4 \leq +2,147,483,647$$

corresponding to -2^{31} to $(2^{31} - 1)$. Although the allowed range of integer values for I_4 may be large, it is still not adequate for many purposes. For example, the upper limit of I_4 is less than $13! = 6,227,020,800$. As a result, it may not be possible to carry out certain types of calculations that involve factorials in terms of integers. We shall see examples later of some of the ways to circumvent this difficulty.

For most calculations in physics, the use of integers alone is too restrictive. Floating numbers broaden the range of values that can be stored in the computer memory by allocating a part of a word as the exponent of each number. That is, each number now has a sign, a fraction part or mantissa, and an exponent. Since the computer memory is made of bits, some arrangements must be made to represent a number in this way. The usual convention is to use four bytes for a single-precision number. Among the 32 bits in such a word, 1 bit is devoted to the sign, 8 bits are assigned to the exponent, and 23 bits are left for the mantissa. In this way, numbers with absolute values in the range from approximately 1.2×10^{-38} to 3.4×10^{38} can be represented. In double precision, eight bytes are used for each word. Among the 64 bits in such a word, 1 bit is for the sign, 11 bits for the exponent, and 52 bits for the mantissa. Floating numbers with absolute values approximately in the range from 2.2×10^{-308} to 1.8×10^{308} may be represented through such an arrangement.

If we get a number whose absolute value is too small to be represented in a computer, an *underflow* condition is created. This happens, for example, when we get a single precision floating number with absolute value less than 10^{-38}. The usual course of action on such occasions is to replace the number by zero. If a number is produced with an absolute value much larger than $\cdot 10^{38}$, an *overflow* condition is created. In this case, the usual response of the computer is to suspend the calculation, as an error will be created if we replace the number by anything else. Generally, one should check for possible underflows and, in particular, overflows in a program where such conditions are likely to occur and take the appropriate response.

The maximum number of significant figures that can be achieved in a floating number calculation is ultimately limited by the number of bits assigned to the mantissa. In single precision, the 23 bits gives a maximum of roughly seven significant figures and in double precision one can get up to 15 significant figures. Since the maximum number of significant figures is limited, *truncation* errors become a part of any floating number calculation. One important consideration in numerical calculations is to choose an algorithm that minimizes the truncation errors. If this is not done, the numerical errors may accumulate and become so large that no significant figures are left in the final results. At the same time, there is no use in trying to design an algorithm with a precision beyond the limitations set by truncation errors. We shall see different aspects of these considerations in almost all the examples in the remaining chapters.

§1-3 Examples of integer calculations

As a first example of integer calculations, let us consider a simple way to find a list of prime numbers. A prime number may be defined as an integer that is divisible only by unity and the number itself. For our purpose, we shall describe it in the following way. Let $\{p_i\}$ be a list of prime numbers arranged in ascending order according to size. For the convenience of discussion, we shall assume that the list starts with 2 and the largest prime number in the list is p_{max}. It is possible to find out whether an integer $n \leq p_{max}$ is a prime number or not by dividing n with each member in the list. What do we mean by the statement that an integer n is divisible by another integer m? Here, we can take it to be that the ratio n/m is also an integer.

Example of calculating prime numbers. Before we design an algorithm to search for prime numbers, let us further examine the question of how integer calculations are carried out on a computer. In ordinary calculations, when an integer n is divided by another integer m, the result is an integer if m is divisible by n; otherwise the result is a real number. For example, $6/2 = 3$ results in an integer, whereas $5/2 = 2.5$ results in a real number. Since integers and real numbers are not represented in the same way inside a computer, the rules to handle these two types of calculation are slightly different. In the integer mode, the result remains an integer. Thus $5/2 = 2$, with the fractional part truncated. By the same token, $7/8 = 0$. If one wants the result 2.5 from dividing 5 by 2, it is necessary to carry out the calculation in terms of floating numbers. This may be done by first changing the numerator 5 to the floating number 5.0 and the denominator 2 to 2.0. The need for these additional steps can be a nuisance at times. On the other hand, we can also turn it into an advantage, as we shall see below.

Because of the way computers carry out integer calculations, we can decide whether an integer is divisible by another integer in the following way. Let us assume that, in the integer mode, the result of n divided by m is the integer k. That is,

$$\frac{n}{m} \longrightarrow k$$

If n is divisible by m, then we have the situation that

$$k * m = n$$

If n is not divisible by m, a part of the quotient is lost due to truncation and

$$k * m < n$$

We shall make use of this property to construct a simple algorithm to generate a list of prime numbers.

Let us assume that we have already found the first n_p prime numbers, p_1, p_2, p_3, ..., p_{max}, arranged in ascending order according to size. To find the next prime number, we start with an integer $N = p_{max} + 1$, where p_{max} is the largest prime number known so far. We can test whether N is prime by dividing it by all the members in the list of prime numbers already found, p_1, p_2, ... If N is divisible by any of the existing prime numbers, it is not a prime number. We shall therefore increase the value of N by 1 and repeat the test. This process goes on until we reach a value of N that is not divisible by any of the existing prime

> **Box 1-1 Program** PRIME
> **Generator for the first K prime numbers**
>
> Initialization:
> (a) Set K as the total number of prime numbers required.
> (b) Set N_BEG$\geq$ 2 as the number of prime numbers to start with.
> (c) Input the first N_BEG prime numbers explicitly.
> (d) Set the number of existing prime numbers to be N_BEG.
> 1. Store the last prime number in the list as NEW.
> 2. Construct a new integer by adding 2 to NEW.
> 3. Test if the integer NEW is prime by dividing it with all the existing prime numbers $\leq$ NEW/2.
>   (a) If divisible, try a new integer by adding 2 to NEW and repeat the test.
>   (b) If not divisible, it is a new prime number.
>   (i) Add the new one to the list.
>   (ii) Augment the number of members in the list by one.
> 4. Repeat steps 1 to 4 to find the next prime number in the list.
> 5. Stop if the total number is K.

numbers. Such a number is therefore a new prime number and may be added to our list as the new member with the largest value. Except for a few refinements to be described next, this is the algorithm outlined in Box 1-1.

A few improvements may be added to the method to make it more efficient. First, we recognize that 2 is a prime number. As a result, all the other even numbers are not prime numbers. For this reason, we can increase N by 2 each time, rather than 1 as given in the previous paragraph. In this way, the speed of the search is increased by a factor of 2. By the same argument, there is no need to include 2 in the list of prime numbers for our tests, as we have already excluded all the even numbers from our considerations. A second improvement is that we do not need to go through the entire list of existing prime numbers in our test. Obviously, if N is not divisible by any member up to $p_i > N/2$, it will not be divisible by any of the prime numbers in the list that are larger than p_i.

To start off the calculation, we need to have at least one prime number in the list. Since 1 is not suitable, our list starts with 2. However, to make use of the improvement to the algorithm given in the previous paragraph, we must have an odd prime number as the last member of the list. For this reason, we must add the number 3 to our starting list. With this list of two prime numbers, all the rest of the prime numbers can be generated.

Our method is not optimal for producing large prime numbers. As the list grows, it takes longer and longer to find the next prime number. In addition, the limitation imposed by the largest integer we can store in a computer word also poses a problem for the method. Our calculation must therefore terminate at some large values of $p_{\max}$, much smaller than sizes of interest to mathematicians. Methods of testing for prime numbers other than the simple one presented here must be used if one is interested in large primes (see, for example, "The Search for Prime Numbers" by C. Pomerance on page 136 of the December 1982 issue of *Scientific American*).

Decomposition into prime numbers. For many calculations, such as the coefficients for associated Legendre polynomials given later in §4-2 and the values of Clebsch-Gordan coefficients in §4-6, the final results often appear in the form of ratios of integers. As an illustration, consider first the binomial coefficient

$$\binom{n}{m} = \frac{n!}{m!(n-m)!} \qquad (1\text{-}1)$$

where the factorial of an integer is defined as the product

$$n! = \prod_{k=1}^{n} k = n(n-1)(n-2)\cdots(2)(1) \qquad (1\text{-}2)$$

A simple way to calculate a binomial coefficient is to evaluate the numerator and the denominator separately and then take the quotient. This method works only for small values of n and m, as the range of integers we can store in the computer memory is limited. For example, we have seen earlier that the largest integer that can be stored in a four-byte word is less than 13!. On the other hand, the largest binomial coefficient with $n = 13$ is only

$$\binom{13}{6} = 1716$$

Obviously, we can make use here of a method that is capable of canceling common divisors between the numerator and denominator at early stages of the calculation.

One simple method of keeping track of possible common factors among a group of integers is to decompose each into a product of prime numbers, as outlined in Box 1-2. Let p_i for $i = 1$ to d be the first d prime numbers. The value of d depends on the magnitude of the largest integer we wish to decompose. In general, if the next prime number beyond our list of d members is p_{next}, then the largest integer we can decompose using the list of prime numbers is restricted to $n_{\max} < p_{\text{next}}$. Any integer n smaller than p_{next} may be expressed as the product of prime numbers in the following way:

$$n = \prod_{i=1}^{d} p_i^{L_{n,i}} \qquad (1\text{-}3)$$

where $L_{n,i}$ is a list, or an *array*, of the powers of prime numbers associated with the number n. For example, the number 728 may be expressed as the following product:

$$728 = 2^3 \times 7^1 \times 13^1$$

This may be represented by an array of d elements in the following way:

$$\{L_{728,i}\} = \{3, 0, 0, 1, 0, 1, 0, 0, \ldots\}$$

with the powers of the first six prime numbers, 2, 3, 5, 7, 11, and 13, equal to, respectively, 3, 0, 0, 1, 0, and 1.

Box 1-2 Subroutine PRM_DCMP(NUM,K_LST,ND)

Decomposition of an integer N into a product of prime numbers

List of arguments:
  NUM: integer to be decomposed.
  K_LST: Array for the powers of prime number for NUM.
  ND: Dimension of K_LST in the calling program.

Initialization:
  (a) Store a list of prime numbers in $\{p_i\}$ in the calling program.
  (b) Zero the array K_LST.
  (c) Let N equal to NUM.

1. Start the decomposition. For $i = 1$ to d, check if N/p_i is an integer.
  (a) If so,
    (a) Let $N = N/p_i$,
    (b) Increase K_LST(i) by 1.
  (b) If not, go to the next p_i in the list.
2. Check if $N = 1$.
  (a) If not, the decomposition failed. Output an error message.
  (b) If $N = 1$, return K_LST as the list of the powers of prime numbers.

In such a representation, the product of any two integers n and m is given by the sum of their respective prime number arrays $\{L_{n,i}\}$ and $\{L_{m,i}\}$:

$$n \times m = \prod_{i=1}^{d} p_i^{L_{n,i}+L_{m,i}}$$

And the ratio of the same two integers is given by their differences in the powers:

$$\frac{n}{m} = \prod_{i=1}^{d} p_i^{L_{n,i}-L_{m,i}}$$

If an element in the array becomes negative, the prime number appears in the denominator, with a power equal to the absolute value of the element. For example, the ratio

$$\frac{4356}{728} = \frac{2^2 \times 3^2 \times 11^2}{2^3 \times 7^1 \times 13^1} = \frac{3^2 \times 11^2}{2^1 \times 7^1 \times 13^1}$$

may be represented as $\{-1, 2, 0, -1, 2, -1, 0, 0, \ldots\}$.

Given a prime number decomposition of a list of integers $1, 2, \ldots, N$, we can construct a list of the prime number decomposition of the factorials of integers up to that of N. By definition,

$$n! = \prod_{i=1}^{d} p_i^{\sum_{k=1}^{n} L_{k,i}}$$

where $L_{k,i}$, for $i = 1, 2, \ldots, d$, defined in (1-3), are the powers of prime numbers to decompose an integer k. The decomposition of a list of factorials may be constructed in the following way. If $\{F_{n,i}\}$ is the array representing the prime number decomposition of $n!$, that is

$$n! \equiv \prod_{i=1}^{d} p_i^{F_{k,i}} = \prod_{i=1}^{d} p_i^{\sum_{k=1}^{n} L_{k,i}}$$

we have the recursion relation

$$F_{n,i} = L_{n,i} + F_{n-1,i} \qquad \text{for} \qquad i = 1, 2, \ldots, d \qquad (1\text{-}4)$$

For the convenience of calculating $F_{n,i}$, we shall define

$$F_{1,i} = 0 \qquad \text{for} \qquad i = 1, 2, \ldots, d$$

as our list of prime numbers starts with 2.

We are now in a position to evaluate the binomial coefficients $\binom{n}{m}$ defined in (1-1). Before finding a way to calculate the expression, we must first realize that there are special cases not explicitly covered by (1-1). For n and m positive integers, we have

$$\binom{n}{m} = 0 \qquad \text{for} \qquad m > n$$

$$\binom{n}{0} = 1 \qquad \binom{0}{m} = \delta_{m,0}$$

$$\binom{n}{-m} = 0 \qquad \binom{-n}{m} = (-1)^m \binom{n+m-1}{m} \qquad (1\text{-}5)$$

These results, as well as those for other special cases, may be found in Riordan (*Combinatorial Identities*, Wiley, New York, 1968). In a well-written computer program, we must take care of such special cases by identifying them and calculating them separately. Alternatively, if the special cases occur only rarely and are not worth the extra effort to program them, the computer program must have the means of detecting and excluding them explicitly by, for example, returning an error message. For this reason, the algorithm outlined in Box 1-3 starts with instructions to take care of the special cases before going into the calculations

---

Box 1-3 Program BINOMIAL
Binomial coefficients using prime number decomposition

Subprogram used:
  PRM_DCMP: Prime number decomposition of an integer (Box 1-3).
Initialization:
  (a) Store a list of prime number decomposition of integers in $\{L_{n,i}\}$.
  (b) Store a list of prime number decomposition of factorials in $\{F_{n,i}\}$.
1. Input integers n and m.
2. Check for special cases:
   (a) For $n > 0$, $m > n$; $n = 0$, $m > 0$; and for $m < 0$, return 0.
   (b) For $m = n$, $n > 0$; and $m = n$, return 1.
   (c) For $n < 0$ and $m > 0$, replace with $(-1)^m \binom{m-n-1}{m}$.
3. The only remaining possibility is $n > m$ with $n > 0$ and $m > 0$. The integer decomposition of $\binom{n}{m}$ is given by $B_i = F_{n,i} - F_{m,i} - F_{n-m,i}$.
4. Calculate the value of the binomial coefficient $b_{n,m} = \prod_{i=1}^{d} p_i^{B_i}$.
5. Return $b_{n,m}$.

---

involving the method of prime number decomposition described in Box 1-2. Note that, for a given size computer word, the value of a binomial coefficient that can be stored as a list of the powers of prime numbers is much larger than the values of the coefficient itself in integer.

The method of decomposition into powers of prime numbers is convenient in many calculations involving products and ratios of integers. On the other hand, if there are any additions or subtractions in the intermediate stages, it is usually necessary to reconstruct first the numbers themselves from the powers of prime numbers. To carry out the actual calculations, some refinements of this basic method are necessary for the following reasons. First, we can easily run into limitations imposed by the largest integer that can be represented in the computer. Second, we can improve the efficiency of the method slightly. It is possible to achieve both aims by taking out first the common factors in the numerators and denominators before the summation or subtraction. Again, this may be done using the prime number representations of the numbers involved. The principles behind the method are illustrated by the following calculation:

$$\frac{N_n}{N_d} \pm \frac{M_n}{M_d} = \frac{N_n \times M_d \pm M_n \times N_d}{N_d \times M_d} \qquad (1\text{-}6)$$

For simplicity, we shall assume that N_n, N_d, M_n, and M_d are positive integers. Instead of carrying out the calculations on the right side of the expression directly, we shall take out first the common factors between N_n and M_n, as well as those between N_d and M_d. In this way, the remaining factors are smaller in size, and their values may be reconstructed from their prime number decomposition lists so as to take their sum or difference. The algorithm for carrying out such a calculation in prime number representation is described in Box 1-4, and the method is used later in §4-5 for calculating Clebsch-Gordan coefficients.

Box 1-4 Program ADD_PRM
Addition of two rational fractions n and m
using prime number decomposition

Subprogram used:
　　PRM_DCMP: Prime number decomposition of an integer (Box 1-2).
Initialization:
　　Store a list of prime numbers.
1. Input the numerator and denominator of n and m.
　　(a) Construct a prime number representation list for n using PRM_DCMP.
　　(b) Construct a prime number representation list for m using PRM_DCMP.
　　(c) Determine the sign between n and m.
2. Sum n and m.
　　(a) Separate the prime number decomposition of the numerator (positive powers) and denominator (negative powers) for n and m.
　　(b) Take away factors common to the two numerators and the two denominators.
　　(c) Find the values of the remainder using (1-6).
3. Reconstruct the final result.

The method of using logarithms. For many calculations, it is more convenient to use a floating number representation even though the final values involved are integers or rational numbers. This occurs, for example, when the results are to be used in another part of the calculation that is in floating number representation. In addition, a floating number representation becomes the only choice if the values involved are too large to be stored as integers in the computer. Some loss in accuracy may take place, but this may not be a serious problem since other parts of the calculation are carried out in terms of floating numbers anyway.

As an example, we shall calculate again the binomial coefficients given by (1-1). Instead of integers, we shall evaluate all the factorials involved as floating numbers in terms of their logarithms. Logarithms are especially useful for more elaborate calculations involving factorials. For our interest here, we can take the logarithm of both sides of (1-1) and obtain the result

$$\ln\binom{n}{m} = \ln(n!) - \ln(m!) - \ln(\{n-m\}!) \qquad (1\text{-}7)$$

On the other hand, the logarithm is a relatively slow function to evaluate. To speed up the calculation, we can store, once and for all, a table of the logarithms of $k!$ for $k = 0, 1, 2, \ldots, N$. This can be done fairly efficiently using the recursion relation

$$\ln(k!) = \ln k + \ln(\{k-1\}!) \tag{1-8}$$

As we shall see later in (4-145),

$$0! = 1 \tag{1-9}$$

As a result, the recursion relation starts with

$$\ln(0!) = 0$$

If we store $\log(k!)$ as an array $L_G(k)$, the calculation of (1-7) reduces to

$$\ln\binom{n}{m} = L_G(n) - L_G(m) - L_G(n-m) \equiv c \tag{1-10}$$

To find the value of $\binom{n}{m}$, we need to take the inverse of the logarithm,

$$\binom{n}{m} = \exp c$$

In this approach, the main part of the computer time in calculating a binomial coefficient is to evaluate the exponential function at the end. The savings come from having an existing table of the logarithms of the factorials. Because of this, the only other calculations involved are two subtractions, as can be seen in (1-10). The algorithm is outlined in Box 1-5.

Box 1-5
Binomial coefficient using a table of $\ln(k!)$

1. Construct a table of the logarithms of factorials using the recursion relation (1-8).
2. Input n and m.
3. Calculate the logarithm of $\binom{n}{m}$ using (1-10).
4. Take the exponential of the logarithm of $\binom{n}{m}$ and return the result.

---

§1-4 Algorithm for computation

The examples given in the previous section demonstrate that there can be differences in the way of solving a problem on a computer from that normally done by hand. Given a problem in physics, how do we go about using a computer to help us in obtaining a solution? The first thing to realize is that, no matter how powerful, computers cannot solve for us all the problems occurring in physics. It is possible to find out whether computers can be of assistance by carrying out some exploration. In general, whether computers can be helpful in a particular situation depends critically on whether we can find a suitable method to reduce the problem to one in which a computer can be used. Numerical methods are often the first that come to mind. This is due, in part, to the fact that most computers are designed for this purpose and in part because we have more extensive experience with them on computers. As a result, it is often far easier to obtain a numerical solution to a specific problem than an algebraic one. For example, it is relatively straightforward to find the numerical values for almost all definite integrals. In general, the same is not true for the analytic solutions of integrals with or without using a computer.

One common criticism of numerical solutions to a problem is that the results often appear in the form of a long table of numerical values, and it is difficult to grasp the meaning of such a large collection of numbers. The numerical solution of a differential equation is one example that comes immediately into our minds. However, this is not a very meaningful criticism. In the first place, the final output of a numerical solution does not have to be a table of numbers. We can either plot the results in the form of graphs or use them to carry out other calculations that have a more direct bearing on the answer for the problem we are seeking. For example, the solution of a differential equation may be the wave function of a quantum mechanical problem. Instead of stopping at the numerical values of the wave function, we can use them to calculate matrix elements of operators corresponding to physical observables, such as energy, momentum, and scattering cross sections. These results may be compared with the measured values of experiments. The long tables of numbers that represent the wave function serve only as intermediate results. They may be essential to debug the computer program but are hardly the final aims of the calculation.

In addition to the ease of obtaining numerical solutions, one should not ignore the fact that computers are also powerful tools for a large variety of algebraic operations. We shall give a brief introduction to algebraic computation in Chapter 10. Even for numerical solutions, it is well known that a little algebraic manipulation can often simplify the problem by a large extent. As a result, the combination of algebraic and numerical methods on a computer can often be a very effective tool for solving physical problems.

What is a good method to use in solving problems? The specific answer to this question depends on the problem, but some general observations may be made here. We have seen earlier that computers are constructed to carry out certain basic functions, such as addition, subtraction, multiplication, and division, at high speeds. Depending on the need, one may incorporate some of the "higher" functions, such as taking the square root, into the hardware of the computer. Regardless of such details, the machine is only able to carry out a very limited number of functions, we must find a way to map a problem from the way we like to think about it to the way the machine operates. This is the role of an *algorithm* (see, for example, the article on "Algorithms" in the April 1977 issue of *Scientific American* by D.E. Knuth). An algorithm is, however, not a computer program — it is usually not even stated in one of the computer languages. In fact, it is desirable to put the statements of an algorithm in a form that is independent of a particular computer language. In this way we have a choice of languages to use in writing the program. To a large extent, the role of an algorithm is to help the human user to formulate the problem in such a way that a computer may be used to carry out the mechanical steps involved. For this reason, an algorithm should be stated in a manner that is easy for the user to understand. At the same time, it must be unambiguous so that a computer program can be based on it. In this sense, it is like to road map that helps the programmer to go from the input of the problem to the output desired.

Another way to see the logical flow of a calculation to be carried out on a computer is to construct a flow chart. Usually, a flow chart gives the details of all the logical steps in a program. As illustration, an example of a flow chart is given in the next section (Fig. 1-1). If a flow chart contains enough information, it becomes very useful for a reader in understanding a program. In contrast, the statements in an algorithm do not normally contain such details. This is necessary so that the complete algorithm for a problem may be stated in a concise way to enable a reader to obtain easily an overall view of the method. From the way in which they are commonly used, we can perhaps distinguish a flow chart from the statements in an algorithm by saying that the former helps a person to understand the program, whereas the latter helps the programmer to understand the problem.

Although the statements of an algorithm are independent of the computer language used in writing the program, the choice of an algorithm for solving a problem very often depends on the language used. A simple example will be of interest here. Suppose we want to calculate the factorial of a positive integer n. Instead of (1-2), a factorial is also given by the recursion relation

$$n! = n \times (n-1)! \qquad (1\text{-}11)$$

starting from $0! = 1$ given earlier in (1-9). We, therefore, have a choice of two methods of calculating the factorial. The first is to take the product of $1, 2, \ldots, n$, as given by (1-2). The second is to use (1-11). For the recursive calculation, a program in the following form may be constructed:

```
SUBPROGRAM N_FACT(n)
    IF n < 0
        RETURN(0)
    ELSE IF n = 0
        RETURN(1)
    ELSE
        RETURN(n × N_FACT(n − 1))
    ENDIF
END
```

We can think of several reasons why this recursive method is preferable. However, FORTRAN 77 and older versions of the language are not constructed in a way that allows recursive calls of a subprogram within the subprogram itself. If, for some other reason, we have to use such a language, we must then choose a nonrecursive algorithm instead.

Requirements of an algorithm. A good algorithm is one that is accurate, efficient, and simple. The reason that accuracy becomes one of the issues derives, in part, from the truncation errors mentioned earlier and, in part, from the fact that it is often necessary to make approximations in order to solve a problem. For example, a definite integral may be replaced by a sum over the approximate values of the integrand in a finite number of small intervals. Among the methods that can be used for this approximation, clearly we must choose one that can yield an adequate accuracy for the problem of interest.

The efficiency consideration of an algorithm should include both the time to carry out the calculation as well as the time to develop the program. This includes both the time to write the problem and to debug it. For this reason, one may, on occasion, use a method that may result in slightly longer execution time so that the programming can be done more easily. In other words, if a realistic cost is to be assigned to a calculation, we must take into consideration both computational time as well as program development time. The relative importance of the two factors depends on how often the program will be used. For many projects in physics, it is more likely that the programming development time will dominate. For this reason, it may be more advantageous some of the time to take a less elegant approach so as to be able to achieve one's final goal of solving the problem of physics in a faster way.

A simple algorithm usually helps in reducing the amount of time required to develop a computer program. However, this is not the only reason for its use. It is often necessary to read the programs later, either to find out what one did some time ago or for someone else to extend the calculations. A simple algorithm, coupled with good programming style, can be very helpful in this respect.

Portability of computer programs. Although all digital computers are based on the same principles, there are still some differences between computers designed by different manufacturers. The differences may come from the emphases put on different functions. Alternatively, they may come as a result of technological improvements in the components. In most cases, the user is sheltered from the variations in the computer architecture by writing the programs in a high-level language and letting the compiler take care of the differences. However, this is not always possible. For example, different computers may have different allowances for the largest floating number they can store. As a result, a program that runs well on a computer that allows a larger floating number may cause an overflow error on a computer with a smaller maximum. It may also happen, whether by intention or otherwise, that the program takes advantage of the special features of a particular computer, such as certain graphics capabilities. Such programs are usually restricted to the computers for which they are designed.

One important consideration in computing these days is that a program may have to run on a variety of different computers. This may happen because a number of different computers are available to the user. As a result, it may be desirable to develop the programs on a smaller machine, such as a workstation, but to carry out the actual computations on a faster one, such as a supercomputer. Alternatively, one may want to share the programs developed with one's collaborators and colleagues elsewhere. Regardless of the circumstance, it is preferable to have programs that can be used on a variety of "platforms." That is, a program should be portable from one computer to another. With the fast technological development in the computing industry, it is also likely that we may have to change our computer from time to time. As a result, portability to future machines is also a concern for any program that takes a large amount of effort to develop.

Since only high-level computer languages are more or less standard among different computers, the first requirement for a program to be portable is that it must be written in a high-level language. In this respect, the FORTRAN language is the best. Except for special features added on by different compilers, the language is universal. In other words, if one stays within the specifications of some widely accepted standard, such as FORTRAN 77, there is very little difficulty in compiling and running a program on different machines. In almost all the other high-level languages, the degree of standardization is not observed to the same extent and, except for very simple programs, there are often problems of compatibility when running them on different machines.

Another issue of portability involves intrinsic functions and subroutine libraries. As emphasized earlier, it is a good practice to make use of existing software in one's program so as to speed up the development. To run such programs on another computer, it is necessary that the same libraries be also present. Portability for programs requiring library functions also means that we must stay with those that are in wide use.

Standard subroutine libraries. As mentioned in §1-1, several subroutine packages are fairly standard in the scientific community. The term "standard" carries a slightly different meaning here. If a set of subroutines that are identical in their use exists in many different machines, programs making use of them are portable among these machines. For subroutines that are available in the form of source codes, there is no problem in this respect. However, this is not often the rule. In the first place, there may be different considerations on different machines because of differences in the architecture. Subroutines in a good library often try to incorporate some of these considerations into the codes. As a result, different machines may require slightly different versions of the same program. It is therefore simpler to supply the object codes of the subroutines so that they can only be used with the machine for which they are designed. This may not be the only reason. Many of these libraries are commercial, in the sense that the cost of developing and maintaining these codes must be recovered from the user together with a profit. As a result, the source codes become proprietary and are not usually distributed. Portability of programs that make use of these commercial packages therefore requires that a copy of the library be purchased for the machine we wish to run.

A number of public domain subroutine libraries are available in the form of FORTRAN source codes. These include EISPACK, which consists of an extensive group of subroutines for eigenvalue programs. A detailed explanation of the package may be found in *Matrix Eigensystem Routines – EISPACK Guide* (Springer-Verlag 1976) by B.T. Smith and others. A similar package, LINPACK, concentrates on solving linear equations, inverting and decomposing matrices. A *LINPACK User's Guide* (SIAM 1979) by J.J. Dongarra is available. Many similar packages are also available through electronic mail, such as NETLIB. Similarly, RedLib, a network for distributing software for algebraic computation based on Reduce, is maintained by the Rand Corporation. A data base, eLib, that covers several libraries is maintained by Konrad-Zuse-Zentrum fuer Informationstechnik Berlin.

Two groups maintain subroutine libraries that are used widely for scientific applications, International Mathematical and Statistical Libraries (IMSL) and National Algorithms Group (NAG). The coverage in both groups is fairly broad and is expanding all the time. It is also quite easy to make use of these libraries because of the detailed documentation associated with each subroutine. For the NAG library, a good starting point is the book *The NAG Library* (Claredon Press, Oxford, 1986) by J.P. Phillips.

Some of the newer books on numerical methods are accompanied by programs on magnetic media. For example, all the subroutines in *Numerical Recipes* (Cambridge, 1986) by W.H. Press and others, are available on computer diskettes. They are specially useful for learning the numerical methods covered in these books.

§1-5 A calculation for the value of π

As an example of computational algorithm, let us calculate the value of π. There are several ways to do this. The first is to realize that $\sin \pi/2 = 1$. As a result, we can make use of the inverse relation

$$\sin^{-1} 1 = \frac{\pi}{2}$$

to obtain the value of π. (The more direct route of using the inverse of $\sin \pi = 0$ does not work since the arc sine function in the intrinsic library usually returns values only in the range $[-\frac{\pi}{2}, +\frac{\pi}{2}]$.) The use of trigonometric functions is a good way, for example, to set the value of π in a program to the same precision as the rest of the calculations we wish to carry out on the computer. However, the method is not of interest here, as it depends on the way the arc sine function is evaluated on the computer.

As an illustration of several important aspects in programming a computer, we shall calculate the value of π using a more elementary approach. Here we encounter the first consideration in solving a problem with a computer: we must select a suitable method. For the value of π, we shall make use of one of the series expansions. There are, in fact, quite a few such series from which we may choose. For example, we can use one of the following two expansions:

$$\frac{\pi}{2\sqrt{2}} = 1 + \frac{1}{3} - \frac{1}{5} - \frac{1}{7} + \frac{1}{9} + \frac{1}{11} - \cdots \qquad (1\text{-}12)$$

$$\frac{\pi}{4} = 1 - \frac{1}{3} + \frac{1}{5} - \frac{1}{7} + \frac{1}{9} - \frac{1}{11} + \cdots = \sum_{n=1}^{\infty} \frac{(-1)^n}{2n+1} \qquad (1\text{-}13)$$

We shall reject (1-12) in favor of (1-13) since the former involves the factor $\sqrt{2}$. As a result, the accuracy one can obtain for the value of π also depends on how well we know $\sqrt{2}$.

It is fairly straightforward to program (1-13). All we need to do is a simple loop that alternates between adding to and subtracting from the sum of the inverse of the next odd integer. On the other hand, if we examine the number of terms required to achieve a given accuracy in this approach, we shall be surprised. For example, if we wish to achieve an accuracy of 10^{-5}, the last term to be included in the summation must be much smaller than 10^{-5}. This means something on the order of 250,000 terms! Here we learn the lesson that a simple, innocent looking expression can be time consuming to evaluate. It is worthwhile to examine our expression a little closer to see if some analytical work can save us a lot of computer time. This is particularly important if the calculation is likely to take a long time. For our simple example here, we can see that, if we take the difference between two adjacent terms, the relation given by (1-13) reduces to the form

$$\frac{\pi}{4} = 1 - 2\left(\frac{1}{3 \times 5} + \frac{1}{7 \times 9} + \frac{1}{11 \times 13} + \cdots\right) \qquad (1\text{-}14)$$

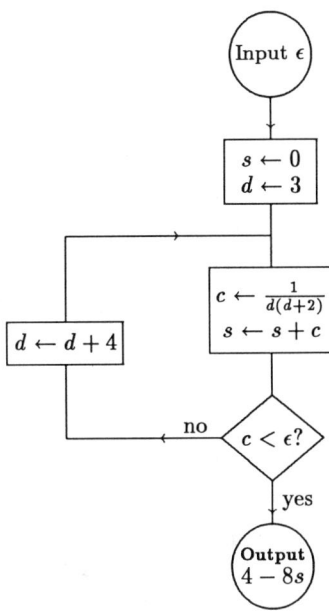

Fig. 1-1 A flow chart for calculating the value of π using (1-14).

As an analytical expression, this is identical to (1-13). However, as a way to calculate π, there is a major difference, as we shall see next.

The steps to carry out the calculation with (1-14) on a computer are shown in the form of an flow chart in Fig. 1-1. The program is only slightly more involved than the one using (1-13). However, the number of terms required to achieve the same accuracy of 10^{-5} is now decreased to 62,500, a factor of 4 reduction. Note that the saving is more than the factor of 2 expected from contracting two terms to one in going from (1-13) to (1-14).

As far as the numerical calculation is concerned, (1-13) represents an example of a poor approach, as successive terms in the sum alternate in sign. For this reason, the calculation approaches the asymptotic value of $\pi/4$ through a zigzag path, as shown by the dashed curve in Fig. 1-2. In contrast, the summation on the right side of (1-14) approaches the limit smoothly, as shown by the solid curve. In general, it is desirable to avoid the kind of fluctuations in the intermediate results given by (1-13). In more realistic cases, the improvement in computation speed by a slight change in the algorithm may not be as obvious as in our example, but the savings can often be much greater.

What is the best accuracy we can achieve in calculating the value of π by the method illustrated? In single precision, we have seen earlier that the limitation is

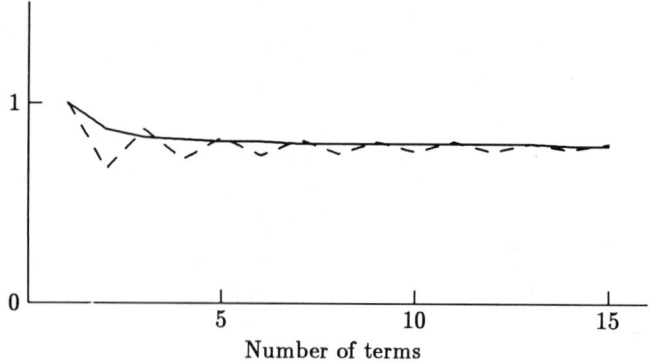

Fig. 1-2 Convergence of two different series to the value of $\pi/4$. The dashed curve is obtained with (1-13) and the solid curve with (1-14).

around seven significant figures, because of the fact that only 23 bits are assigned to store the mantissa. In practice, it is difficult to achieve this degree of accuracy in a long calculation due to truncation errors. It is more likely that, in reality, only five significant figures are obtained. As a result, there is very little value in going much beyond the 62,500 terms we have used earlier for (1-14). For a more detailed discussion of the calculation of π see, for example, the article on "Ramamujan and Pi" in the February 1988 issue of *Scientific American*, p. 112, by J.M. Borwein and P.B. Borwein.

We have mentioned earlier that computational efficiency is an important consideration, but simplicity in the algorithm is also an essential feature. Often, these two considerations may be in conflict with each other. As a result, the choice of an algorithm requires some balance between these two requirements. The difference between the two approaches of calculating the value of π is an elementary example of this problem often encountered in computational physics.

Problems

1-1 If a is any one of the five integers 3, 4, 5, 6, and 7, and b is any one of the three integers 2, 3, and 4, what are the 15 values of $c = a/b$ if the calculation is carried out in integer mode on a computer?

1-2 If a and b are two real numbers, design an algorithm to calculate $c = a*b$ so that the product c is an integer and has the value equal to the nearest integer to the product $a*b$. The purpose of this function is equivalent to that of NINT($a*b$) in the intrinsic function library of a FORTRAN compiler.

1-3 Design an algorithm using the function SIN in the FORTRAN intrinsic function library to calculate $\sin x$ for x from 0° to 90° at every 10°. Note that the SIN function takes arguments in radians.

1-4 Describe a way to calculate the square root of a complex number using a function that can only take the square root of real numbers.

1-5 On some machines, each computer word is made of 36 binary bits. Find the range of integers it can store. Design a way to store floating numbers and state the range of values that can be represented.

1-6 Imagine a simple computer which has 64 storage locations and one register (a special storage area), and can perform only the following 11 operations: input, output, addition, subtraction, multiplication, division, store, if-zero, if-positive, go-to, and stop. Assuming that all the operations are for real numbers, design an assembly language for this machine and write a program with it to calculate the values of y for $y = x^2$ for a set of ten equally spaced values of x in the range $x = 0$ to 1.

1-7 Give two different ways to find on a computer the roots of the quadratic equation $ax^2 + bx + c = 0$ for a set of input values of the three coefficients a, b, and c.

1-8 Design an algorithm for carrying out an indefinite integral

$$I = \int P_n(x)\,dx$$

where the integrand is a power series of the form

$$P_n(x) = a_0 + a_1 x + a_2 x^2 + \cdots + a_n x^n$$

Write a FORTRAN program to carry out this algebraic calculation.

1-9 Construct a function that returns the binomial coefficient $\binom{n}{m}$ of (1-1) using a table of $\ell n(k!)$ as outlined in Box 1-5.

1-10 In the ancient Roman system of notation, the letter I is used to stand for numerical value 1, V for 5, X for 10, L for 50, C for 100, D for 500, and M for 1000. If a letter is followed by one of equal or lesser value, the two values are added. On the other hand, if one of the above letters is followed by one of greater value, it is subtracted from the one following. Thus MCMLXXIV means 1974. (a) Design an algorithm to convert Roman numerals to integers and vice versa. (b) Think of a way to add two integers expressed in Roman numerals without first converting them to their corresponding numerical values.

1-11 Very often we are in need of two real numbers a and b subject to the condition $a^2 + b^2 = 1$. As a result, there is only one degree of freedom among them. If both a and b are positive (and less than 1), there are two ways to take this degree of freedom. The first is to define $a = \sin\theta$ and $b = \cos\theta$ and consider θ as the independent quantity. The second is to define $b = (1 - a^2)^{1/2}$ and consider a as the independent quantity. Use a random number generator to generate a set of values of θ in the first case and a in the second. Check the relative efficiency of the two methods by comparing the amount of time to calculate, for example, 5000 different values in each case.

Chapter 2
Integration and Differentiation

Integration and differentiation enter into such a large variety of physical problems that it is hardly necessary to give an introduction to them here. In this chapter, we shall discuss some of the more standard numerical methods to integrate a function. In addition, we shall also use these mathematical operations as the means to illustrate some of the problems faced by numerical calculations and to appreciate some of the reasons why specialized methods are developed to evaluate definite integrals.

Numerical integration has been an active field in mathematics even before the introduction of computers. This comes from the simple fact that not all integrations can be carried out analytically, and numerical methods, however tedious without the help of a computer, become the only way to solve a problem. Furthermore, the results of many common integrals, such as those involved in defining the error function erf(x) and gamma function $\Gamma(x)$, do not exist as analytical functions and their values must be found numerically. This is quite different from differentiation. For most of the functions encountered in physics, differentiations can usually be carried out by analytical means. However, there are exceptions. On many occasions, we encounter functions that are given numerically. In this case, the functional form of $f(x)$ is not available to us and the derivatives of $f(x)$ can only be obtained using numerical methods. Furthermore, numerical differentiation is a good way to be introduced to the wide world of finite differences, a topic we shall come across frequently in later chapters.

§2-1 Numerical integration

The most common type of integration encountered in physics may be represented by the following operation:

$$I_{[a,b]} = \int_a^b f(x)\, dx \tag{2-1}$$

Since the upper and lower limits of the integration are specified, such an integral is known as a definite integral. Without losing any generality, we can assume that the integrand $f(x)$ is greater than or equal to zero in the interval $x = [a, b]$. In this case, the integral $I_{[a,b]}$ may be interpreted as the area bound above by $f(x)$, below by the x-axis, on the left by the coordinate $x = a$, and on the right by $x = b$. This is shown in Fig. 2-1 by the dotted area. This form of the integral is known in mathematics as the (definite) Riemann integral and is the only form with which we shall be dealing here. In Chapter 10, we shall return for some discussion of indefinite integrals, such as the simple example of

$$\int \sin(x)\,dx = -\cos(x)$$

Instead of numerical integration, our interest there will be concerned with algebraic manipulations.

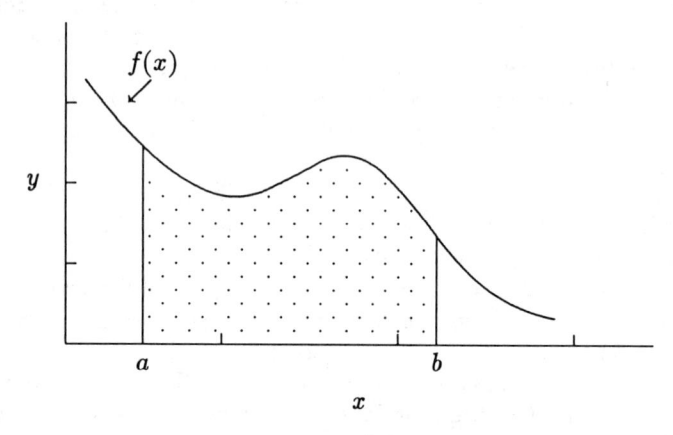

Fig. 2-1 Illustration of a definite integral $\int_a^b f(x)\,dx$ as the area between $f(x)$ and the x-axis in the region between $x = a$ and $x = b$.

The use of numerical integration in physics may be illustrated by the example of the distance traveled by a particle moving along the x-direction with velocity v. If v is constant, the distance d covered in the interval between times $t = a$ and $t = b$ is simply given by

$$d = v \times (b - a)$$

Alternatively, the velocity may be a function of time t. For example, if the particle is acted upon by a constant force so that it is accelerating at a fixed rate of α, the velocity is a linearly increasing function of time t,

$$v(t) = \alpha\,t$$

The distance traveled in the same time interval $t = [a, b]$ is now the familiar result

$$d = \int_a^b v(t)\,dt = \int_a^b \alpha\,t\,dt = \tfrac{1}{2}\alpha t^2 \Big|_a^b = \tfrac{1}{2}\alpha\,(b^2 - a^2)$$

Table 2-1 Distance traveled in ten minutes.

Time (minutes)	Velocity (m/s)	Distance traveled in 1 minute (m)	Total distance traveled so far (m)
1	26	1,560	1,560
2	27	1,620	3,180
3	25	1,500	4,680
4	24	1,440	6,120
5	23	1,380	7,500
6	22	1,320	8,820
7	24	1,440	10,260
8	26	1,560	11,820
9	25	1,500	13,320
10	24	1,440	14,760
Total distance traveled in 10 minutes =			14,760 m

As long as the velocity is expressed in terms of an analytical function of time, the distance can usually be found by integrating over time in a similar manner.

A more common situation in physics perhaps is one where the velocity is not available as an analytical function. As an example, think of a car traveling along a highway. Let us pose the problem of finding the distance traveled in a given amount of time, for example, 10 minutes, without making use of an odometer. One way to do this is to sample the speed of the car regularly, for example, once every minute. As a result, we have the velocity as a function of time in the form of a table of 10 pairs of numerical values. If the car is moving along an empty highway, the speed is essentially a constant, for example, 25 m/s (corresponding to 90 km or about 55 miles an hour). In this case, the distance traveled in 10 minutes is simply 25 m/s × 10 minutes = 15 km. On the other hand, if the highway is congested, the speed varies from one minute to another. We can still find out roughly the distance covered by assuming that the speed is essentially constant within the 1-minute time interval between two successive samplings. The total distance traveled is then the sum of the 10 distances, each covered in 1 minute, as shown in Table 2-1. As long as the speed does not vary too much within a minute, our method of finding the distance is accurate enough for most purposes. On the other hand, it may not yield very good results if the road is very crowded. To accommodate the larger variations of the speed with time, we may have to increase the frequency of measuring the velocity by decreasing the time interval between two consecutive samplings.

Evaluation of an integral by numerical methods is often called numerical quadrature for reasons that will become obvious in §2-4. The basic approach follows from the definition of an integral given earlier: namely to find the area beneath the curve. If we have a constant for the integrand

$$f(x) = H$$

the area is simple $H \times (b-a)$. The value of the integral is then

$$I_{[a,b]} = H \times (b-a)$$

the same as earlier in finding the distance traveled by a particle moving with constant velocity. For the more general situation, better methods are needed. We shall be mostly concerned with one-dimensional integrals, meaning those with only one variable of integration. However, most of the methods we shall be discussing may be extended to multidimensional ones. We shall do this toward the end of this chapter.

The simple example used above illustrates the principle behind numerical integration. In general, our goal is to find the value of the integral from the values of the integrand calculated at different parts of the interval $x = [a, b]$. To be efficient, we shall attempt to minimize the number of times we have to evaluate the integrand, as this is often the most time-consuming part of the calculation. On the other hand, a faithful representation of the behavior of the integrand may require a large number of values evaluated in the integration domain. Since these two requirements are in conflict with each other, different methods are used to meet the variety of situations we encounter in practice.

§2-2 Rectangular and trapezoidal rules

In this section, we shall discuss two simple algorithms for numerical integration. In both cases, the basic idea is centered around the concept described in the previous section: the value of the definite integral of a function $f(x)$ in the interval $[a, b]$ is given by the area bounded by the x-axis, $f(x)$, $x = a$, and $x = b$.

Rectangular rule. The simplest way to calculate the area underneath a curve is to divide the interval between $x = a$ and $x = b$ into a number of smaller ones. For simplicity, we shall begin by taking all the subintervals to be of equal size, each with width h, and there are N of them. As a result, we have the relation

$$Nh = b - a$$

The area underneath the curve we wish to calculate is now divided into N narrow strips, each of width h. If h is sufficiently small, we may approximate each slice of the total area to be rectangular in shape with width h and height given by a reasonable representative value of $f(x)$ in the subinterval.

This is essentially the method we adopted in calculating the distance traveled by a car along a highway in the previous section. In the limit $h \to 0$, equivalent to the situation that the velocity of the car is known at every instant of time, there is only one value of $f(x)$ in each slice and there is no ambiguity in choosing the representative value for the slice. However, in a numerical calculation, efficiency considerations require us to take as large a value of h as possible. As a result, the subdivision must be chosen with some care.

One way to evaluate an integral is to replace the value of the integrand in a subinterval by an average. This idea may be put more quantitatively in the following way. Let the ith subinterval start at $x = x_i$ and end at $x = x_{i+1}$. By definition,
$$h = x_{i+1} - x_i$$
(with $x_0 = a$ and $x_N = b$). An average value of $f(x)$ in subinterval $x = [x_i, x_{i+1}]$ is given by
$$\overline{f}_i = \frac{1}{h} \int_{x_i}^{x_{i+1}} f(x)\,dx$$
If all the values of $\overline{f}_i$ are available for each of the N subintervals, we have the result
$$I_{[a,b]} = \int_a^b f(x)\,dx = h \sum_{i=1}^{N} \overline{f}_i$$
Furthermore, if all the values of $\overline{f}_i$ are exact, we obtain a very accurate value of the integral $I_{[a,b]}$ in this way. However, this is possible only if $f(x)$ is a very simple function, such as a constant. In general, it is not easy to obtain $\overline{f}_i$ exactly and an efficient way is needed to make approximations.

For a slowly varying function, a good estimate of its average in a small subinterval is given by its value in the middle of the range,
$$\overline{f}_i \approx f(x_{i+1/2})$$
where we have used the notation
$$x_{i+1/2} \equiv \tfrac{1}{2}(x_i + x_{i+1}) \tag{2-2}$$
On those occasions where there is no ambiguity, it is also convenient to use the shorthand notation
$$f_i \equiv f(x_i) \tag{2-3}$$
That is, f_i is the value of $f(x)$ evaluated at $x = x_i$. The integral of $f(x)$ in interval $[a, b]$ is then
$$I_{[a,b]} = h \sum_{i=1}^{N} f_{i-1/2} \tag{2-4}$$
Note that, since we have taken $x_0 = a$, the estimate of $f(x)$ in the first interval $[a, a+h]$ is denoted as $f_{1/2} \equiv f(x_{1/2})$.

The method of numerical integration given by (2-4) is known as the rectangular rule. It is easy to show that the error in this method due to finite step size decreases as N the number of intervals is increased. The contribution of a given subinterval $[x_{i-1}, x_i]$ to the entire integral is given by

$$I_{[x_{i-1},x_i]} = \int_{x_{i-1}}^{x_i} f(x)\, dx \approx h f_{i-1/2} \tag{2-5}$$

We can check the accuracy of this approximation within a subinterval by expanding the function $f(x)$ in terms of a Taylor series around the middle point $x_{i-1/2}$,

$$\begin{aligned} f(x) = f_{i-1/2} &+ \frac{1}{1!} f'(x_{i-1/2})(x - x_{i-1/2}) \\ &+ \frac{1}{2!} f''(x_{i-1/2})(x - x_{i-1/2})^2 \\ &+ \frac{1}{3!} f^{(3)}(x_{i-1/2})(x - x_{i-1/2})^3 + \cdots \end{aligned} \tag{2-6}$$

where we have used $f'(x_{i-1/2})$, $f''(x_{i-1/2})$, and $f^{(3)}(x_{i-1/2})$ to denote, respectively, the first, second, and third derivatives of $f(x)$ evaluated at $x = x_{i-1/2}$.

Using (2-6), the exact value of the integral in the subinterval may be expressed in the form

$$\begin{aligned} \int_{x_{i-1}}^{x_i} f(x)\, dx = f_{i-1/2} \int_{x_{i-1}}^{x_i} dx &+ \frac{1}{1!} f'(x_{i-1/2}) \int_{x_{i-1}}^{x_i} (x - x_{i-1/2})\, dx \\ &+ \frac{1}{2!} f''(x_{i-1/2}) \int_{x_{i-1}}^{x_i} (x - x_{i-1/2})^2\, dx \\ &+ \frac{1}{3!} f^{(3)}(x_{i-1/2}) \int_{x_{i-1}}^{x_i} (x - x_{i-1/2})^3\, dx \\ &+ \cdots \end{aligned} \tag{2-7}$$

The first term is the approximation used in (2-5). The second term vanishes because of the integral associated with it. The leading order error in the rectangular rule for a single subinterval of width h is therefore given by the third term,

$$\begin{aligned} \Delta I_{[x_{i-1},x_i]} &\equiv \int_{x_{i-1}}^{x_i} f(x)\, dx - h f_{i-1/2} \\ &= \frac{1}{2!} f''(x_{i-1/2}) \int_{x_{i-1}}^{x_i} (x - x_{i-1/2})^2\, dx \\ &= \frac{h^3}{24} f''(x_{i-1/2}) \end{aligned}$$

The total error in the entire interval $[a, b]$ is the sum of the contributions from all N subintervals,

$$\Delta I_{[a,b]} = \int_a^b f(x)\, dx - I_{[a,b]} \approx \frac{b-a}{24} h^2 f''(\xi) = \frac{1}{24} \frac{(b-a)^3}{N^2} f''(\xi)$$

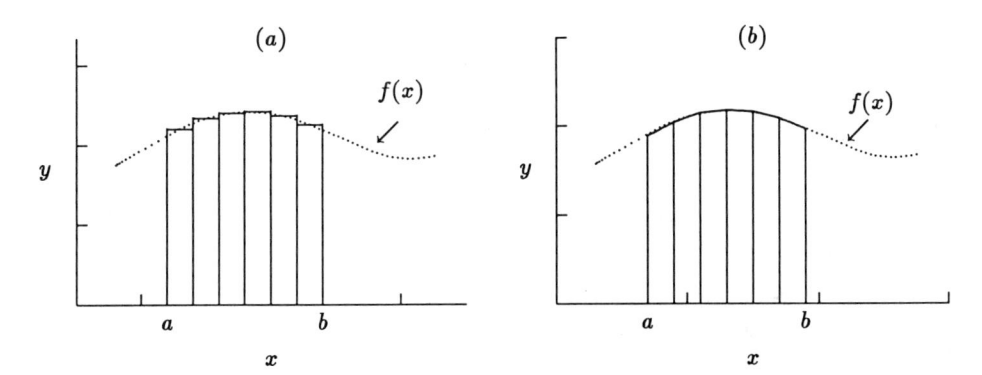

Fig. 2-2 Schematic illustrations of rectangular rule (a) and trapezoidal rule (b). In the former case, the contribution from each subinterval is approximated by a rectangle of width h and height equal to the value of the integrand evaluated at the middle point of the subinterval. In the latter case, the area of a subinterval is approximated by a trapezoid, formed by joining the four points $(x_{i-1}, 0)$, (x_{i-1}, f_{i-1}), (x_i, f_i), and $(x_i, 0)$ with straight lines.

where we have made use of the fact that $Nh = (b - a)$ and have taken $f''(\xi)$ to be the average value of the second derivative of $f(x)$ in $[a, b]$.

Trapezoidal rule. As an alternative to the rectangular rule, we can approximate the area underneath $f(x)$ within each subinterval by a trapezoid with the four corners at coordinates $(x_{i-1}, 0)$, (x_{i-1}, f_{i-1}), (x_i, f_i), and $(x_i, 0)$. The area of this trapezoid is given by

$$I_{[x_{i-1}, x_i]} = \frac{h}{2}(f_{i-1} + f_i) \tag{2-8}$$

Schematically, the difference between rectangular and trapezoidal rules is shown in Fig. 2-2.

For the entire interval $[a, b]$, the integral is again given by the sum over all the slices,

$$I_{[a,b]} = \frac{1}{2} h \sum_{i=1}^{N} (f_{i-1} + f_i) = h\left(\tfrac{1}{2}f_0 + f_1 + f_2 + \cdots + f_{N-1} + \tfrac{1}{2}f_N\right) \tag{2-9}$$

This is a very reasonable result, which we could have anticipated from the beginning. Compared with the other points, the contributions of the end points, f_0 and f_N, are only half as important, as they are at the two ends of the interval. The result follows also from the Euler-Maclaurin summation formula for integrals (Abramowitz and Segun, *Handbook of Mathematical Functions*, Dover, New York, 1965),

$$\int_a^b f(x)\, dx = h\left\{\tfrac{1}{2}f_0 + f_1 + f_2 + \cdots + f_{N-1} + \tfrac{1}{2}f_N\right\}$$

> **Box 2-1 Subroutine** TRAPZ_INT(A,B,KEY,RSLT,H)
> **Trapezoidal rule integration**
> (Constructed for Romberg integration of Box 3-7)
>
> Arguments:
>   A: Lower integration limit.
>   B: Upper integration limit.
>   KEY $= 0$: Initiate a new integration.
> > 0 : Subsequent iterations.
>   RSLT: Value of the integral.
>   H: Integration step size.
>
> Initialization:
>   Define the integrand as an external function.
>
> 1. First call (KEY=0):
>   (a) Calculate the integral using two subintervals.
>   (i) Define the number of subintervals to be 2.
>   (ii) Define SUM to be the sum of the average value of the integrand from the two end points plus the contribution from the midpoint of the interval.
>   (iii) Define the step size H to be half of the distance from A to B.
>   (b) Change the value of KEY to 1.
>   (c) Return the product of H and SUM.
> 2. Subsequent calls (KEY$>$ 0):
>   (a) Calculate the contributions from points halfway between old ones.
>   (i) Start from the middle of the first subinterval.
>   (ii) Calculate the value of the integrand at this point and add the value to SUM.
>   (iii) Move to the middle of the next subinterval and repeat steps (i) and (ii).
>   (b) Double the number of subintervals and halve the step size H.
>   (c) Return H times SUM.

$$-\frac{B_2}{2!}h^2(f'_N - f'_0) - \cdots - \frac{B_{2k}}{(2k)!}h^{2k}(f_N^{(2k-1)} - f_0^{(2k-1)}) - \cdots \tag{2-10}$$

The coefficients B_{2k} are the Bernoulli numbers, with

$$B_1 = -\tfrac{1}{2}$$

and all the other odd orders vanish. The first few even ones have the values

$$B_0 = 1 \qquad B_2 = \frac{1}{6} \qquad B_4 = -\frac{1}{30} \qquad B_6 = \frac{1}{42}$$

More generally, they are given by the generating function,

$$\frac{t}{e^t - 1} = \sum_{k=0}^{\infty} B_k \frac{t^k}{k!} \tag{2-11}$$

Explicit values of B_k up to B_{60} can be found on page 810 of Abramowitz and Segun. Note that, among other things, (2-11) demonstrates that the errors of the trapezoidal rule are given by the even powers of h.

The advantage of the trapezoidal rule is that the values of $f(x)$ are evaluated only at the *grid* or *mesh* points, that is, at $x = x_0, x_1, \ldots, x_N$. As can be seen from Problem 2-1, the errors due to finite step sizes associated with this method are expected to be a factor of 2 larger than those of the rectangular rule. On the other hand, the simplicity of the trapezoidal rule lends itself to further improvements. We shall see later in §3-5 that the method may be used in conjunction with extrapolation techniques to construct a very efficient algorithm for numerical integration known as the Romberg integration. The algorithm for the trapezoidal rule is outlined in Box 2-1.

§2-3 Simpson's rule

In the previous section, we saw that the errors associated with both the rectangular and trapezoidal rules are on the order of $1/N^2$. To improve the accuracy without increasing very much the amount of computation involved, we can use Simpson's rule. In the Taylor series expansion of the integral given in (2-7), we saw that terms involving odd-order derivatives of $f(x)$ vanish if the limits of the integration are taken symmetrically around the point where we make the expansion. To make use of this property in a convenient way, let us make a Taylor series expansion of the area underneath $f(x)$ in two adjacent subintervals. If the expansion is made around the value $x = x_i$, the middle point of the two subintervals involved, we obtain the result

$$\int_{x_{i-1}}^{x_{i+1}} f(x)\,dx = f_i \int_{x_{i-1}}^{x_{i+1}} dx + \frac{1}{1!} f'(x_i) \int_{x_{i-1}}^{x_{i+1}} (x - x_i)\,dx$$

$$+ \frac{1}{2!} f''(x_i) \int_{x_{i-1}}^{x_{i+1}} (x - x_i)^2\,dx$$

$$+ \frac{1}{3!} f^{(3)}(x_i) \int_{x_{i-1}}^{x_{i+1}} (x - x_i)^3\,dx + \cdots$$

$$= f_i \times 2h + 0 + \frac{1}{2!} f_i'' \times \frac{2}{3} h^3 + 0 + O(h^5 f_i^{(4)}) \qquad (2\text{-}12)$$

If we include into the numerical integration terms involving up to the second-order derivative of $f(x)$, we have a result that is more accurate by two orders in the power of h than what we can obtain with the rectangular and the trapezoidal rule.

This may be achieved in the following way. Using the method of finite difference, to be introduced later in §2-7, we can approximate the second-order derivative of $f(x)$ as

$$f''(x_i) \equiv \frac{d^2}{dx^2} f(x)\bigg|_{x=x_i} \approx \frac{1}{h^2}(f_{i-1} - 2f_i + f_{i-1}) \qquad (2\text{-}13)$$

On substituting this result into (2-12), the integral in the subinterval $[x_{i-1}, x_{i+1}]$ may be approximated as

$$I_{[x_{i-1},x_{i+1}]} = 2hf_i + \tfrac{1}{3}h(f_{i-1} - 2f_i + f_{i-1}) + O(h^5 f_i^{(4)})$$
$$= h(\tfrac{1}{3}f_{i-1} + \tfrac{4}{3}f_i + \tfrac{1}{3}f_{i-1}) + O(h^5 f_i^{(4)}) \quad (2\text{-}14)$$

On summing over all the slices in the interval $[a, b]$, we have the result

$$I_{[a,b]} = \tfrac{1}{3}h(f_0 + 4f_1 + 2f_2 + 4f_3 + \cdots + 2f_{N-2} + 4f_{N-1} + f_N) \quad (2\text{-}15)$$

In the expression, we have assumed that the number of intervals N is even for reasons that will become clear in the next paragraph. In this way of dividing the integration range, we find that the contributions from the odd grid points are more important than those from even ones by a factor of 2. This is illustrated by the diagram in Fig. 2-3. The difference in the weight comes from our effort to correct the simple first-order results of more elementary methods by including the contributions from second-order derivatives of $f(x)$. Again, the end points, f_0 and f_N, contribute with a weight that is only half that for the even points.

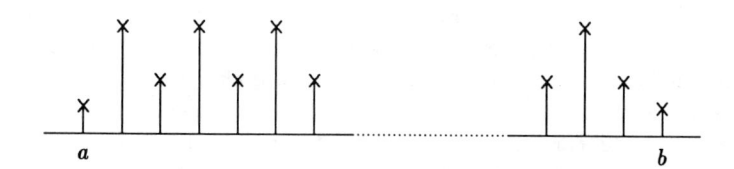

Fig. 2-3 Weighting factors for the contributions from different points in Simpson's rule integration. To reduce numerical errors, contributions from the odd points are treated as twice as important as those coming from even points, and the contributions from the two end points at $x = a$ and $x = b$ weigh only half as much as the even ones.

In addition to improvements in the accuracy, the Simpson's rule given by (2-15) also has the advantage that it leads naturally to a *strategy* for evaluating an integral to the desired accuracy. The importance of this point may be illustrated in the following way. In general, we want to get the maximum accuracy possible from a numerical calculation. This is not always possible since the resulting computation may be too long. On the other hand, there is no point in trying to achieve higher accuracies than that required for the problem in hand. For example, if we wish to compare the result of an integration with the measured values, there is seldom any point in knowing the integral much better than the number of significant figures in the data themselves.

In general, the accuracy of a numerical integration improves with the increasing number of subintervals N used. A "good" algorithm for evaluating an integral is then one in which the number of intervals used is just large enough for the accuracy required. An easy way to achieve this is to use an iterative approach. At the start of the calculation, some small but reasonable value of N is adopted. In the subsequent steps of the calculation, the value of N is increased gradually in each step. Each time we raise the number of subintervals, the accuracy of the calculation is improved. The process continues until the desired accuracy is reached.

Two important considerations must be made in taking such an approach. The first is to know when sufficient accuracy is reached. In principle, we can find the error in a numerical calculation by making, for example, an estimate of the size of the contributions from the leading order term ignored in (2-12). However, this is not always easy to do since it involves a knowledge of the average value of some high-order derivatives, such as $f^{(4)}(\xi)$ in (2-12). A simpler, but less sure, way is to compare the results of two consecutive iterations. If the difference in the calculated values of the integral $I_{[a,b]}$ in two successive integrations with different N is smaller than some preset tolerance ϵ, it is likely that the required accuracy is reached.

The second consideration is to find a way whereby we can apply the iterative approach with a minimum amount of calculation. For the Simpson's rule, the strategy may be implemented in the following way. Consider the case of integrating a function $f(x)$ between $x = a$ and $x = b$ up to some accuracy ϵ. From (2-15), we see that the contributions to the integral may be divided into three parts, the end zone, the odd points, and the even points,

$$S_d(N) \equiv f_0 + f_N$$
$$S_o(N) \equiv f_1 + f_3 + f_5 + \cdots + f_{N-3} + f_{N-1}$$
$$S_e(N) \equiv f_2 + f_4 + f_6 + \cdots + f_{N-4} + f_{N-2}$$

where $f_k = f(a + kh)$ and $h = (b-a)/N$. In terms of S_d, S_o, and S_e, (2-15) may be written in the form

$$I_{[a,b]}(N) = \tfrac{1}{3}h\bigl(S_d(N) + 2S_e(N) + 4S_o(N)\bigr) \qquad (2\text{-}16)$$

We shall see that the value of the integral for a new N may be obtained in terms of S_d, S_o, and S_e with a minimum of calculations.

To make the calculation into an iterative one, we shall start with $N = N_1$, where N_1 is some small but reasonable number of intervals. For example, a value like 6 is adequate if the function $f(x)$ is smooth in the interval $[a, b]$. This constitutes the first step in our iterative calculation. For the next iteration we shall double the number of subintervals and halve the step size h by taking $N = N_2$ with $N_2 = 2N_1$. The advantage of changing the number of subintervals in this way is that, for the new N value, the contributions from the two end points, stored as S_d, are unchanged, and all the even grid points are exactly the ones already calculated in the previous iteration. That is, for the new iteration, the value of S_e

> **Box 2-2   Program SIMPSON**
> **Simpson's rule for numerical integration**
>
> Initialization:
>   (a) Set ϵ as the tolerance for error.
>   (b) Set a maximum for the number of iterations allowed.
>   (c) Define the integrand as a function.
>   (d) Set N equal to the initial number of points.
>   (e) Zero the iteration counter.
> 1. Input a as the lower limit of integration and b as the upper limit.
>   Skip the calculation if $a \geq b$.
> 2. Make an initial calculation of the integral with N points:
>   (a) Calculate the contributions from end points.
>   (b) Sum over the contributions from odd points.
>   (c) Sum over the contributions from even points.
>   (d) Calculate the value of the integral from these three terms according to (2-16).
> 3. Double the number of points and calculate the value of the integral again.
>   (a) Sum the contributions from previous even and odd points as the new even points.
>   (b) Calculate the contribution from the new odd points.
>   (c) Calculate the value of the integral from the contributions of the end points, the new even points, and the new odd points using (2-16).
>   (d) Increase the iteration counter by 1.
> 4. Compare the new result with the previous one.
>   (a) If the difference is greater than ϵ and if the number of iterations is less than the maximum allowed, go back to step 3.
>   (b) If the difference is smaller or equal to ϵ, return the result.

is the sum of S_o and S_e of the previous iteration. The only calculations required are the values of the function at the (new) odd grid points. Thus, we have the result

$$S_e(N_2) = S_e(N_1) + S_o(N_1) \qquad S_o(N_2) = \sum_{i=1}^{N_2/2} f(a + (2i-1)h)$$

where $h = (b-a)/N_2$. We can now apply (2-16) to obtain the new value of $I_{[a,b]}$ for $N = N_2$. If the new result differs from that obtained in the previous iteration by an amount less than ϵ, we shall assume that the calculation has converged. Otherwise, the number of intervals is doubled again by making $N = N_3$ with $N_3 = 2N_2$. The process is repeated until convergence is achieved or a maximum for the number of iterations is reached. An outline of the algorithm is given in Box 2-2.

A simple example will serve to illustrate the power of Simpson's rule. Consider the integral

$$I = \int_0^{\pi/2} \sin x \, dx$$

The result is well known: $\cos(0) = 1$. To obtain the value numerically, we shall start with $N = 6$. Besides the two end points at 0 and $\pi/2$, this gives us three odd points at $x_i = \pi/12$, $\pi/4$, and $5\pi/12$ and two even points at $x_i = \pi/6$ and $\pi/3$. In the first iteration, the integrand must be evaluated at these seven points. The size of the integration interval h is given in Table 2-2 together with the value of the integral obtained.

Table 2-2 Example of the accuracy of Simpson's rule for different N.

N	$\int_0^{\pi/2} \sin x\, dx$	h	$h^4/60$
6	1.0000262	0.262	8×10^{-5}
12	1.0000015	0.131	5×10^{-6}
24	0.9999999	0.065	3×10^{-7}
48	0.9999999	0.033	2×10^{-8}
exact value	1.0000000	—	—

For this simple example, we can also use (2-12) to estimate the error expected for a given N. In a given subinterval the first nonvanishing term beyond the second-order derivatives already included in the Simpson's rule is

$$\Delta_i(N) = \frac{1}{4!} f^{(4)}(x_i) \times \frac{2}{5} h^5$$

For the integral as a whole, we must multiply the result by N,

$$\Delta(N) = \frac{1}{4!} f^{(4)}(\xi) \times \frac{2}{5} h^4 (b - a)$$

where $f^{(4)}(\xi)$ is some average value of the fourth-order derivative of $f(x)$ in $[a, b]$. Since the even-order derivatives of a sine function are also sine functions, we obtain the condition $|f^{(4)}(x_i)| \leq 1$. For the example in hand, $(b - a) = \pi/2$. We can therefore take $(b - a) f^{(4)}(\xi)$ to be on the order of unity for our error estimate. This gives us the result

$$\Delta(N) \approx \frac{h^4}{60}$$

The value is listed in the last column of Table 2-2 for different size h used.

The results of the numerical integration are listed in the second column. When these values are compared with the known exact value of unity, we find that our error estimates for this case are of the correct orders of magnitude. If we double the number of subintervals and take $N = 12$, we expect an accuracy of one part in a million. The tabulated values also show that, for a single precision calculation, there is no point to go beyond $N = 24$, as the size of the error associated with the numerical method becomes comparable with the truncation errors of the computer.

There are several variants of Simpson's rule. For example, it is possible to change the relative weight of the contributions from different grid points and achieve an even higher-order accuracy than that of (2-15). Efficiency can also be improved by using extrapolation techniques, as we shall see later in §3-5 in the discussion of the Romberg method of integration.

§2-4 Gaussian quadrature

The trapezoidal and Simpson's rules discussed in the previous two sections may be regarded as methods that approximate the integrand by second- and third-order polynomials, respectively, in each of the subintervals of h in width. The idea of a polynomial approximation may also be applied to the interval $[a, b]$ as a whole. This leads us to the idea behind Gaussian quadrature, the method of Gauss for numerical integration.

There are several different forms of this method. We shall describe here a representative one, the Gauss-Legendre integration. In general, a function $f(x)$ in a given interval $[a, b]$ may be expanded in terms of a complete set of polynomials $P_\ell(x)$,

$$f(x) = \sum_{\ell=0}^{n} \alpha_\ell P_\ell(x) \qquad (2\text{-}17)$$

The definition of the expansion coefficients α_ℓ depends on the polynomial $P_\ell(x)$ used and we shall see examples later. For convenience, we shall use the subscript ℓ to indicate the order of each polynomial, meaning the highest power of the argument x in $P_\ell(x)$. The order of $f(x)$ is then given by n, the upper limit of the summation in (2-17).

In general, it is more convenient to use a set of orthogonal polynomials to expand a function. In principle, any orthogonal polynomial can be used for making the expansion of (2-17). The particular method we shall discuss below makes use of Legendre polynomials, given later in §4-2. The orthogonal relation between these polynomials is given by

$$\int_{-1}^{+1} P_k(x) P_\ell(x)\, dx = \frac{2}{2\ell + 1} \delta_{k,\ell} \qquad (2\text{-}18)$$

Since Legendre polynomials are defined only in the interval $[-1, +1]$, it is necessary for us to change the interval of integration for an arbitrary integral from $[a, b]$ to $[-1, +1]$ by making the substitution

$$x \longrightarrow \frac{b-a}{2} x + \frac{b+a}{2} \qquad (2\text{-}19)$$

The integral given, for example, by (2-1) is now written in the form

$$I_{[a,b]} = \int_a^b f(x)\, dx \longrightarrow \frac{b-a}{2} \int_{-1}^{+1} f\left(\tfrac{1}{2}(b-a)x + \tfrac{1}{2}(b+a)\right) dx$$

For this reason, it is possible to limit our discussions below to the integral

$$I_{[-1,+1]} = \int_{-1}^{+1} f(x)\, dx \tag{2-20}$$

There is no loss of generality in restricting the integration limits to ± 1.

Let us consider the case that the integrand may be approximated by a function of polynomial order $(2n-1)$,

$$f(x) \approx p_{2n-1}(x) \tag{2-21}$$

In other words, if we expand $f(x)$ in terms of Legendre polynomials using the form given by (2-17), it is sufficient to use $P_\ell(x)$ up to order $\ell = (2n-1)$. If this is a good approximation, it is possible to express the integral of $f(x)$ in term of $2n$ parameters, in a way similar to (2-8) for a first-order polynomial approximation of $f(x)$ and to (2-14) for a second-order polynomial approximation. Gauss's method provides us with a way to find the most accurate approximation for the integral $I_{[-1,+1]}$ in term of $2n$ parameters. In other words, we wish to write the integral in the form

$$I_{[-1,+1]} = \sum_{i=1}^{n} \omega_i f(x_i) \tag{2-22}$$

where both x_i, the abscissas, and ω_i, the weight factors, are to be determined in such a way as to get the best accuracy possible for the amount of calculations involved. As an example, the integral

$$\int_{-1}^{+1} e^x\, dx$$

using a fourth-order Gauss-Legendre quadrature is shown in Fig. 2-4.

Abscissa and weight factors. The method of Gaussian quadrature provides us with a general approach to select the abscissa and weight factors required for (2-22). To simplify the discussion, we shall assume for the time being that the integrand is a polynomial of order $(2n-1)$; that is, (2-21) is an exact identity rather than an approximation. By polynomial division, we can decompose $p_{2n-1}(x)$ into a sum of two terms,

$$p_{2n-1}(x) = p_{n-1}(x) P_n(x) + q_{n-1}(x) \tag{2-23}$$

where $P_n(x)$ is the Legendre polynomial of order n. The other two quantities on the right side, $p_{n-1}(x)$ and $q_{n-1}(x)$, are polynomials of order $(n-1)$.

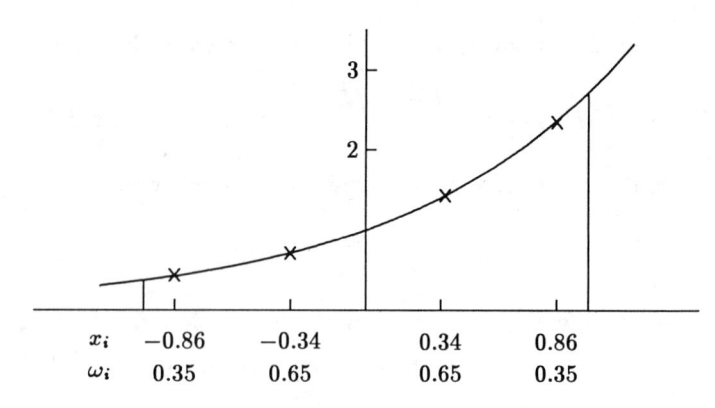

Fig. 2-4 Fourth-order Gauss-Legendre integration for $\int_{-1}^{1} \exp x \, dx$. Only four values of the integrand at $x_i = \pm 0.3399810$ and ± 0.8611363 are required to achieve an accuracy of seven significant figures.

The first term on the right side of (2-23) vanishes on integrating from -1 to $+1$. This arises from the orthogonal property of the polynomials in the following way. By construction, $p_{n-1}(x)$ is a polynomial of order $(n-1)$ and, as a result, may be expanded in terms of Legendre polynomial up to order $(n-1)$,

$$p_{n-1}(x) = \sum_{k=0}^{n-1} a_k P_k(x) \tag{2-24}$$

Since the expression does not contain $P_n(x)$, we obtain the result

$$\int_{-1}^{+1} p_{n-1}(x) P_n(x) \, dx = \sum_{k=0}^{n-1} \int_{-1}^{+1} a_k P_k(x) P_n(x) \, dx = 0$$

where we have interchanged the order of integration and summation and made use of the orthogonal property of $P_\ell(x)$ given in (2-18). From this, we obtain the equality

$$\int_{-1}^{+1} p_{2n-1}(x) \, dx = \int_{-1}^{+1} q_{n-1}(x) \, dx$$

Furthermore, as a polynomial of order n, the function $P_n(x)$ has n zeros in the interval $[-1, +1]$. Let these zero points be at $x = x_1, x_2, \ldots, x_n$. At these points, we have the relation

$$p_{2n-1}(x_i) = q_{n-1}(x_i) \tag{2-25}$$

as can be seen from (2-23).

We can now expand the polynomial $q_{n-1}(x)$ also in terms of Legendre polynomials in a similar fashion as we have done in (2-24):

$$q_{n-1}(x) = \sum_{k=0}^{n-1} \omega_k P_k(x)$$

where ω_k is the weight for polynomial $P_k(x)$. Combining this result with (2-25), we obtain the relation

$$p_{2n-1}(x_i) = \sum_{k=0}^{n-1} \omega_k P_k(x_i)$$

at the roots of $P_n(x)$. This may be written in a matrix notation

$$p_i = \sum_{k=0}^{n-1} P_{ik} \omega_k \qquad (2\text{-}26)$$

where we have used the symbols

$$p_i \equiv p_{2n-1}(x_i) \qquad\qquad P_{ik} \equiv P_k(x_i)$$

Note that $P_k(x_i)$ are the values of Legendre polynomials of order up to $k = (n-1)$ evaluated at the zeros of $P_n(x)$, and the values of x_i are independent of the polynomial $p_{2n-1}(x)$.

The relation given by (2-26) may be inverted to find the value of ω_k in terms of p_i:

$$\omega_k = \sum_{i=1}^{n} p_i \{\boldsymbol{P}^{-1}\}_{ik} \qquad (2\text{-}27)$$

where $\boldsymbol{P}^{-1}$ is the inverse of the matrix $\{P_{ik}\}$:

$$\sum_k \{\boldsymbol{P}^{-1}\}_{ik} P_{kj} = \delta_{i,j}$$

We shall see next how this expansion can help us to evaluate the integral of $f(x)$.

Since we are assuming that (2-21) is exact, we have the result

$$I_{[-1,+1]} = \int_{-1}^{+1} f(x)\,dx = \int_{-1}^{+1} p_{2n-1}(x)\,dx$$

$$= \sum_{k=0}^{n-1} \omega_k \int_{-1}^{+1} P_k(x)\,dx \qquad (2\text{-}28)$$

For $k = 0$, we have $P_0(x) = 1$, and the only nonvanishing integral among the n terms in the sum is for $k = 0$:

$$\int_{-1}^{+1} P_k(x)\,dx = \frac{2}{2k+1}\delta_{k,0} = 2$$

arising from the orthogonal property of the Legendre polynomials. The final result of (2-28) is then

$$I_{[-1,+1]} = 2\omega_0 = 2\sum_{k=1}^{n} p_k \{\boldsymbol{P}^{-1}\}_{k0} \qquad (2\text{-}29)$$

Table 2-3 Abscissa and weight factors for Gauss-Legendre integration.

Order	x_i	ω_i
$n = 4$	±0.339981043584856	0.652145154862546
	±0.861136311594053	0.347854845137454
$n = 5$	0.000000000000000	0.568888888888889
	±0.538469310105683	0.478628670499366
	±0.906179845938664	0.236926885056189
$n = 12$	±0.125233408511469	0.249147045813403
	±0.367831498998180	0.233492536538355
	±0.587317954286617	0.203167426723066
	±0.769902674194305	0.160078328543346
	±0.904117256370475	0.106939325995318
	±0.981560634246719	0.047175336386512

The last equality comes from (2-27).

For the purpose of numerical integration, (2-29) may be written in a more familiar form. Recalling that

$$p_i \equiv p_{2n-1}(x_i) \approx f(x_i)$$

the integral of $f(x)$ reduces to the form of (2-22)

$$I_{[-1,+1]} = \sum_{i=1}^{n} \omega_i f(x_i)$$

where ω_i is the weight of $f(x_i)$ at the zeros of $P_n(x)$ and is given by

$$\omega_i = 2\{\boldsymbol{P}^{-1}\}_{i0}$$

We shall not be concerned here with the method to invert the polynomial matrix $\boldsymbol{P} = \{P_k(x_i)\}$, as it will be discussed later in §5-2. The result is known and may be put in the form

$$\omega_i = \frac{2}{(1-x_i^2)[P_n'(x_i)]^2} \qquad (2\text{-}30)$$

where

$$P_n'(x_i) = \frac{d}{dx} P_n(x)\bigg|_{x=x_i}$$

The values of $P_n'(x_i)$ may be calculated using the methods given in §4-2. Extensive tables of both x_i and ω_i for Gaussian quadrature of different orders may be found in Abramowitz and Segun. Examples for orders $n = 4$, 5, and 12 are given in Table 2-3 for illustration.

Table 2-4 Example of Gauss-Legendre integration for $\int_{-1}^{+1} \exp(x)dx$.

Order	x_i	ω_i	$\exp(x_i)$	$\omega_i \exp(x_i)$
$n = 4$	−0.8611363	0.3478549	0.4226815	0.1470318
	−0.3399810	0.6521451	0.7117838	0.4641864
	0.3399810	0.6521451	1.4049209	0.9162124
	0.8611363	0.3478549	2.3658476	0.8229715
			$\sum_i \omega_i \exp(x_i) =$	2.3504021
$n = 5$	−0.9061798	0.2369269	0.4040649	0.0957338
	−0.5384693	0.4786287	0.5836409	0.2793473
	0.0000000	0.5688889	1.0000000	0.5688889
	0.5384693	0.4786287	1.7133822	0.8200738
	0.9061798	0.2369269	2.4748502	0.5863585
			$\sum_i \omega_i \exp(x_i) =$	2.3504023
			Exact value =	2.350402387

Example of Gaussian-Legendre integration. As an example of Gauss-Legendre integration, let us carry out the integral

$$I = \int_{-1}^{+1} e^x \, dx \tag{2-31}$$

The values of the various quantities that enter into the calculation are summarized in Table 2-4. The abscissas and weight factors are taken from Table 2-3, and the values of the integrand at each of the points x_i are given in columns 2 and 3 of the table. Compared with the exact value of

$$\int_{-1}^{+1} e^x \, dx = e^1 - e^{-1} = 2.350402387$$

we find that, even for $n = 4$ approximation, the result obtained using a Gauss-Legendre integration is accurate up to seven significant figures.

In principle, the exponential function is an infinite series in terms of its argument x,

$$e^x = 1 + \frac{x}{1!} + \frac{x^2}{2!} + \frac{x^3}{3!} + \cdots$$

When we use $n = 4$ for the Gauss-Legendre integration, we are approximating the function by a series up to order $(2n - 1) = 7$. In this limit, terms involving powers higher than x^8 are ignored. As a result, the accuracy may be improved by using instead order $n = 5$, whereby terms up to order x^9 are also included. As we can see from the second part of Table 2-4, the calculated value is now 2.3504023,

---

Box 2-3   Program Gauss_Legendre

Gauss-Legendre quadrature for $\int_{-a}^{+a} e^x dx$

Initialization:
  (a) Convert the integral to the integration interval $[-1, +1]$.
  (b) Input the order, abscissas, and weights.
1. Input the value of a. Stop if $a \leq 0$.
2. Calculate the contribution to the integral from each point:
  (a) Calculate the value of the integrand at each point.
  (b) Taking the product with the weight factor.
  (c) Multiply the result by a as required in (2-32).
  (d) Sum the contributions.
3. Output the sum.

---

essentially the same as the best value that can be achieved with a single precision calculation.

To carry out the same integral for limits other than ±1, we can make the following change of variable:

$$\int_{-a}^{+a} e^x dx = a \int_{-1}^{+1} e^{ay} dy \tag{2-32}$$

The integral now takes on the form of (2-31) and may therefore be carried out in the same way. The algorithm is outlined in Box 2-3. As expected, the accuracy of the calculation deteriorates as a increases beyond 1.

An interesting exercise is to carry out the integral $\int_{-1}^{+1} x^8 dx$. With $n = 4$, the result is 0.21061227, a 5% difference compared with the exact value of 2/9. The reason for the relatively large error is quite obvious — the method approximates the integrand as a polynomial of order $(2n - 1) = 7$. On the other hand, if we use the $n = 5$ Gauss-Legendre integration, a result of 0.22222221 is obtained, essentially identical to the exact value in single precision. This illustrates an important limitation of Gaussian quadrature: the order of the polynomial required to achieve a given accuracy depends very much on the nature of the integrand. In general, it is not easy to know what is the highest-order polynomial approximation required for a given function and accuracy. Furthermore, the method itself does not provide us with a natural way to estimate the error. As a result, it is not possible for us to know the reliability of the value obtained without carrying out some independent checks.

§2-5 Monte Carlo integration

All the methods of numerical integration we have discussed so far involve the evaluation of the integrand $f(x)$ at definite points along the abscissa. For rectangular, trapezoidal, and Simpson's rules, $f(x)$ are calculated at $(N+1)$ evenly spaced points, $x_0, x_1, \ldots, x_N$, along the x-axis in the interval $[a, b]$. In the case of Gauss-Legendre integration, the points x_i are chosen to be the roots of Legendre polynomials of a given order in the interval $[-1, +1]$. In this section, we shall take a different philosophy and use sampling techniques to choose the points at which the values of $f(x)$ are evaluated. The advantage of this approach is the convenience it offers for multidimensional integrals and certain types of improper integrals to which we shall return in the next section. For one-dimensional integrals with well-behaved integrands, which we shall continue to use as the example in this section, the method is not as efficient as those we have seen so far.

For the discussions in this section, it is convenient to take the integration interval to be $[0, 1]$. The choice is connected with the fact that most generators of random numbers with a uniform distribution are designed for this range. For integrations with lower limit a different from 0 or upper limit b different from 1, it is a simple matter to make the transformation

$$x \longrightarrow \frac{1}{b-a}(x-a)$$

so that the actual numerical calculation is carried out in the interval $[0, 1]$. This is similar to that given earlier in (2-19), except there the transformation is for the range of integration to $[-1, +1]$.

The Monte Carlo integration method is based on the idea that, instead of selecting the mesh points ahead of time, we can take a random sample of the possible values of the integration variable x. The integrand is then evaluated at these points. If our sampling is truly random, the points $x_1, x_2, \ldots, x_N$ cover the interval $[0, 1]$ with equal probability. For N number of points, the average distance between two adjacent points is

$$h = 1/(N-1) \xrightarrow[\text{large } N]{} 1/N \qquad (2\text{-}33)$$

The value of the integral is then

$$I_{[0,1]} = \int_0^1 f(x)\, dx = \lim_{N \to \infty} \frac{1}{N} \sum_{i=1}^N f(x_i) \qquad (2\text{-}34)$$

To select the points along the x-axis, we can use a generator for evenly distributed random numbers, such as the ones discussed in §7-1.

If our sampling of the x-axis is truly random, the number of points selected in a given region $[x - \tfrac{1}{2}\Delta x, x + \tfrac{1}{2}\Delta x]$ is independent of the value of x. In other words, if we divide the interval $[0, 1]$ into an arbitrary number of equal-size subintervals, the number of points in each subinterval will be equal to each other within statistical fluctuations. Elementary notions of statistics tell us that this can only be true if we take a large sample, as shown by an example of 5000 random points given in Fig. 2-5.

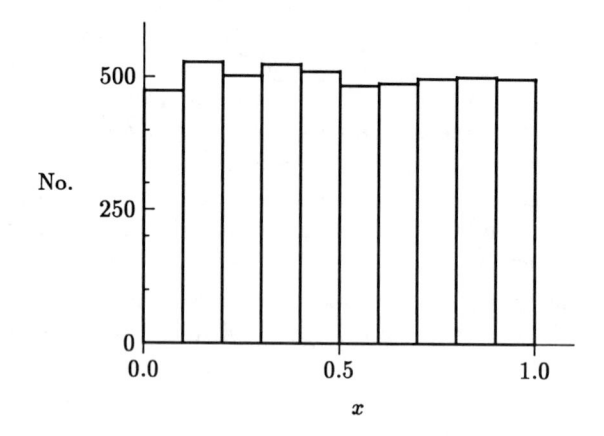

Fig. 2-5 Histogram showing the distribution of 5000 uniform random numbers in the interval $[0,1]$. The subtractive method of Box 7-2 is used to produce the random number.

An iterative approach. One difficulty inherent in Monte Carlo methods is to know when N is large enough for (2-34) to be valid. For our purpose here, we can take a practical approach and use an iterative strategy similar to what we did earlier for Simpson's rule. We start by taking a sample of N_1 points and evaluate the integral using (2-34). The sampling is repeated for another N_1 different points and the integral is evaluated also for these N_1 points. If the two values of the integral obtained with two different sets of N_1 points are each equal within some tolerance ϵ, it is likely that we have a sample that is large enough. To improve the accuracy further, we can sum over the two values obtained and use the average as the final result for the integral. On the other hand, if the difference is too large, we can group both sets of results together as a single one of $N_2 = 2N_1$ points and take another sample of N_2 different points. The results are compared with each other again. The process is repeated until sufficient accuracy is achieved.

In the example for Simpson's rule, we have started with a small number of points ($N = 6$). Since our method here depends on a random sample of evenly distributed points, it is not possible to attain such a distribution with a small N. A cautious person may want to start with a much larger sample, for example, $N_1 \sim$

> **Box 2-4   Program ONE_D**
> **Normal probability function $A(x)$ by Monte Carlo integration**
>
> Subprogram used:
>   RSUB: Subtractive random number generator (Box 7-2)
>
> Initialization:
>   (a) Set ϵ as the maximum error tolerated.
>   (b) Set a limit for the maximum number of iterations.
>   (c) Store the conversion constant $\sqrt{2/\pi}$.
>   (d) Define the integrand $\exp{-t^2/2}$ as a function.
>   (e) Set a starting number of points for the integration and store the value as N.
>   (f) Zero the iteration counter and SUM.
>
> 1. Input:
>   (a) Upper limit of the integral.
>   (b) Seed for the random number generator.
> 2. Calculate the integral with N random points:
>   (a) Generate a uniform random number in $[0,1]$ and use it as the values of t.
>   (b) Calculate the value of the integrand at t and add the value to SUM.
> 3. Convert the value of SUM to the value of the integral using (2-34) and (2-35).
> 4. Repeat steps 2 and 3 with a different set of N random points.
> 5. Compare the two values of the integral obtained.
>   (a) If the difference is larger than ϵ, then:
>   (i) Take the average of the two values.
>   (ii) Double the value of N.
>   (iii) Increase the iteration counter by 1.
>   (iv) Go back to step 4 if the number of iterations is less than the maximum.
>   (b) If the difference is less or equal to ϵ then
>   (i) Take the average of the two values.
>   (ii) Output the result.

100. To see the rate of convergence in a Monte Carlo integration as we increase the sample size, we shall use as an example the normal probability function

$$A(x) = \frac{1}{\sqrt{2\pi}} \int_{-x}^{+x} e^{-t^2/2} dt \tag{2-35}$$

discussed later in §6-1. For simplicity, we shall consider first the case of $x = 1$. Since the integrand is symmetric with respect to $x = 0$, we can change the integral into the form

$$A(x=1) = \sqrt{\frac{2}{\pi}} \int_0^1 e^{-t^2/2} dt$$

This makes it possible for us to use a uniformly distributed random number generator in the interval $[0,1]$ without any conversion. The value is equal to the area

Table 2-5 Convergence of a Monte Carlo integration for $A(x) = \frac{2}{\sqrt{\pi}} \int_0^1 e^{-t^2/2} dt$

Iteration	No. of points	$A(x = 1)$
1	40	0.6931776
2	80	0.6801362
3	160	0.6781273
4	320	0.6887547
5	640	0.6889585
6	1,280	0.6852926
7	2,560	0.6829402
8	5,120	0.6838834
9	10,240	0.6835121
10	20,480	0.6828257
11	40,960	0.6829104
12	81,920	0.6824411
13	163,840	0.6826339
14	327,680	0.6826127
15	655,360	0.6827372
Exact value		0.6826895

underneath a normal distribution curve within one standard deviation on either side of the mean and is given in tables of statistics to be

$$A(x = 1) = 2\left(\frac{1}{\sqrt{2\pi}} \int_{-\infty}^{1} e^{-t^2/2} dt - \frac{1}{2}\right) = 0.6826895$$

The results of numerical integration using the Monte Carlo method discussed above are given in Table 2-5 for different numbers of points sampled. For purposes of illustration, we shall start with a small value of $N = 40$ points. In each successive iteration, the number of points is doubled. At the fourteenth iteration, a total of 655,360 points is used to reach a difference of one part in 10^5. The algorithm is outlined in Box 2-4.

For values of $A(x)$ other than $x = 1$, we can make the following substitution, similar to what we did earlier in (2-32):

$$A(a) = \sqrt{\frac{2}{\pi}} \int_0^a e^{-t^2/2} dt = a\sqrt{\frac{2}{\pi}} \int_0^1 e^{-a^2 y^2/2} dy \qquad (2\text{-}36)$$

where $t = ay$. In general, more points are needed to reach the same level of accuracy if a is much larger than 1.

One important premise of Monte Carlo integrations is that we can take a truly random sample of points along the abscissa. In using a computer, we inevitably use some algorithm to generate the random numbers that are taken to be the points where the integrand is evaluated. As we shall see in more detail in §7-1, these are only *pseudo*-random numbers and must be tested to see if they are truly random for the purpose at hand. For the simple one-dimensional case we have illustrated here, the requirement is that they be evenly distributed in the interval [0, 1], as shown earlier in Fig. 2-5. This is usually not difficult to achieve. However, the main interest of Monte Carlo methods is for the more complicated cases, such as multidimensional integrals. In these cases, the random numbers must be uniformly distributed in the multidimensional volume in which the integral is evaluated. In general, it is not as easy to find out whether this criterion is satisfied or not. For this reason, careful tests must be conducted.

In general, there are two time-consuming parts of a Monte Carlo integration, generating the random numbers and evaluating the integrand. In the one-dimensional example we have seen above, most of the time is, in fact, spent in generating the random numbers. As a result, the method is not efficient, compared with methods that are based on a fixed grid of points. For this reason, Monte Carlo integration methods are favored only in cases where other methods, such as those described in earlier sections, are unsuitable. We shall see examples of such integrals in the next section.

§2-6 Multidimensional integrals and improper integrals

In the previous sections, we have used one-dimensional integrals to illustrate the principles behind numerical integration. The examples employed are usually those where the integral can also be carried out analytically. This is done mainly for the reason that we may have an exact result to compare with. In practice, such integrals are not the main concern of numerical methods. Our interest is primarily in cases where analytical solutions are too difficult or impossible to obtain. This happens frequently in the case of multidimensional integrals and improper integrals. Because of the large variety of such integrals, it is not possible to discuss each type in detail. Instead, a general discussion is given to extend the basic methods of the previous sections to situations commonly encountered in practice for multidimensional and improper integrals.

Multidimensional integrals. All the methods we have treated so far can be generalized, at least in principle, to integrations over several variables. Let us illustrate this point by considering a two-dimensional integral of the form

$$I = \int_c^d \int_a^b f(x,y)\,dx\,dy \tag{2-37}$$

One way to carry out this integration is to divide the interval $[a, b]$ along the x-axis into M small subintervals of width h each. Similarly, the interval $[c, d]$ along the y-axis may be divided into N subdivisions of width k each. Instead of summing over a large number of small areas, as in the one-dimensional case, we are now summing over a number of small volumes. This is shown schematically in Fig. 2-6. In general, we have $h \neq k$. Furthermore, unless the behavior of $f(x,y)$ is similar along both x- and y-directions, the number of subdivisions in one of the two directions does not have to be equal to that of the other one.

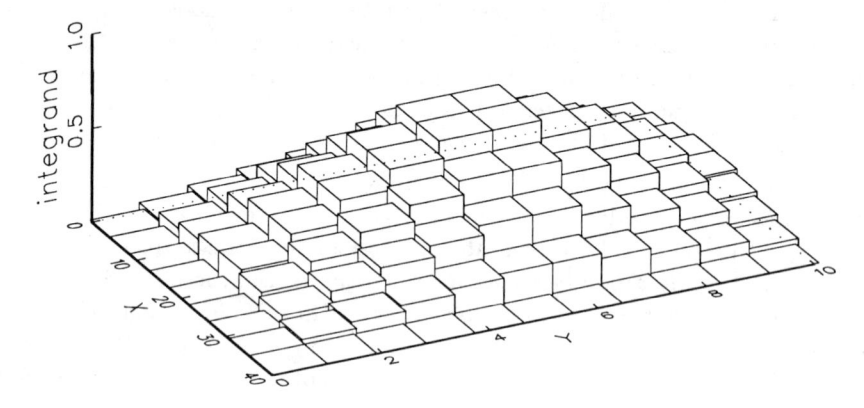

Fig. 2-6 Approximation of an integral of two variables by $(M \times N)$ small columns, each with length h, width k, and height $f(x_i, y_j)$ equal to the average value of the integrand $f(x, y)$ in the rectangular area $(h \times k)$.

The xy-plane is now divided into $(M \times N)$ number of rectangular grids. If the area of each rectangle $(h \times k)$ is sufficiently small, we may approximate the value of $f(x,y)$ in such a rectangular area to be a constant, equal to $f(x_i, y_j)$, the value at the center of the grid. The integral I is then given by the sum of the products of $f(x_i, y_j)$ and the area of each grid,

$$I = \lim_{\substack{M \to \infty \\ N \to \infty}} \sum_{i=1}^{M} \sum_{j=1}^{N} f(x_i, y_j) hk \qquad (2\text{-}38)$$

This is nothing more than an extension of the rectangular rule of §2-2 to the case of two independent variables.

We can improve the accuracy of (2-38) by expanding the value $f(x, y)$ within a rectangular grid in terms of a Taylor series in both x and y. This is similar to what we did earlier in (2-6). If we further incorporate the first (partial) derivatives in both x and y as corrections, in the same way we did in (2-8) for x alone, we obtain the equivalent of the trapezoidal rule for two-dimensional integrals. Similarly, we can also make improvements using second-order derivatives, and an extension of the Simpson's rule is obtained.

To apply the Gauss-Legendre integration to the case of two integration variables, we need a somewhat different approach from the one-dimensional case. One possible way is to carry out the integral in (2-37) in two stages. First, we integrate $f(x,y)$ over x for a given value of y.

$$g(y) = \int_a^b f(x,y)\,dx \qquad (2\text{-}39)$$

Once we have the values of $g(y)$ for all the values of y required in a Gaussian quadrature calculation, we can go to the second stage and carry out the integration over x

$$I = \int_c^d g(y)\,dy \qquad (2\text{-}40)$$

Since, in the numerical integration of (2-40), we need only the values of $g(y)$ at a finite, discrete set of points along the y-axis, there is no difficulty in applying the Gauss-Legendre integration of (2-22) to the integral in (2-39) for a given value of y. However, the calculation must be carried out once for each value of y required in (2-40). Once this is done for all the grid points along the y-axis, we can evaluate I by integrating $g(y)$ over y. Again, the integral is over a single variable and (2-22) may be used. For example, if we decide to use a $(2n-1)$-degree polynomial approximation for $g(y)$, we need only the values of $g(y)$ evaluated at $y_1, y_2, \ldots, y_n$, the zeros of the nth-order Legendre polynomial. For each $g(y_i)$, we need to evaluate once the integral on the right side of (2-39), which, in turn, may be approximated with a polynomial of degree $(2m-1)$.

Monte Carlo methods. In any of the methods described above, the amount of computation required in a two-dimensional integral is proportional to the number of mesh points used. For (2-38), this number is $(M \times N)$, where M is the number of subintervals along the x-axis and N is the corresponding number along the y-axis. To simplify the discussion, let us take $M = N$. In this case, the amount of computation for a two-dimensional integral is proportional to N^2. More generally, for integrals with n integration variables, the amount is proportional to N^n, a number that grows very rapidly with n. Regardless of minor improvements one can make, methods that are based on a fixed mesh become impractical very quickly if the number of integration variables is large.

For multidimensional integrals, Monte Carlo methods may be more economical. Consider as an example the case of a three-dimensional integral:

$$I = \int_e^f \int_c^d \int_a^b f(x,y,z)\,dx\,dy\,dz \qquad (2\text{-}41)$$

Again, we can replace the integral by a sum over the products of values of the integrand in a cell and the volume of the cell:

$$I \approx \sum_{i=1}^N f(x_i, y_i, z_i)\,\Delta v \qquad (2\text{-}42)$$

where N is the number of points in the three-dimensional space where the integrand is evaluated and

$$\Delta v = \frac{(b-a)(d-c)(f-e)}{N}$$

is the average volume of each cell in the three-dimensional space.

Instead of evaluating the integrand in the three-dimensional space at a fixed set of mesh points, we can take a random sample in the cube $[a,b] \times [c,d] \times [e,f]$. As a practical matter, the location of each such point in the cube is represented by three random numbers, x_i, y_i, and z_i. As we saw in the previous section, if the total number of points sampled is large enough, and our random numbers are not biased in any way, the collection of N random points (x_i, y_i, z_i) will eventually fill the cube uniformly. As a result, we may expect that the approximation of an integral by the sum in (2-42) to be a good one. Except for the number of random numbers required to determine the location of a point, such an approach is essentially independent of the number of integration variables involved.

The number of points required for a Monte Carlo integration depends on the accuracy required and the nature of the integrand. Since most of the integrands encountered in physics are fairly smooth, sufficient accuracy of an integral can often be achieved by a relatively small sample. Furthermore, the same iterative strategy as used in the previous section for a one-dimensional case may also be applied to multidimensional integrals. That is, we make a comparison of the results of two different samples of N points each. If the difference is smaller than the tolerance, the value is accepted. Otherwise, the sample size is doubled. The process is repeated until the desired accuracy is reached. In this way our algorithm naturally includes an estimate of the accuracy of the result as well.

As an example, we shall consider the problem of calculating the total amount of charge Q in a cubic volume of length 2 on each size in some arbitrary units. For the charge distribution inside this volume, we shall take

$$\rho(x,y,z) = \exp(-r)/8\pi$$

where $r = \sqrt{x^2 + y^2 + z^2}$. The form is very close to a Yukawa or screened charge distribution used in a variety of applications. The peak of the distribution is at the center of the cube where $r = 0$. The total amount of charge inside this volume is given by the integral:

$$Q = \int_{-1}^{+1} \int_{-1}^{+1} \int_{-1}^{+1} \frac{1}{8\pi} e^{-r} \, dx \, dy \, dz$$

We shall make use of the symmetry of the problem and evaluate only one-eighth of the space by changing the integral into the form

$$Q = \frac{1}{\pi} \int_0^{+1} \int_0^{+1} \int_0^{+1} e^{-r} \, dx \, dy \, dz$$

This is also convenient for making use of evenly distributed random numbers in the interval $[0,1]$ for the Monte Carlo calculation. The result, $Q \approx 0.4/\pi$, is reached with $N = 8000$ points to an accuracy of 0.0002. Obviously, the efficiency of the method depends a great deal on the amount of time required to generate the random numbers. In fact, since the integrand is a simple function, a relatively larger fraction of the computer time is expanded in generating the random numbers than for the rest of the calculations. The algorithm used is outlined in Box 2-5.

Box 2-5   Program THREE_D

Example of three-dimensional Monte Carlo integration

Finding the total charge in a cubic box with exponential density

Subprogram used:
  RSUB: Subtractive random number generator (Box 7-2)

Initialization:
  (a) Set ϵ as the maximum error to be tolerated.
  (b) Set a limit for the number of iterations.
  (c) Set N as the starting number of points for the integration.
  (d) Zero the iteration counter and SUM.

1. Define the integrand as a function $\rho(x, y, z)$.
2. Input:
   (a) Value of a, the scale factor for the charge distribution.
   (b) Seed for the random number generator.
3. Calculate the integral with N random points:
   (a) Generate three uniform random numbers in $[0, 1]$ and use them as the values of x, y, and z.
   (b) Calculate the integrand at (x, y, z) and add the value to SUM.
4. Convert the value of SUM to the value of the integral using (2-42).
5. Repeat steps 3 and 4 with a different set of N random points.
6. Compare the two values of the integral obtained.
   (a) If the difference is larger than ϵ, then:
       (i) Take the average of the two values.
       (ii) Double the value of N.
       (iii) Increase the iteration counter by 1.
       (iv) Go back to step 5 if the number of iterations is less than the maximum.
   (b) If the difference is less or equal to ϵ, then:
       (i) Take the average of the two values.
       (ii) Output the result.

As a side interest, we can use the following calculation to check if the computer program to find the charge in the cubic volume is correct. For this purpose, we see that the method may be extended to arbitrary integration limits by applying transformation similar to that used in (2-36). The number of samplings required to reach the same degree of accuracy is, however, proportional to the volume involved in the integration. On the other hand, the charge density $\rho(x,y,z)$ drops off very quickly with distance from the center of the cube. As a result, very little contributions to the integral can come from regions where the absolute values of x, y, and z are large. For this reason, we shall take the value 10 as the "infinity" in this problem. In other words, we shall carry out the integration only in the interval $[-10, +10]$ for each variable. The result may be compared with the exact value, obtained by the following calculation:

$$\frac{1}{8\pi}\int_{-\infty}^{+\infty}\int_{-\infty}^{+\infty}\int_{-\infty}^{+\infty} e^{-r}\,dxdydz = \frac{1}{8\pi}\int_0^{+\infty}\int_0^{\pi}\int_0^{2\pi} e^{-r}r^2\sin\theta\,d\phi d\theta dr$$

$$= \frac{1}{2}\int_0^{+\infty} e^{-r}r^2\,dr$$

$$= 1$$

To reach the same accuracy of 10^{-3} as we have done earlier, the number of samplings must be increased by a factor of 10^3!

Improper integrals. In calculus, an *improper integral* is defined as one with either the range of integration being infinite or the integrand containing infinities within the integration range. Obviously, the singularities must be of the form that they do not make contributions so that the integral itself becomes infinite. For example, the integrand of the integral

$$I = \int_0^1 \frac{dx}{\sqrt{1-x^2}} \tag{2-43}$$

has a singularity at the upper limit. However, the integral itself exists. Since

$$\int \frac{dx}{\sqrt{1-x^2}} = \sin^{-1} x$$

we have the result $I = \pi/2$. On the other hand, the integral

$$I(y) = \int_0^y \frac{dx}{x} \tag{2-44}$$

does not exist (meaning, it does not have a finite value), since

$$\int \frac{dx}{x} = \ln(x)$$

At the lower limit of 0, the integral $I(y)$ is undefined, as $\ln(0) = -\infty$. We shall call the type of singularity at $x = 0$ in the integrand of (2-43) an *integrable singularity* and the corresponding one in (2-44) a *nonintegrable singularity*.

The use of numerical methods to evaluate improper integrals has certain obvious advantages over analytical methods. However, if an integral does not have a finite value, numerical techniques must also fail. We shall not go into the conditions under which an integral exists. Our interest here is only in the problem of how to find the values of integrals that do not have nonintegrable singularities.

Special problems of numerical integration. Before discussing some of the numerical methods for carrying out improper integrals, it is worthwhile to take note of some of the special problems in numerical calculations involving infinities. For example, it is well known that

$$\frac{\sin x}{x} \xrightarrow[x \to 0]{} 1$$

If we wish to evaluate the ratio of $\sin x$ over x for a range of x values, the normal procedure is to calculate $\sin(x)$ and x separately and then take the quotient. This method fails at the singular point, since the denominator is zero. The problem is a nuisance but not a fundamental difficulty. All that is required is to check whether the denominator is zero before the division. In fact, it is good programming practice to anticipate for the possibilities of overflows and install the equivalent of L'Hospital rule in the program where necessary. For our example, we can avoid any overflow by putting in the condition that if $|x| \leq \epsilon$ let $\sin(x)/x = 1$. The choice for the value of some small, positive number ϵ as the practical criterion for "zero" depends on the computer used and accuracy required in the calculation.

A second difficulty in numerical integration occurs in *infinite integrals* where one or both integration limits are at infinity. In mathematics, infinity is a quantity that is larger than any finite value. We have seen in the Chapter 1 that the range of values that can be stored in a computer word is limited. As a result, it may be tempting to define infinity as the absolute value of the largest number that can be represented by a computer word. This is not very useful for our purpose here. A more practical way is based on the fact that we can only calculate an integral to some finite accuracy ϵ. An infinity in the integration limits for any integral that is finite may therefore be taken as the value beyond which the sum of contributions to the integral is much less than ϵ.

Consider the following integral:

$$I(b) = \int_0^b e^{-x^2} dx$$

If the upper limit b is at infinity, it is an infinite integral. The exact value for $I(b = \infty)$ is $\sqrt{\pi}/2 = 0.886227$. On the other hand, we find that at $b = 3$ the value of the integral is already 0.886207. Thus the value 3 may be taken as infinity for this integral if we are willing to accept an error of $\epsilon = 10^{-4}$. On the other hand, for the integral

$$I = \int_0^\infty \frac{\sin x}{x} dx$$

the value of the integrand fluctuates rapidly between positive and negative values for large x. As a result, we need a value of the upper limit on the order $x \gg 1/\epsilon$ before we can obtain a result by numerical integration that is close to the exact value of $\pi/2$.

A third difficulty with infinite integrals is that some methods, such as the Gauss-Legendre method, may not work. The root of the problem lies with the fact that the polynomials are defined only in the interval $[-1, +1]$. We can, in principle, transform the integration variable such that the domain coincides with $[-1, +1]$, as we have done in (2-19). However, the accuracy may turn out to be rather poor. The proper alternative is to use polynomials defined in the appropriate intervals. For example, for the interval $[0, \infty]$, we can use the Laguerre integration method in which the integral is approximated by the relation

$$\int_0^\infty f(x)\,dx = \sum_{i=1}^n \omega_i\, e^{x_i} g(x_i) \qquad (2\text{-}45)$$

in a similar way as in (2-22) for the Gauss-Legendre method. Here, instead of Legendre polynomials, x_i are the zeros of Laguerre polynomials (see §4-4). It is also possible to make use of the Hermite integration method

$$\int_{-\infty}^\infty f(x)\,dx = \sum_{i=1}^n \omega_i\, e^{x_i^2} g(x_i) \qquad (2\text{-}46)$$

where x_i are the zeros of Hermite polynomials (see §4-1). The values of the abscissas and the weight factors for different degrees of both types of quadrature are given in Abramowitz and Segun.

Numerical methods for improper integrals. Let us now turn our attention to integrals containing integrable singularities inside the domain. Most of the numerical methods we have discussed so far may be used also for this type of integral, provided some minor modifications are made to avoid the singular points. Since singularities cause overflows in the calculation, it is essential that they do not coincide with any of the mesh points. For most of the methods, it is necessary to know ahead of time the locations of the singular point and avoid them in defining the mesh. The exception is the Monte Carlo integration method. The advantage comes from the fact that, in a random sample of the integration domain, the probability of encountering one of the singularities is small. Furthermore, in the event that such a singular point does occur in the sampling, we can sacrifice a little of the accuracy by discarding the point. In this way, it is possible to essentially ignore the existence of such singularities in the calculation. For this reason the Monte Carlo method is the most convenient one to use for improper integrals.

As an example, let us consider again the integral

$$I(b) = \int_0^b \frac{dx}{\sqrt{1-x^2}}\,dx \qquad (2\text{-}47)$$

The integrable singularity occurs at $x = 1$. To carry out a Monte Carlo calculation for this integral, we shall assume that our random numbers are distributed uniformly in the interval $[0, b]$. Unless we obtain a random number $x_i = 1$ (that is, x differs from 1 by some small number ϵ), the existence of the singularity does not give us any difficulty in practice. If the value of one of the randomly selected x_i is essentially 1, the random number is discarded and another is selected to take its place. Since it is only one member of a large sample, the accuracy of the final result is hardly affected by such an approximation.

In some cases, an algorithm based on a fixed mesh of points may be preferred. For our discussion here, let us consider the trapezoidal rule and use the integral of (2-47) for $b = 1$ as our prototype. We cannot use (2-9) here because f_0, the value of the integrand at the upper limit of the integral, is infinite. Since the integral itself exists, the fault must be with our approach. It is not hard to design an alternative algorithm such that the value of the integrand at the upper limit of the integration is not needed. This leads us to the second Euler-Maclaurin summation formula,

$$\int_a^b f(x)\,dx = h\Big\{f_{1/2} + f_{3/2} + f_{5/2} + \cdots + f_{N-3/2} + f_{N-1/2}\Big\}$$
$$+ \frac{B_2}{4}h^2\{f'_N - f'_0\} + \cdots + \frac{B_{2k}}{(2k)!}h^{2k}\{1 - 2^{-2k+1}\}\{f_N^{(2k-1)} - f_0^{(2k-1)}\} - \cdots$$
(2-48)

This result may be obtained from (2-10) by subtracting from it twice that of a similar expression except that the step size is $h/2$. In contrast with the trapezoidal rule given by (2-9), we have here a method to approximate an integral by the sum of the values of the integrand in the middle of each subinterval. One small point of caution must be observed in applying (2-48). Since the formula involves middle points (half-integer grid points), we cannot extend the method to an iterative one by simply doubling the number of intervals as we did earlier. In each iteration, the values calculated in the previous iteration are those for the new integer grid points, rather than at half-integer points required by the method. It is easy to get around this point — instead of doubling the number of grid points in each successive iteration, we can triple the number (or take any other odd multiples).

In general, it is not advisable to use any of the Gaussian quadrature formulas for integrals with singular values in the integrand. One of the premises of Gauss's approach is that the function can be approximated by a polynomial of finite order, and this, in turn, implies that the function is fairly smooth in the interval $[a, b]$. Functions with singularities, except perhaps those occurring at the ends of the interval, are usually not smooth, and it may not be appropriate to approximate them by finite polynomial series.

§2-7 Numerical differentiation

The derivative of a function $f(x)$ at a point x is defined in terms of limiting values. Let x and $(x+h)$ be two nearby points separated by some small distance h and the values of the function at these two points be represented, respectively, as $f(x)$ and $f(x+h)$. In the limit that h goes to zero, the derivative of $f(x)$ at x is defined as

$$f'(x) = \lim_{h \to 0} \frac{f(x+h) - f(x)}{h} \tag{2-49}$$

Graphically, we can think of $f'(x)$ as the slope, or tangent, of $f(x)$ at x. In general, we do not have any difficulties in differentiating the functions normally encountered in physics. As a result, there has not been the need to make use of elaborate numerical methods to calculate derivatives. A concise account of some of the necessary cautions may be found in "Numerical Calculation of Derivatives" by Press and Teukolsky in the January/February 1991 issue of *Computers in Physics*.

Numerical differentiation becomes important in recent years as a result of the increased reliance on computers to solve problems. For example, if a function exists only as tabulated values at discrete intervals along the abscissa, such as data taken by instruments for different input values, the derivatives of the function must be calculated numerically. A more important role of derivatives in numerical calculations is in solving differential equations, to which we shall come in Chapter 8. Derivatives are also the convenient vehicle to introduce certain basic concepts in numerical techniques in general, and we shall carry out some of these discussions in this section.

Since derivatives of a function exist only as limiting values, such as that given by (2-49), it is usually impossible in numerical work to deal with them directly. Instead, we use *finite differences*, meaning small differences in the function at nearby points. A straightforward generalization of (2-49) is to make use of the difference

$$\Delta f(x) \equiv f(x+h) - f(x) \tag{2-50}$$

and equate $f'(x)$ with the quotient $\Delta f(x)/h$ for finite h. The quantity $\Delta f(x)$ is known as the *forward difference* of the function $f(x)$ at x, since it is the difference of $f(x)$ at x and a point ahead of by a distance h. This is not the only way to define the value of the difference of a function at two adjacent points. Two other differences can also be used,

$$\nabla f(x) = f(x) - f(x-h) \tag{2-51}$$

$$\delta f(x) = f(x+h/2) - f(x-h/2) \tag{2-52}$$

The former, $\nabla f(x)$, is known as the *backward difference*, and the latter, $\delta f(x)$, is known as the *central difference*. Graphically, the three types of differences are compared in Fig. 2-7. In many discussions, it is useful to consider Δ, ∇, and

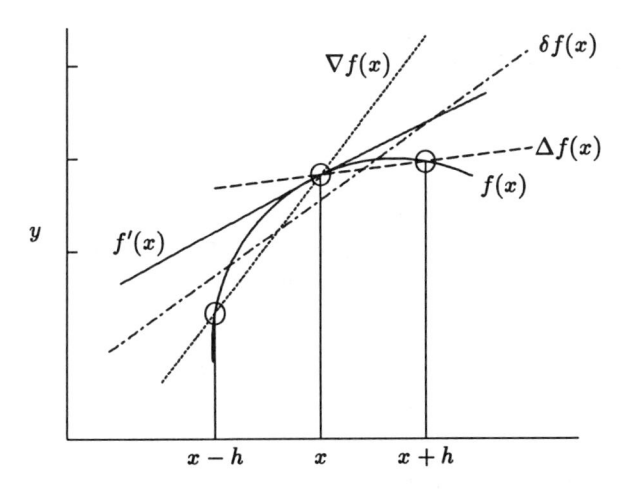

Fig. 2-7 Schematic illustration of the various finite difference approximations to the value of the first-order derivative of a function $f(x)$. The dashed line represents an attempt to substitute the derivative of $f(x)$ at $x = x_i$ by the forward difference $\Delta f(x)$. The dotted line is the backward difference $\nabla f(x)$. The dash-dot line is the central difference $\delta f(x)$. The exact value of $f'(x)$ is the tangent to $f(x)$ at x and is shown by the solid line.

δ as operators that, when acting on a function $f(x)$, generate, respectively, the forward, backward, and central differences given in (2-50) to (2-52).

The most natural way to compute the first-order derivative for $f(x)$ at $x = x_i$ is to use the central difference. In the notation of (2-3), this may be expressed in the form

$$f'_k = \frac{f_{k+1/2} - f_{k-1/2}}{h} \tag{2-53}$$

Mathematically, the relation is correct only in the limit $h \to 0$, as we have seen in (2-44). Here, we shall assume that h is sufficiently small that the relation holds for the numerical accuracy required. As usual, the expression implies that the abscissa is divided into grids of equal width h. Often it may be undesirable to use half-intervals. In such cases, we can take the central difference over two adjacent subintervals

$$f'_k = \frac{f_{k+1} - f_{k-1}}{2h} \tag{2-54}$$

A generalization of the idea behind (2-53) gives us the central difference form of the second-order derivative of $f(x)$:

$$f''_k = \frac{f'_{k+1/2} - f'_{k-1/2}}{h} = \frac{f_{k+1} - 2f_k + f_{k-1}}{h^2} \tag{2-55}$$

a form we used earlier in (2-13).

For many applications, the central difference is preferred over forward and backward differences. The reason may be seen from a Taylor series expansion of $f(x+h)$ at x,

$$f(x+h) = f(x) + \frac{h}{1!}f'(x) + \frac{h^2}{2!}f''(x) + \frac{h^3}{3!}f^{(3)}(x) + \cdots \qquad (2\text{-}56)$$

To obtain the forward difference approximation for $f'(x)$, we can subtract $f(x)$ from both sides of the equation and rearrange the terms. This gives us the result

$$f'(x) = \frac{1}{h}\{f(x+h) - f(x)\} - \frac{h}{2}f''(x) + \cdots$$

We see that the error in approximating $f'(x)$ by $\Delta f(x)/h$ is of the order of h times $f''(x)$, the second-order derivative of $f(x)$ at x. If the function is sufficiently smooth, $f''(x)$ is small in value. Furthermore, if h is also small, we expect the approximation to be a good one.

Similarly, a Taylor series expansion of $f(x-h)$ at x gives us the expression

$$f(x-h) = f(x) - \frac{h}{1!}f'(x) + \frac{h^2}{2!}f''(x) - \frac{h^3}{3!}f^{(3)}(x) + \cdots \qquad (2\text{-}57)$$

From this we obtain the error of expressing $f'(x)$ in terms of the backward difference. As expected, it is of the same order as using the forward difference. On the other hand, by taking the difference between (2-56) and (2-57), we obtain the expression

$$f(x+h) - f(x-h) = 2hf'(x) + 0 + 2\frac{h^3}{3!}f^{(3)}(x) + \cdots$$

All the terms involving odd powers of h vanish on the right side. From this we obtain the first-order derivative of $f(x)$ in the form

$$f'(x) = \frac{1}{2h}\{f(x+h) - f(x-h)\} - \frac{h^2}{3!}f'''(x) + \cdots \qquad (2\text{-}58)$$

The first term on the right side may be identified with the central difference given by (2-54). The error in approximating $f'(x)$ by the central difference here is one order higher in h than forward and backward differences. For small h, we see that central differences are better approximations to the derivatives of function.

In practice, it is difficult to achieve good accuracy in numerical differentiation as compared, for example, with numerical integration. The main reason comes from the fact that we are taking the ratio of two differences, for example, $\{f(x+h/2) - f(x-h/2)\}$ and $\{(x+h/2) - (x-h/2)\}$ as we saw earlier in (2-53). Each of these two differences is between two quantities of very similar values. Underlying our discussions, there is the implicit assumption that the function, whose derivative we are seeking, is a smooth one. If this is true, the values of the function at two nearby points must be very close to each other. The value of the derivative is then related to the small difference between two numbers that are large by comparison. Consider the case of a function that exists only as tabulated values obtained by measuring a physical quantity y at different values x_i. If the measurements are made with n significant figures, and the variation of y as a function of x is a smooth one, the difference between two adjacent values y_{i+1} and y_i will be smaller than either y_{i+1} and y_i. As a result, the number of significant figures in $\Delta y_i \equiv (y_{i+1} - y_i)$ is less than n. The same argument applies to the difference between x_{i+1} and x_i. The net result is that the number of significant figures we can have for any finite differences is smaller than n.

In addition to practical difficulties, there are also intrinsic limitations in replacing derivatives by finite differences. This is best illustrated by an example. Consider the exponential function

$$f(x) = e^{i\omega x}$$

commonly found in the solution of wave equations and Fourier analyses. Here, the symbol i stands for the imaginary number, with $i^2 = -1$, and ω represents the frequency of oscillation. Analytically, the first-order derivative of the function is well known,

$$f'(x) = i\omega e^{i\omega x} \qquad (2\text{-}59)$$

To find the finite difference for $f(x)$, let us divide the abscissa x into equally spaced intervals of width h each and, as usual, use the symbol

$$x_k = x_0 + kh$$

to represent the location of the kth grid points from x_0 along the x-axis. In the notation of (2-3), the value of $f(x)$ at $x = x_k$ is

$$f_k = e^{i\omega x_0} e^{i\omega kh}$$

The value of the numerical derivative using a forward difference approximation is given by

$$f'(x_k) \approx \frac{1}{h}\Delta f_k = \frac{f_{k+1} - f_k}{h} = e^{i\omega x_0} e^{i\omega kh} \frac{e^{i\omega h} - 1}{h}$$

Similarly, the backward and central difference results are, respectively,

$$f'(x_k) \approx \frac{1}{h}\nabla f_k = \frac{f_k - f_{k-1}}{h} = e^{i\omega x_0} e^{i\omega kh} \frac{1 - e^{-i\omega h}}{h}$$

$$f'(x_k) \approx \frac{1}{2h}\delta f_k = \frac{f_{k+1} - f_{k-1}}{2h} = e^{i\omega x_0} e^{i\omega kh} \frac{e^{i\omega h} - e^{-i\omega h}}{2h}$$

To check the accuracy of various finite difference approximations, let us take the ratio of each with the exact value given by (2-59). The three ratios are, respectively,

$$e^{i\omega h/2}\frac{\sin\frac{1}{2}\omega h}{\frac{1}{2}\omega h} \qquad e^{-i\omega h/2}\frac{\sin\frac{1}{2}\omega h}{\frac{1}{2}\omega h} \qquad e^{i\omega h}\frac{\sin\omega h}{\omega h}$$

For any finite size h, the absolute values of these three ratios are always less than unity, unless $\omega \to 0$. In other words, the numerical results constantly underestimate the derivative except for $\omega = 0$. The reason is easy to understand. In all three cases, we have essentially used a linear approximation to the tangent of $f(x)$ at x, whereas $f(x)$ is a function that oscillates with a frequency related to ω.

It is not difficult to make some improvement on the accuracy for derivatives in numerical calculations. In the case of forward and backward difference approximations, we have been using essentially only two points of $f(x)$ to calculate the derivative. For the central difference approximation, we have gone one step further and made a three-point approximation by using f_{k-1}, f_k, and f_{k+1}, even though f_k itself does not appear explicitly in the final result. We can also adopt a five-point approximation by making use of a Taylor series expansion for $f(x \pm 2h)$, in a way analogous to what we did for $f(x \pm h)$ in (2-57) and (2-58). The result is

$$f'_k = \frac{1}{12h}(f_{k-2} - 8f_{k-1} + 8f_{k+1} - f_{k+2}) + \frac{h^4}{30}f_k^{(5)} + \cdots \qquad (2\text{-}60)$$

The error is now on the order of h^4, an improvement of two orders in h over that of (2-58). Similarly, the central difference result for the second-order derivative becomes

$$f''_k = \frac{1}{12h^2}(-f_{k-2} + 16f_{k-1} - 30f_k + 16f_{k+1} - f_{k+2}) + \frac{h^4}{90}f^{(6)}(x) + \cdots \qquad (2\text{-}61)$$

The improvement is again two orders in h over that given by (2-55).

Finite difference methods are used extensively in solving differential equations. There we replace the relations between derivatives with the relations between finite differences. As a result, numerical errors and their propagations become even more important. For a solution to be stable, meaning that the error does not take over, it is necessary that approximations from different parts compensate each other to as large an extent as possible. We shall see examples of such considerations later in Chapter 8.

Problems

2-1 Show that the error associated with the trapezoidal rule of integration for the definite integral

$$I_{[a,b]} = \int_a^b f(x)\, dx$$

is $-(b-a)h^2 f''(\xi)/12$, where $f''(\xi)$ is the average value of the second-order derivative of $f(x)$ in the interval $x = [a, b]$.

2-2 Compare the efficiencies of calculating the normal probability function

$$A(x) = \frac{1}{\sqrt{2\pi}} \int_{-x}^{+x} e^{-t^2} dt$$

using the trapezoidal rule, Simpson's rule, and a Monte Carlo method.

2-3 Use a Gauss-Legendre quadrature of order $n = 4$ to evaluate the integral

$$I(a) = \int_{-a}^{+a} x^9 \, dx$$

for an arbitrarily large value of a. The exact value for this integral is zero. Use the same method to calculate the integral

$$I(b) = \int_{-a}^{+a} x^8 \, dx$$

The exact value is $2a^9/9$. Compare the accuracies obtained for the two integrals and explain why there is a large difference.

2-4 For $0 < n < 1$, the following integral has the analytical solution

$$\int_0^\infty \frac{x^{n-1}}{1+x} dx = \frac{\pi}{\sin n\pi}$$

Use the trapezoidal or Simpson's rule to evaluate the integral. Compare the efficiency and accuracy of carrying out the calculation by splitting the integral into two parts

$$\int_0^\infty \frac{x^{n-1}}{1+x} dx = \int_0^a \frac{x^{n-1}}{1+x} dx + \int_a^\infty \frac{x^{n-1}}{1+x} dx$$

For sufficiently large values of a, the second integral on the right side may be approximated by the result

$$\int_a^\infty \frac{x^{n-1}}{1+x} dx \approx \int_a^\infty x^{n-2} dx = \frac{a^{n-1}}{1-n}$$

When does such a substitution improve the efficiency of the integration?

2-5 The Laguerre and Hermite polynomials are defined, respectively, in the ranges $[0,\infty]$ and $[-\infty,\infty]$ by (4-112) and (4-15). Derive the numerical quadrature formulas analogous to (2-22) using Laguerre and Hermite polynomials instead of Legendre polynomials. Find the roots and weights for order $n = 4$ and check the results with those given in Abramowitz and Segun.

2-6 Use a Taylor series expansion to derive the central difference result given in (2-55) for the second-order derivative of $f(x)$ at $x = x_k$,

$$f''(x_k) = \frac{f_{k+1} - 2f_k + f_{k-1}}{h^2}$$

Calculate the leading order correction term in powers of h.

2-7 Use the same method as for Problem 2-6 and derive the five-point approximate result

$$f''(x_k) = \frac{1}{12h^2}(-f_{k-2} + 16f_{k-1} - 30f_k + 16f_{k+1} - f_{k+2}) + \frac{h^4}{90}f^{(6)}(x) + \cdots$$

given earlier in (2-61).

2-8 Use a Monte Carlo method to carry out the integral

$$I = \int_0^1 \frac{dx}{1+x^2}$$

The result may be tested with the help of the relation

$$\int_0^1 \frac{dx}{1+x^2} = \tan^{-1} x \Big|_0^1 = \frac{\pi}{4}$$

Is this a good way to obtain the value of π or just a way to test the random number generator used in the calculation?

2-9 Use a Monte Carlo approach to evaluate the integral

$$I(q) = \int_0^\infty \frac{e^{-qx}}{\sqrt{x}} dx$$

for $q = 0.5$, 1, and 2. Compare the results with the exact value of $\sqrt{\pi/q}$. It is also possible to use (2-48) for the numerical integration. Compare the accuracies with the Monte Carlo method for the same number of points.

Chapter 3
Interpolation and Extrapolation

A function is a mathematical expression that describes the relation between variables. For simplicity, let us consider a single-valued function with only one independent variable x. If y is a function of x, we can associate a value of y for every one of x. Often the dependence of y on x is given by some known analytical form but this does not always have to be true. For example, we can measure the distances $d_0, d_1, \ldots, d_N$ traveled by a car from some fixed reference point at times $t_0, t_1, \ldots, t_N$, respectively. In this case, the relation between d and independent variable t is given by N pairs of numbers $(t_0, d_0), (t_1, d_1), \ldots, (t_N, d_N)$. Unless we are provided with some additional information, the functional relation between distance d and time t is unknown to us. What happens if we wish to find out the value of d at some time t that is not one of those measured? If t is within the range of time $[t_0, t_N]$ where measurements have been made, the value of d may be approximated by interpolation techniques. If t is outside the range, the corresponding technique is called extrapolation. Since we do not have any knowledge of the functional dependence of d on t, other than the table of $(N+1)$ pairs of values, the calculated result is necessarily an approximate one. As in any other approximation schemes, it is important that the method we use to make the calculations also supply us with an estimate of the error associated with the result. Interpolation and extrapolation techniques are also useful in cases where the functional relations are known but are too time consuming to evaluate. We shall see several examples of this kind.

§3-1 Polynomial interpolation

In interpolation, we are interested in finding the most likely value of y for a given x using a table of the values of y at a finite number of points $x_0, x_1, \ldots, x_N$. This is illustrated by Fig. 3-1. In general, we do not have any knowledge of the functional relation between y and x. Furthermore, we do not usually have any

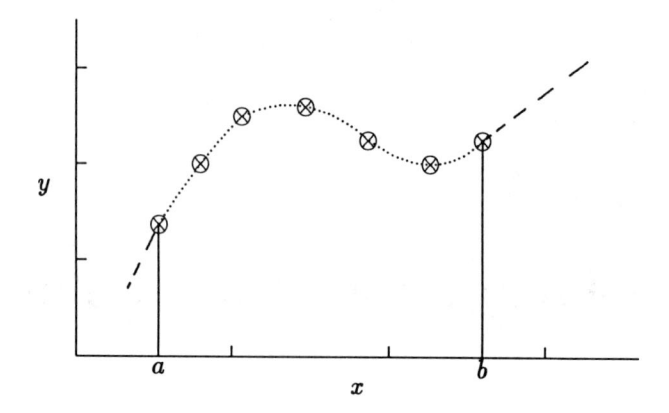

Fig. 3-1 Interpolation and extrapolation for a function $y = f(x)$ from a table of values. The known values of y at a number of points are shown as $\otimes$. The values at other points may be obtained by interpolation if $a < x < b$, such as those indicated by the dotted curve, and by extrapolation if x is outside $[a, b]$, such as those indicated by the dashed curve.

control over the values of x_i where the values of y are known. For this reason, it is not always possible to assume that they are equally spaced.

In general, a continuous function $f(x)$ in a finite interval $x = [a, b]$ can always be fitted by a polynomial $P(x)$. This is known as the Weierstrass convergence theorem, a proof of which can be found in Courant and Hilbert (*Methods of Mathematical Physics*, Interscience Publishers, New York, 1962), vol. II, p. 273-74. Our interest here is to find a polynomial approximation for a function $f(x)$, given to us in the form of $(N + 1)$ pairs of numbers $\{x_i, f(x_i)\}$ for $i = 0, 1, \ldots, N$. This task may be achieved by constructing a polynomial of the form

$$P(x) = \sum_{k=0}^{N} p_k(x) f(x_k)$$

It is clear that if

$$p_k(x) = \frac{(x - x_0)(x - x_1) \cdots (x - x_{k-1})(x - x_{k+1}) \cdots (x - x_N)}{(x_k - x_0)(x_k - x_1) \cdots (x_k - x_{k-1})(x_k - x_{k+1}) \cdots (x_k - x_N)}$$

we have the result that $P(x_i) = f(x_i)$ for $i = 0, 1, \ldots, N$. Such a scheme is known as the Lagrange polynomial interpolation of the function $f(x)$. The polynomial $p_k(x)$ is of degree N, since the numerator is a product of N terms each of which has the form $(x - x_i)$, for $i = 0$ to N except for $i = k$. We shall assume that no two values of x_i are equal to each other, otherwise there will be singularities in the $p_k(x)$ constructed this way.

Neville's algorithm. A polynomial of degree m may also be built in the form of a power series with $(m+1)$ coefficients, $a_0, a_1, \ldots, a_m$:

$$P_m(x) = \sum_{k=0}^{m} a_k x^k \qquad (3\text{-}1)$$

If our function $f(x)$ is known at $(N+1)$ points, it is possible to construct a polynomial of this form, with degree N, in such a way that $P_N(x)$ goes through each of the $(N+1)$ points where the values of the function are known. However, for the purpose of computer calculations, the simple form of (3-1), sometimes known as the Lagrange method, is not the most suitable. Furthermore, it is not easy in such an approach to make an estimate of the errors associated with the calculated result.

In interpolation, our interest is purely local. For this reason, the polynomial needs only to fit the known values of the function in the small region in the vicinity of x we are concerned with. This is different from curve fitting where, as we shall see in Chapter 6, a good overall representation of the function in the entire interval of $x = [a, b]$ is required. For this reason, there is often no point in making use of too many points in an interpolation. It complicates the calculations without necessarily improving the reliability of the final result.

Consider the following example of finding the approximate value of a function at x using the known values at five nearby points. Let the abscissas of these five points be x_1, x_2, x_3, x_4, and x_5. The corresponding values of y at these five points are, respectively, $f(x_1)$, $f(x_2)$, $f(x_3)$, $f(x_4)$, and $f(x_5)$. Although the method to be outlined below is more general, we shall assume, for the sake of simplifying the explanation, that x_3 is the nearest point to x. Furthermore, the value of x lies between x_3 and x_4.

The most naive estimate for the value y at x is given by the known value of $f(x)$ that is closest to x. In our case, we have assumed that this is the point x_3. The zeroth-order approximation to y is then $f(x_3)$. Let us represent this approximate value as f_3. More generally, we shall adopt the notation

$$f_i = f(x_i) \qquad (3\text{-}2)$$

for $i = 1, 2, \ldots, 5$. To improve upon such a lowest-order estimate, we note that x lies between x_3 and x_4. It is therefore possible to obtain a more accurate value for y by making a linear interpolation between the known values of $f(x)$ at these two points. This gives us a new approximate value, which we shall represent as

$$f_{34} = \frac{1}{x_3 - x_4} \{(x - x_4)f_3 - (x - x_3)f_4\} \qquad (3\text{-}3)$$

Again, we shall use the notation in a more general way and treat f_{ij} as the linearly interpolated value between f_i and f_j. It is easy to see that f_{34} is a better approximation to the value of y at $x = x_3$ than f_3, since it makes use of the information provided at two different points.

A further improvement may be achieved by making a quadratic interpolation. In addition to f_3 and f_4, we shall also make use of f_2 as a part of the input to the calculation. There are several ways to do this. The most convenient one for the purpose of programming a computer is to adopt *Neville's algorithm* and make a linear interpolation between the values of f_{34} and f_{23}. This gives us the result

$$f_{234} = \frac{1}{x_2 - x_4}\{(x - x_4)f_{23} - (x - x_2)f_{34}\} \qquad (3\text{-}4)$$

This is very similar in spirit to what we have done in (3-3). The advantage in this approach comes from the fact that (3-3) and (3-4) involve the same types of operation and, as a result, may be put into a recursive form, as we shall see in (3-6).

Table 3-1 Neville's method of interpolation from five known values.

x_1	f_1				
		f_{12}			
x_2	f_2		f_{123}		
		f_{23}		f_{1234}	
x_3	f_3		f_{234}		f_{12345}
		f_{34}		f_{2345}	
x_4	f_4		f_{345}		
		f_{45}			
x_5	f_5				

It is clear that to make further improvement it is necessary to make a cubic approximation. This may be done through a linear interpolation between f_{234} and f_{345}. Let us represent the result as f_{2345}, following the system of notation we adopted earlier. Similarly, by a linear interpolation between f_{123} and f_{234}, we have the value of f_{1234}. Using these two values, we can carry out the next and final step for our interpolation with five pairs of input values. The result is

$$f_{12345} = \frac{1}{x_1 - x_5}\{(x - x_5)f_{1234} - (x - x_1)f_{2345}\} \qquad (3\text{-}5)$$

We can put (3-3) to (3-5) in the form of a recursion relation,

$$f_{ij\ldots k\ell} = \frac{1}{x_i - x_\ell}\{(x - x_\ell)f_{ij\ldots k} - (x - x_i)f_{j\ldots k\ell}\} \qquad (3\text{-}6)$$

That is, the value of $f_{ij\ldots k\ell}$ is the result of a linear interpolation of two similar quantities that are nearest to it but with one subscript less. The quantities $f_{ij\ldots k\ell}$, with m subscripts, are polynomials of x with leading order $(m-1)$. The starting point of this set of relations is the f_i given in (3-2). Note that, in calculating the values of $f_{ij...m}$ with m subscripts, we need x, x_i, x_m, together with the values of $f_{i,j,...}$ and $f_{jk...}$ having one less subscript, as can be seen from examining (3-6).

The successive approximations to $f(x)$ in terms of $f_{ij...k\ell}$ may be represented by the diagram shown in Table 3-1. In general, for an interpolation using N input values, the highest polynomial order we can use is $(N-1)$. This, in turn, means that the maximum number of subscripts for $f_{ij...k\ell}$ is N. The essence of the method is to make small adjustments in each step to the calculated value of y by incorporating one more piece of input information. In the process, the polynomial degree is also increased by 1. The uncertainty in the value obtained may be estimated from the difference between two successful steps, given essentially by the difference between two adjacent columns in Table 3-1. Thus, the difference between f_{2345} (or f_{1234}) and f_{12345} provides us with an error estimate of the final result.

Table 3-2 Interpolation of erf(x) for $x = 0.52$ from five given values at $x = 0.3$, 0.4, 0.5, 0.6, and 0.7.

x_i	f_i	f_{ij}	f_{ijk}	$f_{ijk\ell}$	$f_{ijk\ell m}$
0.3	0.3286268				
		0.5481111			
0.4	0.4283924		0.5380024		
		0.5389214		0.5379062	
0.5	0.5204999		0.5378712		0.5378986
		0.5371711		0.5378923	
0.6	0.6038561		0.5379240		
		0.5447000			
0.7	0.6778012				

A simple numerical illustration will give us a better feeling of the way the algorithm works in practice. Let us try to find the value of the error function

$$\text{erf}(x) = \frac{2}{\sqrt{\pi}} \int_0^x e^{-t^2} dt \tag{3-7}$$

near the point $x = 0.5$ by interpolation. As input, we shall use the five values of erf(x) at $x = 0.3$, 0.4, 0.5, 0.6, and 0.7. Theses are given in the second column of Table 3-2. If we wish to find the value of erf(x) at $x = 0.52$, we can calculate the four linear terms f_{ij} associated with this value of x using (3-6), and these are listed in the third column. The fourth column gives the three values of f_{ijk} in the second-order approximation, the fifth column gives the two values in the third-order, and so on. The final value for a fourth-order approximation is $f_{12345} = 0.5378986$ and it is given in the last column. This is the interpolated value of erf(x) for $x = 0.52$

calculated with the Neville's algorithm using the five input given in column 2. The error of the interpolation process is estimated to be around 7×10^{-6}, as can be seen from the differences between the values in the fifth and sixth columns. For smooth functions normally encountered in physics, the actual accuracy is usually better than that indicated by the error estimate. For this reason, it should not be a surprise for us to find that, for our error function example, the calculated result happens to agree with the exact value up to the seven significant figures listed.

The example also serves as a good illustration of the applications of interpolation techniques. Since the error function is defined only in term of an integral, the values are cumbersome to calculate and are usually found by looking up a table. However, the number of entries we can find in a mathematical table is limited, and it happens frequently that the list does not contain exactly the value of the dependent variable we wish to have. There are two ways to solve the problem. The first is to write a numerical integration program for the function using one of the techniques given in Chapter 2. When this is not the convenient thing to do, we can take the alternative route and use an interpolation procedure to find the values required. As we can see from the example, very accurate results can be obtained using just a few nearby values taken from the table.

Improvements on the algorithm. The recursion relation for $f_{ij\ldots k\ell}$ given by (3-6) is not the ideal one for computation. The reason is that we are making small corrections by taking weighted averages between two relatively large numbers. Better accuracies may be obtained if, at each stage, we deal directly with the small corrections themselves. For this purpose, we shall define two related types of differences. The first is between two values of $f_{ij\ldots k\ell}$ with the last subscript differing by unity,

$$D_{i,\ell} \equiv f_{i,i+1,\ldots,i+\ell} - f_{i,i+1,\ldots,i+\ell-1} \tag{3-8}$$

For the lowest order of $\ell = 1$, we take

$$D_{i,1} = f_i$$

In terms of the quantities given in Table 3-1, we are moving downward along the diagonal, as $f_{i,i+1,\ldots,i+\ell-1}$ is one row above and one column to the left of $f_{i,i+1,\ldots,i+\ell}$. In the absence of a better name, we shall call $D_{i,\ell}$ the "downward difference." There is no point in indicating the intervening subscripts of f from which $D_{i\ell}$ are constructed, as they are always increasing in value by unity each time we move one index to the right. Similarly, we can also define an "upward difference" by taking the difference between two values of $f_{i,i+1,\ldots,i+m}$, with the first subscript differing by unity:

$$U_{i,\ell} \equiv f_{i,i+1,\ldots,i+\ell} - f_{i+1,i+2,\ldots,i+\ell} \tag{3-9}$$

The starting point for this set of differences is $U_{i,1} = f_i$. We can work out the recursion relations between $D_{i,\ell}$ and $U_{i,\ell}$ by substituting $f_{ij...k\ell}$ given in (3-6) into (3-8) and (3-9). The results are

$$D_{i,\ell+1} = \frac{x_i - x}{x_i - x_{i+\ell+1}} \{D_{i+1,\ell} - U_{i,\ell}\}$$

$$U_{i,\ell+1} = \frac{x_{i+\ell+1} - x}{x_i - x_{i+\ell+1}} \{D_{i+1,\ell} - U_{i,\ell}\} \qquad (3\text{-}10)$$

The derivation is left as an exercise (Problem 3-3).

Box 3-1 Subroutine NEVILLE(X_IN,F_IN,N,X,FX,DF)
Neville's algorithm for interpolation
Find the value of $f(x)$ from an input table of $\{x_i, f(x_i)\}$

Argument list:
X_IN(I): Input abscissa.
F_IN(I): value of $f(x)$ at $x =$X_IN(I).
N: Number of input points.
X: Value of x to be calculated.
FX: Returned value of $f(x)$.
DF: Estimated error for FX.

Initialization:
In the calling program:
(a) Store a table of the values of $\{x_i, f(x_i)\}$ for $i = 1$ to N.
(b) Input the value of x.

1. Define U and D according to (3-10) and set up an index IDX to indicate the table entry nearest to x.
2. Make an initial estimate using only the table entry nearest to x. Determine also if x is above (UP = .TRUE.) or below x_{IDX} (UP = .FALSE.).
3. Improving the estimate by higher-order interpolations:
    (a) Let the order $L = 1$.
    (b) If UP is .TRUE.:
        (i) Decrease IDX by 1.
        (ii) If IDX≥ 1, increase the estimate by U_{IDX}.
        (iii) Otherwise, set IDX=1 and use D_{IDX} instead.
        (iv) Change UP to .FALSE. and go to step (d).
    (c) If UP is .FALSE.:
        (i) Increase the estimate by D_{IDX} if IDX $> (N - L)$.
        (ii) Otherwise, set IDX equal to $(N - L)$ and use D_{IDX}.
        (iii) Change UP to .TRUE..
    (d) Increase L by 1 and go back to step (b) until L reaches the maximum.
4. Return the interpolated value and the last improvement as the estimate of error.

Since both $D_{i,\ell}$ and $U_{i,\ell}$ are the "corrections" to the previous order of approximation, we must make a choice as to which of these two quantities to adopt at each order. This can be done by following the way we make use of the successive orders of $f_{ij\ldots k\ell}$. Earlier, we have assumed that the value of x is near the value of x_i in the middle of the table of known values. For the convenience of discussion, let us give the name i_x to this value of the subscript. At the starting point of the interpolation, we have $i_x = i$. In the zeroth-order approximation, we take $f(x) = f_i$.

If $x > x_i$, the next-order approximation comes from $f_{i-1,i}$, following the way the subscripts are defined here. Since the difference between $f_{i-1,i}$ and f_i is given by $U_{i-1,2}$, we shall use $U_{i-1,2}$ to generate the next-order correction in the present scheme. The difference between $f_{i-1,i}$ and $f_{i-1,i,i+1}$ is given by $D_{i-1,3}$, and we shall make use of $D_{i-1,3}$ next. Continuing in this way with the help of Table 3-1, we come quickly to the conclusion that, for $x > x_i$, we start from $f_i = D_{i,1}$ and the successive corrections are $U_{i-1,2}$, $D_{i-1,3}$, $U_{i-2,4}$, $D_{i-2,5}$, and so on. Similarly, a scheme to obtain the successive corrections to the interpolated value of $f(x)$ for $x > x_i$ may be constructed in the following way. We start with f_i and alternate between using U and D for each higher order. Everytime we use U, we decrease the value of the first subscript i_x by 1 and increase the value of the second subscript by an equal amount.

Table 3-3

Interpolated values of erf(x) using the same five inputs as Table 3-2.

x	Tabulated value	Interpolated value	Estimated error
0.46	0.4846554	0.4846556	9.7×10^{-6}
0.47	0.4937451	0.4937451	8.4×10^{-6}
0.48	0.5027497	0.5027497	6.2×10^{-6}
0.49	0.5116683	0.5116683	3.4×10^{-6}
0.50	0.5204999	input	—
0.51	0.5292436	0.5292436	3.4×10^{-6}
0.52	0.5378986	0.5378986	6.2×10^{-6}
0.53	0.5464641	0.5464640	8.4×10^{-6}
0.54	0.5549393	0.5549392	9.7×10^{-6}

A similar approach can also be obtained for $x < x_i$. In this case we start with f_i and alternate between using D and U, with the value of i_x decreased by 1 everytime we encounter an U. A summary of Neville's algorithm in terms of the upward and downward differences is given in Box 3-1. As an example, we shall calculate the values of erf(x) for $x = 0.46$ to 0.54 in steps of 0.01, and the results are compared with exact values in Table 3-3.

§3-2 Interpolation using rational functions

In general, the use of polynomials in approximations works best if the functional relation between the dependent and independent variables is a smooth one. Furthermore, there should not be any singularities either within the range of interest or near one of the end points. Unfortunately, these conditions are not met by a large class of functions, such as $y(x) = \tan x$ near $x = \pi$ and $y(x) = \ln x$ for very small values of x. In these cases, it is more advantageous to make use of rational functions consisting of ratios of two polynomials. In fact, for the purpose of interpolation, rational functions are often superior even when the function is well behaved.

Just as in the case of polynomial approximations, there are many different ways to construct an approximation to a function based on rational functions. The more popular approach is to use the Padé approximation. However, for computational purposes, a slightly different approach is preferred, as suggested by Stoer and Bulirsch (*Introduction to Numerical Analysis*, English translation by R. Bartels, W. Gautschi, and C. Witzgall, Springer-Verlag, New York, 1980). Among other advantages, the method also provides with us an error estimate for the results. For pedagogical reasons, however, we shall begin the discussion with a summary of the Padé approximation.

Padé approximation. The idea behind the Padé approximation may be illustrated by the following example. It is well known that the trigonometry function $\cos x$ may be expanded in terms of a power series:

$$\cos x = 1 - \frac{1}{2!}x^2 + \frac{1}{4!}x^4 - \frac{1}{6!}x^6 + \frac{1}{8!}x^8 - \cdots \qquad (3\text{-}11)$$

If $x \leq 1$, the series is convergent. An approximate form of the same function that is identical to (3-11) up to the x^8 term may be written as a rational function in the following way:

$$\cos x \approx \frac{1 + p_1 x^2 + p_2 x^4}{1 + q_1 x^2 + q_2 x^4} \qquad (3\text{-}12)$$

This is known as the *Padé approximation*. In using a series expansion to approximate a function, the form given by (3-12) is often more compact and, for many purposes, more convenient to use than (3-11).

The values of the four coefficients p_1, p_2, q_1, and q_2 may be found by equating (3-12) to (3-11) up to the x^8 term. Using the binomial expansion

$$\frac{1}{1+\delta} = 1 - \delta + \delta^2 - \delta^3 + \delta^4 - \cdots$$

the denominator on the right side of (3-12) may be written as

$$\frac{1}{1+q_1x^2+q_2x^4} = 1 - (q_1x^2+q_2x^4) + (q_1x^2+q_2x^4)^2$$
$$- (q_1x^2+q_2x^4)^3 + (q_1x^2+q_2x^4)^4 - \cdots$$

$$\begin{aligned}
= 1 &- q_1x^2 \\
&- q_2x^4 + q_1^2x^4 \\
&+ 2q_1q_2x^6 - q_1^3x^6 \\
&+ q_2^2x^8 \quad -3q_1^2q_2x^8 \quad +q_1^4x^8 \\
&\qquad\qquad -3q_1q_2^2x^{10} + 4q_1^3q_2x^{10} - \cdots \\
&\qquad\qquad -q_2^3x^{12} \quad +6q_1^2q_2^2x^{12} - \cdots \\
&\qquad\qquad\qquad\qquad +4q_1q_2^3x^{14} - \cdots \\
&\qquad\qquad\qquad\qquad +q_2^4x^{16} \quad -\cdots
\end{aligned} \qquad (3\text{-}13)$$

When we multiply $(1 + p_1x^2 + p_2x^4)$ to the final form of (3-13) and compare the coefficients of x^2, x^4, x^6, and x^8 with those on the right side of (3-11), we obtain a set of four relations:

$$q_1 - p_1 = \tfrac{1}{2}$$
$$q_1^2 - p_1q_1 + p_2 - q_2 = \tfrac{1}{24}$$
$$q_1^3 - p_1q_1^2 + p_2q_1 - 2q_1q_2 + p_1q_2 = \tfrac{1}{720}$$
$$q_1^4 - p_1q_1^3 + p_2q_1^2 - 3q_1^2q_2 + 2p_1q_1q_2 - p_2q_2 + q_2^2 = \tfrac{1}{40{,}320}$$

By substituting the first of the four into the second, the first and second into the third, and so on, this set of equations may be simplified into

$$q_1 - p_1 = \tfrac{1}{2} \qquad\qquad \tfrac{1}{2}q_1 + p_2 - q_2 = \tfrac{1}{24}$$
$$\tfrac{1}{24}q_1 - \tfrac{1}{2}q_2 = \tfrac{1}{720} \qquad\qquad \tfrac{1}{720}q_1 - \tfrac{1}{24}q_2 = \tfrac{1}{40{,}320}$$

The values of the four coefficients obtained in this way are

$$p_1 = -\frac{115}{252} \qquad p_2 = \frac{313}{15{,}120} \qquad q_1 = \frac{11}{252} \qquad q_2 = \frac{13}{15{,}120}$$

Using these values, the Padé approximation of $\cos x$ for $|x| < 1$ takes the form

$$\cos x \approx \frac{15{,}120 - 6900x^2 + 313x^4}{15{,}120 + 660x^2 + 13x^4} \qquad (3\text{-}14)$$

A test of the accuracy of this expression may be found from the fact that the right side of (3-14) vanishes when

$$x^2 = \frac{6900}{2 \times 313} \pm \left\{ \left(\frac{6900}{2 \times 313}\right)^2 - \frac{15120}{313} \right\}^{1/2}$$

or $x = \pm 1.5708259$, if we take the negative sign in the expression. This is to be compared with the exact value of $x = \pm\frac{1}{2}\pi = \pm 1.5707963$ for the zero of $\cos x$ in this range of arguments. The positive sign gives a value of x that is too large for the approximation of (3-14) to be valid and we shall ignore it here.

The steps involved in arriving at the values of the coefficients p_1, p_2, q_1, and q_2 are somewhat tedious. However, they are very much "mechanical" in nature in the sense that the instructions for carrying out the steps are very simple. Since the actual algebraic calculations are long, it is ideal for a computer to take them over. We shall see in Chapter 10 that all the required calculations here may be carried out by any of the standard algebraic computation packages with only a minimum amount of input.

The general form of the Padé approximation to a function $f(x)$ is to use the ratio

$$f(x) \approx \frac{P_n(x)}{Q_m(x)} \qquad (3\text{-}15)$$

where $P_n(x)$ and $Q_m(x)$ are polynomials of degrees n and m, respectively. In the example given above, we have used only even powers of x for both $P_n(x)$ and $Q_m(x)$. This is done because $\cos x$ is an even function, as can be seen from (3-11). The same principle applies also in the case of odd functions. For example, we can use

$$\sin x \approx x \frac{1 + p_1 x^2 + p_2 x^4}{1 + q_1 x^2}$$

to produce an approximation up to terms involving x^7 for

$$\sin x = x - \frac{1}{3!}x^3 + \frac{1}{5!}x^5 - \frac{1}{7!}x^7 + \cdots$$

More generally, for a function that is neither even nor odd, both $P_n(x)$ and $Q_m(x)$ must involve even as well as odd powers of the argument. Explicitly, we must write (3-15) in the form

$$f(x) \approx \frac{p_0 + p_1 x + p_2 x^2 + \cdots + p_n x^n}{1 + q_1 x + q_2 x^2 + \cdots + q_m x^m}$$

where we have used unity for the first term in $Q_m(x)$ without sacrificing anything.

As illustrations of the accuracies that can be achieved with rational function approximations, we give here the Padé approximations for three familiar functions in the range $0 \leq x \leq 1$.

$$e^{-x} \approx \frac{1.00000\,00007 - 0.47593\,58618x + 0.08849\,21370x^2 - 0.00656\,58101x^3}{1 + 0.52406\,42207x + 0.11255\,48636x^2 + 0.01063\,37905x^3}$$

$$\frac{\tan^{-1} x}{x} \approx \frac{0.99999\,99992 + 1.13037\,54276x^2 + 0.28700\,44785x^4 + 0.00894\,72229x^6}{1 + 1.46370\,86496x^2 + 0.57490\,98994x^4 + 0.05067\,70959x^6}$$

$$\ln \tfrac{1}{2}(1+x) \approx \frac{-0.69314\,71773 + 0.06774\,12133x + 0.52975\,01385x^2 + 0.09565\,58162x^3}{1 + 1.34496\,44663x + 0.45477\,29177x^2 + 0.02868\,18192x^3}$$

(3-16)

The maximum errors for these three approximations are quoted in Fröberg (*Numerical Mathematics*, Benjamin/Cummings, Menlo Park, California, 1985) to be, respectively, 7.34×10^{-10}, 7.80×10^{-10}, and 3.29×10^{-9}. Note also that the constant term in the numerator in each expression is slightly different from the exact value of the function at $x = 0$. This is done so that the overall accuracy in the entire range may be improved.

The method of Stoer and Bulirsch. We have seen earlier that, for numerical calculations, it is desirable for the method to provide us with an estimate of the error associated with the result. Furthermore, it is also useful if there is a natural way to extend the approximation to high orders when the accuracy is inadequate. Neville's algorithm for polynomial interpolation is a good example of such a procedure. In addition to simplicity, the recursion relations given by (3-6) and (3-10) supply us with estimates of the truncation errors and allow the possibility to incorporate higher-order approximations into the calculation with very little additional work. A similar approach for rational function interpolation is found in Stoer and Bulirsch.

Let us assume that the function $f(x)$ is given to us in the interval $x = [a, b]$ in terms of $(N+1)$ pairs of values (x_0, f_0), $(x_1, f_1), \ldots, (x_N, f_N)$. For convenience, we shall arrange the values in ascending order according to x, that is, $x_0 < x_1 < \cdots < x_N$. For such a table, we can write the rational function for interpolation in the following way:

$$R_{\mu\nu s}(x) = \frac{P_{\mu\nu s}(x)}{Q_{\mu\nu s}(x)} = \frac{p_{0;\mu\nu s} + p_{1;\mu\nu s}x + p_{2;\mu\nu s}x^2 + \cdots + p_{\mu;\mu\nu s}x^\mu}{1 + q_{1;\mu\nu s}x + q_{2;\mu\nu s}x^2 + \cdots + q_{\nu;\mu\nu s}x^\nu} \quad (3\text{-}17)$$

where μ is the degree of the polynomial in the numerator and ν is that for the polynomial in the denominator. The third subscript s is needed to distinguish the different independent rational functions that can be constructed. Some further explanation for this label is required.

Since the constant term in the denominator is set to unity, the function $R_{\mu\nu s}(x)$ has only $(\mu + \nu + 1)$ parameters. It is therefore possible to make $R_{\mu\nu s}(x)$

equal to $f(x)$ at $(\mu + \nu + 1)$ number of points. We shall choose $R_{\mu\nu s}(x)$ in such a way that

$$R_{\mu\nu s}(x_i) = f_i \qquad \text{at} \qquad x = x_s, x_{s+1}, \ldots, x_{s+\mu+\nu}$$

In other words, the function $\{R_{\mu\nu s}(x) - f(x)\}$ has $(\mu + \nu + 1)$ nodes and they occur at $x = x_s, x_{s+1}, \ldots, x_{s+\mu+\nu}$. As a result, rational functions having the same degrees μ and ν but differing in the location of their first node may be distinguished by the label s. Furthermore, the numerator polynomial $P_{\mu\nu s}(x)$ and denominator polynomial $Q_{\mu\nu s}(x)$ for a given μ, ν, and s are related to each other, since, in order to have roots at given locations, it is impossible to change one of the two in a nontrivial way without having to modify the other. For this reason, coefficients $p_{k;\mu\nu s}$ of the numerator polynomial $P_{\mu\nu s}(x)$, and $q_{k;\mu\nu s}$ of the denominator polynomial $Q_{\mu\nu s}(x)$, are labeled by four indexes, k, the power of x it is associated with, as well as μ, ν, and s.

One strength of the Stoer-Bulirsch algorithm is that the rational functions $R_{\mu\nu s}(x)$ may be constructed in a recursive manner. For both μ and ν equal to zero, the rational function is simply a constant. As a result, we can have N such functions, each representing $f(x)$ at one of the N points given to us. That is,

$$R_{00i}(x) = \frac{p_{0;00i}}{1} = f_i \qquad (3\text{-}18)$$

Similarly, each of the $\mu = 0$ and $\nu = 1$ rational functions,

$$R_{01i}(x) = \frac{p_{0;01i}}{1 + q_{1;01i}x} \qquad (3\text{-}19)$$

can be made equal to $f(x)$ at two places, $x = x_i$ and $x = x_{i+1}$. This may be achieved by requiring them to satisfy the following relations:

$$R_{01i}(x_i) = \frac{p_{0;01i}}{1 + q_{1;01i}x_i} = f_i$$

$$R_{01i}(x_{i+1}) = \frac{p_{0;01i}}{1 + q_{1;01i}x_{i+1}} = f_{i+1}$$

These two equations may be solved for the two unknown coefficients $p_{0;01i}$ and $q_{1;01i}$. The results are

$$q_{1;01i} = \frac{f_i - f_{i+1}}{x_{i+1}f_{i+1} - x_i f_i}$$

$$p_{0;01i} = f_i + x_i f_i \frac{f_i - f_{i+1}}{x_{i+1}f_{i+1} - x_i f_i} \qquad (3\text{-}20)$$

In a similar way, a $\mu = 1$ and $\nu = 0$ rational function takes on the form

$$R_{10i}(x) = \frac{p_{0;10i} + p_{1;10i}x}{1}$$

with

$$p_{0;10i} = \frac{f_i x_{i+1} - f_{i+1} x_i}{x_{i+1} - x_i} \qquad p_{1;10i} = \frac{f_{i+1} - f_i}{x_{i+1} - x_i}$$

The two functions $R_{01i}(x)$ and $R_{10i}(x)$ are not independent of each other, since both are generated from the same two pieces of input, f_i and f_{i+1}. As we shall see later, the interpolation algorithm requires only one of them and, as a practical algorithm, only $R_{\mu\nu s}$ with $\mu \leq \nu$ is used.

Using (3-20), we can write the rational function for $(\mu\nu s) = (01i)$ of (3-19) in the following form:

$$R_{01i}(x) = \frac{f_i + x_i f_i \frac{f_i - f_{i+1}}{x_{i+1} f_{i+1} - x_i f_i}}{1 + \frac{f_i - f_{i+1}}{x_{i+1} f_{i+1} - x_i f_i} x} = \frac{(x_{i+1} - x_i) f_i f_{i+1}}{f_i(x - x_i) - f_{i+1}(x - x_{i+1})}$$

This result may be expressed in terms of $R_{00i}(x)$ with the help of (3-18):

$$R_{01i}(x) = \frac{\alpha_i - \alpha_{i+1}}{\frac{\alpha_i}{R_{00(i+1)}(x)} - \frac{\alpha_{i+1}}{R_{00i}(x)}}$$

where we have adopted the notation

$$\alpha_i \equiv x - x_i \tag{3-21}$$

We shall soon see that this is just the first member of a general recursion relation among $R_{\mu\nu s}$ for $\nu \geq 1$,

$$R_{0\nu s}(x) = \frac{\alpha_s - \alpha_{s+\nu}}{\frac{\alpha_s}{R_{0(\nu-1)(s+1)}(x)} - \frac{\alpha_{s+\nu}}{R_{0(\nu-1)s}(x)}} \tag{3-22}$$

A similar relation generates $R_{\mu 0 s}(x)$ from $R_{(\mu-1)0s}(x)$ and $R_{(\mu-1)0(s+)}(x)$. We shall not give it here, as it is not needed in any of the later calculations.

To build a general recursion relation for the rational functions $R_{\mu\nu s}(x)$, we must examine first how the polynomials $P_{\mu\nu s}(x)$ and $Q_{\mu\nu s}(x)$ of different degrees are related to each other. It is clear from the definitions of μ and ν that, in going from $(\mu - 1, \nu)$ to (μ, ν), the numerator polynomial $P_{\mu\nu s}(x)$ increases its degree by 1 from $P_{(\mu-1)\nu s}(x)$. The degree of the denominator polynomial $Q_{\mu\nu s}(x)$, however, remains unchanged, but the coefficients $q_{i;\mu\nu s}$ are in general different from $q_{i;(\mu-1)\nu s}$ for the polynomial $Q_{(\mu-1)\nu s}(x)$. It is easy to see that the following method of generating both polynomials from $(\mu - 1, \nu)$ to (μ, ν) fulfills this requirement:

$$P_{\mu\nu s}(x) = \alpha_s q_{\nu;(\mu-1)\nu s} P_{(\mu-1)\nu(s+1)}(x) - \alpha_{s+\mu+\nu} q_{\nu;(\mu-1)\nu(s+1)} P_{(\mu-1)\nu s}(x)$$

$$Q_{\mu\nu s}(x) = \alpha_s q_{\nu;(\mu-1)\nu s} Q_{(\mu-1)\nu(s+1)}(x) - \alpha_{s+\mu+\nu} q_{\nu;(\mu-1)\nu(s+1)} Q_{(\mu-1)\nu s}(x)$$

$$\tag{3-23}$$

By (3-21), we have $\alpha_s = (x - x_s)$ and $\alpha_{s+\mu+\nu} = (x - x_{s+\mu+\nu})$. The multiplication of α_s to $P_{(\mu-1)\nu(s+1)}(x)$ and $\alpha_{s+\mu+\nu}$ to $P_{(\mu-1)\nu s}(x)$ promotes both terms to polynomials of degree μ. However, the polynomial degree of $Q_{\mu\nu s}(x)$ is not increased by this operation since the leading order terms cancel each other on the right side.

A proper proof of the relations given by (3-23) will also require us to show that the rational function, formed by the ratio of the two polynomials, has the value f_i at $x = x_i$ for $i = s, (s+1), \ldots, (s+\mu+\nu)$. In other words,

$$P_{\mu\nu s}(x) - f(x)Q_{\mu\nu s}(x) = 0$$

at $x = x_s, x_{s+1}, \ldots x_{s+\mu+\nu}$ if

$$P_{(\mu-1)\nu s}(x) - f(x)Q_{(\mu-1)\nu s}(x) = 0$$

at $x = x_s, x_{s+1}, \ldots, x_{s+\mu+\nu-1}$. This can be seen by replacing $P_{\mu\nu s}(x)$ and $Q_{\mu\nu s}(x)$ with their explicit values and rearranging the terms into the form

$$P_{\mu\nu s}(x) - f(x)Q_{\mu\nu s}(x) = \alpha_s q_{\nu;(\mu-1)\nu s}\{P_{(\mu-1)\nu(s+1)}(x) - f(x)Q_{(\mu-1)\nu(s+1)}(x)\}$$
$$- \alpha_{s+\mu+\nu} q_{\nu;(\mu-1)\nu(s+1)}\{P_{(\mu-1)\nu s}(x) - f(x)Q_{(\mu-1)\nu s}(x)\}$$

If $P_{(\mu-1)\nu(s+1)}(x) - f(x)Q_{(\mu-1)\nu(s+1)}(x)$ has roots at $x = x_{s+1}, x_{s+2}, \ldots, x_{s+\mu+\nu}$, then the product of $\alpha_s = (x - x_s)$ with it produces an additional root at $x = x_s$. The first term of the above expression therefore satisfies the requirement that there are the correct number of roots and they are at the right locations. The same is true for the second term, except the additional root provided by $\alpha_{s+\mu+\nu} = (x - x_{s+\mu+\nu})$ is at $x = x_{s+\mu+\nu}$. The difference between the two polynomials inside the braces provides the other $(s+\mu+\nu-1)$ roots.

Using the same arguments, we can show that the relations between the polynomials in going from $(\mu, \nu-1)$ to (μ, ν) are

$$P_{\mu\nu s}(x) = \alpha_s p_{\mu;\mu(\nu-1)s} P_{\mu(\nu-1)(s+1)}(x) - \alpha_{s+\mu+\nu} p_{\mu;\mu(\nu-1)(s+1)} P_{\mu(\nu-1)s}(x)$$

$$Q_{\mu\nu s}(x) = \alpha_s p_{\mu;\mu(\nu-1)s} Q_{\mu(\nu-1)(s+1)}(x) - \alpha_{s+\mu+\nu} p_{\mu;\mu(\nu-1)(s+1)} Q_{\mu(\nu-1)s}(x)$$

This, together with (3-23), may be used to demonstrate that the recursion relations between $R_{\mu\nu s}(x)$ are in the forms

$$R_{\mu\nu s}(x) = R_{(\mu-1)\nu(s+1)}(x) + \frac{R_{(\mu-1)\nu(s+1)}(x) - R_{(\mu-1)\nu s}(x)}{\frac{\alpha_s}{\alpha_{s+\mu+\nu}}\left\{1 - \frac{R_{(\mu-1)\nu(s+1)}(x) - R_{(\mu-1)\nu s}(x)}{R_{(\mu-1)\nu(s+1)}(x) - R_{(\mu-1)(\nu-1)(s+1)}(x)}\right\} - 1}$$

(3-24)

$$R_{\mu\nu s}(x) = R_{\mu(\nu-1)(s+1)}(x) + \frac{R_{\mu(\nu-1)(s+1)}(x) - R_{\mu(\nu-1)s}(x)}{\frac{\alpha_s}{\alpha_{s+\mu+\nu}}\left\{1 - \frac{R_{\mu(\nu-1)(s+1)}(x) - R_{\mu(\nu-1)s}(x)}{R_{\mu(\nu-1)(s+1)}(x) - R_{(\mu-1)(\nu-1)(s+1)}(x)}\right\} - 1}$$

(3-25)

We shall see later that further simplifications can be made to these relations for the purpose of interpolation.

We can satisfy ourselves that the relations given by (3-24) and (3-25) are correct by examining first the difference between two rational functions differing by one degree either in the numerator or in the denominator. From the definition of $R_{\mu\nu s}(x)$ given by (3-17), we obtain the result

$$R_{(\mu-1)\nu s}(x) - R_{(\mu-1)(\nu-1)(s+1)}(x)$$
$$= \frac{P_{(\mu-1)\nu s}(x)Q_{(\mu-1)(\nu-1)(s+1)}(x) - P_{(\mu-1)(\nu-1)(s+1)}(x)Q_{(\mu-1)\nu s}(x)}{Q_{(\mu-1)\nu s}(x)Q_{(\mu-1)(\nu-1)(s+1)}(x)}$$

The numerator is a polynomial of degree not exceeding $(\mu + \nu + 1)$ and vanishes at $x = x_{s+1}, x_{s+2}, \ldots, x_{s+\mu+\nu-1}$. As a result, we can write the same difference between the two polynomials in the form

$$R_{(\mu-1)\nu s}(x) - R_{(\mu-1)(\nu-1)(s+1)}(x) = \kappa \frac{(x-x_{s+1})(x-x_{s+2})\cdots(x-x_{s+\mu+\nu-1})}{Q_{(\mu-1)\nu s}(x)Q_{(\mu-1)(\nu-1)(s+1)}(x)}$$

where the constant κ may be shown to have the form

$$\kappa = -p_{\mu-1;(\mu-1)(\nu-1)(s+1)}\, q_{\nu;(\mu-1)\nu s}$$

Similarly, it is possible to show that

$$R_{(\mu-1)\nu(s+1)}(x) - R_{(\mu-1)(\nu-1)(s+1)}(x)$$
$$= \kappa' \frac{(x-x_{s+1})(x-x_{s+2})\cdots(x-x_{s+\mu+\nu-1})}{Q_{(\mu-1)\nu(s+1)}(x)Q_{(\mu-1)(\nu-1)(s+1)}(x)}$$

with $\kappa' = -p_{(\mu-1);(\mu-1)(\nu-1)(s+1)}\, q_{\nu;(\mu-1)\nu(s+1)}$. Next, we can use (3-23) to write the rational function in terms of numerator and denominator polynomials of one degree lower:

$$R_{\mu\nu s}(x) = \frac{\alpha_s q_{\nu;(\mu-1)\nu s} P_{(\mu-1)\nu(s+1)}(x) - \alpha_{s+\mu+\nu}q_{\nu;(\mu-1)\nu(s+1)} P_{(\mu-1)\nu s}(x)}{\alpha_s q_{\nu;(\mu-1)\nu s} Q_{(\mu-1)\nu(s+1)}(x) - \alpha_{s+\mu+\nu}q_{\nu;(\mu-1)\nu(s+1)} Q_{(\mu-1)\nu s}(x)}$$

On multiplying both the numerator and denominator of the right side by

$$\frac{-p_{\mu-1;(\mu-1)(\nu-1)(s+1)}(x-x_{s+1})(x-x_{s+2})\cdots(x-x_{s+\mu+\nu-1})}{Q_{(\mu-1)\nu(s+1)}(x)Q_{(\mu-1)\nu s}(x)Q_{(\mu-1)(\nu-1)(s+1)}(x)}$$

and rearranging terms, we obtain (3-24). The relation given by (3-25) may be proved in the same way. The special case of (3-22) follows from (3-25) for $\mu = 0$, if we define $R_{(\mu=-1)(\nu-1)(s+1)}(x) = 0$.

The recommended procedure given by Stoer and Bulirsch for interpolation is to use rational functions either with $\nu = \mu$ or $\nu = (\mu + 1)$. If there are $(N+1)$

pairs of values of $\{x_i, f(x_i)\}$ available to us, we can start by defining first the $(N+1)$ zero-degree rational functions as

$$R_{00i}(x) = f_i \qquad \text{for} \quad i = 0, 1, \ldots, N$$

From these, we can generate N first-degree polynomials $R_{01i}(x)$, for $i = 1, 2, \ldots, N$, using (3-25). This is followed by using (3-24) to construct $R_{11i}(x)$ for $i = 1, 2, \ldots, (N-1)$. The process ends with a single polynomial for the highest order, $R_{N/2,N/2,1}(x)$ if N is even, and $R_{(N-1)/2,(N+1)/2,1}(x)$ if N is odd. Note that, in order to generate $R_{\mu(\nu=\mu)s}(x)$ with (3-24), we need as input only $R_{(\mu-1)(\nu=\mu)s}(x)$, $R_{(\mu-1)(\nu=\mu)(s+1)}(x)$, and $R_{(\mu-1)(\nu=\mu-1)(s+1)}(x)$. Similarly, (3-25) uses only $R_{\mu(\nu=\mu)s}(x)$, $R_{\mu(\nu=\mu)(s+1)}(x)$, and $R_{(\mu-1)(\nu=\mu)(s+1)}(x)$ to produce $R_{\mu(\nu=\mu+1)s}(x)$. The series of rational functions with either $\nu = \mu$ and $\nu = (\mu+1)$ is a complete set of relations, and no other rational functions are needed to propagate the series.

If we are restricting ourselves to rational functions with $\nu = \mu$ or $\nu = (\mu+1)$, it is possible to simplify the notation somewhat by using a single index,

$$m = \mu + \nu$$

to label the polynomial degrees. It is obvious that m is even for $\nu = \mu$ and odd for $\nu = \mu + 1$. The two recursion formulas, (3-24) and (3-25), may be combined into a single one,

$$R_{ms}(x) = R_{(m-1)(s+1)}(x) + \frac{R_{(m-1)(s+1)}(x) - R_{(m-1)s}(x)}{\frac{\alpha_s}{\alpha_{m+s}}\left\{1 - \frac{R_{(m-1)(s+1)}(x) - R_{(m-1)s}(x)}{R_{(m-1)(s+1)}(x) - R_{(m-2)(s+1)}(x)}\right\} - 1} \tag{3-26}$$

with the understanding that $R_{0s} = f_s$ and $R_{ms}(x) = 0$ for $m < 0$.

Recursion relation between differences. The recursion relation for rational functions given by (3-26) is analogous to that given by (3-6) for polynomials. As we have seen in the previous section, it is advantageous for computational purposes to use instead the recursion relations for the differences between rational functions of different degrees, as we have done in (3-10) for Neville's algorithm. For this purpose, we shall again define two difference functions:

$$\Delta_{ms} \equiv R_{ms}(x) - R_{(m-1)s}(x) \tag{3-27}$$

$$\Theta_{ms} \equiv R_{ms}(x) - R_{(m-1)(s+1)}(x) \tag{3-28}$$

To find the recursion relations between Δ_{ms} and Θ_{ms}, we can start with (3-26) and obtain the result

$$\Theta_{ms} = \frac{R_{(m-1)(s+1)}(x) - R_{(m-1)s}(x)}{\frac{\alpha_s}{\alpha_{m+s}}\left\{1 - \frac{R_{(m-1)(s+1)}(x) - R_{(m-1)s}(x)}{R_{(m-1)(s+1)}(x) - R_{(m-2)(s+1)}(x)}\right\} - 1} \tag{3-29}$$

Note that, by taking the difference between (3-27) and (3-28) for $(m+1)$ instead of m, we have the identity

$$\Delta_{(m+1)s} - \Theta_{(m+1)s} = R_{m(s+1)}(x) - R_{ms}(x) \qquad (3\text{-}30)$$

Similarly, if we change s to $(s+1)$ in (3-27) before taking the difference, we have

$$\Delta_{m(s+1)} - \Theta_{ms} = R_{m(s+1)}(x) - R_{ms}(x) \qquad (3\text{-}31)$$

On substituting these two equations into the right side of (3-29), we obtain one of the two recursion relations for Δ_{ms} and Θ_{ms},

$$\Theta_{(m+1)s} = \frac{(\Delta_{m(s+1)} - \Theta_{ms})\Delta_{m(s+1)}}{\frac{\alpha_s}{\alpha_{m+s+1}}\Theta_{ms} - \Delta_{m(s+1)}} \qquad (3\text{-}32)$$

Box 3-2 Program RAT_INT

Rational function interpolation using Neville's algorithm

Subprogram used:
  RAT_NEV: Subroutine to carry out the calculations.
Initialization:
  Input $\{x_i, f(x_i)\}$ for $i = 0$ to N.
1. Input the value of x.
2. Use RAT_NEV to carry out the interpolation for the value of $f(x)$:
   (a) Lowest-order approximation.
       (i) Set up index IDX to indicate the table entry nearest to x.
       (ii) Define Δ_{0s} and Θ_{0s} to be equal to $f(x_s)$.
       (iii) Let $f(x)$ be equal to the nearest input value.
       (iv) Use a logical variable UP to indicate whether $x < x_i$ (.TRUE.) or not (.FALSE.).
   (b) Improve the estimate using higher-order approximations:
       (i) Let the order $L = 1$.
       (ii) Calculate Δ and Θ according to (3-32) and (3-33).
       (iii) If UP is .TRUE.:
            Decrease IDX by 1.
            If IDX≥ 1, increase the estimate of $f(x)$ using Θ.
            Otherwise, set IDX=1 and use Δ.
            Change UP to .FALSE. and go to step (v).
       (iv) If UP is .FALSE.:
            Increase the estimate by Δ if IDX$> (N-L)$.
            Otherwise, set IDX$= (N-L)$ and use Θ.
            Change UP to .TRUE..
       (v) If L is less than maximum order, increase L by 1 and go back to step (iii).
   (c) Return the interpolated value and the last improvement as the error estimate.
3. Output the result together with the error estimate.

The other may be found by noting the fact that, from (3-30) and (3-31), we have the expression

$$\Delta_{(m+1)s} - \Theta_{(m+1)s} = \Delta_{m(s+1)} - \Theta_{ms}$$

Using this equivalence, we can write

$$\Delta_{(m+1)s} = \Theta_{(m+1)s} + \Delta_{m(s+1)} - \Theta_{ms}$$

$$= \frac{(\Delta_{m(s+1)} - \Theta_{ms})\Delta_{m(s+1)}}{\frac{\alpha_s}{\alpha_{m+s+1}}\Theta_{ms} - \Delta_{m(s+1)}} + (\Delta_{m(s+1)} - \Theta_{ms})$$

$$= \frac{\frac{\alpha_s}{\alpha_{m+s+1}}\Theta_{ms}(\Delta_{m(s+1)} - \Theta_{ms})}{\frac{\alpha_s}{\alpha_{m+s+1}}\Theta_{ms} - \Delta_{m(s+1)}} \quad (3\text{-}33)$$

These two relations are analogous to those of (3-10) for polynomials. By starting with

$$\Delta_{0s} = \Theta_{0s} = f_s$$

we have now a method for rational function interpolation that is very similar to Neville's algorithm for polynomial interpolation. This is true also for the strategy to make a choice between Δ_{ms} and Θ_{ms} to improve the interpolated result at each stage. The algorithm is summarized in Box 3-2.

Table 3-4 Rational function interpolation for $\tan x$ near $x = \pi/2$.

Input		Interpolation			
x	$\tan x$	x	Exact	Rational	Polynomial
1.1	1.9647597	—	—	—	—
1.2	2.5721516	1.15	2.2344969	$2.2344921\,(2 \times 10^{-4})$	$2.0980797\,(2 \times 10^{-2})$
1.3	3.6021024	1.25	3.0095697	$3.0095730\,(2 \times 10^{-4})$	$3.0862589\,(1 \times 10^{-1})$
1.4	5.7978837	1.35	4.4552218	$4.4552164\,(6 \times 10^{-4})$	$4.3438010\,(1 \times 10^{-1})$
1.5	14.1014200	1.45	8.2380928	$8.2381449\,(3 \times 10^{-3})$	$8.7133064\,(2 \times 10^{-1})$

As an application, we shall use the method of rational function interpolation to calculate the values of $\tan x$ for x near the singular point at $x = \pi/2$. The input values are chosen to be at $x = 1.1, 1.2, 1.3, 1.4,$ and 1.5. Since the value of the tangent rises very fast as $x \to \pi/2$, the spacing of 0.1 between two adjacent input x values is rather large. This can be seen from values given in the first two columns of Table 3-4. We have made this choice on purpose so as to demonstrate the power of rational function interpolation. For the purpose of comparison, we have included also the results obtained with polynomial interpolation, as well as the exact values. We see from the table that the results obtained by using rational functions are always superior to those with polynomials. Later in §3-5, we shall once again see the power of using rational polynomials in extrapolations with singularities nearby.

§3-3 Continued fraction

A natural extension of the techniques to approximate a function by rational polynomials is to use continued fractions. In this case, the value f of a function for some value of its argument x is given in the form

$$f = b_0 + \cfrac{a_1}{b_1 + \cfrac{a_2}{b_2 + \cfrac{a_3}{b_3 + \cdots}}} \tag{3-34}$$

The coefficients $a_1, a_2, \ldots$, are called partial numerators and the coefficients $b_0, b_1, \ldots$, are known as partial denominators. In many mathematical tables, (3-34) is often written in a more compact form as

$$f = b_0 + \frac{a_1}{b_1 +} \frac{a_2}{b_2 +} \frac{a_3}{b_3 +} \cdots \tag{3-35}$$

In addition to its role in interpolation, a continued fraction is also a convenient way to evaluate certain functions, and the coefficients for many commonly used ones are available (see, M. Abramowitz, and I.A. Segun, editors, *Handbook of Mathematical Functions*, Dover, New York, 1965). We shall give here a short discussion of the method used to obtain these coefficients and to evaluate functions given in terms of continued fractions. As a practical application, we shall make use of the continued fraction form to calculate the values of the incomplete gamma function $\Gamma(a, x)$ of §4-5.

Consider a function $f(x)$ given to us in terms of its values $f_0, f_1, f_2, \ldots, f_k, \ldots$, at respectively $x = x_0, x_1, x_2, \ldots, x_k, \ldots$. Our interest here is the same as that of the previous two sections: to construct a polynomial approximation for $f(x)$ using the values given to us. Let $v_0(x), v_1(x), \ldots$, be a set of polynomials related to each other through the expression

$$v_k(x) = v_k(x_k) + \frac{x - x_k}{v_{k+1}(x)} \tag{3-36}$$

where x_k is the point where the value of $f(x)$ is f_k. As the starting point of this recursion relation, we shall let $v_0(x)$ be equal to $f(x)$ itself,

$$f(x) = v_0(x) \tag{3-37}$$

Applying (3-36) repetitively to the right side of (3-37), we obtain

$$f(x) = v_0(x) = v_0(x_0) + \frac{x - x_0}{v_1(x)}$$

$$= v_0(x_0) + \frac{x - x_0}{v_1(x_1) + \frac{x - x_1}{v_2(x)}}$$

$$= v_0(x_0) + \frac{x - x_0}{v_1(x_1) + \frac{x - x_1}{v_2(x_2) + \frac{x - x_3}{v_3(x)}}}$$

$$= \cdots \qquad (3\text{-}38)$$

When we identify $v_i(x_i)$ with b_i and $(x - x_i)$ with a_{i+1}, we obtain the standard form of a continued fraction given by (3-34). Our next task is to find the values of the coefficients $v_0(x_0)$, $v_1(x_1)$, ..., $v_k(x_k)$, ..., in terms of f_1, f_2, ..., f_k, ..., that are given to us as input.

It is obvious from (3-36) that $v_k(x)$ depends also on the values of $x_0, x_1, \ldots, x_{k-1}$. In other words, $v_k(x)$ may be expressed as a function of $x_0, x_1, \ldots, x_{k-1}$, as well as x, in the form

$$v_k(x) = \phi_k(x_0, x_1, \ldots, x_{k-1}, x)$$

We shall come back later to define the function $\phi_k(x_0, x_1, \ldots, x_{k-1}, x)$. In the meantime, since $b_k = v_k(x_k)$, we have also the relation

$$b_k = \phi_k(x_0, x_1, \ldots, x_{k-1}, x_k) \qquad (3\text{-}39)$$

The defining equation for $v_k(x)$ given by (3-36) may be rewritten in the form

$$v_{k+1}(x) = \frac{x - x_k}{v_k(x) - v_k(x_k)}$$

This gives us a recursion relation for the function $\phi_k(x_0, x_1, \ldots, x_{k-1}, x)$:

$$\phi_{k+1}(x_0, x_1, \ldots, x_k, x) = \frac{x - x_k}{\phi_k(x_0, x_1, \ldots, x_{k-1}, x) - \phi_k(x_0, x_1, \ldots, x_{k-1}, x_k)} \qquad (3\text{-}40)$$

To have a starting point for this set of relations, we shall define

$$\phi_1(x_0, x) = \frac{x - x_0}{f(x) - f_0} \qquad (3\text{-}41)$$

We are now in a position to construct a difference table of the following form:

$x_0 \quad f_0$

$x_1 \quad f_1 \quad \phi_1(x_0, x_1) = b_1$

$x_2 \quad f_2 \quad \phi_1(x_0, x_2) \quad \phi_2(x_0, x_1, x_2) = b_2$

$x_3 \quad f_3 \quad \phi_1(x_0, x_3) \quad \phi_2(x_0, x_1, x_3) \quad \phi_3(x_0, x_1, x_2, x_3) = b_3$

$\vdots \quad \vdots \quad \vdots \qquad\qquad \vdots \qquad\qquad \vdots$

The last quantity in each line gives us the value of the coefficient b_k. For the purpose of distinguishing from the reciprocal differences to be introduced below, the quantity $\phi_k(x_0, x_1, \ldots, x_k)$ is called the *inverted difference*.

Thiele's continued fraction expansion. The use of inverted difference to find the coefficients of a continued fraction is not convenient for many purposes, as $\phi_k(x_0, x_1, \ldots, x_{k-1}, x)$ is symmetric only in its last two arguments. This can be seen, for example, by writing out $\phi_2(x_0, x_1, x_2)$ explicitly in terms of f_k and x_k. Using (3-40) and (3-41), we have

$$\phi_2(x_0, x_1, x_2) = \frac{x_2 - x_1}{\phi_1(x_0, x_2) - \phi_1(x_0, x_1)} = \frac{x_2 - x_1}{\frac{x_2 - x_0}{f_2 - f_0} - \frac{x_1 - x_0}{f_1 - f_0}}$$

It is quite clear from examining the expression that the function is invariant if we interchange subscripts 1 and 2, but not so if we interchange 0 and 1 or 0 and 2.

A fully symmetric function $\rho_k(x_0, x_1, \ldots, x_k)$ may be constructed from the inverted difference $\phi_k(x_0, x_1, \ldots, x_k)$ through the recursion relation

$$\rho_k(x_0, x_1, \ldots, x_k) = \phi_k(x_0, x_1, \ldots, x_k) + \rho_{k-2}(x_0, x_1, \ldots, x_{k-2}) \qquad (3\text{-}42)$$

This is known as the *reciprocal differences*. The first two of this set of functions have the explicit forms

$$\rho_0(x_0) = \phi_0(x_0) = f_0$$

$$\rho_1(x_0, x_1) = \phi_1(x_0, x_1) = \frac{x_1 - x_0}{f_1 - f_0} \qquad (3\text{-}43)$$

To show that $\rho_k(x_0, x_1, \ldots, x_k)$ is symmetric in all its arguments, let us write out $\rho_2(x_0, x_1, x_2)$ explicitly:

$$\rho_2(x_0, x_1, x_2) = \phi_2(x_0, x_1, x_2) + \rho_0(x_0) = \phi_2(x_0, x_1, x_2) + \phi_0(x_0)$$

$$= \frac{x_0 f_0 (f_1 - f_2) + x_1 f_1 (f_2 - f_0) + x_2 f_2 (f_0 - f_1)}{x_0 (f_1 - f_2) + x_1 (f_2 - f_0) + x_2 (f_0 - f_1)}$$

It is obvious by inspection that the expression is invariant under any interchange among the three subscripts 1, 2, and 3.

We can make use of (3-40) and (3-42) to derive the following recursion relation for the reciprocal differences:

$$\rho_k(x_0, x_1, \ldots, x_{k-2}, x_{k-1}, x_k) = \rho_{k-2}(x_0, x_1, \ldots, x_{k-2})$$
$$+ \frac{x_k - x_{k-1}}{\rho_{k-1}(x_0, x_1, \ldots, x_{k-2}, x_k) - \rho_{k-1}(x_0, x_1, \ldots, x_{k-2}, x_{k-1})}$$
$$(3\text{-}44)$$

The relation between different orders may be put into the form of a difference table

$$
\begin{array}{ccccc}
x_0 & f_0 & & & \\
 & & \rho_1(x_0, x_1) & & \\
x_1 & f_1 & & \rho_2(x_0, x_1, x_2) & \\
 & & \rho_1(x_1, x_2) & & \phi_3(x_0, x_1, x_2, x_3) \\
x_2 & f_2 & & \rho_2(x_1, x_2, x_3) & \\
 & & \rho_1(x_2, x_3) & & \vdots \\
x_3 & f_3 & & \vdots & \\
\vdots & \vdots & \vdots & \vdots & \vdots
\end{array}
$$

In terms of $\rho_k(x_0, x_1, \ldots, x_k)$, the partial denominators b_k are given by

$$b_k = \rho_k(x_0, x_1, \ldots, x_k) - \rho_{k-2}(x_0, x_1, \ldots, x_{k-2})$$

The first two of these coefficients are

$$b_0 = \rho_0(x_0) = f_0 \qquad b_1 = \rho_1(x_0, x_1)$$

Formally, these are the same results obtained using inverted differences.

The main reason for using reciprocal differences is that we can take advantage of the limiting situation when all the arguments $x_1, x_2, \ldots, x_k$ approach the same value x. In this limit, we find from (3-43) that

$$\rho_1(x) = \lim_{\substack{x_0 \to x \\ x_1 \to x}} \rho_1(x_0, x_1) = \lim_{\substack{x_0 \to x \\ x_1 \to x}} \frac{x_1 - x_0}{\rho_0(x_1) - \rho(x_0)} = \left(\frac{d\rho_0(x)}{dx}\right)^{-1}$$

More generally, we can write

$$\rho_k(x) = \lim_{\substack{x_0 \to x \\ x_1 \to x \\ \cdots \\ x_k \to x}} \rho_k(x_0, x_1, \ldots, x_k) \qquad (3\text{-}45)$$

In terms of $\rho_k(x)$, we obtain the following recursion relation from (3-44),

$$\rho_k(x) = \rho_{k-2}(x) + \frac{k}{\rho'_{k-1}(x)} \qquad (3\text{-}46)$$

where

$$\rho'_k(x) \equiv \frac{d\rho_k(x)}{dx}$$

The factor k in the second term on the right side of (3-46) comes from the fact that

$$\lim_{x_k \to x} \frac{x_k - x}{\rho_{k-1}(x, x, \ldots, x, x_k) - \rho_{k-1}(x, x, \ldots, x, x)} = k\left(\frac{d\rho_{k-1}(x, x, \ldots, x, x)}{dx}\right)^{-1}$$

as there are k number of arguments of x in $\rho_{k-1}(x, x, \ldots, x, x)$ and only one x_k in the arguments of $\rho_{k-1}(x, x, \ldots, x, x_k)$. If all the arguments $x_1, x_2, \ldots$ in (3-38) take on the same value x_0, the continued fraction expression assumes the form

$$f(x) = \phi_0(x_0) + \cfrac{x - x_0}{\phi_1(x_0) + \cfrac{x - x_0}{\phi_2(x_0) + \cfrac{x - x_0}{\phi_3(x_0) + \cdots}}} \tag{3-47}$$

The values of $\phi_k(x_0)$ are generated from the recursion relation of (3-46) with

$$\phi_k(x) = \frac{k}{\rho'_{k-1}(x)} \qquad \rho_{-2}(x) = \rho_{-1}(x) = 0 \qquad \phi_0(x) = \rho_0(x) = f(x)$$

As can be seen from (3-47), only the values of $\phi_k(x)$ for $x = x_0$ appear in these relations.

Before we embark on the design of an algorithm to calculate a function given in the form of a continued fraction, it is useful to see an example of how the coefficients $\phi_k(x_0)$ are obtained. Consider the arc tangent function

$$f(x) = \tan^{-1} x$$

near the point $x_0 = 1$. Starting from $k = 0$, we have

$$\phi_0(x_0) = \tan^{-1} x_0 = \frac{\pi}{4}$$

Since $\rho_0(x) = f(x)$ and

$$\frac{d}{dx} \tan^{-1} x = \frac{1}{1 + x^2}$$

we have

$$\phi_1(x) = \frac{1}{\rho'_0(x)} = 1 + x^2$$

In this way, we obtain the $k = 1$ terms:

$$\phi_1(x_0) = 2$$

$$\rho_1(x) = \rho_{-1}(x) + \frac{1}{\rho'_0(x)} = 1 + x^2$$

From $\rho_1(x)$, we can generate the $k = 2$ terms:

$$\phi_2(x) = \frac{2}{\rho'_1(x)} = \frac{1}{x}$$

$$\rho_2(x) = \rho_0(x) + \phi_2(x) = \tan^{-1} x + \frac{1}{x}$$

The first one of these two equations gives

$$\phi_2(x_0) = 1$$

The $k = 3$ terms are

$$\phi_3(x) = \frac{3}{\rho_2'(x)} = \frac{3}{\frac{1}{1+x^2} - \frac{1}{x^2}} = -3x^2(1+x^2)$$

$$\rho_3(x) = \rho_1(x) + \phi_3(x) = (1+x^2) - 3x^2(1+x^2) = 1 - 2x^2 - 3x^4$$

From this, we obtain the results

$$\phi_3(x_0) = -6$$

and

$$\phi_4(x) = \frac{4}{\rho_3'(x)} = \frac{-1}{x+3x^2}$$

$$\rho_4(x) = \rho_2(x) + \phi_4(x) = \tan^{-1} x + \frac{1}{x} - \frac{1}{x+3x^2}$$

Using the fact that

$$\phi_4(x_0) = -\frac{1}{4}$$

we obtain the continued fraction expression for $\tan^{-1} x$ in the form

$$f(x) = \frac{\pi}{4} + \cfrac{x-1}{2 + \cfrac{x-1}{1 + \cfrac{x-1}{-6 + \cfrac{x-1}{-\frac{1}{4} + \frac{x-1}{\cdots}}}}}$$

In practice, this is not the most efficient way to evaluate $\tan^{-1} x$, as the expression converges slowly. A better form, given on page 81 of Abramowitz and Segun, is

$$\tan^{-1} x = \frac{x}{1+} \frac{x^2}{3+} \frac{4x^2}{5+} \frac{9x^2}{7+} \frac{16x^2}{9+}$$

The derivation of this result and similar ones for other functions require generalizations of (3-47). These are given in Hildebrand (*Introduction to Numerical Analysis*, McGraw-Hill, New York, 1956) and we shall not repeat them here.

Method of evaluation. The form of a continued fraction, as it stands in the way given by (3-35), must be evaluated from right to left. This is different from the way a function is usually calculated when it is expanded in terms of a series. In the latter case, it is evaluated from left to right, one term at a time, until convergence is achieved. As a result, some new techniques are needed to arrive at an efficient algorithm for evaluating a function given to us in the form of a continued fraction.

The following theorems, given on page 19 of Abramowitz and Segun, are useful for calculating the values of a continued fraction:

(I) If a_i and b_i are positive, then

$$f_{2n} < f_{2n+2} \qquad \text{and} \qquad f_{2n-1} > f_{2n+1} \qquad (3\text{-}48)$$

where

$$f_n = \frac{A_n}{B_n} = b_0 + \frac{a_1}{b_1+} \frac{a_2}{b_2+} \cdots \frac{a_n}{b_n} \qquad (3\text{-}49)$$

(II) The quantities A_n and B_n obey the recursion relations

$$A_n = b_n A_{n-1} + a_n A_{n-2}$$

$$B_n = b_n B_{n-1} + a_n B_{n-2} \qquad (3\text{-}50)$$

with

$$B_{-1} = 0 \qquad A_{-1} = B_0 = 1 \qquad A_0 = b_0 \qquad (3\text{-}51)$$

The proof of these two theorems may be carried out by induction but we shall not do it here. Our interest lies in the fact that A_n and B_n may be constructed using (3-50), with coefficients a_i and b_i obtained either from a table or derived using methods discussed above. From the ratio of these two quantities, we can calculate f_n using (3-49). Once this is done, these two steps are repeated for $(n+1)$, and so on.

The accuracy of a calculation may be estimated from the differences in the results for two successive values of n. However, because of item (I) above, it is better if we compare f_n with f_{n+2}, rather than with f_{n+1}. If the difference is smaller than the accuracy required, we can say that convergence is achieved.

As an application, we shall try to calculate the incomplete gamma function described later in §4-5 and used in §6-5 to evaluate $Q(\chi^2|\nu)$, the probability integral for a χ^2 distribution. The continued fraction form of the function is given on page 263 of Abramowitz and Segun:

$$\Gamma(a,x) = \int_x^\infty e^{-t} t^{a-1} dt = e^{-x} x^a \left\{ \frac{1}{x+} \frac{1-a}{1+} \frac{1}{x+} \frac{2-a}{1+} \frac{2}{x+} \cdots \right\} \qquad (3\text{-}52)$$

Our interest here lies only in evaluating the function in this form. As a practical matter, (3-52) is usually used only for $x < (a+1)$ where the convergence is fast. For other values of x, a series form given in §4-5 is preferred.

Let us start by constructing a table of the partial numerators a_i and denominators b_i from the continued fraction of (3-52):

i	0	1	2	3	4	5	6	7	$\cdots$
a_i		1	$1-a$	1	$2-a$	2	$3-a$	3	$\cdots$
b_i	0	x	1	x	1	x	1	x	$\cdots$

Using (3-50), we find that

$$A_1 = b_1 A_0 + a_1 A_{-1} = 1 \qquad B_1 = b_1 B_0 + a_1 B_{-1} = x \qquad (3\text{-}53)$$

From the table, we see that, for $i > 1$,

$$a_i = \begin{cases} i/2 - a & \text{for } i = \text{even} \\ (i-1)/2 & \text{for } i = \text{odd} \end{cases} \qquad b_i = \begin{cases} 1 & \text{for } i = \text{even} \\ x & \text{for } i = \text{odd} \end{cases} \qquad (3\text{-}54)$$

This allows us to simplify the recursion relations (3-50) to the form

$$A_{2n} = A_{2n-1} + (n-a)A_{2n-2} \qquad B_{2n} = B_{2n-1} + (n-a)B_{2n-2}$$
$$A_{2n+1} = xA_{2n} + nA_{2n-1} \qquad B_{2n+1} = xB_{2n} + nB_{2n-1} \qquad (3\text{-}55)$$

The only other refinement required here is to prevent the values of A_n and B_n from becoming too large. Since we are only interested in their ratio, the magnitudes of the individual coefficients are, in principle, of no concern to us. However, in computation, large values of either quantities can cause floating number overflows, even though the ratios between them remain finite. To prevent such overflows without changing their ratio, we can divide both A_{2n} and B_{2n} by a constant. For convenience, we can use the value of B_{2n-1} as this constant.

Box 3-3 Function GAMMA_CF(a, x)
Continued fraction approximation to
incomplete gamma function $\Gamma(a, x)$

Initialization:
  (a) Set the maximum number of steps to be 100 and accuracy to be 5×10^{-7}.
  (b) Let $A_0 = 0$, $B_1 = A_0 = 1$, $B_1 = x$, $f = 1.0$ according to (3-51).
1. For $n = 1$ to a maximum of 100, carry out the following steps:
  (a) Calculate A_{2n} and B_{2n} using (3-50) and divide the results by f.
  (b) Divide A_{2n-1} and B_{2n-1} also by f and calculate A_{2n+1} and B_{2n+1} using (3-50).
  (c) If B_{2n+1} does not vanish,
      (i) Set $f = B_{2n+1}$.
      (ii) Calculate $f_{2n+1} = A_{2n+1}/B_{2n+1}$.
  (d) Go to the next n if the difference from f_{2n-1} is greater than the accuracy required.
2. If the desired accuracy if reached, return $\Gamma(a, x) = e^{-x} x^a f_{2n+1}$.

The calculation is carried out in the following way. Starting from $n = 1$, both terms of order n and $(n+1)$ are treated in a single step. That is, in each step, we calculate first A_{2n} and B_{2n} and then A_{2n+1} and B_{2n+1}. At the end of each step, the value of f_{2n+1}, calculated using (3-49), is compared with that of f_{2n-1} to see if convergence is achieved. The reason for taking this approach comes from the fact that convergence is guaranteed among f_n only for n differing by 2, as stated in (3-48) above. The method is outlined in Box 3-3.

Table 3-5
Examples of incomplete gamma function $\Gamma(a,x)$ by continued fraction.

a	x	$\Gamma(a,x)$	$\Gamma(a)$	ν	χ^2	$Q(\chi^2\|\nu)$
0.5	0.5	0.56241876	$\sqrt{\pi}$	1	1	0.31731
1.0	0.5	0.60653061	1	2	1	0.60653
1.5	0.5	0.71009004	$\frac{1}{2}\sqrt{\pi}$	3	1	0.80125
2.0	0.5	0.90979600	1	4	1	0.90980
2.5	0.5	1.2795792	$\frac{3}{4}\sqrt{\pi}$	5	1	0.96257

For our example, we can check the value of $\Gamma(a,x)$ obtained in terms of the probability integral for χ^2 distribution

$$Q(\chi^2|\nu) = \frac{1}{2^{\nu/2}\Gamma(\nu/2)} \int_{\chi^2}^{\infty} e^{-t/2} t^{\nu/2-1} dt = \frac{\Gamma(a,x)}{\Gamma(a)} \qquad (3\text{-}56)$$

where $a = \frac{1}{2}\nu$, $x = \frac{1}{2}\chi^2$, and $\Gamma(a)$ is the ordinary gamma function described in §4-5. The values of the probability integral are given in standard mathematical tables. For the calculated results shown in Table 3-5, the values of $Q(\chi^2|\nu)$ obtained are identical to those tabulated up to the number of significant figures shown.

§3-4 Fourier transform

Fourier coefficients. A periodic function is one that repeats itself after a fixed interval. For example, if

$$f(x) = f(2L + x) \qquad (3\text{-}57)$$

the function $f(x)$ has a period of $2L$. One important property of such a function is that it can be expressed in terms of a Fourier series,

$$f(x) = \tfrac{1}{2}a_0 + \sum_{m=1}^{\infty}\left(a_m \cos\frac{m\pi}{L}x + b_m \sin\frac{m\pi}{L}x\right) \qquad (3\text{-}58)$$

Using the orthogonality relations between trigonometric functions,

$$\int_0^{2\pi} \cos mx \cos nx\, dx = \begin{cases} 0 & \text{if } m \neq n \\ \pi & \text{if } m = n \neq 0 \\ 2\pi & \text{if } m = n = 0 \end{cases}$$

$$\int_0^{2\pi} \sin mx \sin nx\, dx = \begin{cases} 0 & \text{if } m \neq n, \text{ or } m = n = 0 \\ \pi & \text{if } m = n \neq 0 \end{cases}$$

$$\int_0^{2\pi} \cos mx \sin nx\, dx = 0 \qquad \text{all } m, n$$

the expansion coefficients a_m and b_m of (3-58) may be obtained from the following integrals:

$$a_m = \frac{1}{L} \int_0^{2L} f(x) \cos \frac{m\pi}{L} x \, dx$$

$$b_m = \frac{1}{L} \int_0^{2L} f(x) \sin \frac{m\pi}{L} x \, dx \tag{3-59}$$

Alternatively, we can express the sine and cosine functions in terms of exponential functions,

$$\cos \frac{m\pi}{L} x = \frac{1}{2}\left(e^{i\frac{m\pi}{L}x} + e^{-i\frac{m\pi}{L}x}\right)$$

$$\sin \frac{m\pi}{L} x = \frac{1}{2i}\left(e^{i\frac{m\pi}{L}x} - e^{-i\frac{m\pi}{L}x}\right)$$

where $i^2 = -1$. On substituting these relations into (3-58), we obtain the result

$$f(x) = \sum_{\ell=-\infty}^{\infty} g_\ell e^{i\frac{\ell\pi}{L}x} \tag{3-60}$$

with $g_0 = \frac{1}{2} a_0$ and, for $n > 0$,

$$g_n = \tfrac{1}{2}(a_n - ib_n) \qquad g_{-n} = \tfrac{1}{2}(a_n + ib_n)$$

The factor $e^{i\ell x}$ is a complex number with absolute value equal to unity. Two such factors, e^{imx} and e^{inx} with $m \neq n$, differ from each other by only a phase angle. For this reason, the right side of (3-60) is often called a phase polynomial.

Since the relation given by either (3-58) or (3-60) is exact, a complete set of Fourier coefficients $\{a_m, b_m\}$, for $m = 0, 1, \ldots, \infty$ (with $b_0 = 0$), or $\{g_\ell\}$ for $\ell = -\infty$ to $+\infty$, provides an alternative way to specify the function $f(x)$. On the other hand, if the functional form of $f(x)$ is unknown and the relation between x and $f(x)$ is given to us in the form of $(N+1)$ pairs of values (x_i, f_i) for $i = 0, 1, \ldots, N$, we can determine the values of at most $(N+1)$ coefficients. Since the function is periodic, all the coefficients may be determined from the information on the function within a single cycle.

For some calculations, it is more convenient to consider, instead of $[0, 2L]$, the interval $[-L, L]$. For example, if we wish to obtain the values of the coefficients g_ℓ of (3-60) from $f(x)$ without going through the trigonometric functions, as we did earlier, we can make use of the orthogonal relation between exponential functions,

$$\frac{1}{2\pi}\int_{-\pi}^{\pi} (e^{imx})^* e^{inx} dx = \frac{1}{2L}\int_{-L}^{L} (e^{i\frac{m\pi}{L}x})^* e^{i\frac{n\pi}{L}x} dx = \delta_{m,n} \tag{3-61}$$

where the Kronecker delta $\delta_{m,n}$ is defined as

$$\delta_{m,n} \equiv \begin{cases} 1 & \text{if } m = n \\ 0 & \text{if } m \neq n \end{cases} \tag{3-62}$$

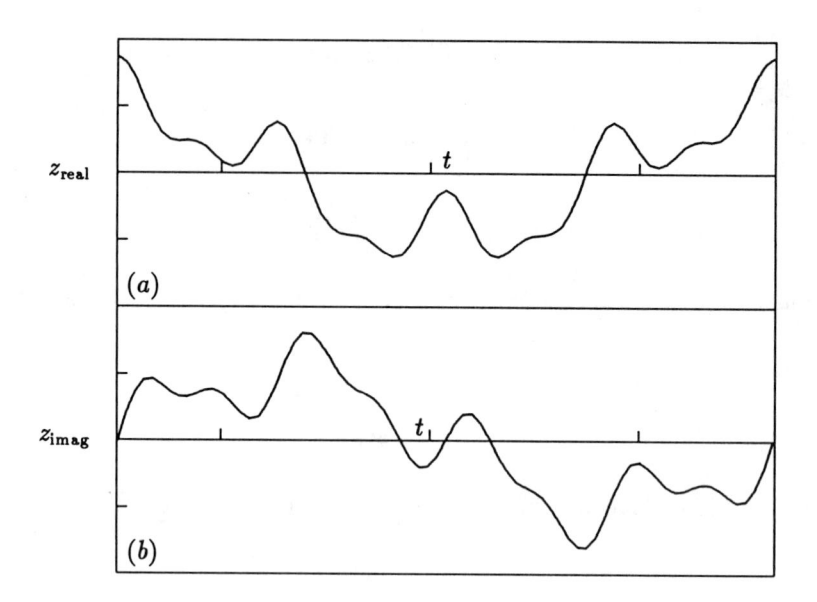

Fig. 3-2 Real and imaginary parts of $z = \sum_{\ell=1}^{3} a_\ell \exp\{i\omega_\ell t\}$ for $(a_\ell, \omega_\ell) =$ (1.0, 1.0), (0.5, 2.0), and (0.25, 8.0). Using the Fourier transform, the values of the three amplitudes and frequencies may be obtained from those of z.

On multiplying both sides of (3-60) with $\exp\{-i\ell\pi x/L\}$ and integrating between the limits $\pm L$, we obtain

$$g_\ell = \frac{1}{2L} \int_{-L}^{L} f(x) e^{-i\frac{\ell\pi}{L}x} dx \qquad (3\text{-}63)$$

This is analogous to (3-59) for the coefficients of a Fourier series in terms of sine and cosine functions. The relation between a function $f(x)$ and its Fourier coefficients is illustrated by Fig. 3-2.

Fourier transform. If we take the limit $L \to \infty$, the summation index m in (3-58) becomes a continuous variable. For later convenience, we shall write it as ω. Similarly, the summation over ℓ becomes an integration over ω and (3-60) takes on the form

$$f(x) = \frac{1}{\sqrt{2\pi}} \int_{-\infty}^{\infty} g(\omega) e^{i\omega x} d\omega \qquad (3\text{-}64)$$

In the place of the expansion of the coefficients g_ℓ, we now have the function $g(\omega)$.

A normalization factor of $1/\sqrt{2\pi}$ is introduced in (3-64) for later convenience. Instead of the orthogonality relation of (3-61), we now have the integral

$$\frac{1}{2\pi} \int_{-\infty}^{\infty} \left(e^{irx}\right)^* e^{isx} dx = \frac{1}{2\pi} \int_{-\infty}^{\infty} e^{i(s-r)x} dx = \delta(s-r) \qquad (3\text{-}65)$$

The function $\delta(s-r)$ is the Dirac delta function and has the property

$$\int_{-\infty}^{\infty} f(x)\delta(x-x_0)\,dx = f(x_0)$$

Using this property, we find that

$$g(\omega) = \frac{1}{\sqrt{2\pi}} \int_{-\infty}^{\infty} f(x) e^{-i\omega x}\,dx \qquad (3\text{-}66)$$

in parallel with (3-63). The function $g(\omega)$ is known as the inverse Fourier transform of $f(x)$. By inserting the factor $1/\sqrt{2\pi}$ in (3-64), the two functions $f(x)$ and $g(\omega)$ are symmetrical with respect to each other, except for a change of sign in the argument of the exponential function in the integrands. Instead of exponential functions, the relation between $f(x)$ and $g(\omega)$ may also be defined in terms of sines and cosines. This is left as an exercise.

The use of Fourier transformation is usually introduced in conjunction with the study of waveforms. Given two waves of the same frequency, the difference between their shapes may be stated in terms of the differences in the sizes of their Fourier coefficients $\{a_i, b_i\}$ or $\{g_\ell\}$. For example, the musical note middle C, which we shall take to be 256 Hz here, sounds quite different depending on whether it is coming from a piano or a violin. One way to "quantify" the difference is to note that a musical note is not a pure sine wave. For a note to sound musical to our ears, it must be a superposition of many waves at frequencies that are multiples, or harmonics, of the fundamental one. For the 256-Hz fundamental frequency we have adopted for our middle C note, the harmonics are at frequencies of 512 Hz, 768 Hz, 1024 Hz, and so on. That is, the amplitude of a note at some given instant of time x is given by

$$\psi(x) = \sum_{\ell=1}^{N} a_\ell \sin(\ell 2\pi \nu x)$$

where ν is the fundamental frequency, 256 Hz for our middle C example here. The intensity of each harmonic is proportional to the square of the expansion coefficient a_ℓ. Different musical instruments have different relative intensities for the harmonics.

Because of this, it is possible, in principle, to synthesize the musical notes from any instrument by providing the proper mixture of different harmonics. This is very similar in spirit to the inverse of the mathematical operation of a Fourier transformation. However, to build a good synthesizer, a large number of pure sine-wave generators are needed to "approximate" the note produced by an actual musical instrument to the accuracy that our ears can detect — usually frequencies in the range from 20 Hz to 20 kHz. In practice, additional complications come from the fact that the set of $\{a_\ell\}$ for a given instrument changes with the fundamental frequency as well as loudness. As a result, large quantities of input information are required to specify the sound of a given instrument.

In quantum mechanics, Fourier transform is used, for example, to change a wave function in coordinate representation to momentum representation. In other words, if $f(x)$ represents the wave function as a function of x, then $g(\omega)$ of (3-66) describes the behavior of the same wave function as a function of different momentum ω.

Perhaps the widest application of Fourier transform these days is in "treating" data. For example, if a sample of data contains noise because of imperfections in the measuring instrument, we can use Fourier transform techniques to filter out some of the noise. For this reason Fourier transform is used in image processing (see, for example, T.M. Cannon and B.R. Hunt, *Scientific American* [October 1981], 214). Because of its broad applications, Fourier transform has become one of the important calculations carried out on computers. (For a historical review as well as a description of some of the modern application of Fourier transform, see R.N. Bracewell, *Scientific American* [June 1989], 86). The advantage of such applications is also greatly enhanced by the development of fast Fourier transformation (FFT) algorithms. In fact, the speed with which a computer can perform FFT is one measure of the power of a modern machine.

Discrete Fourier transform. Our interest here is to calculate the Fourier coefficients $\{a_m\}$ and $\{b_m\}$ of (3-58) or $\{g_\ell\}$ of (3-38) with the function $f(x)$ given to us in terms of $(N+1)$ known values $f_0, f_1, \ldots, f_N$, at $x = x_0, x_1, \ldots, x_N$. For simplicity, we shall consider only the case where the input points are equally spaced along x at distance h between two consecutive points. Our problem is similar to that of the first two sections of this chapter. Instead of polynomials and rational functions, we are, in essence, using trigonometric functions and the equivalent exponential functions to carry out the "interpolations" here. For this reason, Fourier transform for a discrete set of points is sometimes known as trigonometric interpolation.

Although we do not need any information outside the region $x = [x_0, x_N]$ in carrying out the calculations, we shall nevertheless assume that $f(x)$ is a periodic function described by (3-57). In terms of the evenly spaced grid points along the x-axis, the same condition may be stated as

$$f(x_i) = f(x_{i+N+1})$$

We shall take x_0 as the starting point of the cycle with which we are concerned. The corresponding point of the next cycle is x_{N+1}. The indexing scheme used here is slightly different from that adopted elsewhere in this volume. For example, we find that

$$2L = x_{N+1} - x_0 = (N+1)h$$

rather than $2L = (x_N - x_0)$. This is done so that the number of subintervals of width h is $(N+1)$ rather than N. We shall soon see that such a system is more convenient for the method of calculation we wish to adopt. Furthermore, for most of our discussions below we shall use the exponential function form of (3-60). In practical applications, sine and cosine functions may be more useful, and the discussions below apply equally well to them.

We can now make use of (3-60) to construct a set of equations relating the unknown Fourier coefficients $\{g_\ell\}$ with the input quantities $\{f_i\}$ for the function $f(x)$. For simplicity, we shall take $x_0 = 0$ and, as a result of the assumption of evenly spaced points, we have the relation $x_k = kh$. On applying (3-60) for $x = x_k$, we obtain the result

$$f_k = f(x_k) = \sum_{\ell=0}^{N} g_\ell e^{i\ell\pi x_k/L} \tag{3-67}$$

The summation goes from 0 to N here. This comes from the fact that we have only $(N+1)$ pieces of input information, $f_0, f_1, \ldots, f_N$, and, consequently, we can determine at most $(N+1)$ coefficients $g_0, g_1, \ldots, g_N$.

To simplify the notation, we shall write

$$\alpha \equiv e^{i\pi h/L} = e^{i2\pi/(N+1)} \tag{3-68}$$

since $2L = (N+1)h$. For each of the $(N+1)$ values of $f(x)$ given to us, we have an equation of the form of (3-67). In terms of α, these $(N+1)$ equations may be put into the form

$$\begin{aligned}
g_0 + g_1 + g_2 + g_3 + \cdots + g_N &= f_0 \\
g_0 + \alpha g_1 + \alpha^2 g_2 + \alpha^3 g_3 + \cdots + \alpha^N g_N &= f_1 \\
g_0 + \alpha^2 g_1 + \alpha^4 g_2 + \alpha^6 g_3 + \cdots + \alpha^{2N} g_N &= f_2 \\
g_0 + \alpha^3 g_1 + \alpha^6 g_2 + \alpha^9 g_3 + \cdots + \alpha^{3N} g_N &= f_3 \\
&\vdots \\
g_0 + \alpha^N g_1 + \alpha^{2N} g_2 + \alpha^{3N} g_3 + \cdots + \alpha^{NN} g_N &= f_N
\end{aligned}$$

In matrix notation, they may be expressed as

$$Ag = f \tag{3-69}$$

where

$$A = \begin{pmatrix} 1 & 1 & 1 & 1 & \cdots & 1 \\ 1 & \alpha & \alpha^2 & \alpha^3 & \cdots & \alpha^N \\ 1 & \alpha^2 & \alpha^4 & \alpha^6 & \cdots & \alpha^{2N} \\ \vdots & \vdots & \vdots & \vdots & \ddots & \vdots \\ 1 & \alpha^N & \alpha^{2N} & \alpha^{3N} & \cdots & \alpha^{NN} \end{pmatrix} \quad g = \begin{pmatrix} g_0 \\ g_1 \\ g_2 \\ \vdots \\ g_N \end{pmatrix} \quad f = \begin{pmatrix} f_0 \\ f_1 \\ f_2 \\ \vdots \\ f_N \end{pmatrix}$$

Our aim here is to solve this equation and obtain the values of the elements of g.

Formally, the solution of (3-69) may be written as

$$g = Bf \qquad (3\text{-}70)$$

where the matrix $B = A^{-1}$. The fact that B is the inverse of A may be expressed in terms of their matrix elements in the following way:

$$(AB)_{rs} = \sum_{t=0}^{N} A_{rt} B_{ts} = \delta_{r,s} \qquad (3\text{-}71)$$

where A_{rt} is the matrix element of A in row r and column t, and B_{ts} is the matrix element of B in row t and column s. For later convenience, we shall number the $(N+1)$ rows and columns of both A and B starting from 0 and ending with N.

Note that the matrix A has a special feature in that its elements are in the form

$$A_{rs} = (\alpha)^{rs} \qquad \begin{aligned} r &= 0, 1, 2, \ldots, N \\ s &= 0, 1, 2, \ldots, N \end{aligned} \qquad (3\text{-}72)$$

That is, all the matrix elements are the integer powers of the factor α defined in (3-68), with the power given by the product of row and column numbers in the particular way we number the rows and columns. Because of this feature, elements of the inverse matrix B have the form

$$B_{rs} = \frac{1}{N+1} \alpha^{-rs} \qquad (3\text{-}73)$$

That is, B_{rs} is proportional to the inverse of A_{rs}. It is worthwhile emphasizing again that the convenient forms of (3-72) and (3-73) are, in part, the results of the labeling scheme we have adopted.

To show that a matrix with elements given by (3-73) above is the inverse of A, we need to demonstrate that the product of A and B is a unit matrix, as required by (3-71). By explicit calculation, we see that

$$\sum_{t=0}^{N} A_{rt} B_{ts} = \frac{1}{N+1} \sum_{t=0}^{N} \alpha^{rt} \alpha^{-ts} = \frac{1}{N+1} \sum_{t=0}^{N} \alpha^{(r-s)t} = \delta_{r,s} \qquad (3\text{-}74)$$

The final result comes from the following arguments. For $r = s$, each term in the sum is $\alpha^0 = 1$ and there are $(N+1)$ terms in the sum. Consequently, the result is unity.

For $r \neq s$, we shall show that the sum vanishes. Since both r and s are integers, their difference must also be an integer. Let us begin by assuming that $(r - s) = 1$. The sum then has the form

$$S = \sum_{t=0}^{N} \alpha^t \qquad (3\text{-}75)$$

Recalling that $\alpha = e^{i2\pi/(N+1)}$, each of the $(N+1)$ terms in the sum is a point on the unit circle in the complex plane. The phase angle between two adjacent points is a constant equal to $2\pi/(N+1)$. This is illustrated in Fig. 3-3 for the case of $N = 5$. In general, if $(N+1)$ is even, each point has exactly one counterpart with the opposite sign, and the contributions to the sum from these two points cancel each other. For example, at the point j the term is α^j. It is a complex number opposite in sign with the contribution at the point $j+(N+1)/2$, having the value $\alpha^{j+(N+1)/2} = \alpha^j e^{i\pi} = -\alpha^j$. If $(N+1)$ is odd, the cancellation is between a larger set of points. This can be demonstrated to one's satisfaction, for example, by evaluating the case of $(N+1) = 3$ explicitly.

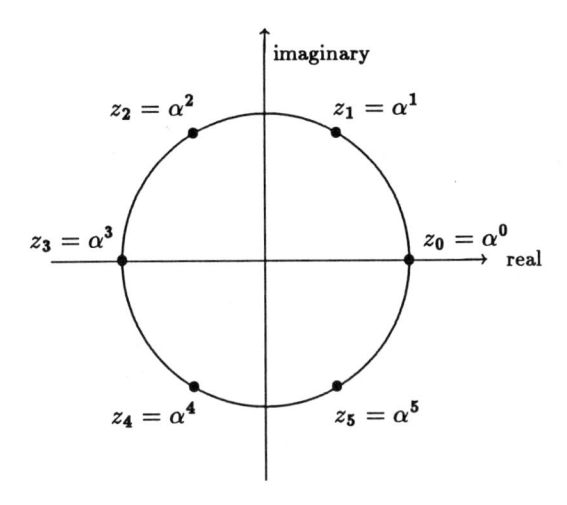

Fig. 3-3 The value of each term in the sum on the right side of (3-75) for $(N+1) = 6$. The imaginary parts of points z_1 and z_5 cancel each other. The same happens for points z_2 and z_4. The real parts of all six points sum to zero.

For $(r-s) = k > 1$, the values of the various terms in the sum of (3-74) are still distributed evenly around the unit circle except that they go around on the complex plane k times. The cancellation between the contributions from different points, however, remains in effect. For $r < s$, the same arguments apply here also, except that now the points go around the circle in the clockwise direction as t increases from 0 to N in (3-75).

With the elements of the inverse matrix B given by (3-73), we can calculate the values of the Fourier coefficients g_ℓ using (3-70):

$$g_\ell = \sum_{s=0}^{N} B_{\ell s} f_s = \frac{1}{N+1}\left(f_0 + \alpha^{-\ell} f_1 + \alpha^{-2\ell} f_2 + \alpha^{-3\ell} f_3 + \cdots + \alpha^{-N\ell} f_N\right)$$

(3-76)

This is very similar in form to (3-67). In fact, we can also express f_k in terms of g_ℓ in an analogous way:

$$f_\ell = g_0 + \alpha^\ell g_1 + \alpha^{2\ell} g_2 + 2 \cdots + \alpha^{N\ell} g_N$$

The similarity is not a coincidence. It is possible to write the pair of relations (3-69) and (3-70) in the form

$$\boldsymbol{f} = \boldsymbol{A}\boldsymbol{g} \qquad \boldsymbol{g} = \boldsymbol{A}^{-1}\boldsymbol{f} \qquad (3\text{-}77)$$

They are essentially the same relations as given by (3-64) and (3-66) except, here, the Fourier transformation is for a set of discrete points rather than for a continuous function. Furthermore, since the matrix elements of $\boldsymbol{A}$ and its inverse differ only by the sign in the powers of α, it is possible to use a single computer program to carry out both types of transformation, Fourier transform from $\boldsymbol{g}$ to $\boldsymbol{f}$ using the matrix $\boldsymbol{A}$ (positive powers of α) and inverse Fourier transform using the matrix $\boldsymbol{B} = \boldsymbol{A}^{-1}$ (negative powers of α).

In principle, we have completed our primary goal of carrying out a discrete Fourier transform by expressing the value of the coefficients g_ℓ in terms of the input quantities f_i. The only trouble is that there are, in general, $(N+1)$ Fourier coefficients to be calculated, and each requires somewhere between one to two times $(N+1)$ basic operations, that is, additions, subtractions, multiplications, divisions, and so on. As a result, the total amount of computation required to obtain all the coefficients is proportional to $(N+1)^2$. In any realistic applications, $(N+1)$ can be a very large number, perhaps on the order of 10^6. As a result, the total amount of computation required can become quite prohibitive. For this reason, algorithms for fast Fourier transform were developed in the 1960s and they are more efficient. An example of such an approach is given below.

Fast Fourier transform. Fast Fourier transform (FFT) algorithms usually work best when the number of available points $(N+1)$ is equal to some integer powers of 2. For the convenience of discussion, let us consider the case of $(N+1) = 2^\eta$ evenly spaced input points. Specifically, we shall illustrate the method with $\eta = 3$ (and $N+1 = 8$). The method itself is, of course, designed for much larger numbers of elements. From (3-68), we see that

$$\alpha^{N+1} = e^{i2\pi} = 1$$

Similarly, we have

$$\alpha^{(N+1)/2} = -1 \qquad \alpha^{(N+1)/4} = i \qquad \alpha^{(N+1)/2+\ell} = -\alpha^\ell$$

As a result, the matrix A in (3-69) takes on a particularly simple form

$$A = \begin{pmatrix} 1 & 1 & 1 & 1 & 1 & 1 & 1 & 1 \\ 1 & \alpha & i & i\alpha & -1 & -\alpha & -i & -i\alpha \\ 1 & i & -1 & -i & 1 & i & -1 & -i \\ 1 & i\alpha & -i & \alpha & -1 & -i\alpha & i & -\alpha \\ 1 & -1 & 1 & -1 & 1 & -1 & 1 & -1 \\ 1 & -\alpha & i & -i\alpha & -1 & \alpha & -i & i\alpha \\ 1 & -i & -1 & i & 1 & -i & -1 & i \\ 1 & -i\alpha & -i & -\alpha & -1 & i\alpha & i & \alpha \end{pmatrix}$$

There are many "symmetries" in this matrix. For example, the elements in the first four rows are similar to those in the second four rows, except that the odd elements (start the counting of the elements in each row from zero) have the opposite signs.

These symmetries carry over to B, the inverse of A. For example, for the $(N+1) = 8$ case we are dealing with here,

$$B = \frac{1}{8} \begin{pmatrix} 1 & 1 & 1 & 1 & 1 & 1 & 1 & 1 \\ 1 & -i\alpha & -i & -\alpha & -1 & i\alpha & i & \alpha \\ 1 & -i & -1 & i & 1 & -i & -1 & i \\ 1 & -\alpha & i & -i\alpha & -1 & \alpha & -i & i\alpha \\ 1 & -1 & 1 & -1 & 1 & -1 & 1 & -1 \\ 1 & i\alpha & -i & \alpha & -1 & -i\alpha & i & -\alpha \\ 1 & i & -1 & -i & 1 & i & -1 & -i \\ 1 & \alpha & i & i\alpha & -1 & -\alpha & -i & -i\alpha \end{pmatrix}$$

where we have made use of the fact that $\alpha^{-\ell} = \alpha^{N+1-\ell}$. By making use of the symmetries in B, methods of FFT can reduce the number of operations required to obtain all $(N+1)$ Fourier coefficients by a large extent, usually from $(N+1)^2$ to $(N+1)\log_2(N+1)$. This is a very significant factor, especially when $(N+1)$ is large.

Before we attempt to derive an algorithm for the general case, it is instructive to work out the $(N+1) = 8$ case explicitly. Because of the symmetries in the matrix B we have seen in the previous paragraph, we shall carry out the calculations in pairs, with the two members in each pair separated by $(N+1)/2$ in the values of their indexes. For the $(N+1) = 8$ example we are working on here, we shall work on g_0 together with g_4, and g_1 with g_5, g_2 with g_6, and finally g_3 with g_5. For the first pair, we find from (3-76) that

$$8g_0 = f_0 + f_1 + f_2 + f_3 + f_4 + f_5 + f_6 + f_7$$
$$8g_4 = f_0 - f_1 + f_2 - f_3 + f_4 - f_5 + f_6 - f_7$$

This can be put in a more symmetric form by defining

$$f_{10} = f_0 + f_2 + f_4 + f_6 \qquad f_{11} = f_1 + f_3 + f_5 + f_7$$

Using these intermediate quantities, we have the results

$$8g_0 = f_{10} + f_{11} \qquad 8g_4 = f_{10} - f_{11}$$

Similar grouping can also be constructed for the other three pairs, (1,5), (2,6), and (3,7). The complete list, including the (0,4) pair above, is given in Table 3-6.

Table 3-6 Reduction of g_ℓ to sums of two terms.

$8g_0 =$	$f_{10} + f_{11}$	$f_{10} =$	$f_0 + f_2 + f_4 + f_6$
$8g_4 =$	$f_{10} - f_{11}$	$f_{11} =$	$f_1 + f_3 + f_5 + f_7$
$8g_1 =$	$f_{12} - i\alpha f_{13}$	$f_{12} =$	$f_0 - if_2 - f_4 + if_6$
$8g_5 =$	$f_{12} + i\alpha f_{13}$	$f_{13} =$	$f_1 - if_3 - f_5 + if_7$
$8g_2 =$	$f_{14} - i\, f_{15}$	$f_{14} =$	$f_0 - f_2 + f_4 - f_6$
$8g_6 =$	$f_{14} + i\, f_{15}$	$f_{15} =$	$f_1 - f_3 + f_5 - f_7$
$8g_3 =$	$f_{16} - \alpha\, f_{17}$	$f_{16} =$	$f_0 + if_2 - f_4 - if_6$
$8g_7 =$	$f_{16} + \alpha\, f_{17}$	$f_{17} =$	$f_1 + if_3 - f_5 - if_7$

The intermediate quantities f_{1k}, for $k = 0, 1, \ldots, 7$, in the table again display a symmetry between the pairs of elements (0,4), (1,5), (2,6), and (3,7). In fact, a similar table to Table 3-6 can be constructed for f_{1k} (see Table 3-7).

Table 3-7 Linear combinations of f_k for calculating g_ℓ.

$f_{10} =$	$f_{20} + f_{21}$	$f_{20} =$	$f_0 + f_4$
$f_{14} =$	$f_{20} - f_{21}$	$f_{21} =$	$f_2 + f_6$
$f_{11} =$	$f_{22} + f_{23}$	$f_{22} =$	$f_1 + f_5$
$f_{15} =$	$f_{22} - f_{23}$	$f_{23} =$	$f_3 + f_7$
$f_{12} =$	$f_{24} - if_{25}$	$f_{24} =$	$f_0 - f_4$
$f_{16} =$	$f_{24} + if_{25}$	$f_{25} =$	$f_2 - f_6$
$f_{13} =$	$f_{26} - if_{27}$	$f_{26} =$	$f_1 - f_5$
$f_{17} =$	$f_{26} + if_{27}$	$f_{27} =$	$f_3 - f_7$

In an actual calculation, we start with the construction of the eight $f_{2\ell}$ from pairs of input f_k as given in Table 3-7. Next we calculate the eight $f_{1\ell}$ from pairs of f_{2k} just obtained. The final step involves the calculation of the eight g_ℓ from pairs of f_{1k}. The total number of operations is therefore $(N+1)\log_2(N+1) = 8 \times 3 = 24$, rather than $(N+1)^2 = 64$. The reduction is not very significant here, since we are only working with a small example. However, the calculation does provide us with an illustration of the basic principle of FFT.

FFT algorithm. To apply the method outlined above as a practical algorithm, we need a way to find the *phase factor* between each pair of elements so that it may be programmed for an arbitrary 2^η number of elements. To derive the phase factors, let us rewrite the two sets of equations in (3-77), one for Fourier transform and the other for inverse Fourier transform, in the form

$$\psi_{\eta,\ell} = \sum_{s=0}^{N} \beta^{\ell s} \phi_s \tag{3-78}$$

where, according to the definitions laid down in (3-69) and (3-70), we have

	Fourier transform	Inverse Fourier transform
$\psi_{\eta,\ell} =$	$(N+1)g_\ell$	f_ℓ
$\phi_s =$	f_s	g_s
$\beta =$	$\alpha^{-1} = e^{-i2\pi/(N+1)}$	$\alpha = e^{i2\pi/(N+1)}$

The presence of a second subscript to $\psi_{\eta,\ell}$ is necessary, as we shall see later, in using the same symbol to represent also the intermediate results in the calculation. The new index labels which of the η steps the results belong to. For the convenience of later discussions, we shall label this index starting from η and decreasing by unity for each step.

For the phase factor $\beta^{\ell s}$, it is useful to recall that, since $\beta = \alpha^{\pm 1}$, we have

$$\beta^{N+1} = +1 \qquad \beta^{(N+1)/2} = -1 \qquad \beta^{\pm(N+1)/4} = \pm i$$

The terms on the right side of (3-78) may be separated into an even group involving ϕ_{2s} and an odd group involving ϕ_{2s+1}:

$$\psi_{\eta,\ell} = \sum_{s=0}^{[N/2]} \beta^{\ell(2s)} \phi_{2s} + \sum_{s=0}^{[N/2]} \beta^{\ell(2s+1)} \phi_{2s+1}$$

$$= \sum_{s=0}^{[N/2]} \beta^{\ell(2s)} \phi_{2s} + \beta^\ell \sum_{s=0}^{[N/2]} \beta^{\ell(2s)} \phi_{2s+1} \tag{3-79a}$$

where for simplicity we have used $[N/2]$ to represent the integer part of the result of N divided by 2. Since $(N+1) = 2^\eta$, we have $[N/2] = (N-1)/2$, and since the summation starts with index 0, there are $(N+1)/2$ or $2^{\eta-1}$ number of terms in each of the sums above.

Similarly, for $\ell < (N+1)/2$, we have

$$\psi_{\eta,\ell+\frac{N+1}{2}} = \sum_{s=0}^{[N/2]} \beta^{(\ell+\frac{N+1}{2})2s}\phi_{2s} + \sum_{s=0}^{[N/2]} \beta^{(\ell+\frac{N+1}{2})(2s+1)}\phi_{2s+1}$$

$$= \sum_{s=0}^{[N/2]} \left(\beta^{N+1}\right)^s \beta^{\ell(2s)}\phi_{2s} + \beta^{\frac{N+1}{2}}\beta^\ell \sum_{s=0}^{[N/2]} \left(\beta^{N+1}\right)^s \beta^{\ell(2s)}\phi_{2s+1}$$

$$= \sum_{s=0}^{[N/2]} \beta^{\ell(2s)}\phi_{2s} - \beta^\ell \sum_{s=0}^{[N/2]} \beta^{\ell(2s)}\phi_{2s+1} \tag{3-79b}$$

This result is the same as that for $\psi_{\eta,\ell}$ in (3-79a) except that the sign for the second term in the final form is different. Each pair of terms, $\psi_{\eta,\ell}$ and $\psi_{\eta,\ell+\frac{n+1}{2}}$, for $\ell = 0, 1, \ldots, [N/2]$, is formed of the even and odd sums of two other terms,

$$\psi_{\eta,\ell} = \psi_{\eta-1,s} + P_{\eta-1,\ell}\psi_{\eta-1,s+\frac{N+1}{2}}$$

$$\psi_{\eta,\ell+\frac{n+1}{2}} = \psi_{\eta-1,s} - P_{\eta-1,\ell}\psi_{\eta-1,s+\frac{N+1}{2}}$$

where

$$\psi_{\eta-1,\ell} \equiv \sum_{s=0}^{[N/2]} \beta^{\ell(2s)}\phi_{2s} \qquad \psi_{\eta-1,s+\frac{N+1}{2}} \equiv \sum_{s=0}^{[N/2]} \beta^{\ell(2s)}\phi_{2s+1}$$

and the phase factor

$$P_{\eta-1,\ell} \equiv \beta^\ell$$

This is exactly what we have shown earlier in Table 3-6 for the $(N+1) = 8$ example. For later convenience, we shall adopt the order of putting the $2^{\eta-1}$ even sums ahead of the odd sums, instead of the chronological order given in the table.

The relations given in (3-79) are recursive. For example, it is possible also to express $\psi_{\eta-1,\ell}$ as a sum of two terms:

$$\psi_{\eta-1,\ell} = \sum_{s=0}^{[N/2]} \beta^{2\ell s}\phi_{2s}$$

$$= \sum_{s=0}^{[N/4]} \beta^{2\ell 2s}\phi_{4s} + \sum_{s=0}^{[N/4]} \beta^{2\ell(2s+1)}\phi_{2(2s+1)}$$

$$= \sum_{s=0}^{[N/4]} \beta^{4\ell s}\phi_{4s} + \beta^{2\ell} \sum_{s=0}^{[N/4]} \beta^{4\ell s}\phi_{4s+2}$$

$$\equiv \psi_{\eta-2,\ell} + P_{\eta-2,\ell}\psi_{\eta-2,\ell+\frac{N+1}{4}}$$

It is obvious that
$$\psi_{\eta-1,\ell+\frac{N+1}{4}} = \psi_{\eta-2,\ell} - P_{\eta-2,\ell}\psi_{\eta-2,\ell+\frac{N+1}{4}} \tag{3-80}$$
since
$$\psi_{\eta-1,\ell+\frac{N+1}{4}} = \sum_{s=0}^{[N/4]} \beta^{4s\frac{N+1}{4}} \beta^{2\ell 2s} \phi_{4s} + \sum_{s=0}^{[N/4]} \beta^{(4s+2)\frac{N+1}{4}} \beta^{2\ell(2s+1)} \phi_{2(2s+1)}$$
and
$$P_{\eta-2,\ell} = \left(\beta^2\right)^\ell$$

There are altogether four different groups of $\psi_{\eta-2,\ell}$ here, corresponding to sums of terms of the forms ϕ_{4s}, ϕ_{4s+2}, ϕ_{4s+1}, and ϕ_{4s+3}, for $s = 0, 1, \ldots, [N/4]$. In each group there are $2^{\eta-2}$ elements, distinguished from each other by the label $\ell = 0$, $1, \ldots, [N/4]$.

In the next step, the number of groups is doubled and the number of elements in each group is reduced to half of the number in each group in the previous step. The process continues until step η, where there are 2^η groups of one element each, the individual ϕ_ℓ in the input. That is,
$$\psi_{0,0}^k = \phi_\ell$$
where we have adopted temporarily a superscript k on the left side of the equation indicating to which of the 2^η groups the element belongs.

In principle, this completes the set of recursion relations for carrying out FFT. However, we still do not have an easy way in each step to program the linear combination of two $\psi_{r,s}$ to form the two new $\psi_{r+1,t}$ for the next step. To achieve this, we need to define a system to order the elements in such a way that it is convenient to form the groups.

Bit-reversed order. The order we shall adopt is related to the degree of "evenness" of the index and is known as the *bit-reversed order*. Before we see the advantages of such a system in FFT calculations, let us first define more precisely what is meant by an integer that is more "even" than another.

In a binary representation, an even number has the last bit equal to 0 and an odd number has the last bit equal to 1. Following this idea, we can judge how "even" an integer is by the number and positions of the 0's in the binary representation. Thus, the index $b000 \cdots 000 = 0$ is the most "even." (We have preceded each number with a letter b to indicate that it is written in binary representation, as we did earlier in §1-2.) The next most even has a 1 in the first bit and all the rest of the bits are zero: $b100 \cdots 000 = 2^{\eta-1}$. Here, η is the total number of binary bits required to represent the set of integers under consideration. This is followed by $b010 \cdots 000 = 2^{\eta-2}$, $b110 \cdots 000 = 2^{\eta-2} + 2^{\eta-1}$, $b001 \cdots 000 = 2^{\eta-3}$, and so on. If we reverse the order of the binary bits, the

Table 3-8 Order of input for FFT.

r	Subscript ℓ	Subscript $\ell + 2^{\eta-1}$	Binary ℓ	Binary $\ell + 2^{\eta-1}$	Bit-reversed ℓ	Bit-reversed $\ell + 2^{\eta-1}$
	0	$2^{\eta-1}$	$000\cdots000$	$100\cdots000$	$000\cdots000$	$000\cdots001$
2	$2^{\eta-2}$	$2^{\eta-2}+2^{\eta-1}$	$010\cdots000$	$110\cdots000$	$000\cdots010$	$000\cdots011$
3	$2^{\eta-3}$	$2^{\eta-3}+2^{\eta-1}$	$001\cdots000$	$101\cdots000$	$000\cdots100$	$000\cdots101$
	$2^{\eta-3}+2^{\eta-2}$	$2^{\eta-3}+2^{\eta-2}+2^{\eta-1}$	$011\cdots000$	$111\cdots000$	$000\cdots110$	$000\cdots111$
⋮	⋮	⋮	⋮	⋮	⋮	⋮
r	$2^{\eta-r}$	$2^{\eta-r}+2^{\eta-1}$	$000\cdots10\cdots$ $\cdot00\cdots000$	$100\cdots10\cdots$ $\cdot00\cdots000$	$000\cdots00\cdots$ $\cdot01\cdots000$	$000\cdots00\cdots$ $\cdot01\cdots001$
	$2^{\eta-r}+2^{\eta-2}$	$2^{\eta-r}+2^{\eta-2}+2^{\eta-1}$	$010\cdots10\cdots$ $\cdot00\cdots000$	$110\cdots10\cdots$ $\cdot00\cdots000$	$000\cdots00\cdots$ $\cdot01\cdots010$	$000\cdots00\cdots$ $\cdot01\cdots011$
⋮	⋮	⋮	⋮	⋮	⋮	⋮
	$2^{\eta-r}+2^{\eta-r+1}$ $+\cdots+2^{\eta-2}$	$2^{\eta-r}+2^{\eta-r+1}$ $+\cdots+2^{\eta-2}+2^{\eta-1}$	$011\cdots10\cdots$ $\cdot00\cdots000$	$111\cdots10\cdots$ $\cdot00\cdots000$	$000\cdots00\cdots$ $\cdot01\cdots110$	$000\cdots00\cdots$ $\cdot01\cdots111$
⋮	⋮	⋮	⋮	⋮	⋮	⋮
	$2^{\eta-1}-1$	$2^\eta - 1$	$011\cdots111$	$111\cdots111$	$111\cdots110$	$111\cdots111$

first four members of this list become $b000\cdots000$, $b000\cdots001$, $b000\cdots010$, and $b000\cdots011$, exactly the numbers 0, 1, 2, and 3, respectively, in decimal notation. In other words, the bit-reversed order is a list of numbers arranged in ascending order if we read the bits in the binary representation backward, as illustrated in Table 3-8.

A good way of getting a feeling for the bit-reversed order is to generate such a list. We can start by writing the members in the binary representation. If there are $(N+1)$ elements in the list, each element may be represented by an index of $\eta = \log_2(N+1)$ binary bits. In this scheme, the first member in the list is simply $b000\cdots000$, with all bits zero. The second member in the list has the index $b100\cdots000$, with all the bits zero except the first. If we reverse the order of the binary bits, the result is the number 1. Similarly, the third member in the bit-reversed order list is $b010\cdots000$, corresponding to the number 2 when the order of bits is reversed. This can go on until we come to the last member, which is $b111\cdots111$ $(=2^{\eta-1})$, as show in Table 3-8.

We now have a natural way to generate a list of 2^η numbers in the bit-reversed order. The first is 0 and this constitutes a group of one element in the list we wish to construct. The next most even number is $M_v = 2^{\eta-1}$. We shall include this number into our list by adding M_v to every member in the existing list. In the process, the length of the list is doubled, since an equal number of elements as before is now added to the list. We can double the length of the list again by reducing the value of M_v by a factor of 2 and adding it to all existing members in the list we have constructed so far. After repeating this operation η times, the value of M_v is reduced to 1 and the total number of elements in the list becomes 2^η. This completes the bit-reversed order of a list of $(N+1)$ members. The algorithm is outlined in Box 3-4.

Box 3-4 Program BIT_REV

Generate a list of 2^η items in bit-reversed order

1. Input η.
2. Start with one element.
   (a) Set this element to zero and put it into the list.
   (b) Define $M_v = 2^\eta$.
3. For k from 1 to η, carry out the following steps:
   (a) Divide M_v by 2.
   (b) Generate new elements by adding M_v to all the existing elements.
4. Output the list.

The advantage of arranging the elements according to how even the subscript is may be seen by working out the step before the last one in our FFT algorithm. The number of groups at this stage is $(N+1)/2$ and there are two elements in each group. The elements of the most even group have the form

$$\psi_{1,\ell} = \sum_{s=0}^{1} \beta^{2^{\eta-1}\ell s} \phi_{2^{\eta-1}s} = \phi_0 \pm \phi_{2^{\eta-1}}$$

where the $\pm$ sign depends on whether ℓ is even or odd. Since there are two members in this group, we shall label them $\ell = 0$ and 1. The next group of two elements is formed from linear combinations of $\phi_{2^{\eta-2}}$ and $\phi_{2^{\eta-2}+2^{\eta-1}}$:

$$\psi_{1,\ell} = \sum_{s=0}^{1} \beta^{2^{\eta-1}(\ell-2)s} \phi_{2^{\eta-2}+2^{\eta-1}s} = \phi_{2^{\eta-2}} \pm \phi_{2^{\eta-2}+2^{\eta-1}}$$

We shall label these two $\ell = 2$ and 3. It is easy to check that the phase factor between the two elements on the right side of each of the two equations is either $+1$ or -1 and that the values of the subscripts the two elements are separated by is $2^{\eta-1}$. Other linear combinations of two elements may be taken in this way for all the $2^{\eta-1}$ groups in this step, and the resulting order of the elements is also shown in Table 3-8.

Before describing an algorithm to take advantage of the bit-reversed order for FFT, let us see why the arrangement is useful. In a system consisting of 2^η elements in the bit-reversed order, the first element is the zeroth element and the second is $\ell = 2^{\eta-1}$. Since these are the two most even elements, the sum constructed from these two elements is the most even of any two elements. The pair of elements following these two are the next most even and they form the next most even sum of two elements. By going through the list of input elements arranged in bit-reversed order in this way, we can form all the sums of two elements in order of decreasing evenness. This gives us half the sums we need in the first of η steps of FFT. The other half of the sums comes by taking the differences between pairs of input elements. Again, the bit-reversed order gives us these sums ordered according to the order we have defined.

Storage considerations. There is one more improvement we can make to the method before implementing it as an algorithm. We have implicitly assumed that the ordered list we generate in each step is stored in a new array of $(N+1)$ elements. To carry out the complete transformation in this way, we will need a total of η arrays, each of length $(N+1)$, to store the results of each step. Since we wish to handle cases with large numbers of elements, the method proves to be uneconomical in terms of storage locations. We can solve this problem by defining two arrays each of length $(N+1)$, one for the input list and one for the output list. At each of the η steps, we move the output list of the previous step into the input list and store the calculated values in the output list. The problem with this method is that the number of operations is increased by almost $(N+1)\log_2(N+1)$, a significant fraction compared with the total number of operations required to carry out the entire FFT. On the other hand, the storage requirement is now reduced to $2(N+1)$ locations.

It is possible to do even better in terms of storage locations. Since each pair of elements in the input list is used only to form the sum and difference of a pair of elements (together with a phase factor), they are not needed in any of the subsequent calculations. As a result, we can store the two output quantities at the end of each group of calculations into the two storage locations no longer needed. For the convenience of the next step, we shall put the result obtained by summing in the array ahead of that obtained by taking the difference. In this way, a single array is adequate for both input and output. The only trouble with this method is that the output list is only partially ordered. This can be seen from Table 3-9, where we have displayed the output from each of the three steps in the eight-element example used earlier. In step 0, the list of f_k is arranged in the bit-reversed order on input and is given in the first column. At the end of step 1, the linear combinations of two f_k's are in the order as shown in the second column. If we number the most "symmetric" pair as 0 and arrange the rest of the elements according to the symmetry we have adopted, the order of the entries in the second column is 0, 2, 1, 3, 4, 6, 5, and 7. It is essentially in ascending order except that, within each group of four elements, there is a reversal between the second and third members. However, the partial order is sufficiently simple

Table 3-9 Order to carry out FFT for eight elements.

s	$\psi_{3,s}$	$\psi_{2,s}$	$\psi_{1,s}$
0	$0+4$	$(0+4)+(2+6)$	$\{(0+4)+(2+6)\}+\{(1+5)+(3+7)\} = g_0$
4	$0-4$	$(0-4)+(2-6)$	$\{(0-4)+(2-6)\}+\{(1-5)+(3-7)\} = g_1$
2	$2+6$	$(0+4)-(2+6)$	$\{(0+4)-(2+6)\}+\{(1+5)-(3+7)\} = g_2$
6	$2-6$	$(0-4)-(2-6)$	$\{(0-4)-(2-6)\}+\{(1-5)-(3-7)\} = g_3$
1	$1+5$	$(1+5)+(3+7)$	$\{(0+4)+(2+6)\}-\{(1+5)+(3+7)\} = g_4$
5	$1-5$	$(1-5)+(3-7)$	$\{(0-4)+(2-6)\}-\{(1-5)+(3-7)\} = g_5$
3	$3+7$	$(1+5)-(3+7)$	$\{(0+4)-(2+6)\}-\{(1+5)-(3+7)\} = g_6$
7	$3-7$	$(1-5)-(3-7)$	$\{(0-4)-(2-6)\}-\{(1-5)-(3-7)\} = g_7$

and, as we shall see later, it is not difficult to recognize and compensate for it in a computer program.

The same type of departure from an ordered list happens again in the output of the third step. The order is now 0, 4, 2, 6, 1, 5, 3, and 7. For our eight-element example, this is the bit-reversed order from the input list. Since this is the last step in this example, it is not surprising that the order is exactly the opposite to the input one. As the input to step 1 is arranged according to bit-reversed order, it is not surprising that the output at the last step is in the ascending order for the Fourier coefficients g_ℓ we want. In this way, an efficient algorithm to carry out FFT is obtained by starting from a list of input in bit-reversed order.

The calculations involved may now be summarized in the following way. We assume that the input for the calculation is arranged in the bit-reversed order. From this list, we can take each pair of input elements in turn and form sums and differences. These are stored back into the two locations from which the input pair was taken. When this operation is carried to the end of the list, the first of the η steps is completed. Because of the partial order in the results, the next step involves the formation of sums and differences of a pair of members separated by one element. Here, we must also incorporate the phase factor

$$P_{k,\ell} = \left(\beta^m\right)^\ell \qquad \text{with} \qquad m = \eta - k \qquad (3\text{-}81)$$

where $\ell = 0, 1, 2, \ldots, 2^{k+1}$ labels the pair and $k = 0, 1, 2, \ldots, (\eta - 1)$ labels the step, as can be seen by generalizing (3-80) to step k. Again, these results are stored in the locations left vacant by the pair used as the input for this step. The process continues until step η, where a complete list of the output in the usual ascending order is produced. This is shown by the $\eta = 3$ step in our $(N+1) = 8$ example earlier. In each step, the pair of input elements with which we wish to work is separated by the distance $(2^{k-1} - 1)$ elements, where $k = 1, 2, \ldots, \eta$ is the step number. That is, in step $k = 1$, the input list is in the bit-reversed order and all the pairs are located adjacent to each other. In step $k = 2$, they are separated by one element, step $k = 3$ by three elements, step $k = 4$ by seven elements, and

> **Box 3-5 Program FFT**
> **Fast Fourier Transform**
>
> 1. Input the list of complex amplitudes.
> 2. Arrange the input amplitudes into bit-reversed order:
>   (a) Generate a bit-reversed order list using Box 3-4.
>   (b) Store the results temporarily in the array for the output.
>   (c) Replace each element of the array by the input element to which it is pointing.
> 3. Initialization for taking sums and differences of two elements:
>   (a) Define the number of elements in each group to be 1.
>   (b) Set $\alpha = \exp\{-2\pi/(2N+1)\}$.
>   (c) Set the base of phase factor equal $n_b = (2N+1)$.
> 4. Carry out the η steps of taking sums and differences of two elements:
>   (a) Set the spacing between elements in the same group.
>   (b) Double the number of elements in each group from the previous step.
>   (c) Reduce n_b to half and calculate the basic phase factor α^{n_b}.
>   (d) Go through all the groups:
>   (i) Form sums and differences in each group.
>   (ii) Store the sum and difference.
> 5. Output the Fourier coefficients.

so on. In the final step, the separation is $(2^{\eta-1}-1)$; each element is combined with another one halfway down the list. This is shown explicitly by the $\eta = 3$ example given in Table 3-9. For simplicity, only the signs, but not the phase factors, of the terms are given. Note also that, since the phase factor depends only on the order of the members in a group, it is more efficient to carry out the transformation for the same pair of members in all the groups first and then proceed to the next pair of members.

The algorithm outlined in Box 3-5 assumes that one wishes to preserve the input list and a separate array is used for the output. To this end, we can use the output array for the dual purpose of generating the list of bit-reversed indexes as well as the storage for intermediate results in the calculation. This is not always desirable. Since FFT is designed to handle cases where the number of elements is large, one may not be able to afford a second array of $(N+1)$ elements. It is possible to modify the algorithm and to use a single array. However, this will require a slightly modified method to order the input. If we can interchange the elements in the input array in a clever way so that it can be reordered into the bit-reversed pattern, all the subsequent calculations can be performed within the same array, with the output delivered to the array used originally for the input. This, as well as other more specialized FFT calculations, may be found in Press and others (*Numerical Recipes*, Cambridge University Press, Cambridge, 1986), as well as many of the standard subroutine libraries such as NAG and IMSL.

§3-5 Extrapolation

Extrapolation is the process of approximating the values of a function $f(x)$ outside the interval $x = [x_0, x_N]$ in which there are known values. It is inherently a very delicate operation, since the available information says very little about the behavior of $f(x)$ outside the interval. For example, it is very common in physics to describe a minimum in the potential by a parabola, as shown schematically in Fig. 3-4. For our purpose here, we can approximate the function around $x = x_0$ in the form

$$f(x) \approx a_0 + a_1(x - x_0) + a_2(x - x_0)^2$$

where a_0 gives the value of $f(x)$ at the minimum. The value of a_1 is related to the asymmetry of $f(x)$ around the point and a_2, the curvature. With the values of $f(x)$ available at three or more points around x_0, we can have a fairly good description of the function in the neighborhood of x_0 using, for example, one of the interpolation routines given in earlier sections. On the other hand, in the absence of additional information, we have no way of predicting that, for the example shown in Fig. 3-4, there is a maximum beyond $x = x_5$ and the way the function increases in value for $x < x_1$. The example amplifies the fact that extrapolation can only be carried out reliably in regions close to the interval where values of $f(x)$ are known.

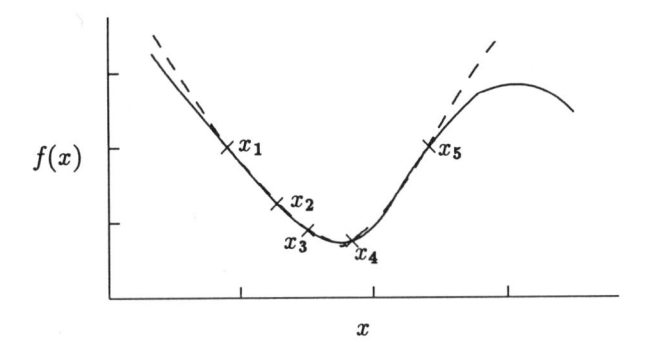

Fig. 3-4 Example illustrating the danger of extrapolation. A second-degree polynomial fit to the vicinity of a minimum of $f(x)$, shown as a dashed curve, fails to predict the maximum beyond $x = x_5$ and gives a poor representation of $f(x)$ for $x \leq x_1$.

Since extrapolation is also based on an approximate form of the unknown function $f(x)$, the starting point of any method to carry out the calculations is essentially the same as that for interpolation methods. Consequently, it is possible to make some minor adjustments to the algorithms for interpolation and adapt the same techniques for extrapolations as well. For this reason, we have taken care of the possibility of x being outside the range of the input values in the algorithm for

Neville's interpolation described in Box 3-1. Had we restricted the value of x to the central region, where the accuracy is the best, a somewhat simpler algorithm could have been adopted.

As an example, we shall consider again the function $\tan x$, used earlier in Table 3-4 for the purpose of illustrating interpolation. The input values used there were taken at $x = 1.1, 1.2, 1.3, 1.4$, and 1.5. Instead of interpolating for the values of $\tan x$ for x within the range $[1.1, 1.5]$, our interest here is to find its values for $x > 1.5$ and $x < 1.1$, using the same techniques of polynomial and rational function approximation to the function. Since there is a singularity of $\tan x$ at $x = \pi/2$, it presents somewhat of a challenge for extrapolation techniques for x near $\pi/2$. The results obtained with both methods are given in Table 3-10 and compared with the exact values. We see that, even at $x = 1.57$, the rational function approximation is able to produce a result with two-significant-figure accuracy, as well as an error estimate that is correct in the order of magnitudes.

Table 3-10 Comparison of rational function and polynomial extrapolation for $f(x) = \tan x$ using the same five input values as in Table 3-4.

x	Exact	Rational	Polynomial
1.00	1.5774077	1.5574998 (4×10^{-3})	5.2353148 (4×10^{0})
1.02	1.6281304	1.6281912 (2×10^{-3})	3.9640360 (3×10^{0})
1.04	1.7036146	1.7036510 (1×10^{-3})	3.0659318 (2×10^{0})
1.06	1.7844248	1.7844439 (8×10^{-4})	2.4730530 (8×10^{-1})
1.51	16.4280917	16.427917 (8×10^{-3})	15.583240 (1×10^{-1})
1.53	24.4984104	24.496742 (7×10^{-2})	19.040157 (5×10^{-1})
1.55	48.0784825	48.063660 (6×10^{-1})	23.236004 (1×10^{0})
1.57	1255.7655915	1237.2043 (4×10^{2})	28.274174 (2×10^{0})

Richardson extrapolation. An important application of extrapolation techniques is found in the method of *Richardson's deferred approach to the limit*. We shall use here an example of numerical integration as an illustration of the method. Later, in Chapter 8, we shall see that the approach is useful also in solving differential equations.

In §2-2, we saw that the errors in numerical integration are proportional to powers of the step size h. One way to improve the accuracy is to reduce the size of h to as small as possible. The use of extrapolation techniques is one possibility to achieve such a goal. Alternatively, we can incorporate correction terms into the method such that the errors are proportional to higher orders in h. We shall deal first with the latter approach.

For this purpose, it is convenient to express the integral in terms of the Euler-Maclaurin summation formula given in (2-10):

$$I \equiv \int_a^b f(x)\,dx$$
$$= h\left\{\tfrac{1}{2}f_0 + f_1 + f_2 + \cdots + f_{N-1} + \tfrac{1}{2}f_N\right\}$$
$$- \frac{B_2}{2!}h^2\left\{f'(b) - f'(a)\right\} - \cdots - \frac{B_{2k}}{(2k)!}h^{2k}\left\{f^{(2k-1)}(b) - f^{(2k-1)}(a)\right\} - \cdots$$
$$= h\left\{\tfrac{1}{2}f_0 + f_1 + f_2 + \cdots + f_{N-1} + \tfrac{1}{2}f_N\right\}$$
$$- \frac{h^2}{12}\left\{f'(b) - f'(a)\right\} + \frac{h^4}{720}\left\{f''(b) - f''(a)\right\} + \cdots \tag{3-82}$$

where we have divided the range $[a,b]$ into N subintervals, each of size h. The first term on the right side is the result given by the trapezoidal rule of integration

$$I(h) \equiv h\left\{\tfrac{1}{2}f_0 + f_1 + f_2 + \cdots + f_{N-1} + \tfrac{1}{2}f_N\right\} \tag{3-83}$$

In the limit of zero step size, all the other terms vanish and we have the equality

$$I = \lim_{h \to 0} I(h) \tag{3-84}$$

In a numerical calculation, it is not possible in practice to reach such a limit, as it will take an infinite number of subintervals and, consequently, an infinite amount of time to evaluate the integrand $f(x)$ at each point. What we can do by using extrapolation techniques is to find the value expected for $h \to 0$, at least in the leading order of error in h.

The Richardson technique is similar to what we did earlier in §3-1 for polynomial interpolation. Recall that, in the Simpson's rule of integration described in §2-3, we were able to eliminate terms of higher orders in h by incorporating into the approximation contributions from higher-order derivatives of $f(x)$. Here, we shall try to achieve the same goal by constructing a difference table in the values of $I(h)$ for different step sizes.

Up to terms involving h^2, (3-82) may be written in the form

$$I_N = I(h) - \frac{1}{12}\{f'(b) - f'(a)\}h^2 \tag{3-85}$$

This expression is not useful as it stands since we do not know the values of the first-order derivative $f'(x)$ at the two limits of the integration. On the other hand, the corresponding value of the integral for step size $h/2$ is given by

$$I_{2N} = I(h/2) - \frac{1}{12}\{f'(b) - f'(a)\}\left(\frac{h}{2}\right)^2 \tag{3-86}$$

Between (3-85) and (3-86), we can eliminate the unknown factor $\{f'(b) - f'(a)\}$ and obtain the result

$$I_{N,2N} = \frac{4}{3}I(h/2) - \frac{1}{3}I(h) = I(h/2) + \frac{1}{3}\{I(h/2) - I(h)\} \qquad (3\text{-}87)$$

This is a better approximation to the value of the integral than that given by $I(h)$ in (3-85), since errors of the order of h^2 have been eliminated.

We can also achieve the same result of eliminating terms of order h^2 using a linear combination I_{2N} and I_{4N}. According to our notation here, these are the approximate values of the integral obtained, respectively, with step sizes $h/2$ and $h/4$. The analogous result to (3-87) takes the form

$$I_{2N,4N} = I(h/4) + \frac{1}{3}\{I(h/4) - I(h/2)\} \qquad (3\text{-}88)$$

The error in both (3-87) and (3-88) is of the order of h^4. This can be seen by deriving these two expressions again starting from (3-82) with terms up to h^4 included:

$$I \approx I_{N,2N} + \frac{1}{720}\{f''(b) - f''(a)\}\frac{1}{4}h^4$$

$$I \approx I_{2N,4N} + \frac{1}{720}\{f''(b) - f''(a)\}\frac{1}{4}\frac{h^4}{2^4}$$

By eliminating terms that depend on the unknown factor $\{f''(b) - f''(a)\}$ from these two equations, we obtain an approximation to the integral that is accurate to order h^4:

$$I_{N,2N,4N} = I_{2N,4N} + \frac{1}{2^4 - 1}\{I_{2N,4N} - I_{N,2N}\}$$

The result is similar to (3-87), except for a further improvement in the accuracy by an order h^2. It is obvious that we can repeat the process and achieve even better accuracies. For example, by making use of $I(h/2)$, $I(h/4)$, and $I(h/6)$, we obtain $I_{2N,4N,6N}$. This, together with $I_{N,2N,4N}$, enables us to eliminate errors of the order of h^6. In each such step, we need the approximate value of the integral given by (3-83) with a step size that is half that of the last one used in the previous step.

The calculations we have carried out in the previous paragraphs may be put into the form of a recursion relation similar to that in (3-6):

$$I_{i,i+2,\ldots,i+m-2,i+m} = I_{i+2,\ldots,i+m-2,i+m}$$
$$+ \frac{1}{2^m - 1}\{I_{i+2,\ldots,i+m-2,i+m} - I_{i,i+2,\ldots,i+m-2}\} \qquad (3\text{-}89)$$

where, to simplify the notation, we have adopted the symbol

$$I_{i,i+2,\ldots,i+m-2,i+m} \equiv I_{2^i N, 2^{i+2} N, \ldots, 2^{i+m-2} N, 2^{i+m} N}$$

> **Box 3-6 Program** RICHARD
> **Integration by extrapolation using**
> **Richardson's deferred approach to limit**
>
> Subprograms used:
> TRAPZ_INT: Trapezoidal rule integration (Box 2-1).
> Initialization:
>   Set the tolerance for error to be $\epsilon = 10^{-5}$.
>   Set the final value of $h^2 = 0$.
>   Set the number of points for the extrapolation to be $m = 5$.
> 1. Input the integration limits.
> 2. Initialize the trapezoidal integration.
> 3. Set up the input to the extrapolation step.
>   (a) Calculate the integral for m different step sizes using TRAPZ_INT.
>   (b) Store the value of the integral and the square of the step size.
> 4. Extrapolate to zero step size.
> 5. Output the result.

Here m indicates that terms up to h^m have been eliminated. The relation between the various order approximations can also be represented by a difference table similar to Table 3-1, but we shall not repeat it here. The algorithm is outlined in Box 3-6.

The reason that only even values of i appear in the expression comes from the fact that only even powers of h are present beyond the first term in the Euler-Maclaurin summation formula. It is also obvious that the coefficients in the recursion relation (3-89) cannot involve any factors that depend on h. For this reason, the Bernoulli numbers multiplying the derivatives of $f(x)$ in (3-82) do not appear in the final form either.

As an example, we shall apply the technique to find the value of the error function of (3-7). The value of the integral for a given step size is calculated using the trapezoidal rule given in Box 2-1. To apply the Richardson's deferred approach to the limit, we need a number of such values, each calculated with a different step size. It is quite adequate to start with a only a few values of $I(h)$ obtained with small numbers of subintervals, such as 2, 4, 8, 16, and 32. From these five "input" values, we can use the recursion relation (3-89) to calculate the four values of I_{02}, I_{24}, I_{46}, and I_{68} and eliminate errors of order h^2. Next, we calculate the three values of I_{024}, I_{246}, and I_{468}. From these values, we can eliminate errors of order h^4. Any dependence on h^6 is taken out when we obtain the two values for I_{0246} and I_{2468}. Our final result for a five-point extrapolation is I_{02468}, which is, in principle, accurate to order h^{10}, as all terms up to h^8 have been eliminated. The values obtained at each stage for erf(x) with $x = 0.5$ are given in Table 3-11. As we can see from the table, it is possible to obtain a good

Table 3-11

Integration using Richardson's deferred approach to the limit for erf($x = 0.5$).

N	I_i	$I_{i,i+2}$	$I_{i,i+2,i+4}$	$I_{i,i+2,i+4,i+6}$	$I_{i,i+2,i+4,i+6,i+8}$
2	0.5158988				
		0.5205060			
4	0.5193542		0.5204999		
		0.5205003		0.5204999	
8	0.5202138		0.5204999		0.5204999
		0.5204999		0.5204999	
16	0.5204284		0.5204999		
		0.5204999			
32	0.5204821				
			Exact value: erf($x = 0.5$)	=	0.5204999

agreement to the exact value for a single-precision calculation with no more than five input points in the extrapolation.

Romberg integration. Another use of extrapolation techniques to carry out a numerical integration is the method of Romberg. The difference from Richardson's deferred approach to the limit may be seen by expressing $I(h)$ as a power series in h^2:

$$I(h) = a_0 + a_1 h^2 + a_2 h^4 + \cdots + a_k h^{2k} + \cdots \qquad (3\text{-}90)$$

To identify the coefficients $a_0, a_1, \ldots$ in the expression, we shall rewrite the Euler-Maclaurin formula of (3-82) in the following form with the help of (3-83):

$$I(h) = I + \frac{h^2}{12}\{f'(b) - f'(a)\} - \frac{h^4}{720}\{f''(b) - f''(a)\} + \cdots + \cdots \qquad (3\text{-}91)$$

where I is the exact value defined by (3-82) and $I(h)$ is the value obtained through numerical integration using (3-83) with a step size h. A comparison of the right sides of (3-90) and (3-91) shows that the exact value of the integral is given by the coefficient a_0:

$$a_0 = I = \int_a^b f(x)\,dx$$

The other coefficients $a_1, a_2, \ldots$ are related to the derivatives of the integrand

$$a_1 = \frac{1}{12}\{f'(b) - f'(a)\} \quad \cdots \quad a_k = \frac{B_{2k}}{(2k)!}\{f^{(2k-1)}(b) - f^{(2k-1)}(a)\}$$

In the limit $h \to 0$, all the terms on the right side vanish except the first, and we recover the result given by (3-84), stating that $I(h)$ is equal to the exact value I as $h \to 0$.

Our interest is in a_0. To extrapolate for this value we need again a number of $I(h)$ calculated with different step sizes:

$$h_1 = \frac{b-a}{n_1} \qquad h_2 = \frac{b-a}{n_2} \qquad h_3 = \frac{b-a}{n_3} \qquad \cdots$$

for $n_1 < n_2 < n_3 < \cdots$. This is the essence of the Romberg method of integration. Conceptually, it is simpler than the Richardson's deferred approach to the limit and the method allows more freedom in choosing the step sizes.

Box 3-7   Program ROMBERG

Romberg integration using Neville's algorithm

to extrapolate to zero step size

Subprograms used:
  NEVILLE: Neville's algorithm for interpolation (Box 3-1).
  TRAPZ_INT: Trapezoidal rule integration (Box 2-1).

Initialization:
  (a) Set the tolerance for error to be $\epsilon = 10^{-5}$, and the final step size for extrapolation $h^2 = 0$.
  (b) Set the number of points for the extrapolation $m = 5$, the maximum number of iterations $I_x = 10$, and the dimension of arrays to be $N_d = I_x + m$.

1. Input the integration limits.
2. Initialize the trapezoidal integration.
3. Set up the input to the extrapolation step:
   (a) Calculate the integral for m different step sizes using the TRAPZ_INT of Box 2-1.
   (b) Store the m values of the integral and squares of the step sizes.
4. Use NEVILLE to extrapolate to $h^2 = 0$.
5. Check the accuracy using the error estimates provided by NEVILLE:
   (a) If the accuracy is not enough,
      (i) Calculate the integral for a smaller step size.
      (ii) Store the results and repeat the extrapolation with the last m values.
   (b) Output the result if the required accuracy is reached.

The algorithm is outlined in Box 3-7. At the start, the integration interval $[a, b]$ is divided into n_1 subintervals, where n_1 is some reasonably small number like 6. We can make use of the trapezoidal rule to calculate the value of $I(h)$ for this step size and the result is stored as $I(h_1)$. Next, the same integral is evaluated with $n_2 = 2n_1$ subintervals and this gives us $I(h_2)$. The process is repeated for a total of m times with $n_3 = 3n_1$, $n_4 = 4n_1$, ..., $n_m = mn_1$. The values of these m pairs of values $\{I(h_1), h_1^2\}$, $\{I(h_2), h_2^2\}$, ..., $\{I(h_m), h_m^2\}$ provide us with the input for extrapolating the value of $I(h)$ to $h^2 = 0$. This step of the calculation may be carried out using the Neville algorithm of Box 3-1.

An estimate of the accuracy of the integration is provided by the uncertainty $\mathcal{E}$ of the extrapolation process. If $\mathcal{E}$ is larger than the error that can be tolerated, the numerical integration is carried out again with $(m+1)n_1$ subintervals, and the extrapolation process is repeated with $\{I(h_{m+1}), h_{m+1}^2\}$ as a part of the input. To keep the calculation simple, we can discard $\{I(h_1), h_1^2\}$ at this stage and use only the last m pairs of values of $I(h)$ and h^2 for the extrapolation. The process of going to smaller and smaller step sizes is repeated until the error is small enough or a maximum number of steps is reached.

Table 3-12 Values of erf(x) obtained using Romberg integration.

x	3-Point extrapolation	4-Point extrapolation	5-Point extrapolation	Exact value
0.3	$0.32862678 \pm (3 \times 10^{-8})$	$0.32862675 \pm (3 \times 10^{-10})$	$0.32862678 \pm (1 \times 10^{-10})$	0.32862676
0.4	$0.42839241 \pm (1 \times 10^{-7})$	$0.42839241 \pm (2 \times 10^{-10})$	$0.42839241 \pm (8 \times 10^{-11})$	0.42839236
0.5	$0.52049989 \pm (3 \times 10^{-7})$	$0.52049994 \pm (5 \times 10^{-10})$	$0.52049994 \pm (1 \times 10^{-10})$	0.52049988
0.6	$0.60385603 \pm (6 \times 10^{-7})$	$0.60385615 \pm (2 \times 10^{-9})$	$0.60385621 \pm (4 \times 10^{-11})$	0.60385609
0.7	$0.67780131 \pm (1 \times 10^{-6})$	$0.67780131 \pm (1 \times 10^{-10})$	$0.67780131 \pm (1 \times 10^{-10})$	0.67780119

As an example, the values of the error function

$$\text{erf}(x) = \frac{2}{\sqrt{\pi}} \int_0^x e^{-t^2} dt$$

are evaluated for $x = 0.3$ to 0.7. Three different sets of results are shown in Table 3-12. The first set, given in column 1, consists of the extrapolated values obtained with three pairs of input calculated with the integration range divided into 6, 12, and 18 subintervals ($m = 3$ and $n_1 = 6$). The second set, given in column 2, is obtained with one additional input of 24 subintervals ($m = 4$) and the last set, given in column 5, is obtained with the addition of one more pair consisting of 30 subintervals ($m = 5$). As we can see from the table, the final results do not differ much from those in the previous column. In each case, the error estimates provided by the extrapolation calculation are given in parentheses.

§3-6 Inverse interpolation

In interpolation, we are interested in finding the value of y at a point x within the interval $[a, b]$ using as input a table of the values of y at a finite number of points $x = x_0, x_1, \ldots, x_N$. On many other occasions, we may, instead, be interested in the opposite question of finding the value of x for a given value of y. For example, we may wish to find out the value of x at which a function $f(x)$ takes on a particular value C. This is the problem of inverse interpolation and is, in

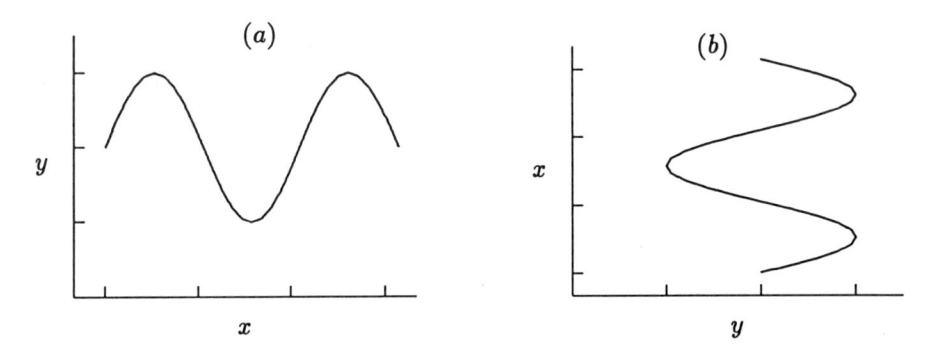

Fig. 3-5 Relation between y as a function of x and its inverse. If there is a local minimum or maximum in y, as shown in (a), there are more than one value of x corresponding to a given y, as shown in (b).

general, different from that of interpolation, as it is not always possible to simply interchange the roles of x and y in the table of input.

There are two main reasons for the difference. First, in interpolation, the value of y as a function of x is assumed to be single valued; otherwise the problem is not well defined. However, for the same function, the value of x as a function of y may not be single valued. This happens, for example, around a minium or a maximum, as illustrated schematically in Fig. 3-5. Second, most of the interpolation techniques approximate the function $y = f(x)$ by a polynomial or a rational function in x. The fact that such an approximation can be made does not guarantee that x may be represented by a polynomial or a rational function in terms of y. For this reason, it is not possible to expect that we can use interpolation techniques to carry out inverse interpolation. Only in simple cases, where $f(x)$ is a smooth, monotonically increasing or decreasing function of x (that is, $df/dx \neq 0$ in the interval) can we be assured of achieving any accuracy in using standard interpolation procedures to carry out an inverse interpolation.

The problem of inverse interpolation may be posed in the following way. Given $f(x)$, we wish to find the value of x for which

$$f(x) - C = 0 \qquad (3\text{-}92)$$

where C is the value of $f(x)$ we wish to locate. To have a unique solution, it is necessary that $f(x)$ not be a constant equal to C in a finite region of x. For this reason, we also need to add the stipulation that $f(x + \delta) - C$ and $f(x + \delta) - C$ are different in sign, at least for some small δ. In this way, the problem is made equivalent to one of finding the roots of (3-92). There are several ways to obtain a solution. One possibility, mentioned in the previous paragraph, is to think in terms of a function $g(y) = x$. If the function $y = f(x)$ is monotonic, it is possible to apply standard interpolation algorithms to find the value of $g(y)$ for a given x. In general, a rational function interpolation is preferred here so as to ensure better results. However, the usefulness of such an approach is limited.

The bisection method. A simple technique that works for a more general class of functions is to use the method of bisection to find the roots for (3-92). Let us define a new function

$$\phi(x) = f(x) - C \tag{3-93}$$

The problem of inverse interpolation for the value of x for which $f(x) = C$ is now reduced to one of searching for the zero of $\phi(x)$ in the range $x = [a, b]$. For the convenience of discussion, we shall assume that there is only one zero of $\phi(x)$ in the interval and that $\phi(a) < 0$ and $\phi(b) > 0$.

To start with, let us find the value of $\phi(x)$ at the midpoint of the interval

$$x_m = \tfrac{1}{2}(a+b)$$

If $\phi(x_m) < 0$, the zero of $\phi(x)$ must be in the interval $[x_m, b]$. On the other hand, if $\phi(x_m) > 0$, the zero must be in the interval $[a, x_m]$. In either case, the range of our search is reduced to half of that of the original one, either $[x_m, b]$ or $[a, x_m]$. The procedure of finding in which half of the interval the zero of $\phi(x)$ lies is now repeated for the new, smaller interval. In this way the search may be continued iteratively, with the interval reduced by a factor of 2 from its previous value in each step. When the size of the interval is smaller than the accuracy of x required, we can stop and adopt the midpoint of the final interval as the value of x desired. The algorithm is summarized in Box 3-8.

Box 3-8 Inverse interpolation using bisection.

Find the only root of $\phi(x)$ in an interval

Initialization:
 (a) Let ϵ be the required accuracy.
 (b) Input a as the lower limit of the range and b as the upper limit.
1. Check whether $\phi(a) < 0$ and $\phi(b) > 0$ (increasing function) or $\phi(a) > 0$ and $\phi(b) < 0$ (decreasing function).
2. Find the midpoint $x_m = \tfrac{1}{2}(a+b)$.
  (a) If $\phi(x_m) > 0$:
    (i) replace b by x_m for an increasing function, or
    (ii) replace a by x_m for a decreasing function.
  (b) If $\phi(x_m) < 0$:
    (i) replace a by x_m for an increasing function, or
    (ii) replace b by x_m for a decreasing function.
  (c) If $\phi(x_m) = 0$, exit. The root of $\phi(x)$ is at x_m.
3. Repeat step 2 if $(b-a) > \epsilon$. Otherwise, return $x = \tfrac{1}{2}(a+b)$.

The major difficulty with this procedure is that, for functions $\phi(x)$ with more than one zero in the interval, we need to know the approximate locations of all the roots. One easy way to solve this particular problem is to make a rough plot of the function using, if necessary, the method of spline described in the next section. The plot or, alternatively, a list of the values from which a plot is made gives us the approximate locations of all the roots with sufficient accuracy to be used as the starting points for the bisection method.

A second, but less serious, difficulty with the bisection method is that the convergence rate is not slow — it takes η bisections to reduce the interval by a factor of 2^η. This may be quite adequate if the need for inverse interpolation is only occasional. For more general applications, it is better to use one of the methods described below.

Newton's method of interpolation. It is useful, partly for historical interest, to derive here the interpolation method of Newton using finite differences. For this purpose, we shall consider only the approach of polynomial interpolation.

Consider the case where we are given $(N+1)$ values of $f(x)$ at $x = x_0, x_1, x_2, \ldots, x_N$. It is possible to construct a polynomial expression of degree N,

$$p(x) = c_0 + \sum_{j=1}^{N} c_j (x - x_0)(x - x_1) \cdots (x - x_{j-1}) \tag{3-94}$$

such that it goes through all the $(N+1)$ given points. That is, it is possible to find a set of values of the $(N+1)$ coefficient $c_0, c_1, \ldots, c_N$, such that

$$f(x) - p(x) = 0 \quad \text{at} \quad x = x_0, x_1, x_2, \ldots, x_N$$

At other values of x, the difference between $p(x)$ and $f(x)$ is given by

$$f(x) - p(x) = \frac{f^{(N+1)}(\xi)}{(N+1)!}(x - x_0)(x - x_1) \cdots (x - x_N) \tag{3-95}$$

where $f^{(m)}(x)$ is the mth derivative of $f(x)$ and ξ is a point in the interval $[x_0, x_N]$. The result follows from the fact that, by definition, the $(N+1)$ derivative of $p(x)$ vanishes and that of the product $(x - x_0)(x - x_1) \cdots (x - x_N)$ on the right side of (3-95) is equal to $(N+1)!$.

The coefficients c_j of $p(x)$ in (3-94) may be found from the *divided differences* of the input values $f(x_0), f(x_1), \ldots, f(x_N)$. From (3-94) and (3-95), we obtain the expression

$$f(x) = c_0 + c_1(x - x_0) + c_2(x - x_0)(x - x_1) + c_3(x - x_0)(x - x_1)(x - x_3)$$
$$+ \cdots + c_N(x - x_0)(x - x_1) \cdots (x - x_{N-1})$$
$$+ \frac{f^{(N+1)}(\xi)}{(N+1)!}(x - x_0)(x - x_1) \cdots (x - x_N) \tag{3-96}$$

Using this, we obtain immediately the result

$$f(x_0) = c_0$$

Let us define the divided difference of $f(x)$ at x and $x = x_0$ as

$$f_{x_0,x} \equiv \frac{f(x) - f(x_0)}{x - x_0}$$

It is obvious from (3-96) that the value of $f_{x_0,x}$ is

$$f_{x_0,x} = c_1 + c_2(x - x_1) + c_3(x - x_1)(x - x_2) + \cdots$$
$$+ c_N(x - x_1)(x - x_2) \cdots (x - x_{N-1}) \tag{3-97}$$

At $x = x_1$, we find that

$$f_{x_0,x_1} = c_1 \tag{3-98}$$

The coefficient c_2 is given by the divided difference between f_{x_0,x_2} and f_{x_0,x_1}. For this purpose, it is convenient to define the second-order divided difference

$$f_{x_0,x_1,x} \equiv \frac{f_{x_1,x} - f_{x_0,x_1}}{x - x_0}$$
$$= c_2 + c_3(x - x_2) + \cdots + c_N(x - x_2) \cdots (x - x_{N-1})$$

Analogous to (3-98), we obtain the result

$$f_{x_0,x_1,x_2} = c_2$$

when we put $x = x_2$ in $f_{x_0,x_1,x}$.

More generally, we can obtain higher-order divided differences through the recursion relation

$$f_{x_i,x_{i+1},x_{i+2},\ldots,c_{k-1},x_k,x} \equiv \frac{f_{x_{i+1},x_{i+2},\ldots,c_{k-1},x_k,x} - f_{x_i,x_{i+1},x_{i+2},\ldots,x_{k-1},x_k}}{x - x_i}$$
$$\tag{3-99}$$

At $x = x_{k+1}$, we obtain the value of coefficient c_{k+1} in terms of the divided difference for order $(k+1)$:

$$f_{x_0,x_1,x_2,\ldots,x_{k-1},x_k,x_{k+1}} = c_{k+1} \tag{3-100}$$

The starting point of this recursion relation is

$$f_x = f(x)$$

Using this relation, we can find all the divided differences required to calculate the coefficients $c_0, c_1, \ldots$ that define the polynomial $p(x)$. This is known as Newton's interpolation formula. An example of $(N + 1) = 6$ is given in Table 3-13.

For input involving equally spaced values of x, that is, $x_k - x_0 = kh$ with h being a constant, the divided differences of (3-99) may be expressed in terms of the forward difference $\Delta f(x)$ defined in (2-50). For example,

$$c_1 = f_{x_0,x_1} = \frac{f(x_1) - f(x_0)}{x_1 - x_0} = \frac{f_{x_1} - f_{x_0}}{x_1 - x_0} = \frac{\Delta f(x_0)}{1!h} \tag{3-101}$$

Table 3-13 Divided differences in Newton's interpolation.

x_0	$f(x_0)$					
		f_{x_0,x_1}				
x_1	$f(x_1)$		f_{x_0,x_1,x_2}			
		f_{x_1,x_2}		f_{x_0,x_1,x_2,x_3}		
x_2	$f(x_2)$		f_{x_1,x_2,x_3}		f_{x_0,x_1,x_2,x_3,x_4}	
		f_{x_2,x_3}		f_{x_1,x_2,x_3,x_4}		$f_{x_0,x_1,x_2,x_3,x_4,x_5}$
x_3	$f(x_3)$		f_{x_2,x_3,x_4}		f_{x_1,x_2,x_3,x_4,x_5}	
		f_{x_3,x_4}		f_{x_2,x_3,x_4,x_5}		
x_4	$f(x_4)$		f_{x_3,x_4,x_5}			
		f_{x_4,x_5}				
x_5	$f(x_5)$					

as can be seen from (3-97). The factorial in the denominator of the final form is included so that the form may be consistent with those we need later. Similarly,

$$c_2 = f_{x_0,x_1,x_2} = \frac{f_{x_1,x_2} - f_{x_0,x_1}}{x_2 - x_0}$$

$$= \frac{1}{2h}\left\{\frac{f(x_2) - f(x_1)}{x_2 - x_1} - \frac{f(x_1) - f(x_0)}{x_1 - x_0}\right\}$$

$$= \frac{1}{2h}\left\{\frac{\Delta f(x_1)}{h} - \frac{\Delta f(x_0)}{h}\right\}$$

$$= \frac{1}{2!h^2}\Delta^2 f(x_0) \tag{3-102}$$

where we have used the notation $\Delta^2 f(x_0) \equiv \Delta f(x_1) - \Delta f(x_0)$. The general case is

$$c_k = f_{x_0,x_1,x_2,\ldots,x_{k-1},x_k} = \frac{1}{k!h^k}\left\{\Delta^{k-1}f(x_1) - \Delta^{k-1}f(x_0)\right\} \equiv \frac{1}{k!h^k}\Delta^k f(x_0)$$

$$\tag{3-103}$$

In terms of these forward differences, Newton's formula takes on a simpler form. For $x = (x_0 + ph)$, (3-96) may be written in the form

$$f(x_0 + ph) = f_0 + \sum_{j=1}^{N} \Delta^j f(x_0) \frac{p(p-1)(p-2)\cdots(p-j+1)}{j!}$$

$$+ f^{(N+1)}(\xi) h^{N+1} \frac{p(p-1)(p-2)\cdots(p-N)}{(N+1)!} \tag{3-104}$$

where, in arriving at the result, we have made use of the relation

$$(x - x_0)(x - x_1)(x - x_2)\cdots(x - x_{k-1}) = ph\,(p-1)h\,(p-2)h\cdots(p-k+1)h$$

$$= h^k p(p-1)(p-2)\cdots(p-k+1)$$

obtained from the fact that $x_k = (x_0 + kh)$.

Bessel's formula for finite difference. A more convenient method for inverse interpolation is named after Bessel, and it may be obtained from that of Newton in the following way. Instead of $\Delta^k f(x_0)$ in (3-103), we can make use of other forward differences. For this purpose, it is worth recalling that the general result of (3-100) does not depend on any particular order of arranging the points x_0, $x_1, \ldots, x_N$. Instead of having x_0 at one end of the interval, we can put it in the middle. The $(N+1)$ input points are now labeled instead as $x_{-N/2}, x_{-N/2+1}, \ldots$, $x_{-1}, x_0, x_1, \ldots, x_{N/2}$, with $x_{-k} = (x_0 - kh)$ and $x_k = (x_0 + kh)$. In the discussions above, we have always made use of the input value $f(x_0), f(x_1), \ldots$, in ascending order according to the value of the subscript, as illustrated in Table 3-13. In the new labeling scheme, the corresponding order is $f(x_{-N/2}), f(x_{-N/2+1}), \ldots$. There is no compelling reason to follow this order. In fact, we shall see that it is more convenient to form divided differences by taking $f(x)$ in the order $f(x_0), f(x_1)$, $f(x_{-1}), f(x_2), f(x_{-2}), f(x_3), f(x_{-3}), \ldots$.

In this new scheme, the coefficient c_1 remains unchanged in form from that given by (3-101). However, the actual input points that enter into the calculation are now taken from the middle of the interval rather than from one end. The expressions for the other coefficients are somewhat different from those in Newton's scheme. For example, in the place of (3-102), we have

$$c_2 = f_{x_{-1},x_0,x_1} = \frac{1}{2!h^2}\Delta^2 f(x_{-1})$$

Similarly,

$$c_3 = f_{x_{-1},x_0,x_1,x_2} = \frac{1}{3!h^3}\Delta^3 f(x_{-1})$$

$$c_4 = f_{x_{-2},x_{-1},x_0,x_1,x_2} = \frac{1}{4!h^4}\Delta^4 f(x_{-2})$$

The function $f(x)$ may now be expressed in the form

$$f(x_0 + ph) = f(x_0) + \frac{p}{1!}\Delta f(x_0) + \frac{p(p-1)}{2!}\Delta^2 f(x_{-1}) + \frac{(p+1)p(p-1)}{3!}\Delta^3 f(x_{-1})$$
$$+ \frac{(p+1)p(p-1)(p-2)}{4!}\Delta^4 f(x_{-2}) + \cdots \qquad (3\text{-}105)$$

instead of Newton's formula (3-103).

On the other hand, Problem 3-8 shows that, by making use of the input values of $f(x)$ in the order $f(x_0), f(x_1), f(x_2), f(x_{-1}), f(x_3), f(x_{-2}), \ldots$, and expanding $f(x)$ around $x = x_1$, a slightly different approximation for $f(x)$ is obtained:

$$f(x_0 + ph) = f(x_1) + \frac{p-1}{1!}\Delta f(x_0) + \frac{p(p-1)}{2!}\Delta^2 f(x_0) + \frac{p(p-1)(p-2)}{3!}\Delta^3 f(x_{-1})$$
$$+ \frac{(p+1)p(p-1)(p-2)}{4!}\Delta^4 f(x_{-1}) + \cdots \qquad (3\text{-}106)$$

The average of (3-105) and (3-106) gives us the Bessel's formula for interpolation

$$f(x_0 + ph) = \frac{1}{2}\{f(x_0) + f(x_1)\} + (p - \tfrac{1}{2})\Delta f(x_0)$$

$$+ \frac{p(p-1)}{4}\{\Delta^2 f(x_{-1}) + \Delta^2 f(x_0)\}$$

$$+ \frac{p(p-1)(2p-1)}{12}\Delta^3 f(x_{-1})$$

$$+ \frac{(p+1)p(p-1)(p-2)}{48}\{\Delta^4 f(x_{-2}) + \Delta^4 f(x_{-1})\}$$

$$+ \cdots \qquad (3\text{-}107)$$

This is not the only way to express $f(x+ph)$ in term of forward differences. In fact, several other interpolation relations can also be derived starting from Newton's formula given by (3-103), but we shall not attempt them here.

The advantage of Bessel's formula is that the coefficients are small in general and vary only slowly with p. Let us see why this is useful in inverse interpolation. For this purpose, it is convenient to rewrite (3-107) in terms of the central difference operator $\hat{\delta}_i$, defined earlier in (2-52) as

$$\hat{\delta}_i f \equiv \hat{\delta}_i^1 f = f(x_{i+1/2}) - f(x_{i-1/2})$$

and

$$\hat{\delta}_i^n = \hat{\delta}_{i+1/2}^{n-1} - \hat{\delta}_{i-1/2}^{n-1}$$

Similarly, we can define an averaging operator $\hat{\mu}$ as

$$\hat{\mu}\phi(x) \equiv \tfrac{1}{2}\{\phi(x - \tfrac{1}{2}h) + \phi(x + \tfrac{1}{2}h)\}$$

In terms of these two operators, (3-107) may be expressed as

$$f(x_0 + ph) = \hat{\mu} f(x_{1/2}) + \frac{p - \tfrac{1}{2}}{1!}\hat{\delta}_{1/2} f + \frac{p(p-1)}{2!}\hat{\mu}\hat{\delta}_{1/2}^2 f + \frac{p(p-1)(p-\tfrac{1}{2})}{3!}\hat{\delta}_{1/2}^3 f$$

$$+ \frac{(p+1)p(p-1)(p-2)}{4!}\hat{\mu}\hat{\delta}_{1/2}^4 f + \cdots \qquad (3\text{-}108)$$

The general form of each term in the series is determined by whether it involves even or odd order central differences, that is, whether it involves $\hat{\delta}_j^k f$ for $k = 2m$ or $(2m+1)$:

$$\begin{cases} \dfrac{p(p^2 - 1^2)(p^2 - 2^2)\cdots\{p^2 - (m-1)^2\}(p-m)}{(2m)!}\hat{\mu}\hat{\delta}_{1/2}^{2m} f & \text{for } k = 2m \\[2mm] \dfrac{p(p^2 - 1^2)(p^2 - 2^2)\cdots\{p^2 - (m-1)^2\}(p-m)(p-\tfrac{1}{2})}{(2m+1)!}\hat{\delta}_{1/2}^{2m+1} f & \text{for } k = 2m+1 \end{cases}$$

The derivation of this relation is given in Hildebrand and we shall not repeat it here.

Table 3-14 Input to a fourth-order Bessel's interpolation formula.

$$
\begin{array}{lllll}
f(x_{-2}) & & & & \\
 & \hat{\delta}_{-3/2}f & & & \\
f(x_{-1}) & & \hat{\delta}^2_{-1}f & & \\
 & \hat{\delta}_{-1/2}f & & \hat{\delta}^3_{-1/2}f & \\
\underline{f(x_0)} & & \underline{\hat{\delta}^2_0 f} & & \underline{\hat{\delta}^4_0 f} \\
 & \underline{\hat{\delta}_{1/2}f} & & \underline{\hat{\delta}^3_{1/2}f} & \\
\underline{f(x_1)} & & \underline{\hat{\delta}^2_1 f} & & \underline{\hat{\delta}^4_1 f} \\
 & \hat{\delta}_{3/2}f & & \hat{\delta}^3_{3/2}f & \\
f(x_2) & & \hat{\delta}^2_2 f & & \\
 & \hat{\delta}_{5/2}f & & & \\
f(x_3) & & & & \\
\end{array}
$$

A method for inverse interpolation. The inverse interpolation problem we are interested in is now reduced to one of finding the value of p in (3-108) for which $f(x_0 + ph) = C$. If we use Bessel's formula involving central differences up to $\hat{\delta}^4_i f$, the input quantities are

$$\hat{\mu}f_{1/2} = \tfrac{1}{2}(f_1 + f_0) \qquad \hat{\delta}_{1/2}f = f(x_1) - f(x_0) \qquad \hat{\mu}\hat{\delta}^2_{1/2}f = \tfrac{1}{2}(\hat{\delta}^2_1 f + \hat{\delta}^2_0 f)$$

$$\hat{\delta}^3_{1/2}f = \hat{\delta}^2_1 f - \hat{\delta}^2_0 f \qquad \hat{\mu}\hat{\delta}^4_{1/2}f = \tfrac{1}{2}(\hat{\delta}^4_1 f + \hat{\delta}^4_0 f) \qquad (3\text{-}109)$$

These are underlined in Table 3-14. Six equidistant values of $f(x_i)$ are required to calculate the central differences involved and they are given in the first column of the table.

Once the five quantities in (3-109) are available, (3-108) may be put in the form of finding the value of p satisfying the equation

$$C \approx \hat{\mu}f_{1/2} + (p - \tfrac{1}{2})\hat{\delta}_{1/2}f + \frac{p(p-1)}{2}\hat{\mu}\hat{\delta}^2_{1/2}f + \frac{p(p-1)(2p-1)}{12}\hat{\delta}^3_{1/2}f$$

$$+ \frac{(p+1)p(p-1)(p-2)}{24}\hat{\mu}\hat{\delta}^4_{1/2}f$$

$$= f_0 + p\hat{\delta}_{1/2}f + \frac{p(p-1)}{2}\hat{\mu}\hat{\delta}^2_{1/2}f + \frac{p(p-1)(2p-1)}{12}\hat{\delta}^3_{1/2}f$$

$$+ \frac{(p+1)p(p-1)(p-2)}{24}\hat{\mu}\hat{\delta}^4_{1/2}f$$

Since the coefficients of all the terms vary only slowly with p, we can rewrite this relation in the following form:

$$p \approx \frac{1}{\hat{\delta}_{1/2}f}\left\{ C - f(x_0) - \frac{p(p-1)}{2}\hat{\mu}\hat{\delta}^2_{1/2}f - \frac{p(p-1)(2p-1)}{12}\hat{\delta}^3_{1/2}f \right.$$

$$\left. - \frac{(p+1)p(p-1)(p-2)}{24}\hat{\mu}\hat{\delta}^4_{1/2}f \right\} \qquad (3\text{-}110)$$

This is not a solution for p, as it appears on both sides of the equation. However, it may be used in an iterative procedure to find the value of p to the desired accuracy.

Table 3-15 A different indexing scheme for central differences (cf. Table 3-14).

$f(x_1)$				
	$\hat{\delta}_1 f$			
$f(x_2)$		$\hat{\delta}_1^2 f$		
	$\hat{\delta}_2 f$		$\hat{\delta}_1^3 f$	
$f(x_3)$		$\hat{\delta}_2^2 f$		$\hat{\delta}_1^4 f$
	$\hat{\delta}_3 f$		$\hat{\delta}_2^3 f$	
$f(x_4)$		$\hat{\delta}_3^2 f$		$\hat{\delta}_2^4 f$
	$\hat{\delta}_4 f$		$\hat{\delta}_3^3 f$	
$f(x_5)$		$\hat{\delta}_4^2 f$		
	$\hat{\delta}_5 f$			
$f(x_6)$				

The lowest order approximation for p from (3-110) is

$$p_0 \approx \frac{1}{\hat{\delta}_{1/2} f}\{C - f(x_0)\}$$

If we write the right side of (3-110) as a function in the following way:

$$\psi(p) \equiv \frac{1}{\hat{\delta}_{1/2} f}\left\{C - f(x_0) - \frac{p(p-1)}{2}\hat{\mu}\hat{\delta}_{1/2}^2 f - \frac{p(p-1)(2p-1)}{12}\hat{\delta}_{1/2}^3 f \right.$$
$$\left. - \frac{(p+1)p(p-1)(p-2)}{24}\hat{\mu}\hat{\delta}_{1/2}^4 f\right\} \quad (3\text{-}111)$$

successive higher order approximations of p may be put into the form of a recursive relation

$$p_{k+1} = \psi(p_k) \quad (3\text{-}112)$$

Note that the possibility of using such an approach is based on the fact that the value of $\psi(p)$ varies only slowly with p.

To carry out an inverse interpolation in practice, we need to construct a difference table of the form of Table 3-14. The five quantities that must be calculated in (3-109) to provide the necessary input to (3-112) come from $\hat{\delta}_{1/2} f$, $\hat{\delta}_1 f$, $\hat{\delta}_0 f$, $\hat{\delta}_{1/2}^3 f$, $\hat{\delta}_1^4 f$, and $\hat{\delta}_0^4 f$. Such an approach may be adequate for hand calculations — to construct a computer program to carry out the work, it is essential that we have a more automated method.

To start with, we need a method to obtain the value of any central difference $\hat{\delta}_m^n f$ from a given list of $f(x_i)$. From Table 3-13, it is easy to see that $(n+1)$ values of $f(x_i)$ are required to calculate the central difference up to order n. If we use the standard method for numbering the subscripts, such as that given in Table 3-14,

Table 3-16 Input table for inverse interpolation of the normal probability function $\frac{1}{\sqrt{2\pi}} \int_{-\infty}^{x} e^{-t^2/2} dt$.

x_i	$f(x_i)$	$\hat{\delta}_i f$	$\hat{\delta}_i^2 f$	$\hat{\delta}_i^3 f$	$\hat{\delta}_i^4 f$
0.0	0.50000000				
		0.07925971			
0.2	0.57925971		−0.00309768		
		0.07616203		−0.00273921	
0.4	0.65542174		−0.00583689		0.00064868
		0.07032514		−0.00209051	
0.6	0.72574688		−0.00792742		0.00082037
		0.06239772		−0.00127016	
0.8	0.78814460		−0.00919758		0.00085319
		0.05320014		0.00041697	
1.0	0.84134474		−0.00961455		0.00075894
		0.04358559		0.00034197	
1.2	0.88493033		−0.00927258		0.00057497
		0.03431301		0.00091694	
1.4	0.91924334		−0.00835564		0.00035030
		0.02595737		0.00126724	
1.6	0.94520071		−0.00708840		0.00013238
		0.01886897		0.00139962	
1.8	0.96406968		−0.00568878		
		0.01318019			
2.0	0.97724987				

the first element of this list is $(m - n/2)$, the next one $(m - n/2 + 1)$, and so on. A simpler approach, as far as programming a computer is concerned, is to use the scheme that each element — the input $f(x_i)$ values as well as finite differences of a given order — is numbered according to the order in which it appears. This is given in Table 3-15. The correspondence with the scheme used in Table 3-14 may be found by comparing the elements at the same locations in the two tables. In this way, the mth finite difference of order n may be calculated from $(n + 1)$ input $f(x_i)$ values starting from the mth element in the list. The algorithm to write a function that returns the finite difference of an arbitrary order is given as a part of Box 3-9.

With a general routine to calculate the central differences, it is now relatively easy to implement Bessel's formula for inverse interpolation. The algorithm is outlined in Box 3-9. As an example, we shall work out the value of x for which the normal probability integral has the value 0.75. That is, we wish to find the value of x satisfying the relation

$$\frac{1}{\sqrt{2\pi}} \int_{-\infty}^{x} e^{-t^2/2} dt = 0.75$$

The input quantities, taken from a standard mathematical table, are listed in the first two columns of Table 3-16, and the maximum error we can tolerate in the inverse interpolation is set to be $\epsilon = 10^{-3}$.

From the table, we see that the value of x we wish to find must occur between 0.6 and 0.8. Since $f(x = 0.6) = 0.72574688$ and $f(x = 0.8) = 0.78814459$, we

> **Box 3-9 Program BES_INTP**
> **Inverse interpolation using Bessel's formula**
> Assuming only one root in the interval
>
> Subprogram used:
>   CENT_DIFF: Calculate the central difference of $f(x)$ for a given order.
> Initialization:
>   (a) Set the accuracy required to be ϵ.
>   (b) Define the interpolation formula (3-111) for p as a function.
> 1. Input the list of $\{x_i, f(x_i)\}$ values.
> 2. Find the size of spacing in x.
> 3. Input y for which the value of x is to be interpolated.
> 4. Start the interpolation:
>   (a) Locate the nearest point in the table.
>   (b) Calculate the five input quantities of (3-109). Use CENT_DFF to obtain the central difference of order n by the following steps:
>   (i) Define the $(n+1)$ zeroth-order differences $\{d_i\}$ to be equal to $\{f_i\}$.
>   (ii) Calculate the next order by taking the difference $(d_{i+1} - d_i)$ and store the result in the location for d_i. For order k, there are $(n+1-k)$ such differences.
>   (iii) Repeat step (ii) until $k = n$ and return the last value of d_1 as the central difference for order n.
> 5. Calculate the value of p using the recursion relation of (3-112).
> 6. Output the result of each iteration.

choose $x_0 = 0.6$. It then follows that

$$f(x_0) = 0.72574688 \qquad \hat{\delta}_{1/2}f = 0.06239772$$

$$\hat{\delta}_0^2 f = -0.00792742 \qquad \hat{\delta}_1^2 f = -0.00919759 \qquad \hat{\delta}_{1/2}^3 f = -0.00127017$$

$$\hat{\delta}_0^4 f = 0.00082034 \qquad \hat{\delta}_1^4 f = 0.00085324$$

From this set of values we obtain

$$p_0 = 0.38868618 \qquad p_1 = 0.37217590 \qquad p_2 = 0.37246385$$

The difference between p_2 and p_1 is less than ϵ. As a result, no further iterations of (3-112) are needed. The final result, $p = 0.372$, corresponds to

$$x = x_0 + ph = 0.6 + 0.372 * 0.2 = 0.674$$

and this is our value for $f(x) = 0.75$.

Coefficients of the interpolation polynomial. There are occasions where the values of the coefficients in the interpolation polynomial are of interest. For simplicity, we shall again examine only the case of polynomial interpolation here. Using (3-1), the relation between $y = f(x)$ as a function of x is approximated by a polynomial of order N,

$$f(x) \approx a_0 + a_1 x + a_2 x^2 + \cdots + a_N x^N \tag{3-113}$$

Our problem here is that, given a table of $(N+1)$ pairs of $\{x_i, f(x_i)\}$, we wish to find the values of the coefficients $a_0, a_1, \ldots, a_N$.

It is obvious from (3-113) that, if we interpolate (or extrapolate) for the value of $f(x)$ at $x = 0$, we obtain a_0,

$$f(x=0) = a_0 \tag{3-114}$$

As we shall see below, this is the basic principle that provides us with a convenient method to extract all the coefficients.

As the starting point of the interpolation, we have a table of $(N+1)$ pairs of values $\{x_0, f(x_0)\}, \{x_1, f(x_1)\}, \ldots, \{x_N, f(x_N)\}$. Using (3-113), we may express the relations between them and the coefficients as a set of $(N+1)$ simultaneous equations

$$\begin{aligned}
f(x_0) &= a_0 + a_1 x_0 + a_2 x_0^2 + \cdots + a_N x_0^N \\
f(x_1) &= a_0 + a_1 x_1 + a_2 x_1^2 + \cdots + a_N x_1^N \\
&\vdots \\
f(x_{N-1}) &= a_0 + a_1 x_{N-1} + a_2 x_{N-1}^2 + \cdots + a_N x_{N-1}^N \\
f(x_N) &= a_0 + a_1 x_N + a_2 x_N^2 + \cdots + a_N x_N^N
\end{aligned} \tag{3-115}$$

Since the value of a_0 is known from (3-114), we can subtract it from $f(x_0)$, $f(x_1), \ldots, f(x_N)$. By dividing the resulting differences in each case by the corresponding value of x, we obtain a set of polynomial expressions of one order lower than those given in (3-115):

$$\begin{aligned}
\frac{f(x_0)-a_0}{x_0} &= a_1 + a_2 x_0 + \cdots + a_N x_0^{N-1} \\
\frac{f(x_1)-a_0}{x_1} &= a_1 + a_2 x_1 + \cdots + a_N x_1^{N-1} \\
&\vdots \\
\frac{f(x_{N-1})-a_0}{x_{N-1}} &= a_1 + a_2 x_{N-1} + \cdots + a_N x_{N-1}^{N-1} \\
\frac{f(x_N)-a_0}{x_N} &= a_1 + a_2 x_N + \cdots + a_N x_N^{N-1}
\end{aligned}$$

Any N of these $(N+1)$ relations may now be used to find the value of a_1 by interpolating (or extrapolating) for $x = 0$ in the same way as we did earlier for the value of a_0. This process is repeated N times until all $(N+1)$ coefficients are found.

Alternatively, we can put (3-115) in matrix notation

$$\begin{vmatrix} 1 & x_0 & x_0^2 & \cdots & x_0^N \\ 1 & x_1 & x_1^2 & \cdots & x_1^N \\ \vdots & \vdots & \vdots & \ddots & \vdots \\ 1 & x_{N-1} & x_{N-1}^2 & \cdots & x_{N-1}^N \\ 1 & x_N & x_N^2 & \cdots & x_N^N \end{vmatrix} \begin{vmatrix} a_0 \\ a_1 \\ \vdots \\ a_{N-1} \\ a_N \end{vmatrix} = \begin{vmatrix} f(x_0) \\ f(x_1) \\ \vdots \\ f(x_{N-1}) \\ f(x_N) \end{vmatrix} \qquad (3\text{-}116)$$

We shall see later in §5-1 that the unknown coefficients $a_0, a_1, \ldots, a_N$ may be found by inverting the determinant made of the powers of 1, $x_0, x_1, \ldots, x_{N+1}$. The particular form of the determinant we have here is known as a Vandermonde determinant and may be inverted more easily than the general case. However, we shall not go into this topic.

§3-7 Cubic spline

A spline is a piece of flexible wood or plastic that can be bent into arbitrary (smooth) shapes. In the days before computers, it was used in tracing a smooth curve between points on a sheet of graph paper. For this reason, the word is now associated with numerical techniques that perform the same function. The basic idea behind most of the methods is to use a simple function to approximate the relation between the dependent and independent variables, similar to what we did in interpolation. For most cases, it is possible to achieve such an aim with a third-degree polynomial of the form

$$f(x) \approx a_0 + a_1 x + a_2 x^2 + a_3 x^3 \qquad (3\text{-}117)$$

and hence the name *cubic spline*.

In physics, we often encounter the situation that a function $f(x)$ is given in the interval $x = [x_0, x_N]$ by a set of $(N+1)$ values $f_0, f_1, \ldots, f_N$ taken, respectively, at $x = x_0, x_1, \ldots, x_N$. In general, the spacing between any two adjacent points x_i and x_{i+1} is not a constant. Our goal in a cubic spline calculation is to find a smooth third-degree polynomial approximation to the unknown function $f(x)$ that makes the best attempt to go through all the given points.

The interpolation polynomials discussed in the first two sections of this chapter do not necessarily satisfy our requirement here, since the aim there was to produce the best approximate values for the function in a small local region. For two adjacent subintervals, the coefficients for the interpolation polynomials may be different from each other because of the different values used to determine the coefficients of the polynomials. As a result, curves constructed from two such polynomials may not join each other smoothly at the boundary. This is adequate for the purpose of interpolation, as we are not interested in the global behavior of the polynomials over the entire region of $x = [x_0, x_N]$.

Mathematically, we can say that two polynomial functions behaving in this way are continuous across the boundary, but their derivatives are not. In principle, a smooth function is one that has derivatives of all orders, and they are continuous everywhere in the region. It is not possible to achieve this requirement using a polynomial of finite degree. For the third-degree polynomial approximation we wish to use here, the best we can do is to have derivatives up to the second order continuous across the boundary.

To construct such a polynomial, let us consider first an arbitrary subinterval $[x_i, x_{i+1}]$. For the approximation to go through f_i and f_{i+1} at the two ends of this subinterval, it is sufficient to use a first-order polynomial of the form

$$f(x) = f_i \lambda_i(x) + f_{i+1} \omega_i(x) \tag{3-118}$$

where

$$\lambda_i(x) = \frac{x_{i+1} - x}{x_{i+1} - x_i}$$

$$\omega_i(x) = 1 - \lambda(x) = \frac{x - x_i}{x_{i+1} - x_i} \tag{3-119}$$

As far as a first-order polynomial is concerned, (3-118) is completely determined, since the two coefficients f_i and f_{i+1} are given by the input quantities. To extend the functional form to a third-degree polynomial, we need some additional information to determine the new coefficients associated with the higher-order polynomials we wish to include. Since all the known values of the function within the subinterval are used up already, these coefficients must be found from the condition that both the first- and second-order derivatives are continuous across the boundaries to the two subintervals on either side.

For later convenience, we shall build the third-order polynomial we are seeking by starting from the right side of (3-118) and considering it as the first term of a series. To this, we shall add terms involving x^2 and x^3. Furthermore, since the first term already has the correct values at the two ends of the subinterval, it is necessary that contributions from the additional terms vanish at these points. One possible way to satisfy these requirements is to use the form

$$f(x)_{[x_i, x_{i+1}]} = f_i \lambda_i(x) + f_{i+1} \omega_i(x)$$
$$+ f_i'' \frac{(x_{i+1} - x_i)^2}{6} \{\lambda_i^3(x) - \lambda_i(x)\}$$
$$+ f_{i+1}'' \frac{(x_{i+1} - x_i)^2}{6} \{\omega_i^3(x) - \omega_i(x)\} \tag{3-120}$$

where the two coefficients f_i'' and f_{i+1}'' will be identified later as the values of the second-order derivatives of $f(x)$ at x_i and x_{i+1}, respectively. The two constant factors of $(x_{i+1} - x_i)^2/6$ are included in the expression for later convenience.

We shall now examine the continuity of the polynomial $f(x)$ and its derivatives across the boundaries to the subinterval on the left at $x = x_i$ and to the subinterval on the right at $x = x_{i+1}$. It is obvious that we have $f_{[x_i, x_{i+1}]}(x_i) = f_i$ and $f_{[x_i, x_{i+1}]}(x_{i+1}) = f_{i+1}$, since both $\{\lambda^3(x) - \lambda(x)\}$ and $\{\omega^3(x) - \omega(x)\}$ vanish at $x = x_i$ and x_{i+1}. Using (3-120), we find that the first-order derivative of $f_{[x_i, x_{i+1}]}(x)$ is

$$\left.\frac{df}{dx}\right|_{[x_i, x_{i+1}]} = \frac{f_{i+1} - f_i}{x_{i+1} - x_i} - f_i'' \frac{x_{i+1} - x_i}{6} \{3\lambda_i^2(x) - 1\}$$

$$+ f_{i+1}'' \frac{x_{i+1} - x_i}{6} \{3\omega_i^2(x) - 1\} \qquad (3\text{-}121)$$

For it to be continuous across the boundary, its value at $x = x_i$ must be equal to that for the subinterval $[x_{i-1}, x_i]$ at the same point. Similarly, its value at $x = x_{i+1}$ must be the same as that for the subinterval $[x_{i+1}, x_{i+2}]$. As we shall see later, these requirements provide us with the two conditions to determine the coefficients f_i'' and f_{i+1}''.

The second-order derivative of $f_{[x_i, x_{i+1}]}(x)$ is

$$\left.\frac{d^2 f}{dx^2}\right|_{[x_i, x_{i+1}]} = f_i'' \lambda_i(x) + f_{i+1}'' \omega_i(x)$$

Since $\lambda(x) = 1$ and $\omega(x) = 0$ at $x = x_i$, the right side is equal to f_i''. As a result, the second-order derivative is continuous across $x = x_i$ if f_i'' is the value of the second-order derivative of the polynomial $f_{[x_{i-1}, x_i]}(x)$ from the subinterval $[x_{i-1}, x_i]$ to the left. Similarly, $\lambda(x) = 0$ and $\omega(x) = 1$ at $x = x_{i+1}$, and the second-order derivative is continuous across $x = x_{i+1}$ if f_{i+1}'' is the value of the second-order derivative of the polynomial $f_{[x_{i+1}, x_{i+2}]}(x)$ from the subinterval $[x_{i+1}, x_{i+2}]$ to the right. From this, we see that the second-order derivatives can be continuous across all the boundaries between subintervals in the entire range of $x = [x_0, x_N]$ by having the same value of f_i'' at the boundary between two adjacent subintervals.

The only calculation remaining now is to ensure that the first-order derivatives from two adjacent subintervals are also equal at to each other at the boundary. Consider now two adjacent subintervals $[x_{i-1}, x_i]$ and $[x_i, x_{i+1}]$. At their boundary $x = x_i$, we have from (3-121) the results

$$\left.\frac{df}{dx}\right|_{[x_i, x_{i+1}]}(x = x_i) = \frac{f_{i+1} - f_i}{x_{i+1} - x_i} - \frac{x_{i+1} - x_i}{3} f_i'' - \frac{x_{i+1} - x_i}{6} f_{i+1}''$$

$$\left.\frac{df}{dx}\right|_{[x_{i-1}, x_i]}(x = x_i) = \frac{f_i - f_{i-1}}{x_i - x_{i-1}} + \frac{x_i - x_{i-1}}{6} f_{i-1}'' + \frac{x_i - x_{i-1}}{3} f_i''$$

For these two quantities to be equal to each other, the following relation must be satisfied:

$$(x_i - x_{i-1}) f_{i-1}'' + 2(x_{i+1} - x_{i-1}) f_i'' + (x_{i+1} - x_i) f_{i+1}''$$

$$= 6 \left(\frac{f_{i+1} - f_i}{x_{i+1} - x_i} - \frac{f_i - f_{i-1}}{x_i - x_{i-1}} \right) \qquad (3\text{-}122)$$

There are three unknown quantities in this equation, f''_{i-1}, f''_i, and f''_{i+1}. Two of the three quantities appear also in a similar equation at $x = x_{i+1}$, obtained from the boundary between subintervals $[x_i, x_{i+1}]$ and $[x_{i+1}, x_{i+2}]$. However, this does not solve our problem of having more unknown quantities than the number of equations, since a new unknown quantity f''_{i+2} enters into our system when we include the additional subintervals into the consideration. It is not difficult to see that we can extend the same consideration to the other boundaries between subintervals in our domain. Each time we include one more boundary point, we have one more equation as well as one more unknown quantity. The net result is that we have $(N-1)$ equations, one for each boundary point between two subintervals, and $(N+1)$ unknown f''_i, one for each point in the region, including the two end points. To obtain a solution, two additional pieces of information must be supplied. These are usually taken as external conditions we wish to impose on the second derivative of $f(x)$.

It is helpful to have (3-122) written out explicitly at the two points of $i = 1$ and $(N-1)$:

$$(x_1 - x_0)f''_0 + 2(x_2 - x_0)f''_1 + (x_2 - x_1)f''_2 = 6\left(\frac{f_2 - f_1}{x_2 - x_1} - \frac{f_1 - f_0}{x_1 - x_0}\right) \quad (3\text{-}123a)$$

$$(x_{N-1} - x_{N-2})f''_{N-2} + 2(x_N - x_{N-2})f''_{N-1} + (x_N - x_{N-1})f''_n$$
$$= 6\left(\frac{f_N - f_{N-1}}{x_N - x_{N-1}} - \frac{f_{N-1} - f_{N-2}}{x_{N-1} - x_{N-2}}\right) \quad (3\text{-}123b)$$

If the two additional input quantities are given to us as the values of f''_0 and f''_1, the only unknown remaining in (3-123a) is f''_2. This gives us the value of f''_2 in terms of the input f''_0 and f''_1. Once we have f''_2, we can use it together with f''_1 to solve for f''_3 using (3-122) by putting $i = 2$. This process may be continued until we come to the end, where $i = (N-1)$. Similarly, if we have the values f''_N and f''_{N-1} instead of f''_0 and f''_1, we can use (3-123b) and solve the problem backward.

Since the function $f(x)$ is unknown, it is difficult in general for us to have an idea what the second-order derivatives should be anywhere in the region of interest. A more realistic approach is for us to make some estimates of the values of f''_0 and f''_N at the two ends of the interval, $x = x_0$ and x_N, respectively. Once this is done, we have enough information to solve the set of $(N-1)$ simultaneous equations given in (3-122). The solution is now slightly more complicated because the boundary conditions are different. Let us first rewrite (3-123) into a form with the unknown second-order derivatives of $f(x)$ only on the left side:

$$2(x_2 - x_0)f''_1 + (x_2 - x_1)f''_2 = 6\left(\frac{f_2 - f_1}{x_2 - x_1} - \frac{f_1 - f_0}{x_1 - x_0}\right) - (x_1 - x_0)f''_0$$

$$(x_{N-1} - x_{N-2})f''_{N-2} + 2(x_N - x_{N-2})f''_{N-1}$$
$$= 6\left(\frac{f_N - f_{N-1}}{x_N - x_{N-1}} - \frac{f_{N-1} - f_{N-2}}{x_{N-1} - x_{N-2}}\right) - (x_N - x_{N-1})f''_N$$

Together with (3-122) for $i = 2, 3, \ldots, (N-2)$, we have a set of $(N-1)$ simultaneous equations, and these may be written in the form of a matrix product:

$$\begin{pmatrix} b_1 & c_1 & 0 & 0 & \cdots & 0 \\ a_2 & b_2 & c_2 & 0 & \cdots & 0 \\ 0 & a_3 & b_3 & c_3 & \cdots & 0 \\ \vdots & \vdots & \vdots & \ddots & \vdots & \vdots \\ 0 & \cdots & 0 & a_{N-2} & b_{N-2} & c_{N-2} \\ 0 & \cdots & 0 & 0 & a_{N-1} & b_{N-1} \end{pmatrix} \begin{pmatrix} f_1'' \\ f_2'' \\ f_3'' \\ \cdots \\ f_{N-2}'' \\ f_{N-1}'' \end{pmatrix} = \begin{pmatrix} d_1 \\ d_2 \\ d_3 \\ \cdots \\ d_{N-2} \\ d_{N-1} \end{pmatrix} \quad (3\text{-}124)$$

where

$$a_i = \begin{cases} 0 & \text{for } i = 1 \\ x_i - x_{i-1} & \text{for } i = 2, 3, \ldots, (N-1) \end{cases}$$

$$b_i = 2(x_{i+1} - x_{i-1})$$

$$c_i = \begin{cases} x_{i+1} - x_i & \text{for } i = 1, 2, 3, \ldots, (N-2) \\ 0 & \text{for } i = (N-1) \end{cases}$$

$$d_i = \begin{cases} 6\left(\dfrac{f_2 - f_1}{x_2 - x_1} - \dfrac{f_1 - f_0}{x_1 - x_0}\right) - (x_1 - x_0)f_0'' & \text{for } i = 1 \\ 6\left(\dfrac{f_{i+1} - f_i}{x_{i+1} - x_i} - \dfrac{f_i - f_{i-1}}{x_i - x_{i-1}}\right) & \text{for } i = 2, 3, \ldots, (N-2) \\ 6\left(\dfrac{f_N - f_{N-1}}{x_N - x_{N-1}} - \dfrac{f_{N-1} - f_{N-2}}{x_{N-1} - x_{N-2}}\right) - (x_N - x_{N-1})f_N'' & \text{for } i = (N-1) \end{cases}$$

(3-125)

Since the main matrix has a tridiagonal form, the solution may be obtained using the method of Gaussian elimination.

Gaussian elimination. There are two steps involved in the approach. The first is to reduce the tridiagonal matrix into an upper diagonal form:

$$\begin{pmatrix} b_1' & c_1' & 0 & 0 & \cdots & 0 \\ 0 & b_2' & c_2' & 0 & \cdots & 0 \\ 0 & 0 & b_3' & c_3' & \cdots & 0 \\ \vdots & \vdots & \vdots & \ddots & \vdots & \vdots \\ 0 & \cdots & 0 & 0 & b_{N-2}' & c_{N-2}' \\ 0 & \cdots & 0 & 0 & 0 & b_{N-1}' \end{pmatrix} \begin{pmatrix} f_1'' \\ f_2'' \\ f_3'' \\ \vdots \\ f_{N-2}'' \\ f_{N-1}'' \end{pmatrix} = \begin{pmatrix} d_1' \\ d_2' \\ d_3' \\ \vdots \\ d_{N-2}' \\ d_{N-1}' \end{pmatrix} \quad (3\text{-}126)$$

The major difference between (3-124) and (3-126) is that the subdiagonal elements, that is, the positions that were occupied by a_i, are now replaced by zeros. If we further normalize all the matrix elements such that the diagonal ones are

$$b_i' = 1$$

> **Box 3-10   Program SPLINE**
> **Cubic spline with Gaussian elimination**
>
> Subprograms used:
>   CUBIC_SPLINE: Calculates the second-order derivatives required for cubic spline.
>   SMOOTH: Generates evenly spaced interpolated values.
> Initialization:
>   (a) Set the limits for the maximum number of input and output points.
>   (b) Define the two second-order derivatives at the boundaries to be zero.
> 1. Input:
>   (a) $\{x_i, f_i\}$ for $i = 0, 1, 2, \ldots, N$.
>   (b) m, the number of evenly spaced output points required.
> 2. Use CUBIC_SPLINE to obtain the second-order derivatives at the internal mesh points.
>   (a) Define the elements of the tridiagonal matrix using (3-125).
>   (b) Reduce the matrix to an upper bidiagonal one using (3-127).
>   (c) Back substitution to obtain y_i'' using (3-128) and (3-129).
> 3. Use the derivatives obtained in SMOOTH to interpolate for a set of evenly spaced points:
>   (a) Find the step size.
>   (b) Set the first output point to be the same as the first input point.
>   (c) Interpolate the value of $f(x)$ for all the points:
>   (i) Find the subinterval in the input to which x belongs.
>   (ii) Calculate the value of $f(x)$ using (3-120).
>   (d) Return x and the interpolated values of $f(x)$.
> 4. Output the values of $f(x)$ for m evenly spaced points between x_0 and x_N.

it is easy to see that the elements of the new and old matrices are related in the following way:

$$c_i' = \begin{cases} \dfrac{c_1}{b_1} & \text{for } i = 1 \\ \dfrac{c_i}{b_i - a_i c_{i-1}'} & \text{for } i = 2, 3, \ldots, (N-2) \end{cases}$$

$$d_i' = \begin{cases} \dfrac{d_1}{b_1} & \text{for } i = 1 \\ \dfrac{d_i - a_i d_{i-1}'}{b_i - a_i c_{i-1}'} & \text{for } i = 2, 3, \ldots, (N-1) \end{cases} \quad (3-127)$$

This completes the first step in solving (3-124).

The second step consists of back substitutions to obtain the solution of an algebraic equation involving an upper diagonal matrix. It is convenient to start from the last row. From of (3-126), we have the simple result

$$f_{N-1}'' = d_{N-1}' \quad (3-128)$$

since b_N' is normalized to unity. This gives us the value of f_{N-1}'' (f_N'' is known and its value is one of the two additional input quantities required to solve the cubic spline problem). The second last row of (3-126) has the form

$$f''_{N-2} + c'_{N-2} f''_{N-1} = d'_{N-2}$$

Again, because $b_{N-1} = 1$, we can make use of the value of f''_{N-1} obtained from (3-128) to solve for f''_{N-2}. The result is

$$f''_{N-2} = d'_{N-2} - c'_{N-2} f''_{N-1}$$

It is obvious that we can continue this process by moving up one row at a time. The general result is

$$f''_i = d'_i - c'_i f''_{i+1} \qquad \text{for} \quad i = (N-2), (N-3), \ldots, 2, 1 \qquad (3\text{-}129)$$

In this way all $(N-1)$ values of f''_i are obtained.

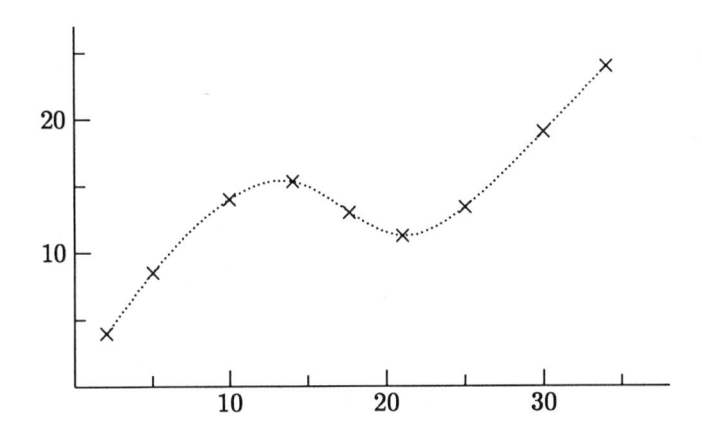

Fig. 3-6 Smoothing a function by cubic spline. The crosses are the values of $f(x)$ supplied and the dots are generated using a cubic spline routine.

Once all the values of f''_i are known, we have a complete set of continuous, third-order polynomial approximations to $f(x)$. Each polynomial has the form given by (3-120). Using such an expression, it is possible for us to calculate the approximate value of $f(x)$ for any x. The algorithm is summarized in Box 3-10. As an example, an arbitrary function, defined by the values at nine input points in the interval $x = [2, 34]$, is shown by the crosses in Fig. 3-6. To draw a smooth curve through these nine points, a number of the intermediate points are calculated using a cubic spline routine based on the algorithm outlined in Box 3-10 and these are shown as dots. As can be seen from the figure, the dots do form a smooth curve joining the input points.

The Gaussian elimination method is not the most sophisticated way to solve a matrix equation. As we shall see later in Chapter 5, other techniques are more convenient to use for the general case. The method used here has the advantage that it is very simple to understand and to implement. Furthermore, it works efficiently for the case in hand. For this reason, it is also used in a variety of other situations where tridiagonal matrices are involved, such as in the numerical solution of differential equations in Chapter 8.

Problems

3-1 If a function $f(x)$ is a polynomial of degree 2, having the form

$$f(x) = a_0 + a_1 x + a_2 x^2$$

show that the second-order difference of Neville's algorithm given by (3-4) is exact.

3-2 In *Computing Methods for Scientists and Engineers* by Fox and Mayers (Clarendon Press, Oxford, 1968), the following table of the values of a function $f(x)$ is given for $x = -0.2$ to 1.2 in steps of 0.2:

$$-0.7328 \quad -0.7071 \quad -0.6528 \quad -0.3981 \quad 0.5721 \quad 3.1165 \quad 8.4372 \quad 18.0797$$

Find the value of x for $f(x) = 0.0$. Compare the results using Neville's method and Bessel's formula.

3-3 Prove the recursion relation of (3-10) by working out the difference between $D_{i+1,\ell}$ and $U_{i,\ell}$. Express the result in terms of $f_{i,i+1,\ldots,i+\ell}$ and $f_{i,i+1,\ldots,i+\ell+1}$.

3-4 Find the Padé approximation for $f(x) = \sin x$ that is accurate up to the x^7 term. Compare the accuracy of the resulting expression with (3-14) for $\cos x$.

3-5 The value of $\sin x$ for $x = 2.5$ is positive, and for $x = 3.5$, the value is negative. Use the bisection method of Box 3-8 on a calculator to find the value of x in the interval $[2.617994, 3.665191]$ for which $\sin x = 0$. Deduce the number of steps required to get an agreement with $x = \pi$ to five significant figures.

3-6 Apply the Fourier transform of (3-58) to a square wave

$$f(x) = \begin{cases} 0 & \text{for } 0 \leq x < 1,\ 2 \leq x < 3,\ \cdots \\ 1 & \text{for } 1 \leq x < 2,\ 3 \leq x < 4,\ \cdots \end{cases}$$

and find the values of the first nine coefficients a_0, a_1, b_1, a_2, b_2, a_3, b_3, a_4, b_4, and a_5 using (3-59). Reconstruct the waveform from the Fourier coefficients obtained and compare the results with the original. The differences come from the higher-order terms ignored.

3-7 Repeat Problem 3-6 with a sawtooth wave form

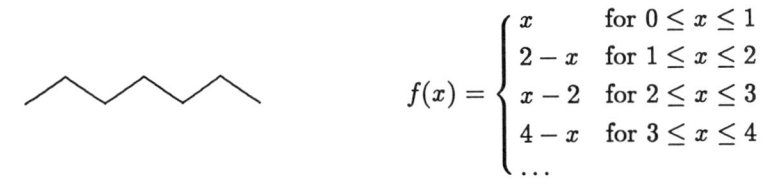

$$f(x) = \begin{cases} x & \text{for } 0 \leq x \leq 1 \\ 2-x & \text{for } 1 \leq x \leq 2 \\ x-2 & \text{for } 2 \leq x \leq 3 \\ 4-x & \text{for } 3 \leq x \leq 4 \\ \cdots \end{cases}$$

3-8 Derive (3-106) using a polynomial approximation of the form

$$f(x) = c_0 + c_1(x-x_0) + c_2(x-x_0)(x-x_1) + c_3(x-x_0)(x-x_1)(x-x_2)$$
$$+ c_4(x-x_0)(x-x_1)(x-x_2)(x-x_{-1}) + \cdots$$

where $x_k = (x_0 + kh)$ for both positive and negative integer values of k and $ph = (x - x_0)$.

3-9 Use the method of bisection of Box 3-8 to find the two values of x in the interval $x = [0, 15]$ for which the function

$$f(x) = 2.5 - 3.0x + 0.2x^2 + 0.015x^3$$

is equal to zero.

3-10 Apply a cubic spline to the periodic function given by the following nine pairs of values:

$x =$	0	$\frac{1}{2}\pi$	π	$\frac{3}{2}\pi$	2π	$\frac{5}{2}\pi$	3π	$\frac{7}{2}\pi$	4π
$f(x) =$	0	1	0	-1	0	1	0	-1	0

Use $f''(x = 0) = f''(x = 4\pi) = 0$ as the two additional conditions required. These sets of points can also be fitted by a sine function. What are the differences between the results of a cubic spline and $\sin x$?

Chapter 4
Special Functions

A large number of mathematical functions have been developed over the years to solve specific problems of interest in physics. In this chapter, we shall examine some of these functions, partly for their intrinsic usefulness and partly as the vehicle to introduce certain computational techniques. Although each function is discussed in relation to a particular type of physical problem, its interest is usually much wider. However, we shall not make any attempt to describe the full range of possible applications of any of the special functions.

§4-1 Hermite polynomials and quantum mechanical harmonic oscillator

A one-dimensional harmonic oscillator is often used to illustrate several fundamental aspects in both classical and quantum mechanics. The reason for its popularity comes from the fact that the potential has the form

$$V_{\text{h.o.}}(x) = \tfrac{1}{2}\mu\omega^2 x^2 \qquad (4\text{-}1)$$

where μ is the mass and ω is the angular frequency of the oscillator. The form is an idealization of the potential experienced by an object trapped near the local minimum, as shown schematically in Fig. 4-1. If x_0 is the location of the minimum, we can expand the potential $V(x)$ in a small region around x_0 in terms of a series,

$$V(x) = c_0 + c_1(x - x_0) + c_2(x - x_0)^2 + c_3(x - x_0)^3 + \cdots$$

The first term c_0 is a constant, which may be removed by redefining the zero point of the potential, as we shall do soon. Since x_0 is the location of a minimum, the first-order derivative of $V(x)$ vanishes at x_0 and, as a result, $c_1 = 0$. If we make the following change of the variables,

$$x \longrightarrow (x - x_0) \qquad V(x) \longrightarrow V(x) - c_0$$

the leading nonvanishing term of $V(x)$ takes on the form of (4-1). In the vicinity of the minimum, the higher-order terms are usually small and they may be regarded as corrections. In this way, the values of the coefficients $c_3, c_4, \ldots$, may be regarded as parameters, describing the departure from a simple quadratic form of the potential. For example, the third-order derivative describes the skewness, meaning how different the slope on one side of the minimum differs from that on the other side. If we ignore such anharmonic effects, we have the result of a simple harmonic oscillator potential.

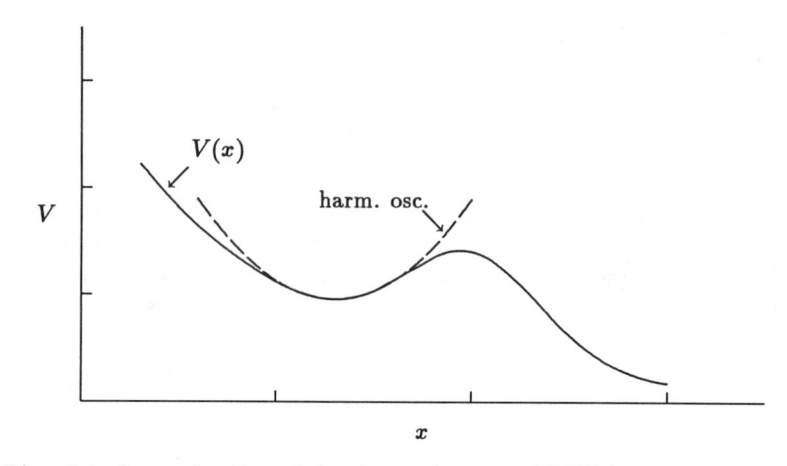

Fig. 4-1 Approximation of the shape of a potential $V(x)$ near a minimum by that of a harmonic oscillator $V_{\text{h.o.}}(x) = \frac{1}{2}\mu\omega^2 x^2$.

One-dimensional harmonic oscillator in quantum mechanics. The particular physical problem we shall be concerned here is the wave function of a one-dimensional harmonic oscillator in quantum mechanics. Many different ways are available in standard texts to find the solution. Here, we shall concentrate on the one that makes use of Hermite polynomials. In addition to the one-dimensional harmonic oscillator, Hermite polynomials are also useful, for example, in a Gram-Charlier expansion of a distribution that is almost a normal one (see, for example, H. Cramer, *Mathematical Methods of Statistics*, Princeton University Press, Princeton, New Jersey, 1946).

The Hamiltonian of a particle of mass μ moving in an harmonic oscillator potential well of the form of (4-1) is given by

$$H = \frac{1}{2\mu}p_x^2 + \frac{1}{2}\mu\omega^2 x^2 \tag{4-2}$$

where p_x is the linear momentum of the particle. To obtain a quantum mechanical solution of the problem, we must first express H in the form of an operator:

$$\hat{H} = -\frac{\hbar^2}{2\mu}\frac{d^2}{dx^2} + \frac{1}{2}\mu\omega^2 x^2 \tag{4-3}$$

where we have made use of the substitution
$$p_x \to \frac{\hbar}{i}\frac{d}{dx}$$
in changing the observables in the system to their corresponding operators. The Hamiltonian is now an operator and may be used to construct the Schrödinger equation,
$$\hat{H}\psi(x) = E\psi(x) \tag{4-4}$$
from which we obtain the eigenvalues and eigenfunctions that describe the system.

Mathematically, the problem is equivalent to one of solving a second-order ordinary differential equation of the form
$$-\frac{\hbar^2}{2\mu^2}\frac{d^2\psi}{dx^2} + \left(\tfrac{1}{2}\mu\omega^2 x^2 - E\right)\psi(x) = 0$$
It is convenient to make a change of the variables here. If we measure energies in units of $\tfrac{1}{2}\hbar\omega$,
$$E \longrightarrow \lambda = \frac{E}{\tfrac{1}{2}\hbar\omega} \tag{4-5}$$
and express lengths in terms of the dimensionless quantity
$$\rho = \sqrt{\frac{\mu\omega}{\hbar}}\,x \tag{4-6}$$
the differential equation is changed into the form
$$\frac{d^2\psi}{d\rho^2} + (\lambda - \rho^2)\psi(\rho) = 0 \tag{4-7}$$
To keep the expression simple, we have dropped an overall factor of $\tfrac{1}{2}\hbar\omega$.

Harmonic oscillator wave function. At large values of ρ, we can ignore the term λ in the coefficient of $\psi(\rho)$ in (4-7) and obtain the approximate relation
$$\frac{d^2\psi}{d\rho^2} - \rho^2\psi(\rho) \approx 0$$
This means that, asymptotically, the wave function $\psi(\rho)$ must assume the form
$$\psi(\rho) \xrightarrow[\rho\to\infty]{} Ae^{+\rho^2/2} + Be^{-\rho^2/2}$$
where A and B are the two unknown coefficients in the solution of a second-order differential equation and they must be determined by boundary conditions. On the other hand, the wave function must be finite as $\rho \to \infty$. Consequently, the first term cannot be present and we conclude that $A = 0$. The asymptotic form of the wave function is then
$$\psi(\rho) \xrightarrow[\rho\to\infty]{} Be^{-\rho^2/2}$$

For small values of ρ, the wave function, in general, differs from an exponential function. For this reason, we shall make use of a trial solution of the following form for (4-7):

$$\psi(\rho) = Be^{-\rho^2/2}H(\rho) \tag{4-8}$$

where $H(\rho)$ is an unknown function of ρ to be determined. The value of B may be found from the normalization requirement for $\psi(\rho)$ and is therefore of no concern to us at the moment.

On substituting (4-8) into (4-7), we obtain the equation that must be satisfied by $H(\rho)$ in the trial solution:

$$\frac{d^2}{d\rho^2}H(\rho) - 2\rho\frac{d}{d\rho}H(\rho) + (\lambda - 1)H(\rho) = 0 \tag{4-9}$$

One way to solve this equation is to use a power series expansion for the unknown function

$$H(\rho) = \sum_{k=0}^{\infty} a_k \rho^{k+\alpha} \tag{4-10}$$

The coefficients α and a_k may be found by inserting this form of $H(\rho)$ into (4-9). The result is

$$\sum_{k=0}^{\infty}\left\{(k+\alpha)(k+\alpha-1)a_k\rho^{k+\alpha-2} - [2(k+\alpha) - (\lambda-1)]a_k\rho^{k+\alpha}\right\} = 0 \tag{4-11}$$

For this equality to be true for all values of ρ, the sum of the coefficients for each power of ρ must vanish.

The lowest power of ρ in (4-11) is $(\alpha - 2)$ and this occurs at $k = 0$. Since the coefficient for this term must also vanish, we obtain as the result

$$\alpha(\alpha - 1)a_0 = 0$$

In general, $a_0 \neq 0$. As a result, we have the condition that either

$$\alpha = 0 \quad \text{or} \quad \alpha = 1$$

Later, we shall see that both possibilities lead to the same set of solutions and we need only take one of them. For $\alpha = 0$, (4-11) may be rewritten into the form with all the coefficients of the same power of ρ grouped together:

$$\sum_{k=0}^{\infty}\{(k+1)(k+2)a_{k+2} - (2k - \lambda + 1)a_k\}\rho^k = 0$$

From the requirement that the coefficients of ρ^k for each value of k must vanish, we obtain the condition to be satisfied by all a_k:

$$a_{k+2} = \frac{2k - \lambda + 1}{(k+1)(k+2)}a_k \tag{4-12}$$

Note that the equation actually represents two independent sets of conditions, one for k even and the other for k odd. This is expected from the fact that $H(\rho)$ is the solution of a second-order differential equation and, hence, there should be two independent sets of solution. These two sets may be distinguished by having one set starting with

$$a_0 \neq 0 \qquad a_1 = 0$$

and the other set with

$$a_0 = 0 \qquad a_1 \neq 0$$

We shall not show the proof here that these two sets of solutions are linearly independent of each other. However, this point is important in demonstrating the fact that we do not need to take the $\alpha = 1$ alternative in (4-10), as a complete set of solutions exists with $\alpha = 0$ alone. In other words, both $\alpha = 0$ and 1 lead to the same two sets of solutions and this gives us the justification to ignore the other possible value of α.

Hermite polynomials. We must now impose the boundary conditions on the solutions. For k large, we can ignore the other factors, λ, 1, and 2, in (4-12), as they are small compared with k. Using this, we obtain the result

$$\frac{a_{k+2}}{a_k} \xrightarrow[\text{large } k]{} \frac{2}{k}$$

This is similar to the coefficients in a power series expansion of the exponential function

$$e^{x^2} = 1 + x^2 + \frac{x^4}{2!} + \frac{x^6}{3!} + \cdots$$

As a result, we find that the wave function has the asymptotical form when we substitute the result for $H(\rho)$ into (4-8):

$$\psi(\rho) \xrightarrow[\rho \to \infty]{} e^{-\rho^2/2} e^{\rho^2} = e^{+\rho^2/2}$$

Since the square of a wave function is proportional to the probability of finding the particle at a given location, such a solution implies that the probability dominates at $r = \infty$. This result does not correspond to the situation of a particle localized around the minimum of a potential well at $r = 0$ in which we are interested. For this reason, the power series of (4-10) for $H(\rho)$ cannot be infinite and must terminate at some finite k value. For this to take place, it is necessary that the energy factor λ of (4-11) be an odd integer, as can be seen from (4-12). That is,

$$\lambda = 2n + 1 \qquad (4\text{-}13)$$

where n is an integer equal to 0, 1, 2, ... Physically, this is the well-known condition stating that the energy of a one-dimensional harmonic oscillator must be an odd multiple of $\frac{1}{2}\hbar\omega$. This can be seen by substituting (4-13) into (4-5).

For each value of n, we have a distinct polynomial $H(\rho)$ with the leading (the highest power) term of ρ equal to n. Let us use a subscript n to label the different polynomials. For a given n, the polynomial $H_n(\rho)$ contains either even or odd powers of ρ, depending on whether n is even or odd. If we take the normalization of each polynomial to be such that the coefficient for the leading term is $a_n = 2^n$, we obtain the Hermite polynomials. The first few have the form

$$H_0(\rho) = 1 \qquad\qquad H_1(\rho) = 2\rho$$
$$H_2(\rho) = 4\rho^2 - 2 \qquad\qquad H_3(\rho) = 8\rho^3 - 12\rho$$
$$H_4(\rho) = 16\rho^4 - 48\rho^2 + 12 \qquad H_5(\rho) = 32\rho^5 - 160\rho^3 + 120\rho \qquad (4\text{-}14)$$

Later, we shall see methods of generating Hermite polynomials of arbitrary orders by using a computer.

The orthogonality relation among Hermite polynomials is given by the integral

$$\int_{-\infty}^{+\infty} e^{-\rho^2} H_n(\rho) H_m(\rho)\, d\rho = 2^n n! \sqrt{\pi}\, \delta_{n,m} \qquad (4\text{-}15)$$

As a result, the wave function of (4-4) for a harmonic oscillator takes on the form

$$\psi_n(\rho) = \frac{1}{\sqrt{2^n n! \sqrt{\pi}}} e^{-\rho^2/2} H_n(\rho) \qquad (4\text{-}16)$$

The eigenvalue corresponding to this wave function is $\lambda = (2n+1)$ in terms of dimensionless energy variable, or $E_n = (n + \frac{1}{2})\hbar\omega$. The shapes of $\phi_0(\rho)$, $\phi_1(\rho)$, $\phi_2(\rho)$, and $\phi_3(\rho)$ are given in Fig. 4-2 for illustration.

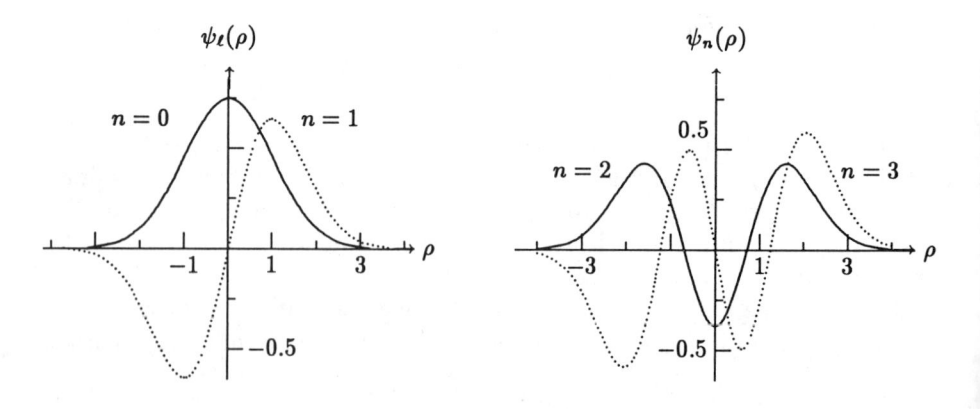

Fig. 4-2 One-dimensional harmonic oscillator wave function $\psi_n(\rho)$ of (4-16) for $n = 0, 1, 2, 3,$ and 4.

Generating function. We shall now turn our attention to Hermite polynomials of an arbitrary order. There are two aspects in which we are particularly interested, the numerical value of $\psi_n(\rho)$ for a given ρ and the algebraic form of $H_n(\rho)$ in terms of a power series in ρ. For both, we need the generating function for $H_n(\rho)$ from which we can derive recursion relations among $H_n(\rho)$ of different degree n.

Consider the following function of two variables ρ and t:

$$g(\rho, t) = e^{-t^2 + 2t\rho} \qquad (4\text{-}17)$$

On making a Taylor series expansion of $g(\rho, t)$ around $t = 0$, we obtain the result

$$g(\rho, t) = \sum_{n=0}^{\infty} h_n(\rho) \frac{t^n}{n!} \qquad (4\text{-}18)$$

where $h_n(\rho)$, the coefficient for $t^n/n!$, may be written in the form

$$h_n(\rho) = \left.\frac{\partial^n}{\partial t^n} g(\rho, t)\right|_{t=0} = e^{\rho^2} \left.\frac{\partial^n}{\partial t^n} e^{-(t-\rho)^2}\right|_{t=0}$$

We shall try to establish that $h_n(\rho)$ and $H_n(\rho)$ are identical to each other by demonstrating that $h_n(\rho)$ satisfies the same differential equation as (4-9) for $H_n(\rho)$. For the function $\exp\{-(t-\rho)^2\}$, we find that

$$\frac{\partial}{\partial(t-\rho)} = \frac{\partial}{\partial t}$$

Using this equivalence, we can write $h_n(\rho)$ as

$$h_n(\rho) = e^{\rho^2} \left.\frac{\partial^n}{\partial(t-\rho)^n} e^{-(t-\rho)^2}\right|_{t=0}$$

$$= (-1)^n e^{\rho^2} \left.\frac{\partial^n}{\partial \rho^n} e^{-(t-\rho)^2}\right|_{t=0}$$

$$= (-1)^n e^{\rho^2} \frac{d^n}{d\rho^n} e^{-\rho^2} \qquad (4\text{-}19)$$

The final form is one of the ways to obtain Hermite polynomials of an arbitrary order n. However, before we can do this, we have to demonstrate that $H_n(\rho) = h_n(\rho)$.

Let us take the partial derivative of $g(\rho, t)$ with respect to t using its exponential form given by (4-17). On expanding the result in terms of $h_n(\rho)$, we obtain

$$\frac{\partial g}{\partial t} = 2(\rho - t) g(\rho, t) = 2(\rho - t) \sum_{n=0}^{\infty} h_n(\rho) \frac{t^n}{n!} \qquad (4\text{-}20)$$

Similarly, the partial derivative of $g(\rho, t)$ with respect to ρ gives the result

$$\frac{\partial g}{\partial \rho} = 2tg(\rho, t) \tag{4-21}$$

By taking the partial derivative of the sum of $\partial g(\rho, t)/\partial t$ and $\partial g(\rho, t)/\partial \rho$ with respect to ρ and eliminating $\partial^2 g(\rho, t)/\partial \rho \partial t$ by replacing it with the partial derivative of (4-21) with respect to t, we obtain the differential equation satisfied by $g(\rho, t)$:

$$\frac{\partial^2 g}{\partial \rho^2} - 2\rho \frac{\partial g}{\partial \rho} + 2t \frac{\partial g}{\partial t} = 0$$

If we replace $g(\rho, t)$ by its form given on the right side of (4-18) and rearrange the terms for different powers of t, we obtain a relation for $h_n(\rho)$:

$$\sum_{n=0}^{\infty} \left\{ \frac{d^2 h_n}{d\rho^2} - 2\rho \frac{dh_n}{d\rho} + 2nh_n(\rho) \right\} \frac{t^n}{n!} = 0$$

For this relation to hold for an arbitrary value of t, the quantity inside the braces must vanish for each power of t. This gives us a differential equation that is identical to (4-9), satisfied by $H_n(\rho)$ for $\lambda = (2n+1)$. This confirms that $h_n(\rho)$ is the same as $H_n(\rho)$ and the $g(\rho, t)$ is the *generating function* for the Hermite polynomials.

Recursion relations. We shall now derive two recursion relations among the Hermite polynomials. For this purpose, it is necessary again to take partial derivatives of $g(\rho, t)$ with respect to t. Using the series expansion form of (4-18), we obtain the result

$$\frac{\partial g}{\partial t} = \sum_{n=0}^{\infty} h_n(\rho) \frac{nt^{n-1}}{n!}$$

On equating this relation with that of (4-20), we obtain the identity

$$\sum_{n=0}^{\infty} \{h_{n+1}(\rho) - 2\rho h_n(\rho) + 2nh_{n-1}(\rho)\} \frac{t^n}{n!} = 0$$

Using the fact that this must be true for an arbitrary value of t, the following recursion relation between Hermite polynomials of different orders is established:

$$H_{n+1}(\rho) = 2\rho H_n(\rho) - 2nH_{n-1}(\rho) \tag{4-22}$$

Similarly, by comparing the partial derivatives of the generating function $g(\rho, t)$ with respect to ρ in terms of its power series form given by (4-18) and that of (4-21), we obtain a relation of $H_n(\rho)$ with its derivative:

$$H'_n(\rho) = 2nH_{n-1}(\rho)$$

> **Box 4-1** Subroutine HERMIT_COEF(K_ORDER,NARY,NDMN)
> **Coefficients of Hermite polynomials by recursion relation (4-24)**
>
> Argument list:
> K_ORDER: Order of the Hermite polynomial.
>   NARY: Two-dimensional array for all the coefficients up to order K_ORDER.
>   NDMN: Dimension of array NARY.
> Initialization:
>   (a) Zero the array NARY.
>   (b) Define coefficients $a_{0,0} = 1$ for H_0, and $a_{1,0} = 0$ and $a_{1,1} = 2$ for H_1.
> 1. Start the propagation from order 2 and continue till K_ORDER.
> 2. Find the nonvanishing coefficients for the order using (4-24).
> 3. Return the array NARY.

Since the low-order Hermite polynomials have a very simple structure, we can use them as the starting point and calculate those of higher orders using these recursion relations.

Both (4-22) and the last equality of (4-19) may be used to generate Hermite polynomials of an arbitrary order n. For large n, it is more convenient to program the recursion relation of (4-22) and find the coefficients in the polynomial expansion of $H_n(\rho)$ with a computer. By definition, $H_n(\rho)$ is a power series in ρ with the highest power n. That is, we can write (4-10) for $\alpha = 0$ in the form

$$H_n(\rho) = \sum_{k=0}^{n} a_{n,k} \rho^k \tag{4-23}$$

where we have explicitly indicated the order of the polynomial for each coefficient by adding n to the subscript of a.

To find the algebraic form of the Hermite polynomial of degree n, we need to calculate the values of the nonvanishing coefficients $a_{n,n}, a_{n,n-2}, \ldots$. On substituting (4-23) into (4-22) and rearranging the terms so that all the coefficients for a given power of ρ are grouped together, we obtain the result

$$\sum_{k=0}^{n+1} \{a_{n+1,k} - 2a_{n,k-1} + 2na_{n-1,k}\} \rho^k = 0$$

For this equality to hold for an arbitrary value of ρ, it is necessary that the coefficients satisfy the relation

$$a_{n+1,k} = 2a_{n,k-1} - 2na_{n-1,k} \tag{4-24}$$

That is, to find the coefficients for order $(n + 1)$ for a given power k, we need as input those for orders n and $(n - 1)$ of the same k. Since $H_0(\rho) = 1$ and

$H_1(\rho) = 2\rho$, we have the starting values

$$a_{0,0} = 1 \qquad a_{1,0} = 0 \qquad a_{1,1} = 2$$

(and $a_{n,k} = 0$ for $k > n$). From these values, it is possible to generate the coefficients $a_{2,0}$ and $a_{2,2}$ for $H_2(\rho)$ ($a_{2,1} = a_{2,k} = 0$ for $k > 2$). Once the coefficients for $H_2(\rho)$ are known, we can use them, together with those of $H_1(\rho)$, to produce those for $H_3(\rho)$, and so on. Since all the coefficients are integers, the method is efficient, limited only by the largest integer that can be stored in the computer. The steps in the calculation are outlined in Box 4-1.

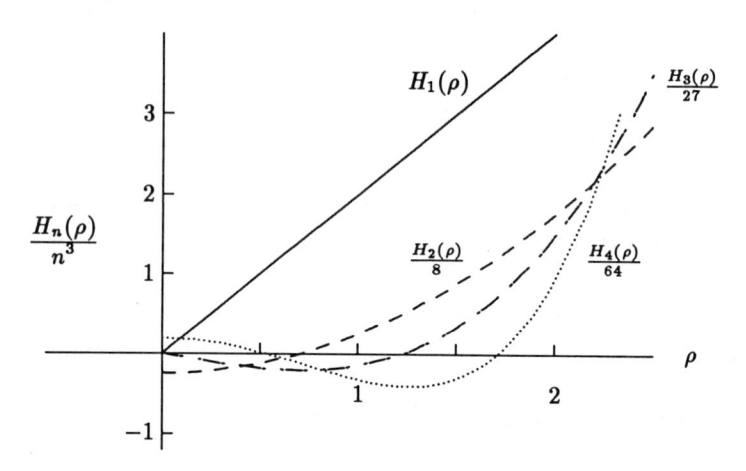

Fig. 4-3 Hermite polynomials $H_n(\rho)$ for $n = 1$ to 4. The plotted values are $H_n(r)/n^3$ so as to bring all four curves to roughly the same range of values.

We can also use (4-22) to calculate the numerical value of $H_n(\rho)$ for a given ρ. Again we start with $H_0(\rho) = 1$ and $H_1(\rho) = 2\rho$. The value of $H_2(\rho)$ is obtained from the expression

$$H_2(\rho) = 2\rho H_1(\rho) - 2H_0(\rho)$$

where the quantities on the right side are numerical values. Next we can apply (4-22) to produce $H_3(\rho)$ using the value of $H_2(\rho)$ for the ρ just obtained and the value of $H_1(\rho)$ we already have. The process is repeated for the next order, and so on, until the desired order is reached. In this way, the value of Hermite polynomials for a given ρ is propagated from those of $H_0(\rho)$ and $H_1(\rho)$, one order at a time, up to any arbitrary high order. The method is simple, but the numerical accuracy is likely to be poor if n is large. A better alternative is to use the method in the previous paragraph to calculate the algebraic form of $H_n(\rho)$ first. The numerical value of $H_n(\rho)$ is then produced by inserting the value of ρ into the algebraic expression just obtained. As an illustration, the forms of the lowest few orders of $H_n(\rho)$ for $\rho < 5$ are plotted in Fig. 4-3.

§4-2 Legendre polynomials and spherical harmonics

In many physical situations, it is possible to make use of spherical symmetry to simplify the problem. One such example is the Helmholtz equation,

$$\nabla^2 \psi(\mathbf{r}) + k^2 \psi(\mathbf{r}) = 0 \tag{4-25}$$

It is a second-order differential equation that appears in a variety of problems in quantum mechanics, electromagnetism, and wave phenomena, among others. If the coefficient k^2 is a constant or only a function of the radial coordinate r, the solution may be expressed as a product of radial and angular parts. To carry out the separation of variables, we shall first rewrite the Laplacian operator

$$\nabla^2 = \frac{\partial^2}{\partial x^2} + \frac{\partial^2}{\partial x^2} + \frac{\partial^2}{\partial x^2}$$

in terms of its components in spherical polar coordinates:

$$\nabla^2 = \frac{1}{r^2}\frac{\partial}{\partial r}\left(r^2 \frac{\partial}{\partial r}\right) + \frac{1}{r^2}\left[\frac{1}{\sin\theta}\frac{\partial}{\partial \theta}\left(\sin\theta \frac{\partial}{\partial \theta}\right) + \frac{1}{\sin^2\theta}\frac{\partial^2}{\partial \phi^2}\right] \tag{4-26}$$

where θ is the polar angle and ϕ is the azimuthal angle, as shown in Fig. 4-4. In quantum mechanics, the angular part of the operator,

$$\frac{1}{\hbar^2}\hat{L}^2 = -\left[\frac{1}{\sin\theta}\frac{\partial}{\partial \theta}\left(\sin\theta \frac{\partial}{\partial \theta}\right) + \frac{1}{\sin^2\theta}\frac{\partial^2}{\partial \phi^2}\right] \tag{4-27}$$

is identified as the negative of the square of the orbital angular momentum operator $\hat{L}^2$ in units of $\hbar^2$. Here $\hbar$ is Planck's constant divided by 2π.

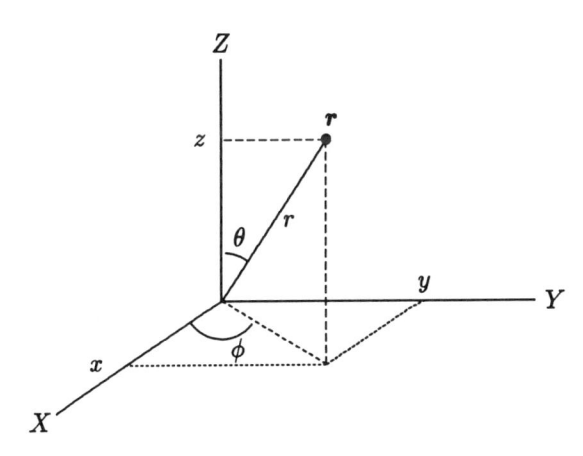

Fig. 4-4 Spherical polar coordinate system. The vector $\mathbf{r}$ is specified by giving its length r, its polar angle θ, and azimuthal angle ϕ.

Separation of variables. For problems with spherical symmetry, we can write the function $\psi(\mathbf{r})$ in (4-25) as a product of radial and angular parts

$$\psi(\mathbf{r}) = R(r)Y(\theta,\phi) \tag{4-28}$$

On substituting this into (4-25) and making use of the form of the Laplacian operator ∇^2 given in (4-26), we obtain

$$\frac{1}{r^2}\frac{\partial}{\partial r}\left(r^2\frac{\partial}{\partial r}\right)R(r)Y(\theta,\phi) + \frac{1}{r^2}\left[\frac{1}{\sin\theta}\frac{\partial}{\partial\theta}\left(\sin\theta\frac{\partial}{\partial\theta}\right) + \frac{1}{\sin^2\theta}\frac{\partial^2}{\partial\phi^2}\right]R(r)Y(\theta,\phi)$$
$$+ k^2 R(r)Y(\theta,\phi) = 0$$

By dividing each term of the equation with $R(r)Y(\theta,\phi)/r^2$ and rearranging the result, we obtain

$$\frac{1}{R(r)}\frac{d}{dr}\left(r^2\frac{d}{dr}\right)R(r) + k^2 r^2 = -\frac{1}{Y(\theta,\phi)}\left[\frac{1}{\sin\theta}\frac{\partial}{\partial\theta}\left(\sin\theta\frac{\partial}{\partial\theta}\right) + \frac{1}{\sin^2\theta}\frac{\partial^2}{\partial\phi^2}\right]Y(\theta,\phi)$$

If the factor k^2 is independent of θ and ϕ, the two sides of the equation have no common variables. The only way to maintain the equality is for each side to be equal to a constant C. As a result, the Helmholtz equation is separated into two equations, one dealing with the radial dependence and the other with angular dependence:

$$\frac{d}{dr}\left(r^2\frac{d}{dr}\right)R(r) + (kr)^2 R(r) = CR(r)$$
$$\left[\frac{1}{\sin\theta}\frac{\partial}{\partial\theta}\left(\sin\theta\frac{\partial}{\partial\theta}\right) + \frac{1}{\sin^2\theta}\frac{\partial^2}{\partial\phi^2}\right]Y(\theta,\phi) = -CY(\theta,\phi) \tag{4-29}$$

Our concern in this section is with the angular part, and we shall delay further discussions of the radial part to the next two sections.

A second separation of variables may be carried out for (4-29). By writing $Y(\theta,\phi)$ as a product of θ- and ϕ-dependent parts,

$$Y(\theta,\phi) = P(\theta)\Phi(\phi)$$

we obtain the following two equations:

$$\frac{1}{P(\theta)\sin\theta}\frac{d}{d\theta}\left(\sin\theta\frac{d}{d\theta}\right)P(\theta) + C + \frac{1}{\sin^2\theta}D = 0$$
$$\frac{1}{\Phi(\phi)}\frac{d^2}{d\phi^2}\Phi(\phi) = D \tag{4-30}$$

where D is constant, in a similar way as C.

Azimuthal angle dependence. Let us deal first with the simpler case of the dependence on the azimuthal angle ϕ. In the usual form of a second-order differential, (4-30) may be written as

$$\frac{d^2\Phi}{d\phi^2} - D\Phi(\phi) = 0$$

The solution of this equation is a familiar one:

$$\Phi(\phi) = Ae^{im\phi} + Be^{-im\phi} \tag{4-31}$$

where $m^2 = -D$. The two constants A and B must be determined by boundary conditions supplied with the problem. On the other hand, since ϕ can only take on values in the range $[0, 2\pi]$, the function $\Phi(\phi)$ must be periodic. For a classical system, it has the property

$$\Phi(\phi + 2\pi) = \Phi(\phi)$$

As a result, m must be an integer. In quantum mechanics, the possibility of a sign change must be allowed if $\Phi(\phi)$ is a part of certain wave functions. As a result,

$$\Phi(\phi + 2\pi) = \pm\Phi(\phi)$$

This comes from the fact that the *parity* of a wave function may be either positive or negative. In this case, m can also be half-integers. However, we shall ignore such a possibility for the moment.

Polar angle dependence and Legendre equation. Let us return to the polar angle dependence of the system given by (4-30),

$$\frac{1}{\sin\theta}\frac{d}{d\theta}\left(\sin\theta\frac{d}{d\theta}\right)P(\theta) + \left(C - \frac{m^2}{\sin^2\theta}\right)P(\theta) = 0$$

This equation may be simplified by writing

$$\eta = \cos\theta$$

with the value of η restricted to the range $[-1, +1]$. The form of (4-30) is now reduced to

$$\frac{d}{d\eta}\left[(1-\eta^2)\frac{dP}{d\eta}\right] + \left(C - \frac{m^2}{1-\eta^2}\right)P(\eta) = 0 \tag{4-32}$$

where P is a function of η and we have made use of the fact that $D = -m^2$.

Let us start by considering the simple case that the system has no ϕ dependence. As a result, $m = 0$. The equation satisfied by P is

$$\frac{d}{d\eta}\left[(1-\eta^2)\frac{dP}{d\eta}\right] + CP(\eta) = 0 \tag{4-33}$$

This is known as the Legendre equation. The more general case $m \neq 0$ is the associated Legendre equation, to which we shall return later.

In the same way as we did earlier for Hermite polynomials, we shall attempt a power series form as the trial solution for $P(\eta)$. We shall see soon that only certain values are possible for the constant C in order for $P(\eta)$ to be a well-behaved function of interest to us. This is very similar to what we saw earlier for Hermite polynomials. Let

$$P(\eta) = \sum_{k=0}^{\infty} a_k \eta^k$$

On substituting this assumption into (4-33), we obtain an equation for the unknown coefficients a_k,

$$\sum_{k=0}^{\infty} \{(k+2)(k+1)a_{k+2} - k(k+1)a_k + Ca_k\}\eta^k = 0$$

For this identity to hold for all the possible values of η in the interval $[-1, 1]$, the quantity inside the braces must vanish. This gives us a relation between a_k for different values of k:

$$a_{k+2} = \frac{k(k+1) - C}{(k+1)(k+2)} a_k \qquad (4\text{-}34)$$

Again, the even and odd powers of η may be separated into two distinct series, one for $a_0 \neq 0$ and $a_1 = 0$ and the other for $a_0 = 0$ and $a_1 \neq 0$.

The power series cannot be infinite since, for large values of k, we have the limiting relation

$$\frac{a_{k+2}}{a_k} \xrightarrow[k\to\infty]{} \frac{k}{k+2}$$

If this is true, $P(\eta)$ diverges for $\eta = \pm 1$. To prevent this from happening, the power series must terminate at some value of k. For this to take place at some value $k = \ell$, it is necessary that

$$C = \ell(\ell + 1)$$

as can be seen from (4-34). This gives us the well-known result that the eigenvalues of the operator $\hat{L}^2$, defined in (4-27), are $\ell(\ell+1)\hbar^2$.

Generating function for Legendre polynomials. Analogous to the parallel situation given by (4-17) for Hermite polynomials, the generating function for the Legendre polynomials has the form

$$g(\eta, t) = \frac{1}{\sqrt{1 - 2\eta t + t^2}} = \sum_{k=0}^{\infty} P_k(\eta) t^k \qquad \text{for} \quad |t| < 1 \qquad (4\text{-}35)$$

We shall first show that $P_k(\eta)$ here satisfies the same differential equation as the Legendre equation of (4-33).

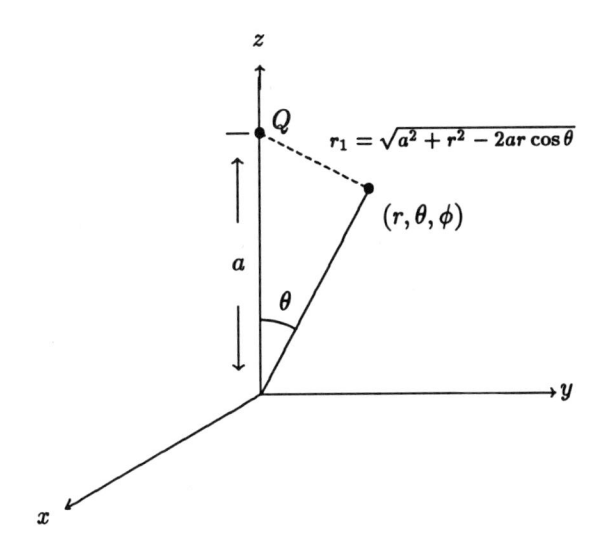

Fig. 4-5 Distance from a point r at (r,θ,ϕ) to a point Q on the z-axis at $z=a$.

The form of (4-35) is familiar from electrostatics. Consider, for example, the potential at a point r due to a charge q located at a point $z = a$ along the z-axis

$$V(r) = c\frac{q}{r_1}$$

where the constant c depends on the system of units used (equal to $1/4\pi\epsilon_0$ in SI and unity in cgs). The distance r_1 between the charge and the point at r, shown in Fig. 4-5, is given by the expression

$$r_1 = \sqrt{a^2 + r^2 - 2ar\cos\theta}$$

where r is the magnitude of the vector r. On letting $a = 1$, $r = t$, and writing $\cos\theta$ as η, we obtain (4-35) as the expansion for r_1^{-1} in terms of a power series in the variable t.

If $g(\eta,t)$ is differentiated with respect to t, we obtain the result

$$\frac{\partial g}{\partial t} = \frac{\eta - t}{(1 - 2\eta t + t^2)^{3/2}} = \sum_{k=0}^{\infty} k P_k(\eta) t^{k-1}$$

This gives us the relation

$$(\eta - t)\sum_{k=0}^{\infty} P_k(\eta) t^k = (1 - 2\eta t + t^2)\sum_{k=0}^{\infty} k P_k(\eta) t^{k-1}$$

For the equality to hold for arbitrary values of t, the coefficients for each power of t must be the same on both sides of the equation. From this requirement we obtain

$$(2k+1)\eta P_k(\eta) = (k+1)P_{k+1}(\eta) + kP_{k-1}(\eta) \tag{4-36}$$

Later, we shall use this equality as the recursion relation to generate the Legendre polynomial of order $(k+1)$, using as input those of orders k and $(k-1)$.

In the same way, we can take the partial derivative of $g(\eta, t)$ with respect to the variable η:

$$\frac{\partial g}{\partial \eta} = \frac{t}{(1 - 2\eta t + t^2)^{3/2}} = \sum_{k=0}^{\infty} P'_k(\eta) t^k$$

and obtain the identity

$$P_k(\eta) + 2\eta P'_k(\eta) = P'_{k+1}(\eta) + P'_{k-1}(\eta) \tag{4-37}$$

To simplify the notation, we have used $P'(\eta) = dP/d\eta$. A similar relation is obtained by differentiating (4-36) with respect to η:

$$(2k+1)P_k(\eta) + (2k+1)\eta P'_k(\eta) = (k+1)P'_{k+1}(\eta) + kP'_{k-1}(\eta) \tag{4-38}$$

Between (4-37) and (4-38), we can eliminate P'_k and obtain the result

$$(2k+1)P_k(\eta) = P'_{k+1}(\eta) - P'_{k-1}(\eta) \tag{4-39}$$

We can use (4-38) and (4-39) to generate the following set of relations:

$$P'_{k+1}(\eta) = (k+1)P_k(\eta) + \eta P'_k(\eta) \tag{4-40}$$

$$P'_{k-1}(\eta) = \eta P'_k(\eta) - kP_k(\eta) \tag{4-41}$$

By replacing $(k-1)$ with k, the last equation is changed to the form

$$P'_k(\eta) = \eta P'_{k+1}(\eta) - (k+1)P_{k+1}(\eta)$$

Using this, we can eliminate $P'_k(\eta)$ with the help of (4-40) and obtain the result

$$(1-\eta^2)P'_{k+1}(\eta) = (k+1)P_k(\eta) - (k+1)\eta P_{k+1}(\eta)$$

or, on replacing $(k+1)$ by k, we have

$$(1-\eta^2)P'_k(\eta) = kP_{k-1}(\eta) - k\eta P_k(\eta)$$

Differentiating both sides of the equation with respect to η and eliminating P'_{k-1} using (4-41), we find that

$$(1-\eta^2)P''_k(\eta) - 2\eta P'_k(\eta) + k(k+1)P_k(\eta) = 0$$

where $P''(\eta) = d^2P/d\eta^2$. It is easy to see that this is the same differential equation as (4-33). This identifies $g(\eta, t)$ of (4-35) as the generating function for Legendre polynomials.

Orthogonality and normalization. Since $P_\ell(\eta)$ is an eigenfunction of the Hermitian operator $\hat{L}^2$, we can make use of theorems in quantum mechanics to show that Legendre polynomials of different order ℓ are orthogonal to each other. The reader is referred to standard texts, such as Merzbacher (*Quantum Mechanics*, 2nd ed., Wiley, New York, 1961), for the proof. The normalization of $P_\ell(\eta)$ is chosen such that

$$\int_{-1}^{+1} P_m(\eta) P_n(\eta)\, d\eta = \frac{2}{2n+1} \delta_{m,n} \qquad (4\text{-}42)$$

Since $P_0(\eta)$ is, by definition, a polynomial of degree zero, it cannot have any dependence on η. With the normalization condition above, it is necessary that

$$P_0(\eta) = 1$$

Similarly, we obtain

$$P_1(\eta) = \eta$$

as it must be a polynomial of degree 1 in η and orthogonal to $P_0(\eta)$.

Table 4-1 Legendre polynomials of low orders.

$P_0(\eta) = 1$	$P_1(\eta) = \eta$
$P_2(\eta) = \frac{1}{2}(3\eta^2 - 1)$	$P_3(\eta) = \frac{1}{2}(5\eta^3 - 3\eta)$
$P_4(\eta) = \frac{1}{8}(35\eta^4 - 30\eta^2 + 3)$	$P_5(\eta) = \frac{1}{8}(63\eta^5 - 70\eta^3 + 15\eta)$

With the values of $P_0(\eta)$ and $P_1(\eta)$ given above as the starting point, we can make use of the recursion relation (4-36) to generate the higher-order polynomials. For this purpose, it is convenient to rewrite the equation as

$$P_\ell(\eta) = \frac{1}{\ell}\{(2\ell - 1)\eta P_{\ell-1}(\eta) - (\ell - 1) P_{\ell-2}(\eta)\} \qquad (4\text{-}43)$$

This is useful, for example, if we wish to find the value of $P_\ell(\eta)$ for a given η from those of $P_{\ell-1}(\eta)$ and $P_{\ell-2}(\eta)$. However, unlike Hermite polynomials, it is not the preferred method to find the coefficients of Legendre polynomials. The difference comes from the fact that the normalization we have chosen for Legendre polynomials produces coefficients that are ratios of polynomials, as can be seen from the explicit forms of the lowest few listed in Table 4-1. As a result, we cannot easily make use of a simple integer representation for the calculations, as we have done earlier for the Hermite polynomials in Box 4-1. For this reason, another method is preferred.

Other ways of representing Legendre polynomials. Let us return to the generating function $g(\eta, t)$ given in (4-35) and consider a power series expansion of the inverse of $\sqrt{1 - (2\eta t - t^2)}$. For this purpose, we shall make use of the binomial series

$$(a + x)^n = a^n + na^{n-1}x + \frac{n(n-1)}{2!}a^{n-2}x^2 + \frac{n(n-1)(n-2)}{3!}a^{n-3}x^3 + \cdots$$

$$= \sum_{k=0}^{n} \binom{n}{k} a^{n-k} x^k \qquad (4\text{-}44)$$

where $\binom{n}{k}$ is the binomial coefficient defined in (1-1). The number of terms in the series is finite if n, the upper limit of the summation, is an integer; otherwise it is an infinite series. For $n = -\frac{1}{2}$ and $a = 1$, it takes on the form

$$\frac{1}{\sqrt{1-x}} = 1 + \frac{1}{2}x + \frac{1 \times 3}{2 \times 4}x^2 + \frac{1 \times 3 \times 5}{2 \times 4 \times 6}x^3 + \frac{1 \times 3 \times 5 \times 7}{2 \times 4 \times 6 \times 8}x^4 + \cdots$$

$$= \sum_{k=0}^{\infty} \frac{(2k-1)!!}{(2k)!!} x^k = \sum_{k=0}^{\infty} \frac{(2k)!}{(2^k k!)^2} x^k \qquad (4\text{-}45)$$

where the double factorial symbol is defined later in (4-149) as

$$k!! = \begin{cases} k(k-2)(k-4)\cdots(4)(2) & \text{for } k \text{ even} \\ k(k-2)(k-4)\cdots(3)(1) & \text{for } k \text{ odd} \end{cases}$$

In arriving at the final form of (4-45), we have made use of the identities

$$(2k)!! = 2^k k!$$

$$(2k - 1)!! = \frac{(2k)!}{(2k)!!} = \frac{(2k)!}{2^k k!}$$

More discussions on the properties of double factorials are given later in §4-4 in conjunction with the gamma function $\Gamma(n)$.

By putting $(2\eta t - t^2)$ in (4-35) as x, we obtain, with the help of (4-45), a binomial series form for the generating function of Legendre polynomials:

$$g(\eta, t) = \sum_{k=0}^{\infty} \frac{(2k)!}{2^{2k}(k!)^2} (2\eta t - t^2)^k = \sum_{k=0}^{\infty} \frac{(2k)!}{2^{2k}(k!)^2} t^k (2\eta - t)^k$$

The factor $(2\eta - t)^k$ on the right side may be expanded by making use of (4-44).

This produces the expression

$$g(\eta, t) = \sum_{k=0}^{\infty} \frac{(2k)!}{2^{2k}(k!)^2} t^k \sum_{j=0}^{k} (-1)^j \binom{k}{j} (2\eta)^{k-j} t^j$$

$$= \sum_{k=0}^{\infty} \sum_{j=0}^{k} (-1)^j \frac{(2k)!}{2^{2k} k! j! (k-j)!} (2\eta)^{k-j} t^{k+j}$$

$$= \sum_{k=0}^{\infty} \sum_{j=0}^{[k/2]} (-1)^j \frac{(2k-2j)!}{2^{2k-2j} j! (k-j)! (k-2j)!} (2\eta)^{k-2j} t^k$$

where, in rearranging the order of summations, we have made use of the identity

$$\sum_{k=0}^{\infty} \sum_{j=0}^{k} \alpha_{k,k-j} = \sum_{m=0}^{\infty} \sum_{n=0}^{[m/2]} \alpha_{n,m-2n} \qquad (4\text{-}46)$$

For j and m integers, the brackets in the summation mean that

$$\left[\frac{m}{2}\right] = \begin{cases} j & \text{if } m = 2j \\ j-1 & \text{if } m = 2j-1 \end{cases}$$

By comparing the coefficients of t^k in (4-46) and (4-35), we obtain an explicit expression for the Legendre polynomials in terms of a power series in η:

$$P_\ell(\eta) = \sum_{k=0}^{[\ell/2]} (-1)^k \frac{(2\ell - 2k)!}{2^\ell k! (\ell - k)! (\ell - 2k)!} \eta^{\ell - 2k} \qquad (4\text{-}47)$$

This may be written in the more familiar form of

$$P_\ell(\eta) = \sum_{k=0}^{\ell} a_{\ell,k} \eta^k$$

where, according to (4-34), the coefficients $a_{\ell,k}$ vanish for odd k, if ℓ is even, and for even k, if ℓ is odd. The nonvanishing coefficients are given by the relation

$$a_{\ell,k} = (-1)^{(\ell-k)/2} \frac{(\ell+k)(\ell+k-1)\cdots(k+2)(k+1)}{2^\ell \ell!} \binom{\ell}{\frac{\ell+k}{2}} \qquad (4\text{-}48)$$

It is more convenient to calculate the values of the coefficients $a_{\ell,k}$ using (4-48) and construct the algebraic form of Legendre polynomials of a given degree with the values of the coefficients obtained in this way. Note that, since k and ℓ must be both odd or both even, the factors $(\ell \pm k)/2$ appearing in the phase factor and the binomial coefficient are integers.

It is useful to have a different form of (4-47) for the purpose of displaying some of the properties of Legendre polynomials. The factor $(2\ell - 2k)!/(\ell - 2k)!$

may be suppressed by differentiating the product of η^ℓ on the right side of (4-47) ℓ times with respect to η. The result is

$$P_\ell(\eta) = \sum_{k=0}^{[\ell/2]} (-1)^k \frac{1}{2^\ell k!(\ell-k)!} \frac{d^\ell}{d\eta^\ell} \eta^{2\ell-2k}$$

Since the coefficients in front do not involve η, we can take the differentiation operators outside the summation. Once we have done this, it is possible to extend the upper limit of the summation from $[\ell/2]$ to ℓ without changing the result. This comes from the fact that terms with $k > [\ell/2]$ vanish on differentiating ℓ times with respect to η. On rearranging the remaining factors in the form of a binomial coefficient, we obtain the Rodrigues formula:

$$\begin{aligned} P_\ell(\eta) &= \frac{1}{2^\ell \ell!} \frac{d^\ell}{d\eta^\ell} \sum_{k=0}^{\ell} (-1)^k \frac{\ell!}{k!(\ell-k)!} \eta^{2\ell-2k} \\ &= \frac{1}{2^\ell \ell!} \frac{d^\ell}{d\eta^\ell} \sum_{k=0}^{\ell} (-1)^k \binom{\ell}{k} (\eta^2)^{\ell-k} \\ &= \frac{1}{2^\ell \ell!} \frac{d^\ell}{d\eta^\ell} \left(\eta^2 - 1\right)^\ell \end{aligned} \qquad (4\text{-}49)$$

Although we shall make very little direct use of this expression, it is the usual form of Legendre polynomials found in standard references.

Associated Legendre polynomials. Let us now return to the more general case of a system that is a function of the polar angle θ as well as the azimuthal angle ϕ. As far as the θ-dependent part is concerned, we no longer have the simplification offered by $m = 0$ in (4-33). Starting from the Rodrigues formula, it is not difficult to see that the function

$$P_{\ell,m}(\eta) = \left(1-\eta^2\right)^{m/2} \frac{d^m}{d\eta^m} P_\ell(\eta) = \frac{1}{2^\ell \ell!} \left(1-\eta^2\right)^{m/2} \frac{d^{\ell+m}}{d\eta^{\ell+m}} \left(\eta^2 - 1\right)^\ell \qquad (4\text{-}50)$$

for $0 \le |m| \le \ell$, satisfies the associated Legendre equation

$$\frac{d}{d\eta}\left[(1-\eta^2)\frac{d}{d\eta}P_{\ell,m}\right] + \left[\ell(\ell+1) - \frac{m^2}{1-\eta^2}\right] P_{\ell,m} = 0$$

This equation may be obtained from (4-32) by replacing C with $\ell(\ell+1)$. The proof can be found, for example, in Arfken (*Mathematical Methods for Physicists*, Academic Press, New York, 1970, page 558) and we shall not give it here. The quantity $P_{\ell,m}(\eta)$ is know as the associated Legendre polynomial.

With this choice, the orthonormal condition for $P_{\ell,m}(\eta)$ for different ℓ but the same m becomes

$$\int_{-1}^{+1} P_{\ell,m}(\eta) P_{\ell',m}(\eta)\, d\eta = \frac{2}{2\ell+1} \frac{(\ell+m)!}{(\ell-m)!} \delta_{\ell,\ell'} \qquad (4\text{-}51)$$

However, the orthogonality condition for associated Legendre polynomials of the same ℓ but different m requires an additional weighting factor. The form is

$$\int_{-1}^{+1} P_{\ell,m}(\eta) P_{\ell,m'}(\eta) \frac{1}{1-\eta^2} d\eta = \frac{(\ell+m)!}{m(\ell-m)!} \delta_{m,m'} \qquad (4\text{-}52)$$

We shall see later that it is more convenient to express the same two conditions for the parallel situation in spherical harmonics. A few of the low-order associated Legendre polynomials for $m > 0$ are given in Table 4-2 as examples. Later, we shall construct a simple algorithm to generate the algebraic form of $P_{\ell,m}(\eta)$ to arbitrary orders starting from (4-48).

Table 4-2 Associated Legendre polynomials $P_{\ell,m}(\eta)$ for $m > 0$ and $\ell \leq 5$.

$P_{1,1}(\eta) = (1-\eta^2)^{1/2}$	$P_{2,1}(\eta) = 3\eta(1-\eta^2)^{1/2}$
$P_{2,2}(\eta) = 3(1-\eta^2)$	$P_{3,1}(\eta) = \frac{3}{2}(5\eta^2-1)(1-\eta)^{1/2}$
$P_{3,2}(\eta) = 15\eta(1-\eta^2)$	$P_{3,3}(\eta) = 15(1-\eta^2)^{3/2}$
$P_{4,1}(\eta) = \frac{5}{2}\eta(7\eta^2-3)(1-\eta^2)^{1/2}$	$P_{4,2}(\eta) = \frac{15}{2}(7\eta^2-1)(1-\eta^2)$
$P_{4,3}(\eta) = 105\eta(1-\eta^2)^{3/2}$	$P_{4,4}(\eta) = 105(1-\eta^2)^2$

It is also possible to show that, instead of (4-50), the expression for $P_{\ell,m}(\eta)$ may also be written as

$$P_{\ell,m}(\eta) = (-1)^m \frac{1}{2^\ell \ell!} \frac{(\ell+m)!}{(\ell-m)!} (1-\eta^2)^{-m/2} \frac{d^{\ell-m}}{d\eta^{\ell-m}} (\eta^2-1)^\ell$$

By comparing this form with that given in (4-50), we find that two associated Legendre polynomials, having the same ℓ but with m value opposite in sign, are related to each other in the following way:

$$P_{\ell,-m}(\eta) = (-1)^m \frac{(\ell-m)!}{(\ell+m)!} P_{\ell,m}(\eta) \qquad (4\text{-}53)$$

The same relation can also be obtained from (4-50) through the use of Leibnitz's formula for the derivatives of a product of two functions:

$$\frac{d^n}{dx^n}\{A(x)B(x)\} = \sum_{k=0}^{n} \binom{n}{k} \left\{\frac{d^{n-k}}{dx^{n-k}} A(x)\right\} \left\{\frac{d^k}{dx^k} B(x)\right\}$$

By taking $A(x) = (\eta-1)^\ell$ and $B(x) = (\eta+1)^\ell$, the relation given in (4-53) is obtained after rearranging the terms.

Spherical harmonics. Since the square of the angular momentum operator $\hat{L}^2$ given in (4-27) is a function of both the polar angle θ and azimuthal angle ϕ, its eigenfunctions are products of $e^{\pm im\phi}$ of (4-31) and associated Legendre polynomials $P_{\ell,m}(\cos\theta)$. These functions are known as the spherical harmonics and they are given by

$$Y_{\ell,m}(\theta,\phi) = (-1)^m \sqrt{\frac{(2\ell+1)}{4\pi}\frac{(\ell-m)!}{(\ell+m)!}}\, e^{im\phi} P_{\ell,m}(\cos\theta)$$

$$= \frac{(-1)^m}{2^\ell \ell!}\sqrt{\frac{(2\ell+1)}{4\pi}\frac{(\ell-m)!}{(\ell+m)!}}\, e^{im\phi}(1-\eta^2)^{m/2}\left(\frac{d}{d\eta}\right)^{\ell+m}(\eta^2-1)^\ell$$

(4-54)

The explicit forms of the lowest few are

$$Y_{0,0}(\theta,\phi) = \sqrt{\frac{1}{4\pi}}$$

$$Y_{1,0}(\theta,\phi) = \sqrt{\frac{3}{4\pi}}\cos\theta$$

$$Y_{1,\pm 1}(\theta,\phi) = \mp\sqrt{\frac{3}{8\pi}}e^{\pm i\phi}\sin\theta$$

$$Y_{2,0}(\theta,\phi) = \sqrt{\frac{5}{16\pi}}(3\cos^2\theta - 1)$$

$$Y_{2,\pm 1}(\theta,\phi) = \mp\sqrt{\frac{15}{8\pi}}e^{\pm i\phi}\cos\theta\sin\theta$$

$$Y_{2,\pm 2}(\theta,\phi) = \sqrt{\frac{15}{32\pi}}e^{\pm 2i\phi}\sin^2\theta$$

$$Y_{3,0}(\theta,\phi) = \sqrt{\frac{7}{16\pi}}(5\cos^3\theta - 3\cos\theta)$$

$$Y_{3,\pm 1}(\theta,\phi) = \mp\sqrt{\frac{21}{64\pi}}e^{\pm i\phi}(5\cos^2\theta - 1)\sin\theta$$

$$Y_{3,\pm 2}(\theta,\phi) = \sqrt{\frac{105}{32\pi}}e^{\pm 2i\phi}\cos\theta\sin^2\theta$$

$$Y_{3,\pm 3}(\theta,\phi) = \mp\sqrt{\frac{35}{64\pi}}e^{\pm 3i\phi}\sin^3\theta \qquad (4\text{-}55)$$

The multiplicative constants for these functions are selected in such a way that spherical harmonics are normalized to unity. The orthogonality condition between those with different values of ℓ and m is given by the following integral:

$$\int_0^{2\pi}\int_0^{\pi} Y^*_{\ell m}(\theta,\phi) Y_{\ell' m'}(\theta,\phi)\sin\theta\, d\theta\, d\phi = \delta_{\ell\ell'}\delta_{mm'} \qquad (4\text{-}56)$$

The complex conjugate of $Y_{\ell m}(\theta, \phi)$ is related to that with the opposite sign of m by the relation

$$Y_{\ell m}^*(\theta, \phi) = (-1)^m Y_{\ell,-m}(\theta, \phi) \tag{4-57}$$

as can be seen from (4-53) and (4-54).

Being the product of $e^{\pm im\phi}$ and associated Legendre polynomials, spherical harmonics are also the eigenfunctions of the z-component of the operator $\hat{L}$ for angular momentum

$$\hat{L}_z Y_{\ell,m}(\theta, \phi) = -i\hbar \frac{\partial}{\partial \phi} Y_{\ell,m}(\theta, \phi) = m\hbar Y_{\ell,m}(\theta, \phi)$$

The other two components of the operator $\hat{L}$ may be written in the form of the angular momentum raising operator $\hat{L}_+$ and lowering operator $\hat{L}_-$, and they have the properties

$$\hat{L}_\pm Y_{\ell,m}(\theta, \phi) = \left(\hat{L}_x \pm i\hat{L}_y\right) Y_{\ell,m}(\theta, \phi)$$

$$= \pm \hbar e^{\pm i\phi} \left(\frac{\partial}{\partial \theta} \pm i \cot\theta \frac{\partial}{\partial \phi} \right) Y_{\ell,m}(\theta, \phi)$$

$$= \hbar \sqrt{(\ell \mp m)(\ell \pm m + 1)}\, Y_{\ell\, m\pm 1}(\theta, \phi) \tag{4-58}$$

For $m = 0$, the spherical harmonics may be expressed in term of Legendre polynomials, as we expect from the definition of $P_\ell(\eta)$:

$$Y_{\ell 0}(\theta) = \sqrt{\frac{2\ell + 1}{4\pi}}\, P_\ell(\cos\theta) \tag{4-59}$$

The same relation can also be obtained from (4-54) by putting $m = 0$. This, together with (4-58), may be used to generate spherical harmonics of other m values.

In terms of $Y_{\ell,m}(\theta, \phi)$, the separation of $\psi(\mathbf{r})$ for a given ℓ and m as a product of radial and angular parts given earlier in (4-28) takes on the form

$$\psi_{\ell m}(\mathbf{r}) = R(r) Y_{\ell,m}(\theta, \phi) \tag{4-60}$$

For this reason, spherical harmonics occur frequently in problems having spherical symmetry.

> **Box 4-2** Subroutine LEG_ALK(L_SIGN,LIST,L_RANK,K_RANK,MD)
> **Coefficients of Legendre polynomials using (4-61)**
>
> Argument list:
>   L_SIGN: Sign of $a_{\ell,k}$.
>   LIST: Array of prime number decomposition of $a_{\ell,k}$.
>   L_RANK: Rank ℓ of $a_{\ell,k}$.
>   K_RANK: Rank k of $a_{\ell,k}$.
>   MD: Dimension of LIST.
>
> Subprogram used:
>   PRM_DCMP: Prime number decomposition of Box 1-2.
>
> 1. Initialize prime number decomposition of integers and factorials using PRM_DCMP.
> 2. Zero the array LIST.
> 3. Check for special cases:
>   (a) Return zero if $(\ell + k)$ odd.
>   (b) Return 1 if both ℓ and k are zero.
> 4. Calculate the nonvanishing coefficients using (4-61):
>   (a) Form the product $(l+k)(l+k-1)\cdots(k+2)(k+1)$.
>   (b) Include contributions from the binomial coefficient and the denominator.
>   (c) Include the phase factor $(-1)^{(\ell-k)/2}$.
> 5. Return the prime number decomposition of $a_{\ell,k}$.

Coefficients of Legendre polynomials and spherical harmonics. We now need a practical method of finding the coefficients of the Legendre polynomials $P_\ell(\eta)$, the associated Legendre polynomials $P_{\ell,m}(\eta)$, and the spherical harmonics $Y_{\ell,m}(\theta,\phi)$. For $P_\ell(\eta)$, it is possible to use the recursion relation (4-36) to produce higher order Legendre polynomials from the starting point of $P_0(\eta) = 1$ and $P_1(\eta) = \eta$). However, a more direct method is to use the explicit expression for the coefficients $a_{\ell,k}$ of $P_{\ell,m}(\eta)$ given by (4-48). That is,

$$P_\ell(\eta) = \sum_{k=0}^{\ell} a_{\ell,k}\eta^k$$

with

$$a_{\ell,k} = \begin{cases} 0 & \text{for } (\ell+k) \text{ odd} \\ (-1)^{(\ell-k)/2}\dfrac{(\ell+k)(\ell+k-1)\cdots(k+2)(k+1)}{2^\ell \ell!}\binom{\ell}{\frac{\ell+k}{2}} & \text{otherwise} \end{cases}$$

(4-61)

The algorithm, given in Box 4-2, is particularly simple if we use the method of decomposition into powers of prime numbers given in §1-2.

For associated Legendre polynomials, we have two indexes, ℓ and m. As a result, the recursion relations are more complicated. Furthermore, the coefficients are ratios of integers because of the normalization condition of (4-52). For these

> **Box 4-3** Subroutine YLM_LEG(KIND,L_RANK,M_RANK)
> **Associated Legendre polynomial**
> (Includes also spherical harmonics of Box 4-4)
>
> Argument list:
>   KIND = 2 for associated Legendre polynomial (= 3 for spherical harmonics).
>   L_RANK: Rank ℓ of $b_{\ell,m;k}$.
>   M_RANK: Rank m of $b_{\ell,m;k}$.
> Subprogram used:
>   LEG_ALK: Coefficients of Legendre polynomials of Box 4-2.
> 1. Set up a symbol to distinguish $P_{\ell,m}$ from $Y_{\ell,m}$ in the output.
> 2. Calculate each of the coefficients $b_{\ell,m;k}$.
>   (a) Use LEG_ALK to generate $a_{\ell,k}$.
>   (b) Convert $a_{\ell,k}$ to $b_{\ell,m;k}$.
>   (i) Use (4-62a) for $m > 0$.
>   (ii) Use (4-62b) for $m < 0$.
> 3. Output the coefficients.

reasons, it is more convenient to make use of the first form of (4-50) for $P_{\ell,m}(\eta)$ with $m \geq 0$. This gives us

$$P_{\ell,m}(\eta) = \left(1-\eta^2\right)^{m/2} \frac{d^m}{d\eta^m} P_\ell(\eta) = \left(1-\eta^2\right)^{m/2} \sum_{k=0}^{\ell-m} b_{\ell,m;k}\, \eta^k$$

where the coefficients $b_{\ell,m;k}$ may be obtained using (4-61):

$$b_{\ell,m;k} = a_{\ell,m+k} \frac{d^m}{d\eta^m} \eta^{k+m} = a_{\ell,m+k}(k+m)(k+m-1)\cdots(k+2)(k+1)$$

$$= \frac{(k+m)!}{k!} a_{\ell,m+k} \qquad (4\text{-}62a)$$

For $m < 0$, we can use (4-53) to relate the coefficients to those with the opposite sign of m. The result is

$$b_{\ell,-m;k} = (-1)^m \frac{(\ell-m)!}{(\ell+m)!} \frac{(k+m)!}{k!} a_{\ell,m+k} \qquad (4\text{-}62b)$$

where m is a positive integer. The algorithm to calculate $b_{\ell,m;k}$, using $a_{\ell,k}$ as the starting point, is given in Box 4-3.

To find the spherical harmonics, we can take the product of the $\exp\{im\phi\}$ factor for the azimuthal angle dependence with the appropriate associated Legendre polynomials. The latter may be obtained using the relation given by the first form of (4-54). In terms of a power series in η, spherical harmonics $Y_{\ell,m}(\theta,\phi)$ may

> **Box 4-4 Subroutine YLM_LEG(KIND,L_RANK,M_RANK)**
> **Spherical harmonics $Y_{\ell,m}$**
> (Includes also associated Legendre polynomial)
>
> Argument list:
>   KIND = 3 for spherical harmonics (= 2 for associated Legendre polynomial).
>   L_RANK: Rank ℓ for $b_{\ell,m;k}$.
>   M_RANK: Rank m for $b_{\ell,m;k}$.
> Subprogram used:
>   LEG_ALK: Coefficients of Legendre polynomials of Box 4-2.
> 1. Set up a symbol to distinguish $P_{\ell,m}$ from $Y_{\ell,m}$ in the output.
> 2. Obtain the coefficient $b_{\ell,m;k}$ of $P_{\ell,m}$ using LEG_ALK.
> 3. Convert $b_{\ell,m;k}$ to $c^2_{\ell,m;k}$ using (4-63).
>   (a) Square $b_{\ell,m;k}$ and multiply by $(2\ell+1)$.
>   (b) Include the factorial factors.
> 4. Return $c^2_{\ell,m;k}$.

be written in the form

$$Y_{\ell,m}(\theta,\phi) = (-1)^m \sqrt{\frac{1}{4\pi}}\, e^{im\phi} \sin^m\theta \sum_{k=0}^{\ell-m} c_{\ell,m;k}\, \eta^k$$

where we have made use of the fact that

$$(1-\eta^2)^{m/2} = (1-\cos^2\theta)^{m/2} = \sin^m\theta$$

The coefficient $c_{\ell,m;k}$ may be calculated from $b_{\ell,m;k}$ of $P_{\ell,m;k}(\eta)$ given in (4-62):

$$c_{\ell,m;k} = \sqrt{\frac{(2\ell+1)(\ell-m)!}{(\ell+m)!}}\, b_{\ell,m;k}$$

As a practical matter, it is easier to calculate the squares of the coefficients:

$$c^2_{\ell,m;k} = \frac{(2\ell+1)(\ell-m)!}{(\ell+m)!}\, b^2_{\ell,m;k} \qquad (4\text{-}63)$$

In this way, $c^2_{\ell,m;k}$ may be expressed in terms of a product of integer powers of prime numbers with the sign of $c_{\ell,m;k}$ following that of $b_{\ell,m;k}$. The algorithm for such a calculation is given in Box 4-4.

§4-3 Spherical Bessel functions

Let us return to the radial part of the Helmholtz equation given by (4-25). If k^2 is a constant in the region of interest, the radial equation (4-28) reduces to the form

$$\frac{d}{dr}\left(r^2 \frac{dR_\ell}{dr}\right) + \{k^2 r^2 - \ell(\ell+1)\} R_\ell(r) = 0 \qquad (4\text{-}64)$$

where we have used $R_\ell(r)$ to represent the radial part of $\psi(r)$ of (4-28) for a given angular momentum ℓ. Since k here has the dimension of inverse length, it is convenient to use, in the place of r, the dimensionless variable

$$\rho = kr$$

The differential equation (4-64) may now be expressed in the form

$$\rho^2 \frac{d^2 R_\ell}{d\rho^2} + 2\rho \frac{dR_\ell}{d\rho} + \{\rho^2 - \ell(\ell+1)\} R_\ell(r) = 0 \qquad (4\text{-}65)$$

It is easy to show that if we make the substitutions

$$W(\rho) = \rho^{1/2} R_\ell(\rho) \qquad \text{and} \qquad \lambda = \ell + \tfrac{1}{2}$$

the equation for $R_\ell(r)$ may be put into the standard form of a Bessel equation for $W(\rho)$

$$\rho^2 \frac{d^2 W}{d\rho^2} + \rho \frac{dW}{d\rho} + (\rho^2 - \lambda^2) W(\rho) = 0 \qquad (4\text{-}66)$$

Since λ is a half-integer here, solutions of (4-65) are related to Bessel functions of half integer ranks. Our main interest here is in (4-65), and we shall only return briefly to (4-66) at the end of this section.

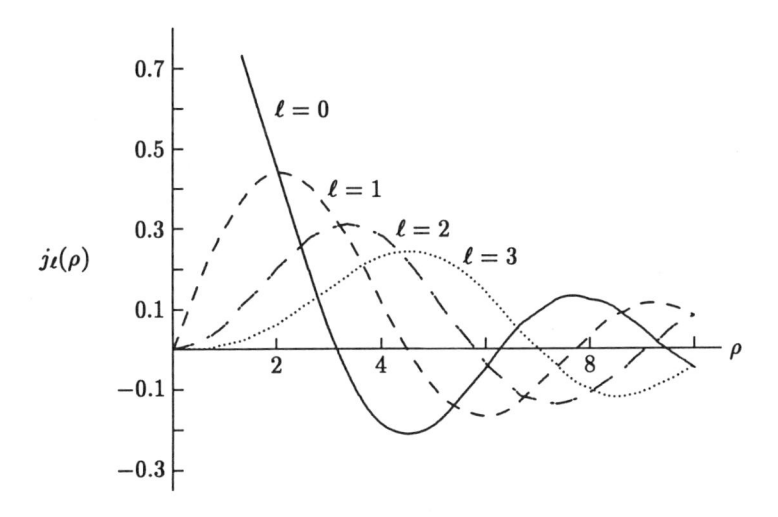

Fig. 4-6 Spherical Bessel functions $j_\ell(\rho)$ for $\ell = 0, 1, 2,$ and 3.

For $\ell = 0$, (4-65) has a particularly simple form:

$$\rho^2 \frac{d^2 R_0}{d\rho^2} + 2\rho \frac{dR_0}{d\rho} + \rho^2 R_0(r) = 0 \tag{4-67}$$

There are two linearly independent solutions for $R_0(\rho)$. The one that is regular at the origin is the spherical Bessel function of rank zero,

$$j_0(\rho) = \frac{\sin \rho}{\rho} \tag{4-68}$$

The other one,

$$n_0(\rho) = -\frac{\cos \rho}{\rho} \tag{4-69}$$

is the spherical Neumann function of rank zero. Particular linear combinations of them,

$$h_0^{(1)}(\rho) = j_0(\rho) + i n_0(\rho) \qquad h_0^{(2)}(\rho) = j_0(\rho) - i n_0(\rho)$$

are known as the spherical Hankel functions of, respectively, the first and second kind.

Recursion relations. It is easy to show that any linear combination of the regular and irregular solutions

$$f_0(\rho) = A j_0(\rho) + B n_0(\rho) \tag{4-70}$$

satisfies (4-65) for $\ell = 0$. As a result, we have the relation

$$f_0'' + \frac{2}{\rho} f_0' + f_0 = 0 \tag{4-71}$$

where we have adopted the notation

$$f' \equiv \frac{df}{d\rho} \qquad f'' \equiv \frac{d^2 f}{d\rho^2} \qquad f^{(n)} \equiv \frac{d^n f}{d\rho^n}$$

Differentiating each term of (4-71) with respect to ρ and rearranging the terms, we obtain

$$f_0''' + \frac{2}{\rho} f_0'' - \frac{2}{\rho^2} f_0' + f_0' = \frac{d^2}{d\rho^2} f_0' + \frac{2}{\rho} \frac{d}{d\rho} f_0' + \left(1 - \frac{2}{\rho^2}\right) f_0' = 0$$

Except for an overall factor of ρ^2, this is the same as (4-65) for $\ell = 1$. As a result, we can identify that $f_0'(\rho)$ is a solution of (4-65) for $\ell = 1$. That is,

$$f_1(\rho) = \frac{d}{d\rho} f_0(\rho) \tag{4-72}$$

More specifically, spherical Bessel and Neumann functions for $\ell = 1$ have the forms

$$j_1(\rho) = \frac{\sin \rho}{\rho^2} - \frac{\cos \rho}{\rho} \qquad n_1(\rho) = -\frac{\cos \rho}{\rho^2} - \frac{\sin \rho}{\rho} \tag{4-73}$$

Chapter 4] Special Functions 173

We shall see later that spherical Bessel and Neumann functions of higher ranks may be generated from those for $\ell = 0$ and 1. For the purpose of illustration, the forms for $\ell \leq 5$ are listed in Table 4-3. Their variations as functions of ρ for $\ell \leq 3$ are given in Fig. 4-6 for spherical Bessel functions and in Fig. 4-7 for spherical Neumann functions.

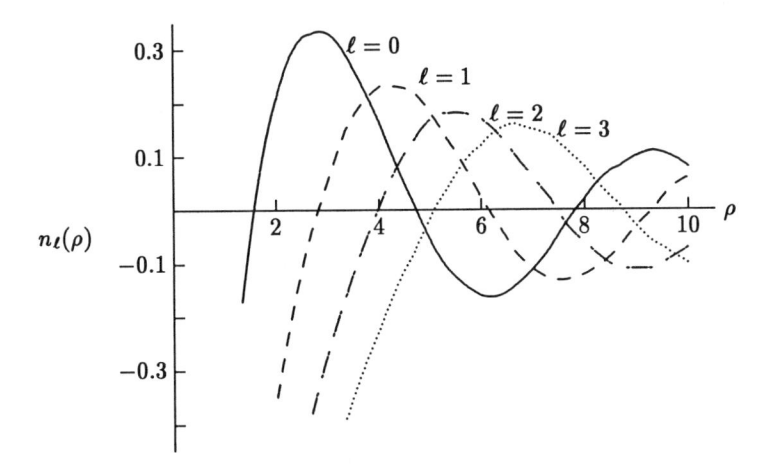

Fig. 4-7 Spherical Neumann functions $n_\ell(\rho)$ for $\ell = 0, 1, 2,$ and 3.

The relation given by (4-72) may be generalized to arbitrary values of ℓ. By induction, we can show that, if $f_n(\rho)$ is a solution of the spherical Bessel equation (4-65) for $\ell = n$, that is,

$$f_n'' + \frac{2}{\rho} f_n' + \left\{ 1 - \frac{n(n+1)}{\rho^2} \right\} f_n = 0 \tag{4-74}$$

then

$$g(\rho) = \rho^n \frac{d}{d\rho} \rho^{-n} f_n(\rho) \tag{4-75}$$

satisfies (4-65) for $\ell = (n+1)$. To derive this relation, we start with the definition of $g(\rho)$ given by (4-75) and obtain

$$g = f_n' - \frac{n}{\rho} f_n$$

$$\frac{dg}{d\rho} = f_n'' - \frac{n}{\rho} f_n' + \frac{n}{\rho^2} f_n$$

$$\frac{d^2 g}{d\rho^2} = f_n''' - \frac{n}{\rho} f_n'' + \frac{2n}{\rho^2} f_n' - \frac{2n}{\rho^3} f_n$$

We can eliminate f_n''' in $d^2g/d\rho^2$ using the relation

$$f_n''' + \frac{2}{\rho}f_n'' + \left(1 - \frac{n^2+n+2}{\rho^2}\right)f_n' + \frac{2n(n+1)}{\rho^3}f_n = 0$$

obtained by differentiating each term of (4-74) with respect to ρ. To this, we add g'' and $2\rho^{-1}g'$ to produce

$$\frac{d^2g}{d\rho^2} + \frac{2}{\rho}\frac{dg}{d\rho} = -\frac{n}{\rho}f_n'' - \left(1 - \frac{n^2+n+2}{\rho^2}\right)f_n' - \frac{2n(n+1)}{\rho^3}f_n$$

Expressing f_n'' in terms of f_n' and f_n using (4-74) and rearranging the terms, we obtain the following relation for g:

$$\frac{d^2g}{d\rho^2} + \frac{2}{\rho}\frac{dg}{d\rho} + \left\{1 - \frac{(n+1)(n+2)}{\rho^2}\right\}g = 0$$

Since this is the same as (4-65) for $\ell = (n+1)$, we deduce that $g(\rho)$ is equivalent to a spherical Bessel function of rank $(n+1)$.

Table 4-3 Spherical Bessel and Neumann functions of $\ell = 0$ to 5.

ℓ	$j_\ell(\rho)$	$n_\ell(\rho)$
0	$\frac{\sin\rho}{\rho}$	$-\frac{\cos\rho}{\rho}$
1	$\frac{\sin\rho}{\rho^2} - \frac{\cos\rho}{\rho}$	$-\frac{\cos\rho}{\rho^2} - \frac{\sin\rho}{\rho}$
2	$\left(\frac{3}{\rho^3} - \frac{1}{\rho}\right)\sin\rho - \frac{3}{\rho^2}\cos\rho$	$\left(-\frac{3}{\rho^3} + \frac{1}{\rho}\right)\cos\rho - \frac{3}{\rho^2}\sin\rho$
3	$\left(\frac{15}{\rho^4} - \frac{6}{\rho^2}\right)\sin\rho + \left(-\frac{15}{\rho^3} + \frac{1}{\rho}\right)\cos\rho$	$\left(-\frac{15}{\rho^4} + \frac{6}{\rho^2}\right)\cos\rho + \left(-\frac{15}{\rho^3} + \frac{1}{\rho}\right)\sin\rho$
4	$\left(\frac{105}{\rho^5} - \frac{45}{\rho^3} + \frac{1}{\rho}\right)\sin\rho + \left(-\frac{105}{\rho^4} + \frac{10}{\rho^2}\right)\cos\rho$	$\left(-\frac{105}{\rho^5} + \frac{45}{\rho^3} - \frac{1}{\rho}\right)\cos\rho + \left(-\frac{105}{\rho^4} + \frac{10}{\rho^2}\right)\sin\rho$
5	$\left(\frac{945}{\rho^6} - \frac{420}{\rho^4} + \frac{15}{\rho^2}\right)\sin\rho + \left(-\frac{945}{\rho^5} + \frac{105}{\rho^3} - \frac{1}{\rho}\right)\cos\rho$	$\left(-\frac{945}{\rho^6} + \frac{420}{\rho^4} - \frac{15}{\rho^2}\right)\cos\rho + \left(-\frac{945}{\rho^5} + \frac{105}{\rho^3} - \frac{1}{\rho}\right)\sin\rho$

We have now established a relation between the solutions of spherical Bessel equation for order n and $(n+1)$. That is,

$$f_{n+1}(\rho) = -\rho^n \frac{d}{d\rho} \rho^{-n} f_n(\rho) \tag{4-76}$$

where, following general practice, we have chosen to insert an overall negative sign on the right side. Using (4-76), we obtain the Rayleigh formulas

$$j_n(\rho) = \rho^n \left(-\frac{1}{\rho}\frac{d}{d\rho}\right)^n \frac{\sin\rho}{\rho}$$

$$n_n(\rho) = -\rho^n \left(-\frac{1}{\rho}\frac{d}{d\rho}\right)^n \frac{\cos\rho}{\rho} \tag{4-77}$$

by starting from the explicit forms of $j_0(\rho)$ and $n_0(\rho)$ given earlier as $f_0(\rho)$.

Since, in arriving at the result of (4-76), the rank n is arbitrary, we can also begin with (4-70) for $n = 0$ and construct the solution for $n = 1$. From $f_1(\rho)$, we can obtain $f_2(\rho)$, and so on. Using similar arguments as those in the previous two paragraphs, it is also possible to show that

$$f_{n-1}(\rho) = \rho^{-(n+1)} \frac{d}{d\rho} \rho^{(n+1)} f_n(\rho) \tag{4-78}$$

By working out the differentiation explicitly on the right sides of (4-76) and (4-78) and eliminating f'_n between them, we obtain the recursion relation

$$(2n+1)f_n(\rho) = \rho\{f_{n+1}(\rho) + f_{n-1}(\rho)\} \tag{4-79}$$

This result may be used to generate the algebraic forms of the spherical Bessel and Neumann functions of arbitrary ranks.

Box 4-5   Program SPH_BES
Coefficients for spherical Bessel function

Initialization:
  (a) Assign arrays for the coefficients $a_{n-1,k}$, $a_{n,k}$, and $a_{n+1,k}$ of $\sin\rho$.
  (b) Assign arrays for the coefficients b_{n-1}, b_n, and b_{n+1} of $\cos\rho$.
  (c) Zero the arrays.
  (d) Define the starting values using (4-82).
  (e) Set $n = 1$.
1. Input the order required.
2. Propagate to $(n+1)$ using (4-81):
  (a) For $k = 0$, let $a_{n+1,0} = -a_{n-1,0}$ and $b_{n+1,0} = -b_{n-1,0}$.
  (b) For $k > 0$, obtain $a_{n+1,k}$ from $a_{n,k}$ and $a_{n-1,k}$ and $b_{n+1,k}$ from $b_{n,k}$ and $b_{n-1,k}$.
3. If $(n+1)$ is less than the required rank,
  (a) Store the values of $a_{n,k}$ and $b_{n,k}$ into the locations of $a_{n-1,k}$ and $b_{n-1,k}$, respectively.
  (b) Store the values of $a_{n+1,k}$ and $b_{n+1,k}$ into the locations of $a_{n,k}$ and $b_{n,k}$, respectively.
  (c) Increase the value of n by 1 and go back to step 2.
4. If $(n+1)$ equals the required rank, output the coefficients for $(n+1)$.

Calculation of spherical Bessel and Neumann functions. It is obvious from the explicit forms of spherical Bessel functions given in Table 4-3 that $j_n(\rho)$ for an arbitrary n consists of a linear combination of $\sin\rho$ and $\cos\rho$. To make use of (4-79) for generating $j_n(\rho)$, we shall start from the expression

$$j_n(\rho) = \sin\rho \sum_{k=0}^{n} \frac{a_{n,k}}{\rho^{k+1}} + \cos\rho \sum_{k=0}^{n} \frac{b_{n,k}}{\rho^{k+1}} \qquad (4\text{-}80)$$

with $a_{n,k} = 0$ for $(n+k)$ equal to an odd integer and $b_{n,k} = 0$ for $(n+k)$ an even integer.

The relation given by (4-79) implies that the coefficients of (4-80) are related in the following way:

$$a_{n+1,k} = (2n+1)a_{n,k-1} - a_{n-1,k}$$
$$b_{n+1,k} = (2n+1)b_{n,k-1} - b_{n-1,k} \qquad (4\text{-}81)$$

with the understanding that

$$a_{n,k} = b_{n,k} = 0 \qquad \text{for} \quad k < 0 \quad \text{and} \quad k > n$$

These may be used as recursion relations to generate $a_{n,\ell}$ and $b_{n,\ell}$ of arbitrary n and k. As starting values, we can use the forms of $j_0(\rho)$ and $j_1(\rho)$ given in Table 4-3. These provide us with the coefficients of the two lowest-order spherical Bessel functions. The nonvanishing ones are

$$a_{0,0} = 1 \qquad a_{1,1} = 1 \qquad b_{1,0} = -1 \qquad (4\text{-}82)$$

Since both $a_{n,k}$ and $b_{n,k}$ are integers, a straightforward propagation procedure, similar to the one used earlier for Hermite polynomials in §4-1, may be employed. This is implemented in the algorithm given in Box 4-5.

Box 4-6   Program VSPH_BES

Value of spherical Bessel function using results of Box 4-5

1. Input the value of ρ.
2. Calculate the values of $\sin\rho$ and $\cos\rho$. Store them as SX and CX.
3. Sum the contributions from the coefficients FS = $\sum_{k=0}^{n} a_{b,k}/\rho^{k+1}$ for $\sin\rho$ and FC = $\sum_{k=0}^{n} b_{b,k}/\rho^{k+1}$ for $\cos\rho$.
4. Return the product (SX*FS + CX*FC).

For technical interests, we shall take a slightly different approach and keep only the coefficients $a_{n,k}$ and $b_{n,k}$ of $j_n(\rho)$ and $a_{n-1,k}$ and $b_{n-1,k}$ of $j_{n-1}(\rho)$ at the end of each step of the propagation. In principle, this is a more flexible method than the one used for the Hermite polynomials, where coefficients of all orders are kept in a two-dimensional array. However, for the low ranks of interest to us in most practical applications, there are very little differences in the two methods. If the interest is in the numerical values, we can start from the algebraic form of (4-80) and make use of the values of the coefficients obtained using the method of Box 4-5. This is done in Box 4-6.

For the spherical Neumann functions $n_n(\rho)$, an identical procedure may be used to calculate the coefficients. The only difference is in the values used as the starting point. Instead of those given in (4-82), the nonvanishing coefficients are

$$a_{0,0} = -1 \qquad a_{1,1} = b_{1,0} = -1$$

as can be seen from Table 4-3.

We can also make use of (4-79) to propagate the numerical value of $j_n(\rho)$. The starting values are those for $j_0(\rho)$ and $j_1(\rho)$ which may be calculated using (4-68) and (4-73). The steps are given in Box 4-7. The same method can also be used to find the value of $n_\ell(\rho)$ by starting from those of $n_0(\rho)$ using (4-69) and $n_1(\rho)$ using (4-73). As usual, it is also possible to obtain the numerical values using the power series forms of $j_n(\rho)$ and $n_n(\rho)$. However, because of the sign, contributions from successive terms have a tendency to cancel each other. This can be seen by examining the examples given in Table 4-3. For this reason, the accuracies in the numerical values obtained in this way are, in general, only comparable with those obtained by propagation. For many purposes, this may not be adequate. It is possible to develop special methods for a particular rank of spherical Bessel or Neumann function that are better in accuracy, especially for small values of ρ. These methods are available in references that specialize on the subject.

Box 4-7 Program PROP_BES
Propagation of the values of spherical Bessel functions

1. Input the rank required.
2. Input the value of ρ.
3. Calculate the starting values:
   (a) Evaluate $j_0(\rho)$ and $j_1(\rho)$ using (4-68) and (4-73).
   (b) Assign $n = 1$.
4. Calculate the value for order $(n+1)$ using (4-79).
5. Increment the value of n by 1.
6. Go back to step 4 until n reaches the order required.
7. Output the calculated value.

Bessel functions of integer ranks. For completeness, we shall end this section with a brief discussion of Bessel functions of integer ranks. The differential equation was given earlier in (4-66):

$$\rho^2 \frac{d^2 W}{d\rho^2} + \rho \frac{dW}{d\rho} + (\rho^2 - \lambda^2) W(\rho) = 0$$

The regular solution, analogous to $j_n(\rho)$, is

$$J_\lambda(\rho) = \sum_{k=0}^{\infty} \frac{(-1)^k}{k!(\lambda+k)!} \left(\frac{\rho}{2}\right)^{\lambda+2k} \tag{4-83}$$

Alternatively, we can also use an integral form,

$$J_\lambda(\rho) = \frac{1}{\pi} \int_0^\pi \cos(\lambda\theta - \rho\sin\theta) \, d\theta$$

to represent the function.

The relation with spherical Bessel functions $j_n(\rho)$ is given by the relation

$$j_n(\rho) = \sqrt{\frac{\pi}{2\rho}} J_{n+1/2}(\rho) \tag{4-84}$$

From this, we obtain an infinite series representation for spherical Bessel functions by substituting (4-83) for $J_\lambda(\rho)$. The value of the factorial of a half-integer quantity that appears in the denominator may be found using the duplication formula of gamma functions (see Problem 4-15):

$$z!(z + \tfrac{1}{2})! = \frac{\pi^{1/2}}{2^{2z+1}}(2z+1)!$$

With this, we obtain the result

$$j_n(\rho) = (2\rho)^n \sum_{k=0}^{\infty} \frac{(-1)^k (n+k)!}{k!(2n+2k+1)!} \rho^{2k}$$

This form is useful, for example, for generating the asymptotic forms of $j_n(\rho)$, used in solving scattering problems in quantum mechanics.

The generating function for Bessel functions is

$$g(\rho, t) = \exp \frac{x}{2}\left(t - \frac{1}{t}\right) = \sum_{\lambda=-\infty}^{\infty} J_\lambda(\rho) t^\lambda$$

Similar to what we did for the Legendre polynomials, we can make use of $g(\rho, t)$ to obtain the recursion relations between Bessel functions:

$$2\lambda J_\lambda(\rho) = \rho \Big\{ J_{\lambda+1}(\rho) + J_{\lambda-1}(\rho) \Big\} \tag{4-85}$$

This is analogous to (4-79) except there it was for spherical Bessel functions. Similarly, we have the relation between the function itself and its derivatives,

$$J_{\lambda+1}(\rho) = -\rho^\lambda \frac{d}{d\rho} \rho^{-\lambda} J_\lambda(\rho) \qquad J_{\lambda-1}(\rho) = \rho^{-\lambda} \frac{d}{d\rho} \rho^\lambda J_\lambda(\rho)$$

These are analogous to (4-76) and (4-78).

The general solution to the Bessel differential equation of (4-66) is a linear combination of the regular solution $J_\lambda(\rho)$ and irregular solution $N_\lambda(\rho)$. The Neumann function $N_\lambda(\rho)$ may be defined as a sum of Bessel functions of positive and negative orders in the following way:

$$N_\lambda(\rho) = \frac{J_\lambda(\rho)\cos(\lambda\pi) - J_{-\lambda}(\rho)}{\sin(\lambda\pi)} \qquad (4\text{-}86)$$

The relation with its spherical counterpart is

$$n_n(\rho) = \sqrt{\frac{\pi}{2\rho}} N_{n+1/2}(\rho)$$

This is similar to the relation between Bessel functions given in (4-84). Several polynomial approximations to $J_0(\rho)$, $J_1(\rho)$, $N_0(\rho)$, and $N_1(\rho)$ are given on pages 369-70 of Abramowitz and Segun (*Handbook of Mathematical Functions*, Dover, New York, 1965). These expressions may be used to obtain the numerical values of these functions. From the values of $\lambda = 0$ and 1, it is also possible to use the recursion relation given in (4-85) to produce the values of higher ranks. For fast and accurate calculations, one should make use of routines available in standard subroutine libraries, such as NAG and IMSL.

§4-4 Laguerre polynomials

We return now to the more general case for the radial part of (4-28). Unlike the previous section, we shall be concerned with the situation in which the factor k^2 is a function of the radial coordinate. This is found, for example, in quantum mechanical problems where there is a potential $V(r)$ acting on the system. In particular, we shall be interested in the case of a central potential, that is, one with radial dependence only. The two most common examples in this category are the isotropic three-dimensional harmonic oscillator potential and the Coulomb potential.

We have already seen the one-dimensional case of a harmonic oscillator potential in §4-1. Here, we have a more general form,

$$V(r) = \tfrac{1}{2}\mu\omega^2 r^2 \qquad (4\text{-}87)$$

where, as we saw earlier, μ is the mass and ω is the angular frequency of the oscillator. It is known as an isotropic harmonic oscillator potential, since there is no angular dependence. Physically, the force in a harmonic oscillator potential increases in strength linearly with the distance r from the center of the well. In contrast, the strength for a Coulomb potential, similar to a gravitational potential, is inversely proportional to the square of the distance. For example, if an electron is outside a nucleus with Z units of positive charges, the potential in cgs units has the form

$$V(r) = -\frac{Ze^2}{r} \tag{4-88}$$

where e is the charge of a proton (equal to the negative of the charge on an electron) and the negative sign indicates it is an attractive potential. If we wish to use SI units instead, we shall regard Z as the number of protons divided by $4\pi\epsilon_0$. Here ϵ_0 is the permittivity of free space.

In both cases, the radial part of the Helmholtz equation of (4-28) reduces to a Laguerre differential equation

$$t\frac{d^2v}{dt^2} + (\alpha + 1 - t)\frac{dv}{dt} + kv = 0 \tag{4-89}$$

Before attempting a solution to the equation, we shall see how this can be done for the two potentials.

Isotropic harmonic oscillator potential. If we generalize the Schrödinger equation for the harmonic oscillator given in (4-4) to three spatial dimensions, we obtain the following second-order differential equation:

$$-\frac{\hbar^2}{2\mu}\nabla^2\psi(\mathbf{r}) + \left(\tfrac{1}{2}\mu\omega^2 r^2 - E\right)\psi(\mathbf{r}) = 0$$

Since the potential has only radial dependence, the angular part of the wave function is given by spherical harmonics, as we saw earlier in (4-60). For this reason, we can write the wave function $\psi(\mathbf{r})$ as a product of $R_\ell(r)$, the radial wave function for angular momentum ℓ, and spherical harmonics $Y_{\ell,m}(\theta,\phi)$:

$$\psi_{\ell m}(\mathbf{r}) = R_\ell(r) Y_{\ell,m}(\theta,\phi)$$

For an isotropic harmonic oscillator potential, $R_\ell(r)$ is the solution of the radial equation

$$\frac{1}{r^2}\frac{d}{dr}\left(r^2 \frac{dR_\ell}{dr}\right) + \left\{\frac{2\mu}{\hbar^2}\left(E - \frac{1}{2}\mu\omega^2 r^2\right) - \frac{\ell(\ell+1)}{r^2}\right\} R_\ell(r) = 0 \tag{4-90}$$

We shall rewrite this equation in the form of a Laguerre differential equation given by (4-89).

Similar to the case of a one-dimensional harmonic oscillator potential discussed in §4-1, it is convenient to express both the energy E and the radial coordinate r in terms of dimensionless quantities. For this purpose, we shall define λ and ρ, in the same way as we did in (4-5) and (4-6), by the following relations:

$$E = \lambda(\tfrac{1}{2}\hbar\omega) \qquad r = \rho\sqrt{\frac{\hbar}{\mu\omega}} \qquad (4\text{-}91)$$

The factor $\sqrt{\hbar/(\mu\omega)}$ has the dimension of length and is often referred to as the harmonic oscillator length parameter. In terms of λ and ρ, the radial equation (4-90) takes on the form

$$\frac{1}{\rho^2}\frac{d}{d\rho}\left(\rho^2\frac{dR_\ell}{d\rho}\right) + \left\{\lambda - \rho^2 - \frac{\ell(\ell+1)}{\rho^2}\right\}R_\ell(\rho) = 0$$

or

$$\frac{d^2R_\ell}{d\rho^2} + \frac{2}{\rho}\frac{dR_\ell}{d\rho} + \left\{\lambda - \rho^2 - \frac{\ell(\ell+1)}{\rho^2}\right\}R_\ell(\rho) = 0 \qquad (4\text{-}92)$$

We shall be using both of these two equivalent relations in our discussions, depending on which is more convenient for the occasion.

At large values of ρ, (4-92) may be approximated as

$$\frac{d^2R_\ell}{d\rho^2} - \rho^2 R_\ell(\rho) \approx 0$$

As a result, the radial wave function must assume the asymptotic shape

$$R_\ell(\rho) \propto e^{-\rho^2/2} \qquad (4\text{-}93)$$

Similarly, at small values of ρ, the radial equation has the form

$$\frac{1}{\rho^2}\frac{d}{d\rho}\left(\rho^2\frac{dR_\ell}{d\rho}\right) - \frac{\ell(\ell+1)}{\rho^2}R_\ell(\rho) \approx 0$$

There are two possible solutions to this equation:

$$R_\ell(\rho) \propto \rho^\ell \qquad\qquad R_\ell(\rho) \propto \rho^{-(\ell+1)}$$

The latter must be rejected as the radial wave function for a harmonic oscillator, since it is irregular at the origin.

The net result of examining the behavior of the solution at small and large values of ρ is that, as the trial solution to (4-90), we shall try

$$R(\rho) = \rho^\ell e^{-\rho^2/2}\phi(\rho) \qquad (4\text{-}94)$$

On substituting this form into (4-92), we obtain the equation that must be satisfied by $\phi(\rho)$:

$$t\frac{d^2\phi}{dt^2} + \left(\ell + \frac{3}{2} - t\right)\frac{d\phi}{dt} + k\phi(t) = 0 \qquad (4\text{-}95)$$

where
$$t = \rho^2 \qquad k = \frac{1}{4}(\lambda - 2\ell - 3) \qquad (4\text{-}96)$$

We see that (4-95) is the same as (4-89) if we let $\alpha = (\ell + \tfrac{1}{2})$.

Coulomb potential and the hydrogenlike atom. For an electron in an atom, the electrostatic attraction provided by the nucleus is given by the Coulomb potential of (4-88). The simplest case is that of a hydrogen atom, as the nucleus has only a single unit of positive charge ($Z = 1$). For other atoms, $Z > 1$, and the neutral atom has more than one electron outside the nucleus. The net effect of the other electrons screens the force acting on an individual electron. This greatly complicates the problem for the wave function of the electron. As a result, the potential given in (4-88) applies only to the idealized situation of a hydrogenlike atom in which all the atomic electrons except one are stripped away. The result is similar to a hydrogen atom except that the number of positive charges in the nucleus is Z instead of one. We shall use such an idealized situation as an example of the applications of Laguerre polynomials. Analogous to the case of a particle in an isotropic harmonic oscillator potential given by (4-90), the radial equation in this case takes on the form

$$\frac{1}{r^2}\frac{d}{dr}\left(r^2 \frac{dR_\ell}{dr}\right) + \left\{\frac{2\mu}{\hbar^2}\left(E + \frac{Ze^2}{r}\right) - \frac{\ell(\ell+1)}{r^2}\right\}R_\ell(r) = 0 \qquad (4\text{-}97)$$

We shall be interested only in the bound states for the electron with energy E being a negative quantity.

It is convenient here to define a dimensionless quantity
$$\rho = \kappa r \qquad (4\text{-}98)$$

For $E < 0$, the quantity κ is a real, positive number:
$$\kappa = 2\sqrt{-\frac{2\mu E}{\hbar^2}}$$

In terms of ρ, the differential equation has the form

$$\frac{1}{\rho^2}\frac{d}{d\rho}\left(\rho^2 \frac{dR_\ell}{d\rho}\right) + \left\{\frac{\eta}{\rho} - \frac{1}{4} - \frac{\ell(\ell+1)}{\rho^2}\right\}R(\rho) = 0 \qquad (4\text{-}99)$$

where
$$\eta = \frac{2\mu Z e^2}{\kappa \hbar^2} = \sqrt{-\frac{\mu Z^2 e^4}{2E\hbar^2}} \qquad (4\text{-}100)$$

is a dimensionless quantity related to the square root of the energy of the system.

Since the Coulomb potential has a long range, the behavior of the radial wave function at large values of ρ is different from that for a short-range harmonic oscillator potential given in (4-93). Instead of $R_\ell(\rho) \to \exp(-\rho^2)$, we find that

$$R_\ell(\rho) \xrightarrow[\rho \to \infty]{} e^{-\rho/2}$$

At short distances, the $\ell(\ell+1)/\rho^2$ term dominates and we have

$$R_\ell(\rho) \xrightarrow[\rho \to 0]{} \rho^\ell$$

For these reasons, we shall take the trial solution for (4-96) as

$$R_\ell(\rho) = \rho^\ell e^{-\rho/2} \phi(\rho) \tag{4-101}$$

On substituting this form of $R_\ell(\rho)$ into (4-99), we obtain the equation that must be satisfied by $\phi(\rho)$,

$$\rho \frac{d^2\phi}{d\rho^2} + (2\ell + 2 - \rho)\frac{d\phi}{d\rho} + (\eta - \ell - 1)\phi(\rho) = 0 \tag{4-102}$$

This is the same equation as (4-89) if we let $\alpha = (2\ell + 1)$ and $k = (\eta - \ell - 1)$.

Solution of the Laguerre equation. Before attempting to solve (4-89), we shall consider first the simpler case with $\alpha = 0$. The differential equation reduces to the form

$$t \frac{d^2\phi}{dt^2} + (1 - t)\frac{d\phi}{dt} + k\phi(\rho) = 0 \tag{4-103}$$

In the same way as in the case of the one-dimensional harmonic oscillator, we shall adopt a trial solution in terms of a power series:

$$\phi(t) = \sum_{j=0} a_j t^j \tag{4-104}$$

where the coefficients a_j are to be determined. On substituting this trial solution into (4-103), we obtain the equation that must be satisfied by these coefficients:

$$\sum_{j=0} \{j(j-1)a_j t^{j-1} + (1-t)j a_j t^{j-1} + k a_j t^j\} = 0$$

For this relation to hold for any arbitrary values of t, the sum of the coefficients for each power of t must vanish. That is,

$$j(j+1)a_{j+1} + (j+1)a_{j+1} - j a_j + k a_j = 0$$

or

$$a_{j+1} = \frac{j - k}{(j+1)^2} a_j \tag{4-105}$$

If $\phi(t)$ is to be used as the radial wave function for a quantum mechanical system, it must be localized in some finite region. As a result, $\phi(t)$ cannot contain arbitrarily high powers of t. For this reason, the series on the right side of (4-104) must terminate at some finite value of j. This, in turn, requires that k must be a positive integer. We shall see later that this provides us with the condition that the eigenvalues of the Schrödinger equation must be discrete for an isotropic harmonic oscillator as well as for a hydrogenlike atom.

Since k of (4-103) enters (4-105), the coefficients a_j in (4-104) depend also on the value of k. For each k, there is a solution of the form

$$L_k(t) = \sum_{j=0}^{k} a_j t^j = 1 - kt + \frac{k-1}{2^2}kt^2 - \frac{(k-2)(k-1)}{2^2 3^2}kt^3$$

$$+ \cdots + (-1)^{k-2}\frac{k^2(k-1)^2}{2!k!}t^{k-2} + (-1)^{k-1}\frac{k^2}{k!}t^{k-1} + (-1)^k \frac{1}{k!}t^k$$

$$= \sum_{j=0}^{k}(-1)^j \frac{k!}{(k-j)!(j!)^2}t^j \qquad (4\text{-}106)$$

where we have chosen to normalize $L_k(t)$ such that

$$a_0 = 1 \qquad (4\text{-}107)$$

This is known as a Laguerre polynomial of degree n. The explicit forms for the lowest few orders are given in Table 4-4 as illustration.

Table 4-4 Laguerre polynomials $L_k(t)$ up to order $k = 6$.

$L_0(t) = 1$

$L_1(t) = 1 - t$

$L_2(t) = \frac{1}{2!}(2 - 4t + t^2)$

$L_3(t) = \frac{1}{3!}(6 - 18t + 9t^2 - t^3)$

$L_4(t) = \frac{1}{4!}(24 - 96t + 72t^2 - 16t^3 + t^4)$

$L_5(t) = \frac{1}{5!}(120 - 600t + 600t^2 - 200t^3 + 25t^4 - t^5)$

$L_6(t) = \frac{1}{6!}(720 - 4320t + 5400t^2 - 2400t^3 + 450t^4 - 36t^5 + t^6)$

The generating function for Laguerre polynomials is given by

$$g(t, z) = \frac{1}{1-z} e^{-tz/(1-z)} = \sum_{k=0}^{\infty} L_k(t) z^k \qquad (4\text{-}108)$$

We can check that $L_k(t)$ defined in this way satisfies (4-103) following a similar procedure as we did earlier for the Hermite polynomials in §4-1. Let us begin by calculating the following sum of the derivatives of $g(t, z)$:

$$S = t\frac{\partial^2 g}{\partial t^2} + (1-t)\frac{\partial g}{\partial t} + z\frac{\partial g}{\partial z}$$

It is possible to show that S vanishes by working out the partial derivatives explicitly using the exponential form $e^{-tz/(1-z)}/(1-z)$ of $g(t,z)$. Similarly, we can apply the same operations to the infinite series form. The result is

$$\sum_{k=0}^{\infty}\left\{t\frac{d^2 L_k}{dt^2} + (1-t)\frac{dL_k}{dt} + kL_k\right\}z^k = 0$$

For this relation to be true for all possible values of z, the sum inside the braces must vanish. That is,

$$t\frac{d^2 L_k}{dt^2} + (1-t)\frac{dL_k}{dt} + kL_k = 0$$

This is identical to (4-103). As a result, we can conclude that $g(z,t)$ of (4-108) is the generating function for Laguerre polynomials.

With $g(t,z)$, we can deduce two recursion relations for Laguerre polynomials,

$$(k+1)L_{k+1}(t) = (2k+1-t)L_k(t) - kL_{k-1}(t) \qquad (4\text{-}109)$$

$$t\frac{dL_k}{dt} = kL_k(t) - kL_{k-1}(t) \qquad (4\text{-}110)$$

in the same way we obtained (4-39) earlier for Legendre polynomials.

It is also possible to express Laguerre polynomials in a differential form

$$L_k(t) = \frac{1}{k!}e^t\frac{d^k}{dt^k}\left\{t^k e^{-t}\right\} \qquad (4\text{-}111)$$

This can be shown to be correct if we carry out all the differentiations in the expression and compare the results with the series form given by (4-106). The orthogonality condition between Laguerre polynomials of different degrees requires a weighting factor of e^{-t} and may be expressed as

$$\int_0^{\infty} e^{-t} L_m(t) L_k(t)\, dt = \delta_{m,k} \qquad (4\text{-}112)$$

Note that it is also possible to adopt a different normalization condition for the Laguerre polynomial than that used in (4-107). For example, many workers prefer the choice of $a_0 = k!$ for $L_k(t)$, as this leads to the result that all the coefficients are integers. Such a choice changes the appearance of, for example, the orthogonality condition and the recursion relations but does not make any difference of substance to our discussions.

Associated Laguerre polynomials.

If we differentiate each term in the Laguerre equation (4-103) once with respect to t, we obtain

$$\frac{d^2 L_k}{dt^2} + t\frac{d^3 L_k}{dt^3} - \frac{dL_k}{dt} + (1-t)\frac{d^2 L_k}{dt^2} + k\frac{dL_k}{dt} = 0$$

On rearranging the terms, this equation may be written as

$$t\frac{d^2}{dt^2}\left(\frac{dL_k}{dt}\right) + \{1 + (1-t)\}\frac{d}{dt}\left(\frac{dL_k}{dt}\right) + (k-1)\frac{dL_k}{dt} = 0$$

Comparing this with (4-89), we find that dL_k/dt is the solution of (4-89) for $\alpha = 1$ and k replaced by $(k-1)$. If we carry out the same type of differentiation p times, we obtain the solution of the Laguerre equation for degree k and $\alpha = p$. In other words,

$$L_k^p(t) = (-1)^p \frac{d^p}{dt^p} L_{k+p}(t) \tag{4-113}$$

The function $L_k^p(t)$ is known as the associated Laguerre polynomial and the forms of the lowest few orders are given in Table 4-5 as illustration. It forms the core of the solutions for the radial equation of an isotropic harmonic oscillator and for a hydrogenlike atom, among others.

Using the series form of $L_k(t)$ given by (4-106) and the definition of $L_k^p(t)$ in (4-113), we obtain a power series expansion for the associated Laguerre polynomials:

$$L_k^p(t) = \sum_{j=0}^{k} (-1)^j \frac{(k+p)!}{(k-j)!(p+j)!j!} t^j$$

$$= \sum_{j=0}^{k} \binom{k+p}{k-j} \frac{(-t)^j}{j!} \qquad \text{for} \quad p > -1 \tag{4-114}$$

Similarly, by starting from (4-111), we obtain a differential form for the same function:

$$L_k^p(t) = \frac{1}{k!} t^{-p} e^t \frac{d^k}{dt^k} \{t^{k+p} e^{-t}\}$$

The orthogonality condition

$$\int_0^\infty t^p e^{-t} L_m^p(t) L_k^p(t)\, dt = \frac{(k+p)!}{k!} \delta_{m,k} \tag{4-115}$$

differs slightly from that for Laguerre polynomials in (4-112) and requires $t^p e^{-t}$ as the weighting factor.

By differentiating the generating function for Laguerre polynomials p times with respect to t, we obtain the generating function for associated Laguerre polynomials

$$g(t, z) = \frac{1}{(1-z)^{p+1}} e^{-tz/(1-z)} = \sum_{k=0}^{\infty} L_k^p(t) z^k \qquad \text{for} \quad |z| < 1 \tag{4-116}$$

Table 4-5 Associated Laguerre polynomials $L^{\ell+1/2}_{(n-\ell)/2}(t)$.

n	ℓ	$L^{\ell+1/2}_{(n-\ell)/2}(t)$	ℓ	$L^{\ell+1/2}_{(n-\ell)/2}(t)$	ℓ	$L^{\ell+1/2}_{(n-\ell)/2}(t)$
0	0	$L^{1/2}_0(t) = 1$				
1	0	$L^{3/2}_0(t) = 1$				
2	0	$L^{1/2}_1(t) = \frac{3}{2} - t$	2	$L^{3/2}_0 = 1$		
3	1	$L^{3/2}_1(t) = \frac{5}{2} - t$	3	$L^{5/2}_0 = 1$		
4	0	$L^{1/2}_2(t) = \frac{15}{8} - \frac{5}{2}t + \frac{1}{2}t^2$	2	$L^{5/2}_1 = \frac{7}{2} - t$	4	$L^{9/2}_0(t) = 1$
5	1	$L^{3/2}_2(t) = \frac{35}{8} - \frac{7}{2}t + \frac{1}{2}t^2$	3	$L^{7/2}_1 = \frac{9}{2} - t$	5	$L^{11/2}_0(t) = 1$

The recursion relations for $L^p_k(t)$ obtained from $g(t,z)$ are

$$(k+1)L^p_{k+1}(t) = (2k + p + 1 - t)L^p_k(t) - (k+p)L^p_{k-1}(t) \tag{4-117}$$

$$t\frac{dL^p_k}{dt} = kL^p_k(t) - (k+p)L^p_{k-1}(t) \tag{4-118}$$

They are similar in form to those for Laguerre polynomials given by (4-109) and (4-110). By subtracting (4-118) from (4-117), we obtain one further recursion relation

$$(k+1)L^p_{k+1}(t) = t\frac{dL^p_k}{dt} + (k+p+1-t)L^p_k(t) \tag{4-119}$$

This is useful in propagating associated Laguerre polynomials to higher k values.

Eigenvalues and radial wave functions of hydrogenlike atoms. We are now in a position to make use of associated Laguerre polynomials as a part of the solution for the radial wave functions of hydrogenlike atoms. Let us start from (4-102) for $\phi(\rho)$:

$$\rho\frac{d^2\phi}{d\rho^2} + (2\ell + 2 - \rho)\frac{d\phi}{d\rho} + (\eta - \ell - 1)\phi = 0$$

This may be compared with the standard form of an associated Laguerre differential equation of (4-89) written in the following manner:

$$t\frac{d^2v}{dt^2} + (p + 1 - t)\frac{dv}{dt} + kv = 0$$

The solution to this differential equation is $L^p_k(t)$, as we saw in (4-113). By making the identifications that $\rho = t$, $k = (\eta - \ell - 1)$, and $p = (2\ell + 1)$, we obtain

$$\phi(\rho) = \frac{1}{\mathcal{N}} L^{2\ell+1}_{\eta-\ell-1}(\rho) \tag{4-120}$$

The only unknown part in the expression is the normalization factor $\mathcal{N}$, to which we shall return later.

Since k must be an integer, only certain values of η are allowed. This, in turn, means that the energies of a hydrogenlike atom are restricted to a discrete set of values. Such an energy spectrum is one of the well-known successes of quantum mechanics in the early days. From the definition of η given in (4-100), we obtain

$$E_n = -\frac{\mu Z^2 e^4}{2n^2 \hbar^2} = -\frac{Z^2 e^2}{2a_0 n^2}$$

where the *principal quantum number*

$$n = \eta = k + \ell + 1$$

is an integer greater or equal to 1, and

$$a_0 = \frac{\hbar^2}{\mu e^2}$$

is the Bohr radius for the hydrogenlike atom with μ as the reduced mass of an electron. In terms of n and a_0, the relation between r and ρ given in (4-98) simplifies to

$$\rho = \frac{2Z}{na_0} r$$

This is the form of ρ commonly found in standard textbooks.

Using (4-120), the radial wave function of (4-101) may now be written in the form

$$R_{n\ell}(\rho) = \frac{1}{\mathcal{N}} \rho^\ell e^{-\rho/2} L_{n-\ell-1}^{2\ell+1}(\rho) \qquad (4\text{-}121)$$

where we have added the principal quantum to the subscript of the wave function to remind ourselves that this is for a particular energy level labeled by the integer n (as well as angular momentum ℓ). Since $R_{n\ell}(r)$ is normalized,

$$\int_0^\infty |R_{n\ell}(r)|^2 r^2 \, dr = 1$$

we have the condition that

$$\mathcal{N}^2 = \left(\frac{na_0}{2Z}\right)^3 \int_0^\infty \rho^{2\ell+2} e^{-\rho} \left|L_{n-\ell-1}^{2\ell+1}(\rho)\right|^2 d\rho \qquad (4\text{-}122)$$

where the constant factor in front of the integral sign comes from the change in the volume element of integration from $r^2 dr$ to $\rho^2 d\rho$. Note that the integral is not the same as the normalization condition for associated Laguerre polynomials given in (4-115).

To carry out the integration, we can make use of the recursion relation for $L_k^p(t)$ given in (4-117) and rewrite it in the following way:

$$t L_m^p(t) = (2m + p + 1) L_m^p(t) - (m+1) L_{m+1}^p(t) - (m+p) L_{m-1}^p(t)$$

When we multiply each term in the expression by $t^p e^{-t} L_m^p(t)$ and integrate with respect to t, the last two terms vanish because $L_m^p(t)$ is orthogonal to $L_{m\pm1}^p(t)$. This leaves us with

$$\int_0^\infty t^{p+1} e^{-t} \{L_m^p(t)\}^2 \, dt = (2m + p + 1) \int_0^\infty t^p e^{-t} \{L_m^p(t)\}^2 \, dt$$

The integral on the right side has the same form as (4-115), and we obtain

$$\int_0^\infty t^{p+1} e^{-t} \{L_m^p(t)\}^2 \, dt = (2m + p + 1) \frac{(m+p)!}{m!} \quad (4\text{-}123)$$

This may be used to carry out the integration in (4-122).

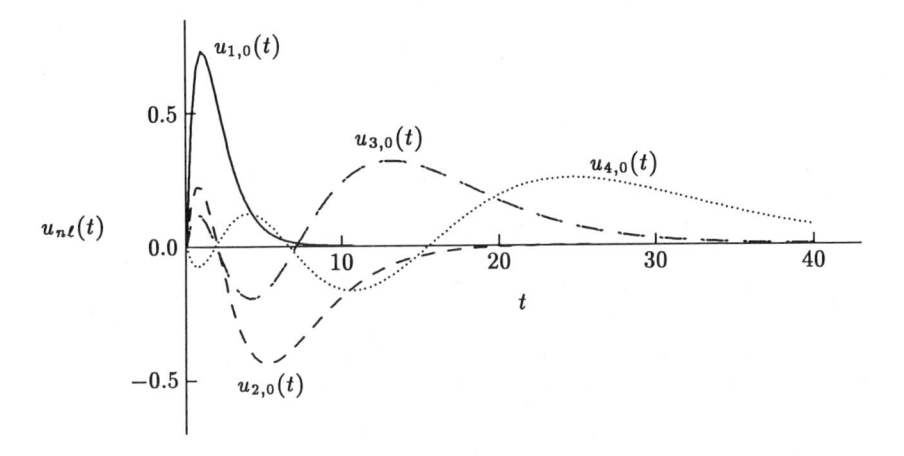

Fig. 4-8 Examples of the radial part of hydrogenlike wave functions $u_{n\ell}(t) = tR_{n\ell}(t)$ for $\ell = 0$ (s-waves) and $n = 1, 2, 3,$ and 4, calculated using Laguerre polynomials.

We can now work out the normalization constant $\mathcal{N}$ for the radial wave functions. On putting (4-123) into (4-122), we have

$$\mathcal{N}^2 = \left(\frac{na_0}{2Z}\right)^3 (2n) \frac{(n+\ell)!}{(n-\ell-1)!}$$

The normalized radial wave function for a hydrogenlike atom is then

$$R_{n\ell}(r) = \left(\frac{2Z}{na_0}\right)^{3/2} \sqrt{\frac{(n-\ell-1)!}{2n(n+\ell)!}} (\kappa r)^\ell e^{-\kappa r/2} L_{n-\ell-1}^{2\ell+1}(\kappa r) \quad (4\text{-}124)$$

Since $\kappa = 2Z/na_0$, the value of $\rho = \kappa r$ is different for states of different principal quantum number n. It is therefore more convenient for many purposes to use a different dimensionless quantity,

$$t = \frac{Z}{a_0} r$$

instead of ρ. Physically, we may interpret t as the length measured in units of the Bohr radius for a hydrogenlike atom consisting of Z units of charge in the nucleus. In terms of t, the radial wave function takes on the form

$$R_{n\ell}(t) = \left(\frac{2}{n}\right)^{3/2} \sqrt{\frac{(n-\ell-1)!}{2n(n+\ell)!}} (2t/n)^\ell e^{-t/n} L^{2\ell+1}_{n-\ell-1}(2t/n) \qquad (4\text{-}125)$$

Since both n and ℓ are integers, $L^{2\ell+1}_{n-\ell-1}(2t/n)$ are given by (4-114). Using these, the radial wave functions $R_{n\ell}(t)$ for low values of n and ℓ may be written down explicitly. The forms of those with $\ell = 0$ and $n \leq 3$ are given here as examples:

$$R_{10}(t) = 2e^{-t}$$

$$R_{20}(t) = \frac{1}{2\sqrt{2}}(2-t)e^{-t/2}$$

$$R_{30}(t) = \frac{2}{81\sqrt{3}}(27 - 18t + 2t^2)e^{-t/3}$$

Their variations as functions of ρ are displayed in Fig. 4-8 in terms of the modified radial wave function $u_{n\ell}(t) = tR_{n\ell}(t)$.

Harmonic oscillator radial wave functions. Using associated Laguerre polynomials, we can also find the energy levels of a particle in an isotropic harmonic oscillator potential well. This may be carried out in a way similar to what we have done above for the hydrogenlike atom. From the fact that k of (4-96) has to be an integer m, we obtain the condition that must satisfied by the parameter λ of (4-91)

$$\lambda = 4m + 2\ell + 3$$

where $m = 0, 1, 2, \ldots$. Since λ is the energy of the particle in units of $\frac{1}{2}\hbar\omega$, we obtain the familiar result

$$E_n = \left(n + \frac{3}{2}\right)\hbar\omega$$

Here $n = (2m + \ell)$ is the principal quantum number, measuring the number of harmonic oscillator quanta above the zero-point energy of $\frac{3}{2}\hbar\omega$.

Except for a normalization factor, the solution to the differential equation for $\phi(t)$ of (4-95) may be obtained by comparing it with the standard form of a Laguerre equation given by (4-89). This gives us the result $\alpha = (\ell + 1/2)$ and $k = (\lambda - 2\ell - 3)/4 = (n - \ell)/2$, or

$$\phi(t) = L^\alpha_k(t) = L^{\ell+1/2}_{(\lambda-2\ell-3)/4}(t) = L^{\ell+1/2}_{(n-\ell)/2}(\rho^2)$$

Since k must be an integer, we find the condition that both n and ℓ must be either even or odd at the same time. For odd ℓ values, α is a half-integer and $L^\alpha_k(t)$ is defined through the generating function (4-116), rather than the Rodrigues

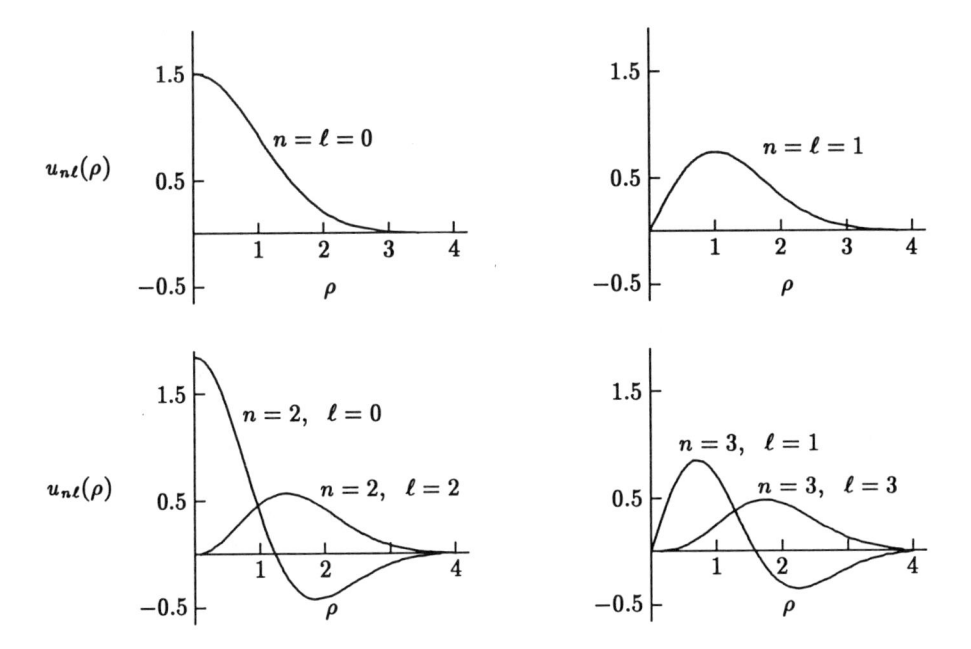

Fig. 4-9 Examples of the modified radial wave functions $u_{n\ell}(\rho) = \rho R_{n\ell}(\rho)$ for a particle in an isotropic three-dimensional harmonic potential well calculated using Laguerre polynomials.

formula given by (4-113). For this reason, the series form of $L_k^p(t)$ given by (4-114) no longer applies. We shall see later how to obtain associated Laguerre polynomials for noninteger p values.

The complete radial wave function for a given n and ℓ is then

$$R_{n\ell}(\rho) = \frac{1}{\mathcal{N}} \rho^\ell e^{-\rho^2/2} L_{(n-\ell)/2}^{\ell+1/2}(\rho^2) \tag{4-126}$$

as can be seen from (4-94). The normalization factor $\mathcal{N}$ is obtained from the requirement that

$$\int_0^\infty |R(r)|^2 r^2 \, dr = 1$$

Using the orthogonality condition of associated Laguerre polynomials given in

(4-115), the square of the normalization factor is given as

$$\mathcal{N}^2 = \left(\frac{\hbar}{\mu\omega}\right)^{3/2} \int_0^\infty \rho^{2\ell} e^{-\rho^2} \left\{L_{(n-\ell)/2}^{\ell+1/2}(\rho^2)\right\}^2 \rho^2 d\rho$$

$$= \frac{1}{2}\left(\frac{\hbar}{\mu\omega}\right)^{3/2} \int_0^\infty t^{\ell+1/2} e^{-t} \left\{L_{(n-\ell)/2}^{\ell+1/2}(t)\right\}^2 dt$$

$$= \frac{1}{2}\left(\frac{\hbar}{\mu\omega}\right)^{3/2} \frac{(\xi+\ell+\frac{1}{2})!}{\xi!}$$

where $\xi = (n-\ell)/2$ is an integer. The radial wave function for a particle in a three-dimensional, isotropic harmonic oscillator well, with principal quantum number n and orbital angular momentum ℓ, can now be put into the form

$$R_{n\ell}(r) = \sqrt{\frac{2^{\ell+2}\nu^{\ell+3/2}(n-\ell)!!}{(n+\ell+1)!!\sqrt{\pi}}} r^\ell e^{-\frac{\nu r^2}{2}} L_{(n-\ell)/2}^{\ell+1/2}(\nu r^2) \qquad (4\text{-}127)$$

where the parameter

$$\nu = \frac{\mu\omega}{\hbar}$$

has the dimension of inverse length squared. In arriving at the result, we have made use of the relations between factorials and double factorials given later in (4-148) to (4-150).

Using the Laguerre polynomials given in Table 4-5, the radial wave functions for $n \le 3$ and $\ell \le 1$ take on the forms

$$R_{00}(r) = \sqrt{\frac{4\nu^{3/2}}{\sqrt{\pi}}} e^{-\nu r^2/2} \qquad\qquad R_{11}(r) = \sqrt{\frac{8\nu^{5/2}}{3\sqrt{\pi}}} r e^{-\nu r^2/2}$$

$$R_{20}(r) = \sqrt{\frac{6\nu^{3/2}}{\sqrt{\pi}}} \left(1 - \frac{2}{3}\nu r^2\right) e^{-\nu r^2/2}$$

$$R_{31}(r) = 2\sqrt{\frac{5\nu^{5/2}}{3\sqrt{\pi}}} r e^{-\nu r^2/2} \left(1 - \frac{2}{5}\nu r^2\right)$$

Plots of these functions are shown in Fig. 4-9 for illustration.

Evaluation of Laguerre polynomials. It is fairly straightforward to evaluate Laguerre polynomials to any order. The general form may be written as a power series in the following way:

$$L_k(t) = \sum_{j=0}^k a_{k,j} t^j \qquad (4\text{-}128)$$

Using (4-106), we find that the coefficient for t^j is

$$a_{k,j} = (-1)^j \frac{k!}{(k-j)!(j!)^2}$$

Similarly, for the associated Laguerre polynomial of integer order (k, p), we have

$$L_k^p(t) = \sum_{j=0}^{k} b_{k,p;j} t^j \qquad (4\text{-}129)$$

with coefficients

$$b_{k,p;j} = (-1)^j \frac{(k+p)!}{(k-j)!(p+j)!j!}$$

as can be see from (4-114).

The relation given by (4-113) and, consequently, (4-114) applies only to integer values of p. As a result, the form for $L_k^p(t)$ in (4-129) cannot be used for p that are half-integers. In its place, we shall make use of the recursion relations of (4-117) and (4-118) obtained from the generating function. Again we shall write the polynomials as a power series:

$$L_k^p(t) = \sum_{j=0}^{k} c_{k,p;j} t^j \qquad (4\text{-}130)$$

To distinguish from (4-129), the coefficients are given the symbol $c_{n,p;j}$ instead of $b_{n,p;j}$.

A recursion relation may be obtained between the coefficients $c_{n,p;j}$. Using (4-119), we have the result

$$(k+1) \sum_{j=0}^{k+1} c_{k+1,p;j} t^j = t \sum_{j=0}^{k} c_{k,p;j} j t^{j-1} + (k+p+1-t) \sum_{j=0}^{k} c_{k,p;j} t^j$$

On rearranging the terms, the same equation may be put into the form

$$\sum_{j=0}^{k+1} \{(k+1)c_{k+1,p;j} - jc_{k,p;j} - (k+p+1)c_{k,p;j} + c_{k,p,j-1}\} t^j = 0$$

where it is understood that $c_{k,p,j} = 0$ for $j > k$. Since the equality holds for arbitrary values of t, the coefficient of t^j must vanish separately for each j. From this we obtain a relation among the coefficients $c_{k,p,j}$ of associated Laguerre polynomials of orders $(k+1)$ and k:

$$c_{k+1,p,j} = \frac{1}{k+1}\{(k+p+j+1)c_{k,p;j} - c_{k,p;j-1}\} \qquad (4\text{-}131)$$

This allows us to calculate, for any given value of p, all the coefficients for order $(k+1)$ from those of order k.

The starting point of the recursion relations are the values of the coefficient for $k = 0$. These may be obtained in the following way. If we put $z = 0$ in (4-116), the generating function takes on the value

$$g(t, 0) = 1$$

Since only the $L_0^p(t)$ term remains inside the summation on the right side in this case, we conclude that

$$L_0^p(t) = 1$$

This is true for arbitrary values of $p \geq 0$. In terms of the coefficients defined in (4-130), we have

$$c_{0,p;j} = \begin{cases} 1 & \text{for } j = 0 \\ 0 & \text{otherwise} \end{cases} \quad (4\text{-}132)$$

This provides us with the necessary starting point for the recursive calculation given by (4-131) to obtain the values of $c_{k,p;j}$ for higher k values. The algorithm is outlined in Box 4-8.

Box 4-8   Program LAG_ASC

Associated Laguerre polynomials $L_n^p(t)$

Coefficients of L_n^p for integer n and arbitrary p

Initialization:
  (a) Zero the array for storing the coefficients.
  (b) Set the starting value according to (4-132).
1. Input n and p.
2. Propagate the coefficients of L_k^p for k from 0 to n.
3. For each L_k^p, calculate all the coefficient for $j = 0$ to k using (4-131).
4. Output the coefficients $c_{n,p,j}$ for $j = 0$ to n.

As a bonus, we find that (4-131) applies to integer values of p as well. Because of this, we can use the same procedure also to generate the coefficient $b_{k,p;j}$. Furthermore, since

$$L_k(t) = L_k^0(t)$$

the same method may also be used to produce the coefficients $a_{k,j}$ by letting $p = 0$. In this way, a single computer program, based on the algorithm of Box 4-8, may be used to produce all the Laguerre polynomials of interest.

For the numerical values of $L_k^p(t)$, we can anticipate that some caution is required in order to obtain good accuracies at high orders. The source of the trouble lies in the fact that the coefficients alternate in sign between successive terms. As a result, cancellations between neighboring terms can take place and reduce the numerical accuracy we can otherwise achieve. The same problem also occurs if we apply the recursion relations directly on the numerical values, as we can see from the forms of (4-109) and (4-117).

§4-5 Error integrals and gamma functions

Integrals of the exponential function appear in a wide variety of physical problems. In this section, we shall give a short account of two such integrals, the error function erf(x) and the gamma function $\Gamma(x)$. A number of other functions are related to these integrals and we shall examine some of them as well.

Error integrals. The error function, or error integral, is defined by the relation

$$\text{erf}(x) = \frac{2}{\sqrt{\pi}} \int_0^x e^{-t^2} dt \tag{4-133}$$

The normalization constant is taken in such a way that

$$\text{erf}(x = \infty) = 1$$

From this condition we obtain a relation with the complementary error function

$$\text{erfc}(x) = \frac{2}{\sqrt{\pi}} \int_x^\infty e^{-t^2} dt = 1 - \text{erf}(x) \tag{4-134}$$

and with the normal probability function

$$P(x) = \frac{1}{\sqrt{2\pi}} \int_{-\infty}^x e^{-t^2/2} dt = \frac{1}{2}\left\{1 + \text{erf}\left(\frac{x}{\sqrt{2}}\right)\right\} \tag{4-135}$$

A function related to $P(x)$ was used earlier in §2-5 as an example for numerical integration. As we shall see in Chapter 6, the error integrals and the related probability integral are needed in many applications involving probability and statistics.

Rational polynomial approximations. In general, it is not possible to use analytical methods to evaluate the error function and functions directly related to it. In Chapter 3, we saw that numerical integration is one possible way to carry out the calculation. An alternative is to use polynomial approximations.

In terms of the inverse of x defined as

$$t = \frac{1}{1 + px} \qquad \text{where} \qquad p = 0.47047$$

the error function may be approximated with an uncertainty of $|\epsilon| \leq 2.5 \times 10^{-5}$ by the expression

$$\text{erf}(x) = 1 - \left(a_1 t + a_2 t^2 + a^3 t^3\right) e^{-x^2} \tag{4-136}$$

where the three coefficients have the values

$$a_1 = 0.3480242 \qquad a_2 = -0.0958798 \qquad a_3 = 0.7478556$$

A more accurate form, with uncertainty $|\epsilon| \leq 1.5 \times 10^{-7}$ is given by the expression

$$\text{erf}(x) = 1 - \left(b_1 t + b_2 t^2 + b_3 t^3 + b_4 t^4 + b_5 t^5\right) e^{-x^2}$$

with

$$t = \frac{1}{1 + 0.3275911x} \qquad b_1 = 0.254829592 \qquad b_2 = -0.284496736$$

$$b_3 = 1.421413741 \qquad b_4 = -1.453152027 \qquad b_5 = 1.061405429$$

Two other methods of approximations are also given on page 299 of Abramowitz and Segun. More elaborate approximations, with even better accuracies, are found in standard subroutine libraries.

Gamma function. There are several ways to define the gamma function $\Gamma(z)$. In terms of an integral, it may be written as

$$\Gamma(z) = \int_0^\infty e^{-t} t^{z-1} dt \qquad (4\text{-}137)$$

The variable z is a complex number in general, but we shall be mainly concerned with cases in which it is a purely real number. In particular, when z is a positive integer n, the function is related to the factorial of the argument

$$\Gamma(n) = (n-1)(n-2)\cdots 1 = (n-1)! \qquad (4\text{-}138)$$

This can be shown from (4-137) using integration by parts. Because of (4-138), $\Gamma(z)$ is also known as the factorial function. The behavior of the gamma function is shown in Fig. 4-10.

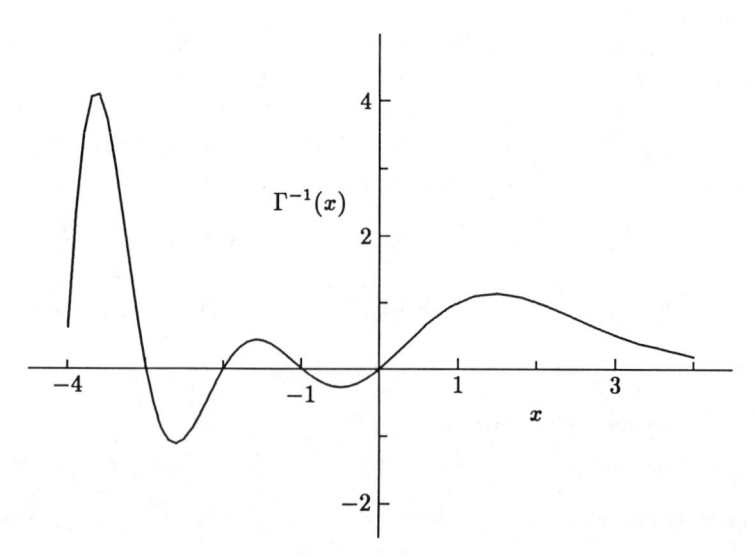

Fig. 4-10 Inverse gamma function, $1/\Gamma(x)$, for real arguments $x = [-4, 4]$.

More generally, for z not equal to zero or a negative integer, the gamma function may be written in the form of a series:

$$\Gamma(z) = \lim_{k \to \infty} \frac{k! k^z}{z(z+1)\cdots(z+k)} \tag{4-139}$$

The equivalence of these two forms may be shown in the following way. Consider the function

$$F(z,k) = \int_0^\infty \left(1 - \frac{t}{k}\right)^k t^{z-1} dt \tag{4-140}$$

In the limit $k \to \infty$, we have the identity

$$\lim_{k \to \infty} \left(1 - \frac{t}{k}\right)^k = 1 - t + \frac{t^2}{2!} - \frac{t^3}{3!} + \frac{t^4}{4!} - \cdots = e^{-t}$$

From this we obtain

$$\Gamma(z) = \lim_{k \to \infty} F(z,k)$$

for $\Gamma(z)$ given by (4-137).

The fact that $F(z,k)$ in the limit $k \to \infty$ is equivalent to $\Gamma(z)$ given by (4-139) may also be shown using integration by parts. For this purpose, it is convenient to make the substitution

$$y = \frac{t}{k}$$

We shall only be interested in $F(x,k)$ for $k \to \infty$. In this limit, the integral of (4-137) is to be taken between 0 and 1 if it is expressed in terms of y instead of t. On integrating over y by parts once, we have the result

$$F(z,k) = k^z \int_0^1 (1-y)^k y^{z-1} dy = k^z \left\{ (1-y)^k \frac{y^z}{z}\bigg|_{y=0}^{y=1} + \frac{k}{z} \int_0^1 (1-y)^{k-1} y^z dy \right\}$$

The first term vanishes and we are left with

$$F(z,k) = k^z \frac{k}{z} \int_0^1 (1-y)^{k-1} y^z dy$$

This is the same as (4-140), except that the power on the first term of the integrand is reduced by 1 and that of the second term increased by 1. On repeating the process k times, the first term becomes unity and we have

$$F(z,k) = k^z \frac{k(k-1)\cdots 1}{z(z+1)\cdots(z+k-1)} \int_0^1 y^{z+k-1} dy$$

$$= k^z \frac{k(k-1)\cdots 1}{z(z+1)\cdots(z+k-1)(z+k)}$$

In the limit $k \to \infty$, the final form is the same as the right side of (4-139).

On substituting $(z+1)$ for z in (4-139), we find that

$$\Gamma(z+1) = \lim_{k \to \infty} \frac{k!k^{z+1}}{(z+1)(z+2)\cdots(z+k+1)}$$

$$= \lim_{k \to \infty} \frac{kz}{z+k+1} \frac{k!k^z}{z(z+1)\cdots(z+k)} = z\Gamma(z) \qquad (4\text{-}141)$$

This gives us a recursion relation, which may be used to show that (4-138) is true. On applying (4-141) for $(z-1)$ times, we end up with

$$\Gamma(z+1) = z(z-1)\cdots(1)\Gamma(1)$$

When we substitute $z=1$ in (4-139), we find that

$$\Gamma(1) = 1$$

This proves the equality given by (4-138).

Another useful form of the gamma function is in terms of an infinite product

$$\frac{1}{\Gamma(z)} = ze^{\gamma z} \prod_{k=1}^{\infty} \left\{ \left(1 + \frac{z}{k}\right) e^{-z/k} \right\} \qquad (4\text{-}142)$$

where the absolute value of z is less than infinity. The Euler's constant γ in the argument of the first exponential function has the value

$$\gamma = \lim_{k \to \infty} \left\{ 1 + \frac{1}{2} + \frac{1}{3} + \frac{1}{4} + \cdots + \frac{1}{k} - \ln k \right\} = 0.5772156649\ldots$$

Using the fact that

$$\prod_{j=1}^{k} \left(1 + \frac{z}{j}\right)^{-1} = \prod_{j=1}^{k} \frac{j}{j+z} = \frac{1 \cdot 2 \cdot 3 \cdots (k-1)k}{(z+1)(z+2)(z+3)\cdots(z+k)}$$

we can rewrite (4-139) in the form

$$\Gamma(z) = \lim_{k \to \infty} \frac{1}{z} \prod_{j=1}^{k} \left(1 + \frac{z}{j}\right)^{-1} k^z$$

or

$$\frac{1}{\Gamma(z)} = z \lim_{k \to \infty} \left\{ e^{-z \ln k} \prod_{j=1}^{k} \left(1 + \frac{z}{j}\right) \right\}$$

where we have made use of the relation $k^z = \exp(z \ln k)$. To convert the argument of the exponential function into the Euler's constant γ, we need to multiply and divide the right side by

$$\exp\left\{ z\left(1 + \frac{1}{2} + \frac{1}{3} + \frac{1}{4} + \cdots + \frac{1}{k}\right) \right\} = \prod_{j=1}^{k} e^{z/j}$$

On taking the limit of $k \to \infty$, we obtain

$$\frac{1}{\Gamma(z)} = ze^{z\gamma} \lim_{k\to\infty} \left\{ \prod_{j=1}^{k} \left(1 + \frac{z}{j}\right) e^{-z/j} \right\}$$

This is the identity of (4-142) we wish to demonstrate in this paragraph.

Using the infinite product form given in (4-142), we find that the product of $\Gamma(z)$ and $\Gamma(-z)$ takes on the value

$$\Gamma(z)\Gamma(-z) = -\left\{ z^2 \prod_{k=1}^{\infty} \left(1 + \frac{z}{k}\right)\left(1 - \frac{z}{k}\right) \right\}^{-1} = -\left\{ z^2 \prod_{k=1}^{\infty} \left(1 - \frac{z^2}{k^2}\right) \right\}^{-1}$$

The right side may be compared with the expansion of $\sin x$ in the following infinity series:

$$\frac{\sin(\pi x)}{\pi x} = \prod_{k=1}^{\infty} \left(1 - \frac{x^2}{k^2}\right)$$

This gives us the identity

$$\Gamma(z)\Gamma(-z) = -\frac{\pi}{z\sin(\pi z)} \tag{4-143a}$$

The same expression may be written as

$$\Gamma(z)\Gamma(1-z) = \frac{\pi}{\sin(\pi z)} \tag{4-143b}$$

obtained by putting the result of (4-142) together with the recursion relation of (4-141).

Polynomial approximations. Several polynomial approximations of $\Gamma(z)$ are available for z not necessarily an integer. Since gamma functions for different arguments are connected by the recursion relation (4-141), all we need is an expression for $0 < z < 1$. A five-term form for $z = x$, where x is a real number, is given by

$$\Gamma(1+x) = 1 + \sum_{k=1}^{5} a_k x^k$$

The values of the coefficients are

$a_1 = -0.5748646$ $a_2 = 0.9512363$ $a_3 = -0.6998588$

$a_4 = 0.4245549$ $a_5 = -0.1010678$

The results calculated with this approximate expression are accurate to $|\epsilon| \leq 5 \times 10^{-5}$. An eight-term form,

$$\Gamma(1+x) = 1 + \sum_{k=1}^{8} b_k x^k \tag{4-144}$$

with

$$b_1 = -0.577191652 \qquad b_2 = 0.988205891 \qquad b_3 = -0.897056937$$
$$b_4 = 0.918206857 \qquad b_5 = -0.756704078 \qquad b_6 = 0.482199394$$
$$b_7 = -0.193527818 \qquad b_8 = 0.035868343$$

has an accuracy of $|\epsilon| \leq 3 \times 10^{-7}$. Both formulas work also in the limit of $x = 0$ and 1, as can be seen by inspection. A series expansion of $1/\Gamma(z)$ for complex argument, consisting of 26 terms, is also given on page 256 of Abramowitz and Segun. Using a combination of the recursion relation $\Gamma(1+z) = z\Gamma(z)$ given in (4-141) and the polynomial approximation of (4-144), a method to calculate $\Gamma(x)$ for $x > 0$ is implemented in the algorithm outlined in Box 4-9.

Box 4-9 Function GAMMA8(x)

Gamma function for real arguments

Using an eight-term approximation of $\Gamma(1+x)$ for $0 \leq x < 1$

Argument list:
  x: A real, nonnegative number.

Initialization:
  Store the values of coefficients $b_1, b_2, \ldots, b_8$ of (4-144).

1. Reject $x \leq 0$.
2. Reduce the argument to the range $[0, 1]$:
   (a) Let $z = (x - 1)$.
   (b) If $0 \leq z < 1$, let $f = 1$ and go to step 3.
   (c) If $z < 0$, use $\Gamma(x) = x^{-1}\Gamma(1+x)$:
       (i) Let $f = x^{-1}$ and $z = (1 + x)$.
       (ii) Go to step 3 to calculate the value of $\Gamma(z)$.
   (d) If $z > 1$, use $\Gamma(1+x) = x\Gamma(x)$ to reduce the argument to ≤ 1.
       (i) Start with $f = 1$.
       (ii) Let $f = fz$ and $z = (z - 1)$.
       (iii) Repeat step (ii) until $z < 1$.
3. Calculate the value of $\Gamma(1+z)$ for $0 \leq z < 1$ using (4-144).
4. Return $\Gamma(x) = f\Gamma(1+z)$.

An efficient way to evaluate the gamma function for large values of the argument is to use Stirling's formula,

$$\Gamma(z) \approx e^{-z} z^{z-\frac{1}{2}} \sqrt{2\pi} \left\{ 1 + \frac{1}{12z} + \frac{1}{288z^2} - \frac{139}{51,840z^3} - \frac{571}{2,488,320z^4} + \cdots \right\}$$

This result may be derived by applying the Euler-Maclaurin summation formula given in (2-25) for the integral form of $\Gamma(z)$ given in (4-137).

Special values. The values of gamma functions for a few special cases are of interest and they may be obtained directly from the properties of $\Gamma(z)$. From the factorial form given in (4-138) and by making use of the recursion relation (4-141), we have

$$(z-1)! = \frac{z!}{z}$$

On letting $z = 1$, we obtain the result

$$0! = 1 \qquad (4\text{-}145)$$

used earlier in §1-3 to calculate the values of binomial coefficients. Similarly, if $z = 0$, we obtain the result

$$(-1)! = \infty$$

Putting this together with the recursion relation of (4-141), we come to the conclusion that the factorial of a negative integer is either $+\infty$ or $-\infty$.

The value of $\Gamma(z)$ for $z = \frac{1}{2}$ may be obtained using (4-143b):

$$\Gamma(z=\tfrac{1}{2})^2 = \frac{\pi}{\sin(\tfrac{1}{2}\pi)} = \pi$$

or

$$\Gamma(\tfrac{1}{2}) = \sqrt{\pi} \qquad (4\text{-}146)$$

On the other hand, by putting $z = \frac{1}{2}$ into (4-143a), we obtain

$$\Gamma(\tfrac{1}{2})\Gamma(-\tfrac{1}{2}) = -\frac{\pi}{\tfrac{1}{2}\sin(\tfrac{1}{2}\pi)} = -2\pi$$

This implies that

$$\Gamma(-\tfrac{1}{2}) = -2\sqrt{\pi} \qquad (4\text{-}147)$$

By a similar maneuver, it is possible to obtain the values of $\Gamma(z)$ for z equal to other ratios of (small) integers. Some of these are listed on page 3 of Abramowitz and Segun.

Double factorial. Using the recursion relation (4-141), we can show that the gamma function for any positive half-integer has the form

$$\Gamma(n+\tfrac{1}{2}) = \frac{2n-1}{2}\Gamma(n-\tfrac{1}{2})$$

$$= \frac{(2n-1)(2n-3)(2n-5)\cdots 3\cdot 1}{2^n}\Gamma(\tfrac{1}{2})$$

$$= \frac{1}{2^n}(2n-1)!!\sqrt{\pi} \qquad (4\text{-}148)$$

where we have used the double factorial symbol in the following sense:

$$k!! \equiv \begin{cases} k(k-2)(k-4)\cdots 3\cdot 1 & \text{for odd } k \\ k(k-2)(k-4)\cdots 4\cdot 2 & \text{for even } k \end{cases} \qquad (4\text{-}149)$$

Note that, for (4-148) to be true for $n = 0$, it is necessary to define $(-1)!! = 1$.

It is also possible to express the factorial of an integer in terms of a double factorial,

$$k! = \left(\frac{2k}{2}\right)!$$
$$= \frac{(2k)(2k-2)(2k-4)\cdots 4 \cdot 2}{2^k} = \frac{1}{2^k}(2k)!! \qquad (4\text{-}150)$$

a result that is useful on many occasions.

Incomplete gamma function. If, instead of ∞, we replace the upper limit of the integral in the definition of the gamma function given in (4-137) by a variable, we obtain the incomplete gamma functions. There are two related forms of this function,

$$\gamma(a, x) = \int_0^x e^{-t} t^{a-1} dt \qquad (4\text{-}151)$$

$$\Gamma(a, x) = \Gamma(a) - \gamma(a, x) = \int_x^\infty e^{-t} t^{a-1} dt \qquad (4\text{-}152)$$

A number of functions involving integrals of exponential functions may be written in terms of these integrals. For example, if we change the integration variable t in (4-151) to $y = t^{1/2}$, the result is proportional to the error function of (4-133):

$$\text{erf}(x) = \frac{1}{\sqrt{\pi}} \gamma(\tfrac{1}{2}, x^2)$$

$$\text{erfc}(x) = \frac{1}{\sqrt{\pi}} \Gamma(\tfrac{1}{2}, x^2) \qquad (4\text{-}153)$$

We shall see later in §6-1 that these functions are also related to the normal probability integral. An example was also given earlier in (4-135).

We have used $\Gamma(a, x)$ in §3-3 as an application of the continued fraction form of a function. A series expansion form for $\gamma(a, x)$ is also available on page 262 of Abramowitz and Segun:

$$\gamma(a, x) = e^{-x} x^a \Gamma(a) \sum_{n=0}^{\infty} \frac{x^n}{\Gamma(a+n+1)} \qquad (4\text{-}154)$$

From the recursion relation (4-141), we have

$$\Gamma(a+n+1) = (a+n)\Gamma(a+n) = (a+n)(a+n-1)\Gamma(a+n-1)$$
$$= (a+n)(a+n-1)\cdots(a)\Gamma(a)$$

Box 4-10 Function SMALL_GAMMA(a, x)

Series expansion of the incomplete gamma function $\gamma(a,x)$ for small x

Argument list:
  a: First argument of $\gamma(a,x)$.
  x: Second argument of $\gamma(a,x)$.
Initialization:
  (a) Set maximum number of terms to be 100.
  (b) Let the accuracy be 5×10^{-7}.
  (c) Let $d = a$.
1. Reject $x < 0$.
2. Return 0 if $x = 0$.
3. For $x > 0$, use (4-155):
  (a) Let the numerator equal 1 and the contribution from first term equal $1/a$.
  (b) Initialize the sum to equal the first term and $d = 1$.
  (c) For $n = 1$ to a maximum of 100, carry out the following steps:
    (i) Increase d by 1 and set $t = t * x/d$ as the contribution of this term.
    (ii) Add t to the sum.
    (iii) Go to step 4 if t is less than the accuracy required.
    (iv) Reject the result if the calculation does not converge.
4. Put in the exponential factor and power of x and return the product.

As a result, the series of (4-154) may also be written in a form that does not involve any gamma functions

$$\gamma(a,x) = e^{-x} x^a \sum_{n=0}^{\infty} \frac{x^n}{(n+a)(n+a-1)\cdots(a)} \qquad (4\text{-}155)$$

For computational purposes, this form is easier to handle, as gamma functions are, in general, more time consuming to evaluate. Explicitly, the various terms in the summation are

$$\sum_{n=0}^{\infty} \frac{x^n}{(n+a)(n+a-1)\cdots(a)} = \frac{1}{a} + \frac{1}{a}\frac{x}{a+1} + \frac{1}{a}\frac{x}{a+1}\frac{x}{a+2} + \cdots$$
$$+ \frac{1}{a}\frac{x}{a+1}\frac{x}{a+2}\cdots\frac{x}{a+n} + \cdots$$

The convergence of the series becomes much faster once we are beyond the term where the additional factor $(a + i)$ in the denominator is larger than x in the numerator. Since the continued fractional form of (3-52) is fast convergent only for $x > (a+1)$, the present form is more useful for $x < (a+1)$. The algorithm to evaluate $\gamma(a,x)$ according to (4-155) is given in Box 4-10.

§4-6 Clebsch-Gordan coefficients

We shall end this chapter with a discussion of the calculation of the most elementary angular momentum coupling coefficients, the Clebsch-Gordan coefficients. In addition to its usefulness in many quantum mechanical studies involving angular momentum, it is also interesting as a problem in computational physics.

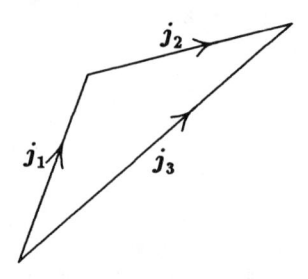

Fig. 4-11 Coupling of two angular momenta, $\boldsymbol{j}_1$ and $\boldsymbol{j}_2$, to total value $\boldsymbol{j}_3$.

When two states, one with angular momentum $\boldsymbol{j}_1$ and the other $\boldsymbol{j}_2$, are coupled together, the resulting angular momentum $\boldsymbol{j}_3$ is the vector sum

$$\boldsymbol{j}_3 = \boldsymbol{j}_1 + \boldsymbol{j}_2$$

The magnitude of the sum is restricted to the range

$$|j_1 - j_2| \leq j_3 \leq j_1 + j_2 \tag{4-156}$$

where j_i is the magnitude of a vector $\boldsymbol{j}_i$. In other words, the three vectors $\boldsymbol{j}_1$, $\boldsymbol{j}_2$, and $\boldsymbol{j}_3$ must form a closed triangle, as shown in Fig. 4-11.

The importance of coupling angular momenta in quantum mechanics comes in the following way. If $\psi_{m_1}^{j_1}$ is the wave function of a state with angular momentum j_1 and projection on the quantization axis m_1, and $\phi_{m_2}^{j_2}$ that of a state with angular momentum j_2 and projection m_2, the wave function of the combined state

$$\Psi = \psi_{m_1}^{j_1} \phi_{m_2}^{j_2}$$

does not, in general, possess a definite angular momentum. The possible range of values is given by (4-156). The product wave function Ψ may be expressed as a linear combination of components, each with a definite value of total angular momentum $\boldsymbol{j}_3$:

$$\Psi = \sum_{j_3 = |j_1 - j_2|}^{j_1 + j_2} \langle j_1 m_1 j_2 m_2 | j_3 m_3 \rangle \left(\psi_{m_1}^{j_1} \times \phi_{m_2}^{j_2} \right)_{m_3}^{j_3} \tag{4-157}$$

The factor $\langle j_1 m_1 j_2 m_2 | j_3 m_3 \rangle$ is the Clebsch-Gordon, or vector addition, coefficient. It expresses the overlap of the product of $\psi_{m_1}^{j_1}$ and $\phi_{m_2}^{j_2}$ with a state formed from the angular momentum coupled function $\left(\psi_{m_1}^{j_1} \times \phi_{m_2}^{j_2}\right)_{m_3}^{j_3}$ that has a definite angular momentum j_3 and projection m_3. Note that, since the projection m_i of a vector j_i is a scalar quantity, the value of m_3 in the above expression must equal the sum of m_1 and m_2. If the Hamiltonian of the system is independent of the total angular momentum, j_3 and m_3 are good quantum numbers and $\left(\psi_{m_1}^{j_1} \times \phi_{m_2}^{j_2}\right)_{m_3}^{j_3}$ is an eigenstate.

Several different ways are used in the literature to write the angular momentum coupling coefficient involving the coupling of (j_1, m_1) and (j_2, m_2) to (j_3, m_3). For example,

$$\langle j_1 m_1 j_2 m_2 | j_3 m_3 \rangle \equiv \langle j_1 j_2 m_1 m_2 | j_1 j_2 j_3 m_3 \rangle \equiv C_{m_1 m_2 m_3}^{j_1 j_2 j_3}$$

We shall use the first form here, since it gives a more direct impression of the overlap of a state with angular momentum ranks (j_3, m_3) with the product of two states, one with angular momentum ranks (j_1, m_1) and the other with (j_2, m_2). Alternatively, we can use the Wigner $3j$-symbol, related to the Clebsch-Gordan coefficients by a simple factor in the following way:

$$\begin{pmatrix} j_1 & j_2 & j_3 \\ m_1 & m_2 & m_3 \end{pmatrix} = \frac{(-1)^{j_1 - j_2 - m_3}}{\sqrt{2j_3 + 1}} \langle j_1 m_1 j_2 m_2 | j_3 \, {-}m_3 \rangle \quad (4\text{-}158)$$

The $3j$-symbol is convenient to express the symmetry relations in the coupling coefficients. For example,

$$\begin{pmatrix} j_1 & j_2 & j_3 \\ m_1 & m_2 & m_3 \end{pmatrix} = \begin{pmatrix} j_2 & j_3 & j_1 \\ m_2 & m_3 & m_1 \end{pmatrix} = \begin{pmatrix} j_3 & j_1 & j_2 \\ m_3 & m_1 & m_2 \end{pmatrix}$$

$$= (-1)^{j_1+j_2+j_3} \begin{pmatrix} j_1 & j_3 & j_2 \\ m_1 & m_3 & m_2 \end{pmatrix} = (-1)^{j_1+j_2+j_3} \begin{pmatrix} j_1 & j_2 & j_3 \\ -m_1 & -m_2 & -m_3 \end{pmatrix}$$

gives the relation between the coefficients under permutations of the arguments.

It is almost universal nowadays to adopt the Condon and Shortley phase convention, which states that, in coupling j_1 and j_2 to the maximum possible angular momentum $(j_1 + j_2)$, and all the projections on the z-axis take on the maximum values allowed, that is, $m_1 = j_1$ and $m_2 = j_2$; the Clebsch-Gordan coefficient

$$\langle j_1 j_1 j_2 j_2 | j_1{+}j_2 \, j_1{+}j_2 \rangle = +1 \quad (4\text{-}159)$$

and the sum

$$\sum_{m_1 m_2} m_1 \langle j_1 m_1 j_2 m_2 | j_3 m_3 \rangle \langle j_1 m_1 j_2 m_2 | (j_3{-}1) m_3 \rangle > 0$$

All the Clebsch-Gordan coefficients are real in this convention, and the explicit values of some of the more useful ones involving low angular ranks are listed in Table 4-6.

The orthogonality relations between two $3j$-coefficients are

$$\sum_{m_1 m_2} \begin{pmatrix} j_1 & j_2 & j_3 \\ m_1 & m_2 & m_3 \end{pmatrix} \begin{pmatrix} j_1 & j_2 & j_3' \\ m_1 & m_2 & m_3' \end{pmatrix} = \Delta(j_1 j_2 j_3) \frac{\delta_{j_3,j_3'} \delta_{m_3,m_3'}}{2j_3 + 1}$$

$$\sum_{m_1 m_2 m_3} \begin{pmatrix} j_1 & j_2 & j_3 \\ m_1 & m_2 & m_3 \end{pmatrix} \begin{pmatrix} j_1 & j_2 & j_3 \\ m_1 & m_2 & m_3 \end{pmatrix} = \Delta(j_1 j_2 j_3)$$

$$\sum_{j_3 m_3} (2j_3 + 1) \begin{pmatrix} j_1 & j_2 & j_3 \\ m_1 & m_2 & m_3 \end{pmatrix} \begin{pmatrix} j_1 & j_2 & j_3 \\ m_1' & m_2' & m_3 \end{pmatrix} = \Delta(j_1 j_2 j_3) \delta_{m_1,m_1'} \delta_{m_2,m_2'} \quad (4\text{-}160)$$

where

$$\Delta(j_1 j_2 j_3) = \begin{cases} 1 & \text{for} \quad |j_1 - j_2| \leq j_3 \leq j_1 + j_2 \\ 0 & \text{otherwise} \end{cases}$$

In terms of Clebsch-Gordan coefficients, the same relations may be written as

$$\sum_{m_1 m_2} \langle j_1 m_1 j_2 m_2 | j_3 m_3 \rangle \langle j_1 m_1 j_2 m_2 | j_3' m_3' \rangle = \Delta(j_1 j_2 j_3) \delta_{j_3,j_3'} \delta_{m_3,m_3'}$$

$$\sum_{m_1 m_2 m_3} \langle j_1 m_1 j_2 m_2 | j_3 m_3 \rangle \langle j_1 m_1 j_2 m_2 | j_3 m_3 \rangle = (2j_3 + 1) \Delta(j_1 j_2 j_3)$$

$$\sum_{j_3 m_3} \langle j_1 m_1 j_2 m_2 | j_3 m_3 \rangle \langle j_1 m_1' j_2 m_2' | j_3 m_3 \rangle = \Delta(j_1 j_2 j_3) \delta_{m_1,m_1'} \delta_{m_2,m_2'}$$

$$(4\text{-}161)$$

Many other symmetry relations can be found in books on angular momentum, such as Brink and Satchler (*Angular Momentum*, 2nd ed., Oxford University Press, Oxford, 1968).

If one or more of the six arguments vanish or are equal to some values such as $\frac{1}{2}$ or 1, the value of the Clebsch-Gordan coefficient may be expressed as a simple function of the remaining arguments. Some examples are given in Table 4-6. For the more general case, a formula due to Racah (*Phys. Rev.* **62** [1942] 438) is often used. For the convenience of programming, it may be written as

$$\langle j_1 m_1 j_2 m_2 | j_3 m_3 \rangle = \delta_{m_1+m_2,m_3}$$

$$\times \sqrt{(2j_3 + 1) \frac{(j_1 + j_2 - j_3)!(j_3 + j_1 - j_2)!(j_2 + j_3 - j_1)!}{(j_1 + j_2 + j_3 + 1)!}}$$

$$\times \sqrt{(j_1 - m_1)!(j_1 + m_1)!(j_2 - m_2)!(j_2 + m_2)!(j_3 - m_3)!(j_3 + m_3)!}$$

$$\times \sum_{\nu_{\min}}^{\nu_{\max}} \frac{(-1)^\nu}{\nu!(j_3 - j_2 + m_1 + \nu)!(j_3 - j_1 - m_2 + \nu)!}$$

$$\times \frac{1}{(j_1 - m_1 - \nu)!(j_2 + m_2 - \nu)!(j_1 + j_2 - j_3 - \nu)!} \quad (4\text{-}162)$$

Table 4-6 The values of some useful Clebsch-Gordan coefficients.

$\langle \tfrac{1}{2} m_s \ell m_\ell | j m \rangle$

j	$m_s = +\tfrac{1}{2}$	$m_s = -\tfrac{1}{2}$
$\ell + \tfrac{1}{2}$	$\sqrt{\dfrac{\ell+\tfrac{1}{2}+m}{2\ell+1}}$	$\sqrt{\dfrac{\ell+\tfrac{1}{2}-m}{2\ell+1}}$
$\ell - \tfrac{1}{2}$	$\sqrt{\dfrac{\ell+\tfrac{1}{2}-m}{2\ell+1}}$	$-\sqrt{\dfrac{\ell+\tfrac{1}{2}+m}{2\ell+1}}$

$\langle 1 m_s \ell m_\ell | j m \rangle$

j	$m_s = +1$	$m_s = 0$	$m_s = -1$
$\ell + 1$	$\sqrt{\dfrac{(\ell+m)(\ell+m+1)}{2(\ell+1)(2\ell+1)}}$	$\sqrt{\dfrac{(\ell-m+1)(\ell+m+1)}{(\ell+1)(2\ell+1)}}$	$\sqrt{\dfrac{(\ell-m)(\ell-m+1)}{2(\ell+1)(2\ell+1)}}$
ℓ	$\sqrt{\dfrac{(\ell+m)(\ell-m+1)}{2\ell(\ell+1)}}$	$\dfrac{-m}{\sqrt{\ell(\ell+1)}}$	$-\sqrt{\dfrac{(\ell-m)(\ell+m+1)}{2\ell(\ell+1)}}$
$\ell - 1$	$\sqrt{\dfrac{(\ell-m)(\ell-m+1)}{2\ell(2\ell+1)}}$	$-\sqrt{\dfrac{(\ell-m)(\ell+m)}{\ell(2\ell+1)}}$	$\sqrt{\dfrac{(\ell+m+1)(\ell+m)}{2\ell(2\ell+1)}}$

$\begin{pmatrix} j & 0 & j' \\ -m & 0 & m' \end{pmatrix} = \langle jmj'-m'|00\rangle = \dfrac{(-1)^{j-m}}{\sqrt{2j+1}}\delta_{j,j'}\delta_{m,m'}$ $\qquad \langle jm00|j'm'\rangle = \delta_{j,j'}\,\delta_{m,m'}$

$\begin{pmatrix} j & 1 & j \\ -m & 0 & m \end{pmatrix} = (-1)^{j-m}\dfrac{m}{\sqrt{j(j+1)(2j+1)}}$

$\begin{pmatrix} j & 2 & j \\ -m & 0 & m \end{pmatrix} = (-1)^{j-m}\dfrac{3m^2 - j(j+1)}{\sqrt{(2j-1)j(j+1)(2j+1)(2j+3)}}$

$\begin{pmatrix} j_1 & j_2 & j_3 \\ 0 & 0 & 0 \end{pmatrix} = \begin{cases} (-1)^g \sqrt{\dfrac{(2g-2j_1)!(2g-2j_2)!(2g-2j_3)!}{(2g+1)!}}\,\dfrac{g!}{(g-j_1)!(g-j_2)!(g-j_3)!} \\ \qquad \text{if} \quad 2g = j_1 + j_2 + j_3 = \text{even} \\ 0 \qquad \text{if} \quad j_1 + j_2 + j_3 = \text{odd} \end{cases}$

The Kronecker delta $\delta_{m_1+m_2,m_3}$ expresses the simple fact that m_3 must be equal to the sum of m_1 and m_2. The first three factorials inside the square root also serve as a check of the triangular relations between the three angular momenta j_1, j_2, and j_3. It is another way of stating the limitations imposed by (4-156) except that, here, the three quantities j_1, j_2, and j_3 are put on a more equal footing. The lower and upper limits of the summation index ν are governed by the requirement that none of the factorials in the denominator inside the summation are those of negative integers. For example, $\nu_{\min}$, the lower limit, is obtained from the second

> **Box 4-11** Function CG_COEF($j_1, j_2, j_3, m_1, m_2, m_3$)
> Clebsch-Gordan coefficient with (4-162)
> Using a table of the logarithm of factorials
>
> List of arguments:
> j_i, m_i, $i = 1, 2$, and 3: Angular momenta and their projections, as defined in (4-157).
>
> Initialize the table of log factorials.
> 1. Angular momentum checks:
>   (a) $m_1 + m_2 = m_3$.
>   (b) $(j_1 + j_2) \geq j_3$ and $(j_1 + j_2 - j_3)$ an integer.
> $(j_3 + j_1) \geq j_2$ and $(j_3 + j_1 - j_2)$ an integer.
> $(j_2 + j_3) \geq j_1$ and $(j_2 + j_3 - j_1)$ an integer.
>   (c) $|m_1| \leq j_1$ and $(j_1 + m_1)$ an integer.
> $|m_2| \leq j_2$ and $(j_2 + m_2)$ an integer.
> $|m_3| \leq j_3$ and $(j_3 + m_3)$ an integer.
> 2. Calculate the overall factor inside the square root.
> 3. Determine the range of summation:
>   (a) The lower limit $\nu_{\min} = 0$, or
>   (i) if $(j_3 - j_2 + m_1) < 0$, then $\nu_{\min} \geq |j_3 - j_2 + m_1|$, or
>   (ii) if $(j_3 - j_1 - m_2) < 0$, then $\nu_{\min} \geq |j_3 - j_1 - m_2|$.
>   (b) The upper limit $\nu_{\max}$ is the minimum of $(j_1 - m_1)$, $(j_2 - m_2)$, and $(j_1 + j_2 - j_3)$.
> 4. Calculate the contributions from the sum:
>   (a) Obtain the phase factor.
>   (b) Calculate the logarithm of the contribution.
>   (c) Convert the logarithm back and add to the sum.
> 5. Return the product of the overall factor and the sum.

and third terms. For both of them to be nonnegative, the value of $\nu_{\min}$ must be large enough such that both $(j_3 - j_2 + m_1 - \nu_{\min})$ and $(j_3 - j_1 - m_2 - \nu_{\min})$ are zero or positive integers. The upper limit of the summation is given by the other four terms. To avoid negative integers, ν cannot be larger than the minimum of $(j_1 - m_1)$, $(j_2 - m_2)$, and $(j_1 + j_2 - j_3)$.

The method to implement (4-162) is outlined in Box 4-11. As discussed in §1-3, we cannot simply calculate the product of the numerator and denominator separately in terms of integers and obtain the result at the end by eliminating common divisors. Except in cases where all six arguments are small, such a method fails due to the limited capacity for storing integers in a computer word. Similarly, the accuracy may be poor if we are not careful in carrying out the calculations in term of floating numbers. For these reasons, the calculation is best done in terms of prime number representations of integers or by using a table of the logarithms of factorials. The former can be quite cumbersome, but the results are useful in calculations requiring exact values. The log factorial method is more

Problems

4-1 Show that the orthogonal property of Hermite polynomials is given by (4-15).

4-2 The algebraic expressions of Hermite polynomials $H_n(\rho)$ are given in (4-14) for n up to 5. Use the method of Box 4-1 to obtain similar expressions for $n = 6$ to 10.

4-3 Write a computer program to evaluate the numerical value of $H_n(\rho)$ for $n = 10$ to 15 and ρ in the interval $[0, 14]$ using the recursion relation of (4-22). For a given value of ρ, the starting point of the calculation is given by $H_0(\rho) = 1$ and $H_1(\rho) = 2\rho$. Compare the accuracies in the results with those calculated using the algebraic expressions obtained in Problem 4-2.

4-4 Show that the relation between spherical harmonics $Y_{\ell,m}(\theta, \phi)$ for $m = 0$ and Legendre polynomial $P_\ell(\cos\theta)$ given by (4-59) follows from the differential forms of $Y_{\ell,m}(\theta, \phi)$, given by (4-54), and $P_\ell(\cos\theta)$, given by (4-49).

4-5 One way to generate spherical harmonics of all possible m values for a given ℓ is to start with $Y_{\ell,m}(\theta, \phi)$ for $m = \pm\ell$ and apply the angular momentum lowering and raising operator $\hat{L}_\pm$ of (4-58). Derive the explicit algebraic forms of $Y_{\ell,m}(\theta, \phi)$ for $m = \pm\ell$ using (4-54) for $\ell = 2$ and show that the results are consistent with those given in (4-55).

4-6 Check that $Y_{2,2}(\theta, \phi)$ of (4-55) is the eigenfunction of $\hat{L}^2$ and $\hat{L}_z$ with the correct eigenvalues by calculating the results of the actions of these two operators on $Y_{2,2}(\theta, \phi)$.

4-7 The numerical values of spherical Bessel function $j_\ell(\rho)$ for very high orders, such as $\ell = 100$, are given in Abramowitz and Segun for $\rho = 1, 2, 5, 10, 50$, and 100. Write a computer program to calculate these values and compare the results with the tabulated ones.

4-8 Evaluate the spherical Bessel function $j_5(\rho)$ for five different values of ρ in the interval $[0, 10]$ using (a) the explicit algebraic form given in Table 4-3 and (b) by propagating from the values of $j_0(\rho)$ and $j_1(\rho)$ for these five arguments. Compare the accuracies of the two methods by checking the results against the tabulated values given in Abramowitz and Segun.

4-9 Make a table of the values of $j_5(\rho)$ for different ρ in the interval $[0, 10]$. Use the method of inverse interpolation in §3-6 to find all the zeros of the function in the interval. What is the main source of errors in the results?

4-10 Calculate the numerical values of $L_n(t)$ for $n = 5$ and 6 for a number of values of t in the range $[0, 5]$. Use the results for $n = 6$ to obtain the value of dL_6/dt by numerical differentiation. Check the relation given by (4-110) using the numerical values obtained.

4-11 Evaluate the expectation value of ρ^2 for the three low-lying states of a hydrogen atom using the numerical values of the radial wave functions calculated with the expressions (4-125). Check the results against those calculated by evaluating the integrals analytically using the algebraic expressions of the wave functions.

4-12 Repeat Problem 4-11 for the isotropic harmonic oscillator using the radial wave functions given by (4-127).

4-13 Verify that the uncertainties in the values of error functions calculated using the approximation of (4-136) are $|\epsilon| \leq 2.5 \times 10^{-5}$ by comparing the results with numerical integration using (4-133). Note that it is necessary to establish first that the error in numerical integration is less than 2.5×10^{-5}.

4-14 The integral representation of $\Gamma(\frac{1}{2})$ is

$$\Gamma(\tfrac{1}{2}) = \int_0^\infty \frac{e^{-t}}{\sqrt{t}}\,dt$$

Choose a method to evaluate this integral numerically. Is this a good way to calculate the value of π?

4-15 Use the properties of $\Gamma(x)$ for half-integer arguments to show that

$$z!(z+\tfrac{1}{2})! = \frac{\pi^{1/2}}{2^{2z+1}}(2z+1)!$$

This result was used earlier in §4-3 to obtain a series form of spherical Bessel functions.

4-16 Verify that the algebraic expression for the Clebsch-Gordan coefficient given by (4-162) is consistent with the phase convention stated by (4-159).

4-17 A state with angular momentum $j = 2$ can have $m = \pm 2, \pm 1$ and 0. Two distinguishable states, both with $j = 2$, can couple to a state with total angular momentum $J = 0, 1, 2, 3$, and 4. Calculate all the Clebsch-Gordan coefficients involved in the coupling. Verify numerically that the Clebsch-Gordan coefficients obtained satisfy the orthogonality conditions of (4-161).

Chapter 5

Matrices

With computers taking over many of the tedious tasks, matrix methods are becoming increasingly more popular in solving problems of physical interest. In this chapter, we shall be mainly concerned with some of the elementary numerical calculations involving matrices. Later, in Chapter 10, we shall see an introduction to certain aspects of algebraic calculations that can be carried out on a computer.

§5-1 System of linear equations

Consider a simple example involving a direct-current circuit, such as that shown in Fig. 5-1. There are three resistors, labeled R_1, R_2 and R_3 and two sources of electromotive force, V_1 and V_2, in the circuit. One possible interest in the problem is to find the three currents, I_1, I_2, and I_3, flowing respectively through the three resistors, R_1, R_2, and R_3.

The solution to the problem may be obtained in the following way. Let us examine first the circuit loop on the left. Since it is a complete loop, the sum of the voltage drops across R_1 and R_2 must be equal to the electromotive force V_1. This gives us one relation among the unknown quantities in the problem:

$$I_1 R_1 + I_2 R_2 = V_1$$

Similarly, the loop on the right gives us the relation

$$-I_2 R_2 + I_3 R_3 = V_2$$

where the negative sign for the first element comes from our choice of considering a current to be positive if it is flowing in the clockwise direction. A third equation among the three unknown quantities is obtained from charge conservation at point A in the diagram:

$$I_1 - I_2 - I_3 = 0$$

These three equations may be written together as a set of three linear equations

$$R_1 I_1 + R_2 I_2 \qquad\quad = V_1$$
$$\qquad\; - R_2 I_2 + R_3 I_3 = V_2$$
$$I_1 - \quad I_2 - \quad I_3 = 0 \tag{5-1}$$

There are several ways to solve this system of equations. The particular method we are interested in here is to make use of matrices.

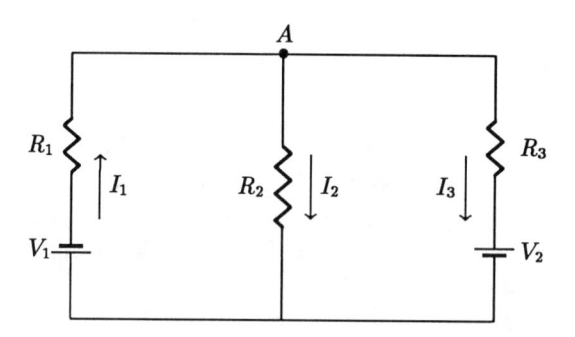

Fig. 5-1 A direct-current circuit consisting of three resistors R_1, R_2, and R_3 and two sources of electromotive force V_1 and V_2. A system of linear equations may be set up to find the current flowing through each of the three resistors.

Matrix notation. A more compact notation is to write the system of linear equations given by (5-1) in terms of matrices,

$$\boldsymbol{RI} = \boldsymbol{V}$$

where $\boldsymbol{R}$ is a (3×3) square matrix,

$$\boldsymbol{R} \equiv \begin{pmatrix} R_1 & R_2 & 0 \\ 0 & -R_2 & R_3 \\ 1 & -1 & -1 \end{pmatrix}$$

That is, it is a matrix with three columns and three rows. The other two matrices, $\boldsymbol{I}$ and $\boldsymbol{V}$, are column matrices with three elements each:

$$\boldsymbol{I} \equiv \begin{pmatrix} I_1 \\ I_2 \\ I_3 \end{pmatrix} \qquad \boldsymbol{V} \equiv \begin{pmatrix} V_1 \\ V_2 \\ 0 \end{pmatrix}$$

In general, a system of n linear equations for n unknown quantities $x_1, x_2, \ldots, x_n$, may be written in the following form:

$$\sum_{j=1}^{n} a_{i,j}\, x_j = y_i \tag{5-2}$$

where $i = 1, 2, \ldots, n$. In matrix notation, the same relations are written in the compact form of

$$AX = Y \tag{5-3}$$

where the matrices A, X, and Y are given by

$$A = \begin{pmatrix} a_{1,1} & a_{1,2} & \cdots & a_{1,n} \\ a_{2,1} & a_{2,2} & \cdots & a_{2,n} \\ \vdots & \vdots & \ddots & \vdots \\ a_{n,1} & a_{n,2} & \cdots & a_{n,n} \end{pmatrix} \qquad X = \begin{pmatrix} x_1 \\ x_2 \\ \vdots \\ x_n \end{pmatrix} \qquad Y = \begin{pmatrix} y_1 \\ y_2 \\ \vdots \\ y_n \end{pmatrix}$$

Sometimes it is convenient to use a shorthand to indicate that the elements of matrix A are $a_{i,j}$ in the following way:

$$A = \{a_{i,j}\}$$

We shall make use of this notation where it is convenient.

When a matrix A is multiplied by a scalar quantity λ, it means that every element of the matrix is multiplied by λ. That is,

$$\lambda A = \begin{pmatrix} \lambda a_{1,1} & \lambda a_{1,2} & \cdots & \lambda a_{1,n} \\ \lambda a_{2,1} & \lambda a_{2,2} & \cdots & \lambda a_{2,n} \\ \vdots & \vdots & \ddots & \vdots \\ \lambda a_{n,1} & \lambda a_{n,2} & \cdots & \lambda a_{n,n} \end{pmatrix}$$

When a matrix A is multiplied by another matrix B to form matrix C,

$$C = AB$$

each element of C is given by a sum of the products of the elements of A and B in the following way:

$$c_{i,k} = \sum_j a_{i,j} b_{j,k}$$

For the multiplication to have meaning, the number of columns of A must equal to the number of rows of B. If the matrix A has n rows and m columns, and B has m rows and ℓ columns, the product matrix C has n rows and ℓ columns. Different types of matrices commonly encountered in physics are listed in Table 5-1.

Table 5-1 Properties of various types of matrices.

Name	Symbol	Matrix element	Property
Null	0	$c_{i,j} = 0$	$0A = A0 = 0$
Unit	1	$c_{i,j} = \delta_{i,j}$	$1A = A1 = A$
Diagonal	D	$c_{i,j} = c_{i,i}\delta_{i,j}$	
Complex conjugate	A^*	$c_{i,j} = a_{i,j}^*$	
Real		$a_{i,j} = a_{i,j}^*$	$A = A^*$
Pure imaginary		$a_{i,j} = -a_{i,j}^*$	$A = -A^*$
Inverse	A^{-1}		$A^{-1}A = AA^{-1} = 1$
Transpose	$\widetilde{A}$	$c_{i,j} = a_{j,i}$	
Symmetric		$a_{i,j} = a_{j,i}$	$A = \widetilde{A}$
Skew-symmetric		$a_{i,j} = -a_{j,i}$	$A = -\widetilde{A}$
Hermitian adjoint	$A^\dagger$	$c_{i,j} = a_{j,i}^*$	
Hermitian		$a_{i,j} = a_{j,i}^*$	$A = A^\dagger$
Skew-Hermitian		$a_{i,j} = -a_{j,i}^*$	$A = -A^\dagger$
Unitary			$A^{-1} = A^\dagger$
Trace	$\mathrm{Tr} A$	$\sum_i a_{i,i}$	
Addition	$A + B = C$	$c_{i,j} = a_{i,j} + b_{i,j}$	
Multiplication	$C = \lambda A$	$c_{i,j} = \lambda a_{i,j}$	
	$C = AB$	$c_{i,j} = \sum_k a_{i,k} b_{k,j}$	

Determinant. The value of the determinant of a square matrix A of order n, that is, one with the number of columns and the number of rows equal to n, is given by

$$\det A = \sum_{i,j,k,\ldots} \epsilon_{i,j,k,\ldots} a_{1,i} a_{2,j} a_{3,k} \cdots \tag{5-4}$$

Here, $\epsilon_{i,j,k,\ldots}$ is the Levi-Civita symbol. It is equal to $+1$ for any even permutations of the n subscripts $\{1, 2, 3, \ldots, n\}$, equal to -1 for odd permutations, and zero if any two indexes are the same. For example, for $n = 3$,

$$\det A = \det \begin{vmatrix} a_{1,1} & a_{1,2} & a_{1,3} \\ a_{2,1} & a_{2,2} & a_{2,3} \\ a_{3,1} & a_{3,2} & a_{3,3} \end{vmatrix}$$

$$= a_{1,1} a_{2,2} a_{3,3} - a_{1,1} a_{2,3} a_{3,2} + a_{1,3} a_{2,1} a_{3,2} - a_{1,2} a_{2,1} a_{3,3}$$
$$+ a_{1,2} a_{2,3} a_{3,1} - a_{1,3} a_{2,2} a_{3,1} \tag{5-5}$$

From this we see that if any two rows, or any two columns, are interchanged the value of the determinant changes sign. By the same token, if any two rows, or any two columns, of a determinant are identical to each other, the value of the determinant vanishes.

In our direct-current circuit example above, if either one or both of the electromotive forces, V_1 and V_2, are not zero, the problem posited by (5-1) has a solution only if the determinant of $\boldsymbol{R}$ does not vanish,

$$\det \boldsymbol{R} \neq 0 \tag{5-6}$$

This is easy to see. Since there are three unknown quantities in our problem, I_1, I_2, and I_3, we must have three independent equations before we can hope to find a solution. In other words, none of the three relations given in (5-1) can be obtained from the other two. For example, if one equation is simply another one multiplied by a constant, the problem does not have a unique solution. In terms of matrices, the same condition appears in the form that if one of the rows is a multiple of another one, the determinant vanishes. We see that (5-6) is equivalent to a statement of linear independence among the members of our system of equations.

One method to evaluate a determinant is based on the Laplace expansion theorem, which states that the value of a determinant is given by the sum of the products of all the elements of an arbitrary row j (or a column) with their corresponding minors:

$$\det \boldsymbol{A} = \sum_i (-1)^{i+j} a_{i,j} M_{i,j} \tag{5-7}$$

The minor $M_{i,j}$ is defined as a determinant obtained from $\boldsymbol{A}$ by removing row i and column j. In the place of minors, one may use cofactors

$$F_{i,j} = (-1)^{i+j} M_{i,j} \tag{5-8}$$

Our $n = 3$ example given in (5-5) may now be written in the form

$$\det \boldsymbol{A} = a_{1,1} M_{1,1} - a_{2,1} M_{2,1} + a_{3,1} M_{3,1}$$
$$= a_{1,1} F_{1,1} + a_{2,1} F_{2,1} + a_{3,1} F_{3,1} \tag{5-9}$$

where

$$M_{1,1} = F_{1,1} = \begin{vmatrix} a_{2,2} & a_{2,3} \\ a_{3,2} & a_{3,3} \end{vmatrix}$$

$$M_{2,1} = -F_{2,1} = \begin{vmatrix} a_{1,2} & a_{1,3} \\ a_{3,2} & a_{3,3} \end{vmatrix}$$

$$M_{3,1} = F_{3,1} = \begin{vmatrix} a_{1,2} & a_{1,3} \\ a_{2,2} & a_{2,3} \end{vmatrix}$$

This method, however, becomes very cumbersome if n is large. For computer calculations, other approaches are preferred.

One property of a determinant is that its value is unchanged if multiples of one row are added to another one. This is also true if we add multiples of one column to another. For example,

$$\det \begin{vmatrix} a_{1,1} & a_{1,2} & \cdots & a_{1,n} \\ a_{2,1} & a_{2,2} & \cdots & a_{2,n} \\ \vdots & \vdots & \ddots & \vdots \\ a_{n,1} & a_{n,2} & \cdots & a_{n,n} \end{vmatrix} = \det \begin{vmatrix} a_{1,1} + \lambda a_{1,2} & a_{1,2} & \cdots & a_{1,n} \\ a_{2,1} + \lambda a_{2,2} & a_{2,2} & \cdots & a_{2,n} \\ \vdots & \vdots & \ddots & \vdots \\ a_{n,1} + \lambda a_{n,2} & a_{n,2} & \cdots & a_{n,n} \end{vmatrix} \qquad (5\text{-}10)$$

This identity may be derived from the Laplace expansion theorem in the following way. If we express the left side of (5-10) in terms of a sum of products of the elements of the first column and their corresponding cofactors, it is easy to see that the following relation is true:

$$\det \begin{vmatrix} a_{1,1} + \lambda a_{1,2} & a_{1,2} & \cdots & a_{1,n} \\ a_{2,1} + \lambda a_{2,2} & a_{2,2} & \cdots & a_{2,n} \\ \vdots & \vdots & \ddots & \vdots \\ a_{n,1} + \lambda a_{n,2} & a_{n,2} & \cdots & a_{n,n} \end{vmatrix}$$

$$= \det \begin{vmatrix} a_{1,1} & a_{1,2} & \cdots & a_{1,n} \\ a_{2,1} & a_{2,2} & \cdots & a_{2,n} \\ \vdots & \vdots & \ddots & \vdots \\ a_{n,1} & a_{n,2} & \cdots & a_{n,n} \end{vmatrix} + \lambda \det \begin{vmatrix} a_{1,2} & a_{1,2} & \cdots & a_{1,n} \\ a_{2,2} & a_{2,2} & \cdots & a_{2,n} \\ \vdots & \vdots & \ddots & \vdots \\ a_{n,2} & a_{n,2} & \cdots & a_{n,n} \end{vmatrix}$$

Since the value of the second determinant on the right side vanishes because the first two columns are identical to each other, we obtain the relation given by (5-10).

Gauss-Jordan elimination method. We can use the property given by (5-10) to transform a determinant such that all the off-diagonal elements vanish. For example, by subtracting $(a_{2,1}/a_{1,1})$ times each of the elements in the first row from the corresponding ones in the second row,

$$a'_{2,j} = a_{2,j} - a_{1,j} \times \frac{a_{2,1}}{a_{1,1}} \qquad \text{for} \qquad j = 1, 2, \ldots, n$$

the $(i, j) = (2, 1)$ element of the resulting determinant vanishes. If this process is carried out for all the rows beyond the first one,

$$a'_{i,j} = a_{i,j} - a_{1,j} \times \frac{a_{i,1}}{a_{1,1}} \qquad \begin{array}{l} \text{for} \quad j = 1, 2, \ldots, n \\ \text{and} \quad i = 2, 3, \ldots, n \end{array} \qquad (5\text{-}11)$$

the result is that the determinant is changed into a form with all the off-diagonal elements in the first column equal zero,

$$\det \begin{vmatrix} a_{1,1} & a_{1,2} & \cdots & a_{1,n} \\ a_{2,1} & a_{2,2} & \cdots & a_{2,n} \\ \vdots & \vdots & \ddots & \vdots \\ a_{n,1} & a_{n,2} & \cdots & a_{n,n} \end{vmatrix} = \det \begin{vmatrix} a_{1,1} & a_{1,2} & \cdots & a_{1,n} \\ 0 & a'_{2,2} & \cdots & a'_{2,n} \\ \vdots & \vdots & \ddots & \vdots \\ 0 & a'_{n,2} & \cdots & a'_{n,n} \end{vmatrix} \qquad (5\text{-}12)$$

Here $a'_{i,j}$ is given by (5-11). More generally, a transformation of the form

$$a'_{i,j} = a_{i,j} - a_{k,j} \times \frac{a_{i,k}}{a_{k,k}} \qquad \text{for} \quad j = k, k+1, \ldots, n$$
$$\text{and} \quad i = j+1, j+2, \ldots, n \qquad (5\text{-}13)$$

reduces the matrix element at location (i, k) to zero. If this is carried out for $k = 1, 2, \ldots, n$, in the same way as we have done in (5-11) for the elements $a_{i,1}$ in the first column, all the elements below the diagonal are reduced to zero without changing the value of the determinant itself. The result is an upper triangular matrix having the same value for the determinant as the original one:

$$\det \begin{vmatrix} a_{1,1} & a_{1,2} & \cdots & a_{1,n} \\ a_{2,1} & a_{2,2} & \cdots & a_{2,n} \\ \vdots & \vdots & \ddots & \vdots \\ a_{n,1} & a_{n,2} & \cdots & a_{n,n} \end{vmatrix} = \det \begin{vmatrix} a_{1,1} & a_{1,2} & a_{1,3} & \cdots & a_{1,n} \\ 0 & a'_{2,2} & a'_{2,3} & \cdots & a'_{2,n} \\ 0 & 0 & a''_{3,3} & \cdots & a''_{3,n} \\ \vdots & \vdots & \vdots & \ddots & \vdots \\ 0 & 0 & 0 & \cdots & a_{n,n}^{(n-1)} \end{vmatrix}$$

As we shall see soon, the value of the determinant may be easily evaluated in this way and the method is known as the Gaussian elimination method.

In principle, we can carry out a back substitution to eliminate all the off-diagonal elements in the upper triangle as well. This is usually done by starting from the nth column and working backward one column at a time. In column n, we can use $a_{n,n}^{(n-1)}$ to reduce all the off-diagonal elements to zero by subtracting suitable multiples of the last row from the other rows. Note that, since all the elements in the last row are zero except the diagonal one, the subtractions do not change the values of the elements in the earlier rows except those in column n. The only effect of this step in the back substitution is to reduce all the off-diagonal elements in column n to zero. Note also that the value of the diagonal element is not affected by any of the back substitution step.

We can now proceed to column $(n-1)$. Again, we have the situation that all the elements in row $(n-1)$ are zero except the diagonal one. As a result, we can subtract suitable multiples of this row from rows $(n-2)$, $(n-3)$, ..., and reduce all the remaining off-diagonal elements in column $(n-1)$ to zero. Similar to what we have done above, no other elements, including the diagonal one in row $(n-1)$, are affected by this set of transformations. It is obvious that we can continue this process until the entire determinant is transformed into a diagonal one, with only nonvanishing elements along the diagonal:

$$\det \begin{vmatrix} a_{1,1} & a_{1,2} & \cdots & a_{1,n} \\ a_{2,1} & a_{2,2} & \cdots & a_{2,n} \\ \vdots & \vdots & \ddots & \vdots \\ a_{n,1} & a_{n,2} & \cdots & a_{n,n} \end{vmatrix} = \det \begin{vmatrix} a_{1,1} & 0 & 0 & \cdots & 0 \\ 0 & a'_{2,2} & 0 & \cdots & 0 \\ 0 & 0 & a''_{3,3} & \cdots & 0 \\ \vdots & \vdots & \vdots & \ddots & \vdots \\ 0 & 0 & 0 & \cdots & a_{n,n}^{(n-1)} \end{vmatrix}$$

The value of such a determinant is simply the product of the diagonal elements,

$$\det \boldsymbol{A} = a_{1,1} a'_{2,2} a''_{3,3} \cdots a_{n,n}^{(n-1)} \tag{5-14}$$

as all the terms except the first in the sum on the right side of (5-4) vanish.

Since the values of the diagonal elements are not changed by the back substitution, the matrix elements that enter on the right side of (5-14) are exactly the same ones as those before the back substitution step. This means that we could have obtained the value of the determinant without actually carrying out the back substitution. In fact, the back substitution is unnecessary if our interest is in the value of the determinant alone.

A slight refinement of the method is necessary before using it as an algorithm for calculating the value of an arbitrary determinant. The basic operation in Gaussian elimination is to reduce the off-diagonal element at position (i,j) to zero by subtracting from it multiples of a diagonal element. The diagonal element used is usually referred to as the *pivot*, for obvious reasons. If the elimination is by columns, as shown in (5-12), the pivot is the diagonal element at position (i,i). On the other hand, if the elimination is to be carried out by rows, the pivot is the diagonal matrix element at position (j,j). This method fails if, for any reason, one of the pivots vanishes. By the same token, the numerical accuracy of the result will be poor if some of the pivots used are much smaller compared with the values of the off-diagonal elements involved in the same calculation. To prevent this from happening, it is advantageous to use as pivot the element with the largest possible absolute value. In practice, this principle is applied in the following way. Consider the situation that the determinant is reduced to the stage that all the off-diagonal matrix elements in the lower half-triangle up to row k are reduced to zero. A search is made among the $(n-k) \times (n-k)$ remaining elements to locate the one with the largest absolute value. If this element is not at the position of the next pivot, the diagonal element at position (k,k), we shall permute the rows and columns of the determinant in such a way so as to bring it to position (k,k). When the refinement of pivoting is included as a part of Gaussian elimination, the method is known as Gauss-Jordan elimination. An algorithm based on it is summarized in Box 5-1.

Box 5-1 Function DETER(A,N,NDMN)
Value of a determinant by Gauss-Jordan elimination

Argument list:
    A: Array for the determinant A.
    N: Order of the determinant.
    NDMN: Dimension of the array A in the calling program.

1. Initialize the value of the determinant to be $\mathcal{D} = 1$.
2. Let $i = 1$.
   (a) Find the element with the largest absolute value among $a_{j,k}$ for $j = i, i+1, \ldots, n$ and $k = i, i+1, \ldots, n$.
      (i) Check if the largest absolute value is zero.
      (ii) If so, return zero for the determinant.
   (b) Move this element to be the position of the diagonal element $a_{i,i}$. Interchange columns and rows of elements if necessary.
      (i) Each time two rows are interchanged, reverse the sign of $\mathcal{D}$.
      (ii) Reverse the sign of $\mathcal{D}$ again if two columns are interchanged.
   (c) Multiply the diagonal matrix element to $\mathcal{D}$.
   (d) Eliminate the rest of the elements in the ith row using (5-13).
3. Increase the value of i by 1 and repeat step 2 until $i = n$.
4. Return $\mathcal{D}$ as the value of the determinant.

Application to solve a system of linear equations. Evaluation of determinants is a key step in solving a system of linear equations. This can be seen from the following consideration. One property of a determinant is that, if all the elements in a row of a determinant are multiplied by a constant, the value of the determinant itself is increased by the same amount. That is,

$$\lambda \det \begin{vmatrix} a_{1,1} & a_{1,2} & \cdots & a_{1,n} \\ a_{2,1} & a_{2,2} & \cdots & a_{2,n} \\ \vdots & \vdots & \ddots & \vdots \\ a_{n,1} & a_{n,2} & \cdots & a_{n,n} \end{vmatrix} = \det \begin{vmatrix} \lambda a_{1,1} & a_{1,2} & \cdots & a_{1,n} \\ \lambda a_{2,1} & a_{2,2} & \cdots & a_{2,n} \\ \vdots & \vdots & \ddots & \vdots \\ \lambda a_{n,1} & a_{n,2} & \cdots & a_{n,n} \end{vmatrix} \quad (5\text{-}15)$$

Using this property, we can transform the first column of the determinant A in (5-3) into the following form:

$$x_1 \det \begin{vmatrix} a_{1,1} & a_{1,2} & \cdots & a_{1,n} \\ a_{2,1} & a_{2,2} & \cdots & a_{2,n} \\ \vdots & \vdots & \ddots & \vdots \\ a_{n,1} & a_{n,2} & \cdots & a_{n,n} \end{vmatrix}$$

$$= \det \begin{vmatrix} a_{1,1}x_1 + a_{1,2}x_2 + \cdots + a_{1,n}x_n & a_{1,2} & \cdots & a_{1,n} \\ a_{2,1}x_1 + a_{2,2}x_2 + \cdots + a_{2,n}x_n & a_{2,2} & \cdots & a_{2,n} \\ \vdots & \vdots & \ddots & \vdots \\ a_{n,1}x_1 + a_{n,2}x_2 + \cdots + a_{n,n}x_n & a_{n,2} & \cdots & a_{n,n} \end{vmatrix}$$

This result is arrived at by replacing λ in (5-15) with x_1. In addition, we have made use of the property of (5-10) and added to the elements of the first column the sum of products of x_j times the elements of the column j, for $j = 2, 3, \ldots, n$. Comparing the result with (5-3), we see that the first column on the right side is equal to y_i for $i = 1, 2, \ldots, n$. This gives us the equality

$$x_1 \det \begin{vmatrix} a_{1,1} & a_{1,2} & \cdots & a_{1,n} \\ a_{2,1} & a_{2,2} & \cdots & a_{2,n} \\ \vdots & \vdots & \ddots & \vdots \\ a_{n,1} & a_{n,2} & \cdots & a_{n,n} \end{vmatrix} = \det \begin{vmatrix} y_1 & a_{1,2} & \cdots & a_{1,n} \\ y_2 & a_{2,2} & \cdots & a_{2,n} \\ \vdots & \vdots & \ddots & \vdots \\ y_n & a_{n,2} & \cdots & a_{n,n} \end{vmatrix}$$

We can turn this relation into one that gives us the value of x_1 in terms of the ratio of two determinants:

$$x_1 = \frac{\det \begin{vmatrix} y_1 & a_{1,2} & \cdots & a_{1,n} \\ y_2 & a_{2,2} & \cdots & a_{2,n} \\ \vdots & \vdots & \ddots & \vdots \\ y_n & a_{n,2} & \cdots & a_{n,n} \end{vmatrix}}{\det \begin{vmatrix} a_{1,1} & a_{1,2} & \cdots & a_{1,n} \\ a_{2,1} & a_{2,2} & \cdots & a_{2,n} \\ \vdots & \vdots & \ddots & \vdots \\ a_{n,1} & a_{n,2} & \cdots & a_{n,n} \end{vmatrix}}$$

In general, we find that

$$x_k = \frac{\det \begin{vmatrix} a_{1,1} & a_{1,2} & \cdots & a_{1,k-1} & y_1 & a_{1,k+1} & \cdots & a_{1,n} \\ a_{2,1} & a_{2,2} & \cdots & a_{2,k-1} & y_2 & a_{2,k+1} & \cdots & a_{2,n} \\ \vdots & \vdots & \ddots & \vdots & \vdots & \vdots & \ddots & \vdots \\ a_{n,1} & a_{n,2} & \cdots & a_{n,k-1} & y_n & a_{n,k+1} & \cdots & a_{n,n} \end{vmatrix}}{\det \begin{vmatrix} a_{1,1} & a_{1,2} & \cdots & a_{1,n} \\ a_{2,1} & a_{2,2} & \cdots & a_{2,n} \\ \vdots & \vdots & \ddots & \vdots \\ a_{n,1} & a_{n,2} & \cdots & a_{n,n} \end{vmatrix}} \tag{5-16}$$

where the denominator is the determinant $\boldsymbol{A}$ itself and the numerator is that of $\boldsymbol{A}$ with the elements in column k replaced by those of $\boldsymbol{Y}$.

Solution of a system of linear equations by elimination. Instead of evaluating determinants, it is also possible to solve a system of linear equations directly by Gaussian elimination. The fact that we can add one row of a determinant to a multiple of another one without changing the value is equivalent to the statement that we can add a multiple of one member of a system of linear equations to another member without affecting the solution. This, in turn, implies that it is possible to obtain the solution for a set of linear equations directly if we carry out the operation of adding columns of determinants on both sides of the equation $AX = Y$ at the same time.

It is more instructive to illustrate the method by a simple example consisting of a system of three linear equations. Consider the case of

$$x_1 + 2x_2 + 3x_3 = 14 \qquad (5\text{-}17a)$$

$$3x_1 + x_2 + 2x_3 = 11 \qquad (5\text{-}17b)$$

$$2x_1 + 3x_2 + x_3 = 11 \qquad (5\text{-}17c)$$

If we subtract three times (5-17a) from (5-17b) and two times (5-17a) from (5-17c), we can eliminate x_1 from the second and third member of the system:

$$x_1 + 2x_2 + 3x_3 = 14 \qquad (5\text{-}18a)$$

$$-5x_2 - 7x_3 = -31 \qquad (5\text{-}18b)$$

$$- x_2 - 5x_3 = -17 \qquad (5\text{-}18c)$$

Again, by subtracting one-fifth times (5-18b) from (5-18c), we can eliminate x_2 from the third member:

$$x_1 + 2x_2 + 3x_3 = 14 \qquad (5\text{-}19a)$$

$$-5x_2 - 7x_3 = -31 \qquad (5\text{-}19b)$$

$$-\frac{18}{5} x_3 = -\frac{54}{5} \qquad (5\text{-}19c)$$

If we write this result in matrix notation, the left side is a product of an upper triangular matrix with a column matrix. From (5-19c), we obtain the solution $x_3 = 3$. We can now carry out the equivalent of back substitutions to obtain the solutions for x_2 and x_3. On inserting the value of x_3 into (5-19b), we have the result $x_2 = 2$. By putting both of these values into (5-19a), we get $x_1 = 1$.

This method is commonly used to solve problems such as those represented by our simple direct-current circuit example in (5-1). A few improvements may be included to make it more suitable for programming a computer to carry out the task. The first is that we can normalize each equation so that the coefficients along the diagonal are unity. This makes it easier to carry out the next elimination. Thus, instead of (5-18), we have

$$x_1 + 2x_2 + 3x_3 = 14 \qquad (5\text{-}20a)$$

$$x_2 + \frac{7}{5} x_3 = \frac{31}{5} \qquad (5\text{-}20b)$$

$$\frac{1}{5} x_2 + x_3 = \frac{17}{5} \qquad (5\text{-}20c)$$

We can now simply subtract one-fifth of (5-20b) from (5-20c) to eliminate x_2 from the third member and normalize the coefficient for x_3 to unity:

$$x_1 + 2x_2 + 3x_3 = 14 \qquad (5\text{-}21a)$$

$$x_2 + \frac{7}{5}x_3 = \frac{31}{5} \qquad (5\text{-}21b)$$

$$x_3 = 3 \qquad (5\text{-}21c)$$

If, for some reason, we wish to keep the value of the determinant on the left side, we must record the normalization constant used at each stage, since the value of a determinant is changed by the amount by which we multiply all the elements of a row.

A second improvement is to eliminate the need of back substitution itself. This can be achieved by a little extra work at each step in the elimination process itself. Instead of only working with the remainder of the equations, we can carry out the elimination of the next unknown quantity from all the equations at the same time. That is, in addition to taking the difference of (5-20c) from (5-20b) in going from (5-20) to (5-21), we also subtract two times (5-20b) from (5-20a). This gives us the result

$$x_1 \qquad + \frac{1}{5}x_3 = \frac{8}{5} \qquad (5\text{-}22a)$$

$$x_2 + \frac{7}{5}x_3 = \frac{31}{5} \qquad (5\text{-}22b)$$

$$x_3 = 3 \qquad (5\text{-}22c)$$

We can now eliminate x_3 from (5-22b) and (5-22a) and obtain the solutions for both x_1 and x_2.

This simple example shows that it is possible to find the solution of a set of n linear equation, such as that represented by (5-3), by taking appropriate linear combinations of two members and eliminating one of the independent variables between them. Several such steps are necessary to reach the stage where only a single independent variable is left in each equation. This is essentially what we have done in finding the value of a determinant by transforming it into a diagonal form. The only difference here is that, in addition to (5-13) for the elements of A, we must apply the transformation to the elements of Y as well. The algorithm is a simple one. However, we should not forget the need for pivoting. To avoid having diagonal elements that are zero or have very small absolute values, it is essential that we search for the element with the largest absolute value in the remaining equations at every stage of the calculation and put it into the diagonal position by permutating rows and columns, similar to what we did earlier in calculating the

> **Box 5-2 Subroutine LIN_EQN(A,Y,N,DET,NDMN)**
> **Solve a system of linear equation $\sum_{j=1}^{n} a_{j,i} x_j = y_i$.**
> Using Gauss-Jordan elimination with pivoting
>
> Argument list:
>   A: Two-dimensional array for the determinant A.
>   Y: Array for the column matrix Y.
>   N: Number of linear equations.
>   DET: Value of the determinant on output.
>   NDMN: Dimension of the arrays A and Y in the calling program.
>
> Initialization:
>   (a) Start with the value of the determinant DET=1.
>   (b) Zero an auxiliary array to keep track of the rows transformed.
>
> 1. Carry out steps 2 to 7 for i equal to 1 to n.
> 2. Find the next pivot:
>   (a) Locate the element with the largest absolute value among rows not yet transformed.
>   (b) If the largest absolute value is zero, return DET=0 to signal a singular case.
> 3. Multiply DET by the value of the pivot found.
> 4. If necessary, move the pivoting element to the diagonal position:
>   (a) Interchange rows of A and Y to put the element at the diagonal position of the column it is found in.
>   (b) Change the sign of DET.
>   (c) Record the interchange using the auxiliary array.
> 5. Divide all elements of A in the row by the pivot so that the diagonal element is now unity. Divide the corresponding element in Y also by the pivot.
> 6. Mark the row as transformed in the auxiliary array.
> 7. Apply the transformation of (5-13) to the rest of the matrix A and Y.
> 8. Return D as the value of the determinant and the array Y as the solution for $\{x_i\}$.

value of determinants. This is the Gauss-Jordan elimination method for solving a system of linear equations and is implemented in the algorithm outlined in Box 5-2.

To check our results, we can substitute the solution obtained for $x_1, x_2, \ldots, x_n$ into the original linear equations and see if the sum of products $\sum_j a_{i,j} x_j$ is equal to y_i within the accuracies required. To do this in a computer program, it is necessary to save the matrix elements of A in a separate array, as the one that goes into the calculation of the solution is usually destroyed in the process of Gauss-Jordan elimination. For the purpose of testing the computer program, it is possible to make use of a random number generator to produce all the input matrix elements $a_{i,j}$ and the components of the vector Y. This approach has the advantage that a large number of test cases can be conducted with relatively little effort.

§5-2 Matrix inversion and *LU*-decomposition

Instead of using determinants, we can also solve a system of linear equations by making use of matrix inversion. If A is a matrix, its inverse A^{-1} is defined by the equation

$$A^{-1}A = AA^{-1} = 1 \qquad (5\text{-}23)$$

Here 1 is the unit matrix. As given in Table 5-1, it is a matrix with all the diagonal elements equal to 1 and all the off-diagonal elements 0. In terms of matrix elements, we can express (5-23) in the following way. Let

$$B = A^{-1}$$

In terms of the matrix elements of B, (5-23) is the same as

$$\sum_k b_{i,k} a_{k,j} = \sum_k a_{i,k} b_{k,j} = \delta_{i,j} \qquad (5\text{-}24)$$

For simplicity, we shall consider only square matrices with the number of rows equal to the number of columns, although many of the considerations below apply also to rectangular matrices with the number of rows not equal to the number of columns.

Using the inverse matrix, we can solve the system of linear equations of (5-3) by multiplying both sides with A^{-1}:

$$A^{-1}AX = A^{-1}Y \qquad (5\text{-}25)$$

From the definition of the inverse given by (5-23), we obtain X in terms of Y:

$$X = A^{-1}Y \qquad (5\text{-}26)$$

In terms of matrix elements, the same relation appears in the form

$$x_i = \sum_j b_{i,j} y_j$$

where $b_{i,j}$ are the elements of A^{-1}. If all the elements $b_{i,j}$ are known, we have a solution of the unknown quantity x_i in terms of the known values of the elements of the column matrix Y.

In principle, we can find the inverse of a matrix by making use of the definition given by (5-24). For a given value of the subscript j, the second form of the definition may be regarded as a system of linear equations for the n unknown quantities $b_{1,j}, b_{2,j}, \ldots, b_{n,j}$:

$$\sum_k a_{i,k} b_{k,j} = \delta_{i,j}$$

This is identical to (5-2) with $b_{k,j}$ taking on the role of the unknown x_i and $\delta_{i,j}$ occupying the position of the known input y_i. On solving this set of equations, we obtain the values of all n elements in column j of $\boldsymbol{A}^{-1}$. For the complete inverse matrix, we must also carry out the same operation for all n columns, with $j = 1, 2, \ldots, n$. As a result, we have to solve a total of n^2 equations to find the n^2 unknown elements $b_{i,j}$ in this approach.

We shall regard this only as a statement that the problem has a solution. The actual calculations involved are much simpler. In practice, the solution requires only a single application of a slightly modified form of the method of Gauss-Jordan elimination for a system of n linear equation, as we shall see next.

Gauss-Jordan elimination. In §5-1, we have used the method of Gauss-Jordan elimination to evaluate the determinant of an arbitrary matrix $\boldsymbol{A}$ by reducing it to a diagonal form. For our discussion here, the operations may be expressed symbolically in terms of an operator,

$$\hat{\mathcal{O}} \boldsymbol{A} = \det |\boldsymbol{A}|\, \mathbf{1} \tag{5-27}$$

where the operator $\hat{\mathcal{O}}$ represents all the steps required to reduce $\boldsymbol{A}$ to a unit matrix multiplied by a constant, given by the value of the determinant of $\boldsymbol{A}$. What happens if the same operations are carried out on a unit matrix? The answer is

$$\hat{\mathcal{O}} \mathbf{1} = \boldsymbol{A}^{-1} \det |\boldsymbol{A}| \tag{5-28}$$

The proof of this relation may be made in the following way. Formally, we can write

$$\boldsymbol{A}^{-1} = \frac{1}{\boldsymbol{A}}$$

If, on the right side of the equation, we apply the same set of operations to both the matrix in the numerator as well as that in the denominator, the ratio is unchanged. In other words,

$$\boldsymbol{A}^{-1} = \frac{\hat{\mathcal{O}} \mathbf{1}}{\hat{\mathcal{O}} \boldsymbol{A}}$$

The result of (5-28) is obtained, since the denominator is transformed into the product of the determinant of $\boldsymbol{A}$ and a unit matrix, as we can see from (5-27).

From a practical point of view, the inverse of a matrix $\boldsymbol{A}$ may be found through the same steps as those used above to reduce it to a unit matrix. To store the steps, we can apply the operations to a unit matrix at the same time. Except for a normalization constant, given by the value of the determinant, $\boldsymbol{A}$ is transformed at the end into a unit matrix, while the unit matrix is transformed into the inverse of $\boldsymbol{A}$. This is essentially what we have done in the previous paragraph in proving the identity of (5-27).

An example using a three-dimensional matrix will serve to illustrate this point. Let $\boldsymbol{U}$ be a unit matrix,

$$\boldsymbol{U} = \mathbf{1}$$

At the start we have matrices

$$A = \begin{pmatrix} a_{1,1} & a_{1,2} & a_{1,3} \\ a_{2,1} & a_{2,2} & a_{2,3} \\ a_{3,1} & a_{3,2} & a_{3,3} \end{pmatrix} \qquad U = \begin{pmatrix} 1 & 0 & 0 \\ 0 & 1 & 0 \\ 0 & 0 & 1 \end{pmatrix} \qquad (5\text{-}29)$$

Using steps similar to those carried out in (5-11), we can reduce the second and third elements of the first row of A to zero. Upon normalizing the diagonal element in the first row to unity, these two matrices are changed to

$$B = \begin{pmatrix} 1 & 0 & 0 \\ \frac{a_{2,1}}{a_{1,1}} & a_{2,2} - a_{2,1}\frac{a_{1,2}}{a_{1,1}} & a_{2,3} - a_{2,1}\frac{a_{1,3}}{a_{1,1}} \\ \frac{a_{3,1}}{a_{1,1}} & a_{3,2} - a_{3,1}\frac{a_{1,2}}{a_{1,1}} & a_{3,3} - a_{3,1}\frac{a_{1,3}}{a_{1,1}} \end{pmatrix} \qquad V = \begin{pmatrix} \frac{1}{a_{1,1}} & -\frac{a_{1,2}}{a_{1,1}} & -\frac{a_{1,3}}{a_{1,1}} \\ 0 & 1 & 0 \\ 0 & 0 & 1 \end{pmatrix}$$

$$(5\text{-}30)$$

Next, we can reduce the first and third elements of the second row of B to zero. This gives us the results

$$C = \begin{pmatrix} 1 & 0 & 0 \\ 0 & 1 & 0 \\ b_{3,1} - b_{3,2}\frac{b_{2,1}}{b_{2,2}} & \frac{b_{2,3}}{b_{2,2}} & b_{3,3} - b_{3,2}\frac{b_{2,3}}{b_{2,2}} \end{pmatrix}$$

$$W = \begin{pmatrix} v_{1,1} - v_{1,2}\frac{b_{2,1}}{b_{2,2}} & \frac{v_{1,2}}{b_{2,2}} & v_{1,3} - v_{1,2}\frac{b_{2,3}}{b_{2,2}} \\ -\frac{v_{2,1}}{b_{2,2}} & \frac{1}{b_{2,2}} & -\frac{b_{2,3}}{b_{2,2}} \\ 0 & 0 & 1 \end{pmatrix}$$

where, to simplify the notation, we have used the matrix elements of B and V of (5-30) instead of those of the original matrix A.

The final step in reducing the left side to a unit matrix requires the first and second elements in the last row of C to be transformed to zeros and the third element normalized to unity. In terms of the matrix elements of C and W, we have the result

$$D = \begin{pmatrix} 1 & 0 & 0 \\ 0 & 1 & 0 \\ 0 & 0 & 1 \end{pmatrix} \qquad X = \begin{pmatrix} w_{1,1} - w_{1,3}\frac{c_{3,1}}{c_{3,3}} & w_{1,2} - w_{1,3}\frac{c_{3,2}}{c_{3,3}} & \frac{w_{1,3}}{c_{3,3}} \\ w_{2,1} - w_{2,3}\frac{c_{3,1}}{c_{3,3}} & w_{2,2} - w_{2,3}\frac{c_{3,2}}{c_{3,3}} & \frac{w_{2,3}}{c_{3,3}} \\ -\frac{c_{3,1}}{c_{3,3}} & -\frac{c_{3,2}}{c_{3,3}} & \frac{1}{c_{3,3}} \end{pmatrix}$$

Since the original matrix is now transformed to D, a unit matrix, the inverse of A is given by the matrix X. In terms of the matrix elements of A, the results may be expressed as

$$A^{-1} = \frac{1}{\det|A|} \begin{pmatrix} a_{2,2}a_{3,3} - a_{2,3}a_{3,2} & a_{1,3}a_{3,2} - a_{1,2}a_{3,3} & a_{1,2}a_{2,3} - a_{1,3}a_{2,2} \\ a_{2,3}a_{3,1} - a_{2,1}a_{3,3} & a_{1,1}a_{3,3} - a_{1,3}a_{3,1} & a_{1,3}a_{2,1} - a_{1,1}a_{2,3} \\ a_{2,1}a_{3,2} - a_{2,2}a_{3,1} & a_{1,2}a_{3,1} - a_{1,1}a_{3,2} & a_{1,1}a_{2,2} - a_{1,2}a_{2,1} \end{pmatrix}$$

$$(5\text{-}31)$$

> **Box 5-3 Subroutine MAT_INV(A,N,DET,NDMN)**
> **Matrix inversion by Gauss-Jordan elimination**
>
> Argument list:
>   A: Input matrix A, stores the inverse of A on output.
>   N: Input the dimension of the matrix.
>   DET: Output value of the determinant.
>   NDMN: Dimension of the array A in the calling program.
>
> Initialization:
>   (a) Set up two auxiliary arrays to record row and column numbers in the original matrix.
>   (b) Let the value of the determinant be DET $= 1$.
>
> 1. Set up the loop to reduce each row starting with $k = 1$.
> 2. Find the pivot by locating the largest element remaining:
>   (a) If the next pivot is zero, the matrix is singular. Terminate the inversion.
>   (b) Otherwise:
>   (i) Store the positions of the row and column of the pivot in the auxiliary arrays.
>   (ii) Update the value of the determinant by multiplying DET with the pivot.
>   (iii) Interchange rows and columns to put the pivot at the diagonal position k.
>   (iv) Reverse the sign of DET for each interchange of rows and columns.
> 3. Change the elements in column k to those of the inverse.
> 4. Apply the transformation to the rest of the matrix except row and column k.
> 5. Normalize the elements of the row k.
> 6. Repeat steps 2 to 5 for the next k value until $k = n$.
> 7. Restore the inverse matrix to the original order using the auxiliary arrays.
> 8. Return the value of DET and the inverse matrix.

We can check that this is, indeed, the inverse matrix by multiplying it with A and finding that a unit matrix emerges.

Two refinements are required for any practical application of the method described above. The first is in reducing the storage required for the elements. The method, as outlined above, requires two $(n \times n)$ arrays to store matrices, for example, A and U of (5-29). However, in each step, the number of newly generated elements of U, as well as their locations in the matrix, corresponds exactly to the number and locations of elements in A reduced to zero. Furthermore, there is really no need to store the unit matrix at the beginning and to have the unit matrix at the end. As a result, the two matrices at all the intermediate stages of the calculations can occupy the same memory locations in the computer. In other words, it is possible to make both matrices coexist in the same array and only a single one needs to be allocated. This complicates the programming somewhat, but is worth doing especially for large matrices.

The second refinement is to apply pivoting, as we did in the previous section. To incorporate this improvement, it is necessary to interchange two columns and two rows of the matrix whenever the element we wish to use as the pivot does not fall on the diagonal. Since interchanging all the elements in two rows or columns is a time-consuming process, methods have been designed to avoid the actual operations, as we did earlier in Box 5-2. However, we shall not carry it out here. We are already using the same two-dimensional array to store two different matrices at the same time — it would make the program even more difficult to understand if we introduce any further complications. For this reason, explicit row and column interchanges are used in the algorithm of Box 5-3, if necessary, at every step of reduction. To restore the inverse matrix to the original order, it is necessary to carry out the same set of interchanges in reverse at the end.

The inverse matrix produced may be checked by multiplying it with the original matrix and seeing if the conditions of (5-24) are satisfied. Again, to test the computer program, we can use a random number generator to produce all the elements of the test matrices. Since the original matrix A is replaced by its inverse at the end of the calculation, a copy of A must be saved for the purpose of checking.

Lower and upper triangular matrices. We have seen in the previous section that, in evaluating a determinant by Gauss-Jordan elimination, the back substitution step is relative easy. The simplification comes from the fact that the matrix is reduced to an upper triangular form, one with zeros as the elements below the diagonal. The same is true if the matrix is reduced to a lower triangular form, with all the elements above the diagonal equal zero. We shall now try to take advantage of the properties of triangular matrices in solving a system of linear equations of the form

$$AX = Y$$

given earlier in (5-3). To achieve this, we shall write the matrix A as a product of a lower triangular matrix L and an upper triangular matrix U,

$$A = LU \tag{5-32}$$

where

$$L = \begin{pmatrix} \ell_{1,1} & 0 & 0 & \cdots & 0 \\ \ell_{2,1} & \ell_{2,2} & 0 & \cdots & 0 \\ \ell_{3,1} & \ell_{3,2} & \ell_{3,3} & \cdots & 0 \\ \vdots & \vdots & \vdots & \ddots & \vdots \\ \ell_{n,1} & \ell_{n,2} & \ell_{n,3} & \cdots & \ell_{n,n} \end{pmatrix} \qquad U = \begin{pmatrix} u_{1,1} & u_{1,2} & u_{1,3} & \cdots & u_{1,n} \\ 0 & u_{2,2} & u_{2,3} & \cdots & u_{2,n} \\ 0 & 0 & u_{1,3} & \cdots & u_{3,n} \\ \vdots & \vdots & \vdots & \ddots & \vdots \\ 0 & 0 & 0 & \cdots & u_{n,n} \end{pmatrix}$$

$$\tag{5-33}$$

We shall see next why such an *LU-decomposition* can help us to solve a system of linear equations represented by (5-3).

In terms of matrices L and U, the relation expressed by (5-3) may be written in the form

$$AX = L(UX) = Y \tag{5-34}$$

The solution to such an equation may be obtained in two stages. First, (UX) may be written as a column matrix Z of dimension n, and its elements are taken to be n unknown quantities, $z_1, z_2, \ldots, z_n$. We can now rewrite (5-34) as

$$LZ = Y \qquad (5\text{-}35)$$

This may be regarded as a system of n linear equations from which we may solve for $z_1, z_2, \ldots, z_n$. Once this is done, we can move to the second stage of the calculation and make use of Z as the right side of the system of linear equations:

$$UX = Z \qquad (5\text{-}36)$$

The solution to this set of equations gives us the elements of the unknown column matrix X, our original aim in solving (5-3).

Let us defer till later the question of how to reduce A into a product of L and U. Here, we shall consider first the problem of solving (5-35) and (5-36). Since each of the two stages of the solution in the previous paragraph involves only a triangular matrix, the calculations are fairly simple. For the first stage, (5-35) may be written in terms of matrix elements in the form

$$LZ = \begin{pmatrix} \ell_{1,1} & 0 & 0 & \cdots & 0 \\ \ell_{2,1} & \ell_{2,2} & 0 & \cdots & 0 \\ \ell_{3,1} & \ell_{3,2} & \ell_{3,3} & \cdots & 0 \\ \vdots & \vdots & \vdots & \ddots & \vdots \\ \ell_{n,1} & \ell_{n,2} & \ell_{n,3} & \cdots & \ell_{n,n} \end{pmatrix} \begin{pmatrix} z_1 \\ z_2 \\ z_3 \\ \vdots \\ z_n \end{pmatrix} = \begin{pmatrix} y_1 \\ y_2 \\ y_3 \\ \vdots \\ y_n \end{pmatrix} \qquad (5\text{-}37)$$

In terms of the known quantities y_i, we obtain from the first row of the matrices the result

$$z_1 = \frac{y_1}{\ell_{1,1}}$$

The value of z_1 obtained may be used to substitute forward to find z_2 in terms of y_2 and z_1

$$z_2 = \frac{1}{\ell_{2,2}}(y_2 - \ell_{2,1} z_1)$$

This process can now be continued to the next row, and so on. On carrying out the forward substitution to the kth row, we obtain the general result

$$z_k = \frac{1}{\ell_{k,k}}\left(y_k - \sum_{j=1}^{k-1} \ell_{k,j} z_j\right) \qquad (5\text{-}38)$$

Once the values of all the elements of Z are found, they can be used in (5-36),

$$UX = \begin{pmatrix} u_{1,1} & u_{1,2} & u_{1,3} & \cdots & u_{1,n} \\ 0 & u_{2,2} & u_{2,3} & \cdots & u_{2,n} \\ 0 & 0 & u_{3,3} & \cdots & u_{3,n} \\ \vdots & \vdots & \vdots & \ddots & \vdots \\ 0 & 0 & 0 & \cdots & u_{n,n} \end{pmatrix} \begin{pmatrix} x_1 \\ x_2 \\ x_3 \\ \vdots \\ x_n \end{pmatrix} = \begin{pmatrix} z_1 \\ z_2 \\ z_3 \\ \vdots \\ z_n \end{pmatrix} \qquad (5\text{-}39)$$

to find the values of the elements of X by back substitution. The detailed calculation is similar to the forward substitution given above, and we shall leave it as an exercise to derive the general formula.

LU-decomposition. From the discussion in the two previous paragraphs, it is clear that, once a matrix is decomposed into the product of a lower and an upper triangular matrix, it is relatively easy to find the solution of X in terms of Y. The only question left is the way to carry out the LU-decomposition. The method is basically provided by (5-32). In terms of the matrix elements of L, U, and A, we have the relation

$$\begin{pmatrix} \ell_{1,1} & 0 & 0 & \cdots & 0 \\ \ell_{2,1} & \ell_{2,2} & 0 & \cdots & 0 \\ \ell_{3,1} & \ell_{3,2} & \ell_{3,3} & \cdots & 0 \\ \vdots & \vdots & \vdots & \ddots & \vdots \\ \ell_{n,1} & \ell_{n,2} & \ell_{n,3} & \cdots & \ell_{n,n} \end{pmatrix} \begin{pmatrix} u_{1,1} & u_{1,2} & u_{1,3} & \cdots & u_{1,n} \\ 0 & u_{2,2} & u_{2,3} & \cdots & u_{2,n} \\ 0 & 0 & u_{3,3} & \cdots & u_{3,n} \\ \vdots & \vdots & \vdots & \ddots & \vdots \\ 0 & 0 & 0 & \cdots & u_{n,n} \end{pmatrix}$$

$$= \begin{pmatrix} a_{1,1} & a_{1,2} & a_{1,3} & \cdots & a_{1,n} \\ a_{2,1} & a_{2,2} & a_{2,3} & \cdots & a_{2,n} \\ a_{3,1} & a_{3,2} & a_{3,3} & \cdots & a_{3,n} \\ \vdots & \vdots & \vdots & \ddots & \vdots \\ a_{n,1} & a_{n,2} & a_{n,3} & \cdots & a_{n,n} \end{pmatrix} \qquad (5\text{-}40)$$

We may regard this matrix equation as a system of n^2 equations. However, the number of unknowns is $(n^2 + n)$, since L and U each has $n(n+1)/2$ nonvanishing elements. This means that we have the freedom of assigning values to n matrix elements among those of L and U.

We shall make use of the freedom in the assignment to simplify the LU-decomposition as much as possible. A common choice is to give the value 1 to all the diagonal elements of L. That is, for $i = 1, 2, \ldots, n$, we have the assignment

$$\ell_{i,i} = 1 \qquad (5\text{-}41)$$

Once this is done, (5-40) may be used to calculate all the other $n(n-1)/2$ elements $\ell_{i,j}$, for $j = 1$ to $(i-1)$ and $i = 1$ to n, as well as the $n(n+1)/2$ elements $u_{i,j}$, for $i = 1$ to n and $j = i$ to n. In principle, we can obtain these matrix elements using one of the methods of solving a system of linear equations we discussed earlier. However, such an approach negates the original purpose of using an LU-decomposition to simplify the solution for a system of linear equations. In fact, we shall soon see that, with the proper method, the LU-decomposition is much faster on a computer than any of the methods described earlier.

Because of the particular shapes of the matrices L and U, solutions of $\ell_{i,j}$ and $u_{i,j}$ in terms of $a_{i,j}$ can be worked out in a simple way. The equality expressed by (5-40) may be stated in terms of individual matrix elements as

$$\sum_m \ell_{i,m} u_{m,j} = a_{i,j} \qquad (5\text{-}42)$$

For $i = j = 1$, all $\ell_{i,m} = 0$ except for $m = 1$. As a result, there is only one term in the sum,
$$\ell_{1,1} u_{1,1} = a_{1,1}$$
Since $\ell_{1,1}$ is chosen to be 1 in (5-41), we obtain at once
$$u_{1,1} = a_{1,1}$$
In fact, this relation applies to all $u_{i,j}$ for $i = 1$ and $j = 1, 2, \ldots, n$. That is, the matrix elements in the first row of U are
$$u_{1,j} = a_{1,j} \tag{5-43}$$
for $j = 1, 2, \ldots, n$. This gives us the solution for all the elements in the first row of U.

At this moment we shall only make use of the value of $u_{1,1}$. For the convenience of pivoting, to which we shall return later, the calculations will be carried out one complete column at a time for all the elements of both U and L. Let us now consider the elements in the first column. There is only one nonvanishing element for U, and this is given by (5-43). For $j = 1$, (5-42) reduces to the form
$$\ell_{i,1} u_{1,1} = a_{i,1} \qquad \text{for} \qquad i = 2, 3, \ldots, n$$
This gives us a solution for all the elements of the first column of L
$$\ell_{i,1} = \frac{1}{u_{1,1}} a_{i,1} \qquad \text{for} \qquad i = 2, 3, \ldots, n \tag{5-44}$$
Note that we have no further use for the elements $a_{j,1}$ for $j = 1$ to n in the first column of the matrix A. As a result, we can make use of their locations in the computer memory to store all the elements in the first columns of both U and L. The only exception is the diagonal element of L. Since it is chosen to be unity, there is no need to have its value kept in the computer. This leaves exactly n nonvanishing elements to be kept, one for U and $(n-1)$ for L. As we shall see later, this is useful in minimizing the storage requirement in carrying out the LU-decomposition.

Once all the elements in the first column are known, we can proceed to solve for those in the second column. First, we shall apply (5-43) to calculate the elements in the rest of the first row of U. Once this is done, we can apply (5-42) for $i = 2$ to produce the equation for all the elements in the second row of U:
$$u_{2,j} = a_{2,j} - \ell_{2,1} u_{1,j} \qquad \text{for} \qquad j = 2, 3, \ldots, n \tag{5-45}$$
Similarly, by applying (5-42) for $j = 2$, we obtain the values for the elements of the second column of L:
$$\ell_{i,2} = \frac{1}{u_{2,2}} (a_{i,2} - \ell_{i,1} u_{1,2}) \qquad \text{for} \qquad i = 3, 4, \ldots, n \tag{5-46}$$

With these results, the values for all the elements in the second column of both L and U are now available.

Let us work out explicitly the equation for the elements of one more column before giving the general result. We have already in our possession (5-43) and (5-45) for the elements in the first two rows of U, and we shall make use of them to obtain the first two elements in column 3. For the elements of U in the third row, we can again invoke (5-42) with $i = 3$. This gives us the result

$$u_{3,j} = a_{3,j} - \left(\ell_{3,1} u_{1,j} + \ell_{3,2} u_{2,j}\right) \tag{5-47}$$

for $j = 3, 4, \ldots, n$. For now, we shall use this equation only to calculate $u_{3,3}$. The elements in the third column of L may be obtained from (5-42) with $j = 3$. That is,

$$\ell_{i,3} = \frac{1}{u_{3,3}} \left\{ a_{i,3} - \left(\ell_{i,1} u_{1,3} + \ell_{i,2} u_{2,3}\right) \right\}$$

for $i = 4, 5, \ldots, n$. Besides $a_{i,3}$, the input required for $\ell_{i,3}$ are $\ell_{i,1}$, already found from (4-44), $\ell_{i,2}$ from (5-46), $u_{1,3}$ from (5-43), and $u_{2,3}$ from (5-45).

The general result is now clear. We shall maintain the procedure of completing the calculation for all the elements one column at a time for both U and L, starting from the first column and proceeding forward one column at a time. By extending the arguments leading to (5-43) for $i = 1$, (5-45) for $i = 2$, and (5-47) for $i = 3$, we can now substitute α for i and k for j in (5-42) and obtain the general result for the elements of U in column k:

$$u_{\alpha,k} = a_{\alpha,k} - \sum_{m=1}^{\alpha-1} \ell_{\alpha,m} u_{m,k} \tag{5-48}$$

for $\alpha = 1, 2, \ldots, k$. For this equation to agree with (5-43) for $k = 1$, the sum must vanish if the upper limit is less than 1.

To calculate $u_{i,k}$, we need as input the values of $\ell_{i,m}$ for $m < k$ (since $m < i$ and $i \leq k$) and $u_{m,k}$ for $m < i$, in addition to that of $a_{i,k}$. All the elements $\ell_{i,j}$ for $j < k$ are already available from the calculations of the previous rows. As for the necessary input from the elements of U, we note that none is needed to obtain $u_{1,k}$. Once $u_{1,k}$ is available, we have enough information to calculate $u_{2,k}$. This, in turn, allows us to obtain $u_{3,k}$, and so on. In this way we always have all the necessary input if the calculations are carried out in the correct order. It is also convenient to complete the calculations for all the elements in a given column first for both U and L before proceeding to the next column. For each column, the calculation starts with first element of U and proceeds down the column one element at a time until we come to the end at the diagonal element.

Once all the elements of U in column k are known, we can go to those of L in the same column. Again, by substituting α for i and k for j in (5-42), we obtain the equation for the elements of L in column k:

$$\ell_{\alpha,k} = \frac{1}{u_{k,k}} \left(a_{\alpha,k} - \sum_{m=1}^{k-1} \ell_{\alpha,m} u_{m,k} \right) \quad \text{for} \quad \alpha = 1, 2, \ldots, (k-1) \quad (5\text{-}49)$$

As in (5-48), it is necessary that the sum on the right side vanish if the upper limit is less than 1 in order for the expression to agree with (5-44). Since $m < k$, all the elements of L required as input to the calculation are those in the earlier columns. Similarly, the required elements from U are in the present column and they have been calculated already, as we saw in the previous paragraph.

So far, we have assumed that it is possible to make an LU-decomposition of a matrix A as long as it is not singular. In fact, the existence of LU-decomposition is more general than that. A proof of the existence theorem for LU-decomposition can be found in Dahlquist and Björck (*Numerical Methods*. English translation by N. Anderson, Prentice Hall, New Jersey, 1974, page 154).

Pivoting. It is important to note here that each step of the calculation requires the result of the previous step. For example, in (5-48), the element $u_{\alpha,k}$ requires all the values of $u_{m,k}$ for $m < \alpha$, as well as the elements of L in the earlier columns. Similarly, for calculating $\ell_{\alpha,k}$ in (5-49), we need the values of all the elements of U in the same column, as well as the elements of L in the previous columns. In calculations of this type, the stability of the solution can be a problem, that is, whether the errors in each step accumulate in such a way that hardly any significant figures are left at the end. Pivoting is one way to improve the accuracy and to stabilize the solution.

In applying the Gauss-Jordan elimination earlier to reduce a determinant to diagonal form, we have selected the element with the largest absolute value in the remaining part of the determinant as the pivot. This was an efficient approach since the pivot appears in the denominator as, for example, in (5-13). In an LU-decomposition, we have an analogous but not identical situation. In (5-49), a diagonal element of U appears in the denominator. As a result, it may be tempting to choose the largest element of U in a column or row as the pivot. Before we decide to do this, two further considerations must be taken into account.

The first is that the value of a particular element of U can be increased or decreased by multiplying a constant factor to the whole row without changing the solution of the system of linear equations. As a result, the criterion for selecting a pivoting element depends not only on the magnitude of the element itself but also on its size relative to the other elements. One way to incorporate the relative size into the selection criterion is to choose the largest element in each row as the weighting factor. We shall see soon how to put this point into practice.

> **Box 5-4   Subroutine** LU_DCMP(A,N,DET,L_ROW,NDMN)
> **LU decomposition of matrix A**
>
> Argument list:
>   A: Matrix A on input. On output, U is in the upper triangle and L in the lower triangle.
>   N: Order of the matrix.
>   DET: Value of the determinant.
>   L_ROW: List of row order.
>   NDMN: Dimension of the arrays.
>
> Initialization:
>   (a) Set the value of the determinant to be DET $= 1$.
>   (b) Zero L_ROW as the original positions of rows.
>   (c) Find the element with the largest absolute value in each row and store it as A_MAX.
>
> 1. Carry out steps 2 to 7 for all n columns starting with $k = 1$.
> 2. Calculate all the elements of U in column k except the diagonal one:
>   (a) Obtain the value of $u_{i,k}$ using (5-48).
>   (b) Store the result in the location for $a_{i,k}$.
> 3. Calculate the elements of L in column k:
>   (a) Form the sum $a_{i,k} - \sum_{m=1}^{k-1} \ell_{i,m} u_{m,k}$ for $i = k$ to n.
>   (b) Store the result in the location for $a_{i,k}$.
>   (c) Search for the largest absolute value in the sum weighted by the largest element of the row. This will be used as the pivot.
>   (d) If the value of the pivot is zero, the matrix is singular. Return DET $= 0$.
>   (e) Check if the pivot is at the diagonal. If not,
>   (i) Interchange rows.
>   (ii) Change the sign of the determinant.
>   (f) Store the original row number of the pivot in L_ROW(k).
>   (g) Multiply the value of the pivot by the determinant.
>   (h) Divide the sums obtained in (a) by $u(k,k)$ to get the elements of L.
> 4. Repeat steps 2 and 3 for the next k value until $k = n$.
> 5. Return DET, L and U stored in A, and the order of the rows in L_ROW.

The second consideration concerns efficiency: to find the pivot without having to do a large amount of work. Since the calculation is sequential, we cannot find the value of the next element of U without having completed all the earlier calculations on both L and U. Thus, the weighting factor for selecting a pivot changes as each element is transformed. As a result, if we wish to find the optimum solution, the selection of the pivot must be carried out essentially from scratch for each element to be reduced. This is a rather time-consuming process and a compromise must be made for the sake of efficiency.

A good way is to select the pivot only from the elements in the column. The choice of calculating by column rather than by row also has the advantage that the upper limit of the summation in (5-49) is the same for all the elements of L in the column. Furthermore, we note that, for column k, the quantity inside the parentheses on the right side of the equation is the same as that of $u_{k,k}$ if it is the diagonal element. As a result, if we limit our choice of the pivot for column k to be among the elements in the lower part of the column, $i \geq k$ where i is the row index, very little additional work is required to find the pivot. In other words, for column k, all the elements $u_{i,k}$ of U up to $i = (k-1)$ are calculated using (5-48). Next we calculate the values inside the parentheses of the right side of (5-49) for $i = k$ to n. Among the latter $(n - k + 1)$ quantities, the one with the largest absolute value, weighted by the element with largest absolute value in row k of A, is taken as the pivot for this column. This element is now put in the diagonal position of column k as the element $u_{k,k}$. To achieve this, it may be necessary to interchange all the elements of two rows. Note also that, since the search for the pivot takes place only among elements $u_{i,k}$ for $i > k$, the interchange of rows does not affect any elements with $i < k$. This is important in the way the elements of U and L are stored, as we shall see in the next paragraph. The only work remaining now is to use the pivot as the denominator for all the quantities inside the parentheses on the right side of (5-49) we calculated earlier and make the quotients as the elements of L for column k.

Storage considerations. We shall now turn to the question of storage. As mentioned earlier, once all the elements of a particular column of U and L are calculated, the values of the elements of the same column in A are no longer needed. Since the calculation is carried out in order, from left to right one column at a time, and for each column, from the top to the bottom one element at a time, the elements of A can be discarded one after another, as the reduction of A to L and U progresses. The array for A may therefore be used to store both U and L. Since one is an upper triangular matrix and the other is a lower triangular matrix, the only conflict arises at the diagonal. This is solved by not storing the diagonal elements of L, which, as we have seen earlier, are taken to be unity. One efficient way to arrange the storage is to use the upper triangle of A, the locations for $a_{i,j}$ for $j \geq i$, for the elements of U and the rest, the locations for $a_{i,j}$ for $j > i$, for the elements of L. The only additional information required is the order of the rows of L, as they may be modified because of the interchanges used for pivoting purposes. It is simplest to put this information in a separate one-dimensional array and to take care of the actual permutations at the stage of making use of the L and U matrices. The algorithm is given in Box 5-4.

* Once the matrices L and U are found, we need to apply them to solve the system of linear equations of interest to us. As we have seen earlier, two separate steps are required. The first is the forward substitution of (5-35) to find the auxiliary vector Z from L and the input vector Y. Once all the elements of column matrix Z are known, (5-36) may be used to solve for the elements of

> **Box 5-5   Subroutine** FBLU_SBS(A,L_ROW,Y,N,NDMN)
> **Forward and back substitution after *LU* decomposition**
>
> Argument list:
> >   A: Upper triangle for U, lower triangle (except diagonal) for L.
> >   L_ROW: Order of rows in L and U.
> >   Y: Input matrix Y.
> >   N: Number of linear equations.
> >   NDMN: Dimension of the arrays.
>
> 1. Forward substitution for $LZ = Y$ of (5-35). Carry out the following steps for all n rows starting from $i = 1$:
>   (a) Use the information in L_ROW to interchange elements of Y to the permuted order.
>   (b) Calculate the elements of Z using (5-38).
>   (c) Store the value of z_i in the location for y_i.
> 2. Back substitution for $UX = Z$ of (5-30). Loop through all n rows starting from $i = n$ and work backward:
>   (a) For $i = n$, let $x_n = z_n$.
>   For $i < n$, let $x_i = \left(z_i - \sum_{m=i+1}^{n} u_{i,m} z_m\right)/u_{i,i}$.
>   (b) Divide the elements by the diagonal element and store them in the locations for y_i.
> 3. Return the array Y as the solution X.

X by back substitution. The algorithm is given in Box 5-5. It is basically a straightforward calculation except for the interchanges made on the rows of L. This is taken care of by permuting the order of the elements of Y before applying (5-36).

The computer program for LU-decomposition may be checked by multiplying L with U and seeing if we can reconstruct the original matrix A to sufficient accuracy. Alternatively, we can combine it with the forward and back substitution step and check the resulting solution of the linear equation in the same way as we did in the previous section.

§5-3   Matrix method to solve the eigenvalue problem

In quantum mechanics, the starting point for many problems is often the time-independent Schrödinger equation,

$$\left(-\frac{\hbar^2}{2\mu}\nabla^2 + V\right)\psi_\alpha = E_\alpha \psi_\alpha \tag{5-50}$$

where the first term on the left side represents the kinetic energy of the system with μ as the reduced mass. The second term comes from the potential energy, with V describing the interaction among different components of the systems.

Both the eigenvalue E_α and the eigenfunction ψ_α are to be found by solving the equation.

There are basically three ways to approach the eigenvalue problem posed by (5-50). The first was given earlier in Chapter 4. When the potential has a simple form, such as in the case of a harmonic oscillator, the solution may be expressed in terms of some known function, such as the Laguerre polynomials in §4-4. In practice, there are only a limited number of cases where such analytic solutions are possible. A second approach is to solve the eigenvalue equation of (5-50) as a differential equation using numerical methods, and we shall discuss this type of approach in Chapter 8. A third way is solve the Schrödinger equation by matrix methods. This is our primary goal for the rest of this chapter.

Basis states. The first step in taking a matrix approach to solve (5-50) is to find a complete set of states $\phi_1, \phi_2, \ldots, \phi_n$ such that any function of interest may be expressed as a linear combination of these basis states. In particular, an eigenvector of (5-50) may be written in the form

$$\psi_\alpha = \sum_{i=1}^{n} C_{\alpha,i}\, \phi_i \tag{5-51}$$

where the coefficients $C_{\alpha,i}$ express the eigenvector ψ_α in terms of ϕ_i. For mathematical convenience, we shall choose the basis states $\{\phi_i\}$ such that each state is normalized to unity and different states are orthogonal to each other. That is,

$$\int \phi_i^* \phi_j \, d\tau = \delta_{i,j} \tag{5-52}$$

where τ represents all the independent variables and

$$\delta_{i,j} = \begin{cases} = 1 & \text{for } i = j \\ = 0 & \text{otherwise} \end{cases}$$

is the Kronecker delta. To accommodate variables that are discrete, such as angular momentum, we can use the bra ($\langle$) and ket ($\rangle$) notation of Dirac and express the integration as

$$\langle \phi_i | \phi_j \rangle = \delta_{i,j}$$

In principle, the total number of basis states n can be infinite; in practice, if n is too large, the problem is not tractable numerically. In addition to restrictions imposed by the size of available memory, numerical accuracy and computational time also limit the size of matrices we can handle. On a fast machine the upper limit of n is on the order of a thousand or so. However, the more practical limits are usually smaller than this number.

To obtain realistic solutions in a restricted space, the choice of the basis to solve a problem by matrix methods requires sound judgment built on reasonable physical grounds. For example, in §4-1 we found that the eigenfunctions of a harmonic oscillator are given by (4-16). For our discussions here, let us write it in the form

$$\phi_m(\rho) = \frac{1}{\sqrt{2^m m! \sqrt{\pi}}} e^{-\rho^2/2} H_m(\rho) \qquad (5\text{-}53)$$

where $H_m(\rho)$ is the Hermite polynomial of degree m and $\rho = x\sqrt{\mu\omega/\hbar}$.

In §4-1, we have also seen that the harmonic oscillator well

$$V(x) = \tfrac{1}{2}\mu\omega^2 x^2$$

is usually an approximation to the shape of a potential well near the minimum of a more realistic one. A better approximation may be achieved by including a quadratic term. The complete potential now has the form

$$V(x) = \tfrac{1}{2}\mu\omega^2 x^2 + \epsilon \hbar \omega \left(\frac{\mu\omega}{\hbar}\right)^2 x^4 \qquad (5\text{-}54)$$

where the constant ϵ is usually a small, dimensionless quantity. For such an *anharmonic oscillator*, the functions $\{\phi_m(\rho)\}$ of (5-53) are no longer the eigenfunctions as

$$\int_{-\infty}^{+\infty} \phi_k(x)\, x^4\, \phi_\ell(x)\, dx \neq 0 \qquad \text{for} \qquad |k-\ell| = 0, 2, 4, \ldots \qquad (5\text{-}55)$$

On the other hand, the collection of $\phi_i(x)$, for $i = 0, 1, \ldots$, remains as a complete set of states in the one-dimensional space of interest to us. As a result, the eigenfunctions of an anharmonic oscillator may be expressed as linear combinations of $\phi_\alpha(x)$ in the same way as we did in (5-51).

For small values of ϵ, each eigenfunction $\psi_\alpha(\rho)$ is likely to be dominated by a single component of ϕ_i. Let us label ψ_α in such a way that the main contribution to it comes from ϕ_α. In this way, the primary effect of the anharmonic terms ϵx^4 is to introduce small admixtures of terms with $i = \alpha \pm 2$, as we can see from (5-55). Other components, with $i = \alpha \pm 4, \alpha \pm 6, \ldots$, are also possible but their contributions are likely to be even smaller. If our interest is in ψ_α, it may be possible to ignore these higher-order terms and approximate the solution using a basis consisting of three states, $\phi_\alpha(x)$, $\phi_{\alpha-2}(x)$, and $\phi_{\alpha+2}(x)$. The problem of the anharmonic oscillator is now reduced to a fairly simple one, consisting of finding a linear combination of these three basis states that satisfy the Schrödinger equation.

The eigenvalue problem. Once a basis is selected for the calculation, the eigenvalue problem is reduced essentially to one of finding the expansion coefficients $C_{\alpha,i}$ of (5-51). This we can see by multiplying both sides of (5-50) by ϕ_j^* and integrating over all the independent variables. In Dirac notation, the result may be written in the form

$$\langle\phi_j|\hat{H}|\psi_\alpha\rangle = E_\alpha\langle\phi_j|\psi_\alpha\rangle \tag{5-56}$$

where $\hat{H}$ is the Hamiltonian operator,

$$\hat{H} = -\frac{\hbar^2}{2\mu}\nabla^2 + V$$

Both the eigenvalue E_α and the corresponding eigenvector ψ_α are not known at this stage of the calculation. Since E_α is a number (rather than an operator), it may be taken outside the integration, as we have done on the right side of (5-56).

We can turn (5-56) into a matrix equation by replacing ψ_α with a linear combination of basis states ϕ_i, using the relation given by (5-51). On applying the orthogonality relation of (5-52), we obtain an algebraic equation for the expansion coefficients:

$$\sum_{i=1}^{n} H_{j,i} C_{\alpha,i} = E_\alpha C_{\alpha,j} \tag{5-57}$$

where $H_{i,j}$ are the elements of the Hamiltonian matrix H, defined by

$$H_{j,i} \equiv \langle\phi_j|\hat{H}|\phi_i\rangle$$

The summation over index i on the right side is eliminated because of the Kronecker delta imposed by the orthogonality relation between the basis states.

The relation given by (5-57) is one of n possible equations we can construct, one for each of the n basis states $\phi_1, \phi_2, \ldots, \phi_n$. The complete set of n linear equations may be written in matrix notation as

$$\begin{pmatrix} H_{1,1} & H_{1,2} & \cdots & H_{1,n} \\ H_{2,1} & H_{2,2} & \cdots & H_{2,n} \\ \vdots & \vdots & \ddots & \vdots \\ H_{n,1} & H_{n,2} & \cdots & H_{n,n} \end{pmatrix} \begin{pmatrix} C_{\alpha,1} \\ C_{\alpha,2} \\ \vdots \\ C_{\alpha,n} \end{pmatrix} = E_\alpha \begin{pmatrix} C_{\alpha,1} \\ C_{\alpha,2} \\ \vdots \\ C_{\alpha,n} \end{pmatrix} \tag{5-58}$$

It is possible to put this set of equations into the standard form of a system of n linear equations by moving all the unknown coefficients $C_{\alpha,i}$ to the left side. The result is

$$\begin{pmatrix} (H_{1,1}-E_\alpha) & H_{1,2} & \cdots & H_{1,n} \\ H_{2,1} & (H_{2,2}-E_\alpha) & \cdots & H_{2,n} \\ \vdots & \vdots & \ddots & \vdots \\ H_{n,1} & H_{n,2} & \cdots & (H_{n,n}-E_\alpha) \end{pmatrix} \begin{pmatrix} C_{\alpha,1} \\ C_{\alpha,2} \\ \vdots \\ C_{\alpha,n} \end{pmatrix} = 0 \tag{5-59}$$

Since the right side is zero, a solution of the equations exits only if

$$\det \begin{vmatrix} (H_{1,1} - E_\alpha) & H_{1,2} & \cdots & H_{1,n} \\ H_{2,1} & (H_{2,2} - E_\alpha) & \cdots & H_{2,n} \\ \vdots & \vdots & \ddots & \vdots \\ H_{n,1} & H_{n,2} & \cdots & (H_{n,n} - E_\alpha) \end{vmatrix} = 0$$

This is known as the characteristic equation and the eigenvalues are the roots of this equation. Once the value of one of the eigenvalues E_α is known, the expansion coefficients $C_{\alpha,1}, C_{\alpha,2}, \ldots, C_{\alpha,1}$ may be found in principle by solving the system of linear equations (5-59).

Matrix diagonalization. A different way of approaching the problem posed by (5-58) is to find a transformation matrix U such that the similarity transformation

$$U^{-1}HU = E \tag{5-60}$$

results in a diagonal matrix E. The process of reducing a matrix H into a diagonal form is called *diagonalization*. Before we find methods to carry out the reduction, let us first see that it actually solves the eigenvalue problem.

Another way of describing the eigenvalue problem of (5-50) is to think hypothetically first in terms of a set of basis states made of the eigenvectors $\{\psi_\alpha\}$. In this basis, the Hamiltonian matrix is diagonal, with all off-diagonal elements equal to zero and the diagonal elements consisting of the eigenvalues. This can be seen by multiplying from the left the complex conjugate of an arbitrary eigenvector ψ_β to both sides of the eigenvalue equation (5-50). On integrating over all the independent variables, the result may be expressed in Dirac notation as

$$\langle \psi_\beta | \hat{H} | \psi_\alpha \rangle = \langle \psi_\beta | E_\alpha | \psi_\alpha \rangle \tag{5-61}$$

Again, we can take E_α outside the matrix element, as it is not an operator. Furthermore, if we assume that the eigenvectors are normalized to unity and orthogonal to each other, the right side is reduced to

$$\langle \psi_\beta | E_\alpha | \psi_\alpha \rangle = E_\alpha \langle \psi_\beta | \psi_\alpha \rangle = E_\alpha \delta_{\alpha,\beta}$$

The left side cannot be simplified further here since $\hat{H}$ is an operator. The relation given by (5-61) now takes on the form

$$\langle \psi_\beta | \hat{H} | \psi_\alpha \rangle = E_\alpha \, \delta_{\alpha,\beta} \tag{5-62}$$

That is, in a basis made of eigenvectors, the Hamiltonian matrix is diagonal with the eigenvalues as the diagonal matrix elements. Such a basis is known as the eigenvector representation. Our goal in matrix diagonalization is to find a transformation that takes us from an arbitrary basis $\{\phi_i\}$ to that made of eigenvectors.

Most of the matrices we encounter in eigenvalue problems are real and symmetric. This comes from the fact that quantum mechanical operators associated with physical observables are Hermitian. As a result, their matrices are also Hermitian. Such matrices can always be put into a form that is real and symmetric. The usual Hamiltonian operator is a member of this class of operators, having the property

$$\boldsymbol{H} = \boldsymbol{H}^\dagger \qquad \text{or} \qquad H_{i,j} = H^*_{j,i}$$

This is very fortunate since it is relatively easy to solve problems involving real, symmetric matrices. However, the power of matrix methods extends far beyond such simple cases, and we shall return later at the end of this chapter to examine briefly a few of the more advanced topics.

The eigenvalue problem we are interested in may also be stated in the following way. We have two different sets of basis states, the set $\{\phi_i\}$ with which it is easy for us to carry out the calculation of matrix elements, and the set of eigenvectors $\{\psi_\alpha\}$ corresponding to the physical states. The relation represented by (5-51) may be described as a transformation from $\{\phi_i\}$ to $\{\psi_\alpha\}$. In terms of matrices, we have the relation

$$\boldsymbol{U}\boldsymbol{\phi} = \boldsymbol{\psi} \tag{5-63}$$

where $\boldsymbol{U}$ is the transformation matrix and

$$\boldsymbol{\phi} = \begin{pmatrix} \phi_1 \\ \phi_2 \\ \phi_3 \\ \vdots \\ \phi_n \end{pmatrix} \qquad \boldsymbol{\psi} = \begin{pmatrix} \psi_1 \\ \psi_2 \\ \psi_3 \\ \vdots \\ \psi_n \end{pmatrix}$$

From (5-51), it is easy to see that the elements of the transformation matrix $\boldsymbol{U}$ are made of the expansion coefficients of ψ_α in terms of ϕ_i:

$$\boldsymbol{U} = \begin{pmatrix} C_{1,1} & C_{1,2} & C_{1,3} & \cdots & C_{1,n} \\ C_{2,1} & C_{2,2} & C_{2,3} & \cdots & C_{2,n} \\ \vdots & \vdots & \vdots & \ddots & \vdots \\ C_{n,1} & C_{n,2} & C_{n,3} & \cdots & C_{n,n} \end{pmatrix}$$

In other words, we have the relation $U_{\alpha,\beta} = C_{\alpha,\beta}$. For this reason, we can also say that $\boldsymbol{U}$ is made of all the eigenvectors in the space.

It remains to be proved that such a transformation brings the Hamiltonian matrix into a diagonal form. Using (5-51), we can write an element of the Hamiltonian matrix in the eigenvector basis in terms of those in the basis formed by $\{\phi_i\}$,

$$\langle \psi_\beta | \hat{H} | \psi_\alpha \rangle = \sum_j C^*_{\beta,j} \sum_i C_{\alpha,i} \langle \phi_j | \hat{H} | \phi_i \rangle$$

Since the expansion coefficients are the elements of the transformation matrix U, the same relation may be expressed in terms of the matrix elements of U:

$$\langle\psi_\beta|\hat{H}|\psi_\alpha\rangle = \sum_j U^*_{\beta,j} \sum_i U_{\alpha,i} \langle\phi_j|\hat{H}|\phi_i\rangle$$

$$= \sum_j U^*_{\beta,j} \sum_i \langle\phi_j|\hat{H}|\phi_i\rangle \widetilde{U}_{i,\alpha} = \left(U^* H \widetilde{U}\right)_{\beta,\alpha}$$

where we have transposed the second U in order to put the expression into the standard form of a matrix multiplication and $\left(U^* H \widetilde{U}\right)_{\beta,\alpha}$ is the element of the matrix $\left(U^* H \widetilde{U}\right)$ at the position (β, α). Using (5-62), we find that

$$\left(U^* H \widetilde{U}\right)_{\beta,\alpha} = E_\alpha \delta_{\alpha,\beta}$$

Since E_α is a real quantity, the same relation can also be written in the form

$$E_\alpha \delta_{\alpha,\beta} = \left\{\left(U^* H \widetilde{U}\right)_{\beta,\alpha}\right\}^* = \left(U^\dagger H U\right)_{\beta,\alpha} \quad (5\text{-}64)$$

where we have made use of the fact that H is Hermitian, $H^\dagger = H$, and the elements of the Hermitian conjugate of a matrix are the complex conjugates of the elements of the transposed matrix,

$$\left(\widetilde{U}\right)^*_{i,j} = \left(U\right)^*_{j,i} = \left(U^\dagger\right)_{i,j}$$

In terms of matrices, (5-64) is equivalent to

$$U^\dagger H U = E$$

where we have used E to represent the diagonal matrix made of eigenvalues that appeared on the right side of (5-62).

This result is not yet in the form of a similarity transformation given in (5-60) until we identify that

$$U^\dagger = U^{-1} \quad (5\text{-}65)$$

This is the property of a unitary matrix, and the transformation induced by a matrix satisfying (5-65) is known as a unitary transformation. It is the property of a Hermitian matrix that it can be brought into a diagonal form by a unitary transformation. Furthermore, the eigenvalues obtained from such a transformation are unique and the eigenfunctions are orthogonal to each. We shall, however, leave the proofs of these points to standard texts on quantum mechanics.

Two-dimensional rotations. The most straightforward method to diagonalize a real, symmetric matrix is the Jacobi method. In this approach, the basic operation consists of a series of rotations among two columns and two rows. Consider first a real, symmetric (2×2) matrix of the form

$$A = \begin{pmatrix} a_{1,1} & a_{1,2} \\ a_{2,1} & a_{2,2} \end{pmatrix}$$

where $a_{2,1} = a_{1,2}$. We shall find a two-dimensional similarity transformation that reduces the two off-diagonal elements to zero.

Because of the unitary requirement of (5-65), there is only one parameter in the transformation. We shall take this to be θ, which may be interpreted as the angle of rotation in the two-dimensional space. It is convenient to write the transformation matrix in terms of trigonometric functions in the following form:

$$T = \begin{pmatrix} \cos\theta & \sin\theta \\ -\sin\theta & \cos\theta \end{pmatrix} \quad (5\text{-}66)$$

Among other advantages, this form also ensures the unitary property of T. Since $T^{-1} = T^\dagger = \widetilde{T}$, a similarity transformation of A using T gives us the result

$$A' = T^{-1}AT = \begin{pmatrix} c & -s \\ s & c \end{pmatrix} \begin{pmatrix} a_{1,1} & a_{1,2} \\ a_{2,1} & a_{2,2} \end{pmatrix} \begin{pmatrix} c & s \\ -s & c \end{pmatrix}$$

$$= \begin{pmatrix} c^2 a_{1,1} + s^2 a_{2,2} - 2cs\,a_{1,2} & cs(a_{1,1} - a_{2,2}) + (c^2 - s^2)a_{1,2} \\ cs(a_{1,1} - a_{2,2}) + (c^2 - s^2)a_{1,2} & s^2 a_{1,1} + c^2 a_{2,2} + 2cs\,a_{1,2} \end{pmatrix}$$

(5-67)

where, in order to write the equation in a compact form, we have used the abbreviations

$$c \equiv \cos\theta \qquad\qquad s \equiv \sin\theta$$

Our goal here is to find a value of θ such that the two off-diagonal matrix elements of A' vanish:

$$a'_{1,2} = a'_{2,1} = 0$$

From (5-67), we see that such a requirement is equivalent to the condition

$$\frac{cs}{c^2 - s^2} = \frac{a_{1,2}}{a_{2,2} - a_{1,1}} \quad (5\text{-}68)$$

In terms of the rotation angle θ, the same condition may be written as

$$\tan 2\theta = \frac{2\cos\theta\sin\theta}{\cos^2\theta - \sin^2\theta} = \frac{2a_{1,2}}{a_{2,2} - a_{1,1}} \quad (5\text{-}69)$$

Although the above equations are expressed in terms of trigonometry functions, the actual relations between the various quantities are algebraic. In other words, the sine, cosine, and tangent functions do not need to appear explicitly in actual calculations. This is important from an efficiency point of view, as trigonometric functions are, in general, time consuming to calculate.

Jacobi method. The same type of rotation as used in (5-67) for a two-dimensional matrix may also be applied to larger matrices. Consider a real, symmetric matrix $\boldsymbol{A}$ of some finite dimension n. If we wish to put to zero a particular pair of off-diagonal matrix elements, for example $a_{k,\ell}$ and $a_{\ell,k}$, we can apply a two-dimensional rotation using a matrix similar to that given by (5-66). The major difference is that, instead of two-dimensional, the transformation matrix $\boldsymbol{T}$ is n-dimensional, as given in (5-70). To bring the complete matrix to a diagonal form, the Jacobi method makes a series of such rotations, each of which annihilates a pair of off-diagonal matrix elements. Let us assume that after $(m-1)$ such rotations the original matrix is transformed into $\boldsymbol{A}(m-1)$ with matrix elements $\{a_{i,j}^{(m-1)}\}$. Since all the transformations are unitary, the matrix $\boldsymbol{A}(m-1)$ remains real and symmetric. Let us further assume that, in the next rotation, the pair of off-diagonal elements we wish to annihilate is $a_{k,\ell}^{(m-1)}$ and $a_{\ell,k}^{(m-1)}$ (with $\ell \neq k$). The transformation matrix to achieve this goal may be written in the form

$$\boldsymbol{T}(m) = \begin{pmatrix} 1 & 0 & 0 & \cdots & 0 & \cdots & 0 & \cdots & 0 \\ 0 & 1 & 0 & \cdots & 0 & \cdots & 0 & \cdots & 0 \\ 0 & 0 & 1 & \cdots & 0 & \cdots & 0 & \cdots & 0 \\ \vdots & \vdots & \vdots & \ddots & \vdots & \ddots & \vdots & \ddots & 0 \\ 0 & 0 & 0 & \cdots & \cos\theta^{(m)} & \cdots & \sin\theta^{(m)} & \cdots & 0 \\ \vdots & \vdots & \vdots & \ddots & \vdots & \ddots & \vdots & \ddots & 0 \\ 0 & 0 & 0 & \cdots & -\sin\theta^{(m)} & \cdots & \cos\theta^{(m)} & \cdots & 0 \\ \vdots & \vdots & \vdots & \ddots & \vdots & \ddots & \vdots & \ddots & 0 \\ 0 & 0 & 0 & \cdots & 0 & \cdots & 0 & \cdots & 1 \end{pmatrix} \quad (5\text{-}70)$$

Similar to (5-69), if we choose the angle $\theta^{(m)}$ such that

$$\tan 2\theta^{(m)} = \frac{2a_{k,\ell}^{(m-1)}}{a_{\ell,\ell}^{(m-1)} - a_{k,k}^{(m-1)}} \quad (5\text{-}71)$$

the elements of the matrix

$$\boldsymbol{A}(m) = \boldsymbol{T}^{-1}(m)\boldsymbol{A}(m-1)\boldsymbol{T}(m) \quad (5\text{-}72)$$

take on the following values.

First, the elements outside columns and rows k and ℓ are unchanged by the transformation, meaning

$$a_{i,j}^{(m)} = a_{i,j}^{(m-1)} \qquad \text{for} \quad \begin{array}{l} i \neq k \neq \ell \\ j \neq k \neq \ell \end{array} \quad (5\text{-}73)$$

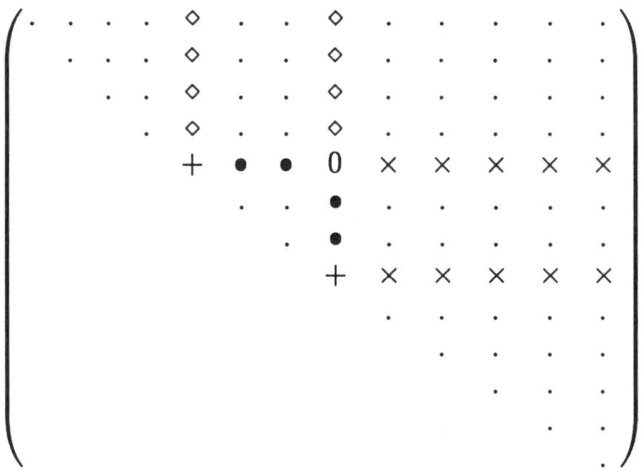

Chapter 5] Matrices 245

$$\begin{pmatrix}
\cdot & \cdot & \cdot & \cdot & \diamond & \cdot & \cdot & \diamond & \cdot & \cdot & \cdot & \cdot & \cdot & \cdot \\
 & \cdot & \cdot & \cdot & \diamond & \cdot & \cdot & \diamond & \cdot & \cdot & \cdot & \cdot & \cdot & \cdot \\
 & & \cdot & \cdot & \diamond & \cdot & \cdot & \diamond & \cdot & \cdot & \cdot & \cdot & \cdot & \cdot \\
 & & & \cdot & \diamond & \cdot & \cdot & \diamond & \cdot & \cdot & \cdot & \cdot & \cdot & \cdot \\
 & & & & + & \bullet & \bullet & 0 & \times & \times & \times & \times & \times & \\
 & & & & & & \cdot & \bullet & \cdot & \cdot & \cdot & \cdot & \cdot & \\
 & & & & & & \cdot & \bullet & \cdot & \cdot & \cdot & \cdot & \cdot & \\
 & & & & & & & + & \times & \times & \times & \times & \times & \\
 & & & & & & & & \cdot & \cdot & \cdot & \cdot & \cdot & \\
 & & & & & & & & & \cdot & \cdot & \cdot & \cdot & \\
 & & & & & & & & & & \cdot & \cdot & \cdot & \\
 & & & & & & & & & & & \cdot & \cdot & \\
\end{pmatrix}$$

Fig. 5-2 Three categories of elements in the Jacobi method of diagonalizing a real, symmetric matrix. The dots (·) indicate elements outside columns and rows k and ℓ that are unchanged by the transformation of (5-70). The plus signs (+) indicate diagonal elements in columns k and ℓ transformed by (5-74). The diamonds (◊), bullets (•), and crosses (×) indicate elements in columns and rows k and ℓ that are transformed according to (5-75). Since the matrix remains symmetric in the transformation, only the upper triangle is shown.

Second, for the matrix elements in columns and rows k and ℓ, we have the analogous situation as (5-67):

$$a_{k,\ell}^{(m)} = a_{\ell,k}^{(m)} = 0$$
$$a_{k,k}^{(m)} = c^2 a_{k,k}^{(m-1)} + s^2 a_{\ell,\ell}^{(m-1)} - 2cs a_{k,\ell}^{(m-1)}$$
$$a_{\ell,\ell}^{(m)} = s^2 a_{k,k}^{(m-1)} + c^2 a_{\ell,\ell}^{(m-1)} + 2cs a_{k,\ell}^{(m-1)} \qquad (5\text{-}74)$$

Here the symbols c and s are cosine and sine functions of the angle $\theta^{(m)}$,

$$c \equiv \cos\theta^{(m)} \qquad\qquad s \equiv \sin\theta^{(m)}$$

Third, since the transformation affects all the elements in columns k and ℓ, as well as in rows k and ℓ, the elements in these columns and rows, other than those already given above, are transformed in the following way:

$$a_{i,k}^{(m)} = a_{k,i}^{(m)} = ca_{i,k}^{(m-1)} - sa_{i,\ell}^{(m-1)}$$
$$a_{i,\ell}^{(m)} = a_{\ell,i}^{(m)} = sa_{i,k}^{(m-1)} + ca_{i,\ell}^{(m-1)} \qquad \text{for} \quad i \neq k \neq \ell \qquad (5\text{-}75)$$

These three types of matrix elements may be distinguished with the aid of the diagram shown in Fig. 5-2.

Because of this third group of transformations in the matrix elements, the Jacobi method must be an iterative one. This we can see from the fact that the off-diagonal matrix elements that vanished in an earlier transformation may become nonzero by a subsequent transformation. The process can, however, be shown to be convergent, since successive transformations increase the absolute values of the diagonal elements and decrease the absolute values of the off-diagonal elements. We shall leave the proof of this point to texts on numerical analysis.

Symbolically, we may represent a Jacobi diagonalization as

$$D = V^{-1}AV \tag{5-76}$$

where the transformation matrix V is the product of a series of two-dimensional rotations, each of which has the form given in (5-70). The total number of two-dimensional rotations required depends on the dimension of the matrix as well as the accuracy we wish to achieve. In principle, the transformation should start with the rotation that annihilates the pair of off-diagonal matrix elements having the largest magnitude and then proceed to the pair with the next largest value. In practice, this is seldom carried out, as the search itself is a time-consuming process. If the matrix dimension is not too large, it is fairly efficient just to carry out the reduction row by row (or column by column), starting from the first off-diagonal element in each row (or column). Since the matrix is symmetric, it is only necessary to examine either the upper or the lower triangle. When we encounter an element whose absolute value is larger than some small number, determined by the accuracy we wish to achieve in the calculation, a rotation is applied to reduce it (and the corresponding one in the other half of the matrix) to zero. After all the $n(n-1)/2$ off-diagonal elements in the matrix have been examined once in this way, we must go back to the beginning and repeat the process if any rotations have been carried out. In general, several iterations are necessary before the magnitudes of all the off-diagonal matrix elements are reduced to less than the error that can be tolerated.

The actual calculations of the elements of the transformation matrix V in (5-76) are very similar to that carried out in (5-74) and (5-75). Let us write the form of V after applying m two-dimensional rotations to A as

$$A(m) = V^{-1}(m)A\,V(m) \tag{5-77}$$

where A is the original matrix we wish to diagonalize. Using (5-72), the same relation may also be expressed in the form

$$A(m) = T^{-1}(m)\{V^{-1}(m-1)AV(m-1)\}T(m)$$

where $T(m)$ is a two-dimensional rotation given by (5-70). Comparing this with (5-77), we obtain the result

$$V(m) = V(m-1)T(m) \tag{5-78}$$

In terms of matrix elements, this is equivalent to

$$v_{i,j}^{(m)} = v_{i,j}^{(m-1)} \quad \text{for} \quad j \neq k \neq \ell$$
$$v_{i,k}^{(m)} = c v_{i,k}^{(m-1)} - s v_{i,\ell}^{(m-1)}$$
$$v_{i,\ell}^{(m)} = s v_{i,k}^{(m-1)} + c v_{i,\ell}^{(m-1)} \tag{5-79}$$

where $i = 1, 2, \ldots, n$. The starting point of the iterative relation given by (5-78) is the unit matrix

$$V(0) \equiv 1$$

with elements $v_{i,j}^{(0)} = \delta_{i,j}$.

Improving efficiency. For computational efficiency, it is advantageous to avoid calculating the trigonometric functions explicitly. We can achieve this through the following manipulation. What we need in (5-74), (5-75), and (5-79) for a rotation are the values of $\cos \theta^{(m)}$ and $\sin \theta^{(m)}$ with the angle $\theta^{(m)}$ given by (5-71). To simplify the notation, we shall use R to represent the inverse of the ratio of matrix elements on the right side of (5-71). That is,

$$R \equiv \frac{a_{\ell,\ell}^{(m-1)} - a_{k,k}^{(m-1)}}{2 a_{k,\ell}^{(m-1)}}$$

Using the trigonometry relation

$$\tan 2\theta^{(m)} = \frac{2 \tan \theta^{(m)}}{1 - \tan^2 \theta^{(m)}}$$

we obtain an equation for $\tan \theta^{(m)}$ in terms of R,

$$\frac{2 \tan \theta^{(m)}}{1 - \tan^2 \theta^{(m)}} = \frac{1}{R}$$

or

$$\tan^2 \theta^{(m)} + 2R \tan \theta^{(m)} - 1 = 0 \tag{5-80}$$

The two standard solutions for such a quadratic equation for $\tan \theta$ are

$$\tan \theta^{(m)} = -R \pm \sqrt{R^2 + 1}$$

Alternatively, we can change (5-80) into a quadratic equation for $\cot \theta$ and obtain the solution

$$\cot \theta^{(m)} = R \pm \sqrt{R^2 + 1}$$

or

$$\tan \theta^{(m)} = \frac{1}{R \pm \sqrt{R^2 + 1}}$$

Any one of these possibilities satisfies (5-71), and a choice among them can be made based on numerical accuracy considerations.

In general, the accuracy of such calculations is best if, in step m, we can manage to zero the off-diagonal matrix element $a_{k,\ell}^{(m-1)}$ with only minimum changes in the other elements. To achieve this, we must take $\cos\theta^{(m)}$ to be as close to 1 as possible and $\sin\theta^{(m)}$ as close to 0 as possible. The advantages of such choices are quite obvious if we examine (5-74) and (5-75). In terms of the angle of rotation, we want the absolute value of $\theta^{(m)}$ to be as close to 0 as possible. The optimum choice is then

$$\tan\theta^{(m)} = \frac{S(R)}{|R| + \sqrt{R^2 + 1}} \tag{5-81}$$

where

$$S(R) = \begin{cases} +1 & \text{if } R > 0 \\ -1 & \text{if } R < 0 \end{cases}$$

From the value of $\tan\theta^{(m)}$, we obtain those for $\cos\theta$ and $\sin\theta$ that actually enter into the calculations for the rotation

$$\cos\theta^{(m)} = \frac{1}{\sqrt{1 + \tan^2\theta^{(m)}}} \qquad \sin\theta^{(m)} = \cos\theta^{(m)} \tan\theta^{(m)} \tag{5-82}$$

In this approach, two square roots are used to replace the slower calculations of an arc tangent, a sine, and a cosine.

To improve the numerical accuracy further, we shall reformulate slightly the three equations of (5-74), (5-75), and (5-79). Analogous to (5-68), we find that

$$\frac{a_{k,\ell}^{(m-1)}}{a_{\ell,\ell}^{(m-1)} - a_{k,k}^{(m-1)}} = \frac{cs}{c^2 - s^2}$$

where we have returned to the use of the abbreviations of $c = \cos\theta^{(m)}$, $s = \sin\theta^{(m)}$, and $t = \tan\theta^{(m)}$. When we substitute this relation into the second member of (5-74) and make use of the fact that $c^2 + s^2 = 1$, we obtain

$$\begin{aligned} a_{k,k}^{(m)} &= a_{k,k}^{(m-1)} + s^2\left(a_{\ell,\ell}^{(m-1)} - a_{k,k}^{(m-1)}\right) - 2cs\, a_{k,\ell}^{(m-1)} \\ &= a_{k,k}^{(m-1)} + \left(s^2 \frac{c^2 - s^2}{cs} - 2cs\right) a_{r,s}^{(m-1)} \\ &= a_{k,k}^{(m-1)} - t\, a_{k,\ell}^{(m-1)} \end{aligned} \tag{5-83}$$

Similarly, the third member of (5-74) may be put into the form

$$a_{\ell,\ell}^{(m)} = a_{\ell,\ell}^{(m-1)} + t\, a_{k,\ell}^{(m-1)} \tag{5-84}$$

The idea here is that the off-diagonal matrix elements are small, in general, especially during the later iterations. Since we have already kept the absolute value of t as small as possible, the above two equations represent small adjustments of the diagonal matrix elements.

The other elements modified by the same two-dimensional rotation are also in columns k and ℓ and in rows k and ℓ. It is useful to rearrange the terms in the first member of (5-75) into the following form:

$$\begin{aligned}a_{i,k}^{(m)} = a_{k,i}^{(m)} &= a_{i,k}^{(m-1)} - s a_{i,\ell}^{(m-1)} - (1-c)a_{i,k}^{(m-1)}\\&= a_{i,k}^{(m-1)} - s\Big(a_{i,\ell}^{(m-1)} + \frac{(1-c)}{s} a_{i,k}^{(m-1)}\Big)\\&= a_{i,k}^{(m-1)} - s\big(a_{i,\ell}^{(m-1)} + \tau a_{i,k}^{(m-1)}\big)\end{aligned} \qquad (5\text{-}85)$$

where

$$\tau \equiv \frac{1-c}{s} = \frac{s}{1+c} = \tan\tfrac{1}{2}\theta^{(m)}$$

Similarly, the second member of (5-75) may be put into the form

$$a_{i,\ell}^{(m)} = a_{\ell,i}^{(m)} = a_{i,\ell}^{(m-1)} + s\big(a_{i,k}^{(m-1)} - \tau a_{i,\ell}^{(m-1)}\big) \qquad (5\text{-}86)$$

Since τ is kept as small as possible, similar to s, the second terms of both (5-85) and (5-86) are small in general. By the same token, the calculations required to obtain the matrix elements of V in the last two members of (5-79) may be put into similar forms:

$$\begin{aligned}v_{i,k}^{(m)} &= v_{i,k}^{(m-1)} - s\big(v_{i,\ell}^{(m-1)} + \tau v_{i,k}^{(m-1)}\big)\\v_{i,\ell}^{(m)} &= v_{i,\ell}^{(m-1)} + s\big(v_{i,k}^{(m-1)} - \tau v_{i,\ell}^{(m-1)}\big)\end{aligned} \qquad (5\text{-}87)$$

These are the calculations carried out in the algorithm for the Jacobi method of matrix diagonalization given in Box 5-6. Many other improvements are possible, but we shall not incorporate them here so as to keep the method relatively simple and still reasonably efficient.

To check the computer program, we start by finding out whether the eigenvectors are normalized to unity and orthogonal to each other, that is, whether they satisfy the relation

$$V^{-1}V = V^\dagger V = 1$$

In terms of matrix elements, this is the same as

$$\sum_k v_{k,i}\, v_{k,j} = \delta_{i,j}$$

These results provide us with an idea of the accuracy achieved in the calculation. A more stringent test is to apply the transformations given by V to the original matrix A and see if (5-76) is satisfied within the numerical accuracy required. This we can do by multiplying the original matrix A on the left by V^{-1} and on the right by V. The result should be a diagonal matrix with the values of the diagonal

Box 5-6 Subroutine JCB_DAG(A,V,E,N,NDMN,ACC)
Jacobi diagonalization of a real, symmetric matrix

Argument list:
  A: Input two-dimensional matrix. Upper triangle used in the calculation. Diagonal elements and lower triangle not changed.
  V: Array for output eigenvectors V.
  E: Array for output eigenvalues E.
  N: Number of rows and columns of the matrices.
  NDMN: Dimension of the arrays A and V in the calling program.
  ACC: Size of off-diagonal matrix element below which it is considered to be 0.

Initialization:
  (a) Set the maximum number of iterations allowed to 50.
  (b) Set E equal to the diagonal matrix elements of A.
  (c) Set V equal to a unit matrix.
  (d) Zero the iteration counter I_t.

1. Increase I_t by 1 and set the requirement for further iteration to .FALSE..
2. Scan all the off-diagonal matrix elements in the upper triangle. If $|a_{i,j}| >$ ACC, then:
   (a) Set the condition for needing further iteration to .TRUE..
   (b) Calculate the parameters t, c, s, and τ given in (5-81), (5-82), and (5-85).
   (c) Modify the diagonal matrix elements using (5-83) and (5-84).
   (d) Put the off-diagonal matrix element $a_{i,j}$ to zero.
   (e) Change the other off-diagonal matrix elements in the upper triangle of columns i and j and rows i and j according to (5-85) and (5-86). Use three separate loops in order not to affect the matrix elements of A along the diagonal and below:
      (i) Elements in columns i and j, represented as diamonds ($\diamond$) in Fig. 5-2 ($a_{k,i}$ and $a_{k,j}$ for $k = 1$ to $i - 1$).
      (ii) The remaining off-diagonal elements in row i between columns i and j, and in column j between rows i and j, represented as bullets ($\bullet$) in Fig. 5-2 ($a_{i,k}$ and $a_{k,j}$ for $k = i + 1$ to $j - 1$).
      (iii) The remaining elements in rows i and j, represented as X's in Fig. 5-2 ($a_{i,k}$ and $a_{j,k}$ for $k = j + 1$ to n).
   (f) Update the matrix V using (5-87).
3. Check for the need of a further iteration:
   (a) If not, return eigenvalues in E and eigenvectors in V.
   (b) If so, check if the maximum number of allowed iterations is exceeded:
      (i) If not, repeat steps 2 and 3.
      (ii) If exceeded, print a warning message and exit.

elements equal to the eigenvalues produced directly by the diagonalization process. Because of the unitary property, there is no need to invert V to obtain V^{-1}. Using the fact that $V^{-1} = \widetilde{V}$, the relation in terms of matrix elements takes on the form

$$(V^{-1}AV)_{i,j} = \sum_{k=1}^{n}\sum_{\ell=1}^{n} v_{k,i} a_{k,\ell} v_{\ell,j}$$

Note the order of the subscripts in the first v. They are the matrix elements of $\widetilde{V}$ and are given by the relation $(\widetilde{V})_{i,j} = v_{j,i}$.

§5-4 Tridiagonalization method of Givens and Householder

The Jacobi method of matrix diagonalization suffers from the fact that the method is an iterative one. In general, the number of iterations required to reduce a symmetric matrix of dimension n to a diagonal one is proportional to n. Since in each iteration the number of off-diagonal elements that has to be annihilated is proportional to n^2, the method requires the order of n^3 operations, thus making it rather impractical for large matrices.

The method of Givens. The main weakness in the Jacobi method lies in the fact that each two-dimensional rotation that annihilates one pair of off-diagonal matrix elements affects the values of other off-diagonal elements in the same column and row. As a result, the value of a pair of matrix elements that were put to zero in an earlier operation may become nonzero again as a result of reducing another pair at a later stage. The method of Givens improves upon the algorithm by eliminating the iterative nature of the calculation. The rotations stay two dimensional, but they are arranged in such a way that the elements already reduced to zero remain as zero. The transformation matrix for each rotation retains the form of (5-69) for the Jacobi method. However, for the convenience of later discussions, we shall now label each rotation matrix T by the indexes of the two columns and two rows it affects:

$$T(k,\ell) = \begin{array}{c} \\ \\ \\ \\ k \\ \\ \ell \\ \\ \\ \end{array}\begin{pmatrix} 1 & 0 & 0 & \cdots & 0 & \cdots & 0 & \cdots & 0 \\ 0 & 1 & 0 & \cdots & 0 & \cdots & 0 & \cdots & 0 \\ 0 & 0 & 1 & \cdots & 0 & \cdots & 0 & \cdots & 0 \\ \vdots & \vdots & \vdots & \ddots & \vdots & \ddots & \vdots & \ddots & 0 \\ 0 & 0 & 0 & \cdots & c & \cdots & s & \cdots & 0 \\ \vdots & \vdots & \vdots & \ddots & \vdots & \ddots & \vdots & \ddots & 0 \\ 0 & 0 & 0 & \cdots & -s & \cdots & c & \cdots & 0 \\ \vdots & \vdots & \vdots & \ddots & \vdots & \ddots & \vdots & \ddots & 0 \\ 0 & 0 & 0 & \cdots & 0 & \cdots & 0 & \cdots & 1 \end{pmatrix}$$

All the diagonal matrix elements of $T(k,\ell)$ equal 1 except those for k and ℓ. They have the values

$$t_{k,k} = t_{\ell,\ell} = c$$

The off-diagonal matrix elements are given by the following relations:

$$t_{i,j} = \begin{cases} s & \text{for } i = k \text{ and } j = \ell \\ -s & \text{for } i = \ell \text{ and } j = k \\ 0 & \text{otherwise} \end{cases}$$

The values of c and s are chosen such that the similarity transformation

$$A' = T^{-1}(k,\ell)AT(k,\ell) \tag{5-88}$$

reduces the off-diagonal element $a_{k-1,\ell}$ to zero (and through symmetry, $a_{\ell,k-1} = 0$) by taking a linear combination of $a_{k-1,\ell}$ with the element $a_{k-1,k}$. The major difference from the Jacobi method here is that the reduction makes use of the superdiagonal element rather than the diagonal element. As a result, it is not possible to bring the matrix to a diagonal form. Instead, the end result is a tridiagonal matrix whose form is shown in Fig. 5-3. As we shall see in the next section, it is a relatively simple matter to diagonalize a tridiagonal matrix.

We shall only try to give an illustration of the method without actually carrying out any calculations with it. In practice, the Givens method is usually used together with the improvements due to Householder. We shall return to the improved version later.

To see the advantage of the Givens method, it is convenient to separate (5-88) into two steps. In the first, let

$$B = T^{-1}(k,\ell)A$$

The only matrix elements in B that are different from those in A are in rows k and ℓ. To complete the transformation of (5-88), we need to carry out the second matrix multiplication

$$A' = BT(k,\ell)$$

Here, the only matrix elements of A' that are different from those of B are in columns k and ℓ. Now if the matrix A is already partially transformed into the form

$$A = \begin{pmatrix} d_1 f_2 \; 0 \; \cdots & 0 & 0 & 0 \cdots 0 & 0 & \cdots & 0 \\ f_2 d_2 f_3 \; \cdots & 0 & 0 & 0 \cdots 0 & 0 & \cdots & 0 \\ 0 \; f_3 d_3 \; \cdots & 0 & 0 & 0 \cdots 0 & 0 & \cdots & 0 \\ \vdots \;\; \vdots \;\; \vdots \;\; \ddots & \vdots & \vdots & \vdots \ddots \vdots & \vdots & \ddots & \vdots \\ 0 \; 0 \; 0 \; \cdots & a_{k-1,k-1} & a_{k-1,k} & \cdots & \cdots & a_{k-1,\ell} & \cdots & a_{k-1,n} \\ \vdots \;\; \vdots \;\; \vdots \;\; \ddots & \vdots & \vdots & \vdots \ddots \vdots & \vdots & \ddots & \vdots \\ 0 \; 0 \; 0 \; \cdots & a_{\ell,k-1} & a_{\ell,k} & \cdots & \cdots & a_{\ell,\ell} & \cdots & a_{\ell,n} \\ \vdots \;\; \vdots \;\; \vdots \;\; \ddots & \vdots & \vdots & \vdots \ddots \vdots & \vdots & \ddots & \vdots \\ 0 \; 0 \; 0 \; \cdots & a_{n,k-1} & a_{n,k} & \cdots & \cdots & \cdots & \cdots & a_{n,n} \end{pmatrix}$$

the action of $T^{-1}(k,\ell)$ changes $a_{\ell,k-1}$ (but not $a_{k-1,\ell}$). We see that this operation does not affect any other matrix elements in column $(k-1)$ other than $a_{k,k-1}$. As a result, $a_{i,k-1}$ remains zero for $i = (k+2)$ to $(\ell-1)$. Similarly, the operation BT reduces $a_{k-1,\ell}$ to zero without affecting any of the matrix elements in row $(k-1)$ other than $a_{k-1,k}$ (and $a_{k-1,\ell}$).

Since the superdiagonal element $a_{k-1,k}$ (and $a_{k,k-1}$) is used in the reduction, they cannot be put to zero by such a rotation. However, if we start from $k = 2$

and reduce all the elements $a_{1,i}$ to zero for $i = 3$ to n and then proceed to $k = 3$, and so on, until we come to $k = n - 1$, the end result is a tridiagonal matrix having the form shown in Fig. 5-3, that is, all the elements are zero except the diagonal ones, which we shall label as $d_1, d_2, \ldots, d_n$, and the superdiagonal ones (those just above the diagonal elements), and the subdiagonal ones (those just below), indicated as $f_2, f_3, \ldots, f_n$. Since there are only $(n - 1)$ superdiagonal matrix elements, we have chosen to number them with subscripts starting from 2. In this scheme, f_1 does not appear.

$$J = \begin{pmatrix} d_1 & f_2 & 0 & 0 & 0 & \cdots & 0 & 0 & 0 \\ f_2 & d_2 & f_3 & 0 & 0 & \cdots & 0 & 0 & 0 \\ 0 & f_3 & d_3 & f_4 & 0 & \cdots & 0 & 0 & 0 \\ 0 & 0 & f_4 & d_4 & f_5 & \cdots & 0 & 0 & 0 \\ \vdots & \vdots & \vdots & \vdots & \vdots & \ddots & \vdots & \vdots & \vdots \\ 0 & 0 & 0 & 0 & 0 & \cdots & d_{n-2} & f_{n-1} & 0 \\ 0 & 0 & 0 & 0 & 0 & \cdots & f_{n-1} & d_{n-1} & f_n \\ 0 & 0 & 0 & 0 & 0 & \cdots & 0 & f_n & d_n \end{pmatrix}$$

Fig. 5-3 A symmetric, tridiagonal matrix with all the elements zero except n diagonal matrix elements $d_1, d_2, \ldots, d_n$, and $(n-1)$ off-diagonal elements $f_2, f_3, \ldots, f_n$, just above and below the diagonal.

In practical terms, each rotation in the method of Givens is similar to that of Jacobi seen in the previous section. The only difference is that, in a Jacobi rotation, we seek to annihilate the off-diagonal matrix element $a_{i,j}$ by taking a linear combination of it with diagonal elements $a_{i,i}$ and $a_{j,j}$, as shown in (5-68). In contrast, in the method of Givens, the cancellation required to reduce $a_{i,j}$ to zero is achieved by using $a_{i,j}$ and $a_{i,i+1}$. As a result, the rotation angle θ of (5-70) or the sine s and cosine c used in the transformation are different in the two methods. The forms of the relation between the matrix elements, however, remain the same as those given by (5-73) to (5-75) and (5-79). On the other hand, the relations in (5-83) and (5-84) depend on the particular values of c and s for a Jacobi transformation and are not valid for a Givens rotation. Furthermore, since the final result of a Givens transformation is a tridiagonal matrix, additional calculations are required to transform it into a diagonal form.

A more efficient method to bring a matrix into a tridiagonal form is that of Householder. Instead of operating on one element at a time, a transformation is carried out in such a way that all the elements in a row $a_{k,\ell}$, for $\ell = k + 2$, $k + 3, \ldots, n$, are reduced to zero. Although the method applies to any Hermitian matrices, we shall restrict our interest here to real, symmetric matrices only.

Inner product and outer product. Before discussing the details of the Householder method to transform a matrix to tridiagonal form, it is useful to differentiate between two types of matrix multiplications, inner and outer products. A column vector with n elements

$$a = \begin{pmatrix} a_1 \\ a_2 \\ a_3 \\ \vdots \\ a_n \end{pmatrix}$$

may be thought as a matrix having only a single column. The transpose of a is a row vector, meaning a matrix consisting of a single row of n elements

$$\tilde{a} = (a_1 \quad a_2 \quad a_3 \quad \cdots \quad a_n)$$

There are two different kinds of matrix product we can form from and they are best illustrated using a column matrix and a row matrix. Let b be a column vector also of dimension n. According to the rule of matrix multiplication, the following product is a scalar quantity,

$$\tilde{a}b = (a_1 \quad a_2 \quad a_3 \quad \cdots \quad a_n) \begin{pmatrix} b_1 \\ b_2 \\ b_3 \\ \vdots \\ b_n \end{pmatrix} = \sum_{i=1}^{n} a_i b_i$$

That is, it is a quantity that is invariant under a transformation of the basis. Such a product of two vectors is known as the scalar or *inner product*. The inner product of $\tilde{a}$ with a,

$$\|a\|^2 = \tilde{a}a = \sum_{i=1}^{n} a_i^2 \tag{5-89}$$

gives the square of the norm of the vector a.

We can form another kind of product between a column matrix and a row matrix. For example, if a is a column matrix with m elements and b is a column matrix of n elements, the product

$$a\tilde{b} = \begin{pmatrix} a_1 \\ a_2 \\ a_3 \\ \vdots \\ a_m \end{pmatrix} (b_1 \quad b_2 \quad b_3 \quad \cdots \quad b_n) = \begin{pmatrix} a_1 b_1 & a_1 b_2 & a_1 b_3 & \cdots & a_1 b_n \\ a_2 b_1 & a_2 b_2 & a_2 b_3 & \cdots & a_2 b_n \\ a_3 b_1 & a_3 b_2 & a_3 b_3 & \cdots & a_3 b_n \\ \vdots & \vdots & \vdots & \ddots & \vdots \\ a_m b_1 & a_m b_2 & a_m b_3 & \cdots & a_m b_n \end{pmatrix}$$

is a rectangular matrix of m rows and n columns. To distinguish from inner products, this type of matrix product is known as the *outer product*.

The method of Householder. Consider the following square matrix of dimension n

$$P = 1 - \beta u \tilde{u} \tag{5-90}$$

where 1 is a unit matrix, β is a constant to be determined later, and u is a column vector having n elements. u is constructed out of another vector a in the following way,

$$u = a \pm \|a\|\epsilon_1 \tag{5-91}$$

where

$$\epsilon_1 = \begin{pmatrix} 1 \\ 0 \\ 0 \\ \vdots \\ 0 \end{pmatrix}$$

is a column matrix with only the first element equal to 1 and the other $(n-1)$ elements equal to 0. In other words,

$$u = \begin{pmatrix} a_1 \pm \|a\| \\ a_2 \\ a_3 \\ \vdots \\ a_n \end{pmatrix}$$

By making a suitable choice of the value of β, it is possible to show that the column vector formed out of the product of P with a is a column vector with only the first element nonzero. This applies to both the plus and the minus signs in (5-91). As we shall see later, the choice of the sign can be used to our advantage when we design an algorithm to transform a matrix to tridiagonal form.

We can express $\tilde{u}$, the transpose of u, in terms of $\tilde{a}$ and $\tilde{\epsilon}_1$. Using this relation, we have the result

$$\begin{aligned} Pa &= (1 - \beta u \tilde{u})a \\ &= a - \beta u(\tilde{a} \pm \|a\|\tilde{\epsilon}_1)a \\ &= a - \beta u(\|a\|^2 \pm \|a\|a_1) \end{aligned} \tag{5-92}$$

Because of the definition of u given by (5-91), it is easy to see that the quantity inside the parentheses is half the square of the norm of u:

$$\|u\|^2 = (\tilde{a} \pm \|a\|\tilde{\epsilon}_1)(a \pm \|a\|\epsilon_1) = 2(\|a\|^2 \pm \|a\|a_1)$$

As a result, if we take

$$\beta = \frac{2}{\|u\|^2} \tag{5-93}$$

we find that
$$Pa = \mp \|a\|\epsilon_1 \tag{5-94}$$

Since all the elements of ϵ_1 are zero except the first (which equals to 1), only the first element of the column vector Pa is different from zero. Similarly, we find that the row vector
$$\tilde{a}\tilde{P} = \mp\|a\|\tilde{\epsilon}_1 \tag{5-95}$$

also has only one nonvanishing element. This particular property of P may be used to transform all the elements of a row or a column of a matrix to a tridiagonal form.

Reduction of one row and one column to tridiagonal form. The property of P given by (5-94) and (5-95) allows us to perform a similarity transformation on an arbitrary real, symmetric matrix and to change one of its columns and rows into a tridiagonal form. To reduce the whole matrix A of dimension n to a diagonal form, it takes, in general, $(n-1)$ such transformations. For the purpose of introducing the method, it is convenient to carry out the transformations starting with the first row and column. Similar to (5-90), the transformation matrix has the form
$$T(1) = 1 - \beta w^{(1)}\tilde{w}^{(1)} \tag{5-96}$$

where the column vector
$$w^{(1)} = \begin{pmatrix} 0 \\ a_{2,1} \pm \alpha \\ a_{3,1} \\ a_{4,1} \\ \ldots \\ a_{n,1} \end{pmatrix} \tag{5-97}$$

is made of the elements of the first column (or row) of A. The only exceptions are that the first element is zero and that the second element contains an additional factor α given by
$$\alpha^2 = \sum_{i=2}^{n} a_{i,1}^2$$

The structure of w is similar to that of u in (5-91), except that the vector is made of all the elements of the first column of the matrix A below the diagonal.

If we choose the value of β in (5-96) in the same way as that given in (5-93) by taking
$$\beta = \frac{2}{\|w^{(1)}\|^2} \tag{5-98}$$

the matrix $T(1)$ is unitary:
$$T^{-1}(1) = \tilde{T}(1) = T(1)$$

This can be shown in the following way. First, we see that the transpose of $T(1)$ is the same as $T(1)$ itself,

$$\widetilde{T}(1) = \widetilde{1} - \beta \widetilde{w^{(1)}\widetilde{w}^{(1)}} = 1 - \beta \widetilde{\widetilde{w}}^{(1)} \widetilde{w}^{(1)} = 1 - \beta w^{(1)} \widetilde{w}^{(1)} = T(1)$$

as $\widetilde{\widetilde{w}}^{(1)}$, the transpose of $\widetilde{w}^{(1)}$, is the same as $w^{(1)}$. Note also that the product of $\widetilde{w}^{(1)}$ with $w^{(1)}$ is an outer product, as this is the type of product used in the definition of $T(1)$ in (5-96). Second, we find that

$$T^2(1) = \left(1 - \beta w^{(1)}\widetilde{w}^{(1)}\right)\left(1 - \beta w^{(1)}\widetilde{w}^{(1)}\right) = 1 - 2\beta w^{(1)}\widetilde{w}^{(1)} + \beta^2 w^{(1)}\widetilde{w}^{(1)} w^{(1)}\widetilde{w}^{(1)}$$

In the last term, we can take first the product of the $\widetilde{w}^{(1)}$ with w in the middle. This is an inner product, since $T^2(1)$ is an inner product by definition. The result is a scalar quantity, the square of the norm of $w^{(1)}$. Because of the choice of β made in (5-98), we have

$$\widetilde{w}^{(1)} w^{(1)} = \frac{2}{\beta}$$

As a result,

$$T^2(1) = 1$$

This, in turn, means that a similarity transformation of A using $T(1)$ can be written in the form $T(1)AT(1)$.

If we think of A as a collection of n column vectors, we find that, by comparing with (5-94), we obtain the result

$$A' = T(1)A$$

$$= \begin{pmatrix} 1 & 0 & 0 & \cdots & 0 \\ 0 & 1-\beta w_2 w_2 & -\beta w_2 w_3 & \cdots & -\beta w_2 w_n \\ 0 & -\beta w_3 w_2 & 1-\beta w_3 w_3 & \cdots & -\beta w_3 w_n \\ 0 & \vdots & \vdots & \ddots & \vdots \\ 0 & -\beta w_n w_2 & -\beta w_n w_3 & \cdots & 1-\beta w_n w_n \end{pmatrix} \begin{pmatrix} a_{1,1} & a_{1,2} & a_{1,3} & \cdots & a_{1,n} \\ a_{2,1} & a_{2,2} & a_{2,3} & \cdots & a_{2,n} \\ a_{3,1} & a_{3,2} & a_{3,3} & \cdots & a_{3,n} \\ \vdots & \vdots & \vdots & \ddots & \vdots \\ a_{n,1} & a_{n,2} & a_{n,3} & \cdots & a_{n,n} \end{pmatrix}$$

$$= \begin{pmatrix} a_{1,1} & a_{1,2} & a_{1,3} & \cdots & a_{1,n} \\ \mp\alpha & a'_{2,1} & a'_{2,3} & \cdots & a'_{2,n} \\ 0 & a'_{3,2} & a'_{3,3} & \cdots & a'_{3,n} \\ \vdots & \vdots & \vdots & \ddots & \vdots \\ 0 & a'_{n,2} & a'_{n,3} & \cdots & a'_{n,n} \end{pmatrix} \tag{5-99}$$

where w_k are the matrix elements of $w^{(1)}$. We shall return later to the forms of the matrix elements $a'_{i,j}$ for both i and j greater than 1. For the moment, the important point is that the elements in the first row of the product matrix A' remain unchanged from those of A. This arises from the fact that the first element of w is zero. Note that the elements of the first column are now in a tridiagonal form.

We can now consider A' as a collection of n row vectors. Using (5-96), we have the result

$$A'' = A'T(1) = \begin{pmatrix} a_{1,1} & \mp\alpha & 0 & \cdots & 0 \\ \mp\alpha & a''_{2,1} & a''_{2,3} & \cdots & a''_{2,n} \\ 0 & a''_{3,2} & a''_{3,3} & \cdots & a''_{3,n} \\ \vdots & \vdots & \vdots & \ddots & \vdots \\ 0 & a''_{n,2} & a''_{n,3} & \cdots & a''_{n,n} \end{pmatrix} \quad (5\text{-}100)$$

Again, the first column of A'' is the same as A' because of the structure of $w^{(1)}$.

The net result obtained from the calculations in the previous two paragraphs is that the transformation $T(1)AT(1)$ reduces the matrix A to the form on the right side of (5-100), with the first column and row in the standard form of a tridiagonal matrix. We can now proceed to apply a second transformation and, instead of (5-97), we shall use a column vector of the form

$$w^{(2)} = \begin{pmatrix} 0 \\ 0 \\ a''_{3,2} \pm \alpha \\ a''_{4,2} \\ a''_{5,2} \\ \cdots \\ a''_{n,2} \end{pmatrix} \qquad \alpha = \sum_{i=3}^{n} (a''_{i,2})^2 \quad (5\text{-}101)$$

The new transformation matrix $T(2)$, made of $w^{(2)}$ in an analogous way as that specified by (5-96), reduces A'' to a form with one more tridiagonal column and row.

We can generalize the procedure to one that transforms a particular row and column of an arbitrary real, symmetric matrix A to tridiagonal form. If we apply such transformations $(m-1)$ times, starting from the first column and row as prescribed in (5-99) and (5-100), the matrix is reduced to the form of a partially tridiagonal matrix

$$A(m-1) = \left(\begin{array}{ccccc|cccc} d_1 & f_2 & 0 & \cdots & 0 & 0 & 0 & 0 & \cdots & 0 \\ f_2 & d_2 & f_3 & \cdots & 0 & 0 & 0 & 0 & \cdots & 0 \\ 0 & f_3 & d_3 & \cdots & 0 & 0 & 0 & 0 & \cdots & 0 \\ \vdots & \vdots & \vdots & \ddots & \vdots & \vdots & \vdots & \vdots & \ddots & \vdots \\ 0 & 0 & 0 & \cdots & f_{m-1} & d_{m-1} & f_m & 0 & \cdots & 0 \\ \hline 0 & 0 & 0 & \cdots & 0 & f_m & a^{(m-1)}_{m,m} & a^{(m-1)}_{m,m+1} & \cdots & a^{(m-1)}_{m,n} \\ 0 & 0 & 0 & \cdots & 0 & 0 & a^{(m-1)}_{m+1,m} & a^{(m-1)}_{m+1,m+1} & \cdots & a^{(m-1)}_{m+1,n} \\ \vdots & \vdots & \vdots & \ddots & \vdots & \vdots & \vdots & \vdots & \ddots & \vdots \\ 0 & 0 & 0 & \cdots & 0 & 0 & a^{(m-1)}_{n,m} & a^{(m-1)}_{n,m+1} & \cdots & a^{(m-1)}_{n,n} \end{array} \right) \quad (5\text{-}102)$$

with $(m-1)$ columns and rows in tridiagonal form. To reduce this matrix to one with one more tridiagonal column and row, we follow (5-96) and use the transformation matrix
$$T(m) = 1 - \beta w^{(m)} \widetilde{w}^{(m)} \tag{5-103}$$
where the elements of the column vector w have the form
$$w_i^{(m)} = \begin{cases} 0 & \text{for } i \leq m \\ a_{m+1,m}^{(m-1)} \pm \alpha & \text{for } i = m+1 \\ a_{m,i}^{(m-1)} & \text{for } m+1 < i \leq n \end{cases} \tag{5-104}$$
with
$$\alpha = \sum_{j=m+1}^{n} \{a_{m,j}^{(m-1)}\}^2$$
Using (5-102) and (5-103), it is easy to show that
$$T(m) A(m-1) T(m) = A(m) \tag{5-105}$$
where $A(m)$ is a matrix with the first m columns and rows in tridiagonal form.

We can take (5-105) as a recursion relation to transform an arbitrary real, symmetric matrix A to a tridiagonal form. The starting point of such a process is $A(0) = A$. The transformation matrix $T(1)$ is given by (5-96), and it generates $A(1)$ with the first column and row in tridiagonal form, as shown on the right side of (5-100). From $A(1)$ we can produce $A(2)$ using $T(2)$ made of $w^{(2)}$ of (5-101), and so on. After $(n-1)$ such similarity transformations, each having the form of (5-103), we arrive at a tridiagonal form of the original A.

It is useful to rewrite (5-105) in a form that is more convenient for practical calculations. Using (5-103), we have
$$A(m) = (1 - \beta w^{(m)} \widetilde{w}^{(m)}) A(m-1)(1 - \beta w^{(m)} \widetilde{w}^{(m)})$$
$$= A(m-1) - \beta A(m-1) w^{(m)} \widetilde{w}^{(m)} - \beta w^{(m)} \widetilde{w}^{(m)} A(m-1)$$
$$\quad + \beta^2 w^{(m)} \widetilde{w}^{(m)} A(m-1) w^{(m)} \widetilde{w}^{(m)}$$
$$= A(m-1) - p\widetilde{w}^{(m)} - w^{(m)} \widetilde{p} + \beta w^{(m)} \widetilde{p} w^{(m)} \widetilde{w}^{(m)} \tag{5-106}$$
where we have defined a column matrix
$$p = \beta A(m-1) w^{(m)} \tag{5-107}$$
Since $A(m)$ is a symmetric matrix, we have
$$\widetilde{p} = \beta \widetilde{w}^{(m)} \widetilde{A}(m-1) = \beta \widetilde{w}^{(m)} A(m-1)$$

The final result may be put in a more convenient form by defining another column matrix

$$q = p - \tfrac{1}{2}\beta(\widetilde{p}w^{(m)})w^{(m)} = p - \tfrac{1}{2}\beta w^{(m)}(\widetilde{p}w^{(m)}) \qquad (5\text{-}108)$$

where $(\widetilde{p}w^{(m)})$ is a scalar product. Using this, the final result of (5-106) may be written as

$$A(m) = A(m-1) - \{p - \tfrac{1}{2}\beta w^{(m)}(\widetilde{p}w^{(m)})\}\widetilde{w}^{(m)} - w^{(m)}\{\widetilde{p} - \tfrac{1}{2}\beta(\widetilde{p}w^{(m)})\widetilde{w}^{(m)}\}$$
$$= A(m-1) - q\widetilde{w}^{(m)} - w^{(m)}\widetilde{q} \qquad (5\text{-}109)$$

This is the form commonly employed in the published programs on Householder tridiagonalization.

It is also possible to start the tridiagonalization process from the last column and row, instead of the first column and row as we did above. In the place of (5-97), we can use a column matrix of the form

$$w^{[n]} = \begin{pmatrix} a_{1,n} \\ a_{2,n} \\ \cdots \\ a_{n-2,n} \\ a_{n-1,n} \pm \alpha \\ 0 \end{pmatrix} \qquad \alpha = \sum_{i=1}^{n-1} a_{i,n}^2$$

Let us name the transformation matrix constructed from this form of $w^{[n]}$ as $T[n]$:

$$T[n] = 1 - \beta w^{[n]}\widetilde{w}^{[n]}$$

It transforms the matrix A into one with column n and row n in a tridiagonal form,

$$A[n] = T[n]AT[n] = \begin{pmatrix} a_{1,1}^{[n]} & a_{2,1}^{[n]} & \cdots & a_{n-1,1}^{[n]} & 0 \\ a_{2,1}^{[n]} & a_{2,2}^{[n]} & \cdots & a_{n-1,2}^{[n]} & 0 \\ \vdots & \vdots & \ddots & \vdots & \vdots \\ a_{1,n-1}^{[n]} & a_{2,n-1}^{[n]} & \cdots & a_{n-1,n-1}^{[n]} & \mp\alpha \\ 0 & 0 & \cdots & \mp\alpha & a_{n,n}^{[n]} \end{pmatrix}$$

The definition of β is the same as that given by (5-98). Next we apply a transformation using $T[n-1]$, constructed in a similar way as $T[n]$ with the two last elements in w equal to zero and using the elements of $A[n]$ for the rest. The result is $A[n-1]$ with columns and rows n and $(n-1)$ in tridiagonal form.

We can proceed in this way to reduce the matrix elements in column and row $(n-2)$, and so on, until the whole matrix is in a tridiagonal form. The final result must be the same as that discussed earlier where the reduction is carried forward starting from column and row 1, rather than backward starting from column and row n. The choice between these two approaches may be made on the grounds of numerical accuracy. It is obvious from the discussions that, if there are large off-diagonal matrix elements in A, it will be better if we try to reduce them to zeros as late as possible. This comes from the fact that cancellation among large numbers has a tendency to reduce the numerical accuracy in a calculation. Since every transformation modifies the values of all the elements in the part of the matrix that is not yet in a tridiagonal form, the absolute magnitudes of the large elements will likely be reduced. As a result, the need for reducing large elements to zero may not materialize in practice. For this reason, it may be advantageous to interchange the columns and rows of A before the tridiagonalization process in such a way that the large off-diagonal matrix elements are, as far as possible, in the later columns and rows.

Most of the published matrix diagonalization programs using the method of tridiagonalization start the reduction from the last column and row. That is, if the dimension of the matrix is n, the first column and row to be reduced is the nth. In this case, the large elements should be put in the leftmost columns and uppermost rows, if possible. Alternatively, if the reduction starts from the first column and row, it is preferable to have the large elements, as far as possible, in the rightmost columns and bottommost rows. To distinguish between these two approaches, we shall use brackets to indicate the steps in transforming A to the tridiagonal form J starting from column and row n, and parentheses for those starting from column and row 1.

To improve the numerical accuracy further, it is also useful to scale the matrix elements that enter into the calculation of $w^{[m]}$. This comes from the observation that some of the off-diagonal matrix elements can be small, especially after the initial few steps in the tridiagonalization process. Since $w^{[m]}$ depends only on the relative size of $a_{m,j}^{[m-1]}$ for $j = 1, 2, \ldots, (m-1)$, the transformation matrix $T^{[m]}$ is not changed if we divide all $a_{m,j}^{[m-1]}$ by a scale factor s. One choice is to take s to be the sum of the absolute values of $a_{m,j}^{[m-1]}$. This is included as a part of the algorithm given in Box 5-7.

Transformation matrix. So far we have concerned ourselves only with the process of changing the matrix A to a tridiagonal form. Our ultimate goal is to obtain both the eigenvalues and the eigenvectors. For this purpose, it is also necessary to diagonalize the tridiagonal matrix. In the approach used here, the transformation matrix that brings the original matrix A to a diagonal one consists of two parts. The first part reduces A to a tridiagonal matrix J and the second part, to which we shall return in the next section, turns J into a diagonal matrix. Let us define a transformation matrix R that accomplishes the first part. That is,

$$\widetilde{R}AR = J \tag{5-110}$$

From the discussions above, it is clear that, if we carry out the reduction by starting from column and row n,

$$R = T[n]T[n-1]\cdots T[2] \tag{5-111}$$

Alternatively, if we carry out the reduction starting from column and row 1, we have

$$R = T(1)T(2)\cdots T(n-1)$$

Note that there are only $(n-1)$ steps in either approach since the matrix is tridiagonal by the end of step $(n-1)$. The matrix R is not the eigenvector matrix, since we still have to diagonalize the tridiagonal matrix.

Storage considerations. It is a relatively simple matter to obtain the transformation matrix R from the product of $T[n]$, $T[n-1]$, ..., $T[2]$. However, to store R, we need, in principle, a two-dimensional array of size n^2. The computer program will require two such arrays, one to store the matrix A and the other for R. As we shall see in the following discussion, it is possible to reduce the storage requirement to only one such array. The possibility to reduce the storage requirement by almost a factor of 2 is important if we wish to handle cases with large n. In many physical problems, it is often necessary to use as large a matrix as possible, and the reduction is therefore useful.

In the tridiagonalization process, the matrix $T[k]$ is usually discarded at the end of step k. If we wish to construct the transformation matrix R, the information on $T[k]$ must be stored before the next step in the reduction is carried out. At the same time, we also notice that, as each column and row of A is reduced to a tridiagonal form, only two nonzero matrix elements are left. The locations used to store the other off-diagonal matrix elements are no longer needed and may be used to store information related to the transformation matrix R. Consider now step k of the reduction. The crucial information concerning R is contained in the column vector $w^{[k]}$. The number of nonvanishing elements in $w^{[k]}$ is $(k-1)$. This is exactly the same as the number of off-diagonal matrix elements annihilated by $T[k]$. As a result, we have the correct amount of space to store the $w^{[k]}$ in the space "discarded" by the off-diagonal matrix elements of $A[k-1]$, and this is true in every step of the tridiagonalization process.

Note that the space vacated by the discarded matrix elements of A is not suitable to store the transformation matrix R itself. Since the transformation of A into a tridiagonal matrix is carried out row by row (or column by column), there is not enough space to store the form of R at the end of each step. This comes from the simple fact that $T[i]$ is a square matrix of dimension n and, consequently, takes up more space than that vacated by the matrix elements of A annihilated by the transformation. However, the amount of space is just right to store $w^{[k]}$. With $w^{[k]}$, it is possible to reconstruct R at the end of the tridiagonalization step, using the space that was originally occupied by A.

Let us use $R[k]$ to represent the product of the first k of the $(n-1)$ number of $T[i]$. In terms of $R[k]$, (5-111) may be put into the form of a recursion relation

$$R[k] = T[k]R[k-1] \tag{5-112}$$

The starting point is

$$R[1] = 1$$

and the end result is given by

$$R = R[n-1]$$

Since $T[k]$ has the form

$T[k] = 1 - \beta w^{[k]} \widetilde{w}^{[k]}$

$$= \begin{pmatrix} 1-\beta w_1^{[k]}w_1^{[k]} & -w_1^{[k]}w_2^{[k]} & -w_1^{[k]}w_3^{[k]} & \cdots & -w_1^{[k]}w_{k-1}^{[k]} & 0 & \cdots & 0 \\ -w_2^{[k]}w_1^{[k]} & 1-\beta w_2^{[k]}w_2^{[k]} & -w_2^{[k]}w_3^{[k]} & \cdots & -w_2^{[k]}w_{k-1}^{[k]} & 0 & \cdots & 0 \\ -w_3^{[k]}w_1^{[k]} & -w_3^{[k]}w_2^{[k]} & 1-\beta w_3^{[k]}w_3^{[k]} & \cdots & -w_3^{[k]}w_{k-1}^{[k]} & 0 & \cdots & 0 \\ \vdots & \vdots & \vdots & \ddots & \vdots & \vdots & \ddots & \vdots \\ -w_{k-1}^{[k]}w_1^{[k]} & -w_{k-1}^{[k]}w_2^{[k]} & -w_{k-1}^{[k]}w_3^{[k]} & \cdots & 1-\beta w_{k-1}^{[k]}w_{k-1}^{[k]} & 0 & \cdots & 0 \\ 0 & 0 & 0 & \cdots & 0 & 1 & \cdots & 0 \\ \vdots & \vdots & \vdots & \ddots & \vdots & \vdots & \ddots & \vdots \\ 0 & 0 & 0 & \cdots & 0 & 0 & \cdots & 1 \end{pmatrix}$$

it is clear that the essential information of $R[k]$, that is, matrix elements that are neither zero nor one, is confined to the upper-left square corner of dimension $(k-1)$. This means that it is possible to construct R using the same two-dimensional array where the $(n-1)$ column matrices $w^{[k]}$ are stored. The useful information for $R[k]$ may be stored without destroying the matrix elements $w^{[i]}$ for $i = k$, $k+1, \ldots, n$.

In terms of $T[k]$ given by the analogous form to (5-103), we can write $R[k]$ as

$$R[k] = \left(1 - \beta w^{[k]} \widetilde{w}^{[k]}\right) R[k-1] = R[k-1] - \beta R'$$

where

$$R' = w^{[k]} \widetilde{w}^{[k]} R[k-1]$$

In terms of matrix elements

$$r_{i,j}^{[k]} = \sum_{\ell=1}^{k-1} (w^{[k]} \widetilde{w}^{[k]})_{i,\ell} r_{\ell,j}^{[k-1]} = w_i^{[k]} \sum_{\ell=1}^{k-1} w_\ell^{[k]} r_{\ell,j}^{[k-1]} \tag{5-113}$$

where $r_{i,j}^{[k]}$ are the matrix element of $R[k]$ at position (i,j). The algorithm based on such an approach is outlined in Box 5-7, and the transformation matrix R produced is used in the next section to generate the eigenvectors of A.

Box 5-7 Subroutine TRI_DIAG(A,N,D,F,NDMN)
Householder tridiagonalization of a real, symmetric matrix

Argument list:
- A: Input real symmetric matrix A. Output matrix R of (5-110).
- N: Matrix dimension n.
- D: Output diagonal elements of tridiagonal matrix J.
- F: Output superdiagonal matrix elements of J.
- NDMN: Dimension of the arrays in the calling program.

1. If $n = 1$, return $a_{1,1}$ as the eigenvalue and 1.0 as the eigenvector.
   If $n = 2$, calculate the eigenvalues and eigenvector by a Jacobi rotation of (5-67).
2. For $n > 2$, carry out the tridiagonalization steps starting from $i = n$.
   (a) Find the scale factor $s = \sum_{j=1}^{m-1} |a_{m,j}|$. If $s = 0$, set $d_m = a_{m,m}$, $f_m = a_{m,m-1}$, $w^{[m]} = 0$, and skip steps (b) to (h).
   (b) Scale the row of matrix elements and find the norm.
   (c) Calculate $\alpha^2 = \sum_{j=1}^{m-1} a_{m,j}^2$ and take the sign of α to be that of $-a_{m,m-1}$.
   (d) Construct $w^{[m]}$ with $w_j = a_{m,j}$ for $j = 1$ to $(m-2)$, $w_{m-1} = a_{m,m-1} - \alpha$, $w_j = 0$ for $j = m$ to n, and $\|w^{[m]}\|^2 = \alpha^2 - \alpha a_{m,m-1}$.
   (e) Store $\{w_i\}$ in array D.
   (f) Construct column matrix p of (5-107) and store the elements in array F.
   (g) Construct column matrix q of (5-108) and replace p_j by q_j in array F.
   (h) Reduce column m and row m of A to tridiagonal form, using (5-109).
   (i) Store $a_{m,m}$ as d_m and $s * a_{m,m-1}$ as f_m.
   (j) Decrease m by 1 and repeat steps (a) to (i) until $m = 1$.
3. Construct the transformation matrix R of (5-110) from $w^{[m]}$.
   (a) Start with $m = 1$ and $r_{1,1} = 1$.
   (b) Obtain $R[m]$ from $R[m-1]$ and $w^{[m]}$. Calculate only matrix elements $r_{i,j}$ for $i = 1$ to $(m-1)$ and $j = 1$ to $(m-1)$ using (5-113).
   (c) Let $r_{m,m} = 1$ and $r_{m,j} = r_{j,m} = 0$ for $j = 1$ to $(m-1)$.
   (d) Repeat steps (b) and (c) until $m = n$.
4. Return diagonal matrix elements d_i in D, superdiagonal matrix elements f_i in F, and transformation matrix R in A.

We can test a computer program for tridiagonalization by checking whether the transformation matrix R produced satisfies the relation

$$\widetilde{R}AR = J$$

given by (5-110). Here J is a tridiagonal, symmetric matrix with diagonal matrix elements $d_1, d_2, \ldots, d_n$ and superdiagonal elements $f_2, f_3, \cdots, f_n$. The advantage of testing the program at such an intermediate stage, rather than waiting till the end of the diagonalization of the next section, is that each section of the calculations can be made correct independent of the others.

§5-5 Eigenvalues and eigenvectors of a tridiagonal matrix

The most straightforward method to find the eigenvalues of a real, symmetric tridiagonal matrix J is to solve the characteristic equation

$$J - \lambda \mathbf{1} = \mathbf{0} \tag{5-114}$$

where $\mathbf{0}$ is the null matrix having the same dimension as J but with all the elements zero. Since the matrix J has the simple form shown in Fig. 5-3, it is fairly easy to find the roots for (5-114).

Sturm sequence of polynomials. Consider first the case of dimension $n = 2$. The characteristic equation may be written as a second-order polynomial in λ,

$$p_2(\lambda) = \det \begin{vmatrix} d_1 - \lambda & f_2 \\ f_2 & d_2 - \lambda \end{vmatrix} = (d_1 - \lambda)(d_2 - \lambda) - f_2^2 \tag{5-115}$$

For $n = 3$, the characteristic polynomial for a tridiagonal matrix has the form

$$p_3(\lambda) = \det \begin{vmatrix} d_1 - \lambda & f_2 & 0 \\ f_2 & d_2 - \lambda & f_3 \\ 0 & f_3 & d_3 - \lambda \end{vmatrix} = (d_3 - \lambda)p_2(\lambda) - f_3^2 p_1(\lambda)$$

where we have defined

$$p_1(\lambda) = d_1 - \lambda \tag{5-116}$$

The second-order polynomial $p_2(\lambda)$ has the form given in (5-115). In fact, if we also define a polynomial of order zero,

$$p_0 = 1$$

we can express $p_2(\lambda)$ of (5-115) in the following form:

$$p_2(\lambda) = (d_2 - \lambda)p_1(\lambda) - f_2^2 p_0 \tag{5-117}$$

By continuing with this line of argument, it is not difficult to see that, for tridiagonal matrices, the characteristic polynomials are given by the following recursion relation

$$p_r(\lambda) = (d_r - \lambda)p_{r-1}(\lambda) - f_r^2 p_{r-2}(\lambda) \tag{5-118}$$

where $2 \leq r \leq n$. For a tridiagonal matrix of dimension n, the characteristic equation is then given by the following polynomial of order n in λ:

$$p_n(\lambda) = (d_n - \lambda)p_{n-1}(\lambda) - f_n^2 p_{n-2}(\lambda) = 0 \tag{5-119}$$

A set of polynomials, $p_0(\lambda)$, $p_1(\lambda)$, $p_2(\lambda)$, ..., $p_n(\lambda)$, satisfying the recursion relation (5-118) is said to form a Sturm sequence. There are many interesting properties of such a sequence of polynomials, and we shall make use of them in designing an algorithm to find the eigenvalues of a tridiagonal matrix.

For all practical purposes, we need to consider only tridiagonal matrices without any of the superdiagonal (and subdiagonal) elements $f_2, f_3, \ldots, f_n$ equal to zero. This can be seen from the following argument. If any one of the matrix elements f_i is zero, the matrix can be split into two separate submatrices. For example, if $f_m = 0$, we have one matrix of dimension $(m-1)$, consisting of diagonal elements $d_1, d_2, \ldots, d_{m-1}$ and superdiagonal (and subdiagonal) elements $f_2, f_3, \ldots, f_{m-1}$, and another one of dimension $(n-m+1)$ consisting of diagonal elements $d_m, d_{m+1}, \ldots, d_n$ and superdiagonal (and subdiagonal) elements $f_{m+1}, f_{m+2}, \ldots, f_n$. Similarly, the basis states of the tridiagonal matrix also split into two groups, one consisting of states with indexes $1, 2, \ldots, (m-1)$ and the other consisting of states $m, (m+1), \ldots, n$. Since there are no matrix elements joining these two blocks of states, any transformations among one group do not affect the other. As a result, the eigenvalues of one submatrix are independent of those in the other.

A simple example is the case of a tridiagonal matrix with dimension $n = 4$,

$$J = \begin{pmatrix} d_1 & f_2 & 0 & 0 \\ f_2 & d_2 & f_3 & 0 \\ 0 & f_3 & d_3 & f_4 \\ 0 & 0 & f_4 & d_4 \end{pmatrix}$$

If $f_3 = 0$, the matrix is equivalent to two $n = 2$ matrices. Transformations among columns (and rows) 1 and 2 do not affect the matrix elements d_3, d_4, and f_4. Conversely, the characteristic equation of J cannot be separated into two or more equations unless some of the superdiagonal (and subdiagonal) matrix elements vanish.

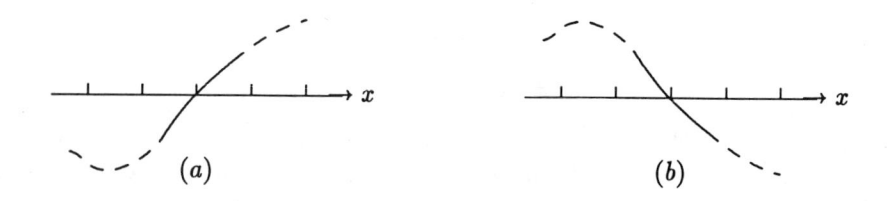

Fig. 5-4 Behavior of a polynomial in an interval where there is only one root. The value of the polynomial must change sign, either from negative to positive as shown in case (a), or from positive to negative as shown in case (b).

Bisection method. A simple way to find the roots of a polynomial is to use the method of bisection, as we did earlier in §3-6 for inverse interpolation. If, by some other means, we know that there is one and only one root of the polynomial $p_r(\lambda)$ in the interval $[v_\ell, v_h]$, the value of the polynomial at $\lambda = v_\ell$ must have a different sign from that at $\lambda = v_h$, as illustrated in Fig. 5-4. Without losing any generality, we can assume that $p_r(v_\ell) < 0$ and $p_r(v_h) > 0$, the same way as illustrated in (a) of Fig. 5-4. We can now divide the interval into two equal parts and calculate the value of $p_r(\lambda)$ at the center of the interval. Let

$$v = \tfrac{1}{2}(v_\ell + v_h) \tag{5-120}$$

If $p_r(v) > 0$, the root is in the lower half of the interval. On the other hand, if $p_r(v) < 0$, the root lies in the upper half of the interval. In the former case, we can reduce the interval to half by replacing the upper limit v_h with v. Similarly, in the latter case, we can achieve the same goal by replacing the lower limit v_ℓ with v. The process of bisection may be repeated as many times as we wish, until the difference between v_h and v_ℓ is less than the accuracy of the root we wish to find. At this point we can take the value of the root as the average between v_h and v_ℓ. The method is quite efficient, particularly if we want only a few of the roots of a polynomial of degree n. However, for this method to work, we must have a way to determine the interval within which each root lies.

For a polynomial of degree 1, there is only root,

$$\lambda^{(1)} = d_1$$

as can be seen from (5-116). The two roots of $p_2(\lambda)$ can be found by solving the quadratic equation on the right side of (5-117):

$$\lambda_1^{(2)} = \tfrac{1}{2}(d_1 + d_2) - \tfrac{1}{2}(d_1 - d_2)\sqrt{1 + \left(\frac{2f_2}{d_1 - d_2}\right)^2}$$

$$\lambda_2^{(2)} = \tfrac{1}{2}(d_1 + d_2) + \tfrac{1}{2}(d_1 - d_2)\sqrt{1 + \left(\frac{2f_2}{d_1 - d_2}\right)^2} \tag{5-121}$$

Since f_2 is assumed to be nonzero, we have the situation that, for $d_1 > d_2$, and both $p_1(\lambda)$ and $p_2(\lambda)$ belong to the same Sturm sequence,

$$\lambda_1^{(2)} < \lambda^{(1)} < \lambda_2^{(2)} \tag{5-122}$$

as can be seen by inspection. (For $d_2 > d_1$, the same inequality holds except that the direction is reversed.)

The property expressed by (5-122) can be shown to be a general one for the polynomials in a Sturm sequence. That is, if $\lambda_i^{(r)}$, for $i = 1, 2, \ldots$, are the roots of $p_r(\lambda)$ and $\lambda_i^{(r+1)}$, for $i = 1, 2, \ldots$, are the roots of $p_{r+1}(\lambda)$, the roots of the two polynomials satisfy the inequality

$$\lambda_1^{(r+1)} < \lambda_1^{(r)} < \lambda_2^{(r+1)} < \lambda_2^{(r)} < \lambda_3^{(r+1)} < \cdots < \lambda_m^{(r+1)} < \lambda_m^{(r)} < \lambda_{m+1}^{(r+1)} <$$

$$\cdots < \lambda_r^{(r+1)} < \lambda_r^{(r)} < \lambda_{r+1}^{(r+1)} \tag{5-123}$$

These relations imply also that two such polynomials, differing in degree by 1, and with $f_r \neq 0$, cannot have any common roots. We can see this from the recursion relation (5-118). Let us assume for the moment that λ_q is a root for both $p_{r+1}(\lambda)$ and $p_r(\lambda)$. For this to be true, it must also be the root of $p_{r-1}(\lambda)$ to satisfy (5-118). By the same token, if λ_q is the root of $p_r(\lambda)$ and $p_{r-1}(\lambda)$, it must also be a root of $p_{r-2}(\lambda)$. When we continue this line of reasoning down to $r = 2$, we reach an absurdity that λ_q is also the root of p_0. Since, by definition, p_0 is a constant and does not have a root, our original assumption must be wrong, and we must conclude that $p_{r+1}(\lambda)$ and $p_r(\lambda)$ cannot share a common root.

To prove the inequality relations of (5-123), we need also to show that there is only one root of $p_r(\lambda)$ between any two adjacent roots of $p_{r+1}(\lambda)$. We shall leave this part of the arguments to books on numerical analysis, such as Wilkinson (*The Algebraic Eigenvalue Problem*, Oxford University Press, Oxford, 1965). Schematically, the relation given by (5-123) is shown in Fig. 5-5.

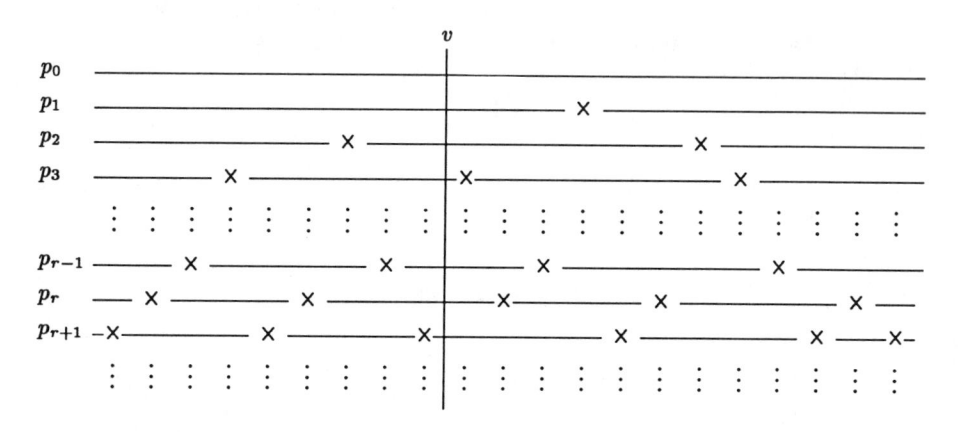

Fig. 5-5 Relation between the roots of the polynomials in a Sturm sequence.

The inequality of (5-123) provides us with a way to find the lower and upper limits of all the roots so as to define the range of our search for each. For this purpose, it is possible to show that all the eigenvalues E_i of a tridiagonal matrix J are bounded by the relation

$$|E_i| \leq \lambda_{\max} \tag{5-124}$$

where

$$\lambda_{\max} = \max\{|f_i| + |d_i| + |f_{i+1}|\} \qquad \text{for} \qquad i = 1, 2, \ldots, n$$

That is, $\lambda_{\max}$ is given by the maximum value among sums of the absolute values of the three elements in a given row. Since the first row has only two nonvanishing elements, d_1 and f_2, the value of f_1 in the above expression is 0. Similarly, $f_{n+1} = 0$, as there are also only two nonvanishing elements in the last row. The quantity $\lambda_{\max}$ is called the maximum or infinite subordinate matrix norm of J. A proof of this relation involves inequalities among matrix norms and can be found in Stoer and Bulirsch (*Introduction to Numerical Analysis* (English translation by R. Bartels, W. Gautschi, and C. Witzgall, Springer-Verlag, New York, 1980). Our interest here lies mainly in the possibility of making use of $\lambda_{\max}$ to provide us with the necessary range of values to start the bisection method for finding the roots of the characteristic equation of (5-119).

Since the largest possible eigenvalue is less than or equal to $\lambda_{\max}$, the roots of any of the polynomials in a Sturm sequence are less than $\lambda_{\max}$. As a result, for $v = \lambda_{\max} + \epsilon$, where ϵ is a small, positive quantity, we have the situation

$$p_r(v) \begin{cases} > 0 & \text{for } r = \text{even} \\ < 0 & \text{for } r = \text{odd} \end{cases} \tag{5-125}$$

This relation implies that, for $v > \lambda_{\max}^{(n)}$, the sign changes between two polynomials differ by 1 in the degree. It is easy to convince ourselves that this is true by considering the limiting case of $v \to +\infty$. In this case, every polynomial is dominated by the first term of (5-118) and

$$p_r(v) \xrightarrow[v \to \infty]{} (-v)^r \tag{5-126}$$

Since there are no roots between $v = +\infty$ and $v = (\lambda_{\max} + \epsilon)$ for any of the polynomials, the signs of $p_r(v)$ cannot change and we obtain (5-125). As we decrease the value of v from the upper limit of $\lambda_{\max}$ until it is just less than the largest eigenvalue $\lambda_n^{(n)}$, the sign of $p_n(v)$ changes. This comes from the fact that it passes through a root. However, the signs of the other polynomials are not changed, as the value of v is still larger than any of the roots of the other polynomials. As a result, the values of $p_n(x)$ and $p_{n-1}(x)$ have the same sign for x in the vicinity of this particular value of v.

Upper and lower limits of eigenvalues. Let us use s to record the number of agreements in sign at any given value of v between any two polynomials differing in degree by 1. When $v = \lambda_{\max} + \epsilon$, no two adjacent polynomials have the same sign, as can be seen from (5-126). In this case, we have $s = 0$. This value of s does not change until we come to $v = \lambda_n^{(n)} - \epsilon$. For small positive values of ϵ, we have $s = 1$ since $p_n(v)$ and $p_{n-1}(v)$ now have the same sign. The value of s remains 1 until v becomes less than $\lambda_{n-1}^{(n)}$, the second largest eigenvalue. This can be seen for the following reasons. As we decrease the value of v from $\lambda_n^{(n)}$, it will eventually become less than $\lambda_{n-1}^{(n-1)}$, the largest root of $p_{n-1}(\lambda)$. At this value, the sign of $p_{n-1}(v)$ changes but not that of $p_n(v)$ or any of the other polynomials, since $\lambda_{n-1}^{(n-1)}$ cannot be the root of any other polynomials. The sign of $p_{n-1}(v)$ now disagrees with that of $p_n(v)$ but is in agreement with that of $p_{n-2}(v)$. As a result, s remains 1. When $v = (\lambda_{n-1}^{(n)} - \epsilon)$, the sign of $p_n(v)$ changes and is now in agreement with that of $p_{n-1}(v)$. Since none of the other polynomials have a change of sign here, we have $s = 2$. If we continue this line of argument, we will come to the conclusion that the value of s is also equal to the number of eigenvalues greater than v. For example, when v is in the interval between $-\infty$ and $(\lambda_1^{(n)} - \epsilon)$, the smallest root of $p_n(\lambda)$ minus a small positive quantity, all the polynomials are positive and the number of sign agreement is $s = n$.

With both upper and lower limits of all the eigenvalues given by (5-124), and the connection between s and the number of eigenvalues given above, we now have a practical way to apply the method of bisection to find the eigenvalues of a tridiagonal matrix. For the convenience of the following discussions, we shall label all the eigenvalues in descending order according to their values. That is,

$$E_1 > E_2 > \cdots > E_n$$

Furthermore, for the time being we shall ignore the possibility of degeneracies, that is, two or more eigenvalues equal in value. However, we shall return to this practical question later when we construct an algorithm for diagonalizing a tridiagonal matrix.

Let us assume that we are interested in the ith largest eigenvalue E_i. In terms of the roots of the polynomial $p_n(v)$, it is equivalent to find the ith largest root $\lambda_i^{(n)}$. From (5-124), we find that the value must lie within the interval $[-\lambda_{\max}, +\lambda_{\max}]$. Using the bisection method, we have

$$v_\ell = -\lambda_{\max} \qquad\qquad v_h = \lambda_{\max}$$

Next, we assign v as the average of v_ℓ and v_h, as we did earlier in (5-120). In addition to calculating the value of $p_n(v)$, we check also, for $r = 0$ to $(n-1)$, whether $p_{r+1}(v)$ agrees in sign with $p_r(v)$ and store the number of agreements as s. There is very little additional work involved in carrying out such a counting if we use the recursion relation (5-118) to evaluate $p_n(v)$. For $s \geq i$, we know that E_i is in the upper half of the interval, and for $s < i$, we find that E_i is in the lower half. The interval is now reduced to half, and the process is repeated until the interval is less than or equal to the required accuracy. This is, essentially, the same procedure as that discussed earlier for the simple case of a single root in a given interval, except that the value of s is used here to determine which half of the interval to use for the next step.

The only complication in the method comes when one of the polynomials becomes zero for some value of v. Let us assume that this takes place at $p_r(v)$. From (5-123), we know that both $p_{r-1}(v)$ and $p_{r+1}(v)$ cannot vanish. A moment of reflection with the help of Fig. 5-5 will convince us that $p_r(v) = 0$ should be considered as a disagreement in sign with $p_{r-1}(v)$ and an agreement in sign with $p_{r+1}(v)$. This completes the basic method for finding the eigenvalues of a tridiagonal matrix using the method of bisection. Because of the possibility of selecting any one of the eigenvalues to search for, the method is most useful if only a few of the eigenvalues are required.

> **Box 5-8 Subroutine** BISECT(D,F,V,N,NDMN,ACC)
> **Eigenvalues of a tridiagonal matrix by bisection**
>
> Argument list:
> - D: Input array for the diagonal matrix elements.
> - F: Input array for the superdiagonal elements.
> - V: Output array of eigenvalues.
> - N: Matrix dimension.
> - NDMN: Dimension of the arrays D, F, and V in the calling program.
> - ACC: Accuracy required.
>
> Initialization:
>   (a) Set up three auxiliary arrays for the square of the superdiagonal elements and the upper and lower limits of each eigenvalue.
>   (b) Form the squares of the superdiagonal matrix elements.
>
> 1. Find the limits of eigenvalues:
>   (a) Calculate the infinite subordinate matrix norm λ_{max} of (5-124).
>   (b) Store λ_{max} as the upper limits and $-\lambda_{max}$ as the lower limits of all the eigenvalues.
> 2. Find the eigenvalues by bisection:
>   (a) Get the interval from the upper and lower limits stored.
>   (b) Calculate the value at the middle point.
>   (c) Count the number of sign agreements:
>   (i) Let p be the value of the polynomial for this order and p_{-1} the previous order.
>   (ii) If $p * p_{-1} > 0$, no sign change,
>   (iii) If $p = 0$, the sign is considered to be different, and
>   (iv) If $p_{-1} = 0$, the sign is considered to be in agreement.
>   (d) Bisect the interval.
>   (e) Update the limits for the other eigenvalues.
>   (f) Check if the difference between the upper and lower limits for the present eigenvalue is smaller than the accuracy required.
>   (i) If so, store the average of the two limits as the eigenvalue.
>   (ii) If not, go back to step (b) using the new interval.
> 3. Return the eigenvalues.

A small improvement may be added to the method. For a given value of v, we know that there are s eigenvalues above v and $(n-s)$ eigenvalues below v. As a result, the value of v becomes the lower bounds for all E_i with $i \leq s$. Similarly, v becomes the upper bounds of the $(n-s)$ eigenvalues with subscript $i < s$. As a result, the value of v used in calculating any one of the eigenvalues can also serve either as the upper or lower limits of all the other unknown E_i, depending on whether s is greater or less than i. It is, therefore, possible to keep a list of the upper and lower limits of all the eigenvalues known so far in the calculation. Each time the value of s is calculated for a particular v, we can check the list and update it if the v is a better estimate of the upper or lower limit of any one of the unknown eigenvalues. This improves the efficiency of the method since the starting range of search for any eigenvalue decreases as more and more eigenvalues are found. The algorithm is summarized in Box 5-8.

It is relatively easy to test a computer program based on the bisection method to find the eigenvalues of a tridiagonal matrix. All we need to do is to substitute the eigenvalues found back into the characteristic equation of (5-119) and see if they are, indeed, the roots of $p_n(v)$ within the specified numerical accuracy. That is, we can check how closely the value $p_n(E_i)$ equals zero for all the eigenvalues found.

QR- and QL-transformations. The strength of the bisection method is in its simplicity. However, as mentioned earlier, it is efficient only in cases where a few eigenvalues are needed. Furthermore, the method does not lend itself easily to calculating the corresponding eigenvectors. For these reasons, the QL- and QR-algorithms are the preferred methods for diagonalizing tridiagonal matrices.

The principle of QL- and QR-methods applies to a wider class of matrices than the tridiagonal one of interest to us here. For this reason, we shall begin by considering symmetric matrices in general. It can be shown that any symmetric matrix $\boldsymbol{A}$ can always be reduced to a product of two matrices,

$$\boldsymbol{A} = \boldsymbol{QR} \tag{5-127}$$

where $\boldsymbol{Q}$ is a unitary matrix, with the property $\boldsymbol{Q}^{-1} = \boldsymbol{Q}^\dagger$, and $\boldsymbol{R}$ is an upper triangular matrix having nonvanishing matrix elements only in the upper triangle. This is similar in spirit to the LR-decomposition used in §5-2 to solve a system of linear equations.

Using the unitary property of $\boldsymbol{Q}$, we obtain from (5-127) the relation

$$\boldsymbol{Q}^\dagger \boldsymbol{A} = \boldsymbol{R}$$

Alternatively, we can define a new matrix $\boldsymbol{B}$ such that

$$\boldsymbol{B} \equiv \boldsymbol{RQ} = \boldsymbol{Q}^\dagger \boldsymbol{A} \boldsymbol{Q} \tag{5-128}$$

Since $\boldsymbol{B}$ is the result of applying a unitary transformation on $\boldsymbol{A}$, its eigenvalues are identical to those of $\boldsymbol{A}$.

In general, $\boldsymbol{B}$ is not a triangular matrix, since multiplying the triangular matrix $\boldsymbol{R}$ by $\boldsymbol{Q}$ produces nonvanishing matrix elements in the other half of the triangle. However, if we define

$$\boldsymbol{A}(1) \equiv \boldsymbol{A}$$

and apply a series of unitary transformations given by the recursion relation

$$\boldsymbol{A}(k+1) = \boldsymbol{Q}^\dagger(k)\boldsymbol{A}(k)\boldsymbol{Q}(k) \tag{5-129}$$

for $k = 1, 2, \ldots$, using unitary matrices $\boldsymbol{Q}(1)$, $\boldsymbol{Q}(2), \ldots$, satisfying (5-127), it is possible to prove that

$$\boldsymbol{A}(k) \xrightarrow[k \to \infty]{} \boldsymbol{D} \tag{5-130}$$

If A is a real, symmetric matrix, D is a diagonal matrix and, as a result, the diagonal matrix elements of D are the eigenvalues of A. In the more general case of A being a nonsymmetric matrix, the diagonal matrix elements of D remain as the eigenvalues of A, but the matrix D is triangular instead, with vanishing elements only above the diagonal. This theorem is not obvious and readers interested for a proof should consult Wilkinson or Stoer and Bulirsch.

Similar to (5-127), we can also decompose A into a product of unitary matrix Q and a lower triangular matrix L with nonvanishing matrix elements only in the lower half,

$$A = QL \qquad (5\text{-}131)$$

This is known as the QL-decomposition. Since there is nothing special about the upper half-triangle of a matrix, compared with the lower half, the result of (5-130) applies here also. We have now a choice between using the QR-decomposition of (5-127) and the QL-decomposition of (5-131) in diagonalizing a matrix. The selection may be made based on numerical accuracy considerations. As we shall see later, if the largest elements are in the lower-right corner of the matrix, the QL-decomposition is more likely to generate better results. On the other hand, if the largest elements are in the upper-left corner, the QR-method is preferred.

As a general method for matrix diagonalization, the QL- and QR-methods are not very useful because of the iterative nature of the procedure, as can be seen from (5-130). On the other hand, for tridiagonal matrices and band matrices, that is, matrices with nonvanishing elements along and near the diagonal, these two methods are likely to be the more efficient ones existing these days. As practical methods for diagonalizing real symmetric matrices, it is therefore most efficient to apply the QL- or QR-algorithm after the matrices are reduced to a tridiagonal form using, for example, the method of Householder given in the previous section. Furthermore, we shall choose the QL approach here, following the most commonly used method originated by Bowdler and others, (J.H. Wilkinson and C. Reinsch, *Linear Algebra*, Handbook for Automatic Computation, vol. II, edited by F.L. Bauer and others, Springer-Verlag, Berlin, 1971, page 227). For the remainder of this section, we shall therefore concentrate on the application of the QL-algorithm to tridiagonal matrices, in particular those obtained from real, symmetric matrices. Later, in §5-7 and §5-8, we shall return briefly to the more general case of a complex, symmetric matrix.

Shift of the diagonal elements. Before we embark on the detailed steps of carrying out a QL-decomposition, it is necessary to realize a special point concerning the question of convergence for the algorithm. In general, the rate an off-diagonal matrix element $a_{i,j}$ goes to zero in (5-130) is proportional to the ratio of the absolute values of the two eigenvalues λ_i and λ_j,

$$\text{rate}\{a_{i,j}^{(k)} \to 0\} \sim \left|\frac{\lambda_i}{\lambda_j}\right|^k \qquad (5\text{-}132)$$

As a result, the convergence of a QL-algorithm can be slow if two eigenvalues happen to be very close to each other in their absolute values. To prevent this from happening, we can shift all the diagonal values by a constant η_k. That is, instead of A, we work with the matrix

$$A' = A - \eta_k 1 \tag{5-133}$$

The value of the parameter η_k will be determined later.

The transformation (5-129) is not affected by such a shift. We can see this in the following way. In step k, we have the relation

$$A'(k) = Q(k)L(k)$$

instead of (5-131). Similar to (5-129), we have the recursion relation

$$A'(k+1) = L(k)Q(k) = Q^\dagger(k)A'(k)Q(k) \tag{5-134}$$

Since

$$Q^\dagger(k)1Q(k) = Q^\dagger(k)Q(k) = 1$$

The result of (5-134) is equivalent to

$$A(k+1) - \eta_k 1 = L(k)Q(k)Q^\dagger(k)A'(k)Q(k) = Q^\dagger(k)A(k)Q(k) - \eta_k 1$$

From this we see that the relation

$$A(k+1) = Q^\dagger(k)A(k)Q(k) \tag{5-135}$$

remains the same as (5-129), except that a QL-decomposition is used here.

Although the shift of (5-133) does not affect the transformation itself, the eigenvalues of A' are different and equal to $(\lambda_i - \eta_k)$, rather than λ_i, the eigenvalues of A. As a result, the relation given by (5-132) becomes

$$\text{rate}\{a_{i,j}^{(k)} \to 0\} \sim \left|\frac{\lambda_i - \eta_k}{\lambda_j - \eta_k}\right|^k \tag{5-136}$$

If $|\lambda_i|$ is not exactly equal to $|\lambda_j|$, we can choose the value of η_k to be as close to that of λ_i as possible and, as a result, the ratio on the right side of (5-136) can be made small. The net result is that $a_{i,j}$ goes to zero in far fewer number of steps than without the shift.

What happens if we have degenerate eigenvalues? First, this cannot take place in a tridiagonal matrix unless one of the superdiagonal elements vanishes, as we saw earlier. As an illustration, let us consider again a two-dimensional case. The two eigenvalues of the matrix

$$C_2 \equiv \begin{pmatrix} d_1 & f_2 \\ f_2 & d_2 \end{pmatrix}$$

are, as we saw in (5-121),

$$\lambda_i = \tfrac{1}{2}(d_1 + d_2) \pm \tfrac{1}{2}(d_1 - d_2)\sqrt{1 + \left(\frac{2f_2}{d_1 - d_2}\right)^2}$$

where the subscript i takes on values of either 1 or 2, depending on which of the two possible values we wish to use. The difference between them is given by

$$\Delta \equiv |d_1 - d_2| = (d_1 - d_2)\sqrt{1 + \left(\frac{2f_2}{d_1 - d_2}\right)^2} = \sqrt{(d_1 - d_2)^2 + 4f_2^2}$$

and cannot be zero if $f_2 \neq 0$.

In practical terms, when the numerical accuracy of the computer cannot distinguish $(|d_1| + f_2)$ from $|d_1|$ or $(|d_2| + f_2)$ from $|d_2|$, the presence of a nonzero f_2 is immaterial. As a result, for the general case of an n-dimensional tridiagonal matrix, we can use the criterion

$$[d] + f_i = [d] \qquad (5\text{-}137)$$

to determine whether f_i may be treated as zero. Here, $[d]$ is the larger of $|d_i|$ and $|d_{i+1}|$. When this condition is true, the tridiagonal matrix may be separated into two distinct blocks, each of which may be handled as a separate matrix. This is essentially the same as we saw earlier in relation with the bisection method. The net conclusion we draw from this discussion is that degenerate eigenvalues are not a problem in QL- and QR-methods. However, a practical algorithm for diagonalizing a tridiagonal matrix must test for the possible existence of degeneracies and make allowance for them. We shall return to this point later when we come to the actual algorithm.

The unitary matrix Q. We shall now turn our attention to the question of finding the unitary matrix $Q(k)$ in (5-131) in step k that gives us $A(k)$ when it is multiplied to a lower triangular matrix $L(k)$. For this purpose, it is more convenient to consider instead its Hermitian adjoint $Q^\dagger(k)$ that changes $A(k)$ into a lower triangular matrix $L(k)$:

$$Q^\dagger(k)A(k) = L(k)$$

If $A(k)$ is a tridiagonal matrix of dimension n, this particular step may be achieved by $(n-1)$ rotations, each of which reduces one of the $(n-1)$ superdiagonal matrix elements of $A(k)$ to zero. Although the method applies to a wider class of matrices, we shall be mainly concerned with real, symmetric tridiagonal ones. Furthermore, as we have seen earlier, several such passes may be necessary, as each rotation does not preserve the value of the off-diagonal matrix element annihilated earlier.

Let us use $P_i^\dagger(k)$ to denote the rotation that annihilates $f_i^{(k)}$, the superdiagonal element at location $(i-1,i)$ of a tridiagonal matrix J. The calculation involves taking a linear combination of elements in rows $(i-1)$ and i such that the element $(i-1, i)$ of $L(k)$ vanishes. The form of $P_i^\dagger(k)$ that achieves this aim may be expressed in terms of two parameters c_i and s_i in the following form:

$$P_i^\dagger(k) = \begin{pmatrix} 1 & 0 & \cdots & 0 & 0 & 0 & \cdots & 0 \\ 0 & 1 & \cdots & 0 & 0 & 0 & \cdots & 0 \\ \vdots & \vdots & \ddots & \vdots & \vdots & \vdots & \ddots & \vdots \\ 0 & 0 & \cdots & c_i & -s_i & 0 & \cdots & 0 \\ 0 & 0 & \cdots & s_i & c_i & 0 & \cdots & 0 \\ 0 & 0 & \cdots & 0 & 0 & 1 & \cdots & 0 \\ \vdots & \vdots & \ddots & \vdots & \vdots & \vdots & \ddots & \vdots \\ 0 & 0 & \cdots & 0 & 0 & 0 & \cdots & 1 \end{pmatrix} \begin{matrix} \\ \\ \\ (i-1) \\ i \\ \\ \\ \\ \end{matrix} \qquad (5\text{-}138)$$

with columns labeled $(i-1)$ and i above.

The values of c_i and s_i must satisfy the condition that $P_i^\dagger(k)P_i(k) = 1$. In terms of matrix elements, this is equivalent to the requirement that

$$c_i^2 + s_i^2 = 1 \qquad (5\text{-}139)$$

A second condition on these two parameters is that the superdiagonal matrix element at location $(i-1, i)$ vanish. That is,

$$c_i \, \bar{a}_{i,i+1} - s_i \, \bar{a}_{i+1,i+1} = 0 \qquad (5\text{-}140)$$

where we have used bars on top of the matrix elements to indicate that they are not necessarily the same as matrix elements $f_{i+1}^{(k)}$ and $d_{i+1}^{(k)}$ of the tridiagonal matrix $J(k)$. As we shall see later, earlier rotations in the same step k may modify their values. Our main concern at the moment is with the values of the two parameters c_i and s_i that are determined by (5-139) and (5-140). Note also that the rotation induced by $P_i^\dagger(k)$ is quite different from those used in the Jacobi or Givens method, since the off-diagonal matrix elements here are reduced to zero by the action of $P_i(k)$ alone, rather than through a similarity transformation of the form $P^{-1}AP$ used in (5-67) or (5-88).

At the start of step k, we have a tridiagonal matrix $J(k)$, having the form shown in Fig. 5-3. Being a real symmetric matrix, the subdiagonal elements of $J(k)$ are equal to the corresponding superdiagonal ones. The matrix is completely specified by n diagonal elements and $(n-1)$ superdiagonal elements. For $J(k)$, these may be taken as $d_1^{(k)}, d_2^{(k)}, \ldots, d_n^{(k)}$ and $f_2^{(k)}, f_3^{(k)}, \ldots, f_n^{(k)}$. To make use of a shift of the eigenvalues to improve the rate of convergence, we shall apply a rotation to the $J(k)$ with the diagonal matrix elements shifted in value by a constant η_k:

$$J'(k) = \begin{pmatrix} d_1^{(k)} - \eta_k & f_2^{(k)} & 0 & \cdots & 0 & 0 & 0 \\ f_2^{(k)} & d_2^{(k)} - \eta_k & f_3^{(k)} & \cdots & 0 & 0 & 0 \\ 0 & f_3^{(k)} & d_3^{(k)} - \eta_k & \cdots & 0 & 0 & 0 \\ \vdots & \vdots & \vdots & \ddots & \vdots & \vdots & \vdots \\ 0 & 0 & 0 & \cdots & d_{n-2}^{(k)} - \eta_k & f_{n-1}^{(k)} & 0 \\ 0 & 0 & 0 & \cdots & f_{n-1}^{(k)} & d_{n-1}^{(k)} - \eta_k & f_n^{(k)} \\ 0 & 0 & 0 & \cdots & 0 & f_n^{(k)} & d_n^{(k)} - \eta_k \end{pmatrix}$$

The choice of η_k is not an easy one since we have no knowledge of any of the eigenvalues at this point. The procedure usually recommended is the following. If we wish to reduce the superdiagonal element $f_i^{(k)}$ to zero, we can make an approximation and treat the submatrix formed by $d_{i-1}^{(k)}$, $d_i^{(k)}$, and $f_i^{(k)}$ as a matrix of dimension 2. From (5-83) and (5-84), the two eigenvalues of this matrix are

$$\lambda_1 = d_{i-1}^{(k)} - t f_i^{(k)} \qquad \lambda_2 = d_i^{(k)} + t f_i^{(k)} \tag{5-141}$$

where $t \equiv \tan \theta^{(i)}$ is given by (5-81). Alternatively, it can be expressed as

$$t = \frac{1}{|\xi| \pm \sqrt{1 + \xi^2}} \tag{5-142}$$

where

$$\xi = \frac{d_i^{(k)} - d_{i-1}^{(k)}}{2 f_i^{(k)}}$$

The choice of the sign in the denominator on the right side of (5-142) depends on the sign of ξ. If ξ is positive, we take the plus sign and, if ξ is negative, we take the negative sign. By approximating η_k with the value of λ_1 calculated in this way, reasonable convergence rates may be obtained.

We shall start the $(n-1)$ rotations by trying first to put to zero the superdiagonal element $f_n^{(k)}$. For this purpose, the rotation matrix $P(k)$ takes on the form

$$P_n^\dagger(k) = \begin{pmatrix} 1 & 0 & \cdots & 0 & 0 \\ 0 & 1 & \cdots & 0 & 0 \\ \vdots & \vdots & \ddots & \vdots & \vdots \\ 0 & 0 & \cdots & c_n & -s_n \\ 0 & 0 & \cdots & s_n & c_n \end{pmatrix}$$

with

$$c_n = \frac{d_n^{(k)} - \eta_k}{r_n} \qquad s_n = \frac{f_n^{(k)}}{r_n}$$

where
$$r_n = \sqrt{(d_n^{(k)} - \eta_k)^2 + (f_n^{(k)})^2}$$

Let us denote the matrix after this rotation as $L_n(k)$, having the form

$$L_n(k) = \begin{pmatrix} d_1^{(k)} - \eta_k & f_2^{(k)} & 0 & \cdots & 0 & 0 & 0 \\ f_2^{(k)} & d_2^{(k)} - \eta_k & f_3^{(k)} & \cdots & 0 & 0 & 0 \\ 0 & f_3^{(k)} & d_3^{(k)} - \eta_k & \cdots & 0 & 0 & 0 \\ \vdots & \vdots & \vdots & \ddots & \vdots & \vdots & \vdots \\ 0 & 0 & 0 & \cdots & d_{n-2}^{(k)} - \eta_k & f_{n-1}^{(k)} & 0 \\ 0 & 0 & 0 & \cdots & c_n f_{n-1}^{(k)} & p_{n-1} & 0 \\ 0 & 0 & 0 & \cdots & 0 & g_n & p_n \end{pmatrix}$$

where
$$p_{n-1} = c_n(d_{n-1}^{(k)} - \eta_k) - s_n f_n^{(k)}$$
$$p_n = s_n f_n^{(k)} + c_n(d_n^{(k)} - \eta_k)$$
$$g_n = s_n(d_{n-1}^{(k)} - \eta_k) + c_n f_n^{(k)}$$

The matrix $L_n(k)$ is not yet in a proper lower triangular form, since we have only performed one of the $(n-1)$ steps in multiplying $Q^\dagger(k)$ with $J(k)$. For the same reason, it is not yet possible to equate $L_n(k)$ with $J(k+1)$, since we have yet to multiply $L(k)$ by $Q(k)$ from the right.

Before we go on any further with the calculation to transform $J(k)$ into $L(k)$, it is important to recall that our ultimate goal in step k is not $L(k)$ but $J(k+1)$, given by the relation

$$J(k+1) = Q^\dagger(k)J(k)Q(k) = L(k)Q(k)$$

For this purpose, we must multiply $L(k)$ obtained from (5-131) by $Q(k)$. Since $Q(k)$ is also made of a series of $(n-1)$ transformations, it is convenient to carry out each rotation in $Q^\dagger(k)J(k)$ in conjunction with the corresponding calculations required for $L(k)Q(k)$. In this way, it is possible to obtain $J(k+1)$ without having to store each of the $(n-1)$ rotations that make up $Q(k)$. This may be achieved in the following way. First, we note that the matrix element $d_n^{(k+1)}$ is not affected by any of the later rotations $P_{n-1}^\dagger(k)$, $P_{n-2}^\dagger(k)$, ... As a result, we can complete the calculation of $d_n^{(k+1)}$ by multiplying the last row of $L_n(k)$ with the last column of $P_n(k)$. This gives us

$$d_n^{(k+1)} - \eta_k = g_n s_n + p_n c_n$$
$$= s_n^2(d_{n-1}^{(k)} - \eta_k) + c_n^2(d_n^{(k)} - \eta_k) + 2c_n s_n f_n^{(k)}$$
$$= (d_n^{(k)} - \eta_k) + s_n^2(d_n^{(k)} - \eta_k) + s_n^2(d_{n-1}^{(k)} - \eta_k)$$

However, we are not yet in a position to work out the value for $f_n^{(k+1)}$, as the matrix elements in column $(n-1)$ will be modified by a later rotation in the $(n-2, n-1)$ plane. This is not a basic problem, as a procedure can be designed such that some quantities are kept from one rotation to the next so as to carry out the transformation for $f_n^{(k+1)}$ later. In this way, a fairly efficient scheme can be constructed to carry out all the rotations in a step.

We can now derive a set of general formulas to carry out the $(n-1)$ rotations for step k. After rotation $(j+1)$, the partial lower triangular matrix $L_{j+1}(k)$, resulting from applying the rotations $P_n^\dagger(k)$, $P_{n-1}^\dagger(k), \ldots, P_{j+1}^\dagger(k)$, takes on the form

$$L_{j+1}(k) = \begin{pmatrix} d_1^{(k)} - \eta_k & f_2^{(k)} & 0 & \cdots & 0 & 0 & 0 & 0 & 0 & \cdots \\ f_2^{(k)} & d_2^{(k)} - \eta_k & f_3^{(k)} & \cdots & 0 & 0 & 0 & 0 & 0 & \cdots \\ \vdots & \vdots & \vdots & \ddots & \vdots & \vdots & \vdots & \vdots & \vdots & \ddots \\ 0 & 0 & 0 & \cdots & f_{j-1}^{(k)} & d_{j-1}^{(k)} - \eta_k & f_j^{(k)} & 0 & 0 & \cdots \\ \hline 0 & 0 & 0 & \cdots & 0 & g_j & p_j & 0 & 0 & \cdots \\ 0 & 0 & 0 & \cdots & 0 & z_{j+1} & g_{j+1} & p_{j+1} & 0 & \cdots \\ 0 & 0 & 0 & \cdots & 0 & 0 & z_{j+2} & g_{j+2} & p_{j+2} & \cdots \\ \vdots & \vdots & \vdots & \ddots & \vdots & \vdots & \vdots & \vdots & \vdots & \ddots \end{pmatrix} \begin{matrix} \\ \\ \\ (j-1) \\ j \\ \\ \\ \\ \end{matrix}$$

(5-143)

Note that only the superdiagonal elements of $L_{j+1}(k)$ beyond column j are zero, since we have yet to apply the operations $P_1(k)P_2(k)\cdots P_j(k)$ on the right to complete the similarity transformation on J. For the same reason, the diagonal matrix elements of $L_{j+1}(k)$ beyond column j are not yet the eigenvalues of the matrix A, and they are denoted as $p_j, p_{j+1}, \ldots, p_n$. At the same time, there are non-vanishing off-diagonal matrix elements $z_{j+1}, z_{j+2}, \ldots$ just below the subdiagonal elements. As we shall see below, these elements arise because of transformation $P_n^\dagger(k)$, $P_{n-1}^\dagger(k), \ldots, P_{j+1}^\dagger(k)$.

The aim of the next rotation is to annihilate the element $f_j^{(k)}$. For this purpose, the rotation matrix $P_j^\dagger(k)$ takes on the form of (5-138). From the conditions given by (5-139) and (5-140), we have the results

$$c_j = \frac{p_j}{r_j} \qquad s_j = \frac{f_j^{(k)}}{r_j}$$

where

$$r_j = \sqrt{p_j^2 + \left(f_j^{(k)}\right)^2}$$

On applying $P_j^\dagger(k)$ to $L_{j-1}(k)$, we obtain the matrix

$$L_j(k) = \left(\begin{array}{ccccc|cccc} d_1^{(k)}-\eta_k & f_2^{(k)} & 0 & \cdots & 0 & 0 & 0 & 0 & 0 & \cdots \\ f_2^{(k)} & d_2^{(k)}-\eta_k & f_3^{(k)} & \cdots & 0 & 0 & 0 & 0 & 0 & \cdots \\ \vdots & \vdots & \vdots & \ddots & \vdots & \vdots & \vdots & \vdots & \vdots & \ddots \\ 0 & 0 & 0 & \cdots & g_{j-1} p_{j-1} & 0 & 0 & 0 & \cdots \\ \hline 0 & 0 & 0 & \cdots & z_j & g_j' & p_j' & 0 & 0 & \cdots \\ 0 & 0 & 0 & \cdots & 0 & z_{j+1} & g_{j+1} & p_{j+1} & 0 & \cdots \\ 0 & 0 & 0 & \cdots & 0 & 0 & z_{j+2} & g_{j+2} & p_{j+2} & \cdots \\ \vdots & \vdots & \vdots & \ddots & \vdots & \vdots & \vdots & \vdots & \vdots & \ddots \end{array}\right) \begin{array}{c} \\ \\ \\ (j-1) \\ j \\ \\ \\ \end{array}$$

A new nonzero matrix element z_j in row j and column $(j-2)$ is produced from the product of s_j in row j and column $(j-1)$ of $P_j^\dagger(k)$ and $f_{j-1}^{(k)}$ in row $(j-1)$ and column $(j-2)$ of $L_{j+1}(k)$. We shall not be concerned with it here, since the final form of $J(k+1)$ is tridiagonal, and any elements outside the diagonal and subdiagonal of $L(k)$ must vanish when it is multiplied by $Q(k)$.

There are four other changes in the matrix elements between $L_{j+1}(k)$ and $L_j(k)$ (in addition to $z_j = s_j f_{k-1}^{(k)}$, with which we do not need to be concerned):

$$p_{j-1} = c_j(d_{j-1}^{(k)} - \eta_k) - s_j g_j$$

$$g_{j-1} = c_j f_{j-1}^{(k)}$$

$$g_j' = s_j(d_{j-1}^{(k)} - \eta_k) + c_j g_j$$

$$p_j' = s_j f_j^{(k)} + c_j p_j = r_j \tag{5-144}$$

The values of p_{j-1} and g_{j-1} are needed in the next rotation and must be stored temporarily. The values of g_j' and p_j' may be put to use right away to produce the values of $d_j^{(k+1)}$ and $f_j^{(k+1)}$ for the matrix $J(k+1)$. For this purpose, we need to multiply with the matrix for the previous rotation $P_{j+1}(k)$, involving matrix elements c_{j+1} and s_{j+1}, as well as $P_j(k)$ of this rotation. From the matrix element at location $(j, j+1)$ in the product $L_j(k)P_{j+1}(k)$, we obtain

$$f_{j+1}^{(k+1)} = s_{j+1} p_j' = s_{j+1} r_j$$

Similarly, the matrix element at location (j,j) now has the value $c_{j+1} r_j$. To produce the matrix element $d_j^{(k+1)}$, we need to multiply from the right by $P_j(k)$. This gives

$$d_j^{(k+1)} = s_j g_j' + c_j c_{j+1} r_j$$

$$= s_j[s_j(d_{j-1}^{(k)} - \eta_k) + c_j g_j] + c_{j+1} c_j r_j$$
$$= s_j^2(d_{j-1}^{(k)} - \eta_k) + s_j c_j g_j + \zeta \qquad (5\text{-}145)$$

where
$$\zeta = c_{j+1} c_j r_j = c_{j+1} p_j$$

The final forms of both (5-144) and (5-145) do not involve c_{j+1} and s_{j+1}. However, we must store the value of ζ before destroying c_{j+1}. Furthermore, it is also possible to arrange the calculations in such a way that the values of $f_{j+1}^{(k+1)}$ and $d_j^{(k+1)}$ are stored in the same locations as those for $f_{j+1}^{(k)}$ and $d_j^{(k)}$, saving the need of a separate array for $J(k+1)$. The actual algorithm, modeled after that described in an article by Wilkinson and Reinsch, and used also in EISPACK, is given in Box 5-9.

The actual calculations required for step $(k+1)$ in a QL-transformation are summarized by Bowdler and others in the following way:

$$p_m = d_m^{(k)} - \eta_k \qquad c_{m+1} = 1 \qquad s_{m+1} = 0$$

$$\text{for } j=m-1, m-2,\ldots, i+1,i \quad \begin{cases} r_{j+1} = \{p_{j+1}^2 + (f_{j+1}^{(k)})^2\}^{1/2} \\ g_{j+1} = c_{j+2} f_{j+1}^{(k)} \\ h_{j+1} = c_{j+2} p_{j+1} \\ f_{j+2}^{(k+1)} = s_{j+2} r_{j+1} \\ c_{j+1} = p_{j+1}/r_{j+1} \\ s_{j+1} = f_{j+1}^{(k)}/r_{j+1} \\ p_j = c_{j+1}(d_j^{(k)} - \eta_k) - s_{j+1} g_{j+1} \\ d_{j+1}^{(k+1)} = h_{j+1} + s_{j+1}\{c_{j+1} g_{j+1} + s_{j+1}(d_j^{(k)} - \eta_k)\} \end{cases}$$

$$f_{i+1}^{(k+1)} = s_{i+1} p_i \qquad d_i^{(k+1)} = c_{i+1} p_i \qquad (5\text{-}146)$$

Usually, more than one iteration is required to bring a superdiagonal matrix elements f_i to sizes sufficiently small compared with the absolute value of the corresponding diagonal elements $|d_{i-1}|$ and $|d_i|$ that they may be considered to be zero. However, the number of iterations required is seldom large, four to five for small values of i if the matrix dimension is $n = 30$ or so, and much less later (larger values of i) when the matrix has undergone several QL-transformations and is partially diagonal already.

To obtain the eigenvectors, we must start with the transformation matrix R in the previous section that reduces the original real symmetric matrix A to a tridiagonal form J:
$$\tilde{R} A R = J$$

> **Box 5-9** Subroutine TRI_QL(D,F,V,N,NDMN)
> **Eigenvalues and eigenvectors**
> **of a real symmetric matrix by QL algorithm**
>
> Argument list:
> >   D: Input diagonal matrix elements. Eigenvalues on output.
> >   F: Input superdiagonal matrix elements.
> >   V: Input transformation matrix R. Output eigenvector matrix.
> >   N: Matrix dimension.
> >   NDMN: Dimension of arrays in the calling programs.
>
> Initialization:
> >   (a) Set up an upper limit for the number of iterations allowed.
> >   (b) Shift the location of f_i by one, that is, $f_{i-1} = f_i$. Let $f_n = 0$.
> >   (c) Let $i = 1$.
>
> 1. Start the reduction for f_i to zero:
>   (a) Initialize the iteration counter for each f_i.
>   (b) Set up a criterion for testing the value of f_i according to (5-137).
>   (c) Check if any superdiagonal element is too small. If so, break the matrix into two blocks and work within each block.
> 2. Shift the diagonal matrix elements:
>   (a) Construct the shift parameter by diagonalizing the (2×2) matrix formed by d_i, d_{i+1}, and f_i.
>   (b) Use λ_1 of (5-141) as the shift parameter η_k.
>   (c) Store the sum of all the shifts so far.
>   (d) Shift all the diagonal elements not yet in diagonal form.
> 3. Carry out the QL-transformation:
>   (a) Initialize $p = d_m$, $c = 1$, and $s = 0$.
>   (b) Carry out the following steps for $j = (m-1)$ to i in steps of -1:
>   (i) Save the old values of c and s.
>   (ii) Calculate r_{j+1}, g_{j+1}, h_{j+1}, c_j, s_j, p_j, $f_{j+1}^{(k+1)}$, and $d_{j+1}^{(k+1)}$ using (5-146).
>   (c) Update the transformation matrix according to (5-149).
>   (d) Calculate the values of $d_i^{(k+1)}$ and $f_i^{(k+1)}$.
>   (e) Check if $f_i^{(k+1)}$ is essentially zero.
>   (i) If not, increment the iteration counter by 1 and repeat the calculations starting with step 3. Stop if the number of iterations exceeds the maximum value.
>   (ii) If so, restore the sum of all the shifts to d_i. Increment the value of i by 1 and go back to step 2 until $i = n$.
> 4. Return the eigenvalues in D and the eigenvectors in A.

The eigenvector matrix V that transforms A to a diagonal matrix D may be defined as

$$\tilde{V}AV = D \tag{5-147}$$

If we use Q to represent the transformation that reduces a tridiagonal matrix J to a diagonal matrix D, we have the relation

$$V = RQ$$

Since Q is the product of many rotations, each having the form $P(k)$ given by (5-138), V is obtained from a recursion relation in the form

$$V(k) = V(k-1)P(k) \tag{5-148}$$

The starting point is $V(1) = R$, given by the transformation matrix that reduces A to a tridiagonal form. If the transformation coefficients are c_j and s_j in $P(k)$ in step k, the matrix elements of $V(k)$ are obtained from those of $V(k-1)$ in the following way:

$$v_{\ell,j+1}^{(k+1)} = c_j v_{\ell,j+1}^{(k)} + s_j v_{\ell,j}^{(k)} \qquad v_{\ell,j}^{(k+1)} = c_j v_{\ell,j}^{(k)} - s_j v_{\ell,j+1}^{(k)} \tag{5-149}$$

where j and $(j+1)$ are the two rows where the rotation takes place. Both ℓ and k take on values $1, 2, \ldots, n$.

§5-6 Lanczos method of constructing matrices

We have seen in the previous two sections that an efficient way to diagonalize a real, symmetric matrix A is to bring it first to a tridiagonal form. Once this is done, it is a relatively simple matter to obtain the eigenvalues and eigenvectors.

So far, we have not been concerned with the basis states in which the matrix A is constructed. This comes from the fact that the final result is independent of the choice of the basis states. The usual practice in physics is to take a basis that is convenient to calculate the matrix elements of the operator in which we are interested. For some problems it is possible to construct a basis, such that the matrix is tridiagonal to start with. Usually, it takes some extra work to find such a set of states. However, the effort may well be worthwhile, as there are two major advantages in using a tridiagonal basis. The first is the obvious ease in diagonalizing such a matrix. The second is in the amount of computer memory required to store the matrix. As we saw earlier, an n-dimensional tridiagonal matrix requires only n locations to store the diagonal elements $d_1, d_2, \ldots, d_n$, and another $(n-1)$ locations to store the off-diagonal elements $f_2, f_3, \ldots, f_n$. This is to be contrasted with the amount $n(n+1)/2$ required to store a real, symmetric matrix in general. As a result, the tridiagonal basis offers the possibility of handling matrices of much larger dimensions.

For eigenvalue problems it is possible to use the Lanczos method to construct a tridiagonal basis for the Hamiltonian operator. An extensive description of the various aspects of this method can be found in Cullum and Willoughby (*Lanczos Algorithms for Large Symmetric Eigenvalue Computations*, vols. I and II, Birkhäuser, Boston, 1985). Our interest here will be limited to an introduction to the subject.

Construction of a tridiagonal basis. Let us again start with the Schrödinger equation. Consider an arbitrary normalized wave function $|\Phi_1\rangle$. We shall assume that it is not an eigenvector of the Hamiltonian. As a result, the action of the Hamiltonian operator $\hat{H}$ on $|\Phi_1\rangle$ produces a function $|U_1\rangle$ that is different from $|\Phi_1\rangle$,

$$\hat{H}|\Phi_1\rangle = |U_1\rangle \qquad (5\text{-}150)$$

In general, $|U_1\rangle$ is not normalized. Let us use N_1 to represent the normalization constant. That is,

$$\langle U_1|U_1\rangle = N_1^2 \qquad (5\text{-}151)$$

A normalized vector $|\Psi_1\rangle$ may be defined by the following relation:

$$|\Psi_1\rangle = N_1^{-1}|U_1\rangle$$

Since $|\Phi_1\rangle$ is not an eigenvector of the Hamiltonian, $|\Psi_1\rangle$ and $|\Phi_1\rangle$ are not equal to each other.

The difference between them may be expressed by saying that $|\Psi_1\rangle$ is a linear combination of $|\Phi_1\rangle$ and another function $|\Phi_2\rangle$. It is convenient to define $|\Phi_2\rangle$ as a normalized function orthogonal to $|\Phi_1\rangle$. That is,

$$\langle \Phi_2|\Phi_2\rangle = 1 \qquad \langle \Phi_2|\Phi_1\rangle = 0$$

In terms of $|\Phi_1\rangle$ and $|\Phi_2\rangle$, we can write $|\Psi_1\rangle$ as

$$|\Psi_1\rangle = \alpha_1|\Phi_1\rangle + \beta_1|\Phi_2\rangle \qquad (5\text{-}152)$$

with

$$\alpha_1^2 + \beta_1^2 = 1 \qquad (5\text{-}153)$$

On multiplying both sides of (5-150) by $\langle\Phi_1|$ and integrating over all the independent variables, we obtain

$$\langle \Phi_1|\hat{H}|\Phi_1\rangle = N_1\langle\Phi_1|\Psi_1\rangle = \alpha_1 N_1 \qquad (5\text{-}154)$$

Since the value of N_1 is known from (5-151), we obtain the value of α_1 by calculating the diagonal matrix element of $\hat{H}$ for $|\Phi_1\rangle$. From this and (5-152), we obtain

$$|\Phi_2\rangle = \frac{1}{\beta_1}\left(|\Psi_1\rangle - \alpha_1|\Phi_1\rangle\right)$$

The value of β_1 may be found from (5-153) or from the normalization requirement of $|\Phi_2\rangle$. This is basically the same as the Gram-Schmidt orthogonalization method of constructing a set of orthonormal vectors from an arbitrary set. Before we proceed further, it is convenient to define two other quantities:

$$d_1 \equiv \langle\Phi_1|\hat{H}|\Phi_1\rangle = \alpha_1 N_1$$

$$f_2 \equiv \langle\Phi_2|\hat{H}|\Phi_1\rangle = N_1\langle\Phi_2|\Psi_1\rangle = \beta_1 N_1 \qquad (5\text{-}155)$$

We shall see later that these are the diagonal and superdiagonal elements of the tridiagonal matrix we are after.

What happens if we apply the Hamiltonian operator on $|\Phi_2\rangle$? Again, $|\Phi_2\rangle$ cannot be an eigenvector of $\hat{H}$. As a result,

$$\hat{H}|\Phi_2\rangle = N_2|\Psi_2\rangle \tag{5-156}$$

where $|\Psi_2\rangle$ is a normalized state, in the same way as $|\Psi_1\rangle$, and N_2 is a constant. In general, $|\Psi_2\rangle$ cannot be a linear combination of $|\Phi_1\rangle$ and $|\Phi_2\rangle$ alone. We shall come back to the reasons for this later. For now, we shall take this premise for granted and, similar to (5-152), we shall express $|\Psi_2\rangle$ as a linear combination of $|\Phi_1\rangle$, $|\Phi_2\rangle$, and another function, which we shall label as $|\Phi_3\rangle$. That is,

$$|\Psi_2\rangle = \alpha_2|\Phi_1\rangle + \beta_2|\Phi_2\rangle + \gamma_2|\Phi_3\rangle \tag{5-157}$$

In other words, any part of $|\Psi_2\rangle$ that cannot come from a linear combination of $|\Phi_1\rangle$ and $|\Phi_2\rangle$ is represented by the new state $|\Phi_3\rangle$. We shall choose $|\Phi_3\rangle$ to be a normalized function and orthogonal to both $|\Phi_1\rangle$ and $|\Phi_2\rangle$. This gives us the condition

$$\alpha_2^2 + \beta_2^2 + \gamma_2^2 = 1$$

obtained by squaring both sides of (5-157) and integrating over all the independent variables.

From the fact that the Hamiltonian is real and symmetric, we find that

$$\langle\Phi_1|\hat{H}|\Phi_2\rangle = \langle\Phi_2|\hat{H}|\Phi_1\rangle = f_2$$

As we have already done in (5-155), this matrix element is labeled as f_2 since it is the first off-diagonal matrix element of $\hat{H}$. The convention we follow here is the same as that adopted in the previous sections in representing the superdiagonal elements of a tridiagonal matrix. On the other hand, from (5-156) and (5-157), we obtain

$$\langle\Phi_1|\hat{H}|\Phi_2\rangle = N_2\langle\Phi_1|\Psi_2\rangle = N_2\alpha_2$$

This gives us the value of α_2 in terms of N_2 and f_2,

$$\alpha_2 = f_2/N_2$$

Using (5-157), we obtain

$$d_2 \equiv \langle\Phi_2|\hat{H}|\Phi_2\rangle$$

The value of β_2 is then given by

$$\langle\Phi_2|\hat{H}|\Phi_2\rangle = N_2\langle\Phi_2|\Psi_2\rangle = N_2\beta_2$$

or

$$\beta_2 = d_2/N_2$$

Similar to (5-154), we have

$$|\Phi_3\rangle = \frac{1}{\gamma_2}\left(|\Psi_2\rangle - \alpha_2|\Phi_1\rangle - \beta_2|\Phi_2\rangle\right)$$

It is useful to define here

$$f_3 \equiv \langle\Phi_3|\hat{H}|\Phi_2\rangle$$

in the same spirit as we did in (5-155). Note also that the matrix element

$$\langle\Phi_3|\hat{H}|\Phi_1\rangle = 0 \tag{5-158}$$

This comes from the fact that, by the definition given in (5-152), the state produced by the action of $\hat{H}$ on $|\Phi_1\rangle$ is only a linear combination of $|\Phi_1\rangle$ and $|\Phi_2\rangle$, and both of them are orthogonal to $|\Phi_3\rangle$.

We shall continue this line of argument for one more step before giving the general result. When the Hamiltonian operator is applied to $|\Phi_3\rangle$, we obtain the result

$$\hat{H}|\Phi_3\rangle = N_3|\Psi_3\rangle$$

in analogy with (5-156). Similarly, we shall assume that $|\Psi_3\rangle$ is normalized to unity. Because of (5-158) and the fact that the Hamiltonian is Hermitian, the matrix is symmetric. As a result, we have

$$\langle\Phi_1|\hat{H}|\Phi_3\rangle = 0$$

This comes from the fact that, by construction, the only nonzero off-diagonal matrix element of $\hat{H}$ acting on $|\Phi_1\rangle$ is $\langle\Phi_2|\hat{H}|\Phi_1\rangle$. It also implies that $|\Psi_3\rangle$ is orthogonal $|\Phi_1\rangle$. In other words, $|\Psi_3\rangle$ can only be a linear combination of $|\Phi_2\rangle$, $|\Phi_3\rangle$, and a new basis state $|\Phi_4\rangle$:

$$|\Psi_3\rangle = \alpha_3|\Phi_2\rangle + \beta_3|\Phi_3\rangle + \gamma_3|\Phi_4\rangle \tag{5-159}$$

Similar to what we have done earlier, $|\Phi_4\rangle$ is normalized and orthogonal to $|\Phi_2\rangle$ and $|\Phi_3\rangle$. Furthermore, since $|\Psi_3\rangle$ is orthogonal to $|\Phi_1\rangle$, the new state $|\Phi_4\rangle$ must also be orthogonal to $|\Phi_1\rangle$.

The relation represented by (5-159) is general. This can be seen in the following way. If we continue the process of finding new basis vector $|\Phi_1\rangle$, $|\Phi_2\rangle$, ..., until we come to $|\Phi_k\rangle$, we have the result

$$\hat{H}|\Phi_k\rangle = N_k|\Psi_k\rangle \tag{5-160}$$

in analogy to (5-156). It is always possible to express the new state $|\Psi_k\rangle$ produced as a linear combination of three components,

$$|\Psi_k\rangle = \alpha_k|\Phi_{k-1}\rangle + \beta_k|\Phi_k\rangle + \gamma_k|\Phi_{k+1}\rangle \tag{5-161}$$

as we have demonstrated for $|\Psi_3\rangle$ in (5-159). The new vector $|\Phi_{k+1}\rangle$ can be made to be orthogonal to all the known basis vectors, $|\Phi_1\rangle$, $|\Phi_2\rangle$, ..., $|\Phi_k\rangle$. Furthermore, we see that the Hamiltonian matrix is tridiagonal in the basis formed by $|\Phi_1\rangle$, $|\Phi_2\rangle$, ...

An exception to (5-161) happens when the dimension of the Hilbert space n is finite. In this case, the total number of linearly independent basis states we can construct is limited to n. When we reach $k = n$ in our basis state construction, all the required states have been found and no new ones can be generated. As a result,
$$|\Phi_{n+1}\rangle = 0$$
This means that the process of finding more basis states cannot be continued beyond $k = n$.

What happens if $\gamma_k = 0$ at some stage of our construction for the tridiagonal basis states? Obviously, we cannot continue the procedure if this takes place, as it implies that the action of $\hat{H}$ on $|\Psi_k\rangle$ does not contain any new component that is not already in the basis states already found. Except for reasons of poor numerical accuracy, this can occur only if the Hilbert space for the problem consists of two or more independent subspaces. Earlier in §5-5 we saw the inverse problem. There we saw that a tridiagonal matrix can be separated into two or more independent submatrices if one or more of the superdiagonal elements vanish. It is obvious that the two topics, $\gamma_k = 0$ and $f_k = 0$, are closely related to each other. We shall delay any further discussion on this point until we come to the question of actually constructing a tridiagonal basis.

Because of truncation errors in the numerical calculations, it may be difficult in practice to judge when γ_k becomes zero when n is large. However, as we shall see next, the strength of the Lanczos method lies in that, for eigenvectors of interest in physical problems, it may be quite adequate to generate only a small fraction of the total number of the tridiagonal basis states in the space, long before any problem of numerical accuracy can dominate the calculations.

Let us label the nonvanishing elements of a tridiagonal matrix in the same way as in the previous sections. That is, the diagonal elements are
$$d_k = \langle\Phi_k|\hat{H}|\Phi_k\rangle \tag{5-162}$$
and superdiagonal elements are
$$f_k = \langle\Phi_{k-1}|\hat{H}|\Phi_k\rangle \tag{5-163}$$
It is easy to see that, in general,
$$\hat{H}|\Phi_k\rangle = f_k|\Phi_{k-1}\rangle + d_k|\Phi_k\rangle + f_{k+1}|\Phi_{k+1}\rangle \tag{5-164}$$
Since it is relatively a simple matter to diagonalize such a matrix, we can carry out the operation even before all the basis states are found. This is a meaningful thing to do since, in most physical problems, we are interested only in a very small fraction of the total number of eigenstates in the Hilbert space. In fact, the

usual situation is that we are only concerned with a few of the low-lying ones near the ground state. In such cases, if the starting state is well chosen, all the basis states that contribute significantly to the low-lying states emerge at relatively early stages in constructing the tridiagonal basis. Once we have the main components, any additional basis states may not affect very much the eigenvectors of interest. As a result, there cannot be much loss in the accuracy of the final results if we do not include any of the basis states that are yet to be found. In other words, good approximations of the results of interest to us may be obtained without having to complete the construction of all the possible tridiagonal basis states in the Hilbert space. This is one of the strengths of the Lanczos approach. What we need now is a practical way to implement the scheme by having a method that recognizes when a sufficient number of basis states is reached.

Illustration with the ground state energy of a system. As an example, let us consider the problem of finding the ground state energy of a system. Based on physical intuition, we can usually pick a starting state $|\Phi_1\rangle$ for our tridiagonal basis that is close to the ground state wave function. Let the value of the diagonal matrix element be $d_1 = \langle \Phi_1|\hat{H}|\Phi_1\rangle$. In general, d_1 will not be equal to the ground state energy, since $|\Phi_1\rangle$ is not the eigenfunction. In fact, from a variational principle point of view, we know that the expectation value of the Hamiltonian in any state in the same Hilbert space must be higher than that in the ground state. If we represent the true ground state energy as $\mathcal{E}_1$, we expect that $d_1 \geq \mathcal{E}_1$.

We can construct a second basis state $|\Phi_2\rangle$ using (5-150) and (5-152). With $|\Phi_1\rangle$ and $|\Phi_2\rangle$, our "active" space now consists of two basis states. According to our labeling scheme, the diagonal matrix element for state $|\Phi_2\rangle$ is d_2 and the off-diagonal matrix element between the two basis states we have so far is f_2. In this enlarged space, we expect to produce an eigenvector that is a better approximation of the ground state. Let the lower one of the two eigenvalues of this (2×2) matrix, formed of d_1, d_2, and f_2, be λ_1. Since, in general, it is a better approximation to the ground state energy, we expect it to be lower in value than d_1 but most likely still higher than $\mathcal{E}_1$.

We can add more tridiagonal basis states to the calculation and enlarge the active space. Each time we add one more state, we expect the lowest eigenvalue to take on a lower value. Eventually, the lowest eigenvalue will converge to the true ground state energy $\mathcal{E}_1$. This is very similar to what one does in a variational calculation. If the convergence is fast, and this seems to be true in many physical problems, the ground state energy may be obtained in an active space that is only a small fraction of the complete Hilbert space for the problem.

There are two ways to recognize when convergence is achieved. The first is to make use of the lowest eigenvalue achieved in the calculation. If the value is not changed in any significant manner when more tridiagonal basis states are added to the space, it is likely that the calculation has converged. The second is to make use of the eigenvector. When we get to the stage that the additional basis states in the active space do not make any essential contributions to the wave function of the lowest state, it may be safe to assume that any further additions will not change the result in any significant way either, and the calculation has converged. In either case, one must ensure that the result is a genuine indication of convergence rather than a mirage caused by truncation errors. One way to get some insurance against possible erroneous conclusions caused by poor numerical accuracy is to take the calculation some distance beyond convergence and to make sure that the new results are not any different. Alternatively, we can try to start with a different $|\Phi_1\rangle$ and see if the calculation converges to the same result.

The most obvious exception to the practical method described above in recognizing convergence is when the Hilbert space consists of two or more parts that have very small off-diagonal matrix elements connecting the states in the different parts. Under such circumstances, it may happen that the starting state chosen is in one part of the space. In this case, the calculation may converge only to the lowest state in that part of the space, but not necessarily to the ground state of the system as a whole. Such pathological cases can usually be detected on physical grounds, and a different starting state may be used.

Realization of tridiagonal basis states. So far, we have discussed the tridiagonal basis states $|\Phi_i\rangle$ in abstract. In actual calculations, it is necessary for us to express $|\Phi_i\rangle$ in terms of some known functions. For this purpose, it is often more convenient to have a complete set of orthonormal functions $|\phi_j\rangle$, for $j = 1$, $2, \ldots, n$, and each tridiagonal state $|\Phi_i\rangle$ may be expressed as a linear combination of $|\phi_j\rangle$:

$$|\Phi_i\rangle = \sum_{j=1}^{n} c_{i,j} |\phi_j\rangle \qquad (5\text{-}165)$$

For a given set of $|\phi_j\rangle$, the function $|\Phi_i\rangle$ is completely specified by giving the values of all the expansion coefficients $c_{i,j}$.

The choice for the set of functions $|\phi_j\rangle$ should be made both on physical grounds and for mathematical convenience. For example, the Hamiltonian may be separated into two parts:

$$\hat{H} = \mathcal{H}_0 + \mathcal{H}'$$

If the eigenfunctions of $\mathcal{H}_0$ are known, such as those for a particle in a harmonic oscillator well of §4-1, we may use this set of functions as our $|\phi_i\rangle$. In particular, if $\mathcal{H}_0$ forms the dominant part of $\hat{H}$, some of the basis states $|\phi_j\rangle$ may be sufficiently close to the eigenvectors of interest to us that the first few tridiagonal basis states have only a few components. As a result, the low-lying tridiagonal basis states of interest to us may be quite simple, and an expansion of the form of (5-165) may only consist of a few nonvanishing terms.

Once a complete set of $|\phi_j\rangle$ is selected, we can proceed to select the starting tridiagonal basis state $|\Phi_1\rangle$ based on some physical notion we have of the system. Let

$$|\Phi_1\rangle = \sum_{j=1}^{n} c_{1,j}|\phi_j\rangle$$

Once the values of $c_{1,j}$ for $j = 1, 2, \ldots, n$ are chosen, the rest of the calculations are quite mechanical and can be easily put on a computer. For example, we can express $|U_1\rangle$ of (5-150) in terms of $|\phi_j\rangle$ by invoking the closure property for a complete set of states:

$$\begin{aligned} |U_1\rangle = \hat{H}|\Phi_1\rangle &= \sum_{j=1}^{n} c_{1,j} \hat{H}|\phi_j\rangle \\ &= \sum_{j=1}^{n} c_{1,j} \sum_{k=1}^{n} |\phi_k\rangle\langle\phi_k|\hat{H}|\phi_j\rangle \\ &= \sum_{k=1}^{n} g_{1,k}|\phi_k\rangle \end{aligned} \quad (5\text{-}166)$$

where

$$g_{1,k} = \sum_{j=1}^{n} c_{1,j}\langle\phi_k|\hat{H}|\phi_j\rangle = \sum_{j=1}^{n} c_{1,j} H_{k,j}$$

To simplify the notation, we have adopted the shorthand

$$H_{i,j} \equiv \langle\phi_i|\hat{H}|\phi_j\rangle \quad (5\text{-}167)$$

and we shall continue with this practice for the rest of this section.

In terms of $H_{i,j}$, the value of the first diagonal matrix element may be written as

$$d_1 = \langle\Phi_1|\hat{H}|\Phi_1\rangle = \sum_{i,j} c_{1,i} c_{1,j} H_{i,j}$$

Since all the coefficients $c_{1,i}$ are available and all the matrix elements $H_{i,j}$ can be calculated, there is no difficulty in obtaining the value of d_1. Once this is done, we shall proceed to find

$$|\Phi_2\rangle = \sum_{j=1}^{n} c_{2,j}|\phi_j\rangle$$

by calculating the coefficients $c_{2,j}$ for $j = 1, 2, \ldots, n$. Using the relations given by (5-164) and (5-166), we have

$$\hat{H}|\Phi_1\rangle = d_1|\Phi_1\rangle + f_2|\Phi_2\rangle = \sum_{j=1}^{n} g_{1,j}|\phi_j\rangle$$

Since the basis states $|\phi_j\rangle$ are independent of each other, we obtain a relation between the coefficients in the form
$$f_2 \, c_{2,j} = g_{1,j} - d_1 c_{1,j}$$
Using the fact that $|\Phi_2\rangle$ is normalized to unity, $\sum_j c_{2,j}^2 = 1$, we obtain the result,
$$f_2^2 = \sum_{j=1}^n (g_{1,j} - d_1 c_{1,j})^2$$
With f_2, we can calculate the values of the coefficients
$$c_{2,j} = \frac{1}{|f_2|}(g_{1,j} - d_1 c_{1,j}) \tag{5-168}$$
There is an overall ambiguity in sign for all the coefficients $c_{2,1}, c_{2,2}, \ldots, c_{2,n}$ that cannot be determined. This affects the sign of f_2, as we can see from the relation
$$f_2 = \langle \Phi_1 | \hat{H} | \Phi_2 \rangle = \sum_{i,j} c_{1,i} \, c_{2,j} H_{i,j}$$
However, it does not have any effect on the diagonal matrix element
$$d_2 = \langle \Phi_2 | \hat{H} | \Phi_2 \rangle = \sum_{i,j} c_{2,i} \, c_{2,j} H_{i,j}$$
We shall see later that such an overall sign is not important physically.

To derive the equations for the general case, let us assume that we have already found k tridiagonal basis states $|\Phi_1\rangle, |\Phi_2\rangle, \ldots, |\Phi_k\rangle$. This means that all the coefficients $c_{i,j}$ for $i = 1, 2, \ldots, k$ and $j = 1, 2, \ldots, n$ are known. The input quantities required to find the next state,
$$|\Phi_{k+1}\rangle = \sum_{j=1}^n c_{k+1,j} |\phi_j\rangle$$
are the kth diagonal and superdiagonal elements, d_k and f_k, and the coefficients $c_{k-1,j}$ and $c_{k,j}$ for $j = 1, 2, \ldots, n$ related to the tridiagonal basis states $|\Phi_{k-1}\rangle$ and $|\Phi_k\rangle$. Let
$$|U_k\rangle = \hat{H} |\Phi_k\rangle = \sum_{j=1}^n g_{k,j} |\phi_j\rangle \tag{5-169}$$
where, similar to (5-166), we have
$$g_{k,j} = \sum_{\ell=1}^n c_{k,\ell} H_{j,\ell}$$
On the other hand, using (5-164), we have
$$\hat{H} |\Phi_k\rangle = f_k |\Phi_{k-1}\rangle + d_k |\Phi_k\rangle + f_{k+1} |\Phi_{k+1}\rangle$$

At this stage, f_{k+1} and the expansion coefficient $c_{k+1,j}$ are still unknown to us.

From the fact that (5-164) holds for an arbitrary set of basis states, we obtain the following relation using (5-169):

$$f_{k+1} c_{k+1,j} = g_{k,j} - f_k c_{k-1,j} - d_k c_{k,j} \qquad (5\text{-}170)$$

Since $|\Phi_{k+1}\rangle$ is normalized to unity, $\sum_j c_{k+1,j}^2 = 1$, we have

$$f_{k+1}^2 = \sum_{j=1}^{n} \left(g_{k,j} - f_k c_{k-1,j} - d_k c_{k,j} \right)^2$$

This gives us the value of f_{k+1} up to a sign. Using this result, we can calculate the values of the coefficients that express $|\Phi_{k+1}\rangle$ in terms of $|\phi_i\rangle$,

$$c_{k+1,j} = \frac{1}{|f_{k+1}|} \left(g_{k,j} - f_k c_{k-1,j} - d_k c_{k,j} \right) \qquad (5\text{-}171)$$

similar to what we have done in (5-168). With this result, we can calculate d_{k+1} using the relation

$$d_{k+1} = \sum_{i,j} c_{k+1,i} \, c_{k+1,j} H_{i,j} \qquad (5\text{-}172)$$

This completes all the calculations associated with the new tridiagonal basis state $|\Phi_{k+1}\rangle$. We can now proceed to find $|\Phi_{k+2}\rangle$, if we wish, by substituting $(k+1)$ for k in (5-169) to (5-172). In principle, we can repeat the process as many times as we wish until the total number of basis states in the Hilbert space is exhausted. However, as mentioned earlier, this is not the usual aim of the Lanczos method.

As soon as a reasonable number of diagonal and superdiagonal elements of the tridiagonal matrix are obtained, we can use either the bisection or QL-method, given in the previous section, to diagonalize the matrix obtained so far. By comparing the results of diagonalizing matrices of successively larger dimensions, it is possible to reach a conclusion on whether it is necessary to include more tridiagonal basis states into the active space.

Note that in (5-172), if the overall sign of $c_{k+1,j}$ is reversed, the sign of f_{k+1} is changed but not that of d_{k+1}, as noted earlier. As a result, such a sign change does not affect the results of the next tridiagonal basis vector. This can be seen by looking at (5-171). If we change the overall sign of all the coefficients $c_{k-1,j}$ for all possible values of j for the basis vector $|\Phi_{k-1}\rangle$, the sign of f_k is changed, but the sign of the product $f_k c_{k-1,j}$ remains the same. Similarly, if we change the overall sign of $c_{k,j}$ for all possible values of j, both f_k and $g_{k,j}$ change sign, but not d_k. The net result is a sign change for all the coefficients $c_{k+1,j}$, for $j = 1$ to n, corresponding to an overall sign change in the tridiagonal basis state. Since such a sign difference cannot be observed in any measurements, it has no physical consequence.

Anharmonic oscillator example. In §4-1, we have seen that analytical solutions can be obtained for a quantum mechanical particle inside a one-dimensional, harmonic oscillator. The Hamiltonian operator has the form

$$\mathcal{H}_0 = -\frac{\hbar^2}{2\mu}\frac{d^2}{dx^2} + \tfrac{1}{2}\mu\omega^2 x^2$$

In more realistic situations, the potential energy

$$V(x) = \tfrac{1}{2}\mu\omega^2 x^2$$

may be regarded as an approximation to that experienced by a particle in a potential minimum, such as that illustrated in Fig. 4-1. In the same way as illustrated in §5-3, we can improve the approximation by including a symmetric *anharmonic* term in the potential:

$$V_{\text{anh}}(x) = \epsilon\hbar\omega\left(\frac{\mu\omega}{\hbar}\right)^2 x^4$$

Here, ϵ is a dimensionless parameter describing the strength of the correction to the simple harmonic potential, as we saw earlier in (5-54).

Again, we shall write the independent variable x in terms of the dimensionless quantity

$$\rho = \sqrt{\frac{\mu\omega}{\hbar}}\,x$$

The complete Hamiltonian takes on the form

$$\hat{H} = \mathcal{H}_0 + \epsilon\rho^4 \tag{5-173}$$

Similar to (4-7), the Schrödinger equation for the anharmonic oscillator may be reduced to the differential equation:

$$\frac{d^2\psi}{d\rho^2} + (\lambda - \rho^2 - \epsilon\rho^4)\psi(\rho) = 0 \tag{5-174}$$

Because of the additional term in the potential, the equation can no longer be solved by the method outlined in §4-1. If the anharmonic term is small, the solution is often obtained using perturbation techniques, such as the one used as an illustrative example in §5-3. Here we shall use the Lanczos method to obtain the low-lying states. The advantage over the perturbative approach is that the parameter ϵ is not limited to small values; however, the convergence rate of the Lanczos method is faster if ϵ is small.

Let us assume that the potential is still dominated by the harmonic term that is proportional to ρ^2. As a result, it is convenient to use the harmonic oscillator wave functions

$$\psi_m(\rho) = \frac{1}{\sqrt{2^m m!\sqrt{\pi}}}\,e^{-\rho^2/2}H_m(\rho)$$

given in (4-16) as our set of functions ϕ_i. Here $H_m(\rho)$ is the Hermite polynomials of degree m defined in §4-1. Note that $\psi_m(\rho)$ is the eigenfunction of our $\mathcal{H}_0$ here, satisfying the relation
$$\mathcal{H}_0 \psi_m = \left(m + \tfrac{1}{2}\right) \psi_m \tag{5-175}$$
with the energy measured in units of $\hbar\omega$.

To match the notations used in this section with those adopted for §4-1, we shall define
$$\phi_i \equiv \psi_{i-1}(\rho)$$
so that ϕ_1 is the ground state harmonic oscillator wave function $\psi_0(\rho)$, ϕ_2 is that with principal quantum number 1, and so on. For our calculations here, we do not explicitly need the forms of the wave functions. Only the matrix elements
$$H_{i,j} \equiv \langle \phi_i | \hat{H} | \phi_j \rangle = \langle \psi_{i-1} | \mathcal{H}_0 + \epsilon \rho^4 | \psi_{j-1} \rangle \tag{5-176}$$
of (5-167) enter into the various terms. Since the functions ϕ_i are chosen to be the eigenfunctions of $\mathcal{H}_0$, we have the relation
$$\langle \phi_i | \mathcal{H}_0 | \phi_j \rangle = \langle \psi_{i-1} | \mathcal{H}_0 | \psi_{j-1} \rangle = \left(i - \tfrac{1}{2}\right) \delta_{i,j}$$
in units of $\hbar\omega$. This is essentially the same result as (5-175) except that it is written in the notation of this section. In terms of the matrix elements of $\hat{H}$, we have
$$\langle \psi_{i-1} | \mathcal{H}_0 + \epsilon \rho^4 | \psi_{j-1} \rangle = \left(i - \tfrac{1}{2}\right) \delta_{i,j} + \epsilon \langle \psi_{i-1} | \rho^4 | \psi_{j-1} \rangle$$
For $m = \min(i,j) - 1$, the matrix elements of ρ^4 between harmonic oscillator wave functions may be expressed in the form
$$\langle \psi_{i-1} | \rho^4 | \psi_{j-1} \rangle = \begin{cases} \dfrac{3}{2}\left(m^2 + m + \tfrac{1}{2}\right) & \text{for } i = j \\[4pt] \left(m + \dfrac{3}{2}\right)\sqrt{(m+1)(m+2)} & \text{for } i = j \pm 2 \\[4pt] \dfrac{1}{4}\sqrt{(m+1)(m+2)(m+3)(m+4)} & \text{for } i = j \pm 4 \\[4pt] 0 & \text{otherwise} \end{cases}$$
$$\tag{5-177}$$

The derivations of these results are left as an exercise (see Problem 5-6).

A reasonable starting state for constructing the tridiagonal basis is the ground state of the simple harmonic oscillator itself,
$$|\Phi_1\rangle = |\phi_1\rangle \equiv \psi_0(\rho)$$
In terms of the expansion coefficients of (5-165), the same condition is stated as
$$c_{1,j} = \begin{cases} 1 & \text{for } j = 1 \\ 0 & \text{otherwise} \end{cases}$$

The first diagonal matrix element in our tridiagonal basis is then

$$d_1 = \langle \Phi_1 | \hat{H} | \Phi_1 \rangle = \langle \psi_0 | \mathcal{H}_0 + \epsilon \rho^4 | \psi_0 \rangle = \frac{1}{2} + \frac{3}{4}\epsilon$$

measured in units of $\hbar\omega$. To obtain the second tridiagonal basis state $|\Phi_2\rangle$, we use (5-166) to get

$$g_{1,k} = \sum_{j=1}^{n} c_{1,j} H_{k,j} = H_{k,1} = \begin{cases} d_1 & \text{for } k = 1 \\ \dfrac{3}{\sqrt{2}}\epsilon & \text{for } k = 3 \\ \sqrt{\dfrac{3}{2}}\epsilon & \text{for } k = 5 \\ 0 & \text{otherwise} \end{cases}$$

From these, we obtain the values

$$f_2^2 = \epsilon^2 \left(\frac{9}{2} + \frac{3}{2} \right) = 6\epsilon^2$$

This, in turn, gives us the value of $c_{2,j}$ using (5-168):

$$c_{2,j} = \begin{cases} \sqrt{\dfrac{3}{4}} & \text{for } j = 3 \\ \sqrt{\dfrac{1}{4}} & \text{for } j = 5 \\ 0 & \text{otherwise} \end{cases}$$

With coefficients $c_{2,j}$ known, we can calculate d_2 according to (5-172), and so on.

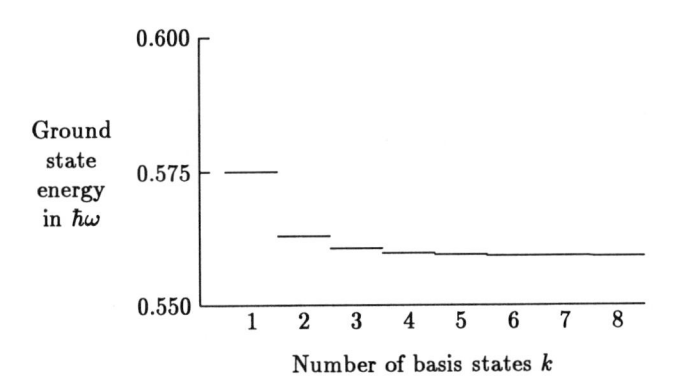

Fig. 5-6 Lowest energy obtained with the Lanczos method for the anharmonic oscillator of (5-137) as a function of the number of tridiagonal basis states. The size of the anharmonic oscillator term is $\epsilon = 0.1$, and convergence is reached with eight basis states. The criterion is set to be a change in the ground state energy of less than 10^{-3} in units of $\hbar\omega$.

The rest of the calculations may be carried out iteratively. For $k \geq 2$, we can calculate d_k by the relation

$$d_k = \sum_{i,j} c_{k,i} c_{k,j} H_{i,j}$$

Similarly, f_k may be obtained using (5-170). This completes the calculations for all the elements in our tridiagonal basis with k states in the active space. We can now diagonalize the matrix we have obtained so far. Since we are interested only in the eigenvalues, the simpler bisection method may be used. Let us label the lowest eigenvalue as E_k, and this is to be compared with E_g, the lowest eigenvalue of a previous iteration. If the difference is larger than the accuracy required, we replace the value of E_g with that of E_k and proceed to obtain $|\Phi_{k+1}\rangle$. This is achieved by calculating $g_{k+1,j}$ using (5-169), f_{k+1} using (5-170), and then $c_{k+1,j}$ using (5-171). The step is completed by finding d_{k+1} using (5-172). For our illustrative example, we shall diagonalize the triangular matrix after each new state is added to active space. If we consider a change in the ground state energy of less than 10^{-3} in units of $\hbar\omega$ as the criterion, the calculation converges at $k = 4$ for a small anharmonic term with $\epsilon = 0.01$ and at $k = 8$ for a larger anharmonic term of $\epsilon = 0.1$. The rate of convergence for the latter case is shown in Fig. 5-6.

§5-7 Nonsymmetric matrices and random phase approximation

In the previous sections, we have been concentrating on real, symmetric matrices, as this is the most common type occurring in physical problems. Most of the methods we have discussed apply to a more general class of eigenvalue problems. In principle, we could have constructed almost any of the algorithms to handle a wider class of matrices, not just real symmetric ones alone. However, this is not a wise direction to take. For the sake of computational efficiency and ease of application, it is better to tailor each program to a particular type of matrix, and we have so far made the choice of specializing in real, symmetric matrices.

Collective excitations in many-body systems. Let us take a small step beyond real, symmetric matrices and consider real, nonsymmetric matrices in this section. As an application in physics, we shall consider the method of random phase approximation (RPA). In many-body systems, there are collective excitations involving a large number of particles acting in a coherent way. In addition to the interest in the phenomenon itself, the study is also one way to understand the underlying interaction between the particles. Since a large number of particles is involved, the amount of computation can be prohibitive if we have to treat the various degrees of freedom of each particle individually. This is especially true for fermions, where the Pauli exclusion principle puts further complications on the problem.

To reduce the problem into a manageable one, we can consider the collective degrees of freedom as the independent variables of the system. Similar to normal coordinates in classical mechanics, calculations may be carried out in terms of these "coordinates" rather than the degrees of freedom of each particle. The random phase approximation is an example of such an approach. The idea originated in plasma oscillations in electron gas and has been generalized to handle a large variety of problems involving collective degrees of freedom in condensed matter physics and nuclear structure problems. We shall describe below the basic ideas involved, the form of nonsymmetric matrices that emerges, and some of the methods used to diagonalize such matrices. For a more complete discussion of the physics, see for example Fetter and Walecka (*Quantum Theory of Many-Particle Systems*, McGraw-Hill, New York, 1971).

Consider the ground state of a system made of a large number of fermions. Following the general practice in the subject, we shall write the wave function of this state as $|0\rangle$. If the particles behave like an ideal gas, that is, without any interaction between them, the lowest energy state is one with all the available low-lying single-particle states filled up to the Fermi energy ϵ_F. Such a many-body state may be labeled in terms of the occupancies of all the single-particle states involved. If we use n_i to represent the number of particles in single-particle state i, with wave function $|\phi_i\rangle$ and eigenvalue ϵ_i, the ground state occupancies are given by

$$n_i = \begin{cases} 1 & \text{for } \epsilon_i \leq \epsilon_F \\ 0 & \text{otherwise} \end{cases} \quad (5\text{-}178)$$

Pictorially, this is shown by the solid line in Fig. 5-7. Excited states in such a system may be formed by promoting particles from the occupied orbits to the empty ones above.

If, instead of an ideal gas, interactions are allowed between the particles. The lowest-energy state of the many-body system in this case may contain components with particles promoted from single-particle states below the Fermi energy to those above. As a result, even in the ground state, the single-particle states just below ϵ_F are only partially occupied and those just above are no longer completely empty. The occupancies of such a correlated ground state may be represented by the dashed curve in Fig. 5-7.

The wave function of a correlated ground state is usually quite complicated in terms of those for single particles. As a result, the excited states of the system are likely to be even more involved, as there are many more possible ways to form such states than the simple case of a Fermi gas system. For simplicity, we shall restrict ourselves to a particular class of excitations, those involving only one particle. These are called one-particle, one-hole ($1p1h$) excitations, as a hole is left behind in the occupied single-particle states because a particle is excited to a state that was empty. For later convenience, we shall use subscripts $i, j, k, \ldots$, to

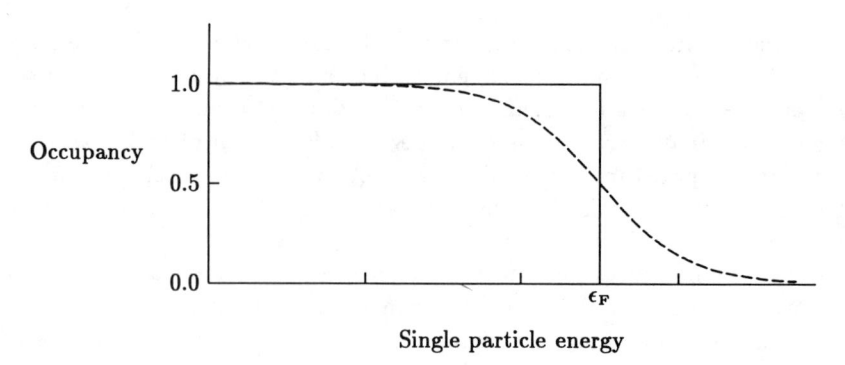

Fig. 5-7 Occupancy of single-particle states for a many-body Fermi system. In an uncorrelated ground state, all the states up to the Fermi energy ϵ_F are occupied and the rest empty. The occupancy is a step function, as shown by the solid curve. More realistically, the occupancy varies smoothly from 1 for single-particle states far below ϵ_F to 0 for those far above, as shown by the dashed curve.

indicate single-particle states below the Fermi energy ϵ_F and subscripts $r, s, t, \ldots,$ for those above.

Let us use $|\nu\rangle$ to represent the wave function of such an excited many-body state and $Q_\nu^\dagger$ as the excitation operator that creates the state from the ground state $|0\rangle$. That is,

$$|\nu\rangle = Q_\nu^\dagger |0\rangle \qquad (5\text{-}179)$$

Since both $|0\rangle$ and $|\nu\rangle$ are eigenstates of the Hamiltonian for the many-body system, we have the relations

$$\hat{H}|0\rangle = E_0|0\rangle \qquad \hat{H}|\nu\rangle = E_\nu|\nu\rangle \qquad (5\text{-}180)$$

where E_0 is the ground state energy and E_ν that of state $|\nu\rangle$.

The conjugate of $Q_\nu^\dagger$ is the de-excitation operator for the excited state $|\nu\rangle$. That is,

$$Q_\nu|\nu\rangle = |0\rangle \qquad (5\text{-}181)$$

Using (5-179), this gives us the relation

$$Q_\nu Q_\nu^\dagger |0\rangle = |0\rangle \qquad (5\text{-}182)$$

In terms of the expectation value in the ground state, this is the same as saying that

$$\langle 0 | Q_\nu Q_\nu^\dagger | 0 \rangle = 1 \qquad (5\text{-}183)$$

What happens if we apply Q_ν to the ground state wave function? Since both $|0\rangle$ and $|\nu\rangle$ are the eigenstates of the Hamiltonian with different eigenvalues, there cannot be any component of $|\nu\rangle$ in $|0\rangle$. As a result,

$$Q_\nu |0\rangle = 0$$

Another way of arriving at the same conclusion is to start from the conjugate of both sides of (5-179)
$$\langle 0|\boldsymbol{Q}_\nu = \langle \nu|$$
From this we obtain the value of the matrix element
$$\langle 0|\boldsymbol{Q}_\nu|0\rangle = \langle \nu|0\rangle = 0 \tag{5-184}$$
We shall make use of these relations in our discussion of random phase approximation.

Tamm-Dancoff approximation. As a starting point, let us consider the simpler case of an uncorrelated ground state, that is, one with occupancies given by (5-178). In this case, the only possible $1p1h$-excitations are those consisting of promoting a particle from one of the states below ϵ_F to one above. In terms of single-particle creation and annihilation operators, the excitation operator $\boldsymbol{Q}_\nu^\dagger$ takes on the form
$$\boldsymbol{Q}_\nu^\dagger = \sum_{ri} X_{ri} a_r^\dagger a_i \tag{5-185}$$
where the operator a_i takes a particle away from single-particle state i (below ϵ_F) and $a_r^\dagger$ puts a particle in single-particle state r (above ϵ_F). The summation index r is over all the empty single-particle states above ϵ_F and the index i is over all the occupied states below. This is known as the Tamm-Dancoff approximation, usually abbreviated as TDA.

Instead of trying to find the wave function of $|\nu\rangle$, we can solve the problem by calculating the excitation operator $\boldsymbol{Q}_\nu^\dagger$. This may be achieved by finding the coefficients X_{ri} on the right side of (5-185). In this approach, all the physical quantities concerning the excited state are expressed with reference to the ground state $|0\rangle$, instead of the wave function of the excited state itself.

To calculate the coefficients X_{ri}, we shall first express the energy of the excited state $|\nu\rangle$ in terms of the excitation operator $\boldsymbol{Q}_\nu^\dagger$ and the ground state wave function $|0\rangle$. That is,
$$\hat{H}|\nu\rangle = \hat{H}\boldsymbol{Q}_\nu^\dagger|0\rangle = E_\nu \boldsymbol{Q}_\nu^\dagger|0\rangle \tag{5-186}$$
The eigenvalue is E_ν here, since $\hat{H}$ is acting on $|\nu\rangle$. On the other hand,
$$\boldsymbol{Q}_\nu^\dagger \hat{H}|0\rangle = E_0 \boldsymbol{Q}_\nu^\dagger|0\rangle$$
In contrast to (5-186), the ground state energy E_0 appears on the right side of the equation, as the Hamiltonian acts on $|0\rangle$. The difference between these two equations gives us
$$[\hat{H}, \boldsymbol{Q}_\nu^\dagger]|0\rangle \equiv (\hat{H}\boldsymbol{Q}_\nu^\dagger - \boldsymbol{Q}_\nu^\dagger\hat{H})|0\rangle = (E_\nu - E_0)\boldsymbol{Q}_\nu^\dagger|0\rangle \tag{5-187}$$
where $[\hat{H}, \boldsymbol{Q}_\nu^\dagger]$ is the commutator between $\hat{H}$ and $\boldsymbol{Q}_\nu^\dagger$.

If we multiply both sides of (5-187) by $\langle 0|a_j^\dagger a_s$, the equation is transformed into one that relates the ground state expectation values of different operators:

$$\langle 0|a_j^\dagger a_s[\hat{H}, Q_\nu^\dagger]|0\rangle = (E_\nu - E_0)\langle 0|a_j^\dagger a_s Q_\nu^\dagger|0\rangle$$

Using the expansion of (5-185), this expression may be written as an equation for the coefficients X_{ri} of the excitation operators $Q_\nu^\dagger$:

$$\sum_{ri}\left\{\langle 0|a_j^\dagger a_s \hat{H} a_r^\dagger a_i|0\rangle - \delta_{r,s}\delta_{i,j}E_0\right\}X_{ri} = (E_\nu - E_0)X_{sj} \qquad (5\text{-}188)$$

In arriving at the result, we have used the relation

$$\langle 0|a_j^\dagger a_s a_r^\dagger a_i|0\rangle = \delta_{r,s}\delta_{i,j}$$

where $\delta_{\alpha,\beta}$ is the Kronecker delta. This comes from the fact that $a_r^\dagger a_i|0\rangle$ removes a particle from single-particle state i and puts it in r. To return to the ground state by the action of the remaining operators $a_j^\dagger$ and a_s, we must remove the particle in r and this requires that $r = s$. Similarly, to put the particle back in state i, it is necessary that $i = j$. The matrix element vanishes if either of these two conditions is not met.

We can put (5-188) into the standard form of an eigenvalue problem in the form

$$\boldsymbol{A}\boldsymbol{X} = (E_\nu - E_0)\boldsymbol{X} \qquad (5\text{-}189)$$

where the matrix elements of $\boldsymbol{A}$ are given by

$$A_{sj,ri} = \langle 0|a_j^\dagger a_s \hat{H} a_r^\dagger a_i|0\rangle - \delta_{r,s}\delta_{i,j}E_0 \qquad (5\text{-}190)$$

and $\boldsymbol{X}$ is a column matrix with elements X_{ri}. The equality of (5-189) has the same form as (5-50) except here, instead of the wave function of a state in terms of the basis states of the Hilbert space, we are looking for the operator $Q_\nu^\dagger$ in terms of the 1p1h-excitation operators $a_r^\dagger a_i$. As shown by (5-189), the eigenvalue in this case is the excitation energy of the state $|\nu\rangle$ relative to the ground state. Since each 1p1h-excitation operator is labeled by two indexes, one for the hole orbit and one for the particle orbit, each term of $Q_\nu^\dagger$ is labeled by two indexes, as shown for instance by X_{sj} on the right side of (5-188). Because of this, each of the two indexes for the matrix elements of $\boldsymbol{A}$ consists of two labels, such as $A_{sj,ri}$ on the left side of (5-190).

To keep the discussion simple, we have restricted the excitations to only one of the particles in the system at a time. In principle, we can also build up excited states by promoting more than one particle at a time and thus creating multiparticle, multihole excitations. In general, it takes more energy to produce such components, and their contributions to the low-lying many-body states decrease with increasing number of particles and holes formed. For this reason, we have ignored such components in our studies for the low-lying excited states.

If the Hamiltonian is real, the matrix elements of A are real for all $1p1h$-excitations in the Tamm-Dancoff approximation. Furthermore, the matrix A is symmetric. As a result, the solution of (5-190) involves diagonalizing a real, symmetric matrix. In the more general case, the Hamiltonian may be complex. However, it must remain Hermitian in order to represent a physical observable. In this case, the Hamiltonian matrix is a complex, symmetric one. We shall see in the next section that there are efficient algorithms to diagonalize such matrices as well.

Random phase approximation. We are now ready to study $1p1h$-excitations built upon a correlated ground state. Although there is no longer a sharp division between the occupied and unoccupied single-particle states, we shall, nevertheless, retain the concept of a Fermi energy. The quantity ϵ_F may be defined as the energy of the highest occupied single-particle state if the states are filled in ascending order according to the single-particle energies. The major difference from TDA discussed above is that, in addition to excitations represented by (5-185), we can also "excite" a particle from a single-particle state above ϵ_F to one below. The new possibility comes from the fact that there are partially occupied states above ϵ_F from which we can remove a particle and partially empty states below into which we can put the particle. The excitation operator of (5-185) now has the more general form

$$Q_\nu^\dagger = \sum_{ri}(X_{ri}a_r^\dagger a_i - Y_{ri}a_i^\dagger a_r) \tag{5-191}$$

where Y_{ri} is the coefficient for removing a particle from orbit r above ϵ_F to orbit i below ϵ_F. This is known as the random phase approximation (RPA). The negative sign in the definition of coefficients Y_{ri} is a convention commonly used in RPA calculations.

The relation expressed by (5-187) remains true for this more general class of excitation operator. However, to arrive at a convenient form of the matrix equation we can use to solve for the values of X_{ri} and Y_{ri}, some additional work is needed. By multiplying $\langle 0|Q_\nu$ on the left to both sides of (5-187), we find that

$$\langle 0|Q_\nu[\hat{H},Q_\nu^\dagger]|0\rangle = (E_\nu - E_0)\langle 0|Q_\nu Q_\nu^\dagger|0\rangle \tag{5-192}$$

We can put the left side into a more symmetric form by subtracting from it the quantity $\langle 0|[\hat{H},Q_\nu^\dagger]Q_\nu|0\rangle$. Because of (5-184),

$$\langle 0|[\hat{H},Q_\nu^\dagger]Q_\nu|0\rangle = 0$$

As a result, the right side of (5-192) does not change. The result is

$$\langle 0|\{Q_\nu[\hat{H},Q_\nu^\dagger] - [\hat{H},q_\nu^\dagger]Q_\nu\}|0\rangle = (E_\nu - E_0)\langle 0|Q_\nu Q_\nu^\dagger|0\rangle$$

This can be put into the form of a commutator between $[\hat{H},Q_\nu^\dagger]$ and Q_ν:

$$\langle 0|[Q_\nu,[\hat{H},Q_\nu^\dagger]]|0\rangle = (E_\nu - E_0)\langle 0|[Q_\nu,Q_\nu^\dagger]|0\rangle \tag{5-193}$$

In fact, by going through a similar set of steps, we can also obtain an expression in the following form:

$$\langle 0|[[Q_\nu, \hat{H}], Q_\nu^\dagger]|0\rangle = (E_\nu - E_0)\langle 0|[Q_\nu, Q_\nu^\dagger]|0\rangle \tag{5-194}$$

Although it is useful in practical calculations to make use of a linear combination of (5-193) and (5-194), we shall only deal with the former for our introduction to the subject here.

The equality expressed by (5-193) holds as long as $Q_\nu^\dagger$ is the creation operator of the excited state $|\nu\rangle$ and satisfies the relations (5-179) and (5-181). That is, it is true only for a given set of the coefficients $\{X_{ri}\}$ and $\{Y_{ri}\}$ related to the state at excitation energy $(E_\nu - E_0)$. To find the values of these two set of coefficients, we can apply variations to Q_ν:

$$\langle 0|[\delta Q_\nu, [\hat{H}, Q_\nu^\dagger]]|0\rangle = (E_\nu - E_0)\langle 0|[\delta Q_\nu, Q_\nu^\dagger]|0\rangle$$

Furthermore, since Q_ν and $Q_\nu^\dagger$ are simply related to each other, there is no point in also varying the coefficients for $Q_\nu^\dagger$. In terms of the two different types of terms given in (5-191), the variational relation may be expressed explicitly as the following set of two equations:

$$\langle 0|[a_j^\dagger a_s, [\hat{H}, Q_\nu^\dagger]]|0\rangle = (E_\nu - E_0)\langle 0|[a_j^\dagger a_s, Q_\nu^\dagger]|0\rangle \tag{5-195}$$

$$\langle 0|[a_s^\dagger a_j, [\hat{H}, Q_\nu^\dagger]]|0\rangle = (E_\nu - E_0)\langle 0|[a_s^\dagger a_j, Q_\nu^\dagger]|0\rangle \tag{5-196}$$

In matrix notation, these may be put into the form

$$\begin{pmatrix} A & B \\ B^* & A^* \end{pmatrix}\begin{pmatrix} X \\ Y \end{pmatrix} = (E_\nu - E_0)\begin{pmatrix} X \\ -Y \end{pmatrix} \tag{5-197}$$

where the elements of the Hermitian submatrix A are

$$A_{sj,ri} = \langle 0|[a_j^\dagger a_s, [\hat{H}, a_r^\dagger a_i]]|0\rangle$$

and the elements of the symmetric submatrix B are

$$B_{sj,ri} = -\langle 0|[a_j^\dagger a_s, [\hat{H}, a_i^\dagger a_r]]|0\rangle$$

The results are obvious for (5-195). To see that (5-197) also satisfies (5-196), we need to invoke that the complex conjugate of a matrix element is given by

$$\langle \psi|\hat{O}|\phi\rangle^* = \langle \phi|\hat{O}^\dagger|\psi\rangle$$

and the relation between single-particle creator $a_r^\dagger$ and its conjugate a_r. This requires some preparation on the techniques of second quantization in quantum mechanics and we shall not get into the question here.

For simplicity, we shall consider only the case of a real Hamiltonian and, as a result, all the matrix elements of A and B are real. For the convenience of discussion, we shall define the following two matrices:

$$H = \begin{pmatrix} A & B \\ -B & -A \end{pmatrix} \qquad Z = \begin{pmatrix} X \\ Y \end{pmatrix} \qquad (5\text{-}198)$$

In terms of these matrices, (5-197) may be written in the form

$$HZ = (E_\nu - E_0)Z \qquad (5\text{-}199)$$

This is similar to (5-190) except that, in the case of TDA, the matrix A is a real, symmetric one. Here, the matrix H is real but nonsymmetric, as can be seen by inspection. Furthermore, if we let all the coefficients Y_{ri} equal zero, we recover (5-189) and (5-190). More detailed discussion of the subject can be found in Fetter and Walecka and Ring and Schuck (*The Nuclear Many-Body Problem*, Springer-Verlag, New York, 1980).

Left and right eigenvectors. Before we start on a discussion of the method to diagonalize a real, nonsymmetric matrix, we shall examine first some of the differences from a real, symmetric matrix. For a Hermitian matrix, both real and complex, the eigenvalues are real. On the other hand, for a real nonsymmetric matrix, some of the eigenvalues can be complex. Complex eigenvalues, if present, appear always in pairs for such matrices, with each member of a pair the conjugate of the other.

So far, we have always written the eigenvalue problem in the form

$$Hv_R = \lambda v_R \qquad (5\text{-}200)$$

where λ is the eigenvalue and v_R the corresponding eigenvector. We shall call eigenvectors, defined in the form of v_R above, right eigenvectors. Alternatively, the eigenvalue equation can also be written in the form with the eigenvector appearing on the left of the matrix H,

$$v_L H = \lambda' v_L \qquad (5\text{-}201)$$

The eigenvectors v_L are known as left eigenvectors. By transposing both sides of the equation, we have

$$\widetilde{v_L H} = \widetilde{H}\widetilde{v}_L = \lambda'\widetilde{v}_L$$

Since the value of the determinant of a matrix is equal to that of its transpose,

$$\det|H| = \det|\widetilde{H}|$$

we find that

$$\det|H - \lambda \mathbf{1}| = \det|\widetilde{H} - \lambda' \mathbf{1}|$$

We see that λ and λ' are the roots of two equivalent characteristic equations. As a result,

$$\lambda = \lambda'$$

That is, the eigenvalues are the same for both left and right eigenvectors.

For a real symmetric matrix $H = \tilde{H}$, we have the result

$$v_L = v_R$$

obtained by comparing (5-200) with (5-201). For this reason, we did not have to make any distinctions between left and right eigenvectors up to now. For a complex Hermitian matrix, $H = H^\dagger$, we have, instead, the relation

$$v_L = v_R^\dagger$$

For a real, nonsymmetric matrix, the situation is slightly more complicated, as we shall see below.

Consider the transformation matrix V_R made of all the right eigenvectors. By definition,

$$V_R^{-1} H V_R = \lambda \tag{5-202}$$

where λ is a diagonal matrix consisting of all the eigenvalues

$$\lambda = \begin{pmatrix} \lambda_1 & 0 & 0 & \cdots & 0 \\ 0 & \lambda_2 & 0 & \cdots & 0 \\ \vdots & \vdots & \vdots & \ddots & \vdots \\ 0 & 0 & 0 & \cdots & \lambda_n \end{pmatrix}$$

Since λ is diagonal, we may also write (5-202) in the form

$$H V_R = V_R \lambda \tag{5-203}$$

Similarly, for the transformation matrix consisting of all the left eigenvectors, we have

$$V_L H = \lambda V_L \tag{5-204}$$

Multiplying V_L from the left to both sides of (5-203), we obtain the result

$$V_L H V_R = V_L V_R \lambda$$

Similarly, we can multiply V_R from the right to both sides of (5-204) and obtain the result

$$V_L H V_R = \lambda V_L V_R \tag{5-205}$$

Comparing the last two equations, we find that

$$V_L V_R \lambda = \lambda V_L V_R$$

For this to be true, it is necessary that the product $V_L V_R$ be a diagonal matrix. We can choose the normalization for each of the eigenvectors v_L and v_R in such a way that

$$V_L V_R = 1 \tag{5-206}$$

That is, the orthogonality relation among the eigenvectors of a real, nonsymmetric matrix is between left and right eigenvectors. In fact, there is no corresponding relation among the left eigenvectors themselves or among the right eigenvectors. Furthermore, if there are degenerate eigenvalues, the left and right eigenvectors may not be orthogonal to each other to start with. In this case, it is necessary to take linear combinations of the degenerate eigenvectors to form the left and right eigenvectors that are orthogonal to each other.

Hessenberg form and QL-transformation. If our interest is in the eigenvalues alone, it is not necessary to diagonalize a matrix completely — it is adequate to bring it to an upper or lower triangular form; that is, a matrix with either all the elements below the diagonal or above the diagonal vanish. The diagonal elements are now the eigenvalues, as can be seen from the roots of the characteristic equation $\det|H - \lambda 1| = 0$ for such a matrix.

For a nonsymmetric matrix H, a Householder transformation of the type of (5-103) cannot reduce it to a tridiagonal form. This is different from a symmetric matrix discussed in §5-4. In general, the most we can hope to achieve here is to reduce to zero all the elements in the lower (or upper) triangle, with the exception of the subdiagonal (or superdiagonal) ones. The transformed matrix takes on the form

$$H' = \begin{pmatrix} h_{1,1} & h_{1,2} & h_{1,3} & \cdots & h_{1,n-1} & h_{1,n} \\ h_{2,1} & h_{2,2} & h_{2,3} & \cdots & h_{2,n-1} & h_{2,n} \\ 0 & h_{3,2} & h_{3,3} & \cdots & h_{3,n-1} & h_{3,n} \\ 0 & 0 & h_{4,3} & \cdots & h_{4,n-1} & h_{4,n} \\ \vdots & \vdots & \vdots & \ddots & \vdots & \vdots \\ 0 & 0 & 0 & \cdots & h_{n,n-1} & h_{n,n} \end{pmatrix}$$

This is known as the upper Hessenberg form. From this, a QL-algorithm may be used to reduce all the subdiagonal elements to zero and transform the matrix into an upper triangular form. Alternatively, we can use a combination of Householder transformation and the QR-algorithm to achieve the same goal. The details of the theory are given in Stoer and Bulirsch and Wilkinson and Reinsch. FORTRAN programs to diagonalize real, nonsymmetric matrices using such an approach can be found in the EISPACK and in Press, and others (*Numerical Recipes*, Cambridge University Press, Cambridge, 1986).

Jacobi method of diagonalizing a real nonsymmetric matrix. To illustrate the differences between diagonalizing a real, nonsymmetric matrix from that of a real, symmetric one as seen in earlier sections, it is instructive for us to adopt the simpler, but less efficient, Jacobi method. Our discussion is based on the paper by Eberlein and Boothroyd that is also reprinted in Wilkinson and Reinsch.

Our aim is to find a matrix T such that a similarity transformation of the form

$$T^{-1}HT = \lambda$$

reduces the real, nonsymmetric matrix H to a lower triangular matrix L. Similar to the Jacobi method for real, symmetric matrices, the transformation matrix T consists of the product of a series of transformations in the form

$$T = T_1 T_2 \cdots T_i \cdots$$

each of which is a rotation in a two-dimensional subspace. Since the matrix is no longer symmetric, a rotation T_i in such a subspace, in general, can transform only one of the two off-diagonal elements to zero. As a result, it is necessary to add a shear, or complex rotation, in the same two-dimensional subspace to ensure convergence to a diagonal matrix at the end.

Each step of the transformation, represented symbolically by T_i, is now a product of two operations, a rotation R_i and a shear S_i. If the rotation is taking place in the subspace consisting of columns and rows k and ℓ with $k < \ell$, the elements of these two transformation matrices are given by

$$r_{k,k} = r_{\ell,\ell} = \cos\theta \qquad s_{k,k} = s_{\ell,\ell} = \cosh p$$
$$r_{k,\ell} = -r_{\ell,k} = -\sin\theta \qquad s_{k,\ell} = s_{\ell,k} = -\sinh p \qquad (5\text{-}207)$$

For $i \neq k \neq \ell$ and $j \neq k \neq \ell$, the corresponding matrix elements are

$$r_{i,j} = \delta_{i,j} \qquad\qquad s_{i,j} = \delta_{i,j} \qquad\qquad (5\text{-}208)$$

Analogous to (5-71), the rotation angle here is selected to be

$$\tan 2\theta = \frac{h_{k,\ell} + h_{\ell,k}}{h_{k,k} - h_{\ell,\ell}}$$

where $h_{i,j}$ are the elements of the real, nonsymmetric matrix whose eigenvalues and eigenvectors we wish to find. The shear parameter p in (5-207) has the value

$$\tanh p = \frac{ed - h}{g + 2(e^2 + d^2)}$$

where

$$e = h_{k,\ell} - h_{\ell,k}$$
$$d = (h_{k,k} - h_{\ell,\ell})\cos 2\theta + (h_{k,\ell} + h_{\ell,k})\sin 2\theta$$
$$g = \sum_{i \neq k,\ell} \left(h_{k,i}^2 + h_{i,k}^2 + h_{\ell,i}^2 + h_{i,\ell}^2 \right) \qquad (5\text{-}209)$$
$$h = \cos 2\theta \sum_{i \neq k,\ell} \left(h_{k,i}h_{\ell,i} - h_{i,k}h_{i,\ell} \right) - \frac{1}{2}\sin 2\theta \sum_{i \neq k,\ell} \left(h_{k,i}^2 + h_{i,k}^2 - h_{\ell,i}^2 - h_{i,\ell}^2 \right)$$

The values are chosen to minimize the (Euclidean) norm of the matrix

$$H(i+1) = (T_1 T_2 \cdots T_i)^{-1} H(T_1 T_2 \cdots T_i)$$

Similar to Jacobi diagonalization of a real, symmetric matrix, the calculation is iterative. Several passes, each going through all the nonvanishing off-diagonal matrix elements $h_{k,\ell}$ and $h_{\ell,k}$ for $k = 1, 2, \ldots, n$ and $\ell = k+1, k+2, \ldots, n$, are necessary before the matrix is reduced to a diagonal form.

Box 5-10 Subroutine JCB_NSYM(A,T,N,NDMN,L_R)
Eigenvalues and eigenvectors of a real, nonsymmetric matrix using a Jacobi-like algorithm

Argument list:
  A: Array for the input matrix.
  T: Array for the output transformation matrix.
  N: Matrix dimension.
  NDMN: Dimension of the arrays in the calling program.
  L_R: Input $+1$ for right eigenvector, -1 for left eigenvector, and 0 if no eigenvector is required. Returns the number of iterations taken.

Initialization:
  (a) Set up two criteria for small numbers, EPS $= 10^{-8}$ and EP $= \sqrt{\text{EPS}}$.
  (b) Let the maximum number of iterations allowed be 50.
  (c) Zero the iteration counter.
  (d) Set up a logical variable MARK to indicate whether further iterations are needed and initialize it to .FALSE..

1. Set T equal to the identity matrix.
2. Check if further iterations are needed:
  (a) No further iteration if the maximum of iterations is exceeded, or
  (b) If MARK is .TRUE., that is, no transformation carried out in the previous iteration, or
  (c) Examine all the off-diagonal matrix elements. Exit if the following three criteria are satisfied: $|h_{i,j} + h_{j,i}| \leq$ EPS, $|h_{i,j} - h_{j,i}| \leq$ EPS, and $|h_{i,i} - h_{j,j}| \leq$ EPS.
3. Carry out a Jacobi-like transformation to H:
  (a) Apply the transformation R_i and S_i according to (5-207).
    (i) Calculate the parameters e, d, g, and h of (5-209).
    (ii) Calculate rotation matrix elements $r_{i,j}$.
    (iii) Calculate the shear matrix elements $s_{i,j}$.
    (iv) Form the transformation matrix $T_i = R_i \times S_i$.
  (b) Skip the transformation if T_i is an identity matrix. Otherwise
    (i) Set MARK to .FALSE..
    (ii) Apply the transformation to H.
    (iii) Update T according to whether the left, right, or no eigenvectors are required.
4. Go back to step 2.
5. Return eigenvalues in A, eigenvectors in T, and number of iterations in L_R.

If all the eigenvalues are real, the transformed matrix $\boldsymbol{\lambda}$ is diagonal

$$\lambda_{k,\ell} = 0 \qquad \text{for } k \neq \ell$$

and the diagonal matrix elements are the eigenvalues. In general, some of the eigenvalues may be complex and, for these components of the matrix $\boldsymbol{\lambda}$, the elements have the form

$$\lambda_{k,\ell} = -\lambda_{\ell,k} \qquad \text{and} \qquad \lambda_{k,k} = \lambda_{\ell,\ell} \qquad \text{for } k \neq \ell$$

where we have used $\lambda_{i,j}$ to represent the matrix elements of $\boldsymbol{\lambda}$. Each complex eigenvalue has the form $\lambda_{k,k} \pm i\lambda_{k,\ell}$. For convenience, we can arrange each pair of such columns (and their corresponding rows), k and ℓ, to be adjacent to each other so that the complex eigenvalues appear as $\lambda_{k,k} \pm i\lambda_{k,k+1}$.

For real eigenvalues, the columns of the transformation matrix $\boldsymbol{T}$ are the right eigenvectors and the rows of $\boldsymbol{T}^{-1}$ are the left eigenvectors. For a complex eigenvalue $\lambda_{k,k} \pm i\lambda_{k,k+1}$, the corresponding eigenvector is also complex. The right eigenvector is $t_{j,k} \pm it_{j,k+1}$ for $j = 1, 2, \ldots, n$, where $t_{j,k}$ and $t_{j,k+1}$ are the matrix elements in columns k and $(k+1)$ of the jth row of $\boldsymbol{T}$. The left eigenvector is a similar linear combination of the matrix elements in rows k and $(k+1)$ of $\boldsymbol{T}^{-1}$.

The relation given by (5-205) may also be used to check the computer program for diagonalization. Because of (5-206), we have

$$V_L H V_R = \boldsymbol{\lambda} \tag{5-210}$$

In other words, we can use each pair of left and right vectors to transform $\boldsymbol{H}$. If the two vectors do not correspond to the same eigenvalue, the result should be zero. On the other hand, for both left and right vectors corresponding to the same eigenvalue, a value λ is expected. To carry out this check, a copy of the original matrix must be saved. At the same time, both the left eigenvector matrix V_L and the right eigenvector matrix V_R must be calculated.

For matrices with relatively small dimensions, algorithms based on the Jacobi method are quite efficient. A simple way to implement it is given in Box 5-10. However, similar to the case of real, symmetric matrices, the amount of computation increases as n^3, where n is the dimension of the matrix. As a result, methods based on Householder and QL-transformations are preferred for matrices with large dimensions. As mentioned earlier, computer programs using these more advanced techniques can be found in standard subroutine libraries.

Normal and defective matrices. It should also be realized that when a matrix has complex eigenvalues, certain properties that are usually associated with matrices with only real eigenvalues may no longer apply. We can see this from the following example.

For a Hermitian matrix, (5-210) may be written as

$$THT^\dagger = \lambda$$

Conversely, we have the relations

$$H = T^\dagger \lambda T \qquad H^\dagger = T^\dagger \lambda^\dagger T$$

By taking the product of H and $H^\dagger$ using these relations, we obtain the results

$$HH^\dagger = T^\dagger \lambda T T^\dagger \lambda^\dagger T = T^\dagger \lambda \lambda^\dagger T$$

Similarly, we have

$$H^\dagger H = T^\dagger \lambda^\dagger \lambda T$$

For a Hermitian matrix, all the eigenvalues are real,

$$\lambda \lambda^\dagger = \lambda^\dagger \lambda$$

and

$$HH^\dagger = H^\dagger H \qquad (5\text{-}211)$$

Matrices obeying such a commutation relation are called *normal* matrices.

In addition to Hermitian matrices, skew-Hermitian and unitary matrices are also normal in the sense of (5-211). In general, real nonsymmetric matrices are not necessarily normal. For a normal matrix, all the eigenvectors are orthogonal to each other and together they span the complete Hilbert space. This means that an arbitrary vector in the space can be expressed as a linear combination of the eigenvectors. For nonnormal matrices, however, the eigenvectors may not span the complete space and, as a result, it may not be possible to make the eigenvectors orthogonal to each other. When this happens, the matrix is known as *defective*. One problem faced by such matrices is that, when there are degenerate eigenvalues, the eigenvectors produced by some diagonalization programs may not be orthogonal among the degenerate states. When this happens in the case of a normal matrix, a Gram-Schmidt procedure may be used to orthogonalize them. For a defective matrix, this is not possible.

In most physical problems, we are dealing with normal matrices. However, this is not always the case. For example, a Hilbert space may contain a number of spurious states. This happens, among others, in problems where our interest is in the excitation energy between particles. In such cases, differences in the kinetic energy as a result of different states of motion of the center of mass of all the particles are irrelevant or spurious. In the absence of any truncation errors, such states, if present, will usually appear as separate eigenstates at the end of the calculation. This is true as long as the set of basis states used for the calculation is complete. Often, for practical reasons, the basis states are truncated according to schemes that do not take the spurious states properly into account. In such cases, we are likely to encounter problems associated with defective matrices. For example, a part of each eigenvector may be made up of spurious states. Usually there is no easy way of knowing the amount of such components and, consequently, the results may not be physically meaningful. The details of such types of calculations depend very much on the physics involved, and it is not useful for us to pursue the subject any further here.

§5-8 Complex matrix, band matrix, and the generalized eigenvalue problem

We shall briefly discuss here a few other topics in eigenvalue problems. On many occasions, we must deal with matrices other than the real, symmetric ones we have been mostly concerned with in this chapter. Most complex matrices we normally encounter are quite easy to handle; it is usually possible to change them into real ones and then diagonalize the resulting real matrices. In fact, this is often the preferred method to handle complex matrices. Band matrices are special types of sparse matrices. For this reason, techniques have been developed such that we can manipulate matrices of this type with dimensions far larger than with matrices in general. Such techniques are useful in solving certain physical problems. The generalized eigenvalue problem occurs in certain physical situations. Although we cannot go into the physics involved in any detail, it is useful to have an introduction to the subject here.

Complex matrices. A complex matrix is one whose elements are complex numbers. For example, the matrix element (k, ℓ) of a complex matrix C has the form

$$c_{k,\ell} = a_{k,\ell} + ib_{k,\ell}$$

where both $a_{k,\ell}$ and $b_{k,\ell}$ are real numbers, and $i^2 = -1$. Using the definition of matrix addition given in Table 5-1, we can write C as the sum of two real matrices A and B in an analogous way,

$$C = A + iB \tag{5-212}$$

Let v be an eigenvector of C with eigenvalue λ, such that

$$Cv = \lambda v \tag{5-213}$$

In general, the elements of v are also complex numbers and, similar to (5-212), we can express it as a sum of two real vectors:

$$v = x + iy \tag{5-214}$$

where the elements of x and y consist of real numbers.

In term of the real matrices A, B, x, and y, the complex eigenvalue problem of (5-213) may be written in the following way:

$$(Ax - By) + i(Ay + Bx) = \lambda(x + iy)$$

The equality must hold for the real and imaginary parts separately. If C is Hermitian, the eigenvalue λ is real. In this case, the relation may be expressed in terms of the following two for real matrices:

$$Ax - By = \lambda x$$

$$Ay + Bx = \lambda y \qquad (5\text{-}215)$$

or

$$\begin{pmatrix} A & -B \\ B & A \end{pmatrix} \begin{pmatrix} x \\ y \end{pmatrix} = \lambda \begin{pmatrix} x \\ y \end{pmatrix} \qquad (5\text{-}216)$$

In other words, the complex eigenvalue problem of (5-213) is equivalent to the following problem for real matrices:

$$C'v' = \lambda v'$$

where

$$C' = \begin{pmatrix} A & -B \\ B & A \end{pmatrix} \qquad v' = \begin{pmatrix} x \\ y \end{pmatrix} \qquad (5\text{-}217)$$

If C is a complex matrix of dimension n, the dimension of the real matrix C' is $2n$. Similarly, v' is a column matrix with twice the number of elements as v.

Since C is Hermitian, its matrix elements have the property

$$c^*_{k,\ell} = c_{\ell,k} \qquad (5\text{-}218)$$

In terms of real numbers, this is equivalent to the relation

$$a_{k,\ell} - ib_{k,\ell} = a_{\ell,k} + ib_{\ell,k}$$

From this, we conclude that submatrix A of (5-217) is a real, symmetric matrix. Similarly, the submatrix B is a skew-symmetric matrix, with elements having the property

$$b_{k,\ell} = -b_{\ell,k}$$

As a result, C' is a real, symmetric matrix. In this way, the eigenvalue problem for a complex, Hermitian matrix reduces to one for a real, symmetric matrix of twice the dimension.

If v of (5-214) is an eigenvector of the complex Hermitian matrix C with (real) eigenvalue λ, then v' of (5-217) is an eigenvector of the real matrix C' with the same eigenvalue. In fact, the eigenvalue equation for λ is also satisfied by the vector

$$u' = \begin{pmatrix} -y \\ x \end{pmatrix}$$

We can see this by writing down the same equation as (5-216) for u'

$$\begin{pmatrix} A & -B \\ B & A \end{pmatrix} \begin{pmatrix} -y \\ x \end{pmatrix} = \lambda \begin{pmatrix} -y \\ x \end{pmatrix}$$

Since this is the same equation as (5-215), we find that v' and u' are linearly independent of each other as far as C' is concerned. As a result, there are two degenerate eigenvectors, v' and u', for each eigenvalue λ. In other words, the number of unique eigenvalues for C' remains the same as that for C. If we rewrite the eigenvectors in terms of complex numbers as

$$v = x + iy \qquad\qquad u = -y + ix$$

we find that

$$u = iv$$

As a result, u is not independent of v, and we find that the number of independent eigenvectors obtained from diagonalizing a complex matrix C of dimension n is exactly the same as diagonalizing an equivalent real matrix C' of dimension $2n$.

In terms of the efficiency of carrying out the calculations on a computer, it takes twice the number of locations to store C' than C. This conclusion is arrived at in the following way. Ignoring for the moment any symmetry properties of the matrices, there are n^2 complex matrix elements in C in general and it takes $2n^2$ locations to store them (n^2 real and n^2 imaginary). In contrast, there are $(2n)^2$ number of real matrix elements in C'. In principle, the factor of 2 increase in the number of matrix elements makes it inefficient to work with C' instead of C. This conclusion can also be reached simply by counting the number of off-diagonal matrix elements that must be reduced to zero. However, this is not the entire story. Since it is usually far less efficient to carry out complex arithmetic on a computer, diagonalization routines written in terms of complex numbers are often slower. In fact, efficient diagonalization packages for complex matrices are often written in terms of real number calculations. For this reason, it is often preferable to use real, symmetric matrix routines to handle complex Hermitian matrices. Many packages for complex matrices are actually written in real number arithmetic.

If the matrix C is not Hermitian, the eigenvalues are complex in general. Furthermore, the symmetry relations between the matrix elements given by (5-218) no longer true. As a result, the matrix C' is, in general, a nonsymmetric one, and methods to "diagonalize" nonsymmetric matrices outlined in the previous section must be used. However, all the other considerations discussed above for Hermitian matrices apply here.

Band matrix. In many physical problems, it is possible to choose the basis states in such a way that the nonvanishing matrix elements occur only along a narrow band around the diagonal. A good example is the case of the anharmonic oscillator of (5-54) and (5-173). If we use the harmonic oscillator wave function $\phi_n(\rho)$ of (4-16) as the basis, the nonzero matrix elements $H_{i,j}$ of the Hamiltonian are found only for $|i-j| = 0, 2$ and 4, as we have seen in (5-177). The Hamiltonian matrix, then, has the form

$$H = \begin{pmatrix} H_{1,1} & 0 & H_{1,3} & 0 & H_{1,5} & 0 & 0 & 0 & 0 & 0 & 0 & \cdots \\ 0 & H_{2,2} & 0 & H_{2,4} & 0 & H_{2,6} & 0 & 0 & 0 & 0 & 0 & \cdots \\ H_{3,1} & 0 & H_{3,3} & 0 & H_{3,5} & 0 & H_{3,7} & 0 & 0 & 0 & 0 & \cdots \\ 0 & H_{4,2} & 0 & H_{4,4} & 0 & H_{4,6} & 0 & H_{4,8} & 0 & 0 & 0 & \cdots \\ H_{5,1} & 0 & H_{5,3} & 0 & H_{5,5} & 0 & H_{5,7} & 0 & H_{5,9} & 0 & 0 & \cdots \\ 0 & H_{6,2} & 0 & H_{6,4} & 0 & H_{6,6} & 0 & H_{6,8} & 0 & H_{6,10} & 0 & \cdots \\ 0 & 0 & H_{7,3} & 0 & H_{7,5} & 0 & H_{7,7} & 0 & H_{7,9} & 0 & H_{7,11} & \cdots \\ \vdots & \vdots & \vdots & \vdots & \vdots & \vdots & \vdots & \vdots & \vdots & \vdots & \vdots & \ddots \end{pmatrix}$$

If the matrix dimension n is large, the number of elements with value zero outweighs the number of nonzero elements. This is an example of a *sparse matrix*, and special methods can be devised to handle such cases. We shall see in later chapters that these types of matrices occur also in solving other numerical problems, such as differential equations.

There are two main areas where we can take advantage of the fact that a matrix is sparse. The first is the amount of storage space. For a band matrix, we can define the width of the band m as

$$H_{k,\ell} = 0 \quad \text{for} \quad |k - \ell| > m$$

If $m \ll n$, where n is the dimension of the matrix, the number of nonzero matrix elements is much less than the number n^2 for a matrix in general. We have already exploited this advantage in §5-5 to diagonalize real, symmetric tridiagonal matrices by storing only the n diagonal elements $\{d_i\}$ and the $(n-1)$ superdiagonal elements $\{f_i\}$. Similar savings can also be achieved for band matrices in general. In this way, matrices with much larger dimensions can be handled on a given computer.

The second advantage in band matrices is that efficient numerical methods are available to solve them. For example, the QL-algorithm is too cumbersome to diagonalize matrices in general but is quite efficient for band matrices of narrow width. We have seen an example of this earlier in §5-5 for the limiting case of a tridiagonal matrix. More generally, the method can be applied to any matrices as long as the width of the band is narrow. Other methods are also available and some of these are discussed in Dahlquist and Björck, Stoer and Bulirsch, and Wilkinson. Computer programs for a large class of band matrices can be found in Wilkinson and Reinsch and in EISPACK.

Generalized eigenvalue problem. The standard form of an eigenvalue problem we have been solving so far may be written as

$$Av = \lambda v$$

where A is an n-dimensional square matrix and the eigenvector v is a column matrix consisting of n elements. The eigenvalue λ is, in general, a complex number unless A is Hermitian. More generally, the eigenvalue problem takes on the form

$$Av = \lambda Bv \tag{5-219}$$

where both A and B are n-dimensional square matrices. As we shall see below, an equivalent way of expressing the generalized eigenvalue problem above is

$$B^{-1}Av = \lambda v \tag{5-220}$$

Before embarking on a brief discussion of the methods to solve (5-219), we shall first see an example of how such problems arise in physics.

Consider again a particle moving in a one-dimensional harmonic oscillator potential well,

$$V = \tfrac{1}{2}\mu\omega^2 x^2 \equiv \tfrac{1}{2}kx^2 \tag{5-221}$$

The kinetic energy of the particle is

$$T = \tfrac{1}{2}\mu\dot{x}^2 \tag{5-222}$$

where $\dot{x} \equiv dx/dt$ is the velocity of the particle and μ is the mass of the particle. The Lagrangian $\mathcal{L}$ for the system is then

$$\mathcal{L} = T - V = \tfrac{1}{2}\bigl(\mu\dot{x}^2 - kx^2\bigr)$$

From this, we obtain the familiar classical equation of motion for a one-dimensional harmonic oscillator:

$$\mu\ddot{x} + kx = 0$$

where $\ddot{x} \equiv d^2x/dt^2$. The dependence of x on time t may be obtained by solving this differential equation under the appropriate boundary conditions.

In the more general case of more than one independent degree of freedom, the kinetic energy involves translational motion as well as angular rotation. In terms of a set of generalized coordinates x_j, we can write T here in a form similar to that given by (5-222):

$$T = \frac{1}{2}\sum_{j,k} T_{j,k}\,\dot{x}_j\dot{x}_k$$

where $T_{j,k}$ is an element of the inertial tensor. Similarly, the potential energy in the more general case has the form

$$V = \frac{1}{2}\sum_{j,k} V_{j,k}\,x_j x_k$$

instead of that given in (5-221). Following standard procedures in classical mechanics, we can construct the Lagrangian from T and V in a similar way as we did in the previous paragraph for the one-dimensional case:

$$\mathcal{L} = \frac{1}{2} \sum_{j,k} \left(T_{j,k}\, \dot{x}_j \dot{x}_k - V_{j,k}\, x_j x_k \right)$$

For each of the generalized coordinates x_k, there is an equation of motion in the form

$$\sum_j \left(T_{j,k}\, \ddot{x}_j + V_{j,k}\, x_j \right) = 0 \tag{5-223}$$

To simplify the solution of this set of equations, we shall assume an oscillatory form for the time dependence. That is,

$$x_j = a_j e^{-i\omega t}$$

where a_j is the complex amplitude for the oscillation of generalized coordinate x_j. The equation of motion of (5-223) becomes

$$\sum_j \left(V_{j,k} a_j - \omega^2 T_{j,k} a_j \right) = 0$$

In terms of matrix notation, this equation may be put into the form

$$\boldsymbol{V}\boldsymbol{a} = \omega^2 \boldsymbol{T}\boldsymbol{a} \tag{5-224}$$

Both $\boldsymbol{V}$ and $\boldsymbol{T}$ are square matrices with dimensions equal to the number of generalized coordinates in the problem. The form of the matrix equation is the same as that which appeared in the generalized eigenvalue problem of (5-219). In fact, (5-224) is a member of a general class of Sturm-Liouville problems that occur in a wide variety of physical situations.

A straightforward and effective way to solve (5-219) is to transform it into the form of (5-220). Since $\boldsymbol{B}$ is given, we can find its inverse $\boldsymbol{B}^{-1}$ and rewrite (5-219) in the form of (5-220). By multiplying together the two square matrices on the left side, we arrive at the standard form of an ordinary eigenvalue problem, and methods such as those discussed in the previous section may be applied.

Alternatively, if $\boldsymbol{B}$ is positive definite, that is, $\boldsymbol{B}$ is Hermitian and $(\boldsymbol{x}^\dagger \boldsymbol{B} \boldsymbol{x}) > 0$ for any vector $\boldsymbol{x}$ in the space, we can decompose $\boldsymbol{B}$ as a product of its "square roots" in the sense

$$\boldsymbol{B} = \boldsymbol{L} \widetilde{\boldsymbol{L}}$$

where $\widetilde{\boldsymbol{L}}$ is the transpose of $\boldsymbol{L}$. This is known as the Cholesky decomposition. It is similar to the $\boldsymbol{LU}$-decomposition of §5-2 except that we are specializing here in symmetric matrices. The inverse of $\boldsymbol{B}$ is then

$$\boldsymbol{B}^{-1} = \widetilde{\boldsymbol{L}}^{-1} \boldsymbol{L}^{-1}$$

This allows us to make the following transformation of the product $B^{-1}A$ on the left side of (5-220):

$$\widetilde{L}B^{-1}A\widetilde{L}^{-1} = \widetilde{L}\widetilde{L}^{-1}L^{-1}A\widetilde{L}^{-1} = L^{-1}A\widetilde{L}^{-1} \equiv G$$

A similar transformation can also be carried out for the generalized eigenvalue problem of (5-220). In either case, the equation is reduced to

$$Gv = \lambda v$$

the common form for the usual eigenvalue problems. Once again, standard diagonalization packages may be used to solve this equation.

The use of Cholesky transformation is more efficient than finding the inverse of B and solving the generalized eigenvalue problem using (5-220). Packages that integrate the transformation together with standard eigenvalue problems are found in Wilkinson and Reinsch and in EISPACK.

Problems

5-1 Find the values of the three currents I_1, I_2, and I_3 for the circuit of Fig. 5-1 using $R_1 = 5\ \Omega$, $R_2 = 10\ \Omega$, $R_3 = 15\ \Omega$, and $V_1 = V_2 = 1.5$ V.

5-2 For a system of three linear equations with A given by (5-29), verify that the solution obtained using the inverse of A, given by (5-25), is identical to that of the Gauss-Jordan elimination method of §5-1.

5-3 Construct an algorithm to obtain the value of a determinant of an arbitrary order n by the method of decomposition into minors. Check the method by writing a computer program based on it. Test the program with the help of a random number generator.

5-4 If all the elements of the two matrices U and Z in (5-39) are known, solve for X by expressing each of the matrix elements, $x_1, x_2, \ldots, x_n$, in terms of $\ell_{i,j}$ and z_k. Compare the relations with those given by (5-38).

5-5 Obtain analytically the eigenvalues and eigenvectors of an anharmonic oscillator in the basis space consisting of three states, ϕ_0, ϕ_2, and ϕ_4, defined by (5-53). Use the potential V given by (5-54) and assume that ϵ is small. Express the results as a power series in ϵ.

5-6 Evaluate the matrix element $\langle \phi_m | \rho^4 | \phi_{m'} \rangle$ using the harmonic oscillator wave functions given by (5-53). Check that the results are consistent with those given by (5-177).

5-7 One way to obtain an approximate solution of the ground state wave function of the anharmonic oscillator of (5-54) is to solve it as an eigenvalue problem in the Hilbert space of three harmonic oscillator states $\phi_0(\rho)$, $\phi_2(\rho)$, and

$\phi_4(\rho)$. Construct the Hamiltonian matrix for $\epsilon = 0.1$ and compare the lowest eigenvalue with that shown in Fig. 5-6, obtained by the Lanczos method.

5-8 Show that the trace of a matrix M is unchanged by a unitary transformation, that is, a similarity transformation $U^{-1}MU$ with the transformation matrix U satisfying (5-65).

5-9 Construct all the polynomials in the Sturm sequence for the characteristic equation of the following tridiagonal matrix:

$$A = \begin{pmatrix} 10 & 5 & 0 \\ 5 & -1 & -1 \\ 0 & -1 & 10 \end{pmatrix}$$

Find the eigenvalues using the bisection method.

5-10 An interesting case in the study of random matrices (see §7-6) is the distribution of the eigenvalues. If the matrix elements of a real, symmetric matrix are made of random numbers with a normal distribution (see §7-1), the distribution of eigenvalues was shown by Wigner [*Ann. Math.* **62** (1955) 548; *SIAM* **9** (1978) 1] to be semicircular on the average. Demonstrate that this is true by constructing a few such random matrices with as large a dimension as feasible (n on the order of a hundred or more) and plotting the eigenvalue distribution in each case.

Chapter 6
Methods of Least Squares

A large part of the activity in physics centers around taking measurements and comparing the results with theoretical expectations. This is a relatively easy task if the data are precise and the theories are well understood. However, such ideal situations occur only rarely. For most of the time, we are more likely to be dealing with cases where the physical phenomena are not clear and the measurements are imperfect. Under such circumstances, it may be necessary to make conjectures or models of the physics. To compensate for our incomplete knowledge, the models may contain a number of parameters that must be determined empirically. The uncertainties in the data, caused either by the limitations of the instruments used or by the nature of the physical quantities themselves, may contrive to make it difficult to determine these parameters from the measured values. If the number of independent pieces of data is larger than the number of free parameters, it is possible to make progress with the help of statistical analyses. Among those commonly used in physics, the method of least squares is by far the most important.

§6-1 Statistical description of data

If we take two separate measurements of a physical object, there is a high probability that the results will be different from each other. As an example, let us think in terms of determining the distance between the curb on the north side of a city block and the one on the south side. There are many possible reasons for the discrepancy between the two results. For example, the measurements may have been carried out by counting the number of paces it takes to walk from one end of the block to the other. Since it is unlikely that the length of the city block is an integer multiple of the length of our pace, we must make some estimates of the remainder in terms of a fraction of a step size. As a result, it may not be possible for us to obtain the distance better than roughly to one-tenth of a pace, on the order of centimeters. In addition, there are other uncertainties in our measurements. For example, it is difficult to control the walk such that all the paces are of exactly the same length.

Other possible sources of discrepancies may also be present and they may have nothing to do with the accuracy of the "apparatus" used for the task. Our city block is not a sharply defined object. As a result, even if we replace our pace with a high-precision measuring device, such as a laser range finder, there will still be differences, perhaps on the order of millimeters, between two measurements. The source of uncertainty in this case is related to the nature of the object we wish to measure. For example, it may not be possible to define the length of a city block to the accuracy of millimeters. In fact, for a city block, there is perhaps no reason to delineate the length to such a precision. In physics, we are usually dealing with quantities that are better defined than our example of a city block. However, this does not mean that everything has a precise value. There are also quantities that cannot be measured precisely, such as those governed by the Heisenberg uncertainty principle.

We can also imagine another possible source of discrepancy between different measurements of the same quantity. If two persons with different strides are used to measure the length of a city block, the results in terms of the number of paces will be different. This is an obvious case of systematic error, resulting from a certain bias in the instruments used. If the sources of systematic error are well known, it may be possible to make corrections for them. However, there are also occasions where this is impossible and the uncertainties associated with systematic errors become an integral part of the data.

Distribution in the measured results. Given the uncertainties associated with any measurements, how can we find the true value of a physical quantity? Let us begin the discussion by assuming that there is such a thing as the true value. In spite of this assumption, there can still be differences in the results of any two measurements. Which one of the two results corresponds to the true value? The most probable situation is that neither is the true value. Furthermore, it is not possible to argue that the average of the two measurements is the true value either. We can arrive at this conclusion in the following way. If we make a third measurement, it is unlikely that the result is exactly equal to the average value of the first two. As a result, the average of all three measurements is different from that of the first two. Since the average value changes with the number of measurements, it is improbable that the average of any two measurements corresponds to the true value. On the other hand, if we make a large number of measurements, the average value will approach some definite value, and additional measurements will not change the average in any significant way. As a result, there is a high probability that the average value in this case is equal to the true value.

The notion that the true value of a quantity is given by the average of a large number of measurements is based on statistics. If the uncertainties associated with the measuring process are random, the values obtained will most likely be scattered around the true value with some definite distribution. In general, it may not be possible to deduce the distribution by theoretical considerations. On the other hand, if enough measurements are carried out under the same conditions, it is possible to map out this distribution to the desired precision. In the absence of any bias in the measurements, such as those caused by systematic errors, the probability of getting a particular measured value is purely statistical. As a result, the distribution reflects the statistical nature of the uncertainties associated with the measurements and the nature of the quantity itself. Furthermore, if all the measured values are scattered randomly around the true value, the average of the measurements must approach the true value when the number of measurements is large.

It is also possible that in many circumstances there may not be a definite value associated with a particular quantity. For example, a diatomic molecule may be thought of as two spheres connected by a spring. At finite temperatures, the two atoms vibrate around the center of mass of the molecule with some finite amplitude. For simplicity, we shall consider the vibration to be along the line joining the two atoms. As a result, the distance between the centers of the two atoms is a function of time. In this case, there is no single value for the distance and we must make some reasonable definition for the "true value" of such a quantity. A good choice in this case is to use the average of values measured for many similar molecules, as this is the quantity that best characterizes the distance between the two atoms at the particular temperature. The distribution of the measured values around the average is therefore a part of the quantity itself and cannot be reduced by any improvements in the measuring apparatus used.

Another common example of such types of quantities is the energy of an excited state in an atom or a nucleus. Because of the finite lifetime Δt associated with the state, the energy has a "natural" width ΔE associated with it that is given by the Heisenberg uncertainty principle $\Delta E \Delta t \approx \hbar$. In this case, there is an uncertainty ΔE associated with the excitation energy even when we use apparatus of infinite precision to carry out the measurements.

Probability distribution. The distribution $p(x)$ of a quantity x may be characterized by a number of parameters, such as where it is located, how broad it is, and what shape it has. If the value obtained in measurement i is x_i, the *mean*, or *centroid*, μ of the distribution is given by the average value in the limit where the number of measurements is infinite:

$$\mu = \lim_{N \to \infty} \frac{1}{N} \sum_{i=1}^{N} x_i \tag{6-1}$$

If the distribution is a continuous one, the probability of finding the value x in the interval dx is given by $p(x)dx$. It is a generally accepted convention to normalize $p(x)$ to unity:

$$\int_a^b p(x)\, dx = 1 \tag{6-2}$$

where a and b are, respectively, the upper and lower limits of the possible values of x. In this case the mean is given by the integral

$$\mu = \int_a^b x\,p(x)\,dx$$

instead of (6-1).

Two other quantities are also often used to indicate the location of a distribution. The *median*, usually denoted as $\mu_{1/2}$, is defined as the value where the probabilities to find x below and above it are the same. That is

$$p(x < \mu_{1/2}) = p(x > \mu_{1/2}) \qquad (6\text{-}3)$$

The *most probable value*, which we shall write as μ_{max}, is given by the value of x where the probability distribution attains the maximum value:

$$p(x = \mu_{max}) \geq p(x \neq \mu_{max}) \qquad (6\text{-}4)$$

In general, these three quantities, μ, $\mu_{1/2}$, and μ_{max}, are not equal to each other, as illustrated in Fig. 6-1 by the distribution

$$p(x)dx = xe^{-x}\,dx \qquad \text{for} \qquad x = [0, \infty] \qquad (6\text{-}5)$$

Mathematically, the mean is the most convenient quantity in many applications, and we shall use it to characterize the location of a distribution for the most part.

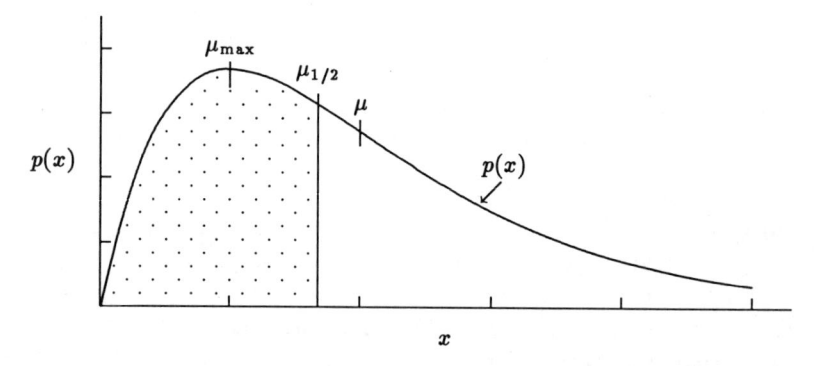

Fig. 6-1 Most probable value μ_{max}, median $\mu_{1/2}$, and mean μ of a distribution. For a normal distribution, the three quantities coincide with each other. This is not true for distributions in general, as illustrated by $p(x)dx = xe^{-x}dx$ shown.

A measure of the deviation of x from the mean is given by the *variance* σ^2. For a discrete distribution, it is defined as

$$\sigma^2 = \lim_{N \to \infty} \frac{1}{N} \sum_{i=1}^{N} (x_i - \mu)^2 \qquad (6\text{-}6)$$

For a continuous distribution in the interval $x = [a, b]$,

$$\sigma^2 = \int_a^b (x - \mu)^2 p(x)\, dx$$

The square root of the variance, σ, is known as the standard deviation. In physics, it is often referred to as the *half-width* of the distribution. Strictly speaking, σ is equal to half of Γ, the full width of the distribution at half maximum, only in the case of a Lorentzian distribution:

$$p(x)dx = \frac{1}{\pi} \frac{\Gamma/2}{(x - \mu)^2 + (\Gamma/2)^2} dx \qquad (6\text{-}7)$$

Such a distribution characterizes, for example, the energy released in the decay of an unstable state. Another quantity related to σ is the *probable error s*. It is defined by the relation

$$\int_{\mu-s}^{\mu+s} p(x)\, dx = \tfrac{1}{2} \qquad (6\text{-}8)$$

and provides the range of the values in which we can find x with a probability of half. For a normal or Gaussian distribution,

$$p(x)dx = \frac{1}{\sigma\sqrt{2\pi}} e^{-(x-\mu)^2/2\sigma^2}\, dx \qquad (6\text{-}9)$$

we have $s = 0.6745\sigma$ and the full width at half maximum is $\Gamma = 2 \times 1.177\sigma$.

To characterize the shape of a distribution in more detail, we can make use of the central moments:

$$\mu_r = \int_a^b (x - \mu)^r p(x)\, dx \qquad (6\text{-}10)$$

The $r = 0$ moment is unity from the normalization condition (6-2), and the first moment, $r = 1$, is zero since we are taking the moments about the mean. The second moment is the same as the variance σ^2 and it provides an idea of how broad the distribution is. An alternate form of the third moment μ_3 is the *skewness*:

$$\gamma_3 = \mu_3/\sigma^3$$

It gives a measure of the asymmetry of the distribution about the mean. Similarly, the fourth moment μ_4 is related to the *excess* or *kurtosis*:

$$\gamma_4 = \mu_4/\sigma^4 - 3$$

For a distribution that is close to normal, a positive γ_4 means it is sharper in the central region than that given by (6-9) and a negative γ_4 means it is flatter. Usually, it is not very meaningful to compare the higher-order moments of two distributions unless their lower-order ones are roughly the same.

The definitions given by (6-1) and (6-2) imply that an infinite number of measurements is required to determine the distribution. Since this is impossible in most cases, it must be postulated in some way. Following the convention used in statistics, we shall call such a hypothetical distribution $p(x)$ the *parent distribution*. The aim, then, of a statistical analysis is to arrive at the parent distribution from a finite sample of the values and to have an estimate of the likely errors, or uncertainties, incurred as a result of our incomplete knowledge of the system. Each measurement we make constitutes one of the infinite numbers required to define the distribution. After a number of samples is taken, we can, for instance, find the mean of the results obtained so far. The value may or may not correspond to the mean of the parent distribution, but it provides an estimate for it. As the number of samples grows, we can be confident that the sample mean is getting closer and closer to the mean of the parent distribution. At the same time, the distribution of our sample results becomes a better and better representation of the parent distribution. Before discussing some of the ways to estimate the parent distribution, we shall find out first what probability distributions are commonly found in nature and see how some of them come about.

Binomial distribution. Consider a sample of N radioactive nuclei. If, in a given time interval dt, the probability for any one nucleus to decay to the ground state is $p\,dt$, what is the probability $P_B(N, n, p)\,dt$ of observing n decays among the N nuclei in the same time interval? For purposes of illustration, we shall assume that the errors associated with our timer and counter for measuring the decays are sufficiently small that we can ignore them. The only source of uncertainty then arises from the probabilistic nature of radioactive decay. Furthermore, the decay of any one nucleus is independent of the states of the other nuclei. As a result, there is no way to predict which of the N nuclei in the sample will decay in the time interval. All we know is that the probability for any one nucleus to decay is p per unit time. The value of p is a property that characterizes the nucleus and is related to its half-life or decay constant.

For our derivation of the probability $P_B(N, n, p)$, it is convenient to think first in terms of a situation in which each of the N nuclei is labeled by a number, 1, 2, ..., N. For nucleus i, the probability to decay in the given time interval is $p\,dt$. Consequently, the probability for two particular nuclei, i and j, to decay in the same time interval is $p^2\,dt$. By the same token, the probability of n particular nuclei to decay in the same time interval is $p^n\,dt$. However, this is not the probability of interest to us — what we want to know is the probability of any n, and only n, of the N nuclei to decay. To arrive at this quantity, several corrections must be made to the result of $p^n\,dt$ obtained above. The first comes from the requirement that none of the remaining $(N - n)$ nuclei have decayed in the mean time. For any one nucleus, the probability not to decay in the time interval is, by definition, $(1 - p)\,dt$. The probability of finding nucleus i decayed and another nucleus k not decayed in the time interval is therefore the product $p(1 - p)\,dt$.

Our thought experiment on radioactive decay can proceed further along the following line. At the beginning of a measurement, we have a sample of N radioactive nuclei in a box. After the given time interval has elapsed, we take out each of them and see whether it has decayed to the ground state or remained in the excited state. The first one we take out of the box can be any of the N nuclei in the box and, since it is immaterial to us which of the N nuclei it is, there are N possible ways to pick the first one. The probability of finding the first nucleus to have decayed is therefore Np. The second one can be any of the remaining $(N-1)$ nuclei, and the probability for this nucleus to be in the ground state is $(N-1)p$. Combining the two arguments, we find that the probability for the first two nuclei to be in the ground state is

$$P(N,2,p)\,dt = \frac{N(N-1)}{2}p^2\,dt$$

The factor of 2 in the denominator on the right side comes from the fact that the two possibilities, one with nucleus i as the first one and nucleus j as the second one, and the other with nucleus j as the first one and nucleus i as the second one, are the same to us. All we need to know is that both the first one and the second one are in the ground state.

The natural extension of this line of argument to the probability for starting with a sample of N excited nuclei and finding the first n nuclei in the ground state at the end of a time interval dt is then

$$P(N,n,p)\,dt = \frac{N(N-1)\cdots(N-n+1)}{n!}p^n\,dt \tag{6-11}$$

Here the factor $n!$ in the denominator on the right side comes from the fact that all we care about is that there are n nuclei in the ground state — the $n \times (n-1) \times \cdots \times 2 \times 1$ different possible orders for the n nuclei to come out of the box are immaterial to us.

The result given by (6-11) is still not the probability $P_B(N,n,p)\,dt$ for which we are looking. This comes from the fact that we have not yet imposed the condition that the other $(N-n)$ nuclei remain in the excited state. Since the probability for $(N-n)$ nuclei not decayed is $(1-p)^{N-n}\,dt$, we obtain the final result that

$$\begin{aligned}P_B(N,n,p)\,dt &= \frac{N(N-1)\cdots(N-n+1)}{n!}p^n(1-p)^{N-n}\,dt \\ &= \frac{N!}{n!(N-n)!}p^n(1-p)^{N-n}\,dt \\ &= \binom{N}{n}p^n(1-p)^{N-n}\,dt\end{aligned} \tag{6-12}$$

where the binomial coefficient $\binom{N}{n}$ was defined earlier in (1-1). This is known as the binomial distribution. The name comes from the binomial theorem

$$(p+q)^n = \sum_{r=0}^{n}\binom{n}{r}p^r q^{n-r}$$

where the coefficient for the $p^r q^{n-r}$ term is given by the binomial coefficient $\binom{n}{r}$. In Problem 6-2, we shall see that the mean and variance of a binomial distribution are, respectively, Np and $Np(1-p)$.

Poisson distribution. For most radioactive transitions, the probability p for a single nucleus to decay per unit time is very small. On the other hand, the sample is usually large, with N far in excess of 10^6. As a result, the product Np remains a finite number. Let us define

$$\mu \equiv Np$$

In the limit that $p \to 0$, $N \to \infty$, and μ remains constant, the binomial distribution approaches a Poisson distribution:

$$P_P(n, \mu) = \frac{\mu^n}{n!} e^{-n} \tag{6-13}$$

In Problem 6-2, we see that both the mean and the variance of this distribution are equal to μ. Since $\sigma^2 = \mu$, the distribution is completely specified by a single parameter, commonly taken to be μ.

The relation between Poisson and binomial distributions may be seen from the following considerations. Let us start from the second form of $P_B(N, n, p)$ given in (6-12) and rewrite it in the form

$$P_B(N, n, p) = \frac{N!}{n!(N-n)!} p^n (1-p)^N (1-p)^{-n}$$

For $p \to 0$, we have $n \ll N$. The first two factors on the right side may be approximated as

$$\frac{N!}{n!(N-n)!} p^n \xrightarrow[p \to 0]{} \frac{1}{n!} N^n p^n = \frac{1}{n!} \mu^n \tag{6-14}$$

where we have made use of the fact $\mu = Np$ for $P_B(N, n, p)$. Similarly the last factor $(1-p)^{-n}$ may be replaced by 1. The third factor may be rewritten in terms of the mean

$$(1-p)^N = \{(1-p)^{1/p}\}^\mu \xrightarrow[p \to 0]{} e^{-\mu} \tag{6-15}$$

This comes from the observation that

$$\left(1 + \frac{1}{m}\right)^m = 1 + m\frac{1}{m} + \frac{m(m-1)}{2!}\left(\frac{1}{m}\right)^2 + \cdots$$

$$+ \frac{m(m-1)\cdots(m-r+1)}{r!}\left(\frac{1}{m}\right)^r + \cdots$$

$$\xrightarrow[m \gg 1]{} 1 + \frac{1}{1} + \frac{1}{2!} + \cdots + \frac{1}{r!} + \cdots = e$$

As a result, on letting $m = -p^{-1}$, we have

$$(1-p)^{1/p} = (1+m^{-1})^{-m} = \frac{1}{\left(1+\frac{1}{m}\right)^m} \longrightarrow e^{-1}$$

When we put together the results of (6-14) and (6-15), the Poisson distribution of (6-13) emerges.

Normal distribution. The normal or Gaussian distribution of (6-9)

$$p(x)dx = \frac{1}{\sigma\sqrt{2\pi}} e^{-(x-\mu)^2/2\sigma^2} dx$$

is specified by two parameters, the mean μ and the variance σ^2. It occurs in a large variety of situations where many independent degrees of freedom are available to the system. The reason for its popularity comes from the central limit theorem in statistics. It states that, if a variable x has a large number of independent degrees of freedom, the distribution of x goes asymptotically to a normal distribution. The theorem is true without much regard to the distribution for each of the degrees of freedom. As an example, consider a number x that is made of the sum of n random numbers. If n is large, the distribution of x has the form of (6-9), essentially independent of the distribution of the n random numbers it is made of. Problem 6-3 provides an illustration of this point.

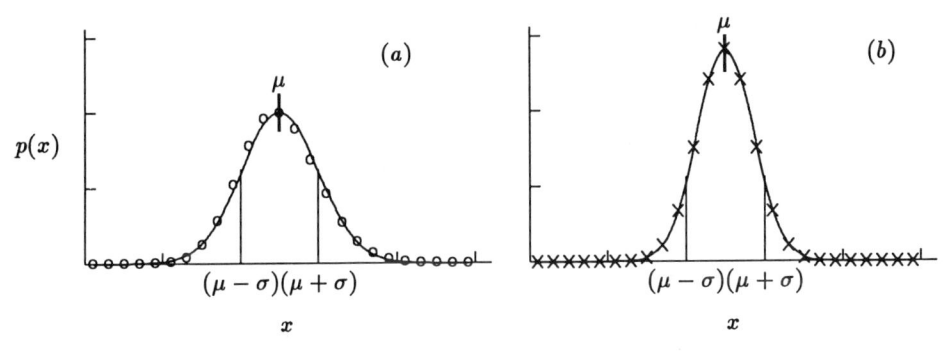

Fig. 6-2 A comparison of Poisson distribution (○) and binomial distribution (×) with normal distributions (solid curve) of the same mean $\mu = 25$. The variance is $\sigma^2 = \mu$ in (a), appropriate for a Poisson distribution, and $\sigma^2 = \mu/2$ in (b), suitable for a binomial distribution with $p = 0.5$ and $N = 50$.

Because of the central limit theorem, the distribution of many quantities approaches that of a normal one when the number of degrees is large. Consider the case of tossing coins. If the coins are unbiased, the probability of any one coin landing head up is exactly $\frac{1}{2}$. If we flip two such coins, the probabilities of having two heads, one head, and no head are given by the binomial distribution with $N = 2$ and $p = (1-p) = 0.5$. The results are, respectively, $P_B(2, 2, 0.5) = 0.25$,

$P_B(2,1,0.5) = 0.5$, and $P_B(2,0,0.5) = 0.25$. Such a distribution is symmetric around the mean of one head, but the shape does not resemble that for a normal distribution. However, if instead of tossing two coins, we deal with a much larger number, such as $N = 50$, the resulting shape will be very close to that of a normal distribution. In the place of coins, we can also illustrate the same point using a large number of independent quantities, each of which has a Poisson distribution. The result is a normal distribution as shown in Fig. 6-2.

Lorentzian distribution. The Lorentzian distribution of (6-7),

$$p(x)dx = \frac{1}{\pi} \frac{\Gamma/2}{(x-\mu)^2 + (\Gamma/2)^2} dx$$

is found in many natural phenomena, such as the power dissipated as a function of frequency across a resistor in an alternating-current circuit consisting of a capacitor, an inductor, and a resistor. The mechanical analog of the same type of situation is the power supplied by a source as a function of frequency to drive a harmonic oscillator.

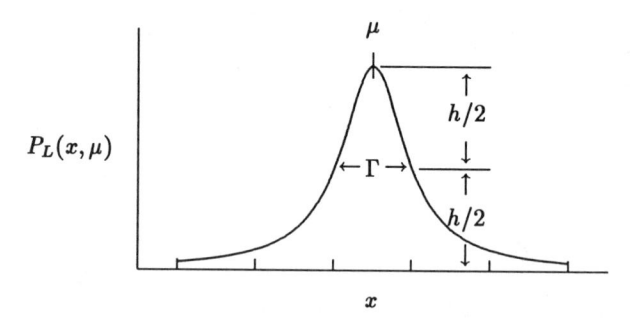

Fig. 6-3 Lorentzian distribution given by (6-7). For this distribution, the full width at half-maximum is equal to twice the standard deviation.

For an excited state of a quantum mechanical system, the Lorentzian distribution gives the *natural line shape* of the energy emitted when the particle decays to a lower state. Because of the Heisenberg uncertainty principle, the energy of an unstable state cannot be infinitely sharp. Instead, it takes on the form shown in Fig. 6-3, with mean equal to μ and full width at half-maximum characterized by Γ. Let the lifetime of the state be $\overline{T}$, meaning the time interval at the end of which only $1/e$ of the original number of excited nuclei remains. For such a state, the width takes on the value

$$\Gamma = \hbar/\overline{T}$$

where $\hbar$ is the Planck's constant h divided by 2π. As mentioned earlier, the standard deviation for a Lorentzian shape distribution is $\frac{1}{2}\Gamma$.

Probability functions. If the probability of finding a quantity with value t in an interval dt is given by $p(t)dt$, the total probability for finding the quantity with a value less or equal to x is given by the integral

$$P(x) = \int_{-\infty}^{x} p(t)\,dt$$

Since the normal distribution is the most common one occurring in nature, the integral

$$P(x) = \frac{1}{\sqrt{2\pi}} \int_{-\infty}^{x} e^{-t^2/2}\,dt \tag{6-16}$$

is useful in a variety of situations and is known sometimes as the *normal probability function*. It is related to the error function of (4-133) by the relation

$$\mathrm{erf}(x) = 2P(\sqrt{2}\,x) - 1$$

as shown earlier in (4-135), and to the incomplete gamma function

$$\gamma(\tfrac{1}{2}, x) = \sqrt{\pi}\{2P(\sqrt{2x}) - 1\}$$

defined in (4-151).

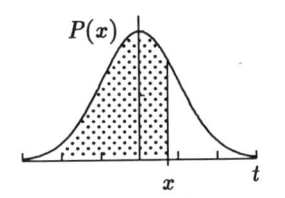

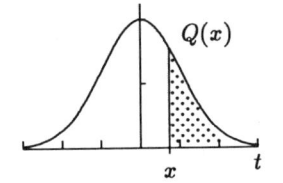

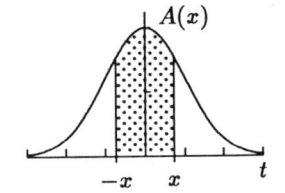

Fig. 6-4 Schematic illustration of the relations between integrals $P(x)$ of (6-16) and integrals $Q(x)$ and $A(x)$ of (6-17).

Two other functions are also useful in probability studies,

$$Q(x) = \frac{1}{\sqrt{2\pi}} \int_{x}^{\infty} e^{-t^2/2}\,dt \qquad A(x) = \frac{1}{\sqrt{2\pi}} \int_{-x}^{x} e^{-t^2/2}\,dt \tag{6-17}$$

Both may be written in terms of $P(x)$

$$P(x) + Q(x) = 1 \qquad P(-x) = Q(x) \qquad A(x) = 2P(x) - 1 \tag{6-18}$$

Graphically, the relations between $P(x)$, $Q(x)$, and $A(x)$ are illustrated in Fig. 6-4.

Earlier in §2-5, we used a method of evaluating $A(x)$ by Monte Carlo integration. An efficient rational polynomial approximation is given in Abramowitz and Segun (*Handbook of Mathematical Functions*. Dover, New York, 1965)

$$P(x) = 1 - \frac{1}{2}\frac{1}{(1 + d_1 x + d_2 x^2 + d_3 x^3 + d_4 x^4 + d_5 x^5 + d_6 x^6)^{16}} + \epsilon(x) \tag{6-19}$$

The six coefficients have the values

$d_1 = 0.0498673470 \qquad d_2 = 0.0211410061 \qquad d_3 = 0.0032776263$
$d_4 = 0.0000380036 \qquad d_5 = 0.0000488906 \qquad d_6 = 0.0000053830$

The error quoted is $|\epsilon(x)| < 1.5 \times 10^{-7}$. Using the relations given in (6-18), the values of $Q(x)$ and $A(x)$ for a given x may be obtained from that of $P(x)$.

§6-2 Uncertainties and their propagation

Since uncertainties are unavoidable in measurements, it is essential that we have some idea of the influence they have on any conclusion we wish to draw from data. Let us begin by considering the case of equating the true value of an object by the average obtained from a finite number of measurements.

Uncertainty associated with finite sample size. Consider a hypothetical experiment to determine the length ℓ of a city block. The values for $N = 25$ measurements are listed in Table 6-1 in units of meters. The average of the 25 results is $\mu = 49.9$ m, and the root mean square deviation from the average is

$$\sigma = \left\{ \frac{1}{N} \sum_{i=1}^{N} (\ell_i - \mu)^2 \right\}^{1/2} = 1.0 \text{ m}$$

The question we wish to ask ourselves is the following: how likely is the average value of 49.9 m to be the true length of the city block? Unless there are some systematic errors, we have no reason to reject the sample average of $\mu = 49.9$ m as the mean of the parent distribution, that is, the mean if the number of measurements N is infinite instead of 25. On the other hand, we have no reason to believe that the sample average is equal to the mean of the parent distribution either.

Table 6-1

Results of 25 measurements of the length ℓ of a hypothetical city block in meters.

Exp. no.	Length	Exp. no.	Length	Exp. no.	Length
1	50.6	2	50.3	3	50.6
4	47.7	5	50.7	6	49.4
7	50.6	8	50.7	9	48.7
10	50.7	11	50.5	12	49.3
13	50.5	14	48.9	15	50.2
16	49.2	17	51.0	18	48.3
19	50.7	20	49.9	21	51.2
22	48.7	23	48.6	24	49.1
25	51.6				
				$\mu = 49.9$	$\sigma = 1.0$

Earlier, we saw some of the possible reasons why the measured results can, in general, be different from each another. We shall assume that there is a large number of independent sources of error, each of which is distributed randomly around some mean μ_i and variance σ_i^2. With this assumption, we can invoke the central limit theorem and claim that the parent distribution of the measured values is a normal one. If the values μ_i and σ_i^2 are known, it is possible to deduce the mean and variance of the parent distribution. Since this is impossible for most cases, we must adopt other methods to estimate the two parameters required to specify the normal parent distribution. One way to achieve this goal is to take the following approach. If the sample of 25 measurements represents a random selection among the members in the parent distribution, the measured values must also distribute normally with mean and variance, μ and σ, that are good estimates of the corresponding values in the parent distribution.

Since we are assuming that the measured values follow a normal distribution, the probability for the mean of the parent distribution to be in the range $[\mu - \sigma, \mu + \sigma]$ is given by the integral $A(x)$ of (6-17) for $x = \sigma$:

$$A(\sigma) = \int_{\mu-\sigma}^{\mu+\sigma} P_G(\mu, \sigma, t)\, dt = 0.68 \qquad (6\text{-}20)$$

In other words, if we perform the experiment of measuring the length of the same city block 100 times, each consisting of 25 measurements, we expect that the average value is between 48.9 m and 50.9 m for 68 of the 100 times. For the remaining 32 times, the average values are expected to be either below 48.9 m or above 50.9 m.

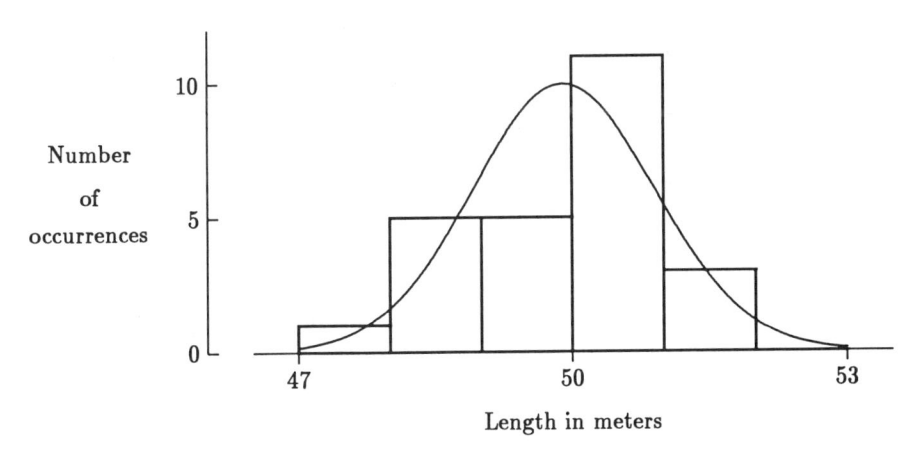

Fig. 6-5 Histogram showing the distribution of the values obtained in 25 measurements of a hypothetical city block listed in Table 6-1. A normal distribution of the same mean and variance (solid curve) is superimposed for comparison.

Before accepting this result, let us check whether it is correct for us to assume that the distribution for the measured values is actually a normal one. For this purpose we shall plot the distribution of the results in the form of a histogram. Since we have only a sample size of $N = 25$, the bin size is chosen to be 1 m. A convenient way to make the plot is include in the first bin all the measured results with $47 \leq \ell < 48$ m, the second bin, those with $48 \leq \ell < 49$ m, and so on. This is shown in Fig. 6-5. There is only one measurement with a value of ℓ in the range $47.0 \leq \ell < 48.0$ m and, as a result, the height of the first bin is 1. There are five measurements with the values of ℓ in the range $48.0 \leq \ell < 49.0$ m and the height of the second bin is 5, and so on. In the figure, we have also superimposed on the histogram a normal distribution of the same mean and variance. For ease of comparison, the normalization of the distribution is 25 instead of 1, as done in (6-9). We notice that in two of the bins, the one for $49.0 \leq \ell < 50.0$ m and the one for $50.0 \leq \ell < 51.0$ m, the deviations from the values expected of a normal distribution curve are quite large.

Let us examine whether these deviations are significant. The value in bin $49.0 \leq \ell < 50.0$ m for a normal distribution with $N = 25$ and $(\mu, \sigma) = (49.9, 1.0)$ may be found from the integral

$$E(a,b) = \frac{N}{\sigma\sqrt{2\pi}} \int_a^b \exp\left\{-\frac{1}{2}\left(\frac{x-c}{\sigma}\right)^2\right\} dx \qquad (6\text{-}21)$$

For $a = 49$ and $b = 50$, the result is 8.9, compared with 5 for the height of the bin. For $a = 50$ and $b = 51$, the result is 8.1 and the height of the bin is 11. The differences are approximately 3 in both cases. Are these differences significant enough that we should question the assumption of a normal distribution for our measured results?

To answer this question, we must first realize that the occupancy in each bin also follows a definite distribution with some mean and variance. For our example, we have only one set of "measured" values, obtained by counting the number of occurrences of the value of ℓ in a given range in Table 6-1. Again, we must assume that this particular set of measured values is only one possible set of 25 measurements among a large number of similar ones in the parent distribution. In other words, if we carry out another set of 25 measurements and plot the results as a histogram in the same way as we have done for our original set, the resulting plot is likely to be different. Now, if we repeat the process of taking 25 measurements many times and plot the histogram each time, we obtain a distribution of the number of occurrences for each of the bins. In this way, we can, in principle, map the distribution for the number of occurrences in each histogram bin. Our interest is to find out the mean and variance for each occupied bin.

In reality, however, we have only one set of measurements and we wish to find out the significance of the set of values we have in hand. With only one value for each bin, the best we can do is to take it as the mean value for the distribution of the bin. Now we have exhausted the only piece of information in our possession. To make progress, we must do something else. A common approach is to assume the Poisson distribution of (6-13) as the parent distribution; that is, the distribution of the height of a particular histogram bin is Poisson in the limit of an infinite number of samples of 25 measurements of our city block. Since the variance of a Poisson distribution is equal to the mean, our distribution is completely specified by the one measured value we have available to us. Furthermore, a Poisson distribution approaches that of a normal distribution when the number of degrees of freedom is large. As a result, we can say that, if the occupancy is n in a particular bin, the standard deviation is $\sqrt{n}$. This means that there is a probability of 68%, as we can see from (6-20), that the true value for this bin is in the range $[n - \sqrt{n}, n + \sqrt{n}]$. In the next section, we shall see that it is also possible to arrive at the same conclusion without having to make the assumption of a Poisson distribution.

Returning to the histogram bin $50 \leq \ell < 51$ m, we have $n = 11$ and $\sqrt{11} = 3.3$. If the parent distribution for the height of this bin is Poisson, the uncertainty associated with the height is estimated to be $\sigma = 3.3$. The difference of the observed value from that expected of (6-21) is of 2.9. Since this is within one standard deviation of the mean, it is probable ($> 68\%$) that the sample result belongs to the same parent distribution as the normal distribution we have assumed. Similarly for the bin $49 \leq \ell < 50$ m, the measured value of n is 5 and the expected value from (6-21) for this bin is 8.8. The difference is therefore outside one standard deviation of $\sqrt{5} = 2.3$. For a normal distribution there is a probability of 32% for this to take place. Among a total of 6 bins in the histogram, we should therefore allow, on the average, roughly two bins to have differences larger than one standard deviation from the mean. For this reason, we cannot treat the deviation from the expected value for this bin as significant. The net conclusion is that, although our histogram does not resemble a bell-shaped distribution, we cannot find any significant departure from one either.

In fact, the results in Table 6-1 are produced using a random number generator with a normal distribution to simulate the uncertainties in the measured values. If we enlarge our sample size from 25 to a much larger number, the resulting histogram more closely resembles a normal distribution, as shown in Fig. 6-6. In the next section we shall develop more quantitative methods to test the similarity of two distributions.

Propagation of error. Consider a quantity y that is a function of the independent variable x:
$$y = f(x)$$
If we vary x by an amount Δx, the value of y changes by an amount Δy. The relation between Δx and Δy may be found from a Taylor series expansion of the value of $f(x \pm \Delta x)$ around that of $f(x)$
$$y \pm \Delta y = f(x \pm \Delta x) = f(x) + \frac{df}{dx}(\pm \Delta x) + \cdots$$

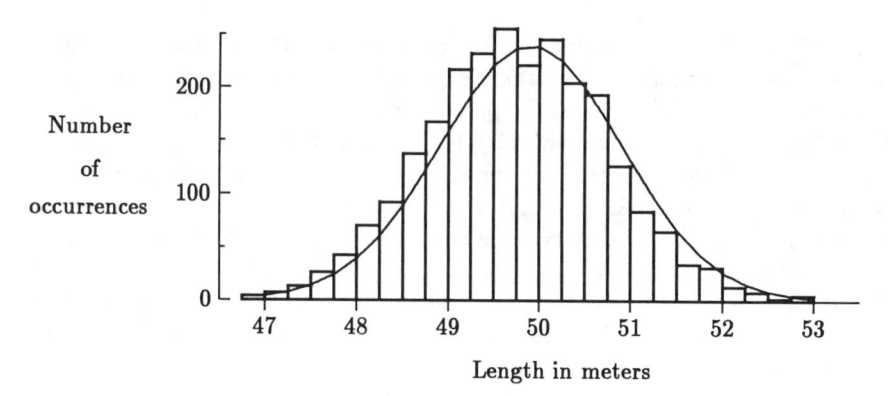

Fig. 6-6 Histogram showing the distribution of the values of 2500 measurements of the type given in Table 6-1. Compared with Fig. 6-5, a much better agreement with a normal distribution of the same mean and variance (solid curve) is obtained. This comes from the fact that the ratio of the uncertainty in each bin, $\sqrt{n}$, to the total number n in each bin decreases as n increases in size.

On truncating the series after the second term, we obtain the result

$$\Delta y = \frac{df}{dx}\Delta x \qquad (6\text{-}22)$$

If y is a function of several independent variables $u, v, w, \ldots$,

$$y = f(u, v, w, \ldots)$$

the relation analogous to (6-22) becomes

$$\Delta y = \frac{\partial f}{\partial u}\Delta u + \frac{\partial f}{\partial v}\Delta v + \frac{\partial f}{\partial w}\Delta w + \cdots \qquad (6\text{-}23)$$

We shall make use of this result to derive the relation between uncertainties in the independent variables and that of a quantity derived from them.

The common practice in physics is to associate the standard deviation σ_x in the distribution of a variable x as the uncertainty. That is,

$$\Delta x = \sigma_x$$

This definition is particularly useful in dealing with experimental data. Since there are many independent sources of random errors associated with a measurement, the values obtained are likely to follow a normal distribution. The assignment of the standard deviation, the square root of the variance, as the uncertainty has a statistical significance and can provide us with an estimate of the likelihood of obtaining values different from the mean, as we have seen earlier.

Using (6-23), we can relate the uncertainties defined in this way in the independent variables $u, v, w, \ldots$, to that of a quantity calculated from them. If there are N measurements of $u, v, w, \ldots$, and the values obtained for them are $u_i, v_i, w_i, \ldots$, in measurement i, the variance in the distribution of y is given by

$$\sigma_y^2 = \lim_{N \to \infty} \frac{1}{N} \sum_{i=1}^{N} \left\{ \frac{\partial f}{\partial u}(u_i - \mu_u) + \frac{\partial f}{\partial v}(v_i - \mu_v) + \frac{\partial f}{\partial w}(w_i - \mu_w) + \cdots \right\}^2$$

$$= \lim_{N \to \infty} \frac{1}{N} \sum_{i=1}^{N} \left\{ \left(\frac{\partial f}{\partial u}\right)^2 (u_i - \mu_u)^2 + \left(\frac{\partial f}{\partial v}\right)^2 (v_i - \mu_v)^2 + \left(\frac{\partial f}{\partial w}\right)^2 (w_i - \mu_w)^2 + \cdots \right.$$

$$\left. + 2\left(\frac{\partial f}{\partial u}\right)\left(\frac{\partial f}{\partial v}\right)(u_i - \mu_u)(v_i - \mu_v) + 2\left(\frac{\partial f}{\partial v}\right)\left(\frac{\partial f}{\partial w}\right)(v_i - \mu_v)(w_i - \mu_w) + \cdots \right\}$$

$$= \left(\frac{\partial f}{\partial u}\right)^2 \sigma_u^2 + \left(\frac{\partial f}{\partial v}\right)^2 \sigma_v^2 + \left(\frac{\partial f}{\partial w}\right)^2 \sigma_w^2 + \cdots$$

$$+ 2\left(\frac{\partial f}{\partial u}\right)\left(\frac{\partial f}{\partial v}\right)\sigma_{uv}^2 + 2\left(\frac{\partial f}{\partial v}\right)\left(\frac{\partial f}{\partial w}\right)\sigma_{vw}^2 + \cdots \qquad (6\text{-}24)$$

where $\mu_u, \mu_v, \mu_w, \ldots$, are the mean values and $\sigma_u^2, \sigma_v^2, \sigma_w^2, \ldots$, the variances of the distributions of $u, v, w, \ldots$, respectively. Since there is more than one independent variable, there may be a correlation between any two of them. This is measured by the covariance

$$\sigma_{uv}^2 = \lim_{N \to \infty} \frac{1}{N} \sum_{i=1}^{N} \{(u_i - \mu_u)(v_i - \mu_v)\} \qquad (6\text{-}25)$$

If u and v are completely uncorrelated, the probability of obtaining a value u_i for u is completely independent of the probability of getting a value v_i for v. In this case,

$$\sigma_{uv}^2 = 0$$

The result comes from the fact that, separately, we have

$$\lim_{N \to \infty} \frac{1}{N} \sum_{i=1}^{N} (u_i - \mu_u) = 0 \qquad \lim_{N \to \infty} \frac{1}{N} \sum_{i=1}^{N} (v_i - \mu_v) = 0$$

A nonvanishing value of σ_{uv}^2 implies that u and v are correlated.

The basic relation between the variances of the dependent and independent variables is given by (6-24). It is useful to see some explicit examples involving elementary functions. For instance, if y is a linear function of two independent variables u and v,

$$y = au \pm bv$$

where a and b are constants, the variance of y is given by

$$\sigma_y^2 = a^2 \sigma_u^2 + b^2 \sigma_v^2 \pm 2ab\, \sigma_{uv}^2 \qquad (6\text{-}26)$$

This is also a good example showing the importance of the covariance σ_{uv}^2. It is not difficult to think of a case where the error in measuring u is always compensated by that in measuring v. In this case, the value of σ_{uv}^2 must be such that σ_y^2 vanishes.

If y is the product of u and v, such as the function

$$y = auv$$

the partial derivatives of y are

$$\frac{\partial y}{\partial u} = av \qquad \frac{\partial y}{\partial v} = au$$

The variance of y is then

$$\sigma_y^2 = (av)^2 \sigma_u^2 + (au)^2 \sigma_v^2 + 2a^2 uv\, \sigma_{uv}^2$$

This is more commonly written in the form

$$\frac{\sigma_y^2}{y^2} = \frac{\sigma_u^2}{u^2} + \frac{\sigma_v^2}{v^2} + 2\frac{\sigma_{uv}^2}{uv} \qquad (6\text{-}27)$$

In this case, any correlation between u and v always increases the uncertainty in y.

For the function

$$y = a\frac{u}{v}$$

the corresponding result is

$$\frac{\sigma_y^2}{y^2} = \frac{\sigma_u^2}{u^2} + \frac{\sigma_v^2}{v^2} - 2\frac{\sigma_{uv}^2}{uv}$$

The difference in the sign of the covariance term from that of (6-27) comes from the fact that

$$\frac{\partial y}{\partial v} = -a\frac{u}{v^2}$$

Derivations of similar relations for other standard functions are given as exercises in Problem 6-5.

§6-3 The method of maximum likelihood

Let us use the decay of an unstable particle as an example to introduce the method of maximum likelihood. Since the decay is purely statistical in nature, we can only predict the probability for it to take place in a given time interval, not the time when a particular particle actually undergoes the transformation. From this we obtain that the number of decays per unit time dN/dt at time t is proportional to $N(t)$, the number of radioactive nuclei present. That is,

$$\frac{dN}{dt} = -\frac{1}{\tau}N(t)$$

where we have written the proportional constant as $1/\tau$ for reasons that will become clear soon. The negative sign on the right side of the expression reflects the fact that each decay decreases the number of radioactive nuclei present.

The solution of this differential equation is the familiar exponential decay law

$$N(t) = N_0 e^{-t/\tau} \tag{6-28}$$

where N_0 is the number of radioactive nuclei at $t = 0$. When $t = \tau$, the number of radioactive nuclei is reduced to $1/e$ of the amount at $t = 0$. For this reason, τ is known as the lifetime or mean life for the decay. The value of τ is a property of the particle and equals the half-life $\tau_{1/2}$ divided by $\ln 2$ ($= 0.693$). From (6-28), we obtain the probability $p(t)$ of observing the particle to decay at time t

$$p(t) = \frac{1}{\tau} e^{-t/\tau} \tag{6-29}$$

where the factor $1/\tau$ on the right side is necessary to normalize $p(t)$ such that the integrated probability is unity in the time interval $[0, \infty]$.

If we wish to measure τ, it is necessary to observe a large number of decays. The need comes from the probabilistic nature of the process. For simplicity we shall assume for now that instrumental errors associated with the measurements are negligible. In the present case, this means that the time at which each of the sample of N unstable particles decays is recorded as precisely as we wish. The data then consist of N values of time, $t_1, t_2, \ldots, t_N$, each giving the time when one of the N particles is observed to decay. The only uncertainty in the results arises because of the statistical nature of the process.

If we take a different set of measurements of a similar collection of the same type of nucleus, the N values of time obtained will most likely be different from those in the set above. However, by definition, the value of τ underlying these N values must be the same in both sets of measurements, since we are, after all, measuring the same type of transition. The method we use to deduce the value of τ from the data must, therefore, be such that the result is, as far as possible, independent of the fluctuations from one set of data to another.

Likelihood function. The probability $\mathcal{L}(\tau)$ for the measured results to come out the way we have recorded above is given by the product of the probabilities (6-29) for the N particles to decay at times $t_1, t_2, \ldots, t_N$:

$$\mathcal{L}(\tau) = \prod_{i=1}^{N} p(t_i) = \prod_{i=1}^{N} \exp\left\{-\frac{t_i}{\tau} - \ln \tau\right\} = \exp\left\{-\frac{1}{\tau}\sum_{i=1}^{N} t_i - N \ln \tau\right\} \quad (6\text{-}30)$$

This expression, however, cannot be used to evaluate $\mathcal{L}(\tau)$, as τ is not yet known. Some estimate for the value of τ must be made before we can proceed with the calculation. An optimum choice is one that maximizes the value of $\mathcal{L}(\tau)$. As we shall see later, if $\mathcal{L}(\tau)$ is a function with a sharply peaked distribution, we do not have much room for any alternate choices. In fact, in the limit that $\mathcal{L}(\tau)$ is a delta function, the value of τ is uniquely determined by such a procedure. However, the limiting case is of no interest to us here, since it implies a completely deterministic situation and there is really no need to construct $\mathcal{L}(\tau)$.

It is easy to see that the peak of the distribution $\mathcal{L}(\tau)$ occurs at the location that satisfies the condition

$$\frac{d\mathcal{L}(\tau)}{d\tau} = \left(\frac{1}{\tau^2}\sum_{i=1}^{N} t_i - \frac{N}{\tau}\right)\mathcal{L}(\tau) = 0$$

From this, we obtain

$$\bar{\tau} = \frac{1}{N}\sum_{i=1}^{N} t_i \quad (6\text{-}31)$$

where we have used a bar over τ to distinguish it from the true value of the lifetime τ. The meaning of (6-31) is that the average of t_i is the value of τ that produces the maximum probability for the function $\mathcal{L}(\tau)$. This way of estimating the value of an unknown quantity is called the method of *maximum likelihood*. For this reason, the function $\mathcal{L}(\tau)$ of (6-30) is called a *likelihood function*. A more formal statement of the method can be found in standard references on probability and mathematical physics, such as Mathews and Walker (*Mathematical Methods of Physics*, Benjamin, New York, 1965). It is important to realize here that this is not the only way to make the choice. In fact, several other criteria are also used in practice. Except for cases where the number of independent pieces of data is small or the uncertainty associated with each piece of data is large, the results obtained using different methods are essentially identical to each other. For this reason we shall concentrate on the method of maximum likelihood.

It is useful to have an estimate of the uncertainty associated with the value of $\bar{\tau}$ determined. If N is large, the shape of $\mathcal{L}(\tau)$ approaches that of a normal distribution of (6-9):

$$\mathcal{L}(\tau) \xrightarrow[N \to \infty]{} \frac{1}{\sigma_\tau \sqrt{2\pi}} e^{-(\tau - \bar{\tau})^2 / 2\sigma_\tau^2} \quad (6\text{-}32)$$

In this form, the uncertainty of τ is given by the standard deviation σ_τ of this distribution. Using (6-32), the derivative of the logarithm of $\mathcal{L}(\tau)$ with respect to τ is

$$\frac{d}{d\tau}\ln\mathcal{L}(\tau) = -\frac{\tau - \bar{\tau}}{\sigma_\tau^2}$$

This gives us the relation

$$\sigma_\tau = \left\{\frac{1}{\bar{\tau} - \tau}\frac{d}{d\tau}\ln\mathcal{L}(\tau)\right\}^{-1/2} \tag{6-33}$$

To obtain an idea of the actual value of σ_τ for a given function, we can put $\mathcal{L}(\tau)$ of (6-30) into (6-33). This gives us σ_τ in terms of τ and N:

$$\sigma_\tau = \left\{\frac{1}{\bar{\tau} - \tau}\left(\frac{1}{\tau^2}\sum_{i=1}^{N}t_i - \frac{N}{\tau}\right)\right\}^{-1/2} = \frac{\tau}{\sqrt{N}} \longrightarrow \frac{\bar{\tau}}{\sqrt{N}} \tag{6-34}$$

This is the same result as we have obtained in the previous section in estimating the uncertainty of the height of a histogram, except there we had to invoke a Poisson distribution.

If N is not large enough for the likelihood function to be approximated by a normal distribution, the final result given by (6-34) may not be valid. In this case, one should make a plot of the likelihood function and see if $\mathcal{L}(\tau)$ has a narrow distribution. Unless $\mathcal{L}(\tau)$ is sharply peaked, the value $\bar{\tau}$ determined may not be very meaningful. This is illustrated schematically in Fig. 6-7.

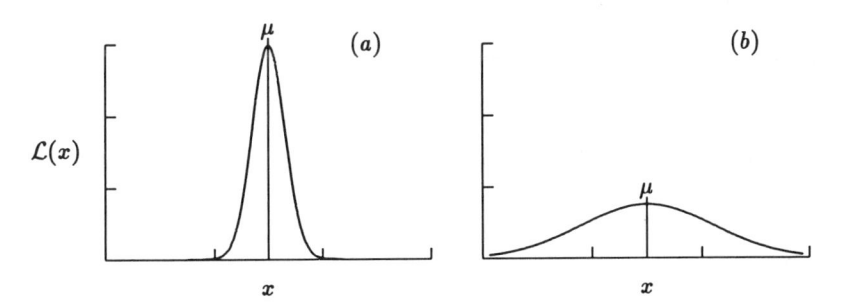

Fig. 6-7 Mean value and its uncertainty. If the likelihood distribution is sharp with a small uncertainty, as that shown in (a) on the left, the mean μ is more effective in characterizing the quantity than that of a flat distribution with a large uncertainty, such as that shown in (b) on the right.

Data points with unequal uncertainties. In the previous example, we assumed that the source of uncertainty in a measurement comes purely from statistical fluctuations. In reality, it is inevitable that we must deal with instrumental uncertainties as well. Because of conditions beyond our control, the accuracy that can be achieved in one measurement may be quite different from that of another. For example, in dealing with a radioactive sample, there are many more decays in the beginning of the experiment and, as a result, we may not be able to measure the time as well as we can do toward the end of the experiment when the activity is lower.

In general, we can associate with any one data point x_i an uncertainty σ_i. Let us find out the influence of σ_i on the values of the quantities we wish to determine from the collection of data. For simplicity, we shall consider a system where the only difference between different measurements comes from random errors in the instrument. Later, in the discussion of nonlinear least-squares fit, we shall return to the more general problem of determining the value of a physical quantity in the presence of both instrumental uncertainties as well as statistical fluctuations.

For each data point, we can assume that the possible values x_i are given by a normal distribution with mean μ and variance σ_i^2:

$$p(x_i)\,dx_i = \frac{1}{\sigma_i\sqrt{2\pi}}e^{-(x_i-\mu)^2/2\sigma_i^2}\,dx_i \qquad (6\text{-}35)$$

Since the only difference between the various data points here is due to instrumental errors, the mean μ is the same for all the data points $x_1, x_2, \ldots, x_N$. However, the variance for each point may be different from the others, as we are considering the more general case in which the uncertainty associated with each data point is independent of the others. We shall further assume that there are many possible sources of instrumental error and, as a result, the central limit theorem applies. In general, it is not an easy question to determine the value of σ_i for each data point as it is related to the particular way the measurement is made. However, we shall not be concerned with this question here. For our purpose, we shall assume that all the values of σ_i are given to us.

We can now construct the likelihood function for the mean μ of the parent distribution under the assumption that each data point follows the distribution given by (6-35). Analogous to (6-30), we have here a product of normal distribution functions:

$$\mathcal{L}(\mu) = \prod_{i=1}^{N}\left\{\frac{1}{\sigma_i\sqrt{2\pi}}e^{-(x_i-\mu)^2/2\sigma_i^2}\right\}$$

$$= \left\{\prod_{i=1}^{N}\frac{1}{\sigma_i\sqrt{2\pi}}\right\}\exp\left\{-\frac{1}{2}\sum_{i=1}^{N}\left(\frac{x_i-\mu}{\sigma_i}\right)^2\right\} \qquad (6\text{-}36)$$

As we have seen earlier, the maximum of this function is given by the condition $d\mathcal{L}(\mu)/d\mu = 0$. This is equivalent to the requirement that

$$\frac{d}{d\mu}\left(\frac{x_i - \mu}{\sigma_i}\right)^2 = 0$$

or

$$\bar{\mu} = \frac{1}{\sum_i \sigma_i^{-2}} \sum_{i=1}^{N} \frac{1}{\sigma_i^2} x_i \tag{6-37}$$

In other words, $\bar{\mu}$, the maximum likelihood value of the mean μ, is given by the average of the data points x_i, weighted by the inverse of the square of the uncertainty of each point. In the limit that the values of σ_i for different points are equal to each other, we obtain the same expression as the average value in the absence of any uncertainties for each data point.

Since $\bar{\mu}$ is a function of x_i, the uncertainty in μ is given by (6-24),

$$\sigma_\mu^2 = \sum_{i=1}^{N}\left\{\sigma_i^2\left(\frac{\partial\mu}{\partial x_i}\right)^2\right\}$$

where we have ignored the covariance between any two data points, as we are assuming that each measurement is independent of the other. From (6-37), we obtain the value of the partial derivative as

$$\frac{\partial}{\partial x_i}\left\{\frac{1}{\sum_j \sigma_j^{-2}} \sum_{k=1}^{N} \frac{1}{\sigma_k^2} x_k\right\} = \frac{1}{\sum_j \sigma_j^{-2}} \frac{1}{\sigma_i^2}$$

This gives us the result

$$\sigma_\mu^2 = \sum_{i=1}^{N}\left\{\sigma_i^2\left(\frac{1}{\sum_j \sigma_j^{-2}} \frac{1}{\sigma_i^2}\right)^2\right\} = \frac{1}{\sum_i \sigma_i^{-2}} \tag{6-38}$$

If all the uncertainties are the same and equal to a constant σ, the expression reduces to

$$\sigma_\mu^2 = \frac{\sigma^2}{N}$$

Thus the uncertainty due to instrument, defined to be σ_μ in the previous section, decreases if more independent measurements are taken in an experiment.

§6-4 The method of least squares

We shall now apply the method of maximum likelihood to a more general situation. If, based on some theoretical grounds, we know that a quantity y is a function of x with m parameters $a_1, a_2, \ldots, a_m$,

$$y = f(a_1, a_2, \ldots, a_m; x) \tag{6-39}$$

what are the most likely values of the parameters that give the best description to a set of measured values of y? This is the familiar problem of fitting a function to a set of data points. If $f(a_1, a_2, \ldots, a_m; x)$ is a linear function in the parameters, the method is known as a linear least-squares fit. In this section, we shall give an introduction to the method by considering the simple case of a straight-line relation between y and x. Later, in §6-6, we shall consider the general case of more than two parameters. The case of fitting with a nonlinear function is treated in §6-7.

Let y_i be the measured value of y at $x = x_i$. Because of uncertainties associated with the process, the value of y_i is likely to be different if we repeat the measurement. If the measurements for y_i are carried out many times, the results are expected to follow a normal distribution:

$$p(y_i)\,dy_i = \frac{1}{\sigma_i\sqrt{2\pi}} e^{-(y_i - f_i)^2/2\sigma_i^2}\,dy_i$$

Again, we are depending on the central limit theorem to arrive at this conclusion. The mean of the distribution $p(y_i)\,dy_i$ is given by

$$f_i = f(a_1, a_2, \ldots, a_m; x_i)$$

The variance σ_i^2 of the distribution comes from a variety of sources, both statistical fluctuations and instrumental uncertainties. In general, however, we do not possess the luxury of having many values of y_i for a given x_i. As a result, $p(y_i)$ is not available to us. On the other hand, if the distribution is a normal one, all we need, in addition to the mean value f_i, is the variance to determine $p(y_i)$. For our interest here, we shall assume that σ_i^2 is provided to us somehow, as there is no way we can deduce its value without being given more information.

The likelihood distribution for the parameters $a_1, a_2, \ldots, a_m$ is given by the product of $p(y_1), p(y_2), \ldots, p(y_N)$. That is,

$$\mathcal{L}(a_1, a_2, \ldots, a_m) = \prod_{i=1}^{N} \left\{ \frac{1}{\sigma_i\sqrt{2\pi}} e^{-(y_i - f_i)^2/2\sigma_i^2} \right\}$$

$$= \left\{ \prod_{i=1}^{N} \frac{1}{\sigma_i\sqrt{2\pi}} \right\} \exp\left\{ -\frac{1}{2} \sum_{i=1}^{N} \left(\frac{y_i - f_i}{\sigma_i} \right)^2 \right\} \tag{6-40}$$

As we shall see soon, the sum in the argument of the exponential function

$$\chi^2 = \sum_{i=1}^{N}\left(\frac{y_i - f_i}{\sigma_i}\right)^2 \tag{6-41}$$

is fundamental to the method of least squares and is given the name *chi-square*.

Similar to what we have done earlier, the maximum of the likelihood distribution $\mathcal{L}(a_1, a_2, \ldots, a_m)$ is obtained by requiring that its partial derivative with respect to each of the parameters vanish. Similar to (6-36), this is equivalent to putting each of the corresponding partial derivatives of χ^2 to zero. That is,

$$\frac{\partial}{\partial a_k}\chi^2 = -2\sum_{i=1}^{N}\left(\frac{y_i - f_i}{\sigma_i}\right)\frac{\partial f_i}{\partial a_k} = 0 \tag{6-42}$$

for $k = 1, 2, \ldots, m$.

Least-squares fit to a straight line. To make progress beyond the relation given by (6-42), we need the explicit functional dependence of f on the parameters $\{a_k\}$. In this section, we shall only consider the simple case of a linear function consisting of two parameters, with $a_1 = a$ and $a_2 = b$. That is,

$$f(a, b; x) = a + bx \tag{6-43}$$

The partial derivatives of $f(a, b; x)$ with respect to parameters a and b are then

$$\frac{\partial f}{\partial a} = 1 \qquad \frac{\partial f}{\partial b} = x$$

The maximum likelihood condition of (6-42) in this case becomes

$$\sum_{i=1}^{N}\left(\frac{y_i - f_i}{\sigma_i^2}\right) = 0 \qquad \sum_{i=1}^{N}\left(\frac{y_i - f_i}{\sigma_i^2}\right)x_i = 0$$

or

$$a\sum_{i=1}^{N}\frac{1}{\sigma_i^2} + b\sum_{i=1}^{N}\frac{x_i}{\sigma_i^2} = \sum_{i=1}^{N}\frac{y_i}{\sigma_i^2} \qquad a\sum_{i=1}^{N}\frac{x_i}{\sigma_i^2} + b\sum_{i=1}^{N}\frac{x_i^2}{\sigma_i^2} = \sum_{i=1}^{N}\frac{x_i y_i}{\sigma_i^2} \tag{6-44}$$

where N is the total number of pieces of data points. The values of a and b are obtained by solving this system of two linear equations. This is a simple example of *linear least-squares* fit to N pieces of data. The proper name for this way of fitting a sample of data points to the relation given by (6-43) is known in statistics

Table 6-2 A sample set of data consisting of the measured values of y for nine different values of x.

i	x_i	y_i	σ_i	i	x_i	y_i	σ_i
1	0.25	0.86	0.27	6	3.64	8.84	0.66
2	1.05	2.18	1.16	7	3.92	8.71	0.98
3	2.25	4.84	1.14	8	4.94	11.98	0.93
4	2.88	5.80	0.93	9	5.92	12.40	0.60
5	2.97	6.99	0.31				

as *linear regression* analysis. We shall, however, follow the more conventional nomenclature in physics.

In matrix form, (6-44) may be written as

$$\begin{pmatrix} \alpha & \beta \\ \gamma & \delta \end{pmatrix} \begin{pmatrix} a \\ b \end{pmatrix} = \begin{pmatrix} \theta \\ \phi \end{pmatrix}$$

with the six known elements given by

$$\alpha = \sum_{i=1}^{N} \frac{1}{\sigma_i^2} \qquad \beta = \sum_{i=1}^{N} \frac{x_i}{\sigma_i^2} \qquad \gamma = \beta$$

$$\delta = \sum_{i=1}^{N} \frac{x_i^2}{\sigma_i^2} \qquad \theta = \sum_{i=1}^{N} \frac{y_i}{\sigma_i^2} \qquad \phi = \sum_{i=1}^{N} \frac{x_i y_i}{\sigma_i^2} \qquad (6\text{-}45)$$

The values of a and b may be obtained by a simple application of the method given by (5-16) using determinants:

$$a = \frac{1}{D} \det \begin{vmatrix} \theta & \beta \\ \phi & \delta \end{vmatrix} = \frac{1}{D} \{\theta\delta - \beta\phi\}$$

$$b = \frac{1}{D} \det \begin{vmatrix} \alpha & \theta \\ \gamma & \phi \end{vmatrix} = \frac{1}{D} \{\alpha\phi - \theta\gamma\} \qquad (6\text{-}46)$$

where the value of the determinant D is given by

$$D = \det \begin{vmatrix} \alpha & \beta \\ \gamma & \delta \end{vmatrix} = \alpha\delta - \beta\gamma = \alpha\delta - \beta^2 \qquad (6\text{-}47)$$

The calculations involved here are quite simple once the six quantities α, β, γ, δ, θ, and ϕ of (6-45) are obtained from the input set of values $\{x_i, y_i\}$ and the uncertainties $\{\sigma_i^2\}$ associated with $\{y_i\}$. As an example, a collection of nine pieces of hypothetical data, together with their uncertainties, is given in Table 6-2. The result of a straight-line fit is shown in Fig. 6-8.

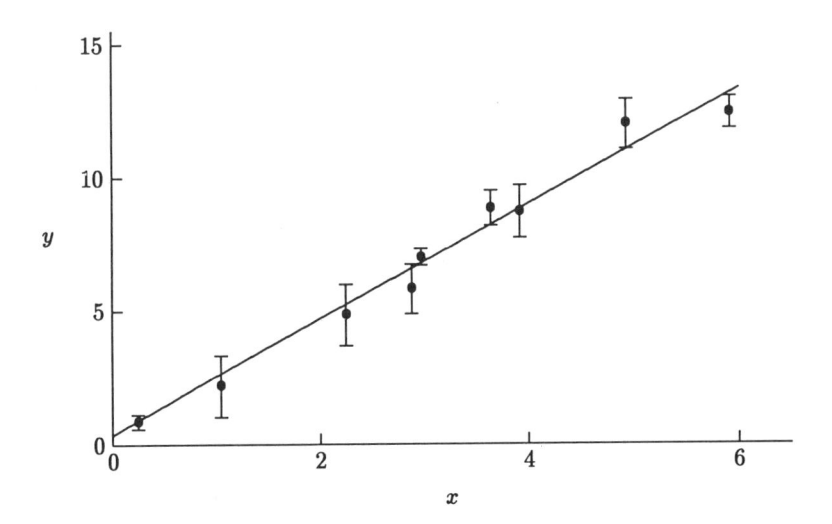

Fig. 6-8 Linear least-squares fit to the nine data points in Table 6-2. The straight line $y = a + bx$ is plotted using the best-fit values of $a = 0.380$ and $b = 2.157$.

Uncertainties in the parameters. In addition to the maximum likelihood values of a and b, we also need to have estimates of the uncertainties associated with them that arise from the uncertainties in y_i. We can accomplish this task with (6-24) by treating both a and b as functions of $y_1, y_2, \ldots, y_N$. For simplicity, we shall assume that the uncertainties in different data points are uncorrelated and the covariance σ_{ij}^2 between any two of them vanishes. As a result, we have

$$\sigma_a^2 = \sum_{i=1}^{N}\left\{\left(\frac{\partial a}{\partial y_i}\right)^2 \sigma_i^2\right\} \qquad \sigma_b^2 = \sum_{i=1}^{N}\left\{\left(\frac{\partial b}{\partial y_i}\right)^2 \sigma_i^2\right\} \qquad (6\text{-}48)$$

Using (6-46), the partial derivatives involved are

$$\frac{\partial a}{\partial y_i} = \frac{1}{D}\left(\frac{1}{\sigma_i^2}\delta - \frac{x_i}{\sigma_i^2}\beta\right) \qquad \frac{\partial b}{\partial y_i} = \frac{1}{D}\left(\frac{x_i}{\sigma_i^2}\alpha - \frac{1}{\sigma_i^2}\gamma\right) \qquad (6\text{-}49)$$

Inserting these results into (6-48), we obtain

$$\sigma_a^2 = \frac{1}{D^2}\sum_{i=1}^{N}\left\{\sigma_i^2\left(\frac{1}{\sigma_i^2}\delta - \frac{x_i}{\sigma_i^2}\beta\right)^2\right\}$$

$$= \frac{1}{D^2}\sum_{i=1}^{N}\left\{\delta^2\frac{1}{\sigma_i^2} + \beta^2\frac{x_i^2}{\sigma_i^2} - 2\delta\beta\frac{x_i}{\sigma_i^2}\right\} = \frac{1}{D^2}\left\{\delta^2\alpha + \delta\beta^2 - 2\delta\beta^2\right\} = \frac{\delta}{D}$$

$$\sigma_b^2 = \frac{1}{D^2}\sum_{i=1}^{N}\left\{\sigma_i^2\left(\frac{x_i}{\sigma_i^2}\alpha - \frac{1}{\sigma_i^2}\gamma\right)^2\right\}$$

$$= \frac{1}{D^2} \sum_{i=1}^{N} \left\{ \alpha^2 \frac{x_i^2}{\sigma_i^2} + \gamma^2 \frac{1}{\sigma_i^2} - 2\alpha\gamma \frac{x_i}{\sigma_i^2} \right\} = \frac{1}{D^2} \left\{ \alpha^2 \delta + \alpha\gamma^2 - 2\alpha\gamma\beta \right\} = \frac{\alpha}{D}$$
(6-50)

where we have made use of the fact that $\beta = \gamma$.

One other quantity that can also provide some idea of the uncertainties in the values of a and b obtained is the covariance between them. This is defined as

$$\sigma_{a,b}^2 = \sum_{i=1}^{N} \left\{ \frac{\partial a}{\partial y_i} \frac{\partial b}{\partial y_i} \sigma_i^2 \right\}$$

Using the forms of $\partial a/\partial y_i$ and $\partial b/\partial y_i$ from (6-49), we find that

$$\sigma_{a,b}^2 = -\frac{\beta}{D}$$
(6-51)

We shall make use of this quantity later. The calculations for σ_a and σ_b are incorporated as part of the algorithm for linear least-squares fit to data outlined in Box 6-1.

Box 6-1   Subroutine LLSQ(NPT,X,Y,SY,A,SA,B,SB)
Linear least-squares fit to a straight line $y = a + b * x$

Argument list:
    NPT: Number of data points.
    X: Array for the values of the independent variable x_i.
    Y: Array for the measured values of the dependent variable y_i.
    SY: Array for σ_i, the uncertainty of y_i.
    A: Output for coefficient a.
    SA: Standard deviation σ_a for a.
    B: Output for coefficient b.
    SB: Standard deviation σ_b for b.

1. Weight each point:
  (a) If $\sigma_i = 0$, the weight is 1.
  (b) If $\sigma_i \neq 0$, the weight is $1/\sigma_i^2$.
2. Calculate the intermediate quantities α, β, δ, θ, and ϕ of (6-45) from the input x_i and y_i with the weighting factor of step 1.
3. Compute the value of the determinant D using (6-47).
4. Calculate the values of a and b using (6-46).
5. Calculate σ_a and σ_b using (6-48).
6. Return a, σ_a, b, σ_b, and $\sigma_{a,b}^2$.

Correlation between y_i and x_i. For our discussion of the method of least-squares fit to a straight-line relation, we have started with the premise that y is given as a function of x in the form of (6-43). As a part of the test of the significance of the fit, we shall also ask the question whether our data can actually be described by such a linear model. As usual in statistical analyses, we cannot hope to obtain a definite yes or no answer to this question. To get an idea of whether the fit is meaningful, we can calculate the correlation coefficient $r_{x,y}$ between $\{y_i\}$ and $\{x_i\}$:

$$r_{x,y} \equiv \frac{\sigma_{x,y}^2}{\sigma_x \sigma_y} = \frac{\sum_{i=1}^{N} \frac{1}{\sigma_i^2}(x_i - \mu_x)(y_i - \mu_y)}{\left\{\sum_{i=1}^{N} \frac{1}{\sigma_i^2}(x_i - \mu_x)^2\right\}^{1/2} \left\{\sum_{i=1}^{N} \frac{1}{\sigma_i^2}(y_i - \mu_y)^2\right\}^{1/2}} \quad (6\text{-}52)$$

where the mean of the data points y_i and x_i is given by

$$\mu_x = \left(\sum_{i=1}^{N} \frac{1}{\sigma_i^2}\right)^{-1} \sum_{i=1}^{N} \frac{1}{\sigma_i^2} x_i = \frac{\beta}{\alpha}$$

$$\mu_y = \left(\sum_{i=1}^{N} \frac{1}{\sigma_i^2}\right)^{-1} \sum_{i=1}^{N} \frac{1}{\sigma_i^2} y_i = \frac{\theta}{\alpha} \quad (6\text{-}53)$$

The denominator on the right side of (6-52) is proportional to the square root of the product of the variances for the distributions of $\{x_i\}$ and $\{y_i\}$:

$$\sigma_x^2 = \left(\sum_{i=1}^{N} \frac{1}{\sigma_i^2}\right)^{-1} \sum_{i=1}^{N} \frac{1}{\sigma_i^2}(x_i - \mu_x)^2$$

$$= \frac{1}{\alpha}\left\{\sum_{i=1}^{N} \frac{1}{\sigma_i^2} x_i^2 - 2\mu_x \sum_{i=1}^{N} \frac{1}{\sigma_i^2} x_i + \mu_x^2 \sum_{i=1}^{N} \frac{1}{\sigma_i^2}\right\} = \frac{\delta}{\alpha} - \frac{\beta^2}{\alpha^2} = \frac{D}{\alpha^2}$$

$$\sigma_y^2 = \left(\sum_{i=1}^{N} \frac{1}{\sigma_i^2}\right)^{-1} \sum_{i=1}^{N} \frac{1}{\sigma_i^2}(y_i - \mu_y)^2 \quad (6\text{-}54)$$

The numerator is related to the covariance between x and y

$$\sigma_{x,y}^2 = \left(\sum_{i=1}^{N} \frac{1}{\sigma_i^2}\right)^{-1} \sum_{i=1}^{N} \frac{1}{\sigma_i^2}(x_i - \mu_x)(y_i - \mu_y)$$

$$= \frac{1}{\alpha}\left\{\sum_{i=1}^{N} \frac{1}{\sigma_i^2} x_i y_i - \mu_x \sum_{i=1}^{N} \frac{1}{\sigma_i^2} y_i - \mu_y \sum_{i=1}^{N} \frac{1}{\sigma_i^2} x_i + \mu_x \mu_y \sum_{i=1}^{N} \frac{1}{\sigma_i^2}\right\}$$

$$= \frac{1}{\alpha}\left\{\sum_{i=1}^{N} \frac{1}{\sigma_i^2} x_i y_i - \mu_x \mu_y \sum_{i=1}^{N} \frac{1}{\sigma_i^2}\right\} = \frac{1}{\alpha}\phi - \frac{\beta \theta}{\alpha^2} = \frac{bD}{\alpha^2} \quad (6\text{-}55)$$

We can now express the square of the correlation coefficient $r_{x,y}$ of (6-52) in terms of the value of σ_x^2 of (6-54) and that of $\sigma_{x,y}^2$ of (6-55):

$$r_{x,y}^2 = \frac{\sigma_{x,y}^4}{\sigma_x^2 \sigma_y^2} = \frac{\frac{bD}{\alpha^2}\sigma_{x,y}^2}{\frac{D}{\alpha^2}\sigma_y^2} = b\frac{\sigma_{x,y}^2}{\sigma_y^2} \quad (6\text{-}56)$$

That is, $r_{x,y}^2$ is related to the slope of y_i as a function of x_i. We shall make use of this result later in §6-6 to construct the correlation coefficient for the more general case.

If $\{x_i\}$ and $\{y_i\}$ are independent of each other, the numerator of the right side of (6-52) vanishes and the correlation coefficient $r_{x,y} = 0$. On the other hand, if y increases linearly as x, we have the result $r_{x,y} = 1$ regardless of the slope (given by b) of a plot of y versus x. Similarly, if y decreases linearly as x, we have the result $r_{x,y} = -1$. For this reason, the absolute value of the correlation coefficient may be used as one of the indicators of whether it is meaningful to carry out a linear least-squares fit of y_i to the form $(a + bx_i)$. More sensitive tests can be constructed from the statistics discussed in the next section.

§6-5 Statistical tests of the results

In carrying out a least-squares fit to a set of data, it is usual to start with a model, based on some reasonable ideas we have on how the independent and dependent variables are related to each other. One of the purposes of the calculation then is to find out whether our model is supported or rejected by the data. Because of the statistical nature of the least-squares analysis and the presence of uncertainties in the data themselves, it is usually impossible for us to reach a definitive answer to the question. Instead, we must make use of tests, or *statistics*, to arrive at an estimate of the probabilities for our model to be correct. A large amount of work in the field of statistics has gone into the design of these tests. We shall describe here three popular ones in physics: Pearson's χ^2-test, Student's t-test, and Fisher's F-test.

χ^2-**test.** In §6-3, we have seen that the method of maximum likelihood produces the most likely values of the parameters a and b in our linear model by minimizing the value of χ^2 defined in (6-41):

$$\chi^2 = \sum_{i=1}^{N}\left(\frac{y_i - f_i}{\sigma_i}\right)^2$$

where σ_i is the uncertainty associated with the measured value y_i, and f_i is the value expected from our model. If the model used is an exact description of the data, we expect $f_i = y_i$ for all the data points. In this case, the value of χ^2

vanishes. In practical situations, it is unlikely for us to obtain $\chi^2 = 0$ in a least-squares calculation, either because of the deficiencies in the model or because of the errors in the measurements, or both. On the other hand, if the fit is a good one, we expect the value of χ^2 to be small. Let us extend this idea into a quantitative test of the quality of the fit.

By definition, χ^2 is the sum of the squares of the differences between y_i and f_i, weighted by the inverse square of the uncertainty associated with y_i. Since it is a sum, its value increases with the total number of data points N. To serve as an indicator for the quality of a fit, such a dependence on N is inconvenient. Our first reaction is to adopt, instead, an average value over N. This is also incorrect. We can see this using the linear model of the previous section as an example. Since there are only two parameters, a and b, an exact fit is obtained if there are only two data points ($N = 2$). Although the value of χ^2 is zero in this case, it is clear that the "fit" is not a meaningful one. This leads us to the concept of the number of *degrees of freedom*, defined as the number of independent measurements, N in our case, minus the number of constraints, m. That is,

$$\nu = N - m \tag{6-57}$$

For our purpose, m is given by the number of free parameters in the model we adopt to fit the data. In the linear case of the previous section, $m = 2$.

A useful measure of how well a model fits the data is given by the value of the χ^2 per degree of freedom, also known as the *reduced chi-square*:

$$\chi_\nu^2 \equiv \frac{\chi^2}{\nu} = \frac{1}{N-n} \sum_{i=1}^{N} \left\{ \frac{1}{\sigma_i^2} (y_i - f_i)^2 \right\} \tag{6-58}$$

We shall see later that the value of χ^2 follows a distribution with the value of the mean equal to ν. As a result, a value of χ_ν^2 of less than or equal to 1 is considered to be a good fit.

To construct a more quantitative criterion, we can compare our model predication with the result of using a set of N random numbers instead of $\{y_i\}$. In other words, if we substitute the measured results with random numbers, what is the probability $P(\chi^2|\nu)$ of obtaining a value that is equal to or larger than χ^2 if the number of degrees of freedom ν remains the same? Clearly, if our fit to the set of data points is not better than that to a set of random numbers, it is not meaningful.

In general, it is more convenient to ask the opposite question: what is the probability

$$Q(\chi^2|\nu) = 1 - P(\chi^2|\nu) \tag{6-59}$$

that our sample of data $\{y_i\}$ is given by the set of $\{f_i\}$ calculated using the parameters obtained from the fitting procedure. For this purpose, let us define a new variable

$$\xi_i = \frac{y_i - f_i}{\sigma_i}$$

The value of χ^2 is then the sum of ξ_i^2. If each y_i is a random number with a normal distribution centered around f_i and having a variance σ_i^2, then ξ_i is a random number with a normal distribution centered around 0 and having a variance of unity. That is, the distribution of ξ_i is given by

$$p(\xi_i)d\xi_i = \frac{1}{\sqrt{2\pi}}e^{-\xi_i^2/2}d\xi_i$$

As we have seen earlier, the normalization of this probability distribution is

$$\int_{-\infty}^{+\infty} p(\xi_i)d\xi_i = 1$$

From these two conditions, we find that the distribution of the variable $\zeta = \xi_i^2$ is given by

$$p(\zeta)\,d\zeta = \frac{1}{\sqrt{2\pi\zeta}}e^{-\zeta/2}d\zeta \qquad (6\text{-}60)$$

with the normalization

$$\int_0^\infty p(\zeta)\,d\zeta = 1$$

This is known as the χ^2-distribution for a single degree of freedom.

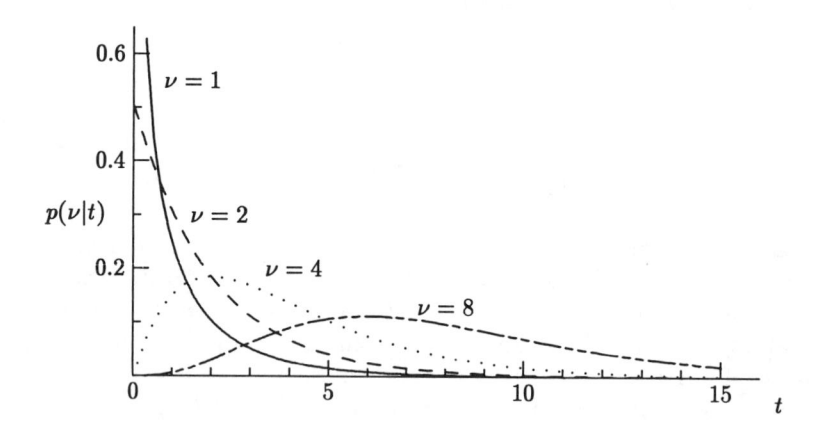

Fig. 6-9 χ^2-distribution $p(\nu|t)$ for $\nu = 1, 2, 4,$ and 8. The shape of $p(\nu|t)$ approaches that of a normal distribution as ν becomes large.

For the purpose of testing our hypothesis, we are more interested in the distribution of the variable

$$t = \sum_{i=1}^\nu \xi_i^2 = \sum_{i=1}^\nu \left(\frac{y_i - f_i}{\sigma_i}\right)^2 \qquad (6\text{-}61)$$

where each quantity ξ_i is an independent random variable with a normal distribution of zero mean and unit variance. The generalization of (6-60) is the χ^2-distribution for ν degrees of freedom,

$$p(t|\nu)\,dt = \frac{1}{2^{\nu/2}\Gamma(\nu/2)} t^{\frac{\nu}{2}-1} e^{-t/2}\,dt \qquad (6\text{-}62)$$

where $\Gamma(x)$ is the gamma function of (4-137). The mean of $p(t|\nu)$ is ν and the variance is 2ν, as can be seen by explicit calculations. The shapes of $p(t|\nu)$ for a few low values of ν are shown in Fig. 6-9.

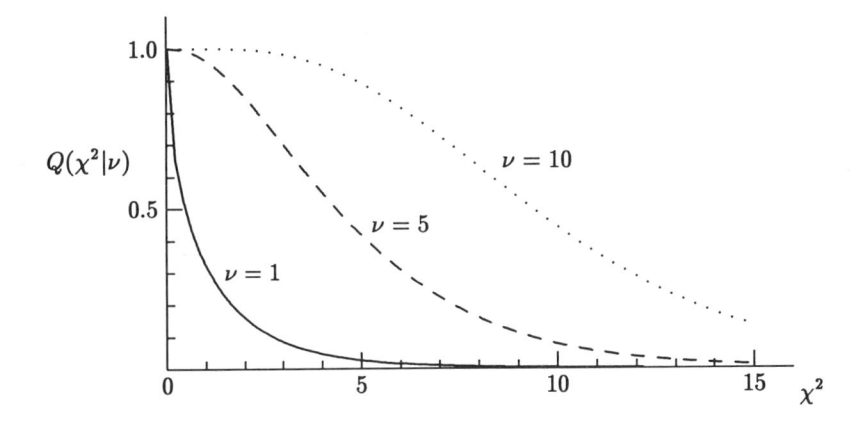

Fig. 6-10 Value of $Q(\chi^2|\nu)$ for $\nu = 1, 5,$ and 10 for $\chi^2 = [0,15]$.

The probability of obtaining a value of t less than or equal to some value χ^2 for a system of ν degrees of freedom is given by the probability integral of the χ^2-distribution:

$$P(\chi^2|\nu) = \int_0^{\chi^2} \frac{1}{2^{\nu/2}\Gamma(\nu/2)} t^{\frac{\nu}{2}-1} e^{-t/2}\,dt \qquad (6\text{-}63)$$

We can identify this quantity as the probability we were looking for earlier to obtain a fit up to some χ^2-value in a least-squares calculation using a function with ν degrees of freedom. Comparing the right side of (6-63) with the incomplete gamma function $\gamma(a,x)$ of (4-151), we find that

$$P(\chi^2|\nu) = \frac{1}{\Gamma(\nu/2)} \gamma(a,x) \qquad (6\text{-}64)$$

with

$$a = \tfrac{1}{2}\nu \qquad\qquad x = \tfrac{1}{2}\chi^2$$

From (4-152) we have $\Gamma(a, x) = \Gamma(a) - \gamma(a, x)$. Using this relation, we obtain

$$Q(\chi^2|\nu) = 1 - \frac{1}{\Gamma(\nu/2)}\gamma(a, x) = \frac{1}{\Gamma(\nu/2)}\Gamma(a, x) \qquad (6\text{-}65)$$

Values of $P(\chi^2|\nu)$ and $Q(\chi^2|\nu)$ are given in standard statistics tables. Alternatively, the values of $Q(\chi^2|\nu)$ may be obtained from those of $\Gamma(x)$ calculated using the algorithm given in Box 4-9 and those of $\gamma(a, x)$ calculated using the algorithm of Box 4-10. An idea of the variation of $Q(\chi^2|\nu)$ as a function of ν and χ^2 is given in Fig. 6-10.

As an example, we shall apply the χ^2-test to the histogram of Fig. 6-5 and see whether the distribution of the data points is, indeed, a normal one. The number of occurrences in each bin h_i is listed in the third column of Table 6-3 and the value n_i expected of a normal distribution is given in the fourth column. Again, we shall assume a Poisson distribution for the possible values in each histogram bin. In this way, we can take that the uncertainty associated with h_i is given by the square root of the number in the bin. The contribution of each bin to the χ^2 is then

$$\chi_i^2 = (h_i - n_i)^2/n_i$$

A small problem occurs in bin 6, where $n_i = 0$. To avoid an unphysical result of infinity as the contribution to the χ^2, we approximate the denominator by unity. The sum of χ_i^2 calculated in this way is 4.4.

Table 6-3 χ^2-test for the histogram in Fig. 6-5 to be a normal distribution.

Bin	Range	Histogram	Normal distribution	χ_i^2
1	47–48	1	0.7	0.1
2	48–49	5	3.9	0.3
3	49–50	5	8.9	3.0
4	50–51	11	8.1	0.8
5	51–52	3	3.0	0.0
6	52–53	0	0.4	0.2
			$\sum \chi_i^2 =$	4.4

Since the normal distribution assumed in the discussion above is adjusted to have the same mean and variance as the data, we have, in effect, imposed the equivalence of two constraints on the system consisting of 6 bins. The number of degrees of freedom is therefore $\nu = (6 - 2) = 4$. For a χ^2-value of 4.4, the corresponding reduced χ^2 is 1.1 in this case. The probability of obtaining a χ^2 larger or equal to 4.4 is $Q(4.4|4) = 0.35$. The result, therefore, supports our assumption of a normal distribution. However, there is also a good chance that the assumption is incorrect. This is, essentially, the same conclusion we reached earlier in §5-2 — the difference is that we have a more quantitative conclusion here.

The χ^2-statistic is widely used in physics to find the probability for a set of data to be in agreement with a model of the expected distribution. However, it is only one of the possible tests and, in many cases, cannot provide a definitive answer without other corroborative evidence. Part of the difficulty in using a χ^2-test comes from the fact that two different sources contribute to the value of the χ^2 at the same time. For a given set of the values of $\{y_i, x_i\}$, there exists a functional relation between them, referred to as the *parent function* in the language of statistics. However, we have no knowledge of this function. In a least-squares calculation, we construct a model by making a guess of this function based on the best information available to us. Any differences between our model and the parent function increase the value of the χ^2. This is the first source contributing to the value of χ^2.

A second source comes from discrepancies between the data and the parent function. For example, some errors may have been introduced into the data, perhaps because of the limited precision we can achieve in the measurement. As a result, even if we make the correct guess for the parent function, we cannot have a perfect fit to the data points and the value of the χ^2 does not vanish. In a χ^2-test, we have no way of distinguishing between these two sources. For this reason, other statistical tests are often applied. This is particularly important when the result from a χ^2-test is ambiguous, as we saw in the example above, and when the conclusion depends heavily on the outcome of the statistical analysis.

Student's *t*-test. Consider again a sample of N pieces of data points, $y_1, y_2, \ldots, y_N$. The mean μ_y and variance σ^2 of the distribution of $\{y_i\}$ are given by

$$\mu_y = \frac{1}{N} \sum_{i=1}^{N} y_i \qquad \sigma^2 = \frac{1}{N-1} \sum_{i=1}^{N} (y_i - \mu_y)^2$$

The Student's *t*-statistic[†] is a test based on the ratio between μ_y and σ. The reason that the denominator in the definition of σ^2 is $(N-1)$ comes from the fact that we are subtracting μ_y from each piece of data. Since μ_y is obtained from the data themselves, the subtraction is equivalent to imposing a constraint on the set of N variables.

If y_i is a random variable with a normal distribution of (unknown) mean μ and the same variance σ^2, then the ratio

$$t = \frac{\mu_y - \mu}{\sigma/\sqrt{N}} \tag{6-66}$$

[†] "Student" is the pen name of W.S. Gosset who published the statistic in 1908.

is also a random variable distributed according to

$$p(t,\nu)\,dt = \frac{dt}{\nu^{1/2} B(\frac{1}{2}, \frac{1}{2}\nu)\left(1 + \frac{t^2}{\nu}\right)^{\frac{1}{2}(\nu+1)}}$$

where $\nu = (N-1)$ is the number of degrees of freedom and $B(\frac{1}{2}, \frac{1}{2}\nu)$ is the beta function

$$B(a,b) = \int_0^1 t^{a-1}(1-t)^{b-1}\,dt = \frac{\Gamma(b)\Gamma(a)}{\Gamma(a+b)} \tag{6-67}$$

with $a = \frac{1}{2}$ and $b = \frac{1}{2}\nu$. More detailed discussion of the Student's t-test and other statistics can be found in Cramer (*Mathematical Methods of Statistics*, Princeton University Press, Princeton, New Jersey, 1946) and in Kendall and Stuart (*The Advanced Theory of Statistics*, vol. 1, Macmillan, New York, 1977). There are several ways to make use of the statistic. The basic definition given by (6-66) provides us with a test to see whether the sample mean μ_y obtained from the data is in agreement with the (unknown) parent mean μ.

Again, we are more interested in the integrated probability of t having a value less than or equal to some value ζ. This is given in terms of the incomplete beta function,

$$I_\zeta(a,b) = \frac{1}{B(a,b)} \int_0^\zeta t^{a-1}(1-t)^{b-1}\,dt \equiv \frac{B_\zeta(a,b)}{B(a,b)} \tag{6-68}$$

The derivation of this result can be found in standard statistics textbooks. The values of $I_\zeta(a,b)$ or, alternatively, the values of

$$A(t|\nu) = 1 - I_{\frac{\nu}{\nu+t^2}}(\tfrac{1}{2}\nu, \tfrac{1}{2}) \tag{6-69}$$

are given in statistics tables. We shall first see how to evaluate this function before applying it to examples in least-squares fits to data in the next section.

Incomplete beta function. The most convenient way to calculate $I_\zeta(a,b)$ on a computer is to use the continued fraction form given on page 944 of Abramowitz and Segun:

$$I_x(a,b) = \frac{x^a(1-x)^b}{aB(a,b)}\left\{\frac{1}{1+}\frac{d_1}{1+}\frac{d_2}{1+}\cdots\right\} \tag{6-70}$$

where the coefficients d_i are given by

$$d_{2m+1} = -\frac{(a+m)(a+b+m)}{(a+2m)(a+2m+1)}x \qquad d_{2m} = \frac{m(b-m)}{(a+2m-1)(a+2m)}x$$

$$\tag{6-71}$$

The expression works best for

$$x < \frac{a-1}{a+b-2} \tag{6-72}$$

For the other values of x, we can substitute $\xi = (1-x)$ for the integral in (6-68) and obtain the relation

$$B_x(a,b) = \int_0^\zeta t^{a-1}(1-t)^{b-1} dt$$

$$= -\int_1^{1-x} (1-\xi)^{a-1} \xi^{b-1} d\xi$$

$$= \int_0^1 (1-\xi)^{a-1} \xi^{b-1} d\xi - \int_0^{1-x} (1-\xi)^{a-1} \xi^{b-1} d\xi$$

$$= B(b,a) - B_{1-x}(b,a)$$

In terms of the incomplete beta function, this is equivalent to the relation

$$I_x(a,b) = 1 - I_{1-x}(b,a) \qquad (6\text{-}73)$$

If we apply the continued fraction form of (6-70) to $I_{1-x}(b,a)$, the condition (6-72) becomes

$$(1-x) < \frac{b-1}{a+b-2}$$

or

$$x > \frac{a-1}{a+b-2}$$

In other words, if x satisfies (6-72), we can use (6-71) to calculate $I_x(a,b)$. Otherwise, we can make use of (6-73) and calculate $I_{1-x}(b,a)$ with (6-71) instead. In this way, the continued fraction form of (6-70) may be employed to calculate all the values of x in the range $[0,1]$.

To evaluate $I_x(a,b)$ using (6-70), we can put the relation into the standard form of (3-50) with partial numerators and denominators:

$$a_i = \begin{cases} 1 & \text{for } i=1 \\ d_{i-1} & \text{for } i>1 \end{cases} \qquad b_i = \begin{cases} 0 & \text{for } i=0 \\ 1 & \text{for } i>0 \end{cases}$$

In terms of d_{2m} and d_{2m+1} given in (6-71), the recursion relation of (3-50) may be written in the form

$$A_{2m} = A_{2m-1} + d_{2m} A_{2m-2} \qquad A_{2m+1} = A_{2m} + d_{2m+1} A_{2m-1}$$

$$B_{2m} = B_{2m-1} + d_{2m} B_{2m-2} \qquad B_{2m+1} = B_{2m} + d_{2m+1} B_{2m-1} \qquad (6\text{-}74)$$

for $m=1,2,\ldots$. The starting point of this particular set of recursion relations is

$$A_0 = B_0 = A_1 = 1 \qquad B_1 = 1 - \frac{a+b}{a+1}x \qquad (6\text{-}75)$$

Note that the indexes for A_i and B_i are shifted by 1 from those used in (3-50). This is done so that we can make convenient use of the relation for d_i given by (6-71). The algorithm to evaluate $I_x(a,b)$ is given in Box 6-2.

> **Box 6-2 Function BETA_I(x,a,b)**
> **Incomplete beta function $I_x(a,b)$**
> Continued fraction approximation for $x < (a-1)/(a+b-2)$
>
> Argument list:
> x: Upper limit of the integral in (6-68).
> a: The factor a in the integrand.
> b: The factor b in the integrand.
>
> Subprogram used:
>   GAMMA_LX: logarithm form of gamma function (cf. Box 4-9).
>
> Initialization:
>   (a) Set the maximum number of terms to be 100, and
>   (b) The accuracy to be 5×10^{-7}.
>
> 1. Check for special values:
>   (a) $I_x(a,b) = 0$ for $x = 0$.
>   (b) $I_x(a,b) = 1$ for $x = 1$.
> 2. Calculate the starting values using (6-75) and set the scale factor $f = 1$.
> 3. For $m = 1$ to a maximum of 100 terms, carry out the following steps:
>   (a) Calculate d_{2m} using (6-71).
>   (b) Calculate A_{2m} and B_{2m} using (6-74).
>   (c) Divide A_{2m} and B_{2m} by f to prevent overflow and underflow.
>   (d) Calculate d_{2m+1} using (6-71) and divide the result by f.
>   (e) Calculate A_{2m+1} and B_{2m+1} using (6-74).
>   (f) If B_{2m+1} does not vanish, let $f = B_{2m+1}$. Otherwise, $f = 1$.
>   (g) Calculate $f_{2m+1} = A_{2m+1}/B_{2m+1}$.
>   (i) Stop the continued fraction if calculation converges.
>   (ii) Otherwise, add more terms by going to the next m value.
> 4. Evaluate $B(a,b)$ using (6-67):
>   (a) Calculate the gamma functions using the logarithmic form of Box 4-9.
>   (b) Obtain $B(a,b)$ from $\exp\{\ln(\Gamma(a)) + \ln(\Gamma(b)) - \ln(\Gamma(a+b))\}$.
> 5. Return $\frac{x^a(1-x)^b}{aB(a,b)} f_{2m+1}$.

Fisher's F-test. The F-statistic is designed to test the probability of two least-squares fits with different χ^2-values being equivalent to each other. For example, in the previous section, we made a two-parameter fit to nine data points. To give a better fit, we may wish to added a third parameter associated with a quadratic term and use the method described in the next section to carry out the calculations. Since we have increased the number of parameters, the value of χ^2 obtained is likely to be smaller. Does this imply that the parent function for the data contains a quadratic term? That is, we wish to find out whether the decrease in the value of χ^2 is significant enough for us to change our model by increasing the number of parameters. This is an aim of the F-test.

The same kind of situation also occurs, for instance, if we carry out two separate sets of measurements for an experiment. If we apply the same least-squares fit to the two sets of results, there is a high probability that the two χ^2-values will be slightly different from each other. If the difference is significant, it may imply that some underlying factors in the experiment have changed between the two measurements. The F-test provides us with a measure of the probability that the two sets are referring to different conditions.

In general, if we have two χ^2-values, χ_1^2 and χ_2^2, corresponding respectively to ν_1 and ν_2 degrees of freedom, the ratio

$$t \equiv \frac{\chi_1^2/\nu_1}{\chi_2^2/\nu_2} \tag{6-76}$$

is distributed according to

$$p(t; \nu_1, \nu_2)\, dt = \frac{\nu_1^{\nu_1/2} \nu_2^{\nu_2/2} t^{(\nu_1/2-1)}}{B(\tfrac{1}{2}\nu_1, \tfrac{1}{2}\nu_2)(\nu_1 t + \nu_2)^{(\nu_1+\nu_2)/2}}\, dt$$

The range of possible values for t is from 0 to ∞. From the definition of t given by (6-76), it is obvious that t and $1/t$ have the same distribution except that the roles of ν_1 and ν_2 are interchanged.

The cumulative probability of t up to some value F is given by the integral

$$P(F|\nu_1, \nu_2) = \frac{\nu_1^{\nu_1/2} \nu_2^{\nu_2/2}}{B(\tfrac{1}{2}\nu_1, \tfrac{1}{2}\nu_2)} \int_0^F t^{(\nu_1/2-1)} (\nu_1 t + \nu_2)^{-(\nu_1+\nu_2)/2}\, dt$$

More commonly, this is put in terms of the incomplete beta function $I_\zeta(a,b)$ in the following way:

$$Q(F|\nu_1, \nu_2) = 1 - P(F|\nu_1, \nu_2) = I_{\frac{\nu_2}{\nu_1 t + \nu_2}}(\tfrac{1}{2}\nu_2, \tfrac{1}{2}\nu_1) \tag{6-77}$$

When $F = 0$, the value of the subscript in the incomplete beta function becomes 1 and

$$Q(0|\nu_1, \nu_2) = I_1(\tfrac{1}{2}\nu_2, \tfrac{1}{2}\nu_1) = 1$$

Similarly, when $F = \infty$, the value of the subscript becomes 0 and

$$Q(\infty|\nu_1, \nu_2) = I_0(\tfrac{1}{2}\nu_2, \tfrac{1}{2}\nu_1) = 0$$

Thus, a value of $Q(F|\nu_1, \nu_2)$ near 0 or 1 implies that the two χ^2-values are quite different from each other. We shall see examples of the applications of the various tests in the examples of the next section.

§6-6 Linear least-squares fit

In this section, we shall apply the method of least squares of §6-4 to functions with an arbitrary number of parameters, $a_1, a_2, \ldots, a_m$, subject only to the condition that the parameters appear linearly in the fitting function. That is, we wish to model the dependence of a quantity y on the variable x by an expression of the form

$$y(x) = \sum_{k=1}^{m} a_k f_k(x) \tag{6-78}$$

where the functions $f_k(x)$ are arbitrary at this moment, but they cannot involve any one of the m parameters $a_1, a_2, \ldots, a_m$. In statistics, this way of analyzing the data is known as *multiple regression*. The more general situation, where the parameters do not necessarily appear linearly in $y(x)$, belongs to the subject of nonlinear least-squares methods, to which we shall return in the next section.

Similar to what we have done earlier, we shall denote as y_i the value of y measured at $x = x_i$. For each y_i, there is an uncertainty σ_i. If the total number of data points is N, the value of χ^2 defined in (6-41) is now given by

$$\chi^2 = \sum_{i=1}^{N} \frac{1}{\sigma_i^2} \left\{ y_i - \sum_{k=1}^{m} a_k f_k(x_i) \right\}^2 \tag{6-79}$$

Since

$$\frac{\partial f_i}{\partial a_\ell} = \left. \frac{\partial y}{\partial a_\ell} \right|_{x=x_i} = f_\ell(x_i)$$

the maximum of the likelihood function $\mathcal{L}(a_1, a_2, \ldots, a_m)$ of (6-40) is given by the conditions

$$\frac{\partial \chi^2}{\partial a_\ell} = -2 \sum_{i=1}^{N} \frac{1}{\sigma_i^2} f_\ell(x_i) \left\{ y_i - \sum_{k=1}^{m} a_k f_k(x_i) \right\} = 0 \tag{6-80}$$

where $\ell = 1, 2, \ldots, m$, and we have made use of the assumption that $\partial f_\ell(x)/\partial a_\ell = 0$ for all $f_\ell(x)$ in (6-78).

Curvature matrix and covariance matrix. The m conditions given by (6-80) may be expressed as a set of m linear equations in the form

$$\sum_{k=1}^{m} a_k \sum_{i=1}^{N} \frac{1}{\sigma_i^2} f_k(x_i) f_\ell(x_i) = \sum_{i=1}^{N} \frac{1}{\sigma_i^2} f_\ell(x_i) y_i \tag{6-81}$$

If we define two matrices, a square matrix $\boldsymbol{F}$ and a column matrix $\boldsymbol{H}$, with elements given respectively by

$$F_{\ell,k} = \sum_{i=1}^{N} \frac{1}{\sigma_i^2} f_k(x_i) f_\ell(x_i) = \frac{1}{2} \frac{\partial \chi^2}{\partial a_\ell \partial a_k} \tag{6-82}$$

$$H_\ell = \sum_{i=1}^{N} \frac{1}{\sigma_i^2} f_\ell(x_i) y_i \tag{6-83}$$

the m linear equations of (6-81) may be expressed as

$$\sum_{k=1}^{m} F_{\ell,k} a_k = H_\ell$$

where $\ell = 1, 2, \ldots, m$. In matrix notation, this set of equations may be written as

$$\boldsymbol{FA} = \boldsymbol{H} \tag{6-84}$$

where $\boldsymbol{F}$ is known as the curvature matrix. The parameters $a_1, a_2, \ldots, a_m$ are the elements of the column matrix $\boldsymbol{A}$. Our aim here is to find the values of the elements of $\boldsymbol{A}$ by solving the system of linear equations.

As in (5-26), the solution of (6-84) for $\boldsymbol{A}$ may be found from the inverse of $\boldsymbol{F}$. Let

$$\boldsymbol{G} = \boldsymbol{F}^{-1}$$

This is known as the covariance matrix. Using $\boldsymbol{G}$, the solution for $\boldsymbol{A}$ is given by

$$\boldsymbol{A} = \boldsymbol{GH} \tag{6-85}$$

In terms of matrix elements, the same result may be written as

$$a_k = \sum_{\ell=1}^{m} G_{k,\ell} H_\ell = \sum_{\ell=1}^{m} G_{k,\ell} \sum_{i=1}^{N} \frac{1}{\sigma_i^2} f_\ell(x_i) y_i \tag{6-86}$$

In this approach, the main calculation in a linear least-squares fit is to obtain the covariance matrix $\boldsymbol{G}$ by inverting the curvature matrix $\boldsymbol{F}$. The algorithm is outlined in Box 6-3.

If, for some reason, a wrong choice is made on the functional form of $y(x)$, two or more of the parameters may, in essence, become simply related to each other. For example, one is a multiple of another. In this case, the curvature matrix becomes singular, the determinant of $\boldsymbol{F}$ vanishes, and the matrix cannot be inverted. There are two ways to correct this problem. The most obvious is to modify our model so that the singular condition does not arise. This may not always be easy to do. The alternative is to use the method of singular value decomposition to solve the matrix inversion problem. This method is discussed in Stoer and Bulirsch (*Introduction to Numerical Analysis*, Springer-Verlag, New York, 1980) and in Press and others (*Numerical Recipes*, Cambridge University Press, Cambridge, 1986). Here we shall stay with the simpler approach and assume that we can take care of any singular situation that arises by changing the functional form of $y(x)$.

> **Box 6-3 Subroutine** MRGS(NPT,N_PARM,X,Y,SY,A,SA)
> **Multiple regression analysis for** $y = \sum_{k=1}^{m} a_k f_k(x)$
>
> Argument list:
>   NPT: Number of data points.
>   N_PARM: Number of parameters.
>   X: Input array for the independent variable $\{x_i\}$.
>   Y: Input array for the dependent variable $\{y_i\}$.
>   SY: Input array for the uncertainties $\{\sigma_i\}$.
>   A: Output array for the coefficients $\{a_j\}$.
>   SA: Output array for the uncertainties in $\{a_j\}$.
>
> Subprograms required:
>   FK: Returns the value of $f_k(x)$ for a given set of $\{a_i\}$.
>   MAT_INV: Matrix inversion of Box 5-3.
>
> 1. Zero the arrays for f_i and $h_{i,j}$.
> 2. From the input values of x_i, y_i, and σ_i, calculate:
>   (a) If $\sigma_i = 0$, let the weighting factor $w_i = 1$, otherwise $w_i = 1/\sigma_i^2$.
>   (b) Store the values of $f_k(x_i)$ in an auxiliary array.
>   (c) Compute H_ℓ using (6-83).
>   (d) Compute the lower half-triangle of $F_{\ell,k}$ using (6-82).
>   (e) Complete the upper triangle of $F_{k,\ell}$ by symmetry.
> 3. Obtain the covariance matrix by inverting $\boldsymbol{F}$ using MAT_INV.
>   Check if the determinant $\boldsymbol{F}$ is singular.
> 4. Compute a_k from the inverse of $\boldsymbol{F}$ using (6-86).
>   Calculate $\{\sigma_{a_i}\}$, the uncertainty of a_i using (6-87).
> 5. Return the values of $\{a_i\}$ and $\{\sigma_{a_i}\}$.

Example of fitting with Legendre polynomial. As an illustration of the method of linear least squares, we shall examine the angular correlation of two γ-rays emitted in the decay of an excited nucleus, one from the decay of an excited state with spin J_0 to an intermediate state with spin J_i, and the other from the intermediate state to the ground state with spin J_f. The angular correlation function $W(\theta)$, which expresses the probability of the second γ-ray to be emitted at angle θ with respect to the first, is given by

$$W(\theta) = 1 + \sum_{k=1}^{k_x} A_{2k} P_{2k}(\cos\theta)$$

where the functions $P_\ell(\cos\theta)$ are the Legendre polynomials of §4-2. The upper limit of summation, k_x, is determined by the spins of the states involved. Because of the symmetry in the problem, only even-order Legendre polynomials can enter the expression. The shape of the angular distribution is characterized by the coefficients A_{2k}, and their values depend on the nature of the two γ-rays emitted and the spins of the three states involved. As a result, angular correlation studies constitute an important tool in determining the spin of a state. For our purpose here, we shall not be concerned with the connection between the coefficients A_{2k} and the types of γ-rays emitted. Our interest is to find the values of the coefficients A_{2k} that give the best fit to the measured correlation function $W(\theta)$.

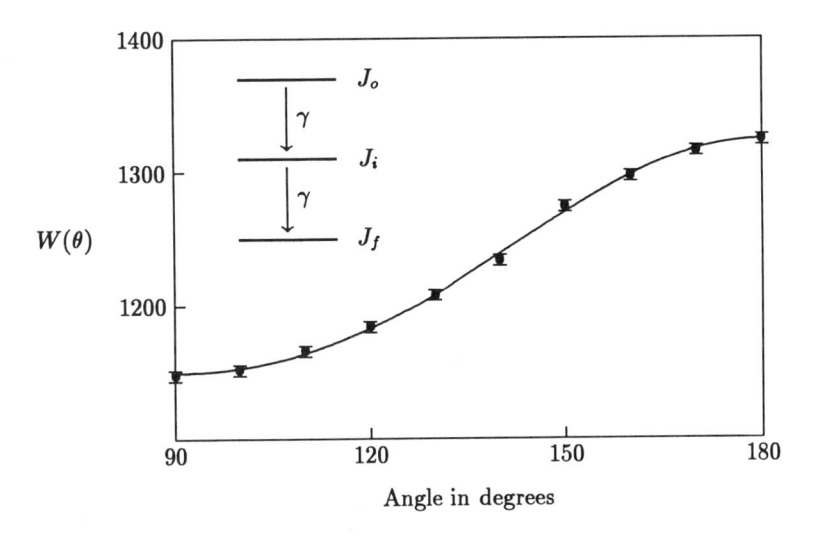

Fig. 6-11 Angular correlation of γ-rays emitted in the decay of ^{60}Ni. The smooth curve is a Legendre polynomial fit using the functional form $C(\theta) = a_0 + a_2 P_2(\cos\theta) + a_4 P_4(\cos\theta)$. The values of the parameters obtained are $a_0 = 1200.3 \pm 1.6$, $a_2 = 110.5 \pm 3.3$, and $a_4 = 12.8 \pm 3.5$. The data are taken from R.M. Steffen, *Adv. Phys.* **4** (1955), 293.

To compare directly with the measured quantities, we shall rewrite the angular correlation function in terms of the number of γ-rays observed at angle θ with respect to another one emitted essentially at the same time,

$$C(\theta) = \sum_{k=0}^{m} a_{2k} P_{2k}(\cos\theta)$$

The experimental data are taken from the decay of ^{60}Ni, measured by R.M. Steffen [*Adv. Phys.* **4** (1955), 293] and the fitted results are shown in Fig. 6-11. Note that, because of the symmetry of even-order Legendre polynomials, the angular distribution of $W(\theta)$ is symmetric around 90°. For this reason, only the backward angles (or the forward angles) are measured in most experiments of this type.

Uncertainties in the parameters. The uncertainties in the parameters a_k determined in a linear least-squares calculation are given by the diagonal matrix elements of the covariance matrix $\boldsymbol{G}$. This can be seen from the following argument. From (6-24), we see that the uncertainty in the value of a_k is given by

$$\sigma_{a_k}^2 = \sum_{i=1}^{N} \sigma_i^2 \left(\frac{\partial a_k}{\partial y_i}\right)^2$$

where we have again ignored the covariance between any two data points. The partial derivative of a_k with respect to y_i may be found from (6-86),

$$\frac{\partial a_k}{\partial y_i} = \sum_{\ell=1}^{m} G_{k,\ell} \frac{1}{\sigma_i^2} f_\ell(x_i)$$

This gives us the result

$$\begin{aligned}
\sigma_{a_k}^2 &= \sum_{i=1}^{N} \sigma_i^2 \sum_{\ell=1}^{m} G_{k,\ell} \frac{1}{\sigma_i^2} f_\ell(x_i) \sum_{j=1}^{m} G_{k,j} \frac{1}{\sigma_i^2} f_j(x_i) \\
&= \sum_{i=1}^{N} \sum_{\ell=1}^{m} \sum_{j=1}^{m} G_{k,\ell} G_{k,j} \frac{1}{\sigma_i^2} f_j(x_i) f_\ell(x_i) \\
&= \sum_{\ell=1}^{m} \sum_{j=1}^{m} G_{k,\ell} G_{k,j} F_{j,\ell} = \sum_{\ell=1}^{m} G_{k,\ell} \delta_{k,\ell} = G_{k,k} \quad (6\text{-}87)
\end{aligned}$$

where we have made use of (6-82) to sum over index i to obtain $F_{j,\ell}$ and the fact that $\boldsymbol{G}$ is the inverse of $\boldsymbol{F}$.

The numerical calculations involved in a linear least-squares fit closely follow the steps leading to (6-85). First, we calculate the matrix elements of $\boldsymbol{H}$ and $\boldsymbol{F}$ from the input values of $\{x_i, y_i, \sigma_i\}$ using (6-82) and (6-83). The inverse of $\boldsymbol{F}$ is obtained using the matrix inversion technique of Box 5-3. The values of the coefficients that give the best fit to the input are then produced from the matrix product of $\boldsymbol{F}^{-1}$ and $\boldsymbol{H}$. The algorithm was outlined earlier in Box 6-3.

Chi-square test of the fit. In carrying out a least-squares fit to a set of data, we usually start with a model based on some ideas we may have concerning the relation between the dependent and independent variables. One indication of whether our model is the correct one is given by the quality of the fit. For this purpose, we can calculate the value of the χ^2. By definition, the minimum in the value of χ^2 for the functional form used occurs when the best-fit set of values for the parameters $\{a_k\}$ are used:

$$\chi_{\min}^2 = \sum_{i=1}^{N} \frac{1}{\sigma_i^2} \left\{y_i - \sum_{k=1}^{m} a_k f_k(x_i)\right\}^2$$

The probability for us to obtain a χ^2-value greater or equal to χ^2_{min} is given by the probability integral for the χ^2-distribution $Q(\chi^2_{min}|\nu)$ of (6-65). Here $\nu = (N-m)$ is the number of degrees of freedom, N is the number of independent data points, and m is the number of free parameters.

As an example, consider the 20 data points in Table 6-4. Let us begin by trying to fit them using the functional form

$$y(a_1, a_2; x) = a_1 + a_2 x$$

Since there are two parameters, the number of degrees of freedom is $\nu = 18$. The value of χ^2 calculated with the best-fit values of $a_1 = 0.115 \pm 0.020$ and $a_2 = 2.282 \pm 0.011$ is $\chi^2_{min} = 168$, giving a reduced χ^2-value of $\chi^2_\nu = 9.3$. The probability of obtaining a χ^2-value of 168 for 18 degrees of freedom is $Q(168|18) = 0.0$. Thus, we can conclude that the fit is not a good one.

Table 6-4 Twenty measured values of y as a function of x.

i	x_i	y_i	y_{fit}	σ_i	i	x_i	y_i	y_{fit}	σ_i
1	2.48	6.04	6.259	0.900	11	4.09	9.29	9.072	1.030
2	2.92	7.57	7.123	0.340	12	2.31	5.89	5.906	0.050
3	1.41	4.03	3.858	0.690	13	2.00	5.47	5.235	0.580
4	0.04	0.16	0.164	0.020	14	2.38	5.94	6.053	0.640
5	0.30	1.61	0.918	0.730	15	3.88	8.76	8.760	0.040
6	0.04	0.49	0.164	0.320	16	0.67	1.74	1.949	0.650
7	4.07	9.14	9.043	0.910	17	3.03	6.80	7.328	0.770
8	1.90	4.60	5.011	0.600	18	0.82	2.18	2.353	0.190
9	0.64	2.51	1.868	0.200	19	3.21	7.94	7.654	1.260
10	1.49	3.79	4.052	1.540	20	0.39	0.96	1.174	0.310

To find a better functional form, we can include a quadratic term in $y(x)$. The fitting function now becomes

$$y(a_1, a_2; x) = a_1 + a_2 x + a_3 x^2$$

With the addition of one more parameter, the number of degrees of freedom is reduced by 1 to 17, and the best-fit values of the parameters are $a_1 = 0.0455 \pm 0.0206$, $a_2 = 2.9656 \pm 0.0566$, and $a_3 = -0.1855 \pm 0.0151$. The value of χ^2 is reduced to $\chi^2_{min} = 17$. Since $Q(17|17) = 0.45$, we may conclude that the new fit is reasonable. The effect of the quadratic term is to put a curvature into the relation between x and y. From Fig. 6-12, we see that, even though the curvature is quite small, the refinement in the fit due to a small x^2-dependent term is quite significant.

The improvement in the fit produced by the quadratic term may prompt us to try to include even higher-order terms into the fitting function $y(x)$ to see if the value of the χ^2 can be further reduced. The addition of a cubic term lowers the χ^2-value only slightly to $\chi^2_{\min} = 16$. However, the smaller value does not necessarily imply a better fit. This can be seen, for example, from the fact that $Q(16|16) = 0.45$, unchanged from that of the quadratic fit. We shall see next that there are also other tests designed to check directly whether an additional parameter is significant.

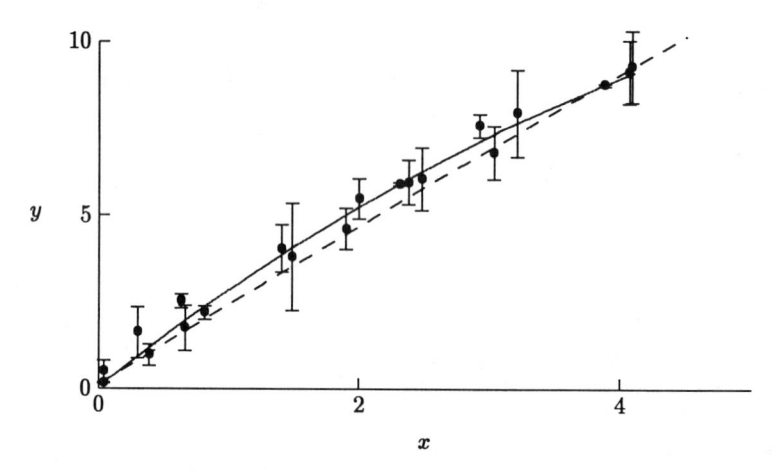

Fig. 6-12 Linear least-squares fit to the 20 data points in Table 6-4. The solid curve is the result of a three-parameter power series $y = a_1 + a_2 x + a_3 x^2$ using the best-fit values of $a_1 = 0.046 \pm 0.021$, $a_2 = 2.966 \pm 0.057$, and $a_3 = -0.186 \pm 0.015$. The dashed curve is a straight-line fit using $y = a + bx$ with $a = 0.115 \pm 0.020$ and $b = 2.282 \pm 0.011$.

Multiple correlation coefficient. Another test of the quality of a fit is based on the correlation coefficient of (6-52). To make a connection with the simple case of a straight-line fit, we shall assume that the first term in the sum on the right side of (6-78) is a constant term, independent of x. This is the equivalent of taking

$$f_1(x) = 1$$

This approach also has the advantage that, when we limit ourselves to two parameters, the form matches that for a straight line used earlier in (6-43). Similar to (6-55), we can calculate the covariance between y and the kth function in the sum on the right side of (6-78) in the following way:

$$\sigma^2_{k,y} \equiv \left(\sum_{i=1}^{N} \frac{1}{\sigma_i^2}\right)^{-1} \sum_{i=1}^{N} \frac{1}{\sigma_i^2} \{f_k(x_i) - \mu_k\}\{y_i - \mu_y\}$$

where μ_k is the mean of $f_k(x)$ and μ_y is the mean of y. These may be obtained in the same way as (6-53):

$$\mu_k = \left(\sum_{i=1}^{N} \frac{1}{\sigma_i^2}\right)^{-1} \sum_{i=1}^{N} \frac{1}{\sigma_i^2} f_k(x_i) \qquad \mu_y = \left(\sum_{i=1}^{N} \frac{1}{\sigma_i^2}\right)^{-1} \sum_{i=1}^{N} \frac{1}{\sigma_i^2} y_i$$

Analogous to (6-56), we can define the square of a *multiple correlation coefficient* as

$$R^2 = \frac{1}{\sigma_y^2} \sum_{k=2}^{m} a_k \sigma_{k,y}^2 \tag{6-88}$$

where the variance of y is given by

$$\sigma_y^2 = \frac{1}{N-1} \sum_{i=1}^{N} (y_i - \mu_y)^2 \tag{6-89}$$

The summation on the right side of (6-88) starts with $k = 2$, as the $k = 1$ term is assumed to be a constant and therefore excluded from the summation.

We use $(N-1)$ rather than N in the denominator on the right side of (6-89) to define σ_y^2 for the following reason. If y_i is a random variable with a normal distribution, then σ_y^2 is a random variable with a χ^2-distribution, as can be seen by comparing (6-89) with (6-61). Since the mean μ_y is obtained by averaging over y_i, there are only $(N-1)$ degrees of freedom left among the N values of $(y_i - \mu_y)$.

The product $R^2 \sigma_y^2$ is also a random variable with a χ^2-distribution, but the number of degrees of freedom is only $(m-1)$. This is evident from the right side of (6-88). On multiplying both sides of (6-88) by σ_y^2, we obtain

$$R^2 \sigma_y^2 = \sum_{k=2}^{m} a_k \sigma_{k,y}^2$$

The right side is now simply a sum over $(m-1)$ squares of random variables. We can decompose σ_y^2 in the following way:

$$\sigma_y^2 = R^2 \sigma_y^2 + (1 - R^2) \sigma_y^2$$

The left side of the equation consists of a random variable with a χ^2-distribution of $(N-1)$ degrees of freedom. Since the first term on the right side is a random variable with a χ^2-distribution of $(m-1)$ degrees of freedom, the second term must be a random variable with $(N-1) - (m-1) = (N-m)$ degrees of freedom. Physically, the first term is a measure of the spread in the dependent and independent variables in the data and the second term provides a feeling of the spread between the fit and the data. Since both quantities follow χ^2-distributions, we can use the F-test of §6-5 to supply a measure of the significance of the fit.

Using the definition of (6-76), we can construct an F-test for the multiple correlation coefficient R in terms of the quantity:

$$F_R = \frac{R^2 \sigma_y^2/(m-1)}{(1-R^2)\sigma_y^2/(N-m)} = \frac{R^2(N-m)}{(1-R^2)(m-1)} \qquad (6\text{-}90)$$

Similar to (6-77), the probability distribution of having a value of F_R larger or equal to some value F is given by the integral $Q(F|\nu_1,\nu_2)$, with $\nu_1 = (m-1)$ and $\nu_2 = (N-m)$. A large value of F_R here implies that the spread between the fit and the data is much smaller than the spread between the dependent and independent variables. This, in turn, is an indication that the fit is a good one.

Test of the need for additional parameters. Earlier, we saw an example of the results of putting additional terms into our fitting function. In general, we can expect a decrease in the value of χ^2 as the number of parameters is increased in the calculation. On the other hand, each additional term also decreases the number of degrees of freedom. As a result, the probability $Q(\chi^2|\nu)$ for obtaining the χ^2-value may not increase, indicating that the new term is not improving the fit. As a more quantitative measure, we can use an F-test to tell us whether the additional term is significant or not.

For the convenience of discussion, we shall treat the additional term as the last one in our function $y(x)$, and there are m terms in total when the new one is included. Let us represent the value of χ^2 obtained with m parameters as $\chi^2(m)$. If we carry out the least-squares calculation without the last term, we obtain, in general, a different value, which we shall label $\chi^2(m-1)$. As we have seen earlier, $\chi^2(m)$ for N data points is a quantity following a χ^2-distribution of $(N-m)$ degrees of freedom, and $\chi^2(m-1)$ is one having $(N-m+1)$ degrees of freedom. The difference between them,

$$\Delta \chi^2 = \chi^2(m-1) - \chi^2(m)$$

is therefore a quantity following a χ^2-distribution of $(N-m+1)-(N-m)=1$ degree of freedom. The ratio

$$F_\chi = \frac{\Delta\chi^2}{\frac{\chi^2(m)}{N-m}} = \frac{\chi^2(m-1) - \chi^2(m)}{\frac{\chi^2(m)}{N-m}} \qquad (6\text{-}91)$$

is then a quantity following an F-distribution, as can be seen by comparing (6-91) with (6-76). If the value of F_χ is small, we can conclude that the last term in our fitting function is not significant in improving the fit. More quantitatively, the probability for having a particular value of F in this case is given by the value of $Q(F|1, N-m)$ of (6-77).

Checking if two samples have the same distribution. In an experiment, it may happen that the measurements are taken over long periods of time. How can we be certain, for example, that nothing in the setup has been changed during the time interval? This is the same problem as the more general one of having several different sets of data, and we wish to know whether they have the same distribution. Consider two such samples with N_1 and N_2 members in each. If both samples are analyzed using the same fitting function $y(x)$ of m parameters, we obtain two sets for values of the parameters. In general, they will not be identical to each other. On the other hand, because of the uncertainties associated with the values determined, it may not be easy to tell whether they are different from each other in a significant way. It is, therefore, useful to devise a statistical test to give us an estimate of the probability of two sets of parameter values describing the same distribution.

Let us represent the value of the kth parameter a_k obtained from the first sample of N_1 data points as $a_k(N_1)$ and that corresponding to the second set with N_2 data points as $a_k(N_2)$. The uncertainties associated with these two values are, respectively, $\sigma_k(N_1)$ and $\sigma_k(N_2)$. In each case, the uncertainty is to be interpreted as the standard deviation of the distribution in the value of the parameter if an infinite number of samples of the N_1, or N_2, data points were available and analyzed in the same way. Similarly, we shall consider that the value of each parameter is given by the mean value of its distribution. The question of whether there are any significant differences between $a_k(N_1)\pm\sigma_k(N_1)$ and $a_k(N_2)\pm\sigma_k(N_2)$ is then equivalent to the question of whether the distributions defined by these two sets of numbers are different from each other. For this purpose, we can construct a variable

$$t = \frac{a_k(N_1) - a_k(N_2)}{\frac{\sigma_k^2(N_1)}{N_1} + \frac{\sigma_k^2(N_2)}{N_2}} \qquad (6\text{-}92)$$

Comparing with the quantity defined by (6-66), we see that the variable t here follows approximately a Student's t-distribution. The number of degrees of freedom is given by

$$\nu = \frac{\left(\frac{\sigma_k^2(N_1)}{N_1} + \frac{\sigma_k^2(N_2)}{N_2}\right)^2}{\frac{1}{N_1-1}\left(\frac{\sigma_k^2(N_1)}{N_1}\right)^2 + \frac{1}{N_2-1}\left(\frac{\sigma_k^2(N_2)}{N_2}\right)^2} \qquad (6\text{-}93)$$

In general, the value of ν here is not an integer. However, this does not cause any difficulty in evaluating the integral $A(t|\nu)$ of (6-69) that provides us with an estimate of the probability.

The possible range of values for t is $[0,\infty]$. At the upper limit of $t = \infty$, it is obvious that $a_k(N_1)$ and $a_k(N_2)$ are quite different from each other. From (6-69), it is easy to see that at $t = \infty$ we have $A(t|\nu) = 1$, as the value of the incomplete beta function vanishes in this limit. A value of $A(t|\nu)$ near 1 means that $a_k(N_1)$ and $a_k(N_2)$ are quite different from each other. On the other hand,

Table 6-5 Results of fitting the data in Table 6-4 as two groups of 10 points each.

Statistic	Set N_1	Set N_2	t	ν	$A(t\|\nu)$
N	10	10			
a_1	0.018 ± 0.024	-0.248 ± 0.274	2.56	6.09	0.96
a_2	3.650 ± 0.306	3.150 ± 0.206	3.59	10.5	1.00
a_3	-0.374 ± 0.104	-0.214 ± 0.036	3.85	7.42	0.99
ν	7	7			
χ^2	8.37	0.83			
$Q(\chi^2\|\nu)$	0.30	1.00			

for $t = 0$, the value of the incomplete beta function becomes 1 and $A(t|\nu) = 0$. As a result, a value of $A(t|\nu)$ near 0 for t and ν defined, respectively, by (6-92) and (6-93), implies that $a_k(N_1)$ and $a_k(N_2)$ are essentially the same.

As an example, we shall divide the 20 data points given in Table 6-4 into two groups, with the first 10 data points in one group and the second 10 in the other. Again, let us use a three-parameter power series

$$y = a_1 + a_2 x + a_3 x^2$$

to fit each of the two groups of ten data points. The results are listed in Table 6-5. We can see from the values of $A(t|\nu)$ obtained that it unlikely for the two sets of parameters to be describing the same distribution. This is not surprising, as the differences between $a_k(N_1)$ and $a_k(N_2)$, for $k = 1, 2$ and 3, are in general larger than the uncertainties associated with each. The suggestion that the two groups are different from each other may also be seen from the values of the χ^2 obtained in the two cases. Such differences are not evident by examining directly the values of the data points in each group.

§6-7 Nonlinear least-squares fit to data

All the methods of least squares we have discussed so far are restricted to fitting functions that are linear in the parameters $a_1, a_2, \ldots, a_m$. That is, the relation between the measured values y_i and the independent variable x_i is given by a function of the form

$$y(x) = f(a_1, a_2, \ldots, a_m; x)$$

satisfying the condition

$$\frac{\partial^2 y}{\partial a_j \partial a_k} = 0 \qquad (6\text{-}94)$$

for all $j = 1, 2, \ldots, m$ and $k = 1, 2, \ldots, m$. [The same is also true for the higher-order partial derivatives of $y(x)$ with respect to the parameters, but we shall not be making any explicit use of this property.] The reason that such a function can simplify the calculations involved comes from the fact that the partial derivatives of $f(a_1, a_2, \ldots, a_m; x)$ with respect to any of the parameters $a_1, a_2, \ldots, a_m$ do not involve any of the parameters. From (6-80), we see that the conditions to minimize the value of χ^2 are given by (6-81) and they are linear in $a_1, a_2, \ldots, a_m$. As a result, the equations may be solved, for example, by matrix inversion, as we did in the previous section.

For fitting functions that are more complicated in form, it is necessary to use instead a nonlinear least-squares method. For example, if we wish to fit a set of data points y_i to a normal distribution with the normalization, mean, and standard deviation as adjustable parameters, we need to use a fitting function of the form

$$f(a_1, a_2, a_3; x) = a_1 \frac{1}{a_3 \sqrt{2\pi}} \exp\left\{-\frac{1}{2}\left(\frac{x - a_2}{a_3}\right)^2\right\} \qquad (6\text{-}95)$$

In this case, $y(x)$ is linear only in the parameter a_1. In contrast, both a_2 and a_3 do not satisfy the condition of (6-94).

It is still possible to find the values of the parameters a_1, a_2, and a_3 using the method of maximum likelihood. In principle, we can follow the same procedure as that used in a linear least-squares calculation and minimize the value of χ^2 by varying the parameters. The result is similar to (6-80). However, since the equations are no longer linear in the unknown quantities a_1, a_2, and a_3, we do not have a general method to obtain their values directly.

An alternative is to go back to the primitive approach of locating the minimum of χ^2 by making a complete survey of the parameter space. Consider first the general case of m parameters. Let us divide the possible range of values for each parameter a_k into n_k equal parts. The complete m-dimensional space is then separated into $(n_1 \times n_2 \times \cdots \times n_m)$ cells. To simplify the discussion, we shall consider $n_1 = n_2 = \cdots = n_m = n$. In this case, there are n^m cells, or grid points, in the space. A complete survey means that we must calculate the value of $\chi^2(\boldsymbol{a})$ at each of the n^m points. By comparing the values of χ^2 obtained, we can locate the minimum in the parameter space. The accuracy of such an approach is limited by the number of subdivisions we can take for each parameter. Furthermore, the amount of computation can be large even for modest values of n and m. Instead of a fixed grid, it is possible to modify the method into an iterative search. This is the principle behind most of the methods of nonlinear least squares, and we shall discuss below a few of the more commonly used ones.

General considerations of the parameter space. For our discussion here, it is convenient to regard χ^2 as a function of m variables $a_1, a_2, \ldots, a_m$, in the form

$$\chi^2(a_1, a_2, \ldots, a_m) = \sum_{i=1}^{N} \frac{1}{\sigma_i^2} \{y_i - f(a_1, a_2, \ldots, a_m; x_i)\}^2 \qquad (6\text{-}96)$$

Our interest is to find the location of its minimum. For $m = 2$, we can visualize χ^2 as a function of a_1 and a_2 by making an analogy to the view of the landscape observed from the top of a mountain. For example, we may identify a_1 with the north-south direction and a_2 with the east-west direction. Depending on the behavior of the function $\chi(a_1, a_2)$, the landscape may be quite flat or it may be full of hills and valleys. To find the lowest point in the area, we shall define a coordinate system with a fixed origin. Any point in the two-dimensional space can then be specified by a vector $\boldsymbol{a}$ from the origin to the point. For $m = 2$, the vector is given by the values of a_1 and a_2, similar to a point on a sheet of graph paper. In general, when m is greater than 2, the vector $\boldsymbol{a}$ represents a set of m numbers

$$\boldsymbol{a} \equiv \{a_1, a_2, \ldots, a_m\}$$

To simplify the discussion, from now on we shall use such a vector notation to denote a set of values of all the parameters.

Most of the nonlinear least-squares methods find the minimum in the parameter space by making use of the local value of the slope of the χ^2-function. The basic idea is a simple one. If the slope is pointing downhill, we know we must move forward in order to reach a lower value. Similarly, if the slope is pointing uphill, we must turn around and move in the opposite direction. A minimum in the area is indicated by the fact that the slope is positive in every direction.

In principle, we can start from any point in the parameter space and locate the minimum by following the slope from one point to another, much as water on land finds its way to the ocean. As a practical procedure, however, we must start from some initial guess of the values of all the m parameters. Let us call this set $\boldsymbol{a}(0)$. Variations of one or more of the m values can be made by adding small increments $\delta\boldsymbol{a}(0)$ to $\boldsymbol{a}(0)$. This generates a new set of values

$$\boldsymbol{a}(1) = \boldsymbol{a}(0) + \delta\boldsymbol{a}(0)$$

By comparing the values of $\chi^2(\boldsymbol{a}(0))$ and $\chi^2(\boldsymbol{a}(1))$, we obtain the slope in the region in between $\boldsymbol{a}(0)$ and $\boldsymbol{a}(1)$. This information provides us with the direction and the magnitude for the changes $\delta\boldsymbol{a}(1)$ in order to find the next set $\boldsymbol{a}(2)$ with the possibility of an even lower $\chi^2(\boldsymbol{a})$. The process is repeated for the next set, and so on. The general approach for step ℓ may be stated as

$$\boldsymbol{a}(\ell+1) = \boldsymbol{a}(\ell) + \delta\boldsymbol{a}(\ell) \tag{6-97}$$

and the end of the search is given by the condition

$$\Delta\chi^2 \equiv \chi^2(\boldsymbol{a}(\ell) + \delta\boldsymbol{a}) - \chi^2(\boldsymbol{a}(\ell)) \geq 0 \qquad \text{for} \quad \delta\boldsymbol{a} \to 0$$

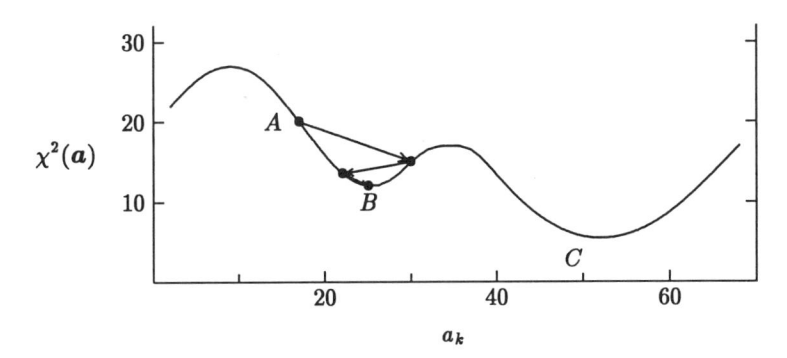

Fig. 6-13 Schematic diagram showing the variation of χ^2 as a function of one of the parameters a_k in a multiparameter, nonlinear least-squares fit. A poor choice of the starting point A, for example, can lead to the search procedure settling in a local minimum in χ^2 in the parameter space at B, rather than the absolute minimum at C.

The major differences among the various nonlinear least-squares methods lie in their ways of making the estimate for the variations $\delta a(\ell)$.

The main problem with the *local* approach outlined above is that there may be more than one minimum in the space. In the case of a linear model, $\chi^2(a)$ is a quadratic function of the parameters a and there is only a single minimum in the m-dimensional space. For a nonlinear model, this is no longer true and there can be a number of local minima in the landscape. For a least-squares fit, we are interested in the absolute minimum of the $\chi^2(a)$ function, as this is the point of maximum likelihood. A local search scheme, however, has no way of finding out what type of minimum one is in. This is illustrated schematically in Fig. 6-13. The only way to identify the absolute minimum in the parameter space for such cases is to perform a complete survey. Since this is usually not feasible, some alternatives must be taken. One way is to repeat the calculation using several different starting points and see if we can reach a different minimum that is lower in value.

Parabolic approximation. For small variations of the parameters, the value of $\chi^2(a(\ell) + \delta a)$ may be expanded in terms of a Taylor series around $\chi^2(a(\ell))$:

$$\chi^2(a(\ell) + \delta a) = \chi^2(a(\ell)) + \sum_{i=1}^{m} \delta a_i \frac{\partial \chi^2}{\partial a_i}\bigg|_{a=a(\ell)}$$

$$+ \frac{1}{2!} \sum_{j=1}^{m} \sum_{i=1}^{m} \delta a_j \delta a_i \frac{\partial^2 \chi^2}{\partial a_j \partial a_i}\bigg|_{a=a(\ell)} + \cdots$$

For $k = 1, 2, \ldots, m$, the changes in $\chi^2(a)$ due to a small variation Δa_k in the parameter a_k are given by

$$\chi^2(a(\ell)+\Delta a_k) - \chi^2(a(\ell)) = \Delta a_k \left\{ \frac{\partial \chi^2}{\partial a_k}\bigg|_{a=a(\ell)} + \sum_{j=1}^{m} \delta a_j \frac{\partial^2 \chi^2}{\partial a_j \partial a_k}\bigg|_{a=a(\ell)} + \cdots \right\}$$

The factor of $\frac{1}{2}$ associated with the second-order derivatives in the previous equation cancels with the factor of 2 coming from the fact that there are two summations involved and both δa_i and δa_j can be equal to Δa_k. We can write the variations of the values of χ^2 due to small changes in a in the form of a derivative:

$$\frac{\Delta \chi^2}{\Delta a_k}\bigg|_{a=a(\ell)} \equiv \frac{\chi^2(a(\ell)+\Delta a_k) - \chi^2(a(\ell))}{\Delta a_k}$$

$$= \frac{\partial \chi^2}{\partial a_k}\bigg|_{a=a(\ell)} + \sum_{j=1}^{m} \delta a_j \frac{\partial^2 \chi^2}{\partial a_j \partial a_k}\bigg|_{a=a(\ell)} + \cdots \quad (6\text{-}98)$$

The minimum of $\chi^2(a)$ is given by the condition

$$\frac{\Delta \chi^2}{\Delta a_k} = 0 \qquad \text{for} \qquad k = 1, 2, \ldots, m$$

If the function $\chi^2(a)$ is approximated by a parabolic form and the series on the right side of (6-98) is truncated after the second term, we obtain a result that is similar to (6-80):

$$\frac{\partial \chi^2}{\partial a_k}\bigg|_{a=a(\ell)} \approx -\sum_{j=1}^{m} \delta a_j \frac{\partial^2 \chi^2}{\partial a_j \partial a_k}\bigg|_{a=a(\ell)} \quad (6\text{-}99)$$

There are m equations of this form for $k = 1, 2, \ldots, m$. In terms of matrices, we can write the m equations as

$$\boldsymbol{\beta} = \boldsymbol{\alpha}\, \delta \boldsymbol{a}$$

where δa_j are the matrix elements of $\delta \boldsymbol{a}$. The matrix elements of $\boldsymbol{\beta}$ and $\boldsymbol{\alpha}$ are given, respectively, by

$$\beta_k = -\frac{1}{2}\frac{\partial \chi^2}{\partial a_k}\bigg|_{a=a(\ell)} \qquad \alpha_{k,j} = \frac{1}{2}\frac{\partial^2 \chi^2}{\partial a_j \partial a_k}\bigg|_{a=a(\ell)} \quad (6\text{-}100)$$

For later convenience, a factor of $\frac{1}{2}$ is included in the definitions of $\boldsymbol{\beta}$ and $\boldsymbol{\alpha}$. We see here also the reason for calling $\boldsymbol{\alpha}$ the "curvature matrix," as it expresses the curvature of $\chi^2(a)$ in the parameter space.

On inverting the curvature matrix $\boldsymbol{\alpha}$, we obtain the solution for δa_k in terms of quantities calculated with $a(\ell)$:

$$\delta \boldsymbol{a} = \boldsymbol{\alpha}^{-1} \boldsymbol{\beta} \quad (6\text{-}101)$$

where α^{-1} is the covariance matrix. This is similar to (6-85) except that, in the linear case, the solution gives directly the location of the minimum. Here, (6-99) is only an approximation. As a result, the solution for δa does not locate the minimum of χ^2 for us, only the direction and an indication of the step size to look for it. As we shall soon see, the choice of step size is an important consideration for fast convergence.

Approximation by linearization. A further simplification may be incorporated into the calculation of the curvature matrix. By taking the second-order partial derivative of $\chi^2(a)$ with respect to any two parameters, a_j and a_k, using the form given by (6-96), we obtain

$$\frac{\partial^2 \chi^2}{\partial a_j \partial a_k} = 2 \sum_{i=1}^{N} \frac{1}{\sigma_i^2} \left\{ \frac{\partial f(a; x_i)}{\partial a_j} \frac{\partial f(a; x_i)}{\partial a_k} - [y_i - f(a; x_i)] \frac{\partial^2 f(a; x_i)}{\partial a_j \partial a_k} \right\} \quad (6\text{-}102)$$

This is essentially the curvature matrix. The calculation can be greatly simplified if we ignore the second term on the right side. This is equivalent to approximating the function $f(a; x)$ as linear in the parameters a and satisfying the condition given by (6-94) as a result. In this approach, elements of the curvature matrix reduce to the simple form

$$\alpha_{k,j} = \frac{1}{2} \frac{\partial^2 \chi^2}{\partial a_j \partial a_k} \approx \sum_{i=1}^{N} \frac{1}{\sigma_i^2} \left\{ \frac{\partial f(a; x_i)}{\partial a_j} \frac{\partial f(a; x_i)}{\partial a_k} \right\} \quad (6\text{-}103)$$

with the factor of $\frac{1}{2}$ in the definition given by (6-100) canceling the factor of 2 on the right side of (6-102). In practice, the second-order partial derivatives $\partial^2 f(a; x_i)/\partial a_j \partial a_k$ in the second term of (6-102) have a tendency to make the numerical calculation unstable. For this reason, also, it is helpful to ignore it.

The method of Marquardt. The parabolic approximation discussed earlier is equivalent to the assumption that the behavior of $\chi^2(a)$ can be described by a quadratic function in a:

$$\chi^2(a) \approx c_0 + \sum_{j,k} c_{j,k} a_j a_k$$

where the coefficients c_0 and $c_{i,j}$ are constants independent of a. This is a familiar way to approximate a function near its minimum, as we have seen, for example, in Fig. 4-1. For the same reason, the approximation is a good one when the search is near a minimum in the parameter space of $\chi^2(a)$. Under such conditions, the calculation converges very quickly.

In regions far away from a minimum, the parabolic approximation is slow since it tends to search in small steps. As an alternative, the gradient method is often used in such regions. In this approach, the step size $\delta a(\ell)$ of (6-97) is proportional to the gradient $\nabla \chi^2(a)$ or the slope of the $\chi^2(a)$ surface at the point $a = a(\ell)$. That is,

$$\delta a \propto -\nabla \chi^2(a(\ell)) \qquad (6\text{-}104)$$

From (6-100), we find that

$$-\nabla \chi^2(a(\ell)) = 2\beta \qquad (6\text{-}105)$$

The gradient method is efficient, since the search follows the path of steepest descent of χ^2 as a function of the parameters. This is especially true in regions far away from a minimum where, in general, the function varies smoothly. To make use of (6-104) to find δa, we need to assign a value to the constant of proportionality. Unfortunately, there is not an easy way to construct one. In general, the method tends to overshoot near a minimum and, as a result, the rate of convergence slows down as we approach a minimum.

In view of the discussion in the previous two paragraphs, the optimum method for a nonlinear least-squares fit is to use a combination of gradient and parabolic approximations. This idea is attributed to Marquardt. Before implementing the method, we need a practical way of determining the step size in the gradient method. A hint can be obtained from a dimensional analysis of the problem. From (6-96), we see that $\chi^2(a)$ is a dimensionless quantity, since σ_i in the denominator has the same dimension as y_i and $f(a; x_i)$ in the numerator. As a result, the matrix elements of β must have the same dimension as the inverse of a, as can be seen from (6-100). However, different parameters can have different dimensions. For example, in (6-95), the distribution may be representing the number of radioactive decays as a function of the energy emitted. In this case, a_1 is dimensionless (counts), whereas a_2 and a_3 are in energy units. From such considerations, we see that the dimension of β_k, the kth element of β, must be the inverse of that of a_k. In terms of matrix elements, (6-101) may be written in the form

$$\delta a_k = \sum_{j=1}^{m} (\alpha^{-1})_{k,j} \beta_j \qquad (6\text{-}106)$$

Using the fact that the product $\alpha^{-1}\alpha$ produces a dimensionless unit matrix, the dimension of the matrix elements of α must be given by

$$[\alpha_{j,k}^{-1}] = [a_j][a_k]$$

where we have used the symbol $[q]$ to represent the dimension of a quantity q.

We have now a way to implement the gradient method. On substituting (6-105) into (6-104), we have

$$\delta a_k = \eta_k \beta_k$$

where the constant of proportionality η_k must have the dimension $[a_k]^2$. The result of the dimensional analysis above tells us that the only quantity in our formulation that has the correct dimension is the inverse of $\alpha_{k,k}$, the kth diagonal matrix element of the curvature matrix. This gives us the result

$$\delta a_k = \frac{1}{\lambda \alpha_{k,k}} \beta_k \qquad (6\text{-}107)$$

where λ is a dimensionless constant of proportionality that must be determined in some other way.

It is possible to find alternate forms of (6-107). The particular choice made above has the advantage that we can incorporate the gradient approximation of (6-107) into a general approach of the parabolic method given by (6-101). For this purpose, we shall define a modified curvature matrix $\boldsymbol{\alpha}'$ whose elements are given by

$$\alpha'_{j,k} = \begin{cases} (1+\lambda)\alpha_{j,j} & \text{for } k = j \\ \alpha_{j,k} & \text{otherwise} \end{cases} \qquad (6\text{-}108)$$

In other words, the diagonal elements of the curvature matrix are increased by a factor $(1+\lambda)$ from the value given by (6-100). When λ is small, $\boldsymbol{\alpha}' = \boldsymbol{\alpha}$ and we have the parabolic approximation together with all its advantage near the minimum. When λ is large, the modified curvature matrix is dominated by the diagonal matrix elements:

$$\alpha'_{j,k} \approx (1+\lambda)\alpha_{j,k}\,\delta_{j,k}$$

where $\delta_{j,k}$ is the Kronecker delta. In this limit, the elements of the covariance matrix may be written as

$$\left(\boldsymbol{\alpha}'^{-1}\right)_{j,k} \approx \frac{1}{(1+\lambda)\alpha_{j,j}} \delta_{j,k}$$

and (6-101) becomes

$$\delta a_k \approx \frac{1}{(1+\lambda)\alpha_{k,k}} \beta_k$$

This is essentially the same result as (6-107) for $\lambda \gg 1$.

The algorithm for carrying out a nonlinear least-squares calculation using the method of Marquardt is outlined in Box 6-4. It starts with an arbitrary set of values $\boldsymbol{a}(\ell)$ for the m parameters and an initial value of λ chosen to be 0.001. The calculations consist of the following steps:

(1) Calculate the value of $\chi^2(\boldsymbol{a})$ with this $\boldsymbol{a}(\ell)$.
(2) Construct $\boldsymbol{\alpha}'$ from $\boldsymbol{\alpha}$ of (6-100) using the value of λ given.
(3) Invert $\boldsymbol{\alpha}'$ and calculate $\delta \boldsymbol{a}(\ell)$ with $\boldsymbol{\alpha}'^{-1}$ in the place of $\boldsymbol{\alpha}^{-1}$ in (6-101).
(5) Calculate $\boldsymbol{a}(\ell+1)$ using (6-97) and then the value of $\chi^2(\boldsymbol{a}(\ell+1))$.
(6) Compare $\chi^2(\boldsymbol{a}(\ell+1))$ with $\chi^2(\boldsymbol{a}(\ell))$:

Box 6-4
Subroutine NLN_FIT(NPT,X,Y,SY,MP,A,SA,AMBDA,CHI2,COV,MD)
Nonlinear least-squares fit with Marquardt method

Argument list:
- NPT: Number of input data points.
- X: Input array of the values of $\{x_i\}$.
- Y: Input array of the values of $\{y_i\}$.
- SY: Input array of the uncertainty in $\{y_i\}$.
- MP: Number of parameters.
- A: Array for the parameters $\{a_i\}$. Input estimates, output final values.
- SA: Array for the uncertainties in a_i.
- AMBDA: Value of λ of (6-107).
- CHI2: Output value of χ^2 for the parameter set.
- COV: Output array for the covariance matrix.
- MD: Dimension of the parameter arrays in the calling program.

Subprograms required:
- FUNC: Returns the value of $f(\boldsymbol{a}; x)$.
- DERIV: Returns dy/da_i for $i = 1$ to m.
- MAT_INV: Matrix inversion of Box 5-3.

1. Check the dimension of the axillary arrays and zero χ^2, β, and covariance.
2. Carry out the following calculations using the input values of $\{x_i, y_i, \sigma_i\}$:
   (a) If $\sigma_i = 0$, let the weighting factor $w_i = 1$; otherwise $w_i = 1/\sigma_i^2$.
   (b) Use FUNC and DERIV to calculate $f(\boldsymbol{a}, x_i)$ and $\partial f/\partial a_k$.
   (c) Compute the matrix elements of β according to (6-100).
   (d) Compute the matrix elements of $\boldsymbol{\alpha}$ according to (6-103).
3. Compute the value of χ^2 for the given $\boldsymbol{a}$ using (6-96).
4. Construct the normalized, modified curvature matrix $\mathcal{A}$ using (6-109).
5. Invert the curvature matrix $\mathcal{A}$.
6. Calculate the next set of the values of $\boldsymbol{a}$:
   (a) Obtain $\boldsymbol{\alpha'}^{-1}$ using (6-110).
   (b) Use (6-106) to get $\delta\boldsymbol{a}$.
   (c) Find a new set of the parameters using (6-97).
7. Evaluate the new $\chi^2(\boldsymbol{a'})$.
8. Compare the new $\chi^2(\boldsymbol{a'})$ with the old $\chi^2(\boldsymbol{a})$.
   (a) If $\chi^2(\boldsymbol{a'}) \geq \chi^2(\boldsymbol{a})$, increase λ by a factor of 10 and go back to step 5.
   (b) If $\chi^2(\boldsymbol{a'}) < \chi^2(\boldsymbol{a})$, decrease the value of λ by a factor of 10 and return:
      (i) The value of $\chi^2(\boldsymbol{a'})$.
      (ii) $\boldsymbol{a'}$ as the new values of the parameters.
      (iii) $\sqrt{\alpha'_{k,k}}$ as the uncertainty of a_k.
      (iv) $\mathcal{A}^{-1}$ together with the diagonal elements of $\boldsymbol{\alpha}$ as the normalization factor to construct the covariance matrix.
9. The calling program determines if further calls to the subroutine are needed by comparing the values of χ^2.

(i) If $\chi^2(a(\ell+1)) \geq \chi^2(a(\ell))$, we are moving away from a minimum. Increase the value of λ by a large factor, such as 10, and go back to step (2).

(ii) If $\chi^2(a(\ell+1)) < \chi^2(a(\ell))$, we are approaching a minimum. Decrease λ by a factor of 10 and go back to step (2) with this new value of λ and increase ℓ by 1.

Convergence is reached when the difference between $\chi^2(a(\ell+1))$ and $\chi^2(a(\ell))$ is smaller than some predetermined value.

Numerical stability considerations. In general, different parameters in a nonlinear function $f(a;x)$ play different roles and, as a result, each may have quite different behavior. This is demonstrated, for example, by the three parameters a_1, a_2, and a_3 of (6-95). For this reason, we expect that the numerical values of the elements of the curvature matrix can be quite different from each other. This is undesirable from the point of view of numerical accuracy. To improve the situation, we can scale the elements of the modified curvature matrix and define a new matrix $\mathcal{A}$ in the following way:

$$\mathcal{A}_{j,k} = \frac{\alpha'_{j,k}}{\sqrt{\alpha_{j,j}\alpha_{k,k}}} \tag{6-109}$$

All the diagonal matrix elements now have the value

$$\mathcal{A}_{j,j} = 1 + \lambda$$

as can be seen by substituting the values of $\alpha'_{j,j}$ from (6-108) into (6-109). Instead of α', the matrix $\mathcal{A}$ is inverted. The relation between the inverse of $\mathcal{A}$ and the covariance matrix is given by

$$\alpha'^{-1}_{j,k} = \frac{\mathcal{A}^{-1}_{j,k}}{\sqrt{\alpha_{j,j}\alpha_{k,k}}} \tag{6-110}$$

In this approach, the diagonal elements of α must be saved before carrying out the transformation of (6-109). They may be regarded as the normalization factors that improve the numerical accuracy in the calculation. No additional storage space is required here, since the matrix α is saved anyway to allow variations on the value of λ as outlined earlier.

As an example, let us consider a hypothetical experiment measuring the radioactivity of an unknown sample as a function of the energy of the radiation emitted. The 50 data points are plotted in Fig. 6-14 as dots, together with the error bar of each point to indicate the uncertainties in the measured number of counts (vertical axis) in each energy interval (horizontal axis). There are two components in the radiation detected, an exponential background coming from a variety of sources of no direct interest to the decay, and the decay of a particular state with a Lorentzian distribution. The counting rate $y(x)$ as a function of energy x takes on the form

$$y(x) = A_1 e^{-\gamma x} + \frac{A_2}{\pi} \frac{\frac{1}{2}\Gamma}{(x-\mu)^2 + (\frac{1}{2}\Gamma)^2} \qquad (6\text{-}111)$$

where A_1 is the strength of the background and γ its decay constant. The Lorentzian distribution is centered at μ with a width Γ and strength A_2. All five quantities, A_1, A_2, γ, Γ, and μ, are treated as unknowns.

We shall now carry out a least-squares fit to find the values of these five quantities from the 50 measurements. For the fitting function, we shall adopt the form

$$f(a_1, a_2, a_3, a_4, a_5; x) = a_1 e^{-a_2 x} + \frac{a_3}{(x-a_4)^2 + a_5}$$

where we have, for mathematical convenience, defined

$$a_3 = A_2 \frac{\Gamma}{2\pi} \qquad a_5 = \left(\tfrac{1}{2}\Gamma\right)^2$$

The other three parameters, $a_1 = A_1$, $a_2 = \gamma$, and $a_4 = \mu$, have a direct relation with their counterparts in (6-111). The change of notation here makes it easier to compare with the rest of the discussion in this section. The five partial derivatives of $f(a; x)$ with respect to the parameters $a_1, a_2, \ldots, a_5$ are

$$\frac{\partial f}{\partial a_1} = e^{-a_2 x} \qquad \frac{\partial f}{\partial a_2} = -a_1 x e^{-a_2 x} \qquad \frac{\partial f}{\partial a_3} = \frac{1}{(x-a_4)^2 + a_5}$$

$$\frac{\partial f}{\partial a_4} = \frac{2 a_3 (x-a_4)}{\{(x-a_4)^2 + a_5\}^2} \qquad \frac{\partial f}{\partial a_5} = \frac{-a_3}{\{(x-a_4)^2 + a_5\}^2}$$

These are required in calculating the elements of $\boldsymbol{\beta}$ and $\boldsymbol{\alpha}$ defined in (6-100) and (6-103).

As initial estimates of the parameters, we can use the value of y_i at $x_i \approx 0$ to give $a_1 \approx 10$ and the rate of the exponential decay to give $a_2 \approx 0.5$. From the location of the Lorentzian peak, we obtain $a_4 = 3$, the height $a_3 \approx 5$, and the width $a_5 = 1$. Using these values together with the recommended value of $\lambda = 0.001$ as the starting point, the calculation converges in four iterations to an accuracy of

$$\frac{\Delta \chi^2}{\chi^2} \leq 0.01$$

where $\Delta \chi^2$ is the absolute value of the difference between the values of χ^2 in two successive iterations. The fitted results are plotted as a smooth curve together with the 50 input data points in Fig. 6-14.

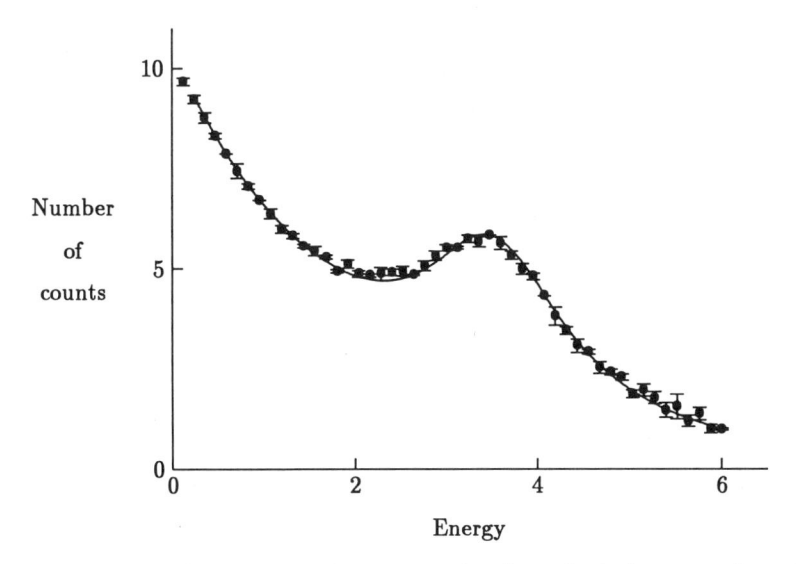

[Figure showing number of counts vs Energy, data with fitted curve showing exponential decay plus peak around energy 3-4]

Fig. 6-14 Nonlinear least-squares fit to a set of 50 hypothetical measured results of a radioactive decay experiment. It is assumed that the counts recorded come from an exponential background and a Lorentzian peak, as given by (6-111). The best-fit parameters are $a_1 = 10.1$, $a_2 = 0.51$, $a_3 = 3.7$, $a_4 = 3.5$, and $a_5 = 0.89$. The χ^2 value of the fit is 67.6.

Uncertainties in the parameters. In the case of a linear least-squares calculation, we have a direct relation between $y(x)$ and the parameters $a_1, a_2, \ldots, a_m$. As a result, there is a clear definition of the uncertainty in any of the parameters a_k in terms of the uncertainties of the input quantities, such as that given by (6-24). For a nonlinear calculation, the final values are obtained as a result of an iterative search, and there is no longer a simple relation between the uncertainties in the input quantities and the final values of the parameters found. The definition of the uncertainties in the parameters found must, therefore, be assigned somewhat arbitrarily. A reasonable choice is to equate the square of the uncertainty in a_k with the kth diagonal element of the inverse of the curvature matrix:

$$\sigma_{a_k} = \left(\alpha^{-1}\right)_{k,k}$$

It can be shown that, with this definition, the value of χ^2 increases approximately by unity if a_k is increased by σ_{a_k}. That is,

$$\chi^2(a_k \pm \sigma_{a_k}) \approx \chi^2(a_k) + 1$$

The proof of this is given in Bevington (*Data Reduction and Error Analysis for the Physical Sciences*, McGraw-Hill, New York, 1969) and we shall not repeat it here.

In carrying out a nonlinear least-squares fit, we need both the functional form of $y(x)$ and the partial derivative of $y(x)$ with respect to each parameter. In principle, we can find the partial derivatives by numerical differentiation. To do this, it is necessary to calculate the value of $y(x)$ at each input point $x = x_i$ for two nearby values of all the parameters. In general, this can be tedious and it is difficult to achieve good numerical accuracy. For this reason, most subroutines for such calculations require, as input, the analytical forms for the partial derivatives. On the other hand, nonlinear least-squares calculations are not restricted to analytical functions. For example, it is possible to express a Hamiltonian in terms of a number of parameters and adjust the values of the parameters so that the eigenvalues give a good fit to the measured energy level spectrum of a quantum mechanical system. In this case, the mathematical operations to find eigenvalues involve matrix diagonalization. As a result, it is not possible to write down in a simple way the functional dependence of the eigenvalues on the parameters. In such cases, numerical differentiation becomes the only way. Such calculations have been carried out in practice; however, they are long and tedious.

Problems

6-1 Calculate the mean μ, medium $\mu_{1/2}$, and most probable value $\mu_{\max}$ for the distribution of (6-5). Compare the results with those obtained for the normal distribution of (6-9).

6-2 Show that the mean and variance of the binomial distribution of (6-12) are

$$\mu = Np \qquad \sigma^2 = Np(1-p)$$

6-3 Use a random number generator with an even distribution in the range $[-1, +1]$ to generate $n = 6$ values and store the sum as x. Collect 1000 such sums and plot their distribution. Compare the results with a normal distribution of the same mean and variance as the x collected. Calculate the χ^2-value. Repeat the calculations with $n = 50$.

6-4 Show that, for a normal distribution given by (6-9), the central moments μ_r of (6-10) are given by

$$\mu_r = \frac{1}{\sigma\sqrt{2\pi}} \int_{-\infty}^{+\infty} (x-\mu)^r e^{-(x-\mu)^2/2\sigma^2}\, dx = \begin{cases} 0 & \text{for } r \text{ odd} \\ \left(\frac{\sigma^2}{2}\right)^m \frac{(2m)!}{m!} & \text{for } r = 2m \end{cases}$$

where $m = 1, 2, \ldots$, is an integer. That is, all the odd central moments vanish and the even ones are related to the variance σ^2.

6-5 Find the relations between the uncertainties in x and y for the following functions: (a) $y = ax^b$, (b) $y = ae^{\pm x}$, (c) $y = a\ln(bx)$, and (d) $y = a/\sin x$.

6-6 Use the propagation of errors given by (6-24) to show that the uncertainty in the average of a quantity decreases proportional to the inverse of the square root of the number of independent measurements taken.

6-7 Obtain the mean and variance of a χ^2-distribution of ν degrees of freedom given by (6-62).

6-8 A good way to test a computer program for linear least-squares calculations is to use a set of inputs produced with the relation

$$y_i = A + Bx_i + \delta y_i$$

For the parameters A and B, adopt the values of 1.0 and 5.0, respectively. Use a random number generator to produce a set of 25 values of x_i in the range $[0, 1]$ and another 25 values for δy_i in the range $[-0.1, +0.1]$. Calculate the value of y_i from x_i and δy_i using the relation given. To each y_i, assign an uncertainty σ_i using a random number generator in the range $[0, 0.1]$. Fit this set of 25 values of (x_i, y_i, σ_i) with the straight-line relation

$$y(x) = a_1 + a_2 x$$

Compare the values of a_1 and a_2 obtained with those of A and B used to generate the input.

6-9 Repeat Problem 6-8 for 100 different sets of 25 input values generated with the same pair of values of A and B, but different random numbers for x_i, δy_i, and σ_i. Plot a histogram of the distribution of the values of a_1 and a_2. Are the results consistent with the uncertainties of a_1 and a_2 produced by the program?

6-10 Repeat Problem 6-8 except fit the input with the function

$$y(x) = a_1 + a_2 x + a_3 x^2$$

Use the tests described in §6-5 and §6-6 to see if the additional quadratic term should be included.

6-11 The following ten sets of values are used as the input data used to produce Fig. 6-11:

$(\cos\theta_i, W(\theta_i), \sigma_i) =$ (0.000, 1148, 4) (-0.174, 1152, 4) (-0.342, 1166, 4)
(-0.500, 1184, 4) (-0.643, 1208, 4) (-0.766, 1234, 4)
(-0.866, 1274, 4) (-0.940, 1297, 4) (-0.985, 1316, 4)
(-1.000, 1324, 4)

where σ_i is the estimated uncertainty in $W(\theta_i)$. Fit the values of $W(\theta)$ with a power series in $\cos^2\theta$

$$W(\theta) = b_0 + b_2 \cos^2\theta + b_4 \cos^4\theta$$

What are the differences from a Legendre polynomial fit of the same order?

6-12 Check the significance of the $\cos^4\theta$ term in Problem 6-11. What about including a $\cos^6\theta$ term?

6-13 The following 20 sets of values

x_i	y_i	σ_i	x_i	y_i	σ_i	x_i	y_i	σ_i	x_i	y_i	σ_i
0.46	0.19	0.05	0.69	0.27	0.06	0.71	0.28	0.05	1.04	0.62	0.01
1.11	0.68	0.05	1.14	0.70	0.07	1.14	0.74	0.08	1.20	0.81	0.09
1.31	0.93	0.10	2.03	2.49	0.03	2.14	2.73	0.04	2.52	3.57	0.01
3.24	3.90	0.07	3.46	3.55	0.03	3.81	2.87	0.03	4.06	2.24	0.01
4.93	0.65	0.10	5.11	0.39	0.07	5.26	0.33	0.05	5.38	0.26	0.08

follow roughly a normal distribution

$$y(x) = a_1 \exp\left\{-\frac{1}{2}\left(\frac{x-a_2}{a_3}\right)^2\right\}$$

with σ_i as the uncertainty of y_i. Use a nonlinear least-squares method to find the values of a_1, a_2, and a_3.

6-14 As an alternative to using a nonlinear least-squares fit, the problem given in Problem 6-13 can also be solved by defining $z = \ln y$ and using a linear least-squares fit for

$$z(x) = b_1 + b_2 x + b_3 x^2$$

Compare the results obtained with the two methods.

Chapter 7
Monte Carlo Calculations

Monte Carlo techniques are useful in solving a variety of problems in physics. Earlier, we made use of such an approach in §2-5 to carry out an integral by evaluating the integrand at a randomly selected set of points in the interval. If the sampling is carried out at a sufficient number of points, the sum of values of the integrand collected at these points provides a good way to approximate the integral. Many physical problems, from simulation of experimental situations to complicated theoretical computations, can take advantage of such a sampling approach. We shall begin this chapter with a discussion on generating random numbers, the starting point of all Monte Carlo calculations.

§7-1 Generation of random numbers

What is a random number? A simple answer is that it is a number chosen at random. More formally, we can say that a random number is a collection of random digits, drawn from $0, 1, 2, \ldots, 9$, each with the probability of 1/10. Both definitions depend on the concept of randomness. Intuitively, all of us have a feeling of what randomness means; however, it is not always easy to put it into words. From a practical point of view, it may be more useful to use the definition, given by Chaitin in the May 1975 issue of *Scientific American* ("Randomness and Mathematical Proof," page 47), which states that "a series of numbers is random if the smallest algorithm capable of specifying it to a computer has about the same number of bits of information as the series itself." There are several other ways to define randomness, but few of them are as able to capture the essence of the term without invoking a large amount of mathematics.

For computational purposes, we seldom deal with truly random numbers. In the first place the random numbers we need are usually generated by an algorithm. In general, we wish to use a simple algorithm so as to be able to produce the numbers quickly. As a result, the numbers obtained cannot be random according to the definition of Chaitin. Furthermore, we may also wish our random numbers to be reproducible on demand, as this is essential, for example, in debugging a computer program that makes use of them. Technically, these are only *pseudo-random* numbers. Pseudo-random numbers that pass a set of randomness tests, such as typical statistical hypotheses of independence, uniformity, and goodness of fit, are often quite adequate for most practical applications that call for random numbers. Following general practice, we shall not make any distinction between pseudo-random numbers and *true* random numbers. It is, however, important to recognize that there are fundamental differences between them. More importantly, pseudo-random numbers that are adequate for one application may not be good enough for another. For this reason, extensive tests should be carried out before using any pseudo-random numbers for a new type of calculation. We shall return to the question of tests for randomness later in this section.

A good random number generator must be fast and simple to use. In addition, it should have the desired statistical properties. For example, because of the limited word length, the total number of different random numbers that can be generated on a computer is finite. In fact, for most generators, the actual range of numbers produced is much smaller than the largest integer that can be stored in a machine word. In addition to the range, the sequence of numbers produced will eventually repeat itself, as we are using an algorithm consisting of a limited amount of instructions. The number of different results before a generator repeats itself is known as the period. Clearly, it is desirable to have as long a period as possible. Another important requirement is that the numbers generated should not be correlated among themselves. There are many different types of correlations that can exit among a sequence of numbers, and it may not be possible to find a generator that can satisfy all the requirements. Fortunately, many applications are sensitive to only a few of these properties and, as a result, generators that are, in principle, imperfect are often useful in practice.

Uniformly distributed random integers. In a Monte Carlo calculation we usually have to take a random sample of some quantity with a definite distribution. As a result, the random numbers we use must have the same distribution. For example, in a one-dimensional Monte Carlo integration, we wish to sample the variable of integration with uniform probability within the integration limits. For this purpose, we need random numbers with a uniform distribution. On the other hand, if we wish to simulate neutrons scattering off a nuclear target, the random numbers must have the angular distribution of neutrons scattering off the same nucleus. We shall see later that it is possible to produce random numbers of any desired distribution starting with a set having a uniform distribution in the range $[0, 1]$. Such random numbers, or uniform deviates in the more formal language, are in turn generated from random integers distributed uniformly in the interval $[0, M]$. Here M is the largest random integer that can be produced by the algorithm. For this reason, we shall start our discussion with methods to generate random integers.

A simple way to obtain a random number is, for example, to read the timer, or system clock, on the computer we are using. Let us assume, for the convenience of discussion here, that the system clock is measured in units of microseconds (μs) and it is set to zero at some arbitrary time, such as 00:00:00 Greenwich Mean Time (GMT) on January 1, 1970, used on many operating systems. A simple function can be written to extract the time t as integers in μs everytime it is called. If t_1 and t_2 are the two values obtained in two successive calls separated by a few seconds, the most significant digits of the two integers will be the same, as they represent the time in units much larger than a second. On the other hand, the three least significant digits, for example, are likely to be random. This is especially true if the timing function is called by some sort of manual control, such as pushing a particular key on the computer keyboard. The source of randomness in this case comes from the fact that it is not possible for us to control our reaction time to be more accurate than the order of 0.1 s. If we obtain a sequence of such values, $t_1, t_2, \ldots$, then

$$r_i = t_i \bmod 1000$$

for $i = 1, 2, \ldots$, is a sequence of random integers, as the mod, or *modulus*, 1000 operation eliminates all the leading significant figures except the last three.

There are several problems in generating random integers using the system clock. In the first place, there can only be 1000 different random numbers in the series using the above method. This comes from the simple fact that our numbers can only be integers in the interval 0 to 1000. In the second place, we cannot use an automated way to read the clock, as any such method will likely produce a series of highly correlated numbers. Finally, we cannot repeat the sequence on those occasions when we need to reproduce the same set of random integers. For these reasons, the only practical use of the system clock as a random number generator is to set the "seed" for some other random number generator, as we shall see later.

Another interesting way to produce random integers is the middle-square method of von Neumann. If we square an integer consisting of several digits, both the most significant digits and the least significant digits are predictable. For example, the square of an integer in the form "13..." has "1" as the leading digit. Similarly, the square of an integer with the last digit "1" must also end with the digit "1." However, it is more difficult to predict the digits in the middle. For example, if we square an integer of three digits,

$$(123)^2 = 15129$$

the result is a five-digit integer. By chopping off the leading and trailing digit, we obtain the number 512. If we square this number and retain only the third, fourth, and fifth digits, we obtain the number 214. In this way, a sequence of random integers can be obtained, each constructed from the square of the previous one with only the middle part of the digits retained. Although this method has been in use for a long time, it is no longer a preferred way to generate random integers these days. Tests have shown that there are many ways such a sequence can develop into a bad direction. For example, if for any reason a member of the sequence becomes zero, all the remaining members of the sequence will obviously be zero.

The linear congruence method. The most popular way to generate random numbers with a uniform distribution is based on the linear congruence method. In this approach, a random integer X_{n+1} is produced from another random integer X_n through the operation

$$X_{n+1} = (aX_n + c) \bmod m \qquad (7\text{-}1)$$

where the modulus, m, is a positive integer. The multiplier a and increment c are also positive integers but their values must be less than m. To start off the sequence of random integers $X_0, X_1, X_2, \ldots$, we need to input a random integer X_0, generally referred to as the *seed* of the random sequence.

The spirit of (7-1) is very similar to that behind the system-clock and middle-square methods discussed above. Instead of taking the square, the seed is multiplied by an integer a, and to prevent a bad sequence from developing, an increment c is added to the product. Similar to the system-clock approach, only the least significant part of the sum (less than m) is retained through the modulus operation.

The quality of random numbers generated by the linear congruence method depends critically on the values of m, a, and c selected. Because of the modulus operation, all the random integers produced by this method must be in the range $[0, m]$. For this reason, we want m to be as large as possible. On the other hand, m^2 cannot be larger than the longest integer the computer can store in its memory. It is also clear that the choices of these three quantities are related with each other. For example, in order to have a long period, it is known that

(1) c must be relative prime to m, meaning that a and m have no common factors between them other than 1.
(2) $(a-1)$ is a multiple of p, if p is a prime factor of m, and
(3) $(a-1)$ is a multiple of 4, if m is a multiple of 4.

For example, a possible set of values for computers with word length of 32 bits is

$$m = 714025 \qquad\qquad a = 1366 \qquad\qquad c = 150889$$

The maximum number of different random numbers that can be generated is 714205 in this case. Other possible combinations are given in Press and others (*Numerical Recipes*, Cambridge University Press, Cambridge, 1986) and detailed discussions of the various considerations that must go into the choices are given in Knuth (*The Art of Computer Programming*, vol. II, 2nd edition, Addison-Wesley, Menlo Park, California, 1981).

Conversion to random numbers. If instead of random integers we wish to have random (floating) numbers, we can convert an integer X_n in the interval $[0, m]$ to a floating number R_n in a given interval $[a, b]$. For most generators, the interval is taken to be $[0, 1]$. Since the largest integer produced by the linear congruence method of (7-1) is m, any random integer X_n may be converted to a random number in the interval $[0, 1]$ by dividing X_n with m. To carry out this procedure on a computer, we must first change both X_n and m into floating numbers and then take the quotient

$$R_n = X_n/m$$

Note that, since X_n in (7-1) is needed to generate the next random integer, it must be preserved.

Because of the wide range of applications of random numbers, computer operating systems and high-level languages are usually equipped with random number generators. The most common one found on 32-bit machines is based on a random integer generator that uses a slight variant of (7-1),

$$X_{n+1} = \frac{aX_n + c}{d} \bmod m \qquad (7\text{-}2)$$

with

$$m = 32768 \qquad a = 1103515245 \qquad c = 12345 \qquad d = 65536$$

The range of random integers generated by this method is $0 \leq X_n < 2^{15}$. There is, however, a small problem if we wish to implement (7-2) using the FORTRAN language. Since the integer a is larger than 2^{30}, there is a high probability that an overflow will be produced on a 32-bit word length computer when it is multiplied by a random integer X_n. One way to circumvent the problem is to code the method in a different language in such a way that it may be called by a FORTRAN program.

Improving randomness by shuffling. One weakness of the linear congruence method is that every random integer is generated from the one produced prior to it. As a result, there is a strong tendency for the random numbers to be sequentially correlated with each other. One consequence of such a correlation may be illustrated by the following example. If we use groups of three random numbers $(R_{3n}, R_{3n+1}, R_{3n+2})$ as the coordinates (x, y, z) of points inside a cube of length 1 (in arbitrary units), there is a possibility that the points will fall mainly on a small number of surfaces inside the cube rather than filling up the volume evenly. A simplified illustration of this point is provided by Problem 7-1.

In spite of these shortcomings, the linear congruence method is widely used as it is an efficient way to generate random numbers. Several remedies have been introduced to overcome its weakness. A simple way to alleviate the correlation tendency is to shuffle the sequence. That is, a large number of random numbers is

> **Box 7-1 Function** RSHFL(ISEED)
> **Improved random number generator by shuffling**
>
> Argument list:
> ISEED: Random number seed on input. Returns the seed for the next random number.
>
> Initialization:
> (a) Select a stack of length $L = 97$.
> (b) Store the stack $S(j)$ with random numbers produced by a linear congruence generator.
> (c) Mark the stack as initialized.
>
> 1. Make sure that the stack is initialized.
> 2. Generate a new random number X.
> 3. Convert X to an integer j in the range $[1, L]$.
> 4. Return $S(j)$ as the random number required.
> 5. Replace $S(j)$ by X.

generated by the linear congruence method and stored in an array. Let the size of the array be L, preferably a prime number. Anytime we need a random number, we take it from the array rather than having it generated directly. However, instead of taking them in sequence, a random integer j in the range $[1, L]$ is generated. The required random number is then taken from the jth location of the array. To prevent the same random number from being used again, the jth location in the array is replaced by a new random number produced by the generator. This is implemented in the algorithm of Box 7-1. For simplicity, we can use the linear congruence generator supplied by the operating system to fill the array, as the most serious problem of correlation is now corrected by the shuffling.

Alternate procedures. Another way to improve the linear congruence method is to construct a new formula that makes use of more than one previous member of the sequence. Instead of X_n alone, as in (7-1), we can use two previous members on the right side of the equation. The simplest algorithm takes on the following form:

$$X_{n+1} = (X_n + X_{n-1}) \bmod m$$

This is, however, a bad choice, since two adjacent members have a tendency to be correlated, as mentioned earlier. To avoid the problem, we can take two previous members separated by a longer distance,

$$X_{n+1} = (X_n + X_{n-k}) \bmod m$$

with k being a number larger than 15 or so. For example, the following choice is given in Knuth:

$$X_{n+1} = (X_{n-24} + X_{n-55}) \bmod m \qquad \text{for} \quad n \geq 55 \qquad (7\text{-}3)$$

This is known as an additive generator. A variant of this that is often used in practice is the subtractive generator,

$$X_{n+1} = (X_{n-55} - X_{n-24}) \bmod m \tag{7-4}$$

Among other advantages, this method can be made completely *portable*. That is, the computer program may be written in a high-level language, such as FORTRAN or C, and may be made to run on any machine that has a compiler for the language. The algorithm is summarized in Box 7-2.

Box 7-2 Function RSUB(ISEED)

Subtraction method of random number generation

Argument list:
  ISEED: Random number seed on input. Returns the seed for the next random number.

Initialization:
  (a) Define three constants $m = 10^9$, $i_s = 21$, and $i_r = 30$.
  (b) Set up a stack S of length $L = 55$ for integers.
  (c) Store the absolute value of the seed in $S(L)$.
  (d) Let $k = 1$ and $j =$ ISEED.
  (e) For $\ell = 1$ to $(L-1)$, carry out the following steps to store the stack:
      (i) Let $\ell_x = i_s * \ell \bmod L$.
      (ii) Store the value k in $S(\ell_x)$.
      (iii) Replace k by $(j - k)$. If $k < 0$, increase j by 10^9.
      (iv) Let $j = S(\ell_x)$
  (f) Randomize the stack by carrying out the following steps three times:
      (i) For $\ell = 1$ to 24, replace $S(\ell)$ by $S(\ell) - S(\ell + 31)$.
      (ii) For $\ell = 25$ to L, replace $S(\ell)$ by $S(\ell) - S(\ell - 24)$.
      (iii) If $S(\ell)$ becomes negative, increase $S(\ell)$ by 10^9.
  (g) Set stack counter $j_s = 0$.
  (h) Mark the stack as initialized.

1. Go back to the initialization step if the input seed is negative. This allows initializing the stack anytime by the calling program.
2. Increase the stack counter j_s by 1.
3. If $j_s > L$,
   (a) Replace the stack by repeating the steps in (f) of the initialization step.
   (b) Set stack counter $j_s = 1$.
4. Return (float) $S(j_s) \times 10^{-9}$ as the uniform random number in $[0, 1]$.

It is also possible to design methods to generate random numbers using special features of the central processing unit of the computer. Two such methods are given by Marsaglia, Narasimhan, and Zaman (*Comp. Phys. Comm.* **60** [1990], 345) and by Chiu and Guu (*Comp. Phys. Comm.* **47** [1987], 129). In general, when carrying out Monte Carlo calculations, it is useful to have more than one random number generator available. As we shall see below, it is not possible to have a complete set of tests for randomness. As a result, there is always a possibility that our Monte Carlo results are the artifact of the random numbers used. For this reason, it is a good practice to run the calculations using random numbers obtained from two or more different methods.

Tests of randomness. How do we know that a given set of numbers is random? This is not an easy question to answer, especially for the *pseudo*-random numbers we are dealing with here. We know that it does not fulfill all the fundamental criteria for randomness. For most practical needs, however, the requirement for randomness is not as stringent as the definitions imply. For example, in a one-dimensional Monte Carlo integration, we are essentially satisfied with a set of random numbers that distribute evenly in the integration interval. On the other hand, if we use the same set of random numbers for a two-dimensional Monte Carlo integration, we have the additional concern of whether a pair of adjacent random numbers is correlated. This comes from the following reason. A point in a two-dimensional space may be specified by two numbers, for example, the x- and y-coordinates with respect to some fixed set of axes. To select a random point in this space, we need two random numbers (x_i, y_i). If there are correlations between a pair of random numbers, the distribution of points sampled by a set of the values of (x_i, y_i) may not fill the area evenly.

In general, there are many pitfalls in constructing a random number generator, as well as a number of ways to make bad use of a good generator. One such case of the latter category is given in Problem 7-3 as illustration. For this reason, testing of random numbers is an important part of any Monte Carlo calculation. All the tests of randomness are based on statistical measures. Since there is no single measure that encompasses all the desirable features, it is difficult, if not impossible, to design a comprehensive test for randomness. A large number of measures are available in the literature, and a good summary of some of the more important ones is given in Knuth. Here we shall only describe three of them: distribution, correlation, and run tests.

Frequency test. Up to now, we have been dealing with random numbers that are supposed to be evenly distributed in the interval $[0,1]$. The first obvious test is to see whether this is, indeed, true. As we did in §2-5, we can divide the range of possible values into a number of equal size bins and count the number of random numbers that fall into each bin. If the random numbers are distributed evenly in the interval, the numbers in each bin should be equal to each other within statistical fluctuations.

As a practical procedure, let the total number of bins be N_{bin}. For uniform random numbers in the range $[0,1]$, it is convenient to choose the width of each bin as

$$w = 1/N_{\text{bin}}$$

For a random number X_j, we can find the bin number k to which it belongs by multiplying X_j with N_{bin} and taking the integer part of the product,

$$k = \text{INT}(X_j \times N_{\text{bin}}) + 1 \tag{7-5}$$

For convenience, we have added 1 to the product so that the bin number starts from 1. The function $\text{INT}(x)$ takes the integer part of a floating number x and returns an integer that is less or equal to x, as we saw earlier in Table 1-1. There is a small probability the k will take on the value $(N_{\text{bin}} + 1)$ in (7-5). This happens under the extremely unlikely occasion where $X_j = 1$ due to round-off errors. However, a good computer program must allow for such a possibility. The easiest way to achieve this is to allocate an extra bin to take care of the possibility of $k = (N_{\text{bin}} + 1)$.

If the total number of random numbers tested is n, the number expected in each bin is $(n \times w)$ on the average. Since the random numbers are assumed to be uncorrelated, the distribution of the possible numbers in each bin follows a Poisson distribution, as we saw in §6-1. The standard deviation of such a distribution is $\sqrt{nw}$, and this provides us with a measure of the average fluctuation in the number we can expect in each bin. A χ^2-test may be applied to the overall distribution of the number of random numbers in all the bins. If M_j is the actual number found in bin j, the χ^2 for the entire distribution is then

$$\chi^2 = \sum_{j=1}^{N_{\text{bin}}} \frac{(M_j - nw)^2}{nw} \tag{7-6}$$

The number of degrees of freedom is one less than the number of bins,

$$\nu = N_{\text{bin}} - 1$$

since we have a constraint that the sum of the occupancies in all the bins must equal to n. Obviously, if the χ^2 has a large value, the sequence of random numbers is not evenly distributed in the interval. On the other hand, a perfectly even distribution, one with $\chi^2 = 0$, is also an indication of departure from randomness. This comes from the fact that fluctuation is an integral part of randomness. If the sequence is truly random, we expect that the departure for any one bin from the average value of nw is given by a normal distribution, as we saw in §6-2. For our present case, the normal distribution centers around zero with a variance of nw. This gives us an expected value for the reduced χ^2 to be $\chi^2_\nu \sim 1$. As a result, a small χ^2-value $[\chi^2 \ll \nu$ or $Q(\chi^2|\nu) \geq 0.99]$ implies a lack of fluctuation and, hence, a sign of departure from randomness as well.

The results of applying a frequency test to random numbers generated from the linear congruence and subtractive methods are given in Table 7-1. To have an idea of the distribution on a fine scale, the two sequences were analyzed using 50 bins ($N_{\text{bin}} = 50$). For good statistics, the total number of random numbers used

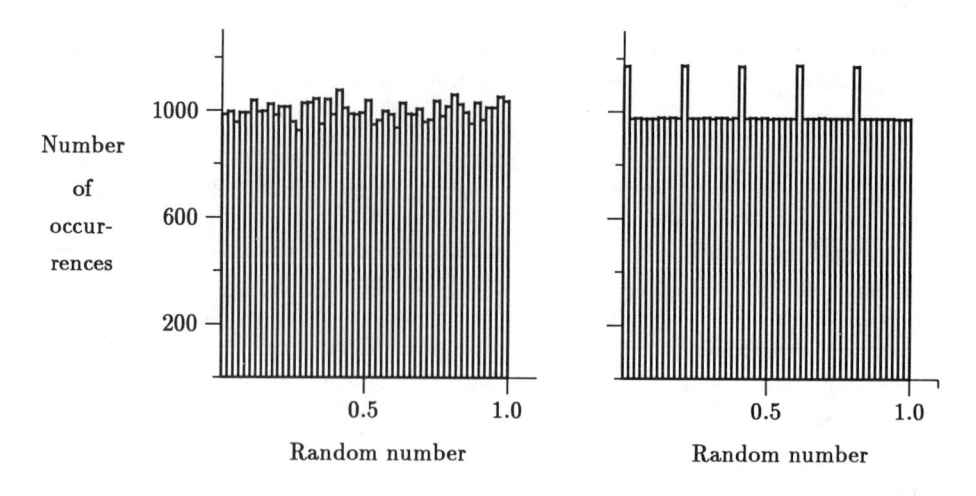

Fig. 7-1 Histograms showing the distribution the number of random numbers of a given size as a function of the size. A total of 50,000 random numbers is involved in each case. On the left, they are produced using the subtraction method and, on the right, the linear congruence method with a "bad" set of parameters.

in each case is $n = 50,000$. This gives us a value of $nw = 1000$. The χ^2-values for both cases are around 50, showing that both sequences are evenly distributed.

For comparison, we include also the result obtained using a "bad" linear congruence generator:

$$X_{j+1} = (137X_j + 187) \bmod 256$$

The χ^2-value in this case is 174, giving a probability $Q(\nu = 49|\chi^2 = 174) = 0.000$ for being an evenly distributed set of numbers. As expected, shuffling of the sequence of random number does not change the χ^2-values for the results of both the good and the bad linear congruence generators. This comes from the simple fact that the frequency test is not concerned with the order in which the random numbers are generated. The algorithm for the test is summarized in Box 7-3.

An interesting feature of the failure of the bad linear congruence generator is shown by comparing the two histograms in Fig. 7-1. On the left we find the histogram for 50,000 random numbers generated with a good generator. Since $nw = 1000$ in each of the 50 bins, the fluctuations are expected to be $\sqrt{1000} = 31.6$ on the average. On the whole, this is roughly what is observed in the plot. On the other hand, the histogram for random numbers produced by the bad generator shows regular peaks that cannot be statistical in nature, a conclusion one can also draw from the large χ^2-value produced.

> **Box 7-3** Subroutine FRQNCY(ISEED,N_TTL,R_GEN)
> **Frequency test of random numbers**
> χ^2-test of even distribution in the interval $[0, 1]$
>
> Argument list:
>   ISEED: Random number seed.
>   N_TTL: Total number of random numbers to be tested.
>   R_GEN: Name of the external function that generates the random numbers.
> Initialization:
>   (a) Set up an array for 50 histogram bins.
>   (b) Zero the bins.
> 1. Carry out the following steps for all the random numbers to be tested:
>   (a) Generate a random number X.
>   (b) Find the bin number X belongs to.
>   (c) Increase the count in that bin by 1.
> 2. Carry out the χ^2-test for an even distribution.
>   (a) Obtain the number of degrees of freedom.
>   (b) Calculate the contribution of each bin to the χ^2 using (7-6).
> 3. Output the χ^2-value, the number of degrees of freedom, and the count in each bin.

Serial correlation test. There are many ways that members of a sequence of random numbers are correlated with each other. The easiest one to test for is linear correlation. The primary aim here is to find out whether there is a tendency for each random number to be followed by another one of a particular type. For example, it will be quite unacceptable if, in a collection, a large random number is always followed by a small one (or a large one). For this purpose, we can define a correlation coefficient between a sequence of n random numbers $X_1, X_2, \ldots, X_n$ as

$$C = \frac{n\left(\sum_{i=1}^n X_{i-1}X_i\right) - \left(\sum_{i=1}^n X_i\right)^2}{n\sum_{i=1}^n X_i^2 - \left(\sum_{i=1}^n X_i\right)^2} \tag{7-7}$$

For simplicity, we can take the X_0 in the first term of the numerator to be equal to X_n.

The coefficient C defined above is slightly different from similar ones, such as that given by (6-52). There, the correlation is between two different sets of variables $\{x_i\}$ and $\{y_i\}$. In contrast, we are taking here a sum over products of the type $X_{i-1}X_i, X_iX_{i+1}, \ldots$, instead of $x_iy_i, x_{i+1}y_{i+1}, \ldots$. As a result, the expected values of the two types of coefficients are different from each other. In the present case, the value of C for an uncorrelated sequence is expected to be in the range $[\mu_n - 2\sigma_n, \mu_n + 2\sigma_n]$ for 95% of the time, with

$$\mu_n = -\frac{1}{n-1} \qquad \sigma_n = \frac{1}{n-1}\sqrt{\frac{n+1}{n(n-3)}}$$

In contrast, ordinary correlation coefficients, such as the one in (6-52), are bound in the range $[-1,+1]$.

Table 7-1 Tests of random number generators.

Test	Frequency χ^2	Serial correlation C	Run-up χ^2
Expected value	$\nu = 49$	$[-0.028, +0.028]$	$\nu = 6$
Linear congruence method			
Bad	174	-0.022	682
After shuffling	171	0.001	842
Good	41	-0.023	4.7
After shuffling	44	-0.003	2.8
Subtraction method	56	-0.003	4.9

The results of a correlation test for the three different random number generators, a bad linear congruence generator, a good linear congruence generator, and a subtraction generator — the same three as those used for the frequency test earlier — are also listed in Table 7-1. In each case, a set of 5000 random numbers is used. The fact that the linear congruence method has a tendency to be correlated is reflected by the relatively large absolute values of both the good and the bad generator. Shuffling among the random numbers destroys such short-range correlations, as can be seen from the much smaller absolute values of C obtained for the results. Furthermore, since the main difference between the good and the bad linear congruence generators is in the distribution of the random numbers produced, there is very little distinction between the results of the two methods as far as the correlation test is concerned. The value of C for the subtraction method is -0.003 in the test carried out, essentially a null value for a set of 5000 numbers. In fact, repeated tests with different sets of 5000 numbers, each set produced by a different seed, give values of C of roughly the same magnitude with both positive and negative signs.

Run test. A more sensitive test of the correlation among random numbers than the coefficient of (7-7) is the length of a run of either increasing or decreasing size. Here, the aim is to check whether there is a tendency for consecutive random numbers in a sequence to be decreasing (run-down test) or increasing (run-up test) in value. To simplify the discussion, we shall only consider the latter.

By the length ℓ_{up} of a run, we mean the number of random numbers in a sequence that is increasing in value. A run starts at X_i if $X_i < X_{i-1}$ and the run continues till (but excluding) X_{i+j} if

$$X_i < X_{i+1} < X_{i+2} < \cdots < X_{i+j-1} > X_{i+j}$$

The length of such a run is defined to be

$$\ell_{\rm up} = j$$

For simplicity, we shall ignore the possibility of two consecutive random numbers being equal to each other in any of our discussions. However, in practical applications such a possibility, however remote, cannot be ignored. For this reason we shall make an approximation in an actual computer program to carry out the calculation and consider all the lesser signs ($<$) above to be lesser or equal signs ($\leq$), even though the statistic is defined in terms of the lesser sign alone.

Box 7-4   Subroutine RUN_UP(ISEED,N_TTL,R_GEN)
Run-up test of a sequence of random numbers

Argument list:
   ISEED: Random number seed.
   N_TTL: Total number of random numbers to be tested.
   R_GEN: Name of the external function that generates the random numbers.
Initialization:
   (a) Set up an array $K_{\rm up}(i)$ of length 7 to store the histogram values.
   (b) Zero the array.
   (c) Zero the counter for the total number of runs.
1. Generate a random number and store it as $X_{\rm old}$.
   Initialize the length of the run to $\ell_{\rm up} = 1$.
2. Generate another random number and store it as $X_{\rm new}$.
3. Compare $X_{\rm new}$ with $X_{\rm old}$.
   (a) If $X_{\rm new} > X_{\rm old}$,
      (i) Increase the run length $\ell_{\rm up}$ by 1.
      (ii) Store $X_{\rm old}$ as $X_{\rm new}$.
      (iii) Go back to step 2.
   (b) If $X_{\rm new} \leq X_{\rm old}$, the run ends:
      (i) Increase run counter $N_{\rm ttl}$ by 1.
      (ii) If $\ell_{\rm up} \leq \ell_{\rm max}$, increment the histogram bin $K_{\rm up}(\ell_{\rm up})$ by 1.
      (iii) If $\ell_{\rm up} > \ell_{\rm max}$, increment the histogram bin $K_{\rm up}(\ell_{\rm max})$ by 1.
      (iv) If the total number of runs $N_{\rm ttl}$ is less than the maximum required, go back to step 1 and start a new independent run. Otherwise go to the next step.
4. Calculate the χ^2-value using (7-9) and output the histogram.

An explicit example with random digits is useful here to illustrate the meaning of a run. For the sequence

$$\cdots 9|138|7|69|7|5|3\ |1\cdots$$

we have $\ell_{\rm up} =$ 3 1 2 1 1 1

according to the definition given in the previous paragraph. Unfortunately, it is not easy to analyze the statistical behavior of such a set of quantities, since the

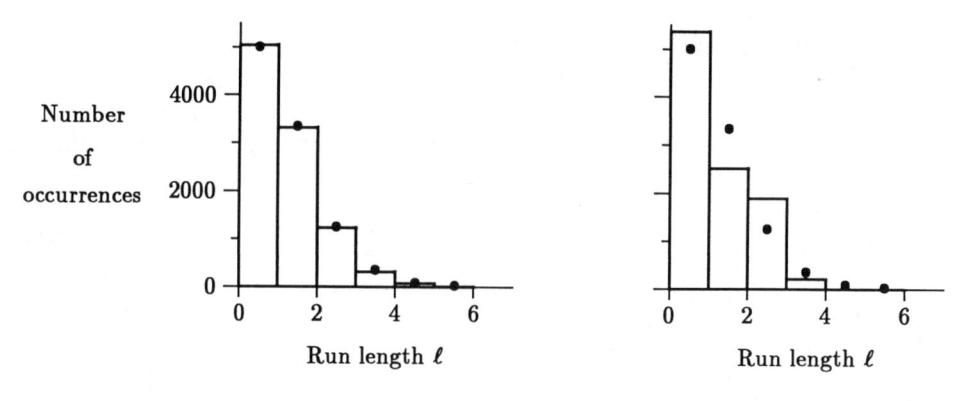

Fig. 7-2 Histograms showing the distributions of the number of runs of length ℓ for two sequences of random numbers. A total of 10,000 runs of increasing size are collected in each case. The results on the left are taken from a random number generator using the subtraction method and those on the right are obtained by a linear congruence method using a "bad" set of parameters. The dots are the expected values of (7-8). The uncertainties expected from a Poisson distribution are too small to be visible in the plot.

end of one run appears as the head of another run. Because of such a correlation between the runs, it is rather difficult to calculate the expected value of the distribution of ℓ_{up}. To simplify the situation, Knuth has suggested that one should disregard the comparison of the last random number of a run with the next one. The resulting runs are now uncorrelated with each other. For the random digit example above, this corresponds to the following method of counting:

$$\ell_{up} = \begin{matrix} \cdots 9|138|\,7\,|69|\,7\,|5|\,3\;\;|1\cdots \\ 2\;\;\odot\;\;2\;\odot\;1\;\odot \end{matrix}$$

where the symbol $\odot$ indicates run of length 1 being discarded. As a result, some of the runs have reduced lengths compared with those in the corresponding correlated case above, such as the first run (length 2 instead of 3). The probability of obtaining a run of length ℓ for a completely uncorrelated sequence is given by

$$p_\ell = \frac{1}{\ell!} - \frac{1}{(\ell+1)!} \qquad (7\text{-}8)$$

That is, in a sample of m runs constructed from n random numbers, there are mp_ℓ runs with length ℓ. Since, in general, each run involves more than one random number, $m < n$. The algorithm to make a run-up test is summarized in Box 7-4.

We can apply a χ^2-test to see if the distribution of the length ℓ corresponds to our expectation of a random sequence. If the number of uncorrelated runs of length ℓ in a sample of random numbers is found to be K_ℓ, the square of the difference between K_ℓ and mp_ℓ, defined in the following way

$$\chi^2 = \sum_{\ell=1}^{L} \frac{1}{mp_\ell}(K_\ell - mp_\ell)^2 \qquad (7\text{-}9)$$

follows a χ^2-distribution. This is the same definition for χ^2 as that of (6-41). The weighting factor σ_i^2 for each point is $(mp_\ell)^{-1}$ here, based on the assumption of a Poisson distribution for the number of runs for a given length ℓ. From (7-8), we see that the probability of observing a long sequence decreases rapidly. Since $6! = 720$, there is very little point in extending L, the upper limit of the summation, to beyond 6. On the other hand, since there is no restriction for the upper limit of the length ℓ, the number of degrees of freedom for the χ^2-distribution is $\nu = L$ (rather than $L-1$). A good sequence of random numbers is expected to have a χ^2-value of approximately ν or a reduced χ^2-value of $\chi_\nu^2 \approx 1$. The actual distributions for two sequences of random numbers produced by two different generators are plotted in Fig. 7-2 as illustration.

The results of a run-up test on the three random number generators are also listed in Table 7-1 for $m = 10,000$. The sensitivity of the test itself is shown by the large differences in the χ^2-values obtained. The reason that shuffling does not improve the results of the bad generator in the run-up statistic, as in the case of serial correlation, may be traced to the fact that the run test measures longer-term correlations than the next-neighbor correlations of the serial correlation test. Since we have chosen, for purposes of illustration, a very poor random number generator, shuffling does not improve the results.

Random numbers of a given distribution. In many calculations, we need random numbers with a distribution that is different from a uniform one. The general approach to obtain a set of random numbers with a given distribution $P(x)$ is to start with a uniform set and change it into the one required. There are several ways to carry out such a transformation and we shall describe here three representative ones. By transformation, we are not necessarily restricted here to those that can be carried out analytically. In fact, on many occasions the required distribution may not even be represented by an analytical function. A summary of the various methods commonly used is found in Abramowitz and Segun (*Handbook of Mathematical Functions*, Dover, New York, 1965). Methods of obtaining many standard distributions can be found in Press and others (1986).

Let us start the discussion by assuming that we have a set of random numbers evenly distributed in the interval $[0,1]$. The probability of finding a number in the interval dx around x is given by

$$p_u(x)\,dx = \begin{cases} dx & \text{for } 0 \leq x \leq 1 \\ 0 & \text{otherwise} \end{cases} \qquad (7\text{-}10)$$

The normalization condition here is

$$\int_{-\infty}^{+\infty} p_u(x)\,dx = 1 \tag{7-11}$$

The subscript u emphasizes the fact that the distribution is uniform. Our present interest is to find a transformation from x to y so that the random number y has the desired distribution $P(y)$. It is convenient for us to assume that $P(y)$ has the same normalization as that for $p_u(x)$ given in (7-11). From the transformation of probability, we have the equality

$$P(y)\,dy = p_u(x)\,dx \tag{7-12}$$

This gives us

$$P(y) = \left|\frac{dx}{dy}\right| p_u(x)$$

where the Jacobian of the transformation is given by the absolute value of the derivative of x with respect to y. This comes from the fact that probabilities cannot be negative.

For some $P(y)$, it is possible to find analytically the transformation from x to y. As an illustration of this approach, consider the exponential distribution

$$P(y)\,dy = e^{-y}\,dy \tag{7-13}$$

To obtain the transformation function $y(x)$, we can start with (7-12). For $p_u(x)\,dx$ given by (7-10), we obtain

$$e^{-y} = x$$

by equating the absolute values of the indefinite integrals of both sides of (7-12). On inverting this relation, we obtain the transformation

$$y(x) = -\ln(x)$$

that changes a set of uniformly distributed random numbers $\{X_i\}$ to a set $\{Y_i\}$ having an exponential distribution. Such a method to carry out the transformation works well in cases where it is possible to carry out the indefinite integral

$$x = F(y) = \int P(y)\,dy$$

and the inverse of $F(y) = x$ can be found. In practice, the relation

$$y = F^{-1}(x) \tag{7-14}$$

may be complicated in form. In such cases it may be more efficient to adopt some other method, such as those discussed below.

Random number with a normal distribution. Besides the uniform distribution, random numbers with a normal distribution are useful in many applications. For the convenience of discussion, let us consider a set centered around zero and having a variance of unity. That is,

$$P(y)\,dy = \frac{1}{\sqrt{2\pi}} e^{-y^2/2}\,dy \tag{7-15}$$

In this case, a simple transformation from a uniform set cannot be worked out easily, as we have done above for an exponential distribution. On the other hand, the transformation between two pairs of such random numbers, (x_1, x_2) and (y_1, y_2), of the following form

$$y_1 = \sqrt{-2\ln x_1}\cos(2\pi x_2) \qquad y_2 = \sqrt{-2\ln x_1}\sin(2\pi x_2) \tag{7-16}$$

can be shown to have the desired property by the following argument. The relation between the joint probabilities $P(y_1, y_2)dy_1 dy_2$ for y_1 and y_2 and $p_u(x_1, x_2)dx_1 dx_2$ for x_1 and x_2, is given by a simple extension of (7-12):

$$P(y_1, y_2)\,dy = p_u(x_1, x_2)\,dx_1 dx_2$$

The Jacobian of the transformation in this case is the determinant

$$\frac{D(x_1, x_2)}{D(y_1, y_2)} = \begin{vmatrix} \frac{\partial x_1}{\partial y_1} & \frac{\partial x_1}{\partial y_2} \\ \frac{\partial x_2}{\partial y_1} & \frac{\partial x_2}{\partial y_2} \end{vmatrix} = -\frac{x_1}{2\pi} = -\frac{1}{2\pi} e^{-(y_1^2 + y_2^2)/2}$$

as can be shown by an explicit calculation. Since this is the product of two normal distributions, one for y_1 and another for y_2, we see that the joint distribution function $P(y_1, y_2)$ is a product of two independent normal distributions. Note that the joint distribution function of x_1 and x_2 is the product of $p_u(x_1)$ and $p_u(x_2)$, each of which is given by (7-10).

For practical applications, the transformation given by (7-16) is too time consuming to carry out because of the trigonometric functions involved. To speed up the calculations, we can regard x_1 and x_2 as the coordinates of a point on a two-dimensional surface. Since x_1 and x_2 are individually distributed uniformly in the interval $[0, 1]$, the point (x_1, x_2) is uniformly distributed in a square of length 1 on each side. If we change (x_1, x_2) into (X_1, X_2) through the transformation

$$X_1 = 2x_1 - 1 \qquad X_2 = 2x_2 - 1$$

and reject those with

$$R^2 \equiv X_1^2 + X_2^2 \geq 1$$

we have a set of points (X_1, X_2) distributed evenly in a circle of radius 1 on a two-dimensional surface. Now if we use $(2\pi z_2)$ to represent the angle between the vector $\boldsymbol{R}$ from the origin at the center of the circle to the point (X_1, X_2) and the horizontal axis, we have the relations

$$\cos(2\pi z_2) = \frac{X_1}{R} \qquad \sin(2\pi z_2) = \frac{X_2}{R}$$

If we further define another variable z_1 as

$$z_1 = R^2$$

we can write down a similar transformation as (7-16)

$$y_1 = \sqrt{-2\ln z_1}\cos(2\pi z_2) = \sqrt{-2\ln R^2}\frac{X_1}{R} = X_1\sqrt{\frac{-2\ln R^2}{R^2}}$$

$$y_2 = \sqrt{-2\ln z_1}\sin(2\pi z_2) = \sqrt{-2\ln R^2}\frac{X_2}{R} = X_2\sqrt{\frac{-2\ln R^2}{R^2}} \qquad (7\text{-}17)$$

The advantage here is that, instead of two trigonometric functions and a square root in (7-16), only a square root is needed.

In this approach, some loss of efficiency is also incurred from discarding pairs of random numbers with $R > 1$. The fraction retained is $\pi/4 \sim 0.79$ on the average, given by the ratio of the area of a unit circle to that of a square of length 2 on each side. Programming for the transformation is given as an exercise in Problem 7-5. An approximate scheme, based on the central limit theorem, is given in Problem 7-4 and is used in many applications.

An important condition for the success of the method is that the random points, given by a pair of random numbers (X_{2i}, X_{2i+1}), must uniformly cover the square area (see Problem 7-1). As a result, any serial correlation between pairs of random numbers can be potentially very damaging. For this reason, one should test the uniform random numbers used as the input for possible correlations, in addition to checking whether the distribution of $\{y_i\}$ is a normal one.

Rejection method. If the desired distribution of random numbers is not given by an analytical function, or if the inverse function for the indefinite integral of $P(x)$ does not exist, the transformation method given by (7-14) does not work. In such cases, the rejection method may be used. The basic idea involved in this method is quite simple. If we throw away those numbers that do not follow the desired distribution, the remainder must be distributed in the way we want. This type of approach was used, in part, in generating random numbers with a normal distribution above.

Consider a two-dimensional surface. As we have seen earlier, the location of each point on this surface may be specified by the values of its x- and y-coordinates with respect to some fixed system of reference. A random point on this surface may be chosen by selecting two random numbers X_1 and X_2, and let the values of the coordinates $x = X_1$ and $y = X_2$. If both X_1 and X_2 are uniformly distributed

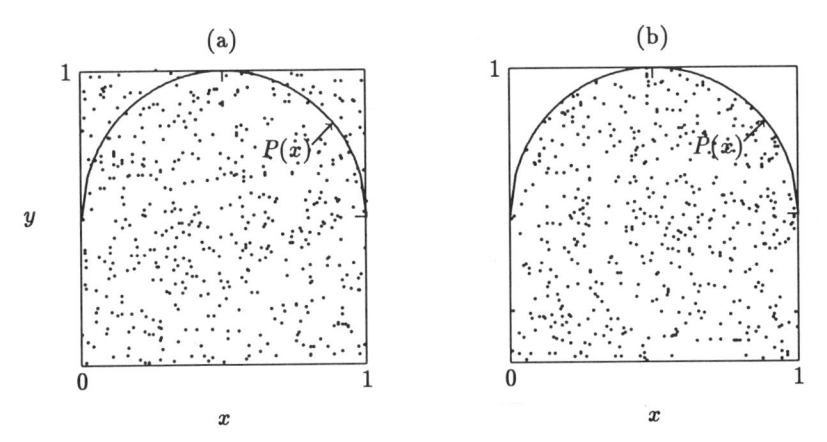

Fig. 7-3 (a) Random points, each specified by two random numbers as its x- and y-coordinates, filling a square area with equal probability. (b) By rejecting points with $y > P(x)$, the remaining random points fill only the area under the curve $P(x)$. As a result, x has the desired distribution $P(x)$.

in the interval $[0, 1]$, the random point has an equal probability of being anywhere in the square area, as shown in Fig. 7-3(a). If a large number of such random points are available, the square area will be evenly covered. On the other hand, if we reject those points with $y > P(x)$, where $P(x)$ is the desired distribution of random numbers, the remaining points fill only the area underneath the curve $P(x)$, as shown in Fig. 7-3(b). Consider now only the distribution of the values of x-coordinates of the points remaining. Since the ratio of the number of points in the small interval dx at $x = a$ and that at $x = b$ is given by $P(x=a)/P(x=b)$, the distribution of x values is given by the probability $P(x)\, dx$. In more practical terms, if we have a collection of N points, represented by their x- and y-coordinates $\{X_1(i), X_2(i)\}$, for $i = 1, 2, \ldots, N$, and we discard all those points [both $X_1(i)$ and $X_2(i)$] with $X_2(i) > P(x = X_1(i))$, the remaining random numbers $X_1(i)$ have the desired distribution $P(x)$.

A simple example may be helpful here to illustrate the principle behind the rejection method. Let us consider the semicircular distribution

$$P(x) = \frac{2}{\pi}\sqrt{1-x^2} \qquad \text{for} \qquad |x| \leq 1 \tag{7-18}$$

For $x > 0$, the distribution covers only a quarter of a circle of unit radius with the center of the circle located at the origin. To produce random numbers having such a distribution with the rejection method, we start from a set of random numbers with a uniform distribution in the interval $[0, 1]$. For each pair of such random numbers, $X_1(i)$ and $X_2(i)$, we shall consider $X_1(i)$ as the x-coordinate and $X_2(i)$ as the y-coordinate of a random point in a square area of unit length on each side. If $\{X_1^2(i) + X_2^2(i)\} \geq 1$, the point is outside the quarter-circle and we reject the point. Another point is produced by selecting a new pair of random numbers $X_1(i+1)$ and $X_2(i+1)$. If $\{X_1^2(i+1) + X_2^2(i+1)\} < 1$, the point is inside the unit circle and we retain $X_1(i+1)$ as one of the random numbers with a semicircular distribution. To show that the accepted sequence of X_1 follows the required distribution, a set of 50,000 such random numbers is plotted in the form of a histogram in Fig. 7-4. The fact that the distribution is semicircular may be seen by comparing the histogram with the quarter-circle superimposed on the diagram in the form of a solid curve. The method is essentially the same as that used earlier to produce random numbers with a circular distribution that serve as the intermediate quantities in generating random numbers with a normal distribution.

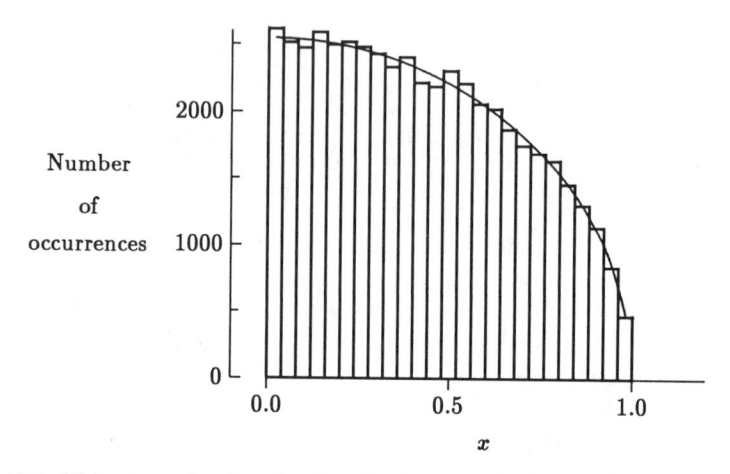

Fig. 7-4 Histogram showing the distribution of 50,000 semicircular random numbers generated by the rejection method. The smooth curve is the analytical form given by (7-18).

Since the quarter-circle occupies most of the area of the square covered by our original random points, only a small fraction of the points is rejected. We can calculate the ratio from the area of a quarter circle ($=\pi/4$) and that of a square (unity). However, since each point requires two random numbers, only $\pi/8$ of all the random numbers generated belong to the distribution $P(x)$. The addition factor of $\frac{1}{2}$ comes from the fact that only one of each pair of uniform random numbers is used (X_1 but not X_2).

In general, if the area underneath the desired distribution $P(x)$ is only a small fraction of the rectangular area covered by pairs of uniform random numbers, it is more efficient to transform the random numbers first to a distribution $g(x)$ satisfying the condition

$$g(x) \geq P(x)$$

before applying the rejection method. The function $g(x)$ chosen must also fulfill the requirement that the transformation method can be applied with ease. For example, if we wish to use the analytical transformation of (7-14), the inverse function $F^{-1}(x)$ must have a simple form so that the necessary calculations can be carried out efficiently. Furthermore, $g(x)$ should be as close to $P(x)$ as possible in order to minimize the number of random numbers to be rejected. For example, for the distribution $P(x)$ shown in Fig. 7-5, the choice of a normal distribution for $g(x)$ is a good one (a random number generator with normal distribution for the y-coordinate and a random number generator with uniform distribution for the x-coordinate). Only the fraction of points between $P(x)$ and the desired normal curve is discarded.

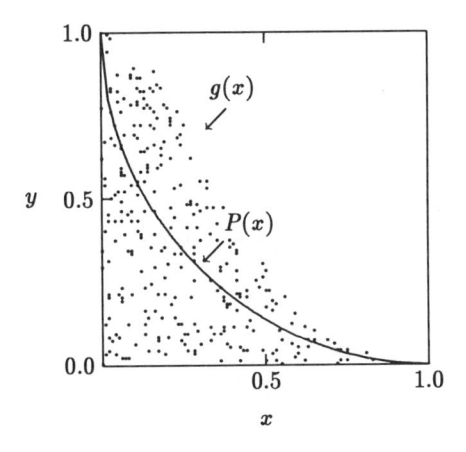

Fig. 7-5 Improving the efficiency of the rejection method for generating random numbers. If the distribution required is $P(x)$, indicated by the solid curve, the starting uniform random numbers are first transformed to a set with distribution $g(x)$, given by a normal distribution here. Since the required distribution $P(x)$ fills a much larger fraction of the area underneath $g(x)$ than the square area, a much smaller fraction of the points has to be rejected if they are first compressed into the distribution $g(x)$.

The Metropolis algorithm. A method to produce random numbers of a given distribution that is especially convenient for problems in statistical mechanics and related applications was given by Metropolis and others (*J. Chem. Phys.* **21** [1953], 1087). Instead of individual random numbers, the interest here is mainly in the distribution of random points in a multidimensional space, such as those representing members of a canonical ensemble in statistical mechanics. If a system consists of N particles, each of which is specified by its location (x, y, z) and momentum (p_x, p_y, p_z), the dimension of the phase space is $d = (3 + 3)N$. Each point in this N-particle space then requires d qualities to label it. To simplify the notation, we shall use a single bold faced letter,

$$\boldsymbol{X} \equiv (x_1, x_2, \ldots, x_d)$$

to represent the location of a point in such a space.

The way a random point in the space is selected in the Metropolis scheme is closely related to the rejection method. For each point, a trial value $\boldsymbol{X_t}$ is proposed. Whether $\boldsymbol{X_t}$ is accepted or rejected as a member of the ensemble depends on the distribution of $\boldsymbol{X}$ we wish to generate. Let us use $P(\boldsymbol{X})$ to represent the required distribution. Furthermore, we assume that at some stage in the generation of random points there are already n points $\boldsymbol{X_1}, \boldsymbol{X_2}, \ldots, \boldsymbol{X_n}$ that are distributed according to $P(\boldsymbol{X})$. The algorithm to find the next point $\boldsymbol{X_{n+1}}$ is given by the following procedure. First, we construct a trial point $\boldsymbol{X_t}$ from the last accepted point $\boldsymbol{X_n}$ by adding to it a set of small changes $\boldsymbol{\delta}$. That is,

$$\boldsymbol{X_t} = \boldsymbol{X_n} + \boldsymbol{\delta} \tag{7-19}$$

The size of $\boldsymbol{\delta}$ depends on the physical problem we wish to solve. It is an important consideration in order to make the algorithm efficient. However, we shall ignore this question for the time being. Similarly, we shall also delay any discussion of how to obtain the starting point $\boldsymbol{X_0}$ until the end of this section.

The decision whether $\boldsymbol{X_t}$ should be accepted as a member of the distribution depends on the ratio

$$r = \frac{P(\boldsymbol{X_t})}{P(\boldsymbol{X_n})} \tag{7-20}$$

If $r \geq 1$, the trial point $\boldsymbol{X_t}$ is accepted. This condition may be stated as

$$\boldsymbol{X_{n+1}} = \boldsymbol{X_t} \qquad \text{for} \qquad r \geq 1 \tag{7-21a}$$

If $r < 1$, the trial point is accepted with probability r. In practical terms, the decision whether to accept the trial point $\boldsymbol{X_t}$ is made by generating a random number x whose value is equally probable anywhere in the interval $[0, 1]$. If $x > r$, the trial point is rejected; otherwise it is accepted. In other words,

$$\boldsymbol{X_{n+1}} = \boldsymbol{X_t} \qquad \text{with probability } r \qquad \text{for} \qquad r < 1 \tag{7-21b}$$

This way of deciding whether to accept $\boldsymbol{X_t}$ is very similar to the spirit of flipping a coin when we have to make an unbiased yes or no decision.

We shall now show that the sequence of random points $\boldsymbol{X_1}, \boldsymbol{X_2}, \ldots,$ generated this way has the desired distribution $P(\boldsymbol{X})$. Let $N(\boldsymbol{X})$ be the density of points in the neighborhood of $\boldsymbol{X}$ and $K(\boldsymbol{X} \to \boldsymbol{Y})$ be the transition probability from $\boldsymbol{X}$ to $\boldsymbol{Y}$. The meaning of $K(\boldsymbol{X} \to \boldsymbol{Y})$ may be interpreted as the probability for the system to move the point $\boldsymbol{X}$ in the phase space to a point at $\boldsymbol{Y}$. At equilibrium, the density of points in the neighborhood of any point must be a constant in time. This, however, does not mean the system is static — it simply implies that transitions out of the region around $\boldsymbol{X}$ are balanced by the transitions into the same region. In other words, the equilibrium condition is given by

$$\sum_i N_e(\boldsymbol{X})K(\boldsymbol{X} \to \boldsymbol{Y_i}) - \sum_i N_e(\boldsymbol{Y_i})K(\boldsymbol{Y_i} \to \boldsymbol{X}) = 0 \tag{7-22}$$

This is, essentially, a statement of the principle of detailed balance. Away from equilibrium, the change in $N(X)$ is governed by the difference between two terms

$$\Delta N(X) = \sum_i \{N(X)K(X \to Y_i) - N(Y_i)K(Y_i \to X)\}$$

$$= \sum_i N(Y_i)K(X \to Y_i)\left\{\frac{N(X)}{N(Y_i)} - \frac{K(Y_i \to X)}{K(X \to Y_i)}\right\} \quad (7\text{-}23)$$

An alternate way to state the condition of equilibrium of (7-22) is then

$$\frac{N_e(X)}{N_e(Y_i)} = \frac{K(Y_i \to X)}{K(X \to Y_i)} \quad (7\text{-}24)$$

The results of (7-24) imply also that the system is stable. For example, if the ratio of $N(X)/N(Y_i)$ is larger than $K(Y_i \to X)/K(X \to Y_i)$, the system moves away from the region around X [that is, $\Delta N(X) > 0$], and vice versa if $N(X)/N(Y_i) < K(Y_i \to X)/K(X \to Y_i)$.

We shall now show that the equilibrium distribution of points is equal to $P(X)$. To do this we note that, from the way the random points are generated, the transition probability $K(X \to Y)$ is a product of two factors,

$$K(X \to Y) = T(X \to Y)A(X \to Y) \quad (7\text{-}25)$$

where $T(X \to Y)$ is the probability that, given the system is at X, the trial value X_t is Y. In other words, $T(X \to Y)$ is determined by the way we select δ in (7-19). For this reason, it must be symmetric with respect to X and Y:

$$T(X \to Y) = T(Y \to X) \quad (7\text{-}26)$$

The other factor, $A(X \to Y)$, in (7-25) is the probability of accepting the trial point and is given by (7-21). In terms of $T(X \to Y)$ and $A(X \to Y)$, the equilibrium condition (7-24) may be stated as

$$\frac{N_e(X)}{N_e(Y)} = \frac{T(Y_i \to X)}{T(X \to Y_i)}\frac{A(Y_i \to X)}{A(X \to Y_i)} = \frac{A(Y_i \to X)}{A(X \to Y_i)} \quad (7\text{-}27)$$

where we made use of (7-26) to cancel out the dependence on $T(X \to Y)$.

To make connection with (7-21), let us consider the case of starting from X and using Y as the trial point. Here, the ratio

$$\frac{P(Y)}{P(X)} > 1 \qquad \text{for} \qquad P(Y) > P(X)$$

In this case, the trial point Y is accepted. In terms of $A(X \to Y)$, we can state the same result by saying that

$$A(X \to Y) = 1 \qquad \text{for} \qquad P(Y) > P(X)$$

At the same time, for $P(Y) > P(X)$, the ratio

$$r = \frac{P(X)}{P(Y)} < 1$$

and

$$A(Y \to X) = r$$

The equilibrium condition (7-27) is then given by

$$\frac{N_e(X)}{N_e(Y)} = \frac{r}{1} = \frac{P(X)}{P(Y)} \tag{7-28}$$

Conversely, the ratio

$$r = \frac{P(Y)}{P(X)} < 1 \qquad \text{for} \qquad P(Y) < P(X)$$

and

$$A(X \to Y) = r = \frac{P(Y)}{P(X)}$$

or $\qquad A(Y \to X) = 1 \qquad\qquad \text{for} \qquad P(Y) < P(X)$

Again, we find that the equilibrium condition is given by (7-28): the density at equilibrium at any point X is proportional to the probability $P(X)$ at the point. This, in turn, demonstrates that the Metropolis algorithm of accepting or rejecting a trial point by (7-21) generates a equilibrium distribution $P(X)$.

The disadvantage of the Metropolis algorithm is that, from the way each trial point is generated by (7-19), the sequence of points X_i, X_{i+1}, ..., is strongly correlated. This is especially a serious problem at the beginning of a sequence since it is difficult to choose a starting point X_0 that is a proper member of the set. As a result, a whole sequence of points that is outside the distribution we wish to have may be produced. One way to avoid the need to find a good starting point is to use an arbitrary one and discard it, together with the first few generated from it. We have used a technique similar to this earlier in the subtraction method of generating uniformly distributed random numbers in Box 7-2. Because of its statistical mechanics origin, the technique of discarding the initial part of a sequence is often called "thermalizing" the distribution. The same basic principle may also be used to reduce the strong correlation between successive random points by discarding a few points between any two that are adopted as members of the physical ensemble. An application of the Metropolis algorithm is given later in §7-5 for the Ising model.

§7-2 Molecular diffusion and Brownian motion

As an elementary application of Monte Carlo methods, we shall study molecular diffusion as a *random walk* problem. In its simplest form, the problem may be described in the following way. At some initial time t_0, we have a concentration of molecules at a point x_0 in space. If the temperature of the system is finite, each molecule possesses some nonzero kinetic energy. As a result, a collection of such molecules will diffuse into the surrounding unless there are barriers preventing them from doing so. The rate of diffusion is determined by two factors, the velocity of each molecule and the probability of collision with other molecules occupying the same space. The former is characterized by the temperature T and the latter by the mean free path $\overline{\ell}$.

A well-known application of the random walk approach to diffusion is Einstein's illustration of Brownian motion. The aim here is to simulate the motion of a particle in a fluid. To simplify the picture, we can view the motion of the Brownian particle as one that takes a step of constant size $\overline{\ell}$ in a time interval τ. If we further confine the motion to one dimension, the motion is limited either to taking a step forward $(+\overline{\ell})$ or a step backward $(-\overline{\ell})$ in each time step. For a given particle, its location at time t is then proportional to the excess of the number of positive steps over that of negative steps (or the other way around). In a time interval Δt, the total number of steps taken is

$$n = \frac{\Delta t}{\tau}$$

If the initial position is taken to be $x = 0$, the location of the particle $x(t)$ at time t is given by

$$x(t) = m\overline{\ell}$$

where m is difference between the number of steps in the positive direction, $(n+m)/2$, and the number of steps in the negative direction, $(n-m)/2$. Since the direction taken in each step, $+$ or $-$, is completely independent of any other steps, the probability for the difference to be m is given by

$$p(n,m) = P_B(n, \tfrac{1}{2}\{n+m\}, \tfrac{1}{2})$$

$$= \binom{n}{\{n+m\}/2}\left(\frac{1}{2}\right)^n = \left(\frac{1}{2}\right)^n \frac{n!}{\left(\frac{n+m}{2}\right)!\left(\frac{n-m}{2}\right)!}$$

where $P_B(n, \{n+m\}/2, \tfrac{1}{2})$ is the binomial distribution, defined earlier in (6-12), for choosing $(n+m)/2$ objects out of n, each with a probability $p = \tfrac{1}{2}$ to be selected.

The distribution $p(n,m)$ is symmetric with respect to $m = 0$, corresponding to the point $(n+m)/2 = n/2$ in $P_B(n, \tfrac{1}{2}\{n+m\}, \tfrac{1}{2})$. This means that it is equally likely to find a particle in the positive direction as in the negative direction.

However, the probability of finding a particle decreases with the distance from $x=0$. If we average over the locations of a large number of Brownian particles, we find that the mean position $\overline{x(t)}$ is zero. This does not mean that all the particles remain at $x = 0$. In fact, we know that the concentration of particles at $x = 0$ must decrease with time. Such a drop is indicated by a linear increase of the variance of the distribution $p(n, m)$ with time:

$$\overline{x^2(t)} = \frac{\overline{\ell}^2}{\tau}t \tag{7-29}$$

This can be seen from the fact that the variance of $P_B(n, \frac{1}{2}\{n+m\}, \frac{1}{2})$ is $n/4$, and n increases linearly with t. The physical reason that the standard deviation of the distribution $\{\overline{x^2(t)}\}^{1/2}$ is proportional to $t^{1/2}$ rather than t comes from collisions with other particles. In the absence of such collisions, a particle will continue its motion along a straight line at constant velocity and the standard deviation of the distribution will increase linearly with time. In our one-dimensional model, each collision has a 50% possibility of turning the Brownian particle backward (and another 50% possibility of making it go forward). In this way, the rate of growth in the standard deviation of the distribution is reduced from that in the absence of collisions.

Such a simple model of Brownian motion may be simulated by a Monte Carlo calculation. Let the location of a particle at time t be $x(t)$. At time $(t + \tau)$, its location is given by

$$x(t + \tau) = x(t) \pm \overline{\ell} \tag{7-30}$$

The sign on the right side is completely random. To simulate such a probabilistic result, we need a random number generator. In principle, we can use one that produces a random number r with a uniform distribution in the interval $[0, 1]$. If $r < 0.5$, we take $x(t + \tau) = x(t) - \overline{\ell}$, and if $x \geq 0.5$, we take $x(t + \tau) = x(t) + \overline{\ell}$. In most practical calculations of this type, however, a much simpler and faster method of using a random bit generator is preferred.

Random bit generator. A random bit generator is one that produces a result of either 1 or 0 with equal probability. For the purpose of discussion, we shall assume that each computer word is 32 bits long, even though the basic method works also for a shorter or longer word length. Among the 32 bits, we shall take an arbitrary one as the random bit of interest to us. Without any loss of generality, we can assume that this is the 19th bit. Again, most other choices among the 32 bits can also serve the purpose.

In the same way as in generating random numbers, we start with an integer seed, s_i, obtained by some other means that need not concern us at the moment. If bit 19 of s_i is 1, the random bit produced by the generator is 1. Similarly, if bit 19 of s_i is 0, the random bit produced is 0. The main part of the random bit generator, similar to random number generators in general, concerns the way to produce another seed, s_{i+1}, from which we obtain the next random bit. For this purpose, we shall adopt the method of *primitive polynomial modulo 2* given in Knuth. One advantage of this method is that we can use the extremely fast Boolean operations AND [IAND(i_1, i_2) in FORTRAN], OR [IOR(i_1, i_2) in FORTRAN], and XOR [IEOR(i_1, i_2) in FORTRAN] to carry out all the calculations. Each one of these functions performs a bit by bit comparison between two computer words, i_1 and i_2, and returns the results as shown in Table 7-2.

Table 7-2 Boolean operations.

| i_1 | 0 | 0 | 1 | 1 | Result |
i_2	0	1	0	1	
IAND(i_1, i_2)	0	0	0	1	$= 1$ if both i_1 and i_2 are 1
IOR(i_1, i_2)	0	1	1	1	$= 1$ if either i_1 or i_2 is 1
IEOR(i_1, i_2)	0	1	1	0	$= 1$ if either i_1 or i_2 is 1 but not both

In this method, the computation of the next seed s_{i+1} is carried out in the following manner. If bit 19 of s_i is zero, s_{i+1} is produced by shifting all the bits of s_i to the left by one binary location. This is equivalent to multiplying s_i by 2, except that the first bit of s_i is discarded. In addition to speed, the shift operator also has the advantage that it does not produce an integer overflow when the first bit of s_i is not zero. The resulting integer seed s_{i+1} is an even number with the last bit zero. This is of no concern to us, since we are only interested in bit 19. The real work comes when bit 19 of s_i is 1. In this case, a *mask* is used to modify various bits in the seed. In the "primitive polynomials modulo 2" method, the mask is made up of a polynomial consisting of a sum of several terms, each one of which is in the form of 2 to an integer power. We shall not go into the proof here of why the use of such a mask ensures that a sequence of random bits generated in this way is uncorrelated. The selection of a mask also depends on which bit in the word is used as the random bit. In other words, both the location of the random bit and the mask are an integral part of selecting the polynomial. A list of several choices can be found in Press and others (1986).

For our selection of bit 19, a suitable polynomial is

$$m = 2^4 + 2^1 + 2^0$$

As mentioned earlier, the mask is needed only in the case where the previous random bit is 1. In this case, the new seed s_{i+1} is obtained from the previous seed s_i by the following steps:

(a) Let $s' = \text{IEOR}(s_i, m)$. That is, compare all 32 bits of s_i one by one with the corresponding bits in mask m. If both of them are 0 or 1, put the corresponding bit in s' to be 0. If one of them is 1 and the other is 0, put the corresponding bit in s' to be 1.

(b) Shift s' left by 1 bit (equivalent to multiply s' by 2).

(c) Add 1 to s'. Store the result as the new seed s_{i+1}.

This completes the generation of a random bit.

Random walk in one dimension. Let us return to our random walk problem. Since the whole process of generating a random bit can be coded in two or three lines, there is no need to make the calculation into a subprogram, as is usually done with random number generators. This makes the calculation more efficient, but we lose the advantage of having the random bit generator as an independent unit. The remainder of the calculation for the random walk problem is also very simple and may be coded almost completely in terms of integers.

The ultimate interest of our simulation is the distribution of Brownian particles along x at time t. Since the step size is fixed to be $\bar{\ell}$, we can measure all the distances using $\bar{\ell}$ as the unit. Similarly, since the time is always in steps of τ, we can use τ as the unit of time. The distribution at any time t is then given by the number of particles at integer number of length units from the origin. An array of length $(2n+1)$ with

$$n = t/\tau$$

may be used to store the number of particles at each possible location.

The actual calculation may be carried out in the following way. Each particle starts its random walk at $t = 0$ from its initial position at $x = 0$. The location at time $t = \tau$ is either $+1$ or -1 in units of $\bar{\ell}$, depending on whether the random bit generated is 1 or 0. Let us represent the result as $x(t = \tau)$. The location at $t = 2\tau$ is obtained using (7-30), with $x(t)$ on the right side given by the location at $t = \tau$. The resulting location gives us the result for $x(t = 2\tau)$. From this, we can obtain the location at the next time interval $t = 3\tau$, and so on. Each time, the random bit generator is used to determine whether the motion is taking a step forward $(+\bar{\ell})$ or backward $(-\bar{\ell})$. After n time steps the position is recorded by increasing the appropriate counter by 1. This completes the motion of one particle. All the steps are then repeated for another particle with a different set of random bits. This is repeated for as many particles as needed to produce good statistics for the final distribution.

As a practical procedure, we can carry out the calculation for some large number, like a thousand particles at a time. After each thousand particles, the contents of all the counters are checked to see if there are enough particles in most of them for a meaningful statistical analysis of the distribution. When this is true, we can proceed to the next step of calculating the mean and variance

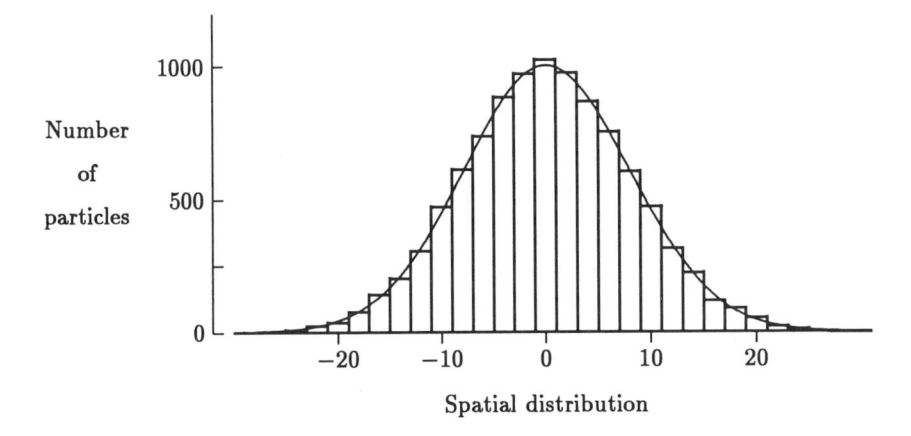

Fig. 7-6 Histogram showing the spatial distribution of Brownian particles after 64 time steps. The solid curve is for a normal distribution having a standard deviation of 8 spatial steps and a normalization equal to the total number of particles involved.

of the distribution. Otherwise, the calculation is repeated for another thousand particles. As mentioned earlier, the distribution gets broader as the number of time steps taken by each particle is increased. For this reason, a larger number of particles is required for studying distributions with longer time spans if we wish to maintain the same accuracy.

The behavior of a collection of one-dimensional Brownian particles is well known. Initially, all the particles are assumed to be located at $x = 0$ in the form of a delta function. The sharp distribution is quickly dispersed, as particles drift to both sides. After a few time steps, the distribution becomes an essentially normal one centered at the origin. A typical result after 64 time steps is shown in Fig. 7-6. The distribution of particles becomes more and more diffused with time, as more and more particles move away from the origin. The shape of the distribution, however, remains normal. To demonstrate the fact that the standard deviation of the distribution increases with the square root of time, the results obtained in a Monte Carlo calculation are displayed in Fig. 7-7.

The analytical solution for the distribution of Brownian particles in one dimension has been known for a long time and the results of our numerical simulation agree well with it. In general, the primary purpose of a Monte Carlo simulation is not to repeat calculations that can be done easily by analytical methods. For our example, we have made several assumptions so as to mimic the physical conditions under which the analytical results are derived. As far as the numerical calculation is concerned, many of the simplifications are not essential. For this reason, it is possible for us to carry out much more realistic simulations of the true physical situation. For example, real molecular motions depend on the probability of colliding with other particles, instead of always moving a constant distance $\bar{\ell}$ in time

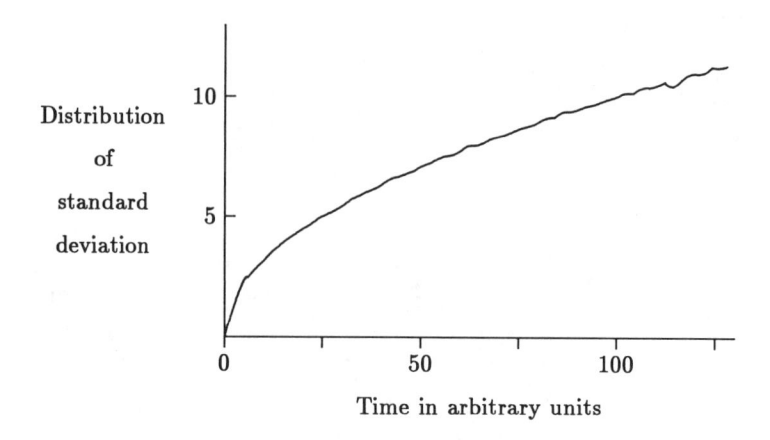

Fig. 7-7 Standard deviation of the spatial distribution of Brownian particles as a function of time. Since the variance grows linearly with time, the standard deviation increases as the square root of time.

τ. Furthermore, the collisions are not necessarily simple hard-sphere scattering. Many things can happen during such a collision, including the possibility that two particles may join together to form a single molecule. Under suitable conditions, a gas may even condense into a liquid. Some of these considerations are not easily incorporated into analytical solutions. For this reason, many of the recent understandings of molecular dynamics are obtained through computer simulations involving Monte Carlo calculations.

There are also some important limitations in simulating molecular motion on a computer. For example, any realistic collection of gas molecules involves on the order of 10^{23} (Avogadro number) particles. Obviously, we cannot hope to follow the motion of each particle when the number of particles is so large, regardless of how simple the motion of each particle may be. As a result, many restrictions are placed on the calculations. Fortunately, the assumptions made for the numerical works are often quite different from those in analytical solutions. As a result, the two methods complement each other in many ways in our attempt to understand a complicated but interesting physical problem.

§7-3 Data simulation and hypothesis testing

In Chapter 6 we discussed several methods of carrying out least-squares fits to data. Our primary aim there was to find the values of the parameters of an expression so as to obtain the best possible description of the data. From a slightly different point of view, the same calculations may also be regarded as a test of our idea, or hypothesis, of how the data should behave. If the expression that we use to fit the data is based on a sound understanding of the physical system, the uncertainties in the parameters obtained will be small, reflecting mainly the uncertainties in the measurements. On the other hand, if we have an incorrect understanding of the system under investigation, our fit will be poor and the value of the χ^2 will be much larger than that resulting from the uncertainties in the measurements alone.

In this way, the goodness of fit may also serve as an indication of how well the expression used in the fit represents the true nature of the physical system we are studying. In cases where the χ^2-values are small and all the uncertainties associated with the individual parameters obtained are negligible, we have the confidence that our hypothesis is correct. For example, for the angular correlation of γ-rays given as an illustration in Fig. 6-11, we can be fairly certain that the distribution has the form made of a sum of three even-order Legendre polynomials, $P_0(\cos\theta)$, $P_2(\cos\theta)$, and $P_4(\cos\theta)$ (as expected of electric quadrupole transitions). On the other hand, if we try to fit the same set of experimental data assuming no angular dependence, the value of χ^2 will be large and we can feel confident in rejecting the hypothesis of an isotropic distribution. In general, the results will not be as clear-cut and a more quantitative criterion is required on which to base on our judgment.

Confidence region and confidence level. A statistical measure of whether a hypothesis should be accepted or rejected is usually stated in terms of the *confidence region*, or confidence interval, at a given *confidence level*. Each of these terms has a specific meaning in statistics. In the place of formal definitions, it is perhaps more useful here to illustrate their meaning by an example.

Consider the case of measuring the mass of a neutrino. In the report of Lubimov and others (*Phys. Lett.* **94B** [1980], 266), the mass μ_ν of an electron antineutrino $\overline{\nu}_e$ is given as 14 eV$\leq \mu_\nu c^2 \leq$ 46 eV at a confidence level of 99%. This means that, if the same experiment to measure the value of $\mu_\nu c^2$ is performed 100 times, we expect, on the average, the results of 99 measurements will be in the region between 14 and 46 eV.[†]

Let us try to see the meaning of this statistical statement in terms of what one can actually do in an experiment. As emphasized several times earlier in Chapter 6, the value q_{exp} obtained in a measurement is not necessarily equal to the true value q_{true} for the quantity we are measuring. In fact, if we repeat the measurement many times, it is likely that the outcome will be slightly different each time. Let us use $p_{\mathrm{exp}}(q)$ to represent the probability distribution of the measured values. Our interest here is to infer from the measured values q_{exp} and their distribution $p_{\mathrm{exp}}(q)$ the true value q_{true} of the quantity q. From a statistical

---

[†]The electron-volt, abbreviated eV, is a convenient unit of mass μ for elementary particles in terms of their rest mass energy μc^2, where c is the speed of light. A rest mass energy of 1 eV is equivalent to a mass of 1.78×10^{-36} kg. The mass of an electron μ_e is 9.11×10^{-31} kg or $\mu_e c^2 = 0.511 \times 10^6$ eV.

point of view, we can take q_{true} as the mean value of $p_{\text{true}}(q)$, obtained after a very large number of measurements, as we did in (6-1). In practice, however, it is not possible to carry out the large number of measurements required to give a reasonable representation of $p_{\exp}(q)$. Under such circumstances, one must try to do the best one can to arrive at an estimate of the true value by some other means.

Before we see how Monte Carlo calculations can help us to acquire some additional confidence in our measured results, let us see first the mathematical relation between confidence level and confidence region. Pretend for the moment that the probability distribution $p_{\text{true}}(q)$ is known. The probability of obtaining a measured value $q_{\exp}$ in the region $[q^\ell, q^h]$ is given by the integral

$$P(q^\ell \le q_{\exp} \le q^h) = \int_{q^\ell}^{q^h} p_{\text{true}}(q)\, dq \qquad (7\text{-}31)$$

In practice, the problem one faces is the inverse. The probability distribution $p_{\text{true}}(q)$ is unknown and cannot be established without carrying out a large number of measurements. If it is not possible to repeat the experiment many times, the best we can do is to have an estimate of q_{true} from the measured value we have in hand. For this purpose, we select first a confidence level, or *confidence coefficient* P, for example 99%, and ask for the (confidence) interval $[q^\ell, q^h]$ in which the integrated probability $P(q^\ell \le q_{\exp} \le q^h)$ is equal to 0.99. For the probability distribution, we can use the measured distribution $p_{\exp}(q)$, if it is available, or make the best estimate we can for $p_{\text{true}}(q)$. Usually, a combination of the two approaches is used. For example, we know that the neutrino mass μ_ν cannot be a negative quantity. As a result, we can restrict the distribution $p(q)$ to the region with $q \ge 0$, as we shall see in an example later.

Alternatively, we can ask the same question in terms of the functional relation between the set of measured values $\{y_i\}$ on some independent variable $\{x_i\}$. Since we do not have a complete knowledge of the relation between y and x, the expression may contain a number of adjustable parameters $\boldsymbol{a} \equiv (a_1, a_2, \ldots, a_m)$ in the form

$$y = f(a_1, a_2, \ldots, a_m; x)$$

This is the type of question we asked ourselves in Chapter 6 on least-squares calculations. If $f(\boldsymbol{a}; x)$ is used to fit to a number of measured values and the parameters $\boldsymbol{a}$ are determined up to some uncertainties $\delta \boldsymbol{a}$, we may wish to know the probability of getting a particular set of results for $\boldsymbol{a}$ in the interval $[\boldsymbol{a}^\ell, \boldsymbol{a}^h]$. Analogous to (7-31), we have

$$P(a_1^\ell \le a_1 \le a_1^h, a_2^\ell \le a_2 \le a_2^h, \ldots, a_m^\ell \le a_m \le a_m^h; x)$$
$$= \int_{a_1^\ell}^{a_1^h} \int_{a_2^\ell}^{a_2^h} \cdots \int_{a_m^\ell}^{a_m^h} f(a_1, a_2, \ldots, a_m; x)\, da_1\, da_2 \cdots da_m$$

The confidence region in this case is a volume V in the m-dimensional parameter space, made of the product $\Pi_{i=1}^{m}(a_i^h - a_i^\ell)$. It is also possible in this case to define the confidence volume in terms of contours of constant χ^2-value. However, we shall not pursue this particular alternative here.

The relation between confidence limit and confidence region becomes somewhat complicated when the distribution of one or more of the parameters extends over into the unphysical region. For example, in an experiment, the measured value of the neutrino rest mass energy may turn out to be 30 eV with an uncertainty of 40 eV. If the probability distribution is a normal one, there is the probability of

$$\frac{1}{40\sqrt{2\pi}} \int_{-\infty}^{0} e^{-\{(x-30)/40\}^2/2} dx = 0.23$$

or 23% that the true value of $\mu_\nu c^2$ is less than zero. A straightforward interpretation of the results means that, if we repeat the experiment 100 times, it is likely that in 23 of the 100 measurements the value of μ_ν will turn out to be negative. In reality, mass is a positive quantity and a negative value should be rejected.

It is clear that a naive statistical approach is unsatisfactory here. Instead, it may be more correct to exclude the unphysical region from our probability distribution. There is no clear guidance in probability theory for the proper procedure to follow in this case. In physics, it is often recommended that, instead of (7-31), the relation between confidence limit and confidence region should be given by

$$P(q^\ell \leq q_{\exp} \leq q^h) = \frac{\int_{q_\ell}^{q^h} p_{\text{true}}(q) \, dq}{\int_{\text{phys. reg.}} p_{\text{true}}(q) \, dq}$$

In this case the denominator on the right side normalizes the probability distribution $q(p)$ to unity within the physical region.

Example of neutrino mass measurement. There are several ways that Monte Carlo calculations may be used to test our hypothesis about the distribution of measured values. Let us again use the neutrino mass measurement as the example. Most of the measurements to date involve the β-decay of tritium (t), a nucleus made of two neutrons and one proton,

$$t \to {}^3\text{He} + e + \overline{\nu}_e$$

The nucleus in the final state is ^{3}He, a light isotope of helium made of two protons and one neutron. In addition, an electron (e) as well as an electron antineutrino ($\overline{\nu}_e$) are also emitted. The total amount of energy released in the decay, or the end-point energy, is 18,578 eV, given by the difference in the rest mass energy of the tritium in the initial state and the sum of the rest mass energies of the ^{3}He (in its ground state), the electron, and the antineutrino in the final state. The energy is shared by the kinetic energy of the electron, the neutrino, and any excitations of the helium atom.

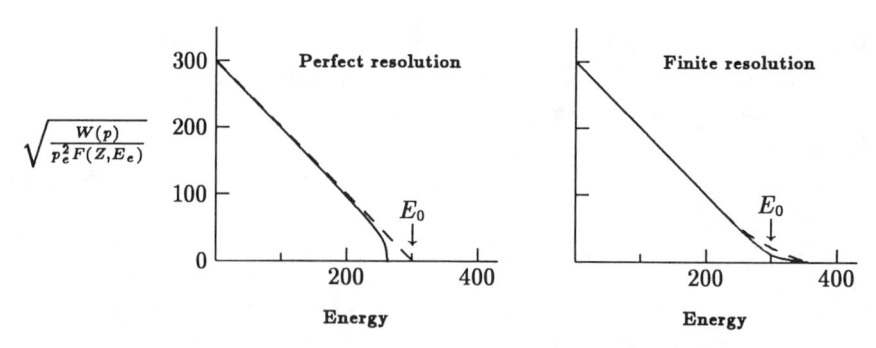

Fig. 7-8 Schematic diagram of a Kurie plot. Variations of the square root of the number of electrons (or positrons) with momentum P_e divided by $P_e^2 F(Z, E_e)$ are shown as a function of E_e, the energy of the electron (or positron) emitted in a β-decay. The solid lines are for the case of a finite neutrino mass μ_ν and the dashed lines are for $\mu_\nu = 0$. With perfect resolution, the plot, as shown in (a), is a straight line intersecting the horizontal axis at the end-point energy E_0 for $\mu_\nu = 0$. With finite resolution of the detector, the observed results in the region near the end-point energy are modified in the way shown in (b).

In the absence of uncertainties due to the apparatus used for the measurement, the number of electrons emitted with energy E_e is given by

$$W(E_e) = CF(Z, E_e)P_e^2(E_0 - E_e)\{(E_0 - E_e)^2 - (\mu_\nu c^2)^2\}^{1/2}$$
$$\xrightarrow[\mu_\nu \to 0]{} CF(Z, E_e)P_e^2(E_0 - E_e)^2 \qquad (7\text{-}32)$$

where C is a constant depending on the nuclear structure of the initial and final states involved in the decay and $F(Z, E_e)$ is the Fermi function that expresses the influence of the Coulomb field of the final nucleus on the electron emitted. The momentum P_e is that of the electron emitted and is related to the kinetic energy of the electron E_e through the relation $P_e^2/2\mu_e \approx E_e$. The quantity E_0, equal to 18578 eV, is the end-point energy. If the decay goes to the ground state of the ^{3}He atom, the maximum energy an electron can have is E_0.

For all the known nuclear β-decays, E_0 is much greater than any reasonable value expected of the rest mass energy of the neutrino. As a result, the expression for $W(E_e)$ is often approximated by the final form of (7-32), with the μ_ν term ignored. In this approximation, a plot of $\sqrt{W(E_e)/(F(Z, E_e)P_e^2)}$ as a function of E_e gives a straight line that intercepts the horizontal axis at electron energy $E_e = E_0$, as shown in Fig. 7-8. This is known as a Kurie plot (for more details, see Wong, *Introductory Nuclear Physics*, Prentice Hall, Englewood Cliffs, New Jersey, 1990). A finite neutrino mass is indicated by a slight departure from the straight line near the end-point energy, as can be seen from (7-32). Schematically, this is shown by the solid curves in Fig. 7-8.

In nuclear reactions, the energies involved are usually much higher than the excitation energies of the neutral atoms involved. As a result, atomic effects do not enter into any of the considerations. This is not true for neutrino mass measurements using tritium β-decay. The value of $\mu_\nu c^2$ is expected to be on the order of 30 eV or less, and there is an excited state of the ^{3}He atom at energy $E^* = 43$ eV above the ground state. In the decay of the tritium nucleus, there is certainly enough energy for the final state of the ^{3}He atom to be in the excited state. When this happens, the total amount of kinetic energy available to the electron and neutrino becomes $(E_0 - E^*)$, instead of E_0. As a result, (7-32) is modified, and the number of electrons emitted with energy E_e becomes a sum of two terms:

$$\mathcal{W}(E_e) = CF(Z, E_e)P_e^2 \Big[p_1(E_0 - E_e)\{(E_0 - E_e)^2 - (\mu_\nu c^2)^2\}^{1/2}$$
$$+ p_2(E_0 - E^* - E_e)\{(E_0 - E^* - E_e)^2 - (\mu_\nu c^2)^2\}^{1/2} \Big]$$
(7-33)

where p_1 and p_2, with the normalization condition $(p_1 + p_2) = 1$, are the probabilities of having, respectively, the ground and excited state of the ^{3}He as the final state. Unfortunately, our understanding of atomic physics at the moment is not sufficient for us to calculate p_1 and p_2 reliably, and this ignorance becomes one of the sources of uncertainties in deducing the value of μ_ν.

Experimentally, one can only detect the electron emitted in the decay, as neutrinos interact only weakly with matter. If our interest is in the mass of the neutrino that forms a part of the decay, it must be deduced indirectly from the shape of the distribution of the number of electrons emitted for a given energy E_e. Since the expected neutrino mass is small, its influence on the electron distribution is noticeable only where $(E_e - E_0) \approx 0$, as can be seen from (7-32). The most important region in a measurement is therefore around electron energy $E_e \sim E_0$. Unfortunately, this is also the place where the number of electrons emitted in the β-decay approaches zero, as can be see from the Kurie plot. For this reason, it is difficult to accumulate enough counts for a sound statistical analysis. Furthermore, any limitations in the measuring equipment tend also to skew the observed distribution in such a way that it can be easily confused with effects due to μ_ν. A typical set of measured results in the region around the end-point energy is shown schematically in Fig. 7-9.

Because of finite instrumental resolution, the number of electrons $N(P)$ observed at momentum P is modified from that given by (7-33). The actual form depends on the particular experimental setup and we shall not be concerned with this question here. [For details, see Lubimov and others, *Phys. Lett.* **94B** (1980), 266, and *Neutrino Mass and Related Topics:* Proceedings of the XVI INS International Symposium, Tokyo, 1988, World Scientific, Singapore, 1988]. In the analysis of Lubimov and coauthors, the following expression is used:

$$N(P_i) = A \sum_k \mathcal{W}(E_k)\{1 + \alpha(P_0 - P_k)\} R_{i,k} + \phi$$

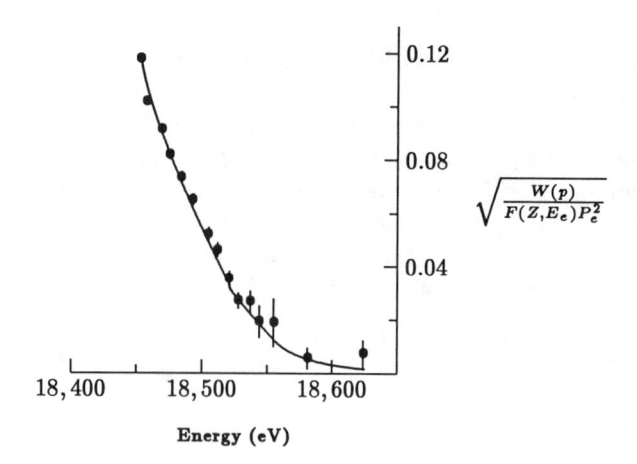

Fig. 7-9 Schematic diagram showing the scatter of data points in a Kurie plot for electrons near the end-point energy E_0 in the β-decay of tritium nuclei. The data are most sensitive to the mass of electron antineutrinos in this region. The solid line represents the Kurie plot if the neutrino has a small finite mass of ~ 30 eV/c^2.

where A is a normalization constant and $\mathcal{W}(E_e)$ is given by (7-33). The other factors are related to the detection equipment: α is a correction constant, $R_{i,k}$ is the resolution function or the probability for an electron emitted with momentum P_k and detected as an electron with momentum P_i, and ϕ is the background. Some of these factors cannot be determined exactly, either experimentally or from our theoretical knowledge. They, together with the value of μ_ν, are treated as parameters to be determined by a least-squares fit to the actual data obtained. Since these parameters reflect only the equipment and the type of sample (consequently, the atomic effects) used, they are independent of the value of neutrino mass of interest to us. Except for statistical fluctuations, these parameters will remain constant if we repeat the experiment using the same type of sample and the same apparatus.

Estimate of confidence region using Monte Carlo calculations. Since the experiment is very difficult to perform, it is not easy to accumulate a large number of measured values. For this reason, Monte Carlo calculations are used to test the reliability of the value of μ_ν obtained. Our interest here is to find the likelihood of obtaining the value of μ_ν deduced from the data. For this purpose, we can simulate on a computer the decay of the same number of tritium nuclei as found in the actual experiment. For each decay, we have no way of knowing the energy E_k and the momentum P_k of the electron emitted other than that they must satisfy energy and momentum conservation laws. In the simulation, we can replace E_k and P_k with random numbers of the correct distributions and subject them to the constraints imposed by the conservation laws. Similarly, the chance of reaching either one of the two possible final ^{3}He atomic states is also random, except that the probability must be equal to the value p_1 for ground state or p_2 for the excited state, with the values of p_1 and p_2 determined by the most reliable calculations available. The electron then goes through the detection apparatus and is recorded as having a momentum P_i, with a probability given by the resolution function $R_{i,k}$ and corrected by the factor $\alpha(P_0 - P_k)$. After the right number of decays are generated, we can add in the background in the form of random numbers having the distribution ϕ. The result is a set of Monte Carlo data. We can treat a set of such values as if it were obtained from an experiment and subject it to the same analysis as the original data. When this is done, a value of μ_ν is obtained. In general, the result will be different from that deduced from the experiment, since the individual pieces of data are not identical to those obtained in the actual experiment. If the results of several different sets of this type of data are accumulated, we have a distribution of the possible values of μ_ν. Figure 7-10 gives a plot of such a distribution using the values given in the article of Lubimov and coauthors. By setting a confidence level of 99%, the mass of electron antineutrino was determined to be in the interval 26 eV$\leq \mu_\nu c^2 \leq$ 46 eV in this way.

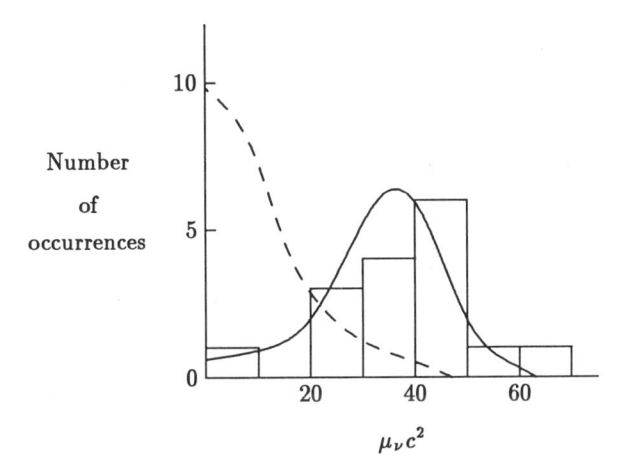

Fig. 7-10 Schematic diagram showing the histograms of the distribution of neutrino mass obtained from a Monte Carlo simulation of the experimental data. The solid curve corresponds to the expected distribution for $\mu_\nu c^2 = 35$ eV and the dashed curve to $\mu_\nu c^2 = 0$. (Adapted from Lubimov and others, *Phys. Lett.* **94B** [1980], 266.)

Subsequent to the 1980 publication, many experiments have been conducted by several different groups around the world to make independent measurements of the value of μ_ν. Because of the tremendous demand on the accuracy in the measurements, there is still no consensus on the actual value of μ_ν. Indirect measurements, such those from the neutrino signals detected from supernova 1987a, put an upper limit on the value that is not inconsistent with that quoted above. However, this question is not of prime interest to us here.

The procedure outlined above for testing our hypothesis about a given physical situation and to establish a confidence region is a general one and can be applied to many similar cases. The approach may be summarized in the following way:

(1) We have a physical system described by a set of parameters a_{true} whose values are unknown to us.

(2) Through measurement we obtain a set of data points D_{exp}.

(3) Using least-squares techniques, we obtain a set of best-fit values for the parameters a_{fit} from D_{exp}.

(4) With a_{fit}, simulations of the experimental data using Monte Carlo techniques are carried out. Each of these simulations gives us a set of "data" $D_{\text{sim}}(i)$.

(5) Least-squares analyses are performed on the simulated data $D_{\text{sim}}(i)$. From each group of simulated results, a set of values $a_{\text{fit}}(i)$ of the parameters is obtained.

(6) From the distribution of $a_{\text{fit}}(i)$ for $i = 1, 2, \ldots$, the confidence interval of a is established for a given confidence limit.

In the above steps, it is understood that the fitting function and the least-squares procedure applied to the simulated data are the same as those used on the true data. Since Monte Carlo calculations are usually far easier to carry out than most experiments, the method becomes a powerful way to provide a meaningful interpretation of the actual data. In particular, the technique is indispensable in situations when the experiment cannot be repeated.

In this section, we have essentially followed the traditional approach to statistical treatment of data. For the most part, these methods were developed before the birth of modern computers. As a result, there is a reluctance to use techniques that require extensive computation. Evidence for such a point of view can be found, for example, in the fact that statistical measures applied to the analyses are usually restricted to those that have analytical results. With the widespread use of computers in laboratories, such limitations have become artificial and many other methods of statistical analyses of data are now feasible. We shall not go into these new methods here since they have yet to filter down to the problems in physics we wish to address. For an introduction to these new methods, the reader is urged to consult, for example, the *Scientific American* article on *Computer-Intensive Methods in Statistics* by Diaconis and Efron that starts on page 116 in the May 1983 issue.

§7-4 Percolation and critical phenomena

An interesting feature in many physical systems is the long-range order induced by short-range interactions. For example, the typical range of forces acting between molecules is on the order of the diameter of a single molecule, and yet collective behavior can take place involving as many as millions of molecules. A good example is ferromagnetism. It is well known that many atoms and molecules possess weak magnetism resulting from the intrinsic spin and orbital motion of the electrons. The magnetism observed in ferromagnetic materials, iron, nickel, and cobalt, is, however, much stronger and requires the alignment of a large number of atoms. In other words, many atoms must act in a coherent or collective manner in order to produce the observed strength. On the other hand, the interaction between the atoms in ferromagnetic materials has only a range on the order of 10 angstroms, not much larger than the diameter of each molecule. This takes place also in many other collective phenomena where the role played by the interaction is a minor one and the physics is determined primarily by statistical considerations. A class of such phenomena is *percolation*, a special type of phase transition that is observed in condensed matter and other many-body systems.

Phase transition. In statistical mechanics, we are usually dealing with the behavior of systems made of large numbers of individual atoms or molecules. The influence of the surroundings is characterized in terms of such parameters as temperature, pressure, and applied magnetic field. Because of the dominance of large numbers, typically on the order of Avogadro number (10^{23}), the properties of these systems are governed to a large extent by statistics. As a result, they usually vary smoothly as functions of the external parameters. The exceptions occur at phase transitions. Typical examples are water changing from liquid to gas (steam), a ferromagnetic material being raised to temperatures above the Curie point and losing its special magnetic property, and a superconductor being heated above its critical temperature and turning into a normal conductor.

The changes in the properties occurring during a phase transition can often be characterized by a single parameter, known as the *order* parameter. For example, for ferromagnetic materials, we may use as the order parameter the magnetic susceptibility

$$\chi_m = \frac{\partial M}{\partial H} \qquad (7\text{-}34)$$

where M is the magnetization and H is the magnetic field. Alternatively, we may use the specific heat

$$C_v = \left(\frac{\partial U}{\partial T}\right)_v \qquad (7\text{-}35)$$

where U is the internal energy and T is the temperature. The term order parameter comes from the observation that quantities, such as χ_m and C_v, characterize the change from an ordered state, such as the ferromagnetic state or the superconducting state of a material, to a disordered one.

Critical phenomena. For some materials, the order parameter itself is discontinuous across a phase transition. This is the case of a first-order phase transition. Our interest here is mainly in second-order phase transitions in which the order parameter itself is continuous at the phase transition but its first derivative is not.

A simple illustration, given by Schulman and Seiden (*Science* **233** [1986] 425), is the spread of a hypothetic disease, "percolitis." Consider a very special group of N individuals living in a completely isolated community. For simplicity, we shall assume that each individual makes one and only one contact each day with every other individual in the community. Furthermore, we shall assume that percolitis has an incubation period of exactly one day and lasts also only for a day. The way the disease is spread is by direct contact between two individuals and the probability of catching the disease from an affected individual is p in each contact with such an individual. At a given time t, measured in units of days, let the number of individuals afflicted with percolitis be $n(t)$. The probability that any one individual in the community is suffering from the disease is given by

$$\rho(t) = \frac{n(t)}{N}$$

The probability for any one not having percolitis at time t is therefore $\{1 - \rho(t)\}$.

Our immediate interest here is the probability for an individual to catch percolitis. In a contact with a diseased individual, the probability not to catch percolitis is $(1-p)$. Among the $(N-1)$ contacts one makes with different individuals in each day, we need only be concerned with those involving individuals afflicted with percolitis. Since there are $n(t)$ such individuals in the community, the probability for not catching the disease is $(1-p)^{n(t)}$. This is then the fraction of the population not afflicted with percolitis on day $(t+1)$. By definition, this is equal to $\{1 - \rho(t+1)\}$. As a result, we have the equation

$$1 - \rho(t+1) = (1-p)^{n(t)} \tag{7-36}$$

The relation may be used to solve for $\rho(t)$. For this purpose, let us define

$$x \equiv Np$$

and, on making use of the following infinite series expansion of the exponential function, as we did earlier in (6-15),

$$e^x = \lim_{s \to \infty} \left(1 + \frac{x}{s}\right)^s$$

the right side of (7-36) may be put into the form

$$(1-p)^{n(t)} = \left(1 - \frac{x}{N}\right)^{n(t)} = \left(1 - \frac{x\rho(t)}{n(t)}\right)^{n(t)} \approx e^{-x\rho(t)}$$

On substituting this result to the right side of (7-36), we have the result

$$\rho(t+1) = 1 - e^{-x\rho(t)} \tag{7-37}$$

If a continuous process is used for contracting the disease rather than our simple step function in t, the same relation will be in the form of a differential equation. For our purpose here, it is not worthwhile to bring in the additional complication in mathematics involved in deriving the more proper result.

We wish now to find out the fraction of the population that will ultimately contract percolitis. This is given by

$$\rho_{max} = \lim_{t \to \infty} \rho(t)$$

At $t = \infty$, (7-37) takes on the form

$$\rho_{max} = 1 - e^{-x\rho_{max}} \qquad (7\text{-}38)$$

There are many ways to solve this equation. It is instructive for us to take a graphical approach. The behavior of the left side of (7-38) as a function of ρ_{max} is simple and given by

$$y_1 = \rho_{max}$$

The behavior of the right side of the equation has the form

$$y_2 = 1 - e^{-x\rho_{max}}$$

The value of ρ_{max} that satisfies both equations depends also on x, related to the probability of catching percolitis on each contact. The forms of y_2 for three typical values of x are shown in Fig. 7-11.

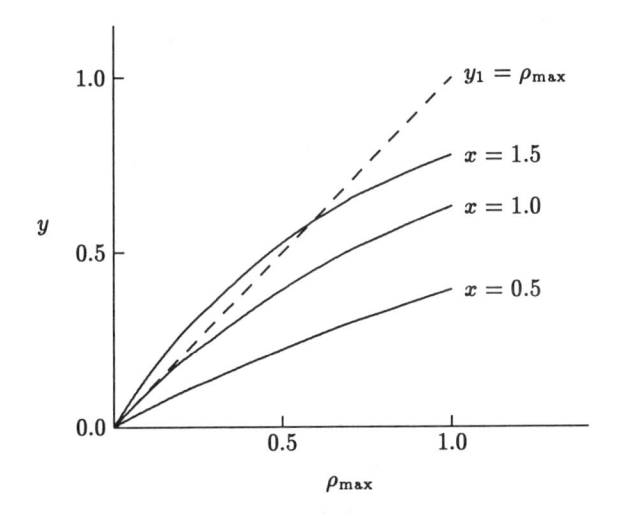

Fig. 7-11 Illustration of a second-order phase transition given by (7-38). The solution to the equation is given by the intersection of $y_1(\rho_{max}) = \rho_{max}$ (dashed line with slope unity) and $y_2(\rho_{max}) = 1 - \exp(-x\rho_{max})$ (solid curves labeled as $x = 0.5$, 1.0, and 1.5). For $x \leq 1$, we have $y_2(\rho_{max}) < y_1(\rho_{max})$ for all values of $\rho_{max} > 0$ and the only solution is $\rho_{max} = 0$. For $x > 1$, a second solution is found for $\rho_{max} > 0$, with a value that increases with the value of x.

For a given x, the solution of (7-38) occurs at the intersection of $y_1(\rho_{\max})$ with $y_2(\rho_{\max})$. If $x \leq 1$, the value of y_2 is always less than y_1, as can be seen in Fig. 7-11. In this case, there is only one possible solution for (7-38) at $\rho_{\max} = 0$. This means that percolitis cannot spread if p is less than $1/N$. That is, it will eventually die out in the community. However, for $x > 1$, the value of y_2 starts off larger than $y_1 = \rho_{\max}$ and falls eventually below the value of y_1 as $\rho_{\max}$ approaches the maximum value of 1. We now have a solution of (7-38) for $\rho_{\max} > 0$. The actual value depends on x, but we are not particularly concerned with the precise answer here. The important thing for us to realize is that we have a critical point at $x = 1$. Let us call this value of x as x_c. Below $x = x_c$, percolitis will eventually die out in the community, and above $x = x_c$, there will ultimately be a stable number in the community suffering from percolitis on a particular day. At $x = x_c$, we have a phase transition. The order parameter here is $\rho_{\max}$ and its value as a function of x is continuous at $x = x_c$. However, the first derivative $\partial \rho_{\max}/\partial x$ is discontinuous at $x = x_c$. The phase transition is, therefore, second order.

The example is a nice and effective way to illustrate the phenomenon of phase transition. In spite of its simplicity, the model is fairly close to reality. In fact, Schulman and Seiden were able to demonstrate by computer simulation that the structure of spiral galaxies can come from a percolation phase transition if the stars are rotating with different velocities. This is a departure from the traditional points of view that are based on the dynamics of the system.

Critical exponents. A critical phenomenon is usually associated with the divergence of some physical quantity. For our percolitis example, the quantity is the relaxation time

$$\Delta(t) \equiv \rho_{\max} - \rho(t)$$

the time for the disease to reach its equilibrium level in the community. It can be shown (L.S. Schulman and P.E. Seiden, *J. Stat. Phys.* **27** [1982] 83) that, as $\rho(t) \to \rho_{\max}$,

$$\Delta(t) \sim \text{constant} \times e^{-t/\xi}$$

where ξ is a time constant, generally known as the *correlation length*. Near the critical point, we have

$$\xi \sim |x - x_c|^{-1}$$

In general, the exponent is not a simple integer as we have here, and the behavior of ξ near the critical point has the form

$$\xi \sim |x - x_c|^{-\nu} \tag{7-39}$$

The quantity ν is known as the critical exponent.

For most of the critical phenomena studied, the second-order phase transition takes place at a particular temperature, T_c, known as the critical temperature.

For different systems, the value of T_c may be different, but the behaviors at temperatures near T_c are often very similar. For example, below T_c, the magnetization of a ferromagnetic material decreases with increasing temperature and vanishes at $T = T_c$. At $T \sim T_c$, the magnetization is given by a simple power law,

$$\mathcal{M} \propto (T_c - T)^\beta$$

Again, the exponent β here is not an integer. For iron, the value is measured at the critical temperature of $T_c = 1044$ K to be

$$\beta = 0.34 \pm 0.02$$

The specific heat at zero magnetic field ($H = 0$), on the other hand, is given by

$$C \propto \begin{cases} (T - T_c)^{-\alpha} & \text{for } T > T_c \\ (T_c - T)^{-\alpha'} & \text{for } T < T_c \end{cases}$$

For iron, it is found that

$$\alpha = \alpha' = 0.12 \pm 0.01$$

Similarly, at $H = 0$, the magnetic susceptibility is given by

$$\chi_m \propto \begin{cases} (T - T_c)^{-\gamma} & \text{for } T > T_c \\ (T_c - T)^{-\gamma'} & \text{for } T < T_c \end{cases}$$

For iron, we have

$$\gamma' \approx \gamma = 1.333 \pm 0.015$$

At the critical temperature, the magnetization is proportional to the magnetic field following the power law

$$\mathcal{M} \propto H^{1/\delta}$$

The measured value for nickel at $T_c = 631.58$ K is known to be

$$\delta = 4.2 \pm 0.1$$

Quantities such as α, β, γ, and δ defined above are known as the *critical exponents* and physical properties that go to zero or infinity following such simple power laws are known as critical phenomena. For a more comprehensive introduction, the interested reader should consult specialized books such as *Modern Theory of Critical Phenomena* by Ma (Benjamin, Reading, Massachusetts, 1976), *Introduction to the Renormalization Group and to Critical Phenomena* by Pfeuty and Toulouse (English translation by G. Barton, Wiley, London, 1977), and Huang (*Statistical Mechanics*, 2nd ed., Wiley, New York, 1987).

One reason that critical phenomena are interesting is the fact that critical exponents are observed to be essentially the same for a variety of materials. In general, these materials may have quite different critical temperatures, but their behaviors are very similar to each other near their critical points. This is known as *universality* and some quantities are more universal than others. The reason why there should be any difference is an intriguing question as well. Since the values of these exponents are not integers, they cannot come from some simple analytical relations. Furthermore, since the behavior near a critical point is singular in nature, standard perturbative methods are not applicable and new methods of investigation must be explored. For this reason, Monte Carlo calculations and computer simulations have become an important tool in the study of critical phenomena.

Percolation. One of the more direct ways to make computer simulations of critical phenomena is through percolation studies. The usual meaning of percolation is the process by which water passes through a porous body, such as a filter. In physics, it is best described by the example of a collection of random, two-dimensional clusters of conductors. Consider two parallel conducting bars, each of length L and separated from each other by a distance equal to their length. Let us divide the area between the top and bottom conductors into $(N \times N)$ small squares. Electricity can flow from one of these squares to its neighbor if both are occupied by conductors. For simplicity, we shall consider that any square not occupied by a conductor is filled with an insulator. A current can flow between two conductors that are located side by side, like this ■■, and on top of each other, like this ▪. However, diagonal passages, such as ▪ and ▪, are not allowed.

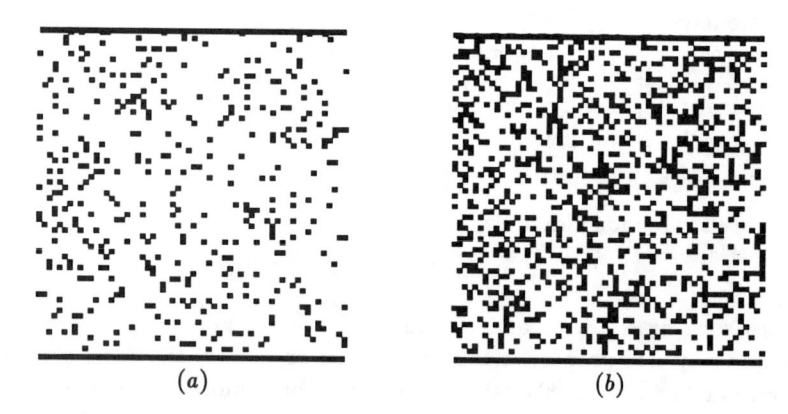

Fig. 7-12 Random conductors, indicated as black squares, filling up a square lattice with probability of $p = 0.1$ in (a) and $p = 0.3$ in (b).

For our present purpose, a group of conductors that are joined together so that charge carriers can flow between them is called a *cluster*. In percolation studies we are interested in the formation and size of such clusters. In particular, we want to know whether a current can flow between the top and bottom conducting bars if the space in between is filled randomly with small square conductors.

Clearly, the answer depends on the number of sites occupied by conductors. In other words, the question is related to the probability p that a square is occupied by a conductor. An illustration is given in Figs. 7-12 and 7-13 for a case of $N = 60$. In Fig. 7-12a, the (60 × 60) squares in between the two bars are filled with probability $p = 0.1$. That is, on the average, one-tenth of the lattice sites are occupied by conductors and the rest are filled with insulators. It is easy to see that most of conductors are in isolated clusters and there is no possibility for a charge carrier to move from the top bar to the bottom bar. One way of classifying such a state is to say that the average size of conducting clusters is small. When p is increased to 0.3, that is, 30% of the sites are occupied on the average, the cluster sizes become much larger. However, no continuous path can be found between the top and bottom, as can be seen by looking at Fig. 7-12b. On the other hand, a careful examination of the $p = 0.6$ result, shown in Fig. 7-13a, reveals that there are several such paths available. In this case, charge carriers can now "percolate" between the top and the bottom bars. When the value of p is increased to 0.9, the whole space is a single large cluster (with 10% holes), as can be seen in Fig. 7-13b.

If the size of our two-dimensional lattice is infinite instead of (60 × 60), the critical probability, or percolation threshold, is known to take place at $p_c = 0.59275$. For finite-size lattices, the transition is not a sharp one and this is one of the limitations of computer simulation, as we shall see in more detail later. The onset of percolation may be regarded as a phase transition. For our random conductor example, the space in between the two conducting bars changes from an insulator to a conductor when there is at least one cluster extending from one bar to the other. Instead of a random collection of conductors, we can imagine the same phenomenon taking place in a more realistic situation in the form of an idealized polymerization process, whereby (noninteracting) molecules coagulate into large molecules. The probability p in this case is analogous to temperature T in critical phenomena we encountered earlier. In this way, many behaviors of a system near the critical point may be examined through percolation studies.

In addition to our simple conducting square example, many other types of lattices can also be simulated on a computer. For example, instead of allowing conduction between only neighboring sites in a square lattice, we can have a triangular lattice. In this case, if we use label i to denote the row number of each site and label j the location of a site in the row, then conduction can take place between an occupied site (i, j) and any one of the three occupied neighboring sites $(i - 1, j)$, $(i, j - 1)$, and $(i - 1, j - 1)$. Instead of two-dimensional lattices, we can also construct three-dimensional ones. For example, our square lattice can be generalized into a cubic one if, in addition to conduction with sites to the immediate left, right, top, and bottom, we include also front and back. In fact, for computer simulation, we can work with arbitrary dimensions, a question of interest in many situations but difficult to carry out in controlled experiments. In this way,

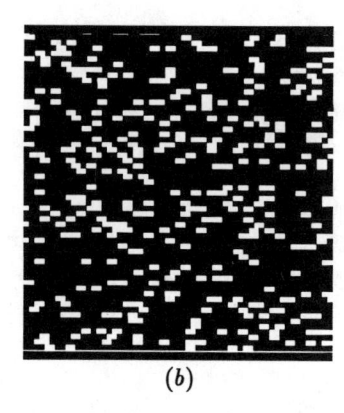

Fig. 7-13 Random conductors, indicated as black squares, filling up a square lattice with probability of $p = 0.6$ in (a) and $p = 0.9$ in (b).

numerical experiments through computer simulations become an important way to understand the underlying physics in critical phenomena and related subjects.

There are some subtle differences between percolation and critical phenomena, but we shall not enter into a discussion of such finer points here. Our main purpose here is to give an introduction to the role of computer simulation in percolation studies, using a square lattice in two dimensions as the example. For a more comprehensive review the reader should consult monographs such as *Introduction to Percolation Theory* by Stauffer (Taylor and Francis, London, 1985) and the series on *Phase Transitions and Critical Phenomena* edited by Domb and Lebowitz (Academic Press, London).

Simulation of a square lattice. Let us examine some of the details in carrying out a computer simulation of percolation using a square lattice of somewhat larger size than our example above. There are three steps involved. The first is to divide the area into a lattice of small squares. For simplicity, we shall take the area to be square with each side divided into N equal parts. The size of N depends on the total number of little squares we wish to have. For our illustrative examples above, a value of $N = 60$ was used. This is sufficiently small so that a small personal computer can handle the calculations involved with ease. On the other hand, we cannot hope to mimic realistic situations involving the order of 10^{23} particles with such a small lattice. We shall soon see that, even for a larger N of 500, finite size effects remain significant.

The second step in the simulation is to put small square conductors randomly with probability p into the $(N \times N)$ lattice sites. This may be done by starting from the top row and going through each of the N sites one by one, from the left to the right. For each square, a random number x with uniform distribution in the interval $[0, 1]$ is generated. If $x \leq p$, a conductor is put into the site, and if $x > p$, the site is filled with an insulator. The complete process of checking and filling all the sites requires $(N \times N)$ random numbers. However, this is a relatively simple part of the calculation.

The third step is to analyze the result. We shall be concerned with only two questions here: the sizes of the clusters and whether there is a contagious path of conductors from the top to the bottom. If the lattice is small, such as the (60×60) example we used in Figs. 7-12 and 7-13, it is possible to check the conduction path by hand. This can be done, for example, by plotting the lattice on a sheet of paper and joining the occupied neighboring sites using a pencil. In more realistic applications, where the lattices are much larger, a better method is needed. In fact, this is the most involved part of the simulation.

We shall describe a method based on the technique published by Hoshen and Kopelman (*Phys. Rev.* **B14** [1976] 3438) whereby the analysis of the result is carried out along with the process of filling the lattice. To group occupied sites into clusters, we shall associate a label with each cluster. All sites belonging to the same cluster bear the same label. For example, the first cluster that appears as we go through the lattice sites is numbered 1, the second 2, and so on. To simplify the checking process, the insulator sites can be assigned the label 0 or some number X that is larger than any possible label to be assigned to an occupied site. In practice, the latter choice is more efficient on a computer, as we shall see soon.

The assignment of a cluster label to a lattice site is carried out as soon as we have decided to fill it with a conductor or an insulator. For a square lattice, it is possible to find out whether a site newly occupied by a conductor is a member of an existing cluster by checking whether the site to its left and just above are conductors. There are three possibilities. The first is that both neighbors are unoccupied. Here, we have the beginning of a new cluster. The counter for the total number of clusters is increased by 1 and the value is assigned as the cluster number of the new site. The second possibility is that either the left or top cell is occupied. In this case, the newly occupied site belongs to the same cluster as the occupied neighbor and, as a result, inherits the same cluster label.

The third possibility is that both neighbors are occupied. Two different situations can occur here. The first is that both neighbors belong to the same cluster. The newly occupied site joins them in the same cluster and takes on the same cluster label. The second situation is that the two neighbors belong to two different clusters. As a result of filling the site in question with a conductor, the two clusters that are separated so far are now joined together to form a single cluster. We shall adopt the rule here that, when two different clusters are joined together to form a single cluster, the lower one of the two cluster labels becomes the label for the union.

At this stage, in principle we can go back and change the labels of all the members of the cluster to the smaller cluster label. This is very inefficient, as can be seen from the following considerations. To be able to change the cluster label of a site, the information must be stored somewhere. This means a large array to record the labels and to keep them up to date. A more efficient way is to associate a counter $L_c(i)$ with each cluster. If cluster i is an isolated cluster, $L_c(i)$ keeps track of the number of members in the cluster. That is, each time a new member is added to cluster i, the counter $L_c(i)$ is increased by 1. When two clusters i and j merge, the value of $L_c(j)$ is added to that of $L_c(i)$ (assuming $j > i$), such that

$$L_c(i) + L_c(j) \to L_c(i)$$

As a result, $L_c(i)$ now reflects the size of the newly merged cluster. Instead of updating the labels of all the former members of cluster j, we shall change $L_c(j)$ to become a pointer, stating the fact that now cluster j no longer exists and all its members are merged with those of cluster i. This can be achieved by, for example, making the assignment

$$L_c(j) = -i$$

In this way, whenever we encounter a cluster label j, we examine the value of its counter $L_c(j)$. If $L_c(j) > 0$, it is a good cluster with $L_c(j)$ members. On the hand, if $L_c(j) < 0$, cluster j is delinquent and we shall refer, instead, to the counter of the cluster whose label is given by $|L_c(j)|$. It is possible that the counter cluster label $L_c(j)$ "points" to is also a pointer. In this case, we shall follow the new pointer to another label, and so on, until we come to a good cluster number. By following the pointers we can ultimately arrive at the good cluster for any occupied site. Since the number of members in each of the good clusters is stored in the counter, we have also the advantage of having the size of each cluster available to us when we finish with the process of filling the lattice.

This method also gives us a natural way to find out whether the top and bottom bars are connected, that is, whether charge carriers can percolate through the lattice. As we saw earlier, if a lattice site is occupied by a conductor, it bears the label of the cluster to which it belongs. If it is occupied by an insulator, the label is X. For each row we can store this information in a one-dimensional array of N integers. For conduction to go from the top (first row) to the bottom (last row), these two rows must share at least one common cluster. Thus, by comparing the array of cluster labels for the first row with a similar array for the last row, taking care to change all the cluster labels in both rows first to those of "good" clusters, we can tell at once whether there can be an electric current flowing between them or not.

Note that, besides the cluster counters $L_c(i)$, only two one-dimensional arrays are needed in this method, one to keep a copy of the occupancy pattern of the first row and one for the row just above the one we are in the process of filling. When a new site is filled with a conductor, we need the information on the cluster label of the site to the left and on top in order to determine the cluster to which the newly occupied site belongs. Let us assume that this information for row i is stored in the array ℓ_k for $k = 1, 2, \ldots, N$. Consider now site j of row $(i+1)$. The cluster label of the neighbor on top is in ℓ_j. This is all the information we need

Box 7-5 Program PER_CNT
Percolation on a Square Lattice

Definitions:
    N: Number of site in a row and number of rows in the lattice.
    X: Label for an empty site.
    n_c: Number of different clusters.
    $\ell(i)$: Cluster number of site i of a row.
    $f(i)$: Cluster number of site i for the first row.
    $L_c(i)$: If > 0, it is the length of cluster i; if < 0, it is a pointer indicating that the cluster is now merged with cluster $y = |L_c(i)|$.

Initialization:
  (a) Input random number seed and occupation probability p.
  (b) Initialize $n_c = 1$ and set the first element of a row $\ell(1) = X$.

Subprograms used:
    CLPS: Eliminate merged clusters from the array.
    ANALYSIS: Analysis of cluster size.
    RSUB: Uniform random number generator (Box 7-2).

1. First row: Fill the site $i = 2$ to $(N+1)$.
  (a) Generate a random number r.
  (b) If $r \leq p$, the site is to be occupied. Let $\ell(i) = c$, with c equal to:
    (i) If the left neighbor is empty, increase n_c by 1, and set $L_c(n_c) = 1$ and $c = n_c$.
    (ii) If occupied, set $c = \ell(i-1)$ and increase $L_c(\ell(i))$ by 1.
  (c) If $r > p$, mark the site as unoccupied by setting $\ell(i) = X$.
2. Keep a copy of $\ell(i)$ in $f(i)$.
3. For rows 2 to N, carry out steps 4 to 7 for $i = 2$ to N in each row.
4. Generate a random number r.
5. If $r > p$, mark the site as unoccupied by setting $\ell(i) = X$.
6. If $r \leq p$, the site is to be occupied. The cluster number c is given by:
  (a) If both the neighbor to the left and above are empty, increase the cluster number n_c by 1 and set $c = n_c$ and $L_c(n_c) = 1$.
  (b) If the neighbor on the left or above is occupied, let y equal to the cluster number of the occupied site.
    (i) If $L_c(y) > 0$, y is a "good" cluster; set $c = y$ and add 1 to $L_c(y)$.
    (ii) If $L_c(y) < 0$, y is "delinquent." Let $y = -L_c(y)$ and return to (i).
  (c) If both the neighbor to the left and above are occupied:
    (i) Find the "good" cluster number y_1 of $\ell(i-1)$ and y_2 of $\ell(i)$ using steps (i) and (ii) of (b) above.
    (ii) If $y_1 = y_2$, let $y = y_1$.
    (iii) If $y_1 \neq y_2$, merge the two clusters by
        Setting $y = \min(y_1, y_2)$ and $z = \max(y_1, y_2)$.
        Letting $L_c(y) = L_c(y) + L_c(z)$ and $L_c(z) = -y$.
7. At any stage, if n_c becomes equal to X, use CLPS to eliminate the merged clusters from the list and update the cluster numbers in $\ell(i)$ and $f(i)$ accordingly.
8. When all N lines are filled, call ANALYSIS to carry out the following steps:
  (a) Group the clusters according to size and output the number in each group.
  (b) Find the number of clusters common to the top and bottom.

for site j of row $(i+1)$, as far as its relation with the previous row is concerned. Once this is used, the information in location ℓ_j is no longer needed and we can use it to store the cluster label of site j for row $(i+1)$. If this rule is followed from the beginning of filling row $(i+1)$, the information concerning the site to the left, site $(j-1)$ in row $(i+1)$, is stored in location ℓ_{j-1}. In other words, as we fill the sites in row $(i+1)$, the information in array ℓ_j is replaced by the cluster label of the new row. In this way, the two pieces of information we need to assign the cluster label to a new site in row $(i+1)$ are both stored in a single array $\{\ell_k\}$.

For the convenience of programming, we shall add a dummy site to the left of the first one in each row and fill this site with an insulator. In this way, the starting cell is the second member of a one-dimensional array and all the comparisons of cluster labels can be carried out without having to pay special attention to whether a cell is the first one in the row and has no left neighbor. The algorithm is summarized in Box 7-5.

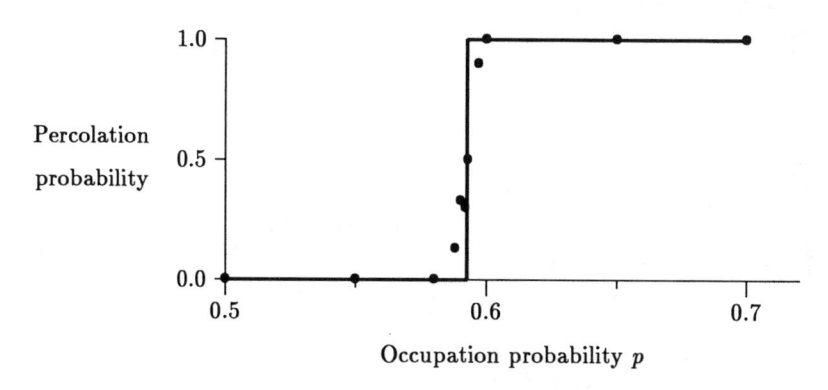

Fig. 7-14 Probability of percolation as a function of the occupation probability p. The dots are the results for a finite lattice of (500 × 500) sites and the solid line is for an infinite lattice, having a percolation threshold $p_c = 0.5927$.

Finite size and other problems. Using the algorithm described above, a percolation study is made on a (500 × 500) square lattice. Since the percolation threshold p_c for an infinite square lattice is known to occur at $p_c = 0.59275$, we restrict our study to occupation probability between 0.5 to 0.7. The calculated results show that the average cluster size increases from 7.5 to 95, as p increases from 0.5 to 0.7. However, there are no dramatic changes around $p = 0.59$. It is perhaps more interesting to examine instead the changes in the size of the largest cluster. For example, at $p = 0.5$, clusters larger than 1000 are very rare. At $p = 0.55$, the average cluster size is 12 and the size of the largest cluster is around 3000. At $p = 0.58$, the corresponding numbers are 20 and 10,000. When we get to $p = 0.59$, large clusters consisting of over 50,000 members begin to appear, even though the average size remains only around 23. At $p = 0.6$, the average size is 24, hardly different from that for $p = 0.59$. On the other hand, the largest clusters now have over 100,000 members. For our (500 × 500) square lattice with 60% occupation ($p = 0.6$), the total number of occupied sites is only 150,000. This means that almost 70% of all the conductors are connected as a single cluster. At $p = 0.7$, the same fraction increases to 98%.

The percolation phenomenon for which we are looking is connected to the question of whether there is a phase transition in the region between $p = 0.59$ and $p = 0.60$, indicated by some sharp changes taking place in the region. The results obtained above do not show any discontinuity in the properties of the system; however, they are sufficiently encouraging for us to take a closer look at our simulation. For this purpose, let us examine whether there is a sudden change as far as percolation is concerned, that is, whether there is some value of p below which it is impossible for a charge carrier to move from the first row of the lattice to the last row and above which there is conduction. Since we are doing a Monte Carlo simulation, a single calculation cannot determine the outcome. This point may be illustrated by the following consideration. At the low occupation probability of $p = 0.002$, there are, on the average, only 500 sites occupied in the lattice. On the other hand, even with such a small number, there is an extremely small, but nonzero, probability that all 500 occupied sites align on top of each other and form a conduction path. The same low probability also exists for no percolation at high p values. This can happen at $p = 0.998$, for example, if all 500 insulators are in the same row. For this reason, the simulation for given p values must be repeated many times, each starting with a different input random number seed. Any conclusion to be drawn from the simulation must be inferred from the distribution of many sets of calculated values.

For simplicity, we shall only carry out ten different runs for each p value. The probability of observing a percolation in the ten run, the fraction of ten runs with percolation, for different values of p between 0.5 and 0.7 is shown in Fig. 7-14. In principle, we should have carried out far larger numbers of runs to obtain better statistics, but this is not the central issue for our illustration here. It is quite clear from the figure that the simulated results do not indicate a sharp transition in percolation as a function of p, as required of a critical phenomenon. This is not a surprise as far as our calculations are concerned. We have seen in the previous paragraph how percolation can take place in samples of finite size even at very low values of p. However, the probability of observing percolation at low p values decreases with increasing size of the lattice. To illustrate this point, let us consider a square lattice of ($\ell \times \ell$) sites. If the probability that any one site is occupied is p, the probability for all the sites in a column being occupied is then p^ℓ and goes to zero (since $p < 1$) in the limit that $\ell \to \infty$. By the same token, a sharp change in a simulation of percolation can only appear for an infinite lattice.

In any computer simulation, it is only possible to work with a finite ℓ. One way to have some feeling for the size of a realistic lattice is to examine the number of atoms in a sample of reasonable size. The order of magnitude is given by the Avogadro number, 10^{23}. If this number of atoms is arranged in the form of a cubic lattice, each side will have more than 10^7 atoms. A realistic simulation may not require ℓ of this size, but certainly a size much larger than anything we can contempt at this moment on a computer. As a result, finite-size effect becomes one of the problems in percolation studies using a computer. For a simple square lattice, a percolation threshold exists only in the limit of infinite lattice, and we must somehow infer the results for an "infinite" lattice from studies made with finite ones. The procedure to carry out this task is more complicated and specialized than we can go into here.

§7-5 The Ising model

In the previous section, we dealt with the formation of clusters from particles that have no interaction with each other. This is evident from the fact that, in the calculations, we did not make any references to the energies involved. Furthermore, in determining whether a site in the lattice should be occupied by a conductor or an insulator we did not pay any attention to the occupancies of the neighbors. Realistic physical systems are usually somewhat different from such idealized situations. For many problems in condensed matter physics and field theory, the interaction between neighboring particles is important. This is true even for particles localized on a lattice. Historically, the best known example of the statistical properties of a system that includes interaction between neighbor particles is perhaps the Ising model.

The model was introduced originally to understand ferromagnetism. Ferromagnetic materials are characterized by the existence of a magnetic moment even in the absence of an external field. On a macroscopic scale, the strength of magnetism is measured in terms of the *spontaneous* magnetic moment, and it decreases smoothly with increasing temperature until a critical value T_c, known as the Curie temperature, is reached. For iron, this occurs at $T_c = 770°$ C. For cobalt and nickel, the corresponding values are 1130° C and 358° C, respectively. A simple model to understand the magnetic property of these materials is to consider all the molecules to be localized in space. For convenience, we shall further constrain the model to be two-dimensional in the form of a square lattice. There is no fundamental difficulty in generalizing computer simulations to the more realistic case of a three-dimensional lattice, but we shall not do it here.

For our square lattice, the only degree of freedom at each lattice site is in the orientation of the molecular spin J. Let us use the symbol $J_{i,j}$ to represent the value of J at site (i,j). To simplify the discussion, we shall only allow two possible orientations, $J_{i,j} = +1$ (up) or -1 (down). An example of such a lattice is shown in Fig. 7-15.

Chapter 7] Monte Carlo Calculations 435

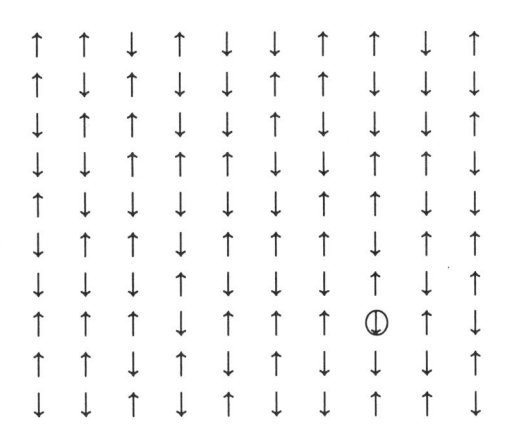

```
↑ ↑ ↓ ↑ ↓ ↓ ↑ ↑ ↓ ↑
↑ ↓ ↑ ↓ ↓ ↑ ↑ ↓ ↓ ↓
↓ ↑ ↑ ↓ ↓ ↑ ↓ ↓ ↓ ↑
↓ ↓ ↑ ↑ ↑ ↓ ↓ ↑ ↑ ↓
↑ ↓ ↓ ↓ ↓ ↓ ↑ ↑ ↓ ↓
↓ ↑ ↑ ↓ ↑ ↑ ↑ ↓ ↑ ↑
↓ ↓ ↓ ↑ ↓ ↓ ↓ ↑ ↓ ↑
↑ ↑ ↑ ↓ ↑ ↑ ↑ ⓛ ↑ ↓
↑ ↑ ↓ ↑ ↓ ↑ ↓ ↓ ↓ ↑
↓ ↓ ↑ ↓ ↑ ↓ ↓ ↑ ↑ ↓
```

Fig. 7-15 Schematic representation of a two-dimensional Ising model lattice. Interaction between spins can take place only between nearest neighbors. The strength is $+\epsilon$ if two adjacent spins are oriented in the same direction and $-\epsilon$ if they are oriented in opposite directions. Thus, for the site marked by a circle, the four nonvanishing interaction strengths are $-\epsilon$, $-\epsilon$, $-\epsilon$, and $+\epsilon$, if we start from the neighbor to the left and move in a clockwise direction.

The model Hamiltonian. In our model, the only "internal" interactions we allow are those between the spins of two nearest neighbors $J_{i,j}$ and $J_{i',j'}$, with $(i'j') = (i\pm 1, j)$ or $(i, j\pm 1)$. The interaction energy involved may be parametrized in the following way. If the spins of both sites are oriented in the same direction, the contribution to the energy of the whole system is $+\epsilon$; and if they are in the opposite direction, the amount is $-\epsilon$. The Hamiltonian for such an interaction takes on the form

$$\hat{H}_{\text{int}} = -\epsilon \sum_{\langle {i,j \atop i',j'} \rangle} \hat{\sigma}_0(i,j)\, \hat{\sigma}_0(i',j') \qquad (7\text{-}40)$$

where $\hat{\sigma}_0(i,j)$ is the operator that gives $+1$ if the spin of the molecule at site (i,j) is up and -1 if the corresponding spin is down. The angle brackets surrounding the summation indexes (i,j) and (i',j') imply that the possible values of (i',j') are restricted to $(i\pm 1, j)$ and $(i, j\pm 1)$. We shall defer till later the introduction of a better indexing scheme for the lattice sites so that this restricted sum can be carried out conveniently in practice.

Although we have limited the orientations of the spins to two definite directions, the model is still a classical one. For a fully quantum mechanical system, the interaction Hamiltonian takes only the form

$$\hat{H}_{\text{int}} = -\epsilon \sum_{\langle {i,j \atop i',j'} \rangle} \hat{\boldsymbol{\sigma}}(i,j) \cdot \hat{\boldsymbol{\sigma}}(i',j')$$

where, for a spin-half system, $\hat{\boldsymbol{\sigma}}(i,j)$ is the Pauli spin matrix operator acting on the spin of the particle at site (i,j). This is known as the Heisenberg model. We shall, however, restrict ourselves to the classical case and consider only the Ising model.

In general, we can also include a term in the Hamiltonian that represents the interaction with an external magnetic field B supplied, for example, by a solenoid. The total Hamiltonian for the system then becomes

$$\hat{H} = -\epsilon \sum_{\langle {}^{i,j}_{i',j'} \rangle} \hat{\sigma}_0(i,j)\hat{\sigma}_0(i',j') - \eta B \sum_{i,j} \hat{\sigma}_0(i,j) \qquad (7\text{-}41)$$

where η is the strength of the interaction with the external B field. For convenience of notation, we shall absorb η into the definition of B and omit it from now on. Both ϵ and B are now in energy units and they may be regarded as the two parameters of the model. The energy of the system for a given configuration $\mathcal{C}$ of the assembly of $(N \times N)$ spins is then

$$E_\mathcal{C} = -\epsilon \sum_{\langle {}^{i,j}_{i',j'} \rangle} J_{i,j} J_{i',j'} - B \sum_{i,j} J_{i,j} \qquad (7\text{-}42)$$

It is a function of the orientation of each of the spins in the configuration.

Since there are $(N \times N)$ spins in our system and each can have two possible orientations, there is a total of $2^{N \times N}$ possible arrangements or *configurations*. The complete set of $2^{N \times N}$ configurations forms a canonical ensemble. If the system is ergodic, the probability $p(\mathcal{C})$ for the system to be in configuration $\mathcal{C}$ is given by

$$p(\mathcal{C}) = \frac{1}{Z(\epsilon,B)} e^{-E_\mathcal{C}/kT} \qquad (7\text{-}43)$$

where $e^{-E_\mathcal{C}/kT}$ is the Boltzmann factor, with $E_\mathcal{C}$ being the energy of the configuration given by (7-42). The other factors in the argument of the exponential function are k, the Boltzmann constant, and T, the temperature.

The normalization of the probability distribution (7-43) is provided by the partition function

$$Z(\epsilon, B) = \sum_\mathcal{C} e^{-E_\mathcal{C}/kT}$$

Following standard statistical mechanics procedures, we shall calculate first the partition function. From $Z(\epsilon, B)$, other quantities of interest to the thermodynamics of the system may be obtained, such as the internal energy

$$\mathcal{E} = kT^2 \frac{\partial}{\partial T} \ln Z(\epsilon, B) = \sum_\mathcal{C} p(\mathcal{C}) E_\mathcal{C} \qquad (7\text{-}44)$$

and magnetization

$$\mathcal{M} = kT \frac{\partial}{\partial B} \ln Z(\epsilon, B) = \sum_\mathcal{C} p(\mathcal{C}) \left(\sum_{i,j} J_{i,j} \right) \qquad (7\text{-}45)$$

From these two quantities, we obtain the specific heat

$$C_B = \frac{\partial \mathcal{E}}{\partial T} = \frac{1}{kT^2}\left\{\sum_C p(C)E_C^2 - \mathcal{E}^2\right\} \qquad (7\text{-}46)$$

and magnetic susceptibility

$$\chi_m = kT\frac{\partial \mathcal{M}}{\partial B} = \sum_C p(C)\left(\sum_{i,j} J_{i,j}\right)^2 - \mathcal{M}^2 \qquad (7\text{-}47)$$

For our illustrative example, we shall be mainly concerned with $\mathcal{E}$ and $\mathcal{M}$, as it is far easier to obtain good numerical accuracies for these two quantities than for C_B and χ_m, which are derivatives of $\mathcal{E}$ and $\mathcal{M}$.

Infinite two-dimensional model. For an infinite two-dimensional system, it is possible to obtain an analytical solution for the Ising model. The results show that, in all thermodynamic functions, a singularity is found at the critical temperature T_c given by

$$2\tanh^2 \frac{2\epsilon}{kT_c} = 1$$

The same result may also be written in the form

$$z \equiv e^{-2\epsilon/kT_c} = \sqrt{2} - 1$$

corresponding to the value

$$\frac{\epsilon}{kT_c} = 0.44$$

In the absence of an external field ($B = 0$), the internal energy has the form

$$\mathcal{E} = -\epsilon \coth\frac{2\epsilon}{kT}\left\{1 + \frac{2}{\pi}K_1(\zeta)\left(2\tanh^2\frac{2\epsilon}{kT} - 1\right)\right\}$$

where

$$\zeta = \frac{2\sinh\frac{2\epsilon}{kT}}{\cosh^2\frac{2\epsilon}{kT}}$$

and $K_1(\zeta)$ is the complete elliptic integral of the first kind

$$K_1(\zeta) = \int_0^{\pi/2} \frac{d\phi}{1 - \zeta^2 \sin^2\phi}$$

At temperatures $T \approx T_c$, the specific heat has the form

$$C_B \approx \frac{2k}{\pi}\left(\frac{2\epsilon}{kT_c}\right)^2\left\{-\ln\left(\left|1 - \frac{T}{T_c}\right|\right) + \ln\left(\frac{kT_c}{2\epsilon}\right) - \left(1 + \frac{\pi}{4}\right)\right\}$$

and the spontaneous magnetization per spin takes on the value

$$\frac{1}{N^2}\mathcal{M} = \begin{cases} \dfrac{(1+z^2)^{1/4}(1-6z^2+z^4)^{1/8}}{\sqrt{1-z^2}} & \text{for } T < T_c \\ 0 & \text{for } T > T_c \end{cases}$$

The calculations leading to these results are somewhat involved and can be found, for example, in Huang.

Computer simulation of a finite square lattice. The advantage of carrying out a computer simulation for the Ising model lies in the wider freedom in choosing the interactions and in the relative ease of extending to three dimensions. The disadvantage is that we have to be contented with a finite system. Even for a modest value of $N = 10$, the total number of possible configurations in a canonical ensemble for a two-dimensional system is $2^{N \times N} = 2^{100} \approx 10^{30}$. For such large numbers, it is not feasible to carry out the calculations by averaging over all the members in the ensemble, as implied by (7-44) to (7-47). The alternative is to use a sampling approach. For this reason, Monte Carlo calculations become the favored method to make computer studies of Ising and related models.

For our square lattice, each configuration is characterized by the orientations, $J_{i,j} = \pm 1$, of the $(N \times N)$ spins. For this reason, it takes N^2 random numbers (random bits is adequate here since each $J_{i,j}$ has only two possible orientations) to construct a member of the ensemble. The statistical weight of a particular configuration in the ensemble is given by (7-43). Since we can only take a relatively small sample of all the possible configurations, the Metropolis algorithm, discussed at the end of §7-1, becomes the most convenient method. In this scheme, we start with a given configuration $\mathcal{C}$. From $\mathcal{C}$ we generate a trial configuration $\mathcal{C}_t$ by going through each of the N^2 spins and check whether the orientation of any one should be flipped. From (7-42), we find that the energy change connected with the reorientation of the spin at site (i,j) with original spin orientation $J_{i,j}$ is given by

$$\Delta E = -2J_{i,j}\{\epsilon(J_{i-1,j} + J_{i+1,j} + J_{i,j-1} + J_{i,j+1}) + B\} \tag{7-48}$$

where the first term, associated with the parameter ϵ, comes from the interaction with the four neighboring sites and the second term, associated with the parameter B, arises from the interaction with an external magnetic field.

In the Metropolis algorithm, the decision whether the trial configuration $\mathcal{C}_t$ should be accepted or rejected depends on the ratio r of (7-20) and is related to the probabilities of having the two configurations $\mathcal{C}_t$ and $\mathcal{C}$. From (7-43) and (7-48), we find that the value of r is given by

$$r = \frac{e^{-E_{\mathcal{C}_t}/kT}}{e^{-E_{\mathcal{C}}/kT}} = e^{-\Delta E/kT} \tag{7-49}$$

For our square lattice, there are only $(2 \times 5) = 10$ possible values for r. The factor 2 comes from the two possible orientations of $J_{i,j}$, and the factor 5 corresponds to the five different possible values, 0, ± 2, and ± 4, for the sum of the relative orientations of the four neighboring spins. As a result, the ten values for a given set of parameters (ϵ, B) can be stored as a (2×5) matrix and we do not have to recalculate any of them each time they are needed. We shall refer to this matrix as the *decision matrix*.

As discussed at the end of §7-1, the direction of the spin at site (i,j) is reversed if $r > 1$. For $r \leq 1$, a uniformly distributed random number x in the interval $[0, 1]$ is generated. If $x \leq r$, the spin orientation is changed. If $x > r$, the trial configuration involving a change of the spin orientation at site (i, j) is rejected, that is, the spin at site (i, j) is unchanged.

To find a new configuration $\mathcal{C}_{n+1}$ from an old $\mathcal{C}_n$ for our square lattice requires a "sweep" over all $(N \times N)$ spins. The decision whether to flip each of the spins is made by the scheme described in the previous paragraph. As we shall see soon, this is the part of the calculation that takes most of the time, and any improvement in the speed, such as putting the decision matrix as a stored array, is important. The steps of the Metropolis scheme to sweep through the lattice are given in Box 7-6.

Box 7-6 Subroutine SWEEP(MS,NP1,NDMN,DS_MTX)

One sweep of the lattice according to the Metropolis algorithm

Argument list:
  MS: Array for the spin orientations of the square lattice.
  NP1: Number of spins in a row (or column) plus 1.
  NDMN: Dimension of the MS array in the calling program.
  DS_MTX: Decision matrix.

Initialization:
  (a) Construct in the calling program the decision matrix of (7-49).
  (b) Set up the periodic boundary condition.

1. Check each site (i, j) of the lattice to see if the spin orientation should be reversed:
   (a) Impose the periodic condition:
      (i) If $i = 1$, use $J_{N,j}$ for $J_{i-1,j}$.
      (ii) If $i = N$, use $J_{1,j}$ for $J_{i+1,j}$.
      (iii) If $j = 1$, use $J_{i,N}$ for $J_{i,j-1}$.
      (iv) If $j = N$, use $J_{i,1}$ for $J_{i,j+1}$.
   (b) Set up the two indexes for the decision matrix.
      (i) Find the value of the spin $J_{i,j}$ and use this to build the index i_s.
      (ii) Find the sum of four neighboring spins $J_{i-1,j} + J_{i+1,j} + J_{i,j-1} + J_{i,j+1}$ and use the value to build index j_s.
   (c) Use i_s and j_s to find the value of the ratio r of (7-49) from the decision matrix.
   (d) If $r > 1$, change the sign of the spin $J_{i,j}$.
   (e) If $r \leq 1$, generate a uniform random number x in $[0, 1]$.
      (i) If $x \leq r$, change the sign of spin $J_{i,j}$.
      (ii) If $x > r$, the sign of $J_{i,j}$ is not changed.
2. Return the new configuration in array MS.

It is convenient to discuss here two other technical points in the calculation. The first is that each random configuration generated by the Metropolis algorithm is strongly correlated with the previous one from which it started. This is undesirable. One way to minimize the correlation is to discard a few configurations between any two that are accepted as members of our sample. This means that several sweeps through the lattice have to be made before we can select a configuration whose physical properties we shall calculate and include in the ensemble averages we wish to study. For the same reason, it is also a good practice to discard the first few configurations generated from the one we use as the starting point. As we saw earlier, this may be regarded as a "thermalization" process, since the starting configuration may not be at equilibrium.

The second point concerns finite-size effects arising from the small value of N that can be used in a realistic computer calculation. To partially compensate for this limitation, we shall impose a periodic boundary condition. In other words, we shall consider that our square lattice is only one of many identical ones in the space, with the top of our lattice bordering the bottom of another. Similarly, the left side is in contact with the right side of another. In practical terms, this means that we make the following replacements of the spins at the borders of our lattice:

$$J_{i-1,j} \to J_{N,j} \quad \text{for} \quad i = 1 \qquad J_{i+1,j} \to J_{1,j} \quad \text{for} \quad i = N$$
$$J_{i,j-1} \to J_{i,N} \quad \text{for} \quad j = 1 \qquad J_{i,j+1} \to J_{i,1} \quad \text{for} \quad j = N$$

Again, this condition can be imposed efficiently by adding one more row to the top of our $(N \times N)$ array, with spin orientations identical to those of the last row, and by adding one more row to the bottom that has the same spin orientations as the first row. Similarly, an additional column is put on the left as well as on the right of our lattice.

Ensemble averages. Since the configurations are selected according to the probability of $p(\mathcal{C}) \sim \exp -\{E_\mathcal{C}/kT\}$, calculations of the ensemble average of physical quantities are equivalent to averaging over the sums of selected configurations. For example, the internal energy of (7-44) becomes

$$\mathcal{E} = \sum_\mathcal{C} p(\mathcal{C}) E_\mathcal{C} \longrightarrow \frac{1}{N_{\text{ens}}} {\sum_\mathcal{C}}' E_\mathcal{C}$$

where N_{ens} is the number of configurations sampled. The prime over the summation indicates the fact that the configurations included have been selected with the appropriate weighting factor $p(\mathcal{C})$. Furthermore, since the "weighting" factor $p(\mathcal{C})$ is already built into the sampling process, it is no longer necessary to put it in explicitly in the calculations.

For each member of the ensemble, all four thermodynamic quantities of interest to us, $\mathcal{E}$, C_B, $\mathcal{M}$, and χ_m, are derived from two terms, the spin-spin term and the spin term. The former depends on the factor

$$t_s(\mathcal{C}) = \sum_{i,j} J_{i,j}(J_{i-1,j} + J_{i+1,j} + J_{i,j-1} + J_{i,j+1}) \tag{7-50}$$

and the latter depends on the factor

$$t_b(\mathcal{C}) = \sum_{i,j} J_{i,j} \tag{7-51}$$

> **Box 7-7 Program ISING**
> **Two-dimensional Ising model**
>
> Calculates the energy, specific heat, magnetization, and susceptibility for a square lattice
>
> Subprograms called:
>   INI_LAT: Initializing the lattice.
>   CALC_EM: Calculate the spin-spin and spin-B contributions for a configuration.
>   OUT_GRP: Output the ensemble average.
>   SWEEP: One Metropolis algorithm sweep through the lattice (Box 7-6).
>   DISPLAY: Display the lattice.
>
> Initial settings:
> N_{therm}: Number of thermalization sweeps ($=20$).
> N_{skip}: Number of sweeps to skip between ensembles ($=5$).
> N_{ens}: Number of ensembles in a group ($=100$).
> N_{group}: Number of groups in the calculation ($=10$).
>
> 1. Input the random number seed, number of spins in each row N, energy parameter ϵ/kT, and external field B/kT.
> 2. Initialize the random number generator and the decision matrix (7-49).
> 3. Construct an initial configuration of $(N \times N)$ lattice using INI_LAT:
>   (a) For each lattice site, generate a uniform random number in $[0,1]$.
>   (b) If $x > 0.5$, the spin orientation is up ($J = +1$).
>   (c) Otherwise, the spin orientation is down ($J = -1$).
> 4. Thermalize the lattice by making N_{therm} calls to SWEEP (Box 7-6).
> 5. Construct an ensemble by carrying out the following steps N_{group} times:
>   (a) Make a sweep through the lattice using SWEEP.
>   (b) Use CALC_EM to calculate and store t_s of (7-50) and t_b of (7-51).
>   (c) Call SWEEP N_{skip} times to reduce correlation between ensemble members.
> 6. Repeat the previous step N_{group} times to construct a collection of ensembles.
> 7. Use OUT_GRP to calculate and output the ensemble averages and standard deviations of $\mathcal{E}$, C_B, $\mathcal{M}$, and χ_m from t_s and t_b using (7-52).

In terms of the values of these two terms for each configuration, the ensemble averages may be expressed as

$$\mathcal{E} = \frac{1}{N_{\text{ens}}} {\sum_{\mathcal{C}}}' \left\{ -\epsilon t_s(\mathcal{C}) - B t_b(\mathcal{C}) \right\}$$

$$C_B = \frac{1}{kT^2 N_{\text{ens}}} {\sum_{\mathcal{C}}}' \left\{ -\epsilon t_s(\mathcal{C}) - B t_b(\mathcal{C}) - \mathcal{E} \right\}^2$$

$$\mathcal{M} = \frac{1}{N_{\text{ens}}} {\sum_{\mathcal{C}}}' t_b(\mathcal{C})$$

$$\chi_m = \frac{1}{N_{\text{ens}}} \sum_{\mathcal{C}}{'} \left\{ t_b(\mathcal{C}) - \mathcal{M} \right\}^2 \tag{7-52}$$

The actual steps of the calculation are given in Box 7-7.

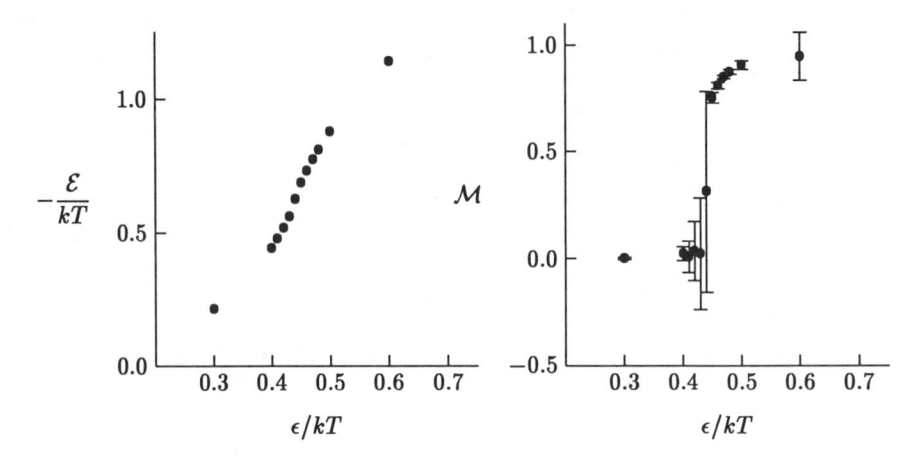

Fig. 7-16 Variations of the internal energy $\mathcal{E}/kT$ and magnetization $\mathcal{M}$ per spin as functions of ϵ/kT for a (60 × 60) square Ising lattice. The external field B is set to zero. Each point is the average of 20 ensembles consisting of 100 samples in each. The fluctuations in the values of $\mathcal{E}$ between different ensembles are small and not shown. The error bars for $\mathcal{M}$ indicate the sizes of one standard deviation in the distribution of values among the 20 ensembles calculated.

The results of a simulation for $\mathcal{E}/kT$ and $\mathcal{M}$ on a square lattice of $N = 60$ are given in Fig. 7-16. Each point in the plot is the average of 20 ensembles, with 100 members in each ensemble. The external magnetic field is held at $B = 0$ so as to mimic the situation of spontaneous magnetization. In each case, the values of the internal energy $\mathcal{E}$ and magnetization $\mathcal{M}$, in units of kT, are plotted as functions of ϵ/kT. Since all the energies are measured in units of kT, the horizontal axes may be interpreted as variations of the temperature for a constant value of the spin-spin interaction strength ϵ (with temperature T decreasing for increasing ϵ/kT).

In the figure, we see that the negative of the internal energy varies smoothly with temperature, without anything resembling a discontinuity to indicate a phase transition. The magnetization, on other hand, is essentially zero below $\epsilon/kT \approx 0.45$ and increases with decreasing temperature (increasing ϵ/kT) beyond this point. The fluctuations in the value of $\mathcal{M}$ are large just below (higher temperatures) the expected place of a critical point, and this makes it difficult to identify whether the results show a phase transition. Part of the reason for the large fluctuations comes from the fact that, since there is no external magnetic field, there is no reference direction with which the magnetic dipole moment associated with each spin can align. As a result, it is equally probable for the spins to be preferentially in the $J_{i,j} = +1$ direction in one member of the ensemble and in the $J_{i,j} = -1$ direction in another. At temperatures below the critical point (large ϵ/kT), the spontaneous magnetic moment provides a reference direction and, as a result, fluctuations in the values are smaller. In spite of the ambiguity caused by fluctuations, there are good indications for a second-order phase transition in the results shown in the figure. This is gratifying, especially in view of the small lattice size used.

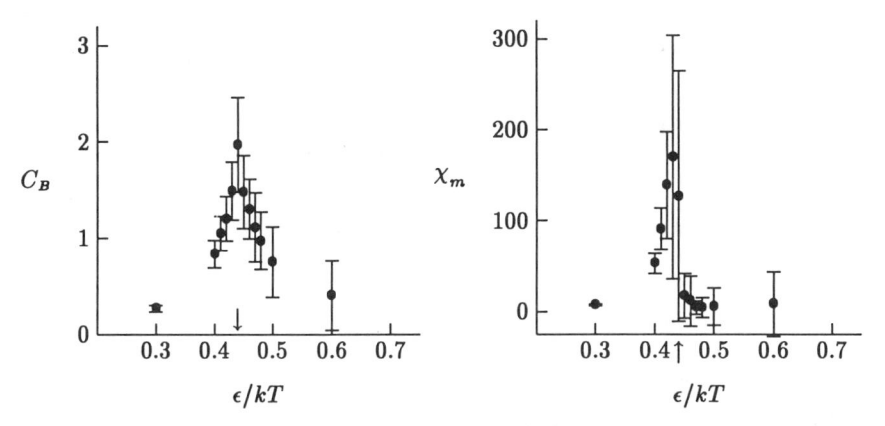

Fig. 7-17 Variations of the specific heat per spin and magnetic susceptibility as functions of ϵ/kT for a (60 × 60) square Ising lattice. The external field B is set to zero. For an infinite system a critical point exists at $\epsilon/kT = 0.44$, marked by an arrow in each plot.

The results for specific heat C_B and magnetic susceptibility χ_m are plotted in Fig. 7-17. As can be seen from (7-52), these two quantities are proportional to the variances of $\mathcal{E}$ and $\mathcal{M}$, respectively. For this reason, we expect large variations in their values from one ensemble to another. Because of our finite lattice and sample sizes, the fluctuations in the simulated results can be quite significant. In spite of this, we see a cusp in each case at roughly the analytical result of $\epsilon/kT = 0.44$, giving an indication of the existence of a possible critical point.

The real interest in a computer simulation of the Ising model is not in the simple two-dimensional case we have shown here. With a little more effort, we can use essentially the same method to carry out a model calculation in three dimensions. In this case, analytical solutions are difficult to obtain. Furthermore, it is much easier to introduce other physical effects in a computer modeling, such as interactions with the next-nearest neighbors. The main shortcoming we have is in interpreting the computer-generated results because of finite size effects, as we saw above. The limitations on the size of N we can use come from the amount of computation involved. For the three-dimensional case, we must sweep through all N^3 spins several times before we can produce a single member of the ensemble. As a result, the computational time increases as N^3. On the other hand, the model is well suited to parallel processing and, no doubt, many interesting results will be forthcoming when new computer architecture becomes widely available.

§7-6 Random matrix ensembles

We have seen earlier in §5-3 that the matrix method is a good way to solve eigenvalue problems in quantum mechanics. In general, the calculation consists of three separate steps: selecting a Hamiltonian, constructing the matrix, and diagonalizing it. The physics input is mainly in choosing the Hamiltonian. For our purpose here, it is convenient to separate the input information in constructing the Hamiltonian into two parts. The first is in the type of Hamiltonian operator we wish to use. This is governed by the general physical principles under which the system is operating, such as symmetry relations and conservation laws. For example, most of the Hamiltonians are symmetric since the systems we wish to study are invariant under time reversal. The second part of the input information concerns the details, such as the strengths of various terms in the Hamiltonian operator. The output of a calculation consists of the eigenvalues and eigenfunctions obtained from diagonalizing the Hamiltonian matrix.

Our interest in the eigenvalue problem here is slightly different from the usual one. Instead of examining the properties of a particular Hamiltonian, we are more concerned with the statistical behavior of an entire class of Hamiltonians. In terms of the two types of input information outlined in the previous paragraph, we are interested in finding out about the general properties of the system, independent of the details of the input parameters. There are many features shared by a variety of quantum mechanical systems — atoms, nuclei, and elementary particles — that have quite different interactions between their constituents. This is similar to critical phenomena, discussed in the previous section, where many properties of a system near a critical point are universal, independent of the particular physical system on which the studies are made.

To keep the discussion simple, we shall ignore the eigenvectors here. In principle, we can also ask questions on the statistical behavior of physical quantities calculated from the eigenvectors, such as the average value of the transition rate from one state to another or the distribution of decay width of a state. We shall, however, not get into such details in this introduction to the topic of random matrix. Some of the classical papers on the subject are reprinted in Porter (*Statistical Theory of Spectra: Fluctuations*, Academic Press, New York, 1965) and a recent review can be found in Brody and others (*Rev. Mod. Phys.* **53** [1982] 385).

In the early 1950s, Wigner proposed that the statistical properties of eigenvalues might be studied by replacing the elements of a Hamiltonian matrix with random numbers. The eigenvalues obtained from such a matrix depend on both the structure of the matrix, resulting from the first type of input discussed earlier, and the values of the matrix elements used as the second type of input. However, since the matrix elements are selected randomly, a different set of random number has an equal probability of being the input. The eigenvalues obtained with different sets of matrix elements will, in all probability, be quite different. In the spirit of Monte Carlo calculations, we shall take a collection of many different sets of random numbers as the input and examine the distributions of the eigenvalues obtained. A collection of such *random matrices*, all constructed in the same way except with different random numbers as the matrix elements, forms an ensemble similar to the case of different systems prepared in the same way form an ensemble in statistical mechanics.

Similar to other studies made on ensembles, there are two possible outcome in the results. The first is that the distributions among the values obtained from different members are broad, something resembling that shown schematically in Fig. 6-7(b). In this case, the ensemble average has little meaning. On the other hand, the distributions may be quite narrow, similar to that shown in Fig. 6-7(a). In this case, the ensemble average implies a property shared by most members of this class of Hamiltonians. As we shall see soon, many properties of random matrix ensembles have fairly narrow distributions and they reveal certain universal characters of quantum mechanical systems that are quite intriguing.

Gaussian orthogonal ensemble. Since most of the quantum mechanical systems we wish to examine are invariant under time reversal, the Hamiltonian is Hermitian and the matrices can be put in a real, symmetric form. For a given Hamiltonian, the value of each matrix element depends on the basis states used in constructing the matrix. On the other hand, the eigenvalues obtained from diagonalizing the matrix are independent of the basis used. If we make a orthogonal transformation of the basis states, the value of each individual matrix element will, in general, be changed. However, the distribution of the matrix elements as a whole will remain the same. To satisfy this particular invariance requirement, it is necessary that the distribution of the values of the matrix elements be given by a normal or Gaussian distribution. For this reason, an ensemble of such matrices is given the name Gaussian orthogonal ensemble or GOE for short.

It can be shown analytically that the distribution or density of eigenvalues $\rho(E)$ for an ensemble of GOE matrices is semicircular in shape, with a radius of 2 in units of the standard deviation. That is,

$$\rho(E) = \frac{1}{2\pi}\sqrt{4 - x^2} \qquad (7\text{-}53)$$

where $x = (E - C)/\sigma \leq 2$, with C and σ, respectively, the mean and the standard deviation of the distribution. If we plot the number of eigenvalues in the range $[E - \frac{1}{2}\Delta E, E + \frac{1}{2}\Delta E]$ as a function of E, the distribution will be similar to that given by (7-53). The numerical result of such an ensemble, made of 50 random

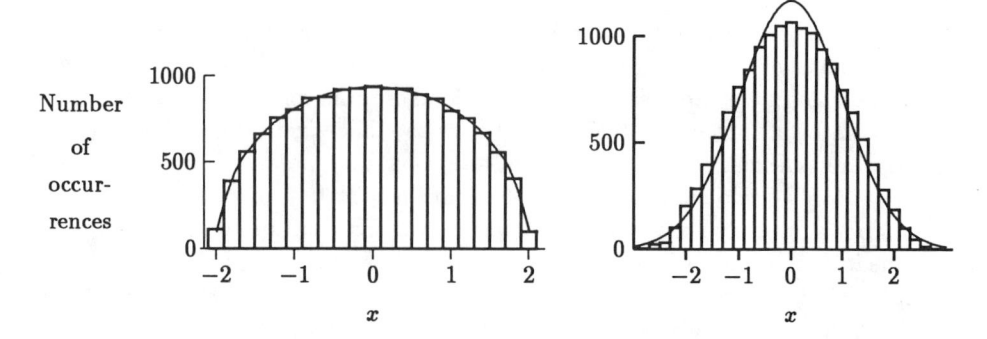

Fig. 7-18 Histograms of eigenvalue distributions. On the left, the plot is for GOE with the solid curve describing a semicircular distribution. On the right, the plot is for TBRE with the solid curve given by a normal distribution. There are 50 members in each ensemble and each random matrix has $N = 294$ eigenvalues. The eigenvalues E_i for each matrix are measured in units of the standard deviation $\sigma = \{N^{-1}\sum(E_i - C)^2\}^{1/2}$ and their distributions are centered around $C = N^{-1}\sum_i E_i$.

matrices each with dimension 294 (there are 294 eigenvalues in each matrix) is shown in on the left of Fig. 7-18.

It is quite easy to carry out a Monte Carlo simulation of the GOE ensemble. In addition to a generator for random numbers with a normal distribution, we need a diagonalization routine for real, symmetric matrices. Before embarking on the calculations, we must decide on the dimension N of the matrix we wish to study. In general, the limitation on N comes from the size of matrices that can be handled on a computer due to limitations imposed by the available memory and the amount of computational time required. The value $N = 294$ used in the examples here comes mainly from the considerations of two-body random ensembles to be described below, and it is not too far from the maximum size of real, symmetric matrices that can be diagonalized without resorting to special techniques or expensive computers.

To construct a real, symmetric random matrix of dimension N, we need $N(N+1)/2$ random numbers as the elements for the upper half of the matrix. (The matrix elements in the lower half are produced by equating to the corresponding ones in the upper half through symmetry requirements.) If the eigenvectors are not needed, the most efficient way to diagonalize such matrices is to use the tridiagonalization method of §5-4 together with the method of bisection of §5-5. To find the distributions of the eigenvalues for an ensemble of similar random matrices, it is necessary to generate many different sets of $N(N+1)/2$ random numbers and diagonalize the resulting matrix for each set.

We do not expect all the eigenvalues of a finite-dimensional matrix to correspond to the energy levels of, for example, a real nucleus. This is most readily seen from the fact that the energy spectrum of a nucleus is unbound at the high-energy end. In other words, as we give more and more excitation energy to the nucleus, we expect to find more and more levels. The practical limit comes when the energy levels become too dense to be resolved individually. To model such a physical system, the matrix dimension must be infinite, a problem not suitable for numerical solution on a computer. Realistically, one can only concentrate on a small, low-lying region of the spectrum. If our Hamiltonian matrix constructed to understand such a region is at all reasonable, the eigenvalue distribution must have a part resembling that of a real physical system. As we shall see below, the GOE does not satisfy this requirement.

Let us use the energy levels in a nucleus as an example. It has been known for a long time that the low-lying energy-level density has an exponential form given by

$$\rho(E) \sim e^{2\sqrt{aE}} \tag{7-54}$$

where a is the level-density parameter. This is known as the Fermi-gas level-density formula, derived by Bethe in 1934 based on a Fermi gas model of the nucleons in a nucleus. It is known to given a reasonable description of the low-lying region of nuclei.

The exponential shape of the level-density formula given by (7-54) is at variance with that of the semicircular form of (7-53). For example, the slope, or first derivative, of the exponential form is increasing with increasing energy. On the other hand, the slope for a semicircular distribution is decreasing with increasing energy. Such a glaring difference casts doubts on whether GOE gives a true representation of physical systems. In fact, it is difficult to find a physical system whose energy-level density resembles that at the low-energy end of a semicircular distribution. For this reason there is interest in finding other types of random matrix ensembles that can give a better representation to realistic systems.

Two-body random ensemble. The GOE is the most general ensemble one can construct for real, symmetric random matrices that are invariant under an orthogonal transformation. For this reason it may be too broad as a model for realistic quantum mechanical systems, as the ensemble contains all the possible matrices satisfying this set of conditions. While it is true that all the physically interesting matrices must fall into this category, there are also many members in the ensemble that do not correspond to realistic systems. In addition to invariance under an orthogonal transformation, physically interesting matrices have many other constraints imposed on them. For example, the interaction is dominated by one- and two-body forces. As we shall see soon, this is an important feature in the eigenvalue distribution. If an ensemble is dominated by the unphysical cases, it is not surprising that the ensemble averages do not reflect the properties of real physical systems.

From the Hamiltonian point of view, the GOE includes all possible interactions between the constituents of the system. Simple statistical arguments, which we shall not reproduce here, lead to the conclusion that, if many-body interactions are present, most of properties of an ensemble will be dominated by these interactions. In reality, however, one- and two-body interactions are usually the most important ones in physical systems. Distributions of eigenvalues for one-body interactions are given by combinatorial considerations and are not of primary interest to us here. For this reason, ensembles of random matrices involving only two-body interactions have been studied by numerical simulations. This gives us the two-body random ensemble, or TBRE for short.

It is not difficult to see that, in a many-body system consisting of m particles interacting via a one-body force, the energy-level density follows a normal distribution. The reason comes from combinatorial arguments and can be found in a review article by Brody and others (1982). The same type of reasoning applies also to the case of two-body interactions. More generally, we can use k to represent the particle rank of the interaction, with $k = 1$ for one-body interactions and $k = 2$ for two-body interactions. Obviously, k must be less or equal to m. It has been shown by Monte Carlo simulation and confirmed by analytical studies that, for $k << m$, the level density follows a normal distribution. However, as we increase k, the density approaches a semicircular distribution when $k \approx m$. For this reason, we expect TBRE to have a normal distribution for the energy-level density with the number of active particles $m >> 2$. The results for a 294-dimensional TBRE with $m = 6$ were given earlier on the right of Fig. 7-18. Since 6 is not very much larger than 2, there are small departures from a normal distribution that are especially noticeable in the central part of the spectra. On the other hand, it is also obvious that, at the low-energy end, the slope of the level density for TBRE increases exponentially with energy, a form that is in agreement with observed energy levels.

Let us see how a TBRE may be constructed. By a k-body interaction, we mean a force that vanishes in a space where the number of active particles m is less than k. For $m > k$, the force acts only on k of the m particles at a time, with the remaining $(m - k)$ particles behaving essentially as spectators. For example, the action of a two-body force in a three-particle space is given by the sum of the interactions between the pairs of particles $(1, 2)$, $(2, 3)$, and $(3, 1)$.

The eigenvalues of an m-body system for a one-body interaction are simply the sum of the single-particle energies of the occupied states. Since the distribution of such a system is well known, we shall concentrate mainly on two-body interactions here. It is convenient to think of a two-body interaction in terms of the changes in the single-particle states of a pair of particles. Consider two particles, one in single-particle state $|r\rangle$ and the other in $|s\rangle$. A two-body interaction is one that can affect the state of both particles at the same time. As a result of the interaction, the single-particle states of the two particles are changed to $|t\rangle$ and

$|u\rangle$. The strength of such a term for a two-body potential V in the Hamiltonian may by represented by the two-body matrix element

$$W_{rstu} \equiv \langle t, u|V|r, s\rangle$$

where $|r, s\rangle$ represents the wave function of a state consisting of two particles, one in single-particle state $|r\rangle$ and the other in $|s\rangle$. Similarly, the wave function $\langle t, u|$ is the conjugate of that of a two-particle state occupying single-particle states $|t\rangle$ and $|u\rangle$. If the values of all such matrix elements are known, the two-body interaction is completely specified in the space of interest to us. For the $m = 3$ example we have used above, the matrix element of a two-body interaction potential between the initial state $|i_1, i_2, i_3\rangle$, where particle 1 is in state i_1, particle 2 in state i_2, and particle 3 in state i_3, and final state $|f_1, f_2, f_3\rangle$ is given by

$$\langle f_1, f_2, f_3|V|i_1, i_2, i_3\rangle = W_{i_1 i_2 f_1 f_2} + W_{i_2, i_3, f_2, f_3} + W_{i_3, i_1, f_3, f_1} \qquad (7\text{-}55)$$

For the convenience of discussion, we have ignored here any symmetries between the particles.

The notation used here is based on second quantization, involving single-particle creation and annihilation operators. For our discussion below, we do not have to rely on any of these more advanced techniques of quantum mechanics. However, it will be very awkward to carry out any realistic calculations without making use of the power offered by second quantization.

In a given basis, we can calculate the Hamiltonian matrix for an m-body system if a complete set of two-body matrix elements is given to us. When the matrix for the m-body system is diagonalized, we obtain the eigenvalues. A collection of m-body eigenvalues for all the possible sets of two-body matrix elements forms a two-body ensemble. If, on the other hand, we replace the two-body matrix elements by random numbers, the collection of m-body eigenvalues for different sets of random two-body matrix elements constitutes a two-body random ensemble.

Since a two-body interaction is invariant with respect to any orthogonal transformations in the two-particle space, the random two-body matrix elements should also follow a normal distribution. In any realistic calculations, it may also be necessary to impose further symmetry requirements, such as those arising from rotational invariance of the Hamiltonian. The construction of matrix elements in the m-particle space can be rather involved, especially if the particles are fermions governed by the Pauli exclusion principle. However, the basic principle of the calculation is a simple one and may be viewed as the process to "propagate" the matrix elements from the two-particle space to the m-particle space, similar to the way we have illustrated in (7-55) for the $m = 3$ case. Various techniques to handle such propagation are discussed in the review article by Brody and others, and we shall not enter into such specialized technical matter here. Except for the method to arrive at the m-particle matrix, the calculations and statistical analyses of the eigenvalues for TBRE are identical as those for GOE. Besides level-density distributions, we shall also make a comparison below of some of the other properties of the two ensembles.

Nearest-neighbor spacing distribution. We have seen earlier that level-density distributions of GOE and TBRE are different from each other and the cause was attributed to the fact that GOE is dominated by many-body interactions. From this we concluded that the exponential rise in the level density observed in nuclei and other quantum mechanical systems is a consequence of the fact that k, the particle rank of the interaction, is much smaller than the number of active particles.

Level density is only one of the properties in the distribution of energy levels of a quantum mechanical system. In fact, we may regard it as somewhat a special one in that it is a *global* property, as it involves the behavior of many levels at the same time. Often, the *local* properties and fluctuations from smooth, or locally averaged, values are of interest. One item in this category is the distribution of the spacing between two adjacent energy levels, generally known as the nearest-neighbor spacing distribution.

In a quantum mechanical system, the energy levels obtained from diagonalizing a single matrix display the phenomenon of *level repulsion*. Let us illustrate this point using a real, symmetric matrix in two dimensions. Such a matrix may be expressed in the following form:

$$\mathcal{M} = \begin{pmatrix} d_1 & f_2 \\ f_2 & d_2 \end{pmatrix}$$

As we have seen earlier in (5-121), the two eigenvalues, λ_1 and λ_2, are the roots of the characteristic equation

$$(d_1 - \lambda)(d_2 - \lambda) - f_2^2 = 0$$

The values of λ_1 and λ_2 are given by

$$\lambda_i = \tfrac{1}{2}\{(d_1 + d_2) \pm \sqrt{(d_1 - d_2)^2 + 4f_2^2}\,\}$$

The "spacing" between the two eigenvalues is then

$$S = |\lambda_2 - \lambda_1| = \sqrt{(d_1 - d_2)^2 + 4f_2^2}$$

As long as the off-diagonal element f_2 of the matrix $\mathcal{M}$ does not vanish, we have the result $S > |d_2 - d_1|$.

For a given absolute value of $(d_2 - d_1)$, the difference between the two eigenvalues increases with the absolute value of f_2, as shown in Fig. 7-19. If we take the view that f_2 is caused by an "external perturbation" on the two-dimensional system, we find that the distance between the two eigenstates increases as the strength of the perturbation is increased; hence the name level repulsion. For matrices with dimension $N > 2$, the situation is more complicated but the basic behavior of the spacing between neighboring eigenvalues remain the same as in the two-dimensional case.

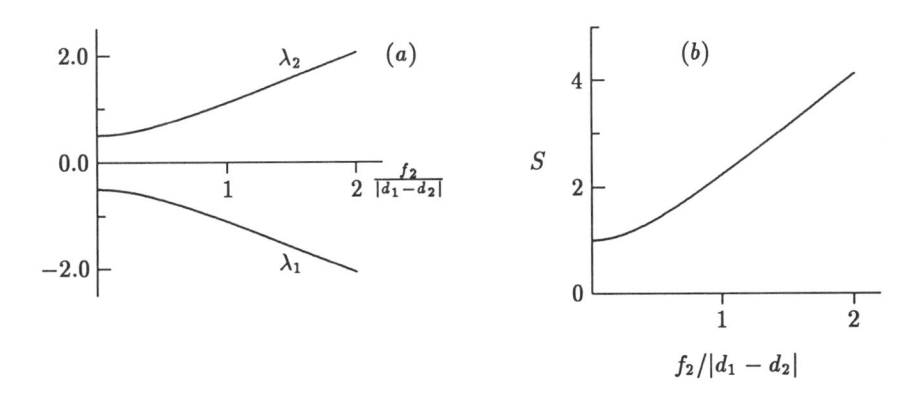

Fig. 7-19 Level repulsion in two-dimensional matrices. As the size of the off-diagonal matrix element f_2 increases, the lower eigenvalue λ_1 decreases in value and the upper eigenvalue λ_2 increases in value, as shown in (a). The distance between the two eigenvalue $S = (\lambda_1 - \lambda_2)$ is shown in (b).

Our two-dimensional example illustrates the reason why small spacings are scarce between eigenvalues of the same matrix. Let us examine the distribution of level spacing more generally. Before we can make any quantitative statements, we need a convenient energy scale for our discussion. For a sequence of n eigenvalues $E_1 < E_2 < \cdots < E_n$, it is convenient to adopt as the unit the average spacing

$$D = \frac{1}{n-1}(E_n - E_1) \qquad (7\text{-}56)$$

The denominator is $(n-1)$ here, as there are only $(n-1)$ spacings between n levels. Later, we shall see a better definition for the term "average spacing," but the form above is adequate for the time being. If the probability of obtaining small spacings, $S/D \ll 1$, is small, then the probability for obtaining large spacings, $S/D \gg 1$, must also be correspondingly small. This is necessary, since the sum of all $(n-1)$ spacings must be equal to $(n-1)D$. When this condition is combined with the idea of level repulsion, we obtain the Wigner surmise for the distribution of the nearest-neighbor spacing:

$$P(S) = \frac{\pi S}{2D^2} \exp\left(-\frac{\pi S^2}{4D^2}\right) \qquad (7\text{-}57)$$

The derivation of this distribution law is given in Brody and others (1982), as well as in Porter (1965).

We are not yet in a position to compare the Wigner surmise with either observations or random matrix results. An underlying assumption in deriving (7-57) is that the average spacing D is constant in the interval between E_1 and E_n. Earlier, we stated that the measured level densities increase exponentially with excitation energy. As a result, a constant average level spacing can only be assumed in a small energy region. On the other hand, we need a large number of spacings for a meaningful statistical analysis. To realize this in a small energy region, we must go to high energies where the density is high. Unfortunately, it is difficult experimentally to resolve the individual levels in such a region. For this reason, a direct check of (7-57) with observed level spacings can be difficult in most cases. Where such comparisons have been made, the Wigner surmise is found to be in good agreement with the observed values (with an important stipulation to which we shall return later).

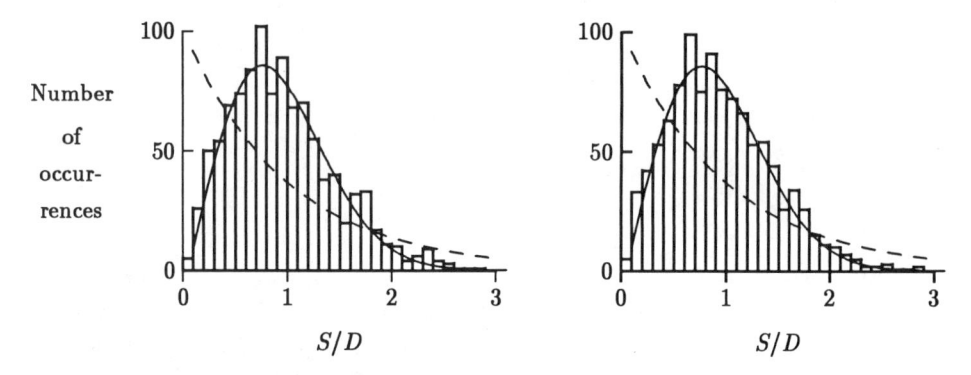

Fig. 7-20 Histograms showing the distributions of the nearest-neighbor spacing S. On the left, the plot is for GOE, and on the right, the plot is for TBRE. The solid curve is the expected result from the Wigner surmise of (7-57). To avoid finite-size effects, only 20 spacings in the middle of the 294 eigenvalues of each of the 50 members of each ensemble are used. The dashed curve shows an exponential distribution, true when level repulsion is absent.

For the ensemble results, we see from the level-density plots given in Fig. 7-18 that there is only a small region in the middle of each spectrum where the level density and, hence, the average level spacing are constant. In particular, we find that the region of constant D is small, especially for TBRE. For this reason, it is possible only to take a narrow central region in each spectrum for our study of the nearest-neighbor spacing distribution. To improve the statistics, we can collect the nearest-neighbor spacings in the same region from all the members of an ensemble. The results shown in Fig. 7-20 are calculated using the central 20 spacings from each of the 50-member ensembles. In contrast with the large difference in the level-density distributions between the two ensembles, we find that both GOE and TBRE have very similar distributions for their nearest-neighbor level spacings.

As we demonstrated with two-dimensional matrices, level repulsion takes place between eigenvalues of the same matrix. Mathematically, we can trace the reason to the fact that the off-diagonal matrix element $f_2 \neq 0$. For the two-dimensional case, the eigenvalues λ_1 and λ_2 are equal to the diagonal matrix elements d_1 and d_2 if the off-diagonal matrix element vanishes. In the more general case of $N > 2$, a matrix can be separated into two different submatrices if all the off-diagonal matrix elements connecting the two parts are zero, as we saw in §5-5 in connection with tridiagonal matrices. In this case, the eigenvalues of one submatrix are unrelated with those of the other, except through the way the two submatrices are constructed. In this case, level repulsion takes places only among the eigenvalues of each submatrix and there is nothing to prevent degeneracy between eigenvalues coming from different submatrices.

Physically, such a separation of a matrix into blocks, with all the off-diagonal matrix elements between blocks vanishing, is evidence of some underlying symmetry. For example, the nuclear Hamiltonian is rotational invariant. In this case, matrix elements between two states having different angular momenta vanish. As a result, the Hamiltonian matrix for states of different angular momenta form separate blocks, and the eigenvalues of different blocks are not related to each other (except for the fact that they are derived from the same Hamiltonian). If we plot the distribution of nearest-neighbor spacing of such a Hamiltonian for states having different angular momenta, it will no longer follow the Wigner surmise of (7-57), as level repulsion operates only within subsets of these states. In fact, if states of a large variety of angular momenta are mixed together in a small energy region, the distribution of the nearest-neighbor spacing is exponential:

$$P(S) = \frac{1}{D} \exp -\frac{S}{D}$$

In contrast to the Wigner surmise, the distribution has a relatively larger fraction of small spacings, as can be seen by comparing the solid and dashed curves in Fig. 7-20.

Correlation coefficient. The spacing distribution is closely related to the correlation coefficient between two adjacent level spacings, S_i and S_{i+1}:

$$C = \frac{n}{n-1} \frac{1}{\sum_{i=1}^{n}(S_i - D)^2} \sum_{i=1}^{n-1}(S_i - D)(S_{i+1} - D) \qquad (7\text{-}58)$$

where the ith nearest-neighbor spacing S_i is given by

$$S_i = E_{i+1} - E_i$$

If the total number of level spacings in the interval is large, it is possible to find the value of C analytically. The result, -0.271, is in agreement with those observed in experimental data and in the central regions of the members of both ensembles, GOE and TBRE, we have been examining.

The negative sign in C means that a small spacing is likely to be followed by a large one. This is essentially a consequence of level repulsion mentioned earlier. The small absolute value of C implies that the spectrum is rather "rigid." In other words, the position of a particular level in one member of an ensemble does not deviate very much from those of the corresponding level in other members of the same ensemble.

The Δ_3 statistic. Another way to see the correlation between level spacings is to make use of the cumulative number of levels $y(E)$, up to some energy E:

$$y(E) = \int_{-\infty}^{E} \rho(x)\,dx$$

where $\rho(x)$ is the density of eigenvalues. If we plot $y(E)$ as a function of E, the result resembles that of a staircase, with $y(E)$ increasing by a constant amount whenever E corresponds to one of the eigenvalues E_i. For the convenience of the discussion here, we shall normalize $\rho(E)$ such that the integrated value over the entire energy interval is n, the number of eigenvalues in the interval. With this normalization, the integral $y(E)$ increases by unity where $E = E_1, E_2, \ldots, E_n$. The results for the central region of a member of GOE and a member of TBRE are shown in Fig. 7-21.

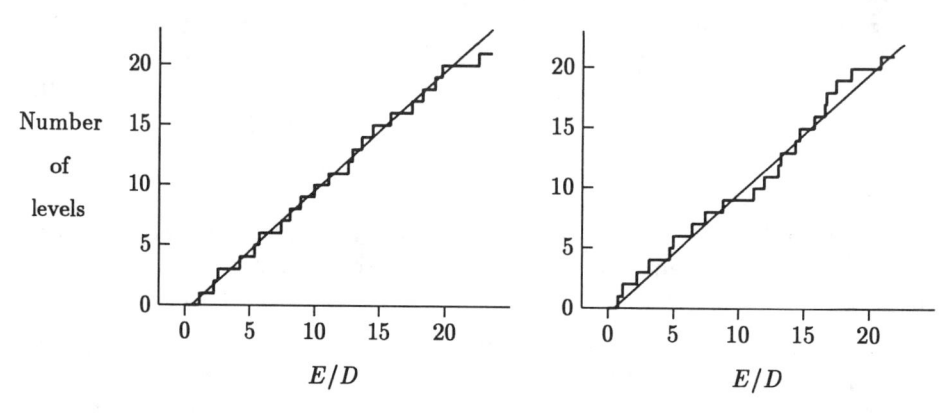

Fig. 7-21 Plots of the cumulative level number as a function of energy. The 20 levels are taken from the eigenvalues in the middle of a member of an ensemble of 294-dimensional random matrices. On the left, the ensemble is GOE and on the right, TBRE. The straight lines with constant slopes are obtained from the best fit to each case. The deviation of the staircase from the straight line is measured in terms of the Δ_3 statistic.

If the energy spectrum is completely rigid, the eigenvalues occur only at constant intervals and the form of $y(E)$ will be that of a set of stairs with constant distance between any two steps. Realistic spectra do not behave in such a regular way. On the other hand, since they are relatively rigid, as we have seen from correlation coefficient studies, departures from a regular structure are small in general and may be regarded as *fluctuations*. As a measure of such fluctuations we can calculate the deviations of $y(E)$ from a straight line of constant slope

$$z(E) = a + bE$$

The values of a and b here are treated as parameters to be found by a least-squares fit of $z(E)$ to $y(E)$. The result is the Δ_3 statistic introduced by Dyson and Mehta (*J. Math. Phys.* **4** [1963] 489):

$$\Delta_3 = \min_{a,b}\left\{\frac{1}{2L}\int_{-L}^{L}[y(E) - a - bE]^2 dE\right\} \quad (7\text{-}59)$$

where the energy interval is taken from $-L$ to L.

For an infinite system, the value of Δ_3 was given by Dyson and Mehta to be

$$\Delta_3 = \frac{1}{\pi^2}\{\ln n - 0.0687\} \pm 0.11 \quad (7\text{-}60)$$

The standard deviation of Δ_3 is 0.11 independent of n, the number of levels in the interval $[-L, +L]$. As a result, it is necessary for n to be on the order of 100 (or $\ln n/\pi^2 \sim 0.5$) before the statistic becomes significant. For our 294-dimensional ensembles, it is not possible to obtain such large values of n without resorting to correcting for finite-size effects. For purposes of illustration here, we shall avoid such a step by giving up the requirement of a statistically meaningful Δ_3 value. In this way, we can taken a smaller value of n such as 20. For the 50-member GOE and TBRE ensembles, the values obtained are 0.30 ± 0.10 and 0.31 ± 0.11, respectively, indistinguishable from the value of 0.30 ± 0.11 obtained using (7-60).

The main conclusion we can draw from random matrix studies is that level-density distributions are different for the two random matrix ensembles, GOE and TBRE. The fluctuations in the eigenvalues, however, seem to be fairly universal to quantum mechanical systems. The implication of such findings to quantum chaos and other statistical studies has not yet been fully explored. Much of the recent progress on the subject has been made with numerical simulation, since analytical work involving eigenvalues is complicated in general.

§7-7 Path integrals in quantum mechanics

In quantum mechanics, the usual way to find the time evolution of a state, characterized by the wave function $|\Psi(t)\rangle$, is to solve the Schrödinger equation

$$i\hbar\frac{\partial}{\partial t}|\Psi(t)\rangle = \hat{H}|\Psi(t)\rangle \quad (7\text{-}61)$$

where, in the nonrelativistic limit, the Hamiltonian operator $\hat{H}$ is made of the sum of kinetic and potential energies. In terms of operators in coordinate space, the Hamiltonian for a particle of mass μ may be written as

$$\hat{H} = -\frac{\hbar^2}{2\mu}\nabla^2 + V(\mathbf{r}, t)$$

From the solution, we obtain the eigenvalues and wave functions. In general, it is difficult to solve (7-61) as a differential equation for any realistic interaction potential V one encounters in physical systems. An alternative in such cases is to use the path integral method. As we shall see below, the method follows closely the analog in classical mechanics in minimizing the "action."

For a free particle,
$$V(\boldsymbol{r}, t) = 0$$

If the wave function at $t = 0$ is a plane wave with momentum $\boldsymbol{p}$,

$$\psi(\boldsymbol{r}, 0) = \frac{1}{(2\pi\hbar)^{3/2}} e^{i(\boldsymbol{p}\cdot\boldsymbol{r})/\hbar} \tag{7-62}$$

the solution of (7-61) at any time t has the familiar form

$$\psi(\boldsymbol{r}, t) = \frac{1}{(2\pi\hbar)^{3/2}} e^{i(\boldsymbol{p}\cdot\boldsymbol{r} - Et)/\hbar} \tag{7-63}$$

where $E = p^2/2\mu$ (and $p = |\boldsymbol{p}|$) is the kinetic energy of the particle.

The bra-ket notation. For the convenience of discussion and to make connection with standard references on the topic, we shall adopt the bra-ket notation here. Let $|\Psi\rangle$ represent a state vector in general. Similar to an ordinary vector $\boldsymbol{v}$ in three-dimensional space, a coordinate system must be set up to specify it further. For example, in a Cartesian system, $\boldsymbol{v}$ is given by the values of its projections on the x-, y-, and z-axes. Alternatively, we may wish to use a spherical coordinate system, and the same vector is described in terms of the values of its three spherical components (v_r, v_θ, v_ϕ). Our state vector $|\Psi\rangle$ is a more general quantity and often requires a continuum of values to identify it. For example, in coordinate representation, $|\Psi\rangle$ is specified by giving its "components" $\psi(\boldsymbol{r}_1)$, $\psi(\boldsymbol{r}_2), \ldots, \psi(\boldsymbol{r}_i), \ldots$, at a continuum of points $\boldsymbol{r}_1, \boldsymbol{r}_2, \ldots, \boldsymbol{r}_i, \ldots$ in coordinate space. Alternatively, we may realize the same state vector in momentum representation by giving its components $\psi(\boldsymbol{p}_1), \psi(\boldsymbol{p}_2), \ldots, \psi(\boldsymbol{p}_i), \ldots$, at a continuum of points $\boldsymbol{p}_1, \boldsymbol{p}_2, \ldots, \boldsymbol{p}_i, \ldots$ in momentum space.

Formally, we can extract a particular component of $|\Psi\rangle$, for example $\psi(\boldsymbol{r})$, by defining a function $|\boldsymbol{r}\rangle$ as the probability amplitude of a particle localized at $\boldsymbol{r}$. Since $|\boldsymbol{r}\rangle$ forms a complete set of states, we have the orthonormal condition

$$\langle \boldsymbol{r} | \boldsymbol{r}' \rangle = \delta(\boldsymbol{r} - \boldsymbol{r}')$$

where $\delta(\boldsymbol{r} - \boldsymbol{r}')$ is the Dirac delta function given in (3-65). The overlap of $|\Psi\rangle$ with $|\boldsymbol{r}\rangle$, or the "projection" of $|\Psi\rangle$ on $|\boldsymbol{r}\rangle$, gives us the $\boldsymbol{r}$-component of $|\Psi\rangle$. That is,

$$\psi(\boldsymbol{r}) = \langle \boldsymbol{r} | \Psi \rangle$$

Similarly, if $|\,p\,\rangle$ represents the wave function of a particle with momentum p, the components of $|\,\Psi\,\rangle$ in momentum representation are given by

$$\psi(p) = \langle p|\Psi\rangle$$

Returning to our free particle example, we see that the right side of (7-62) is nothing but the coordinate representation of a particle with momentum p. As a result, we have the relation

$$\langle r|p\rangle = \frac{1}{(2\pi\hbar)^{3/2}} e^{i(p\cdot r)/\hbar} \qquad (7\text{-}64)$$

In this notation, the transformation between the momentum and coordinate representations of the state vector $|\,\Psi\,\rangle$ may be expressed as

$$\langle p|\Psi\rangle = \iiint d^3r \, \langle p|r\rangle\langle r|\Psi\rangle \qquad (7\text{-}65)$$

This is the same as inserting a complete set of intermediate states $|\,r\,\rangle$ into the "matrix element" $\langle p|\Psi\rangle$.

The free particle. We shall now derive an expression for the wave function of a free particle at time $t \neq 0$. In general, a free particle is represented by a wave packet made of a linear combination of plane waves each with a different momentum. At $t = 0$, the wave function for a particle with momentum p is given by (7-62). For our wave packet, the wave function at time $t = 0$ is then a linear combination of such components weighted by the factor $\langle p|\Psi\rangle$:

$$\psi(r,0) = \langle r|\Psi(t=0)\rangle = \frac{1}{(2\pi\hbar)^{3/2}} \iiint d^3p \, e^{ip\cdot r/\hbar} \langle p|\Psi(t=0)\rangle \qquad (7\text{-}66)$$

This is, essentially, a Fourier transform of the wave function from the coordinate space to the momentum space. For our later needs, it is more convenient to put it into the following form:

$$\langle r|\Psi(t=0)\rangle = \iiint d^3p \langle r|p\rangle\langle p|\Psi(t=0)\rangle$$

where we have made use of (7-64) to change the form of the integrand from that of (7-66).

At time t, the wave function $\psi(r,t)$ for a free particle in coordinate space has the form

$$\langle r|\Psi(t)\rangle = \iiint d^3p \, \langle r|p\rangle e^{-iEt/\hbar} \langle p|\Psi(t=0)\rangle$$

This may be expressed as a relation between the wave function in coordinate space at time t and that at time $t = 0$ by inserting a complete set of states $|r'\rangle$ on the right side,

$$\langle r|\Psi(t)\rangle = \iiint d^3r' \iiint d^3p \langle r|p\rangle e^{-iEt/\hbar} \langle p|r'\rangle \langle r'|\Psi(t=0)\rangle$$

$$= \iiint d^3r' K(r,t;r',0) \langle r'|\Psi(t=0)\rangle \tag{7-67}$$

the same as we have done in (7-65). For later convenience, we have rewritten the final result in terms of

$$K(r,t;r',0) \equiv \iiint d^3p \langle r|p\rangle e^{-iEt/\hbar} \langle p|r'\rangle \tag{7-68}$$

Physically, we may regard $K(r,t;r',0)$ as the transition probability amplitude for a particle at location r at time $t = 0$ to that at location r' at time t. Similarly, the wave function at some other time t' may be expressed in terms of that at $t = 0$ in the form

$$\langle r'|\Psi(t')\rangle = \iiint d^3r\, K(r',t';r,0)\langle r|\Psi(t=0)\rangle \tag{7-69}$$

as there is nothing special about time t or t'.

Comparing (7-69) with (7-67), we can relate the wave functions at two different times t_f and t_i in the following way:

$$\langle r_f|\Psi(t_f)\rangle = \iiint d^3r_i\, K(r_f,t_f;r_i,t_i)\langle r_i|\Psi(t_i)\rangle \tag{7-70}$$

where

$$K(r_f,t_f;r_i,t_i) = \iiint d^3p \langle r_f|p\rangle e^{-iE(t_f-t_i)/\hbar}\langle p|r_i\rangle \tag{7-71}$$

The quantity $K(r_f,t_f;r_i,t_i)$ may also be interpreted as a "propagator" that connects the wave functions of a free particle at two different times. It is also possible to regard it as the overlap of the wave functions $\psi(r_f,t_f)$ and $\psi(r_i,t_i)$,

$$K(r_f,t_f;r_i,t_i) = \langle r_f,t_f|r_i,t_i\rangle \tag{7-72}$$

since the right side of (7-70) may also be regarded as the "expansion" of $\psi(r_f,t_f)$ in terms of the complete set of states $\psi(r_i,t_i)$.

For a free particle, it is possible to carry out the integral on the right side of (7-71) explicitly. The result,

$$K(r_f,t_f;r_i,t_i) = \left\{\frac{\mu}{2\pi i\hbar(t_f-t_i)}\right\}^{3/2} \exp\left\{i\frac{\mu}{2\hbar}\frac{(r_f-r_i)^2}{t_f-t_i}\right\} \tag{7-73}$$

may be verified in the following way. First, we construct a differential equation satisfied by the propagator $K(r,t;r',t')$ for a free particle. This may be done by

substituting (7-71) into the Schrödinger equation (7-61) and using the condition that, at $t' = t$,
$$K(\mathbf{r}, t; \mathbf{r}', t) = \delta(\mathbf{r} - \mathbf{r}')$$
The resulting differential equation has the form
$$i\hbar \frac{\partial}{\partial t} K(\mathbf{r}, t; \mathbf{r}', t') = -\frac{\hbar^2}{2\mu} \nabla^2 K(\mathbf{r}, t; \mathbf{r}', t')$$
where ∇^2 operates only on $\mathbf{r}$ and not on $\mathbf{r}'$. It is obvious that the right side of (7-73) is a solution of such a equation.

Classical action. We shall now show that the argument of the exponential function on the right side of (7-73) is proportional to the classical action. In classical mechanics, the action for the path of a particle that moves from a point $\mathbf{r}_i$ at time t_i to a point $\mathbf{r}_f$ at time t_f is defined as the following integral over time:
$$S(\mathbf{r}_f, t_f; \mathbf{r}_i, t_i) = \int_{t_i}^{t_f} \mathcal{L}\, dt$$
where $\mathcal{L}$ is the Lagrangian. For a free particle, the Lagrangian $\mathcal{L}$ is independent of time and is given by the kinetic energy
$$\mathcal{L} = \tfrac{1}{2}\mu v^2$$
For our example here, the velocity is a constant, which may be found using the relation
$$v = \frac{\mathbf{r}_f - \mathbf{r}_i}{t_f - t_i}$$
As a result, the classical action has the explicit form
$$S(\mathbf{r}_f, t_f; \mathbf{r}_i, t_i) = \tfrac{1}{2}\mu v^2 (t_f - t_i) = \frac{\mu}{2} \frac{(\mathbf{r}_f - \mathbf{r}_i)^2}{t_f - t_i}$$
In terms of $S(\mathbf{r}_f, t_f; \mathbf{r}_i, t_i)$, we can write the propagator of (7-73) as
$$K(\mathbf{r}_f, t_f; \mathbf{r}_i, t_i) = \left\{ \frac{\mu}{2\pi i \hbar (t_f - t_i)} \right\}^{3/2} \exp\{i S(\mathbf{r}_f, t_f; \mathbf{r}_i, t_i)/\hbar\} \quad (7\text{-}74)$$

Although the result is derived for the case of a free particle, the form is actually a more general one. If a particle moves in a potential $V(\mathbf{r})$ that depends only on the coordinates, the propagator is identical to that given in (7-74). The only difference is that we must replace the free-particle Lagrangian in the action $S(\mathbf{r}_f, t_f; \mathbf{r}_i, t_i)$ with the appropriate one that includes the interaction potential. The derivation of the more general case in terms of Hamiltonians can be found in Lee (*Particle Physics and Introduction to Field Theory*, Harwood, Chur, Switzerland, 1981) and Ryder (*Quantum Field Theory*, Cambridge University Press, Cambridge, 1985).

We can put (7-72) in a more convenient form by rewriting it in the usual way for carrying out calculations in quantum mechanics. A solution of the Schrödinger equation may be obtained in the following manner. Ignoring for the moment the fact that $\hat{H}$ is an operator, we may rewrite (7-61) formally as

$$\frac{d\Psi(t)}{\Psi(t)} = -\frac{i}{\hbar}\hat{H}\, dt$$

Since $\hat{H}$ is independent of time in the Schrödinger representation we are using here, it is possible to integrate both sides of the equation between times t_i and t_f and obtain the result

$$\ln \Psi(t_f) - \ln \Psi(t_i) = -\frac{i}{\hbar}\hat{H}(t_f - t_i)$$

This is equivalent to

$$\Psi(t_f) = e^{-i\hat{H}(t_f-t_i)/\hbar}\Psi(t_i)$$

The meaning of an operator in the argument of a mathematical function may be interpreted in terms of a series expansion of the function, as we shall see later in (8-136). In bra-ket notation, the relation above may be written as

$$|\Psi(t_f)\rangle = e^{-i\hat{H}(t_f-t_i)/\hbar}|\Psi(t_i)\rangle \tag{7-75}$$

Using this result, we can express the right side of (7-72) in terms of a matrix element between states $|r_i, t\rangle$ and $|r_f, t\rangle$:

$$K(r_f, t_f; r_i, t_i) = \langle r_f, t_f | e^{-i(t_f-t_i)\hat{H}/\hbar} | r_i, t_i\rangle \tag{7-76}$$

Let $|n\rangle$ be a member of a complete set of eigenstates of $\hat{H}$ with eigenvalue E_n. That is,

$$\hat{H}|n\rangle = E_n|n\rangle$$

and

$$\psi_n(r, t_f) = \langle n | r, t_f\rangle$$

By inserting a complete set of states $|n\rangle$ into the matrix element on the right side of (7-76), we obtain the result

$$K(r_f, t_f; r_i, t_i) = \sum_n \psi_n^*(r_f, t_f)\psi_n(r_i, t_f)\, e^{-i(t_f-t_i)E_n/\hbar} \tag{7-77}$$

In this way, we see that the propagator $K(r_f, t_f; r_i, t_i)$ may be constructed once we have all the eigenfunctions of $\hat{H}$. However, this is not always feasible and the path integral method of Feynman provides us with an alternate method of solving the problem.

Path integral. One way to evaluate $K(\boldsymbol{r}_f, t_f; \boldsymbol{r}_i, t_i)$ is to divide the time interval between t_i and t_f into two parts, one from t_i to t_k and the other from t_k to t_f. Using (7-70), we have the wave functions at times t_k and t_f in coordinate space in the form

$$\langle \boldsymbol{r}_k | \Psi(t_k) \rangle = \iiint d^3 r_i \, K(\boldsymbol{r}_k, t_k; \boldsymbol{r}_i, t_i) \langle \boldsymbol{r}_i | \Psi(t_i) \rangle \tag{7-78}$$

$$\langle \boldsymbol{r}_f | \Psi(t_f) \rangle = \iiint d^3 r_k \, K(\boldsymbol{r}_f, t_f; \boldsymbol{r}_k, t_k i) \langle \boldsymbol{r}_k | \Psi(t_k) \rangle \tag{7-79}$$

It is possible to combine these two results and obtain the wave function $\langle \boldsymbol{r}_f | \Psi(t_f) \rangle$ at time t_f in terms of $\langle \boldsymbol{r}_i | \Psi(t_i) \rangle$ at t_i by substituting the expression for $\langle \boldsymbol{r}_k | \Psi(t_k) \rangle$ given in (7-78) into the right side of (7-79). This gives us the relation

$$\langle \boldsymbol{r}_f | \Psi(t_f) \rangle = \iiint d^3 r_i \iiint d^3 r_k K(\boldsymbol{r}_f, t_f; \boldsymbol{r}_k, t_k) K(\boldsymbol{r}_k, t_k; \boldsymbol{r}_i, t_i) \langle \boldsymbol{r}_i | \Psi(t_i) \rangle$$

Comparing this result with that of (7-70), we obtain the following relation between the propagators:

$$K(\boldsymbol{r}_f, t_f; \boldsymbol{r}_i, t_i) = \iiint d^3 r_k \, K(\boldsymbol{r}_f, t_f; \boldsymbol{r}_k, t_k) K(\boldsymbol{r}_k, t_k; \boldsymbol{r}_i, t_i)$$

Instead of propagators, it is perhaps more instructive to express the same relation in terms of the overlap between two wave functions in the form

$$\langle \boldsymbol{r}_f, t_f | \boldsymbol{r}_i, t_i \rangle = \iiint d^3 r_k \langle \boldsymbol{r}_f, t_f | \boldsymbol{r}_k, t_k \rangle \langle \boldsymbol{r}_k, t_k | \boldsymbol{r}_i, t_i \rangle \tag{7-80}$$

The physical meaning of the last two equations is that the transition from $(\boldsymbol{r}_i, t_i)$ to $(\boldsymbol{r}_f, t_f)$ may be taken as the sum of transitions from $(\boldsymbol{r}_i, t_i)$ to all the possible intermediate points $\boldsymbol{r}_k$ at time t_k, followed by a transition from $\boldsymbol{r}_k$ to $(\boldsymbol{r}_f, t_f)$. This is similar in spirit to the use of Huygens's principle in geometric optics.

By subdividing the time interval $[t_i, t_f]$ into more than two parts, we can split the path into even smaller segments. For latter convenience, we shall divide the time between t_i and t_f into n equal intervals, each of duration δ_t. Taking $t_0 = t_i$ and $t_n = t_f$, the value of t_k at the end of interval k is given by

$$t_k = t_0 + k\delta_t$$

For a particular path, let the location of the particle at time t_k be $\boldsymbol{r}_k$. Analogous to (7-80), the propagator, expressed in the form of the overlap between two wave functions, is now the integral

$$\langle \boldsymbol{r}_f, t_f | \boldsymbol{r}_i, t_i \rangle = \int \cdots \int d^3 r_1 d^3 r_2 \cdots d^3 r_{n-1}$$
$$\times \langle \boldsymbol{r}_f, t_f | \boldsymbol{r}_{n-1}, t_{n-1} \rangle \langle \boldsymbol{r}_{n-1}, t_{n-1} | \boldsymbol{r}_{n-2}, t_{n-2} \rangle \cdots \langle \boldsymbol{r}_1, t_1 | \boldsymbol{r}_i, t_i \rangle$$

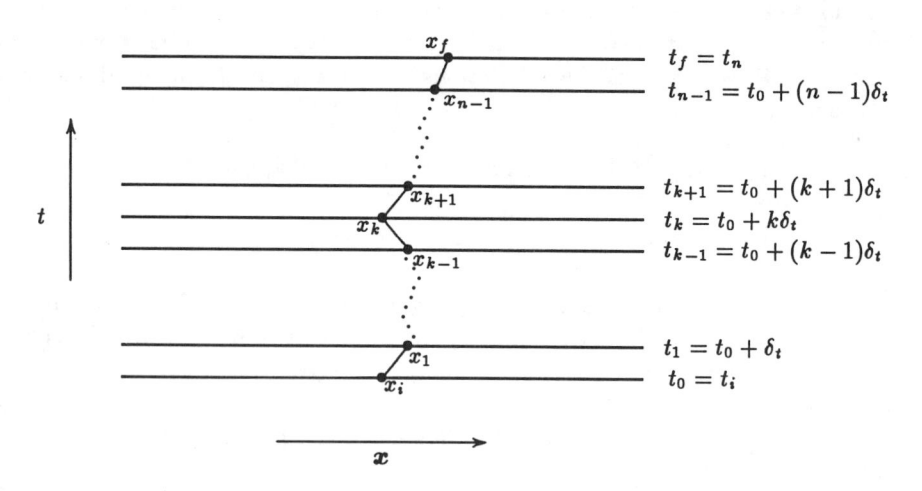

Fig. 7-22 Example of a path from (x_i, t_i) to (x_f, t_f). It goes from (x_i, t_i) to (x_1, t_1), then from (x_1, t_1) to (x_2, t_2), and so on, with the final point at x_f.

In terms of propagators, this may be put in the following way:

$$K(\mathbf{r}_f, t_f; \mathbf{r}_i, t_i) = \int \cdots \int d^3 r_1 d^3 r_2 \cdots d^3 r_{n-1}$$
$$\times K(\mathbf{r}_f, t_f; \mathbf{r}_{n-1}, t_{n-1}) K(\mathbf{r}_{n-1}, t_{n-1}; \mathbf{r}_{n-2}, t_{n-2}) \cdots K(\mathbf{r}_1, t_1; \mathbf{r}_i, t_i)$$

The integrations here are taken over all the possible intermediate points to which the various paths can lead. One such path for the simple case of a single spatial dimension is shown in Fig. 7-22 as illustration. For simplicity, we have approximated the motion from $\mathbf{r}_k$ to $\mathbf{r}_{k+1}$ by a straight line, the shortest distance joining the two points. As a result, each path consists of a series of zigzags made of short straight segments. In the limit that the number of intervals $n \to \infty$, the entire path becomes a smooth curve.

For each segment of the path, (7-74) applies and the propagator is given in terms of the classical action $S(\mathbf{r}_{k+1}, t_{k+1}; \mathbf{r}_k, t_k)$ for the interval from t_k to t_{k+1}:

$$K(\mathbf{r}_{k+1}, t_{k+1}; \mathbf{r}_k, t_k) = \left(\frac{\mu}{2\pi i \hbar \delta_t}\right)^{3/2} \exp\{iS(\mathbf{r}_{k+1}, t_{k+1}; \mathbf{r}_k, t_k)/\hbar\} \quad (7\text{-}81)$$

where

$$S(\mathbf{r}_{k+1}, t_{k+1}; \mathbf{r}_k, t_k) = \int_{t_k}^{t_{k+1}} \mathcal{L} \, dt$$

Substituting this result into (7-81), we obtain the transition probability amplitude from (r_i, t_i) to (r_f, t_f) in terms of the classical action

$$K(r_f, t_f; r_i, t_i) = \frac{1}{Z} \int \cdots \int d^3r_1 d^3r_2 \cdots d^3r_{n-1} \prod_{k=1}^{n-1} \exp\{iS(r_{k+1}, t_{k+1}; r_k, t_k)/\hbar\}$$

$$= \frac{1}{Z} \int \cdots \int d^3r_1 d^3r_2 \cdots d^3r_{n-1}$$
$$\times \exp\{iS(r_n, t_n; r_{n-1}, t_{n-1}; \cdots; r_0, t_0)/\hbar\} \quad (7\text{-}82)$$

where, for convenience, we have used $(r_n, t_n) = (r_f, t_f)$, $(r_0, t_0) = (r_i, t_i)$, and

$$S(r_n, t_n; r_{n-1}, t_{n-1}; \cdots; r_0, t_0) = \sum_{k=0}^{n-1} S(r_{k+1}, t_{k+1}; r_k, t_k)$$

In (7-82), we have used Z to represent the normalization constant, the same symbol as the partition function in statistical mechanics. This is done on purpose because of the close analogy between the two quantities, as we shall see better later.

There are two major advantages to using the path integral method for solving quantum mechanical problems and both are contained in (7-82). The first is that only the classical action enters into the calculation of the propagator. Since $S(r_n, t_n; r_{n-1}, t_{n-1}; \cdots; r_0, t_0)$ does not involve operators and wave functions, it is, in general, much easier to evaluate. The usual quantum mechanical approach of solving the time-dependent Schrödinger equation, such as that used in arriving at (7-77), is applicable only for simple cases. For the more realistic situations, approximations such as perturbation methods are often used. However, if the interaction involved is not weak, perturbation techniques do not apply and the path integral method becomes one of the few available alternatives. The second major advantage is that the action is evaluated in small time intervals. By making the interval δ_t sufficiently small, it is possible to approximate the integral by the product of the average value of the integrand and δ_t, similar to what we have done in the rectangular rule of numerical integration. This greatly simplifies the method and makes it possible to solve problems involving complicated interactions, such as those occurring between quarks in quantum chromodynamics.

The connection between the path integral approach and Monte Carlo calculations is also demonstrated by (7-82). For good accuracy, it is necessary to make the number of intervals n large. This, in turn, means that there are many possible paths. In fact, in the limit that $n \to \infty$, the number of possible paths is also infinite. In the same spirit as that used in the Ising model, it is not necessary to evaluate all the possible paths. A sample is taken and all the physically interesting quantities are evaluated and averaged over the paths sampled. The analogy to statistical mechanics is sufficiently close that the application of the path integral method to quantum field theory has been given the name statistical field theory. As an illustration, we shall carry out below a trivial application of the technique.

Application to a harmonic oscillator. We shall describe a trivial application of the path integral method using Monte Carlo techniques. For simplicity, we shall consider only a single spatial dimension x. Mathematically, (7-77) holds also if we take $t_i = 0$ and $t_f = -it_{\text{total}}$. Furthermore, if the Hamiltonian is time independent, the basis states $\psi_n(x)$ can also be made to be time independent. Under these conditions, (7-77) reduces to the form

$$K(x_f, -it_{\text{total}}; x_i, 0) = \sum_n \psi_n^*(x_f)\psi_n(x_i)e^{-E_n t_{\text{total}}/\hbar}$$

In the limit of large t_{total}, the term with the smallest energy E_n dominates the sum on the right side. As a result, the relation may be approximated as one for the absolute square of the ground state wave function:

$$|\psi_0(x)|^2 \approx e^{-E_0 t_{\text{total}}/\hbar} K(x, -it_{\text{total}}; x, 0) \qquad (7\text{-}83)$$

Thus, by evaluating the propagator $K(x, -it_{\text{total}}; x, 0)$ for some large value of t_{total}, we can obtain the absolute square of the ground state wave function of a system.

For our illustrative example, we shall use the potential

$$V(x) = \tfrac{1}{2}\mu\omega^2 x^2 \qquad (7\text{-}84)$$

given earlier in (4-1) for a particle of mass μ in a harmonic oscillator well with frequency ω. To calculate the propagator using (7-82), we need to sum over the classical action S for a sequence of small time intervals, each of duration δ_t. In each interval, the action is given by

$$S(x_{k+1}, t_{k+1}; x_k, t_k) = \int_{t_k}^{t_{k+1}} \mathcal{L}\, dt$$

In terms of the complex time τ, given by $t = -i\tau$, this integral takes on the form

$$S(x_{k+1}, -i\tau_{k+1}; x_k, -i\tau_k) = -i\int_{\tau_k}^{\tau_{k+1}} \mathcal{L}\, d\tau$$

Since we have restricted ourselves to the nonrelativistic limit and our potential is time independent, the classical Lagrangian may be written in the form

$$\mathcal{L} = \frac{\mu}{2}\left(\frac{dx}{dt}\right)^2 - V(x)$$

On replacing t by $-i\tau$, we have

$$\mathcal{L} = -\frac{\mu}{2}\left(\frac{dx}{d\tau}\right)^2 - V(x)$$

The right side has the same form as the negative of the total energy, with τ taking over the role of time.

If the complex time interval $\delta_\tau = i\delta_t$ is sufficiently short, the classical action may be approximated by the average value in the interval. The result is

$$S(x_{k+1}, -i\tau_{k+1}; x_k, -i\tau_k) = -i \int_{\tau_k}^{\tau_{k+1}} \{-E(x,\tau)\} d\tau \approx i\overline{E}(x_k, \tau_k)\delta_\tau \quad (7\text{-}85)$$

where $\overline{E}(x_k, \tau_k)$ is the average value of the total energy (in terms of complex time τ) in the interval τ_k to τ_{k+1} and δ_τ is the length of the interval ($\delta_\tau = \tau_{k+1} - \tau_k$). The sum of actions over a complete path is then

$$S(x, \tau_n; x_{n-1}, \tau_{n-1}; \ldots; x, 0) = \sum_{k=1}^{n-1} S(x_{k+1}, -i\tau_{k+1}; x_k, -i\tau_k)$$

$$\approx i\delta_\tau \overline{E}(x, x_1, x_2, \ldots, x_{n-1}, x)$$

where

$$\overline{E}(x, x_1, x_2, \ldots, x_{n-1}, x) = \sum_{k=0}^{n-1} \overline{E}(x_k, \tau_k)$$

and $x = x_0$.

Using (7-82), the propagator becomes

$$K(x, -it_{\text{total}}; x, 0) = \frac{1}{Z} \int \cdots \int dx_1 dx_2 \cdots dx_{n-1}$$
$$\times \exp\{iS(x, \tau_n; x_{n-1}, \tau_{n-1}; \cdots; x, 0)/\hbar\}$$
$$= \frac{1}{Z} \int \cdots \int dx_1 dx_2 \cdots dx_{n-1} \exp\{-\frac{\delta_\tau}{\hbar}\overline{E}(x_1, x_2, \ldots, x_{n-1})\}$$

By inserting this result into (7-83), we obtain the absolute square of the ground state wave function in the form

$$|\psi_0(x)|^2 = \frac{e^{-E_0 t_{\text{total}}/\hbar}}{Z} \int \cdots \int dx_1 dx_2 \cdots dx_{n-1} \exp\{-\frac{\delta_\tau}{\hbar}\overline{E}(x, x_1, x_2, \ldots, x_{n-1}, x)\} \quad (7\text{-}86)$$

Since the wave function is normalized to unity,

$$\int_0^\infty |\psi_0(x)|^2 dx = 1$$

the constant Z here is given by the integral

$$Z = e^{-E_0 t_{\text{total}}/\hbar} \int \cdots \int dx\, dx_1 dx_2 \cdots dx_{n-1} \exp\{-\frac{\delta_\tau}{\hbar}\overline{E}(x, x_1, x_2, \ldots, x_{n-1}, x)\}$$

It has the same form as a partition function in statistical mechanics when we make the analogy of $\delta_\tau/\hbar$ with $1/kT$.

Again, it is convenient in our calculations to convert the Schrödinger equation of (7-61) into a dimensionless form by measuring length x in units of $\sqrt{\hbar/(\mu\omega)}$ and time t in units of ω^{-1}. In other words, we define a dimensionless length η and a dimensionless time ξ by the following relations:

$$x = \eta\sqrt{\frac{\hbar}{\mu\omega}} \qquad\qquad t = \xi\frac{1}{\omega}$$

On replacing (x,t) by (η,ξ) in (7-61), we obtain for the harmonic oscillator potential of (7-84) the result

$$i\frac{\partial\psi}{\partial\xi} = -\frac{1}{2}\left\{\frac{\partial^2\psi}{\partial\eta^2} - \eta^2\right\} \qquad (7\text{-}87)$$

The Hamiltonian and all the energies are now measured in units of $\hbar\omega$. In the time interval between τ_k and τ_{k+1}, the contribution to the classical action may be approximated by the average of the sum of kinetic and potential energies in the interval. Since x_k is the location of the particle at time τ_k, the kinetic energy during the small time interval $[\tau_k, \tau_{k+1}]$ is proportional to $(x_{k+1} - x_k)^2$. Furthermore, we are measuring the time in units of ω^{-1}. As a result, the appropriate form of the kinetic energy contribution to the action for the time interval is

$$T_{[\tau_k,\tau_{k=1}]} = \frac{1}{2}\mu\left(\frac{x_{k+1} - x_k}{\tau_{k+1} - \tau_k}\right)^2 = \tfrac{1}{2}\mu\omega^2(x_{k+1} - x_k)^2$$

The classical action of (7-85) now takes on the form

$$S(x_{k+1}, -i\tau_{k+1}; x_k, -i\tau_k) \approx -i\delta_\tau\left\{\frac{\mu}{2}(x_{k+1}-x_k)^2 + \frac{\mu\omega^2}{2}\left(\frac{x_{k+1}+x_k}{2}\right)^2\right\}$$

$$= -i\frac{\hbar\delta_\xi}{2}\left\{(\eta_{k+1}-\eta_k)^2 + \tfrac{1}{4}(\eta_{k+1}+\eta_k)^2\right\}$$

where

$$\delta_\tau = \delta_\xi\frac{1}{\omega}$$

and η_k is the location of the kth point of the path in units of $\sqrt{\hbar/(\mu\omega)}$. In time interval δ_τ, the particle moves from η_k to η_{k+1}. The contribution to the kinetic energy from the kth time interval is then $(\eta_{k+1} - \eta_k)^2$ in units of $\hbar\omega$. For the contribution from the potential energy term, we can take the average value in the interval by assuming the particle to be located at the middle point between n_k, its position at the beginning of the interval, and η_{k+1}, its position at the end of the interval.

The square of the ground state wave function given in (7-86) now takes on the form

$$|\psi_0(\eta)|^2 = \frac{e^{-E_0 t_{\text{total}}/\hbar}}{Z}\int\cdots\int d\eta_1 d\eta_2\cdots d\eta_{n-1}$$

$$\exp\left\{-\tfrac{1}{2}\delta_\xi \overline{E}(\eta,\eta_1,\eta_2,\ldots,\eta_{n-1},\eta)\right\} \qquad (7\text{-}88)$$

with

$$\overline{E}(\eta,\eta_1,\eta_2,\ldots,\eta_{n-1},\eta) = \sum_{k=0}^{n-1}\left\{\left(\eta_{k+1}-\eta_k\right)^2 + \tfrac{1}{4}\left(\eta_{k+1}+\eta_k\right)^2\right\} \tag{7-89}$$

It is advantageous to establish an evenly spaced grid system for η (or x) in the numerical calculation by choosing a constant step size δ_η. All the values of our dimensionless length η can now be specified as integer multiples of the step size

$$\eta_k = m_k \delta_\eta$$

In terms of δ_η and δ_ξ, the energy for a path may be expressed in the form

$$\overline{E}(\eta_1,\eta_2,\ldots,\eta_{n-1}) = \delta_\eta^2 \sum_{k=0}^{n-1}\left\{\left(m_{k+1}-m_k\right)^2 + \tfrac{1}{4}\left(m_{k+1}+m_k\right)^2\right\}$$

As we shall see later, this choice depends also on the step size δ_ξ for time as well.

We are now ready to calculate the value $|\psi_0(\eta)|^2$ using (7-88). The Metropolis algorithm is used to sample possible paths specified by $\eta, \eta_1, \eta_2, \ldots, \eta_n$. Since we are calculating the absolute square of a wave function, rather than some sort of transition amplitude that takes the system from one location to another, the last point of the path η_n must coincide with the starting point η. For a given path, the weighting factor of its contributions to the ensemble average is $\exp\{-\tfrac{1}{2}\delta_\eta \overline{E}(\eta,\eta_1,\eta_2,\ldots,\eta_{n-1},\eta)\}$. Since this factor is included in selecting the path, each that is chosen by the algorithm contributes unity to the (unnormalized) value of $|\psi_0(\eta)|^2$. After a large number of samples is taken, the values of $|\psi_0(\eta)|^2$ for all values of η have to be normalized by the total number of paths sampled.

The efficiency of the calculation may be improved by making the following adjustments. Since the last point in the path must be located at the same point in space as the starting point, a selected path $\{\eta_0, \eta_1, \ldots, \eta_{n-1}, \eta_n = \eta_0\}$ that begins and ends at η_0 may also be regarded as a path that starts at location η_1. This can be done by moving the first point η_0 to position η_n. The result is a path going through the points $\{\eta_1, \ldots, \eta_{n-1}, \eta_0, \eta_{n+1} = \eta_1\}$. The energy is not changed by this move, as can be seen from (7-89). However, we have a "new" path that begins and ends at $\eta = \eta_1$ and contributes to $|\psi_0(\eta)|^2$ at $\eta = \eta_1$. This process may be repeated by moving η_2 to the point η_{n+2} and we have a path for $\eta = \eta_2$. This argument can be continued until we come to η_{n-1}. As a result, each selected path may be counted as n different paths, each contributing to $|\psi_0(\eta)|^2$ at $\eta = \eta_0, \eta_1, \ldots, \eta_{n-1}$ (and adding n to the normalization).

In principle, we can apply the Metropolis algorithm in the same way as we did earlier in the Ising model by going through each point $k = 0, 1, \ldots, (n-1)$, one by one and selecting a new path by varying the value of η_k. Because of the constraint that the path must terminate at the same spatial point as the starting point of the path, that is, $\eta_n = \eta$, a path has to be rejected if it cannot meet this condition. Since this occurs at the end of constructing a new path, it becomes rather wasteful, as a lot of work has already gone into constructing the partial path. A more effective alternative is to select a point k randomly along the path and see if its location should be varied according to the rules of the Metropolis algorithm. The trial paths are constructed by varying η_k only by amounts that are δ times the step size. Using (7-89), the change in energy because of $\eta_k \to \eta_k + \delta\delta_\eta$ or $m_k \to m_k + \delta$ is given by

$$\begin{aligned}\Delta_k \overline{E} &= \tfrac{1}{2}\delta_\eta^2 \{(m_{k+1} - m_k - \delta)^2 - (m_{k+1} - m_k)^2 \\ &\quad + (m_k + \delta - m_{k-1})^2 - (m_k - m_{k-1})^2 \\ &\quad + \tfrac{1}{4}(m_{k+1} + m_k + \delta)^2 - \tfrac{1}{4}(m_{k+1} + m_k)^2 \\ &\quad + \tfrac{1}{4}(m_k + \delta + m_{k-1})^2 - \tfrac{1}{4}(m_k + m_{k-1})^2\} \\ &= \delta_\eta^2 \{\delta^2 + \delta(2m_k - m_{k-1} - m_{k-1}) + \tfrac{1}{4}(\delta^2 + \delta(2m_k + m_{k-1} + m_{k+1}))\}\end{aligned}$$

Similar to (7-49), the trial path is accepted if

$$r = \exp-\left\{\frac{\delta_\xi}{2}\Delta_k \overline{E}\right\} \tag{7-90}$$

is greater than 1. For $r < 1$, it is accepted with a probability r (that is, a uniform random number x in the interval $[0, 1]$ is generated and the trial path is accepted only if $x \leq r$). Since every point along an accepted path contributes to $|\psi_0(\eta)|^2$, we add 1 to $|\psi_0(\eta)|^2$ at $\eta = \eta_k + \delta$ if the new path is accepted and to $|\psi_0(\eta)|^2$ at $\eta = \eta_k$ if the new path is rejected.

For the convenience of calculation, we shall take the absolute value of δ to be a constant; only the sign is arbitrary. Thus, in addition to a random number to determine the point k, we need another random number (a random bit will be adequate) to determine the sign of δ. The obvious choice for the value of $|\delta|$ is the step size for our dimensionless length η. As a result, we have $\delta = \pm 1$, with the sign determined by the random number generated.

The efficiency of the Metropolis algorithm may be improved by decomposing r of (7-90) into a product of kinetic and potential energy parts:

$$r_{\text{k.e.}} = \left(b_{\text{k.e.}}\right)^{i_k} \qquad r_{\text{p.e.}} = \left(b_{\text{p.e.}}\right)^{i_p} \tag{7-91}$$

where

$$\begin{aligned} b_{\text{k.e.}} &= \exp-\{\delta_\eta^2 \delta_\xi\} & i_k &= 1 \pm (2m_k - m_{k+1} - m_{k-1}) \\ b_{\text{p.e.}} &= \exp-\{\tfrac{1}{4}\delta_\eta^2 \delta_\xi\} & i_p &= 1 \pm (2m_k + m_{k+1} + m_{k-1}) \end{aligned}$$

In this way, the values of the most often encountered values of $b_{\text{k.e.}}$ and $b_{\text{p.e.}}$ may be stored as two one-dimensional arrays. For these cases, the calculation of r may now be carried out as a product of the values of $b_{\text{k.e.}}$ and $b_{\text{p.e.}}$ recalled from the

$$|\psi(x)|^2$$

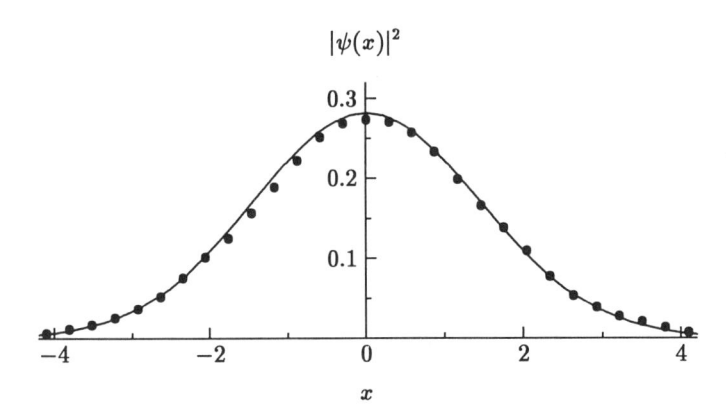

Fig. 7-23 Absolute square of the ground state wave function of a harmonic oscillator calculated using the path integral method. The dots are obtained with a total time of $t_{\text{total}} = 30$, and the entire path is divided into $n = 128$ intervals. The step size is taken to be $\delta_\eta = 1.25 t_{\text{total}}/n$. The solid line represents the exact results of (7-92). The value of $(\hbar/m\omega)^{1/2}$ used is 1.45.

arrays. In this way, explicit calculations of the corresponding factors are need only for values of i_k or i_p that are outside the range of the stored values.

We need a starting path for the Metropolis algorithm. This may be chosen arbitrarily. However, before we can begin taking samples, it is necessary to "thermalize" the initial choice, just as we did earlier for the Ising model. The steps involved in the thermalization process are essentially the same as in selecting a path.

In addition to the initial path, three other quantities must be chosen at the start of the calculation, the total time t_{total}, the number of time intervals n, and the step size δ_η in x. To make the ground state the dominant component, we have assumed that $t_{\text{total}} \to \infty$. In practice, a finite value must be adopted for t_{total} in numerical calculations. As a result, we always have small admixtures of excited states in the value of $|\phi(x)|^2$ obtained. A comparison of the results calculated using different values of t_{total} was given by Lawande, Jensen, and Sahlin (*Comp. Phys.* **3** [1969] 416; **4** [1969] 451). In general, values of $t_{\text{total}} > 20$ (in units of ω^{-1}) were found to be necessary for a harmonic oscillator potential in order to obtain good agreements with the analytical form of

$$\psi_0(\eta) = \pi^{-1/4} e^{-\eta^2/2} \tag{7-92}$$

given earlier in (4-16). For the results shown in Fig. 7-23, a value of $t_{\text{total}} = 30$ was used.

The number of steps for the path to develop from $\tau = 0$ to $\tau = t_{\text{total}}$ depends on the numerical accuracy we wish to achieve in the calculation. For the illustration, a value of 128 was used. The choice depends in part on the step size

δ_η. Since the ground state wave function for a harmonic oscillator is essentially confined to the region $-3 \leq \eta \leq 3$, a sufficient number of η_k must lie within this interval. A value a few times t_{total}/n seems to be satisfactory. The actual steps of the calculations are summarized in Box 7-8.

Box 7-8   Program PATH_INTEGRAL
Path integral calculation
For the square of the ground state wave function of a harmonic oscillator

Subprograms called:
   THE_VAL: Compares the results with theoretical values.
   PROB_NRML: Rational polynomial approximation to the normal probability function.
   M_RATIO: Generate the Metropolis probability ratios of (7-91).
   V_PATH: Sampling the path.
   THERMAL: Thermalization sweeps.
   PATH_PRINT: Output the path.
   RSUB: Uniform random number generator (Box 7-2).

Parameters:
   t_{total}: Total time (=30).
   n: Number of intervals (=128).
   δ_η: Spatial step size (input as $\alpha\delta_\xi = \alpha t_{\text{total}}/n$).
   n_{therm}: Number of thermalization steps (=300).
   n_{var}: Number of random points to select a new path (=300).

Initialization:
   (a) Zero the array for $|\psi(\eta_k)|^2$.
   (b) Generate a set of n random numbers $\{x_k\}$ and use $5x_k$ as the initial path.
   (c) Use M_RATIO to construct the decision table $r_{k.e.}$ and $r_{p.e.}$ of (7-91).

1. Use THERM to thermalize the initial choice by carrying out the following steps n_{therm}^2 times:
   (a) Generate a uniform random number x in $[0, 1]$.
       Use $k = (nx + 1)$ as the point for path variation.
   (b) Generate another uniform random number x' in $[0, 1]$.
       If $x' > 0.5$, the trial variation is $+1$ spatial unit; otherwise -1.
   (c) Determine the change in kinetic and potential energies due to a move of the point η_k by $+1$ or -1 unit.
   (d) Calculate the ratio r of (7-90). Use the stored decision tables if possible; otherwise calculate explicitly using (7-91).
   (e) If $r_k \geq 1$, the trial variation is accepted. If $r_k < 1$,
       (i) Generate another uniform random number x'' in $[0, 1]$.
       (ii) If $x'' \leq r_k$, the trial variation is accepted; otherwise it is rejected.

2. Use V_PATH to vary the path. The steps are identical to those in THERM except that after the last step $|\psi(\eta_k)|^2$ is increased by unity each time.

3. Use THE_VAL to:
   (a) Normalize $|\psi(\eta_k)|^2$.
   (b) Output the value of $|\psi(\eta_k)|^2$.

Lattice gauge calculation. The path integral method is used extensively in field theoretical calculations where perturbation techniques are inappropriate. One of the more interesting applications involves strong interactions between quarks in quantum chromodynamics (QCD). Here, the strength of the interaction at energies accessible to laboratory experiments is too strong for methods of calculation that are based on perturbation expansions. In 1974, Wilson (*Phys. Rev.* **D10** [1974] 2445) suggested the use of a numerical approach that is now known as lattice gauge calculations. The word lattice comes from the fact that the space is divided into grids similar to what we have done in the example above. However, the problem is much more complicated here since we must take the coordinate space to be three dimensional. A modest grid that divides each spatial direction into 16 parts gives us $(16)^3 = 4096$ points. In contrast to the single degree of freedom of $\delta = \pm 1$ at each site in our harmonic oscillator example, there are now many more, given by the product of the different types of quarks and the degrees of freedom each quark can assume, such as color and spin. It is obvious that an exact calculation is impossible and sampling of the multidimensional space becomes the only viable alternative. Although the basic calculations involved are not too different from our simple example above, the time development of such a complex system becomes far more complicated. In fact, the calculations have become one of the challenges of computational physics and computers. Several machines have been designed specially to solve such problems, as this may be the only way to carry out nonperturbative investigations in QCD. This is a fast-developing subject and only the current literature on the topic can keep a person informed of the latest developments.

§7-8 Fractals

The era of computers has generated many new ideas as well as new branches of studies in physics. Among these, fractals have perhaps received the widest attention. From a relatively obscure mathematical curiosity, which may be traced back to its origins in the early 1960s, fractals have grown to be a popular form of computer recreation reported frequently in publications such as *Scientific American*. The number of volumes of books appearing on and around the subject for both specialists and nonspecialists is also growing at an amazing rate. What are fractals? Why are they of interest to physicists? It is perhaps more useful to begin the discussion by examining how the two diagrams shown in Figs. 7-24 and 7-25 are made.

Mandelbrot set and Julia set. Let $z_n = x_n + iy_n$ be a complex number defined by the recursive relation

$$z_n = z_{n-1}^2 + c \qquad (7\text{-}93)$$

where $c = a + ib$ is another complex number. If we start the recursion from some value, such as $z_0 = 0$, and repeat several times the process of generating z_n from z_{n-1} in the way specified by (7-93), we find that there are two possible results. The first is that the magnitude of z_n becomes unbound as n increases. The second is that the magnitude of z_n remains finite regardless of how many times we iterate the relation. The outcome clearly depends on the value of c used. It is therefore possible to divide the set of all complex numbers $\{c\}$ into two groups according to the behavior after a large number of iterations of (7-93). A Mandelbrot set is a group that contains all c such that

$$\lim_{n \to \infty} |z_n| = \text{finite}$$

A graphical display of the set may be carried out by plotting c in the complex plan. In Fig. 7-24, complex numbers $c = (a+ib)$ belonging to a Mandelbrot set are represented by points within the area enclosed by the curve. Points outside the area belong to the other group, and they produce values of z_n whose magnitudes become infinite eventually as we apply (7-93) more and more times.

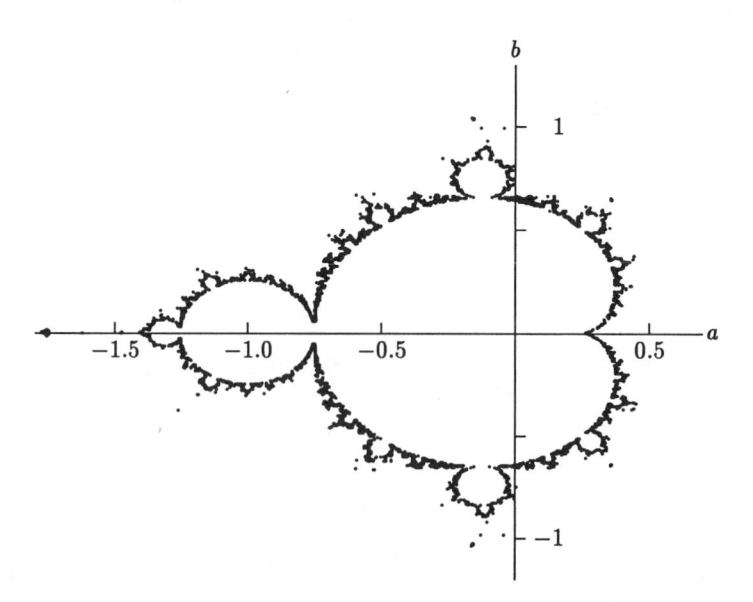

Fig. 7-24 A Mandelbrot set made of complex numbers $c = a + ib$. These numbers, found in the area bound by the curve shown, have the property that the magnitude of z_n, defined by (7-93), remains finite when $n \to \infty$.

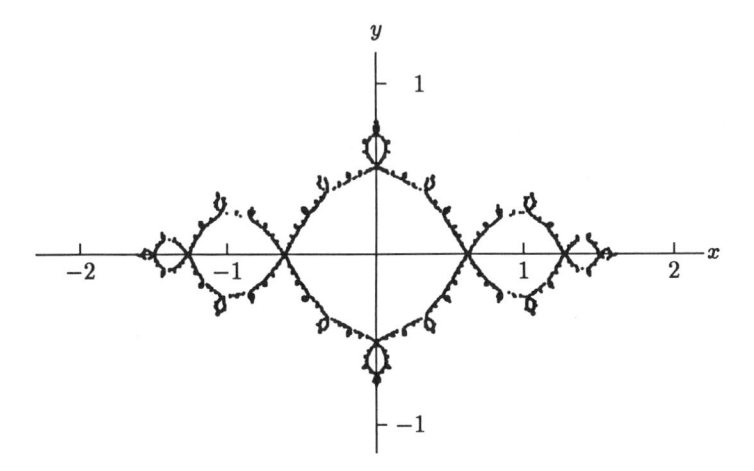

Fig. 7-25 A Julia set containing all the complex numbers $z_0 = x + iy$ for which $|z_n|$, with $z_n = z_{n-1}^2 + c$, remains finite as $n \to \infty$. The value of c used here is -1.

Instead of c, we can also separate the starting complex numbers z_0 into two groups according to the criterion of whether the magnitude of z_n stays finite or not after carrying out the iteration (7-93) an infinite number of times. A collection of z_0 that leads to finite z_n is know as a Julia set. Since the result of iterating (7-93) now depends on the value of c used, there are many possible Julia sets, each corresponding to a different choice of the complex number c. The set enclosed by the curve in Fig. 7-25 corresponds to the choice $c = -1$. Other choices are also possible and many of them display interesting structures.

Coloring a Mandelbrot plot. The usual pictures that one sees of Mandelbrot and Julia plots are very colorful and fascinating, such as those that appeared in the *Scientific American* articles by A.K. Dewdney (**260** [February 1989] 108; [June 1989] 125; and references therein). The rendition of color for these plots is done in the following way. It is sufficient for us to consider here only Mandelbrot sets, as it is quite easy to adapt the same method to Julia sets. For simplicity, we shall treat only the situation of displaying the results on the color monitor of a computer. As we shall see in more detail in Chapter 10, the screen of a computer monitor is divided into a number of pixels. We shall take the total number of available pixels to be $(N_x \times N_y)$. That is, there are N_x dots in a horizontal line and there are N_y vertical lines on a screen. For a color monitor, each pixel can appear in a number of different colors and we shall take the number to be N_c. Typical values of N_x, N_y, and N_c are, respectively, 640, 480, and 16, although higher values are becoming more common.

To display a picture of a Mandelbrot set, we shall consider the screen as the complex plane and each pixel on the screen is a point in the plane. Following the usual convention in displaying complex numbers, we shall use the horizontal direction to represent the real part and the vertical direction for the imaginary part. To put a complex number $c = (a + ib)$ on the screen, we need to select an origin and a scale factor. For Fig. 7-24, we have assumed that there are 640 horizontal pixels on a line and there are 400 lines on the screen. The original of our coordinate system is at pixel (461,236). The vertical axis is then along pixels 461, counting from the left of the screen, and the horizontal axis along line 236, counting from the bottom of the screen. For the convenience of discussion, we shall represent a complex number $c = (a + ib)$ by specifying its coordinates in the complex plane in terms of a pair of numbers (a, b). To display all the numbers belonging to the Mandelbrot set, our real axis has the range of values from -1.8 to 0.7, with each pixel equivalent to $2.5/640 = 0.0039$ in size. The imaginary axis goes from -1.25 to 1.25, with each pixel equivalent to $2.5/480 = 0.0052$.

As far as the color of the display is concerned, the members of a Mandelbrot set are not of interest. In other words, any number whose magnitude never becomes unbound when we iterate according to (7-93) does not produce an interesting effect. The rich display of color that is so impressive in pictures of fractals is obtained from regions bordering a Mandelbrot set. This comes from the following reason. If a point c does not belong to the set, the magnitude of z_n will eventually become very large. The criterion commonly used to indicate large magnitude is $|z_n| > 2$. We shall not attempt a proof here that any z_n with a magnitude greater than 2 will eventually become unbound if we carry out the iteration (7-93) further, and we shall simply take it as the limit for z_n to be unbound. For a Mandelbrot set, different c outside the set reach this limit at different rates. For example, some c may only take $n = 5$ iterations to reach the limit of $|z_n| > 2$, whereas others may take $n = 100$. If we associate a color with each value of n for a complex number c to reach the level of $|z_n| > 2$, a colorful picture is obtained.

In an actual computation, we must choose some value $n_{\max}$ as the largest number of iterations we wish to carry out. For all practical purposes, $n_{\max}$ is the "infinity" for our iteration process. To generate a plot, each complex number c is now subject to the recursive process specified by (7-93), starting with $z_0 = 0$. At the end of each iteration, the magnitude of z_n is checked. If $|z_n| > 2$, the complex number c is not a member of the Mandelbrot set. The iteration process stops and the pixel corresponding to this value of c is given a color by the following procedure. Let us use the symbol n_{esc} to represent the number of iterations taken. Since the number of available colors N_c is usually smaller than $n_{\max}$, we cannot assign a unique color to every possible value of n_{esc}. Instead, we shall associate the pixel with a color I_c, given by the relation

$$I_c = n_{\text{esc}} \bmod N_c \tag{7-94}$$

Here I_c is a number in the range $[0, N_c]$, and each integer value, $0, 1, 2, \ldots$, corresponds to a particular color, such as black, blue, green, $\ldots$. If the value of $|z_n|$ after an iteration is less than or equal to 2, the iteration process is repeated.

If after n_{max} iterations, $|z_n|$ is still less than or equal to 2, we have a member of the Mandelbrot set. The pixel associated with this value of c is colored black. Obviously, our picture is more colorful if we have a larger value of N_c at our disposal. Similarly, our picture will be more interesting, and more accurate, if we are willing to take a larger value of n_{max} for our calculation.

For better efficiency, it is advantageous to carry out the complex arithmetics involved in terms of the real numbers, as we saw earlier in discussions related to complex matrices in §5-8. This can be done by handling the real and imaginary parts of the complex number calculation in (7-93) separately:

$$x_n = x_{n-1}^2 - y_{n-1}^2 + a \qquad y_n = 2x_{n-1}y_{n-1} + b \qquad (7\text{-}95)$$

where we have used x_n and y_n to be, respectively, the real and imaginary parts of z_n. Similarly, a is the real part of c and b the imaginary part. The shape of the output depends also on the range of c we wish to examine. Instead of $a = [-1.8, +0.7]$ and $b = [-1.25, +1.25]$, more interesting plots are obtained by concentrating on small areas just outside the border of a Mandelbrot set. A mathematical fascination of fractals is the amount of detail we can find as we zoom into an increasingly small area.

In making an actual Mandelbrot plot, it is more convenient to think in terms of each individual pixel on the screen and translate it to the corresponding value of complex number c, rather than the other way around. The calculations involved may be thought of in terms of three loops. The first scans the N_y rows in the entire screen one at a time, starting from the bottom. For each row, the second loop examines the N_x pixels one by one. For each c associated with a particular pixel, a third loop handles the iteration given by (7-93), using $z_0 = 0$ as the starting point. Anytime the magnitude of z_n becomes greater than 2, it is output to the screen with a color index I_c given by (7-94). If after n_{max} iterations the magnitude of z_n remains less than or equal to 2, the complex number is considered to be a member of the Mandelbrot set and the pixel is given the color black (blank in Fig. 7-24). The algorithm outlined in Box 7-9 is quite simple, but it may take a long time to run if n_{max} is chosen to be a large number. For the plot in Fig. 7-24, a modest value of $n_{max} = 16$ is used. However, for the more colorful plots of a small part of the border area, much larger numbers of iterations are required. Values of n_{max} in excess of 100 are quite common and this can be very time consuming on a computer. Some of the interesting ranges of c examined are given in the *Scientific American* article by Dewdney cited above.

The construction of a Julia plot is very similar to that of a Mandelbrot plot. The pixels on a computer screen now represent points in the complex plane for all the possible values of z_0, and each plot is for a fixed value of c. The color of each pixel is assigned according to the number of iterations it takes to start with z_0 and end up with a magnitude of z_n that is greater than 2. Instead of the value $c = -1$ used in Fig. 7-25, different Julia plots can be obtained with other values, such as $c = -0.9 + 0.12i$.

> **Box 7-9   Program MANDEL_PLOT**
> **Mandelbrot plot**
>
> Make a Mandelbrot plot using subprograms from Chapter 9
>
> Subprogram used:
>   SET_PXL: Turn the pixel at (IX,IY) to color KOL.
>   ST_LINE: Bring the present pixel position to (IX,IY) with color KOL.
> Parameters in the calculation:
> N_x: Number of pixels in a horizontal line (= 640).
> N_y: Number vertical lines on the screen (= 480).
> 1. Input:
>   (a) The range of complex number c.
> x_{start}: Minimum value of the real part.
> x_{range}: Range of the real part.
> y_{start}: Minimum value of the imaginary part.
> y_{range}: Range of the imaginary part.
>   (b) n_{max}: The maximum number of iterations allowed.
> 2. Calculate the values of c that each pixel corresponds to.
> 3. Initialize the graphics screen.
> 4. Iterate each value of c in the range using (7-93):
>   (a) Imaginary part (vertical direction): For $i_y = 1$ to N_y:
>   (i) Start with the minimum value specified in the input.
>   (ii) Increase by an amount corresponding to a pixel each time i_y increases by 1.
>   (b) Real part (horizontal direction): For $i_x = 1$ to N_x:
>   (i) Start with minimum value specified in the input.
>   (ii) Increase by an amount corresponding to a pixel each time i_x increases by 1.
>   (c) Iterate z_n according to (7-93) up to a maximum of n_{max} times:
>   (i) Start with $x_0 = y_0 = 0$.
>   (ii) Calculate x_n and y_n using (7-95) and let $r^2 = x_n^2 + y_n^2$.
>   (iii) If $r^2 > 4$, use SET_PXL to output the pixel to the screen with color index I_c given by (7-94) and return to step (b).
>   (iv) If $r^2 \leq 4$ after n_{max} iterations, use SET_PXL to output the pixel in black and return to step (b).

Fractals and fractal dimension. The beauty of fractal pictures, however, does not provide us with much insight into the question of what fractals are and the ways in which they are useful in physics. The usual definition of fractals is that they are objects with fractional dimension. To understand this definition, we must first describe what is the meaning of the term "dimension."

Consider an ordinary object, for example, a cube with each side of length ℓ. The volume is equal to ℓ^3. For this reason, it is called a *three-dimensional* object. More formally, we can say the object has dimension $d = 3$. A printed page, on the other hand, is a two-dimensional object ($d = 2$), as the thickness of the paper on which it is printed is immaterial. A straight line is a one-dimensional object ($d = 1$), since it has no width or thickness. For our discussion here, it is convenient to define dimension in a slightly different way. For most objects, the density is a constant. As a result, the mass of a three-dimensional object is proportional to the third power of its linear size. Similarly, the mass of a sheet of paper (of constant thickness and density) is proportional to the square of its linear dimension, and the mass of a piece of wire (of constant cross-sectional area) is proportional to the length. In other words, the mass $\mu(\ell)$ of an object, with linear dimension ℓ, is given by the relation

$$\mu(\ell) \sim \ell^d \tag{7-96}$$

where d is the dimension of the object.

More generally, the same relation can be expressed in terms of a scale change. If we double the linear dimension of a three-dimensional object, the mass is increased by 2^3. In general, the scaling relation may be written as

$$\mu(\lambda\ell) = \lambda^d \mu(\ell) \tag{7-97}$$

This is a better definition for dimension d than (7-96), since it is in the form of a mathematical identity.

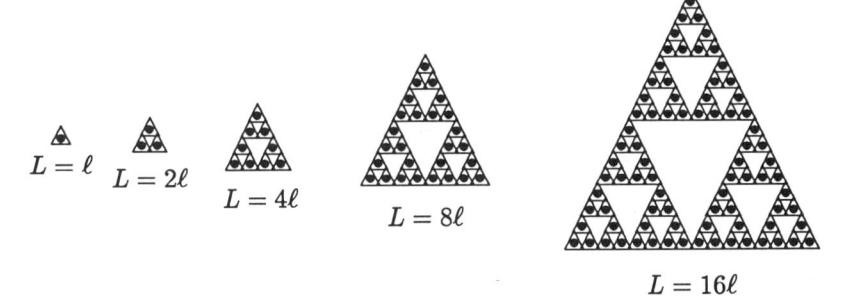

Fig. 7-26 The Sierpinski gasket as an example of self-similar objects. It is constructed by repeatedly applying the process of putting three identical equilateral triangles on top of each other to form one with twice the length of each side.

Almost all the objects we encounter in daily life have integer values for their dimensions. However, there is also a class of objects we can construct that has non-integer dimensions, or fractals. Before we go into the subject of the significance of fractals in physics, we shall first give the example of a Sierpinski gasket to see how objects with fractional dimensions can occur.

Consider an equilateral triangle with all three sides having the same length ℓ. For the ease of argument, we shall take the mass of each such triangle to be unity. Next, we put three such triangles together in such a way that the top corner of triangle 2 touches the lower-left corner of triangle 1, the top corner of triangle 3 touches the lower-right corner of triangle 1, and the lower-right corner of triangle 2 touches the lower-left corner of triangle 3. The result is a triangle with the length of each side $L = 2\ell$, as shown in Fig. 7-26. Since the $L = 2\ell$ triangle is made of three $L = \ell$ triangles, its mass is three units, or

$$\mu(2\ell) = 3\mu(\ell)$$

We can now repeat the process by putting three $L = 2\ell$ triangles together in the same way as we did in forming the $L = 2\ell$ triangles themselves. This results in a triangle with the length of each side $L = 4\ell$. The mass of this new triangle is

$$\mu(L = 4\ell) = 9\mu(\ell)$$

as can be seen by looking at Fig. 7-26. By the same token, the mass of an $L = 8\ell$ triangle is

$$\mu(L = 8\ell) = 27\mu(\ell)$$

The process can be repeated as many times as we wish. Each time the length of the resulting triangle is doubled and the mass is tripled. The scaling relation is then

$$\mu(2^n \ell) = 3^n \mu(\ell)$$

Comparing this with (7-97), we obtain the result

$$\left(2^n\right)^{d_f} = 3^n$$

where

$$d_f = \frac{\ln 3}{\ln 2}$$

The quantity d_f is the fractal dimension of the Sierpinski gasket and is different from d discussed earlier. Where there is the need to differentiate between d and d_f, the former is referred to as the Euclidean dimension, for obvious reasons. Since the Sierpinski gasket is drawn on a sheet of paper, its Euclidean dimension is $d = 2$. The fractal dimension is $d_f = 1.5849\ldots$, and it is not an integer.

Self-similarity of fractals. Another property that characterizes fractals is the existence of self-similarity over many different length scales. Take the Sierpinski gasket as an example. From a distance, a large gasket, for example, with $L = n\ell$ with $n = 2^{100}$, will look like an equilateral triangle made of three other similar triangles, each with sides half the length of the large one. On a closer examination, each of the three constituent triangles is again made of three similar triangles. Such a similarity in form persists for many length scales (100 in this case). Usually, the self-similarity observed in fractals is not exact, unlike our Sierpinski gasket example. In general, the self-similarity is to be understood in terms of statistical measures, as in the case of one handful of fine sand being similar to another handful from the same pile.

In this sense, the borders of Mandelbrot and Julia plots are fractals. As we zoom in closer and closer to examine a smaller and smaller area of the border, more and more details are revealed. Each magnification displays finer details of the area, and the pictures consist of different shapes and are made of different combinations of colors. However, we cannot use any reasonable statistical measures to determine that they are different in any significant way from the pictures of a different magnification.

Many physical objects possess this property of self-similarity. The most obvious example is a crystal. If we break up a large crystal into small pieces, each one will have essentially the same shape as the large crystal. The small crystals may be broken up further and even smaller crystals of similar shapes are obtained. For this reason, the study of fractal growth has become a topic of broad interest in statistical physics. As an application of Monte Carlo techniques, we shall give below one such example in the form of diffusion-limited aggregation.

Example of diffusion-limited aggregation. Aggregation is the process in which particles are collected together to form larger entities, or *aggregates*. Consider a volume of space that is empty to start with. Slowly, particles are released into it that have some form of very short range attraction among themselves. As the particles diffuse throughout the volume because of random thermal motion, there is a finite chance that two of them will come into contact with each other. If the force acting between the two particles is attractive, there is a finite probability that they will stick together and form a larger unit. Since the new unit is larger, there is an even higher probability for another particle to collide and stick with it. The additional particle makes the aggregate even larger and increases the rate of growth further. This process continues as long as the supply of free particles is not exhausted and no instabilities develop as the size gets larger. If there are no preferences as to where a new particle should join the aggregate, the resulting object for a two-dimensional system will look similar to that shown in Fig. 7-27, a pattern that is observed in a variety of situations, such as the path of an electric discharge or the growth of lichen on rocks. On the other hand, if there are preferences where the particles should stick, the results may become objects such as snowflakes. Things formed in this way are fractals. This can be seen from the self-similarity property if any part of it is examined in more detail. It is also possible to define a fractional dimension from such objects, but we shall not be going into the question here.

The diffusion-limited aggregation (DLA) model is designed to carry out computer simulations of such processes. For simplicity, let us consider a case in two (Euclidean) dimensions. Such a surface may be divided into a lattice consisting of $(N \times N)$ sites. At the start, our "aggregate" consists of a single particle in the center of the grid. The diffusion of other particles is simulated by allowing one particle to enter into the space at a time and wander around the lattice in the form of a random walk. If the particle comes within the interaction distance of those in the aggregate, there is a finite probability for it to become a member by sticking to an existing particle.

Fig. 7-27 Pattern generated by a diffusion-limited aggregation algorithm. The result is a fractal and resembles that of a large variety of physical processes.

The simplest DLA model we can construct is to assign unity to the sticking probability if the wandering particle comes into an empty site next to an occupied one. Once this takes place, the wandering particle is stuck and becomes a member of the aggregate (and the site in which it stops becomes an occupied site). The process is now repeated by letting another particle wander into the space. Since the motion of the wandering particle is random, it is possible that it will wander off the area. We shall ignore such "lost" particles. In this way, each particle we introduce into the lattice either becomes a part of the aggregate or wanders off the lattice.

The algorithm is basically a very simple one. To simulate random walk in two-dimensional space, we can extend the basic idea of the one-dimensional random walk in §7-2. One way to do this is to use a random number x that is uniformly distributed in the interval $[0, 1]$. If $x < \frac{1}{4}$, the particle moves up one lattice site; if $\frac{1}{4} \leq x < \frac{1}{2}$, it moves down one lattice site; if $\frac{1}{2} \leq x < \frac{3}{4}$, it moves to the left by one lattice site; and if $x \geq \frac{3}{4}$, it moves to the right by one lattice site. Everytime the particle arrives in a new location, we check whether there is a nearest-neighbor site that is occupied; i.e., whether there is a particle above, below, to the left, or to the right. If so, the particle becomes a part of the aggregate. Otherwise, the random walk resumes again. The process stops either when the particle moves off the area or becomes a part of the aggregate.

In principle, the growth of the aggregate should be independent of the point of launch where new particles are introduced into the lattice. This is true as long as the launching point is sufficiently far away from any part of the aggregate so that the particle has an equal probability of sticking to any available site on the perimeter. On the other hand, it may take a large number of random walk

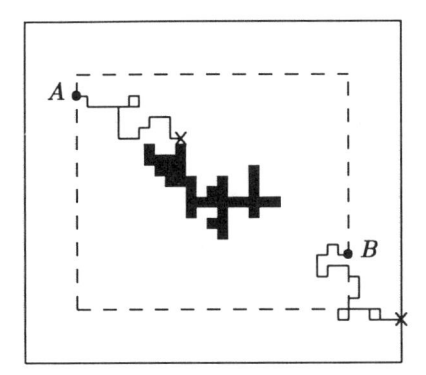

Fig. 7-28 Generating a diffusion-limited aggregation pattern. Each particle starts its wandering through the lattice by random walk from the launching zone, indicated by the dashed border. If a particle gets into a site next to an occupied one, it sticks to the existing aggregate and becomes a member, as shown by the one that starts from the point A. Particle B, on the other hand, wanders off in the wrong direction and reaches the "kill" zone, indicated by the solid line border. Once a particle reaches the kill zone, it is considered to be permanently lost.

steps for a particle to reach the aggregate from a large distance away. From a computational point of view, this is extremely time consuming. For this reason, a *launch zone* that is closer to the existing aggregate is used instead of the border of the lattice. Usually, this consists of a circle around the center of the lattice with a radius R_{launch} that is much larger than the maximum extent of the aggregate. Instead of the border of the lattice, each new particle starts its random walk journey from a random point on this circle. As the aggregate grows in size, the launch radius is increased. For the sake of computational efficiency, it is better to use a square launch zone rather than a circular one. By avoiding any calculations of the circumference and angle associated with a circle, it is possible to confine the operations related to the launching process (other than generating random numbers) to integers. In Fig. 7-28, the launch zone is shown by the dashed lines.

A large amount of computational time can also be saved if we impose a kill zone around the aggregate. Instead of waiting for the wandering particle to go off the lattice before declaring it to be lost, any particle that goes beyond the kill zone is discarded. Again, the kill zone is increased in size as the aggregate grows. Obviously, the kill zone should be at some distance outside the launch zone to allow the possibility for a particle that moves outside the launch zone to move back inside. In Fig. 7-28, the kill zone is indicate by the solid square box. The steps of the DLA algorithm are outlined in Box 7-10.

The calculation stops when enough particles are accumulated in the aggregate. This limit may be set at the beginning of the calculation to be some number

> **Box 7-10   Program DLA**
> **Simulation of diffusion-limited aggregation**
>
> Parameters:
> ℓ_{dist}: Distance of the launch zone from the nearest member.
> k_{dist}: Distance of the kill zone from the launch zone.
> m_{dist}: Size of an artificial occupation zone at the start.
> n_{max}: Maximum size of the aggregate.
> Subprogram called:
>   INIT: Initialize the lattice, and the launch and kill zones, and so on.
>   LNCH: Set up a random point on the launch zone to launch a particle.
>   SET_PXL: Turn the pixel at (IX,IY) to color KOL.
>   RSUB: Uniform random number generator in the interval $[0, 1]$ (Box 7-2).
> Initialization:
>   (a) In INIT:
>   (i) Zero the entire lattice and set the central site to 1.
>   (ii) Set up the launch and kill zones and an artificial occupation boundary.
>   (b) Initialize the graphics screen.
>   (c) Input a seed for the random number generator.
> 1. Launch a particle using LNCH:
>   (a) Generate a uniform random number r_1 in $[0, 4]$ and take the integer part to be i_{xy}.
>   (b) Determine the area where the particle is launched, along the top, bottom, right, or left of the launch zone for $i_{xy} = 0, 1, 2$, or 3, respectively.
>   (c) Generate another random number r_2 and use its value to determine the point along the top, bottom, right or left side of the launch zone where the particle starts.
> 2. Random walk of the particle through the lattice:
>   (a) Generate a uniform random number r_3 in $[0, 4]$, and take the integer part to be j_{xy}.
>   (i) If $j_{xy} = 0, 1, 2$ or 3, move the particle, respectively, up, down, left, or right by one lattice site.
>   (c) If the particle is outside the kill zone, go back to step 1 and launch a new particle.
>   (d) Check for sticking by finding out if a neighboring site is occupied.
>   If so, mark the site of the particle as occupied, and
>   (i) Increase the counter for the size of the aggregate by 1.
>   (ii) Plot the location of the particle using SET_PXL.
>   (iii) Check if the launch and kill zones need to be enlarged.
>   (iv) If the aggregate is smaller than n_{max}, go back to step 1 and launch another particle.
>   (f) Otherwise, go back to step 2 and carry out another random walk step.
> 3. Stop if a sufficient number of particles have accumulated at the aggregate.

N_{max}. In general, it is necessary to choose N_{max} to be much smaller than $(N \times N)$, the total number of sites in the lattice. As in other problems with a finite lattice, the border, or finite size, effect can be important. For example, for the algorithm to work, it is necessary that the kill zone be within the lattice, the launch zone well within the kill zone, and any occupied lattice well within the launch zone. As a result, a fairly large lattice is required to build an aggregate of reasonable size.

It is possible to change the sticking probability from unity for a neighboring site to one having a preference in direction. In this way, different patterns can be generated to simulate different physical situations. It is also of interest to color the particles according to the order of arrival so that one can see how the aggregate grows with time. One can also add to the visual effect by displaying the location of a particle as it wanders throughout the lattice. These are left as exercises for Chapter 10, as it has more to do with the graphical presentation of results than with Monte Carlo calculations.

Problems

7-1 As an example of a bad choice for the parameters of the linear congruence method of (7-1),
$$X_{n+1} = (aX_n + c) \bmod m$$
Knuth gives the values of $m = 256$, $a = 137$, and $c = 187$. Use this generator to select random points in a square surface area of 10 cm on a side. Plot the distribution of points obtained.

7-2 Apply frequency, correlation, and run-up tests to the random numbers generated in Problem 7-1. Which of the three tests provides a quantitative statement for the fact that the random points do not fill the square evenly?

7-3 As an example of a bad way of using random generators, consider a Monte Carlo calculation that requires 250 random numbers as the input. To find the statistical distribution of the results, we shall run each calculation 100 times. To make the 100 calculations independent of each other we need 100 different seeds for the random number generator, and we use each to generate a set of 250 random numbers required for each calculation. There are two ways we can carry out this part of the calculation: (a) Use the random number to generate the 100 seeds first and then use each seed to generate 250 random numbers. (b) Starting with one seed, generate 250 random numbers for the first calculation, and then 250 more random numbers for the next calculation, and so on. What is wrong with approach (a)?

7-4 A popular way to generate random numbers with a normal distribution is to sum n random numbers with a uniform distribution in the interval $[0, 1]$. Apply a frequency test to the results carried out with $n = 3$, 4, and 5. Find out how good a normal distribution can be obtained in this way.

7-5 Start with a uniform random number generator and obtain a sequence of 1000 random numbers with a normal distribution using the transformation technique of (7-16). Repeat the process except now use the transformation of (7-17) instead. Compare the computer time taken to obtain each set.

7-6 Extend the computer simulation of the Ising model of §7-5 to a three-dimensional, cubic lattice of $n = 8$.

7-7 Since the spectrum of eigenvalues is rather "rigid," as we demonstrated in §7-6, there is the possibility of detecting an intruder in a set of eigenvalues. Design a statistical test to find out if a set of eigenvalues has been corrupted by an intruder. Test your method by diagonalizing a real, symmetric matrix of dimension 100 and replace one of the eigenvalues by a random number. Compare your idea with Dyson's F-test (see Brody and coauthors) based on the correlations between the eigenvalues.

7-8 A simple two-dimensional random walk problem may be stated in the following way. Consider a vertical area with infinite array of pegs placed at equal intervals in the form

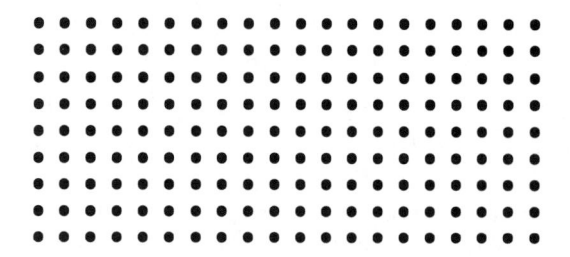

```
• • • • • • • • • • • • • • • • • • • • • • • •
• • • • • • • • • • • • • • • • • • • • • • • •
• • • • • • • • • • • • • • • • • • • • • • • •
• • • • • • • • • • • • • • • • • • • • • • • •
• • • • • • • • • • • • • • • • • • • • • • • •
• • • • • • • • • • • • • • • • • • • • • • • •
• • • • • • • • • • • • • • • • • • • • • • • •
• • • • • • • • • • • • • • • • • • • • • • • •
• • • • • • • • • • • • • • • • • • • • • • • •
```

A particle dropping down from the top may hit one of the pegs and, as a result, scatter randomly either to the left or the right. The particle may, in turn, strike another peg one row down, and so on. Such an arrangement is sometimes referred to as a "probability machine." To simplify the problem, we can assume that the scattering always leads to another peg to the left or to the right one row down. Construct a program to simulate the movement of a particle dropping down the array. Demonstrate through computer simulation that, if all the particles originate from the same point at the top, the resulting distribution of particles at the bottom is a normal one.

7-9 A plausible reason for the spread of the Indo-European languages is due to the introduction of a new method of agriculture (see C. Renfrew, "The Origins of Indo-European Languages," *Scientific American*, **261** [Oct. 1989] 106). Use a two-dimensional random walk to simulate the spread of the farming technique by assuming that the method passes only from parents to children. If each farmer on coming of age moves 18 km in a random direction from the parents farm to establish a new one and the interval between two generations is 25 years, find the average rate for the spread of the technique and, hence, the language spoken by the people.

7-10 A simple study of self-similar diagrams may be carried out with the von Koch snowflake curve. Take a straight line and divide it into three equal segments. Replace the middle segment with two straight lines of the same length as the segment and use them to form the two sides of an equilateral triangle. The complete curve now takes on the form _/_. Repeat the process for each one of the four segments and form a curve with sixteen segments, and so on.

Design an algorithm to carry out this construction on a computer and try it out.

7-11 Modify the algorithm in Box 7-9 to display a Julia set. Construct the plot with $c = (-1 + 0i)$ and $c = (-0.9 + 0.12i)$.

7-12 Instead of mapping out a complete Mandelbrot set as done in Fig. 7-24, concentrate the plot on a small area near the border. As the starting point, some of the ranges suggested in the *Scientific American* articles of Dewdney may be used.

7-13 A colorful plot of the DLA model may be constructed by changing the color of the pixels each time after, say, 50 particles are added to the aggregate. Try this out on a color monitor.

7-14 How can we change the sticking probability in a DLA model so as to generate snowflakelike patterns?

Chapter 8

Ordinary and Partial Differential Equations

Solving differential equations by numerical methods is one of the more developed areas of computational techniques. This arises in part from the needs in a broad range of fields, from engineering to modeling the economy. Besides the method of finite difference — our primary concern here — finite element methods are also used extensively. The subject of numerical methods for solving differential equations is vast and we can give here only a brief account of some of the more common approaches used in problems of interest to physics. There are large numbers of excellent specialized books on both numerical and analytical solutions to various types of differential equations and we shall make reference to them at the appropriate places. Software packages and subroutine libraries are also available to handle some of the more standard forms of differential equations. Although we shall not make any use of them in our introduction to the subject, it is advantageous for anyone planning to carry out extensive calculations to consult these professionally written programs.

§8-1 Types of differential equations

A differential equation expresses the relation of one or more derivatives of the dependent variable in terms of the independent variables. This type of problem appears quite often in physics. For example, the variation of velocity v as a function of time for an object with mass μ falling freely toward the surface of the earth is given in Newtonian mechanics in terms of its derivative with respect to time (that is, acceleration):

$$\mu \frac{dv}{dt} = -\mu g$$

where g is the acceleration due to gravity. The solution of this equation is a trivial one, given by the integral

$$v(t) = v_0 + \int_{t_0}^{t} g\, dt$$

where v_0 is the velocity at time $t = t_0$. More generally, friction may be present in the system. If we assume that such a damping term is proportional to the velocity, the equation for the acceleration now takes on the form

$$\frac{dv}{dt} = \frac{b}{\mu}v - g \qquad (8\text{-}1)$$

where the term bv comes from frictional forces. The solution of this equation may also be represented by an integral:

$$\int_{v(t_0)}^{v(t)} \frac{dv}{\frac{b}{\mu}v - g} = t + c$$

where c is an integration constant to be determined by either the boundary condition or the initial condition of the problem.

If b is a constant, once again, the integration can be easily carried out. In reality, the frictional forces are often a complicated function of the velocity. If this is the case, we can no longer rely on analytical methods to carry out the calculation. Direct numerical integration usually does not help very much here, since the integrand involves the velocity $v(t)$ whose solution we are seeking. On the other hand, it is a relatively simple matter to solve (8-1) directly by numerical methods. Since it involves only total derivatives, this type of differential equation is called an ordinary differential equation, or ODE for short. Furthermore, only the first-order derivative of the dependent variable appears in the equation. Consequently, it is a first-order ODE. We shall see later that solutions to first-order ODEs form the basis for a large class of solutions to differential equations in general.

Ordinary differential equations. As long as the factor b in (8-1) is a constant or a smooth function of t, there is no difficulty in developing methods to "integrate," that is, to obtain a solution for, the equation. On the other hand, differential equations we normally encounter in physical problems are more complicated than first-order ODE. For example, the equation of motion for a simple harmonic oscillator involves second-order derivatives. Consider an oscillator made of a mass μ hanging vertically from one end of a spring with spring constant k. Let the other end of the spring be fixed. If the mass is displaced vertically by a small amount from its equilibrium position and then released, the resulting motion is governed by the equation

$$\mu \frac{d^2\phi}{dt^2} = -k\phi \qquad (8\text{-}2)$$

where $d^2\phi/dt^2$ is the acceleration of the mass along the vertical direction and $-k\phi$ is the restoring force provided by the spring. More generally, we can include a damping term, $-b\,d\phi/dt$, due to friction that is proportional to the velocity, and a driving term $F_0 \cos \omega_d t$ provided by an external force with frequency ω_d. The result is the familiar equation for a forced harmonic oscillator:

$$\frac{d^2\phi}{dt^2} + \frac{b}{\mu}\frac{d\phi}{dt} + \frac{k}{\mu}\phi = \frac{F_0}{\mu}\cos\omega_d t \qquad (8\text{-}3)$$

Both (8-2) and (8-3) are examples of second-order ODE, as the highest-order derivative involved is second-order.

It is always possible to reduce a second-order ODE into a set of two coupled first-order ODEs. For example, if we define

$$\eta(t) = \frac{d\phi}{dt} \qquad (8\text{-}4)$$

we can rewrite (8-3) in the form

$$\frac{d\eta}{dt} + \frac{b}{\mu}\eta + \frac{k}{\mu}\phi = \frac{F_0}{\mu}\cos\omega_d t \qquad (8\text{-}5)$$

Instead of (8-3), we can solve (8-4) and (8-5) together as a set of two coupled equations. In general, it is advantageous to solve a second-order ODE of the form

$$\frac{d^2\phi}{dx^2} + p(x)\frac{d\phi}{dx} = q(x) \qquad (8\text{-}6)$$

as a set of two coupled first-order ODEs:

$$\frac{d\phi}{dx} = v(x) \qquad \frac{dv}{dx} = q(x) - p(x)v(x) \qquad (8\text{-}7)$$

The same approach may also be extended to handle higher-order ODEs. For this reason, most of the numerical methods for solving ODEs are designed for coupled first-order equations.

Partial differential equations. If there is more than one independent variable in the physical problem, the resulting equation often involves partial derivatives of the dependent variables. For example, in the case of the vibrating string shown schematically in Fig. 8-1, the vertical displacement $\phi(x,t)$ is a function of time t as well as the location x along the string. The equation of motion has the well-known form

$$\frac{\partial^2 \phi}{\partial x^2} - \frac{1}{v^2}\frac{\partial^2 \phi}{\partial t^2} = 0 \qquad (8\text{-}8)$$

where v is a quantity having the dimension of velocity and is given by the ratio of tension T and linear density ρ of the string,

$$v^2 = T/\rho$$

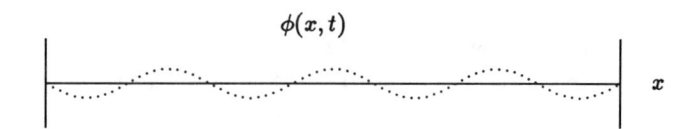

Fig. 8-1 Schematic diagram of a vibrating string. The displacement $\phi(x,t)$ is a function of both x, the location along the string, and t, the time.

The particular form of the partial differential equation (PDE) given by (8-8) is known as a hyperbolic equation.

In general, we can write a two-dimensional (that is, two independent variables) second-order PDE in the form

$$p\frac{\partial^2 \phi}{\partial x^2} + q\frac{\partial^2 \phi}{\partial x \partial y} + r\frac{\partial^2 \phi}{\partial y^2} + s\frac{\partial \phi}{\partial x} + t\frac{\partial \phi}{\partial y} + u\phi + v = 0 \tag{8-9}$$

where p, q, r, s, t, u, and v may be functions of the independent variables x and y, as well as the dependent variable ϕ and its derivatives. If $q^2 < 4pr$, it is called an *elliptic* equation; if $q^2 = 4pr$, it is a *parabolic* equation; and if $q^2 > 4pr$, it is a *hyperbolic* equation.

The wave equation we saw earlier in (8-8) is an example of a hyperbolic PDE. A good example of elliptic equations is the two-dimensional Poisson equation in electrostatics:

$$\frac{\partial^2 \phi}{\partial x^2} + \frac{\partial^2 \phi}{\partial y^2} = -\rho(x,y)$$

It describes the field $\phi(x,y)$ of a charge distribution $\rho(x,y)$ in two dimensions. The standard example of a parabolic equation is the diffusion equation in statistical mechanics:

$$\frac{\partial \phi}{\partial t} = -\frac{\partial}{\partial x}\left(D\frac{\partial \phi}{\partial x}\right)$$

where ϕ may be the concentration of a certain kind of particle and D is its diffusion coefficient.

For the most part, we shall be dealing mainly with linear differential equations; that is, a differential equation involving only linear functions of the dependent variable ϕ and its derivatives. In terms of the prototype given by (8-9), a linear differential equation is one with coefficients p, q, r, $\ldots$, independent of ϕ and any of its derivatives. Later in this chapter, we shall see examples of nonlinear differential equations in which powers, products, and transcendental functions of ϕ and its derivatives may appear. It is much more difficult to design general methods to solve these equations, as questions of stability and convergence of the solution are much harder to analyze.

From the point of view of physics, various types of differential equations may be classified in the following way. In general, elliptic equations occur when we wish to find the equilibrium of a system under a given set of boundary conditions. Parabolic and hyperbolic equations, on the other hand, describe the time development of a system under a given set of initial conditions.

Initial and boundary value problems. Differential equations can also be classified into initial value and boundary value problems. For numerical solutions, the differences between these two categories can often be more important than some of the considerations above. This comes from the fact that an initial value problem propagates the solution forward in time from the values given at the starting point. In contrast, a boundary value problem has known values, which must be satisfied at both the start and the end of the interval. The distinction between temporal and spatial variables is mainly in their physical significance. As far as the numerical solutions are concerned, there are very few differences between them except, perhaps, in the way in which they appear in the equation under certain circumstances. Our primary concern in the differences between initial value and boundary value problems is in the way the constraints are placed on the solution. If all the conditions that must be satisfied by the solution are given at one end of the interval, we have an initial value problem. On the other hand, if the constraints on the solution are applied to both ends of the interval, we have a boundary value problem. For partial differential equations, it may also happen that the conditions for some independent variables are given in the form of initial values and others as boundary conditions. If this happens, we have an initial value boundary problem, a mixture of both types in one problem.

As an example, let us consider the prototype second-order differential equation given by (8-6). The only independent variable is x and we shall be interested in the solution in the domain $x = [a, b]$. Since it is a second-order differential equation, two pieces of information must be supplied by the physics of the problem before we can solve the equation. If we specify them, for example, in terms of the value of $\phi(x)$ and its first-order derivative at $x = a$ (or $x = b$), we have an initial value problem, independent of whether the physical meaning of x is time or distance. On the other hand, the two necessary conditions may be given as the values of $\phi(x)$ at $x = a$ and $x = b$. In this case, we have a boundary value problem. Obviously, quite different numerical methods must be used to solve these two types of problems.

Euler's method for initial value problems. It is useful to illustrate the philosophy behind most numerical methods for solving differential equations using Euler's method. In particular, we shall treat an initial value problem, as it is more straightforward to do so. We shall see later that many improvements can be made to this basic principle of numerical solution. In fact, most of the practical methods used these days have moved so far away from this naive approach that they bear only vague resemblance to the simple steps we shall be carrying out here.

Consider the case of a simple harmonic oscillator with angular frequency

$$\omega_0 = \left(\frac{k}{\mu}\right)^{1/2}$$

where k is the spring constant and μ is the mass, as we had earlier in (8-2). For such an oscillator, the relation given by (8-2) may now be written in the form

$$\frac{d^2\phi}{dt^2} + \omega_0^2 \phi(t) = 0 \qquad (8\text{-}10)$$

Similar to what we did in numerical integrations, the first step in making use of most numerical methods to solve a differential equation is to "discretize" the interval of interest to us into a grid or a mesh consisting of a finite number of points. For example, let us assume that our interest is in the solution of (8-10) in the continuous interval $t = [t_0, t_N]$. For an analytical solution, our aim will be to obtain a function $\phi(t)$ such that the value of the amplitude ϕ is known at any time t in the interval. For a numerical solution, we must be satisfied with the values of $\phi(t)$ at a discrete number of points at $t = t_0, t_1, t_2, \ldots, t_N$. The distance between two consecutive points is

$$h_i = t_{i+1} - t_i$$

and there are N such steps in $[t_0, t_N]$. Only in the limit that all $h_i \to 0$ do we recover the continuous solution.

On a discrete grid of points for the independent variable, we must convert the derivatives of the dependent variables into finite differences before we can find a solution for (8-10). In terms of the central differences introduced in (2-52), the second-order derivative with respect to time at $t = t_i$ may be written in the form

$$\left.\frac{d^2\phi(t)}{dt^2}\right|_{t=t_i} \longrightarrow \frac{1}{h^2}\{\phi(t_{i+1}) - 2\phi(t_i) + \phi(t_{i-1})\} \qquad (8\text{-}11)$$

as we have done in (2-55). For simplicity, we shall take the step size

$$h = t_{i+1} - t_i$$

to be a constant, independent of time. The differential equation (8-10) may now be expressed as the relation between the value of $\phi(t)$ at three different times $t = t_{i-1}, t_i,$ and t_{i+1}:

$$\{\phi(t_{i+1}) - 2\phi(t_i) + \phi(t_{i-1})\} + h^2\omega_0^2\phi(t_i) = 0$$

Using the shorthand notation

$$\phi_i \equiv \phi(t_i)$$

the equation may be put into the form

$$\phi_{i+1} - (2 - h^2\omega_0^2)\phi_i + \phi_{i-1} = 0 \qquad (8\text{-}12)$$

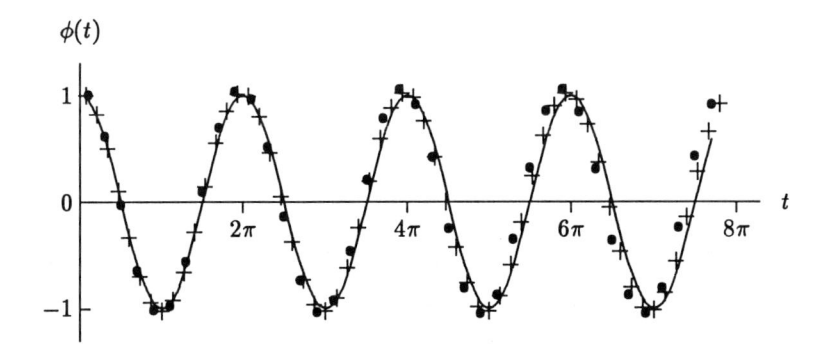

Fig. 8-2 Simple harmonic oscillation. The dots are the results obtained by solving the second-order ordinary differential equation (8-10) using Euler's method and the solid curve is the exact solution, $\phi(t) = \cos\omega_0 t$. The difference in phase results from the poor accuracy in the numerical solution, which comes in part from the large step size of $h = 0.63$ used. When the step size is reduced by 30%, the accuracy of the results (indicated by the symbol +) is improved.

This is called a finite difference equation (FDE), as it relates the differences in the values of $\phi(t)$ at neighboring points along a grid of points for the independent variable.

We can solve (8-12) as an *initial value* problem if sufficient information is available about $\phi(t)$ at the starting time. In terms of the physics, the initial conditions may be supplied in terms of the location and velocity of the mass at $t = 0$. For example, the mass may be displaced at the start by one unit in distance in the positive direction. That is,

$$\phi(t=0) = 1$$

It is then released from this arrangement at $t = 0$ without being given any initial velocity. This may be expressed by the condition

$$\left.\frac{d\phi(t)}{dt}\right|_{t=0} = 0$$

By making use of finite differences, these two initial conditions may be approximated by the following assignments:

$$\phi(t=h) \approx \phi(t=0) = 1 \tag{8-13}$$

where h is the step size in t. The finite difference equation (8-12) can now be expressed as the relation for $\phi(t)$ at $t = t_{i+1} = (i+1)h$ and its values at two previous time steps $t = t_i = ih$ and $t_{i-1} = (i-1)h$:

$$\phi_{i+1} = (2 - h^2\omega_0^2)\phi_i - \phi_{i-1} \tag{8-14}$$

If we label the two starting values provided by (8-13) as ϕ_1 and ϕ_0, respectively, we can calculate ϕ_2 with (8-14) by letting $i = 1$. Once the value of ϕ_2 is available, we can obtain ϕ_3 using ϕ_1 and ϕ_2 as the input, and so on. In this way, the value of $\phi(t)$ at a given time may be propagated forward one step at a time. The results are shown in Fig. 8-2. For the sake of efficiency and accuracy, several improvements can be made to the method and we shall discuss three ways later in §8-2, 8-3 and 8-4.

Example of a boundary value problem. Instead of an oscillator, we may be interested in the sound produced in a hollow pipe. In this case, the dependent variable $\phi(x)$ represents the air pressure along the length x of the cylindrical pipe. The differential equation for $\phi(x)$ has the same form as (8-10),

$$\frac{d^2\phi(x)}{dx^2} + \kappa^2 \phi(x) = 0 \tag{8-15}$$

except that the factor κ^2 here is related to the compressibility of the air inside the pipe. The constraints on the system in this case occur at the two ends of the pipe. For simplicity, we shall think in terms of a pipe with pressure maintained at some constant value p at both ends, $x = 0$ and $x = \ell$. This is a *boundary value* problem, with

$$\phi(x = 0) = p \qquad \phi(x = \ell) = p$$

as the boundary conditions.

It is also possible to express the relation given by (8-15) in the form of a difference equation, in the same way as we did for the initial value problem. Let us divide the interval between $x = 0$ and $x = \ell$ into N equal parts, each of length $h = \ell/N$. The interval $[0, \ell]$ is now represented by a mesh of points with the value of pressure at each point, $x = 0, 1h, 2h, \ldots, Nh = \ell$, represented, respectively, by $\phi_0, \phi_1, \phi_2, \ldots, \phi_{N-1}$, and ϕ_N. Similar to (8-14), we obtain from (8-15) a finite difference equation, relating the values of $\phi(x)$ at three adjacent points $(i+1)$, i, and $(i-1)$:

$$\phi_{i+1} = (2 - h^2 \kappa^2)\phi_i - \phi_{i-1} \tag{8-16}$$

Although the form of this equation is identical to that of (8-14), it is not possible to solve it by propagation. In the case of an initial value problem, we have enough input values at the beginning to propagate the solution forward step by step in time. In the case of a boundary value problem, we do not have enough information to start the calculation and some other way must be found so as to get around the deficiency. All the methods used to find the solution for this type of equation involve some sort of iterative calculations that brings us closer and closer to the final solution.

One approach used to handle a boundary problem of the type of (8-15) is to make a guess for the second condition we must have at the starting point so as to solve the differential equation as an initial value problem. For example, if ϕ_0 is given by the boundary condition, we may make a reasonable estimate of the value of ϕ_1 and, as a result, (8-16) essentially becomes the same as (8-14). Most likely, our initial guess will not be the correct one. For this reason, the value of ϕ_N obtained, when the solution is propagated from ϕ_0 and ϕ_1 to ϕ_N at $x = \ell$, will not satisfy the boundary condition at $x = \ell$. The discrepancy provides us with a way to improve our initial estimate of ϕ_1 and to make a fresh start to solve the problem. The process is repeated several times until solution converges to a value that satisfies the boundary condition for $\phi(x)$ at $x = \ell$. This is the basis of shooting methods, to which we shall return in more detail in §8-5.

Alternatively, we can regard the difference equation (8-16) as a set of algebraic equations. This can be seen by writing the $(N-1)$ equation represented by (8-16) explicitly for $i = 1, 2, \ldots, (N-1)$. The result is

$$-(2 - h^2\kappa^2)\phi_1 + \phi_2 = -p$$
$$\phi_1 - (2 - h^2\kappa^2)\phi_2 + \phi_3 = 0$$
$$\cdots \quad \cdots \quad \cdots \quad \cdots$$
$$\phi_{N-3} - (2 - h^2\kappa^2)\phi_{N-2} + \phi_{N-1} = 0$$
$$\phi_{N-2} - (2 - h^2\kappa^2)\phi_{N-1} = -p$$

In matrix notation, this set of equations may be written as

$$\begin{pmatrix} d & 1 & & & & \\ 1 & d & 1 & & & \\ & 1 & d & 1 & & \\ & & \ddots & \ddots & \ddots & \\ & & & 1 & d & 1 \\ & & & & 1 & d \end{pmatrix} \begin{pmatrix} \phi_1 \\ \phi_2 \\ \phi_3 \\ \vdots \\ \phi_{N-2} \\ \phi_{N-1} \end{pmatrix} = \begin{pmatrix} -p \\ 0 \\ 0 \\ \vdots \\ 0 \\ -p \end{pmatrix}$$

where, to simplify the notation, we have used

$$d \equiv h^2\kappa^2 - 2$$

Furthermore, we have omitted from the square matrix on the left side all the matrix elements that are zero. The problem is now equivalent to one of finding the roots of a set of $(N-1)$ linear algebraic equations. In principle, we can use one of the techniques discussed in §5-2 to find the solution. However, to obtain good accuracy, the step size must be small. Consequently, N may be quite large. For this reason, special techniques have been developed to solve such equations and we shall see some of these in later sections.

Stability and convergence. Besides the obvious requirement of efficiency, a numerical method has to be stable and the solution converges to that of the differential equation. The importance of these two conditions can be seen from the oscillator example we encountered earlier. Since the full solution is obtained by propagating from two initial values, the truncation error in each step can be cumulative and may eventually make the numerical results meaningless.

The main source of *truncation errors* in the numerical solution comes from that introduced in replacing the differential equation by a finite difference equation. Consider first a first-order ODE of the form

$$\frac{d\phi}{dt} = f(\phi, t) \tag{8-17}$$

In the method of Euler used above, a forward difference approximation is used to replace the differential equation by a finite difference equation of the form

$$\frac{1}{h}\{\phi_{i+1} - \phi_i\} = f(\phi_i, t_i) \tag{8-18}$$

We can find out the errors introduced in this approximation by making a Taylor series expansion of $\phi(t)$ at $t = t_{i+1}$ in terms of $\phi(t)$ and its derivatives at $t = t_i$

$$\phi_{i+1} = \phi_i + h\frac{d\phi}{dt}\bigg|_{t=t_i} + \frac{h^2}{2!}\frac{d^2\phi}{dt^2}\bigg|_{t=t_i} + \cdots \tag{8-19}$$

If we replace $d\phi/dt$ at $t = t_i$ on the right side by $f(\phi_i, t_i)$ and compare the resulting expression with (8-18), we find that the leading order difference is proportional to $h^2 d^2\phi(t)/dt^2$ evaluated at $t = t_i$.

Errors of this type, introduced in approximating a differential equation by a finite difference equation, are the most important ones in the numerical solution of differential equations. For this reason, we shall ignore in the discussions in this chapter errors due to finite length of a computer word. As a result, truncation errors may be defined here as the differences between the value $f(\phi_i, t_i)$ given by (8-18) and the exact value of $f(\phi, t)$ at $t = t_i$.

If the function $\phi(t)$ is well behaved, we do not expect the value of $d^2\phi/dt^2$ in (8-19) to become very large anywhere in the range of t of interest to us. As a result, the truncation error of the method goes to zero as $h \to 0$. This is a necessary requirement for any numerical method and is sometimes called the *consistency* condition. However, consistency does not guarantee *convergence*, which is the requirement that the numerical results approach the true solution for the differential equation. There are methods, whose errors (differences from the true solution) actually increase as the step size h is decreased, even though it satisfies the consistency condition. Such methods are called *unstable*, and we shall not elaborate on these methods here. On many occasions, it may not be easy to demonstrate that a solution is stable and convergent without actually carrying out the calculations. This is especially true for nonlinear differential equations where often the condition for consistency can only be demonstrated analytically after making some drastic approximations.

The Taylor series expansion of (8-19) provides us also with a hint of the way to improve the accuracy of our numerical solutions. For example, if we wish to reduce the truncation error in our finite difference equation to order h^3, we must somehow include contributions from the h^2 term into our approximation. There are many ingenious ways to do this, as we can see from the following illustration. Again, let us use the first-order ODE of (8-17) as an example. For this purpose, we can rewrite (8-18) in the form

$$\phi_{i+1} = \phi_i + hf_i \tag{8-20}$$

where $f_i \equiv f(\phi_i, t_i)$. Only the values at mesh point i are used in this approximation to calculate the value of the solution at $(i+1)$. This is known as a one-step process. To reduce the truncation errors, we must somehow include more input information on the right side. In general, we can express the solution at a point $(i+k)$ as a function of the value at the previous k points in the following way:

$$\phi_{i+k} = a_0 \phi_i + a_1 \phi_{i+1} + \cdots + a_{k-1} \phi_{i+k-1}$$
$$+ hF(\phi_i, t_i; \phi_{i+1}, t_{i+1}; \ldots; \phi_{i+k}, t_{i+k}) \tag{8-21}$$

where $a_0, a_1, \ldots, a_{k-1}$ are known coefficients and F is a known function of both the dependent variable ϕ and independent variable t at mesh points i, $i+1, \ldots$, $i+k$. For $k=0$ and $F = f(\phi_i, t_i)$, we recover (8-20). For $k \geq 1$, the method represented by (8-21) is known as a multistep process. In practice, the function F is often replaced by a linear combination of the value of $f(\phi, t)$ at the previous k mesh points, such as

$$F(\phi_i, t_i; \phi_{i+1}, t_{i+1}; \ldots; \phi_{i+k}, t_{i+k}) = \sum_{j=0}^{k} \beta_j f_{i+j}$$

For $k > 0$, this is known as a linear multistep process. If the coefficient β_k does not vanish, the value of ϕ_{i+k} appears also on the right side of the finite difference equation. In this case, the method is an implicit one. For our simple example of (8-20), and its equivalent form for multistep processes in which ϕ_{i+k} appear only on the left side of the equation, the method is known as an explicit one.

We shall end this section by converting the second-order ODE of (8-10),

$$\frac{d^2\phi}{dt^2} + \omega_0^2 \phi(t) = 0$$

into a system of two first-order differential equations and thence to the corresponding finite difference equations. As we did in going from (8-6) to (8-7), let us define

$$y_1(t) = \phi(t) \qquad y_2(t) = \frac{d\phi}{dt} \tag{8-22}$$

The same relation as (8-10) is now expressed in terms of the following two equations:

$$\frac{dy_1}{dt} = y_2(t) \qquad \frac{dy_2}{dt} = -\omega_0^2 \, y_1(t) \qquad (8\text{-}23)$$

In terms of central differences, we can construct the finite difference equations for these two first-order ODEs and put them in the forms

$$\frac{1}{h}\{y_1(t_i + \tfrac{1}{2}h) - y_1(t_i - \tfrac{1}{2}h)\} = y_2(t_i)$$

$$\frac{1}{h}\{y_2(t_i + \tfrac{1}{2}h) - y_2(t_i - \tfrac{1}{2}h)\} = -\omega_0^2 y_1(t_i) \qquad (8\text{-}24)$$

These two equations replace the FDE of (8-12). For the convenience of discussing the more general situation, such as ODEs of order higher than the second one, we shall adopt the notation

$$y(j;k) \equiv y_j(t_k)$$

In terms of $y(j;k)$, the two equations of (8-24) may be expressed as

$$y(1; 2j+1) - y(1; 2j-1) = h y(2; 2j)$$

$$y(2; 2j+1) - y(2; 2j-1) = -h\omega_0^2 \, y(1; 2j) \qquad (8\text{-}25)$$

It is easy to see that these are the same as (8-12) by substituting the values of $y(2; 2j \pm 1)$ obtained from the first equation into the second one.

More generally, we can use a "vector" notation to express the relations given in (8-25). We can think of the vector $\boldsymbol{y}(t)$ as an array of n functions:

$$\boldsymbol{y}(t) \equiv \{y_1(t), y_2(t), \ldots, y_n(t)\} \qquad (8\text{-}26)$$

where n is the number of different dependent functions in the system. For our example of a second-order ODE, we have $n = 2$. Both equations in (8-23) may be represented by the form

$$\frac{d\boldsymbol{y}}{dt} = \boldsymbol{f}(t) \qquad (8\text{-}27)$$

where $\boldsymbol{f}(t) = \{f_1(t), f_2(t)\}$ with

$$f_1(t) = y_2(t) \qquad f_2(t) = -\omega_0^2 y_1(t)$$

The finite difference equation of (8-25) may now be expressed in the form

$$\boldsymbol{y}(t_{k+1}) = h\boldsymbol{f}(k) + \boldsymbol{y}(t_{k-1}) \qquad (8\text{-}28)$$

where we have used $k = 2j$ and $h = (t_{k+1} - t_{k-1})$ to put the equation into a form more convenient for computational purposes.

The only work remaining for us is to express the conditions of (8-13) in a way that we can carry out the calculations as an initial value problem. There is a minor difficulty here, as each of the two equations represented by the shorthand notation of (8-28) requires two values, $f(1)$ and $y(0)$, to get the propagation started. On the surface, it seems that we need a total of four initial values. This is in violation of the fact that we are dealing with a second-order ODE here. We shall leave as an exercise the problem of finding the initial conditions suitable for the present calculation.

The methods used here to solve differential equations are based on *finite differences*. We shall be mainly concentrating on such methods in this chapter, as it is the popular approach for the simple cases normally encountered in physical problems. For a large variety of the more complicated cases, a finite element approach may be more appropriate. In general, computer programs based on finite element methods tend to be aimed at particular types of problems. We shall return to an example of such an approach at the end of the chapter. Some of the numerical methods we shall discuss in this chapter are applicable to both ODEs and PDEs. For this reason, we shall not make clear separations in the discussions for these two categories as is usually done in treatments on analytical methods.

§8-2 Runge-Kutta methods

In Euler's method of solving initial value problems, used as an illustration in the previous section, we have replaced the set of differential equations

$$\frac{d\boldsymbol{y}}{dt} = \boldsymbol{f}(\boldsymbol{y}, t) \qquad (8\text{-}29)$$

given in (8-27) by a set of finite difference equations of the form

$$\boldsymbol{y}(t_{k+1}) = \boldsymbol{y}(t_k) + h\boldsymbol{f}(\boldsymbol{y}(t_k), t_k) \qquad (8\text{-}30)$$

As we recall from (8-26) above, the notation used here is that a bold-faced letter represents a set of n quantities, for example, $\boldsymbol{y}(t) = \{y_1(t), y_2(t), \ldots, y_n(t)\}$. The relation given by (8-29) is therefore for a set of n coupled first-order ODEs in general.

The approximation used in going from (8-29) to (8-30) is the same as that used in (8-20), and the truncation errors are of the order h^2, where h is the step size in the independent variable t. The method is quite simple to implement; all we have to do is to evaluate the functions $\boldsymbol{f}(\boldsymbol{y}, t)$ at the beginning of each interval. This is an example of a one-step explicit method, as the right side involves only the values of t and $\boldsymbol{y}(t)$ at a single previous point. As mentioned in the previous section, we can improve the accuracy of the numerical solution by taking a linear combination of the values at several previous points instead of those at $t = t_k$ alone, as done in (8-30). The aim of Runge-Kutta methods is to achieve this goal without departing from a one-step calculation. The price one pays here is that the function $\boldsymbol{f}(\boldsymbol{y}, t)$ must be evaluated at several different points.

Higher-order approximations. Let us see first how the method works for a simple case. Consider two points, t_k and $(t_k + \alpha h)$, where α is an adjustable parameter to be determined later. Let us use $\boldsymbol{p}$ and $\boldsymbol{q}$ to represent the values of $d\boldsymbol{y}/dt$ at these two points. From (8-29), we obtain the relations

$$\boldsymbol{p} = \boldsymbol{f}(\boldsymbol{y}(t_k), t_k)$$
$$\boldsymbol{q} = \boldsymbol{f}(\boldsymbol{y}(t_k) + \alpha h \boldsymbol{p}, t_k + \alpha h) \qquad (8\text{-}31)$$

Again, the bold-faced symbols imply that there are n different values of p and n different values of q, each pair corresponding to one of the n different functions $f(y_1(t_k),t), f(y_2(t_k),t), \ldots, f(y_n(t_k),t)$, each evaluated at $t = t_k$ and $t = (t_k + \alpha h)$.

A connection with Euler's method may be made by taking $\alpha = \frac{1}{2}$. In this case

$$\boldsymbol{q} = \boldsymbol{f}(\boldsymbol{y}(t_k) + \tfrac{1}{2} h \boldsymbol{p}, t_k + \tfrac{1}{2} h)$$

The central difference value of $d\boldsymbol{y}/dt$ at $t = (t_k + \tfrac{1}{2} h)$ is

$$\left.\frac{d\boldsymbol{y}}{dt}\right|_{t=t_k+\frac{1}{2}h} = \frac{1}{h}\{\boldsymbol{y}(t_{k+1}) - \boldsymbol{y}(t_k)\} = \boldsymbol{f}(\boldsymbol{y}(t_k + \tfrac{1}{2}h), t_k + \tfrac{1}{2}h) \qquad (8\text{-}32)$$

The value of $\boldsymbol{y}(t_k + \tfrac{1}{2}h)$ in the argument of $\boldsymbol{f}(\boldsymbol{y},t)$ may be approximated by those at $t = t_k$ through the use of the first two terms of a Taylor series expansion

$$\boldsymbol{y}(t_k + \tfrac{1}{2}h) = \boldsymbol{y}(t_k) + \frac{h}{2}\left.\frac{d\boldsymbol{y}}{dt}\right|_{t=t_k} + \cdots \approx \boldsymbol{y}(t_k) + \tfrac{1}{2} h \boldsymbol{p}$$

On substituting the final result into (8-32), we find that

$$\frac{1}{h}\{\boldsymbol{y}(t_{k+1}) - \boldsymbol{y}(t_k)\} \approx \boldsymbol{f}(\boldsymbol{y}(t_k) + \tfrac{1}{2} h \boldsymbol{p}, t_k + \tfrac{1}{2} h) = \boldsymbol{q}$$

This is the same result as (8-25) for our oscillator example in the previous section. The truncation error in this case is given by

$$\mathcal{E} = \frac{1}{h}\{\boldsymbol{y}(t_{k+1}) - \boldsymbol{y}(t_k)\} - \boldsymbol{q}(\boldsymbol{y}(t_k) + \tfrac{1}{2} h \boldsymbol{p}, t_k + \tfrac{1}{2} h)$$

and was found to be on the order of h^2 in (8-19) by expanding $\boldsymbol{y}(t_{k+1})$ and $\boldsymbol{q}$ in terms of their values at $t = t_k$. Note also that, although $\boldsymbol{q}$ involves the time $t = t_k + \alpha h$ in the explicit dependence of $\boldsymbol{f}(\boldsymbol{y},t)$ on t, only the values of $\boldsymbol{y}(t)$ at $t = t_k$ are used. Thus, by the definition given in (8-20), the method remains as one step, and there is no difficulty in calculating $\boldsymbol{p}$ and $\boldsymbol{q}$ once we have $\boldsymbol{y}(t_k)$.

A further reduction of the truncation error $\mathcal{E}$ may be achieved by modifying the difference equation of (8-30) and changing it into the form

$$\frac{1}{h}\{\boldsymbol{y}(t_{k+1}) - \boldsymbol{y}(t_k)\} = \beta_1 \boldsymbol{p} + \beta_2 \boldsymbol{q} \qquad (8\text{-}33)$$

We shall now find the values of the three parameters β_1, β_2, and α (required to calculate q) that produce the smallest value of $\mathcal{E}$. A Taylor series expansion of $y(t_{k+1})$ and q gives us the results:

$$y(t_{k+1}) = y(t_k) + h y_t(t_k) + \tfrac{1}{2} h^2 y_{tt}(t_k) + \tfrac{1}{6} h^3 y_{ttt}(t_k) + O(h^4)$$

$$q = f(y(t_k) + \alpha h q, t_k + \alpha h)$$
$$= f(y(t_k), t_k) + \alpha h \{ f_t(y(t_k), t_k) + f_y(y(t_k), t_k) y_t(t_k) \}$$
$$+ \tfrac{1}{2}(\alpha h)^2 \{ f_{tt}(y(t_k), t_k) + 2 f_{ty}(y(t_k), t_k) y_t(t_k)$$
$$+ f_{yy}(y(t_k), t_k)(y_t)^2 \} + O(h^3)$$
$$= y_t + \alpha h y_{tt} + \tfrac{1}{2}(\alpha h)^2 \{ y_{ttt} - f_y y_{tt} \} \quad (8\text{-}34)$$

where, to simplify the notation, we have used subscripts to indicate derivatives of a function. Thus, we have

$$y_t(t) \equiv \frac{dy}{dt} = f(y, t)$$

$$y_{tt}(t) \equiv \frac{d^2 y}{dt^2} = \frac{\partial f}{\partial t} + \sum_{i=1}^{n} \frac{\partial f}{\partial y_i} \frac{\partial y_i}{\partial t} = f_t(t) + f_y(t) y_t(t)$$

$$y_{ttt}(t) \equiv \frac{d^3 y}{dt^3} = \frac{\partial^2 f}{\partial t^2} + \sum_{i=1}^{n} \left\{ 2 \frac{\partial^2 f}{\partial t \partial y_i} \frac{\partial y_i}{\partial t} + \sum_{j=1}^{n} \frac{\partial^2 f}{\partial y_i \partial y_j} \frac{\partial y_i}{\partial t} \frac{\partial y_j}{\partial t} + \frac{\partial f}{\partial y_i} \frac{\partial^2 y_i}{\partial t^2} \right\}$$
$$= f_{tt}(t) + 2 f_{ty}(t) y_t(t) + f_{yy}(t)(y_t)^2 + f_y y_{tt}$$

with $f_t \equiv \partial f / \partial t$ and $f_y \equiv \partial f / \partial y$. Note that the coefficient of the h^2 term in the Taylor series expansion of q in (8-34) is different from y_{ttt} above by the factor $f_y y_{tt}$.

The truncation error in using (8-33) is then

$$\mathcal{E} = \frac{1}{h} \{ y(t_{k+1}) - y(t_k) \} - (\beta_1 p + \beta_2 q)$$
$$= y_t(t_k) + \tfrac{1}{2} h y_{tt}(t_k) + \tfrac{1}{6} h^2 y_{ttt}(t_k) - \beta_1 y_t(t_k)$$
$$- \beta_2 \{ y_t(t_k) + \alpha h y_{tt}(t_k) + \tfrac{1}{2}(\alpha h)^2 [y_{ttt}(t_k) - f_y y_{tt}(t_k)] \} + O(h^3)$$
$$= (1 - \beta_1 - \beta_2) y_t(t_k) + h(\tfrac{1}{2} - \alpha \beta_2) y_{tt}(t_k)$$
$$+ h^2 \{ \tfrac{1}{6} y_{ttt}(t_k) + \tfrac{1}{2} \alpha^2 \beta_2 [y_{ttt}(t_k) - f_y y_{tt}(t_k)] \} + O(h^3)$$

Because of the $f_y y_{tt}(t_k)$ term, it is not possible to eliminate all the h^2 dependence for an arbitrary function $y(t)$. There are, however, several choices available to us to make $\mathcal{E}$ as small as possible. For example, one possibility is to take $\alpha = 1$ and $\beta_1 = \beta_2 = \tfrac{1}{2}$. Alternatively, we can use the *midpoint* method by taking $\alpha = \tfrac{1}{2}$,

$\beta_1 = 0$, and $\beta_2 = 1$. We shall not elaborate on either of these two choices in our illustration here. As far as numerical accuracy is concerned, it is possible to construct formulas with truncation error much less than $O(h^2)$, as we shall see next.

The basic idea of minimizing the truncation error may be used to construct a set of FDEs that is accurate to order h^4. One result commonly used in practical applications is the fourth-order Runge-Kutta formula

$$y(t_{k+1}) = y(t_k) + \tfrac{1}{6}h(p + 2q + 2r + s) \qquad (8\text{-}35)$$

with

$$p = f(y(t_k), t_k) \qquad\qquad q = f(y(t_k) + \tfrac{1}{2}hp, t_k + \tfrac{1}{2}h)$$
$$r = f(y(t_k) + \tfrac{1}{2}hq, t_k + \tfrac{1}{2}h) \qquad s = f(y(t_k) + hr, t_k + h) \qquad (8\text{-}36)$$

We shall not try to reproduce here the tedious proof that the formula is fourth order, that is, all the truncation errors up to and including order h^4 vanish. The form is reminiscent of Simpson's rule for an integral of the form

$$y(t_{k+1}) - y(t_k) = \int_{t_k}^{t_{k+1}} y'(t)\,dt$$

We recall from (2-15) that this method is accurate up to the fourth order. Again, to evaluate p, q, r, and s of (8-36), the only values of $y(t)$ needed are those at $t = t_k$. The algorithm for solving a system of n-coupled first-order ODEs in this way using a fixed step size h is outlined in Box 8-1.

Application to damped harmonic oscillation. If we ignore the driving force term in (8-3), we obtain a second-order differential equation for an harmonic oscillator

$$\mu \frac{d^2\phi}{dt^2} + b\frac{d\phi}{dt} + k\phi = 0 \qquad (8\text{-}37)$$

where $\phi(t)$ is the displacement at time t, μ is the mass, b is the constant of proportionality for the damping term due to friction, and k is the spring constant. In addition to oscillators, this form of second-order ODE appears in a variety of other physical problems. For example, in the case of an alternating-current circuit, we have the differential equation

$$L\frac{d^2Q}{dt^2} + R\frac{dQ}{dt} + \frac{1}{C}Q = 0 \qquad (8\text{-}38)$$

for the charge $Q(t)$ as a function of time in a circuit consisting of three elements connected in series: an inductor with inductance L, a resistor with resistance R, and a capacitor with capacitance C.

Chapter 8] Ordinary and Partial Differential Equations 503

Box 8-1 Subroutine RNG_KTT4(H,BFY,ND_EQN, ND_PTS)
Fourth-order Runge-Kutta method
For a system of n-coupled first-order, initial value ODE

Argument list:
  H: Input step size h.
  BFY: Two-dimensional array for the solution y.
  ND_EQN: Number of equations.
  ND_PTS: Total number of steps in t to be taken.

Subprogram used:
  FVAL: Calculate the value of $f(y,t)$ using $y(t)$ and t as the input.

Initialization:
  In the calling program, store the initial conditions in BFY.

1. Carry out the following calculations for step size h for each of the n ODEs:
   (a) For each t_k, calculate p, q, r, and s of (8-36) by the following steps:
       (i) Call FVAL to calculate p using y_k and t_k.
       (ii) Call FVAL to calculate q using $(y_k + \frac{1}{2}hp)$ and $t = (t_k + \frac{1}{2}h)$.
       (iii) Call FVAL to calculate r using $(y_k + \frac{1}{2}hq)$ and $t = (t_k + \frac{1}{2}h)$.
       (iv) Call FVAL to calculate p using $(y_k + hr)$ and $t = (t_k + h)$.
   (b) Calculate y_{k+1} from the values of y_k, p, q, r, and s using (8-35).
2. Return the values of $y(t)$ in the array BFY.

Equations of the type given by (8-37) and (8-38) may be put into the generic form

$$\frac{d^2\phi}{dt^2} + 2\gamma\frac{d\phi}{dt} + \omega_0^2\phi = 0 \tag{8-39}$$

where, in the case of (8-37), we have

$$2\gamma = \frac{b}{\mu} \qquad \omega_0^2 = \frac{k}{\mu}$$

To solve the equation numerically as a system of coupled first-order ODEs, we shall follow the method used in arriving at (8-22) and define two new independent variables:

$$y_1(t) = \phi(t) \qquad y_2(t) = \frac{d\phi}{dt}$$

In terms of $y_1(t)$ and $y_2(t)$, the second-order ODE of (8-39) is changed into the following two first-order equations:

$$\frac{dy_1}{dt} = y_2 \qquad \frac{dy_2}{dt} = -2\gamma y_2 - \omega_0^2 y_1$$

This can be put into the standard form of (8-29) with

$$f_1(t) = y_2 \qquad f_2(t) = -2\gamma y_2 - \omega_0^2 y_1$$

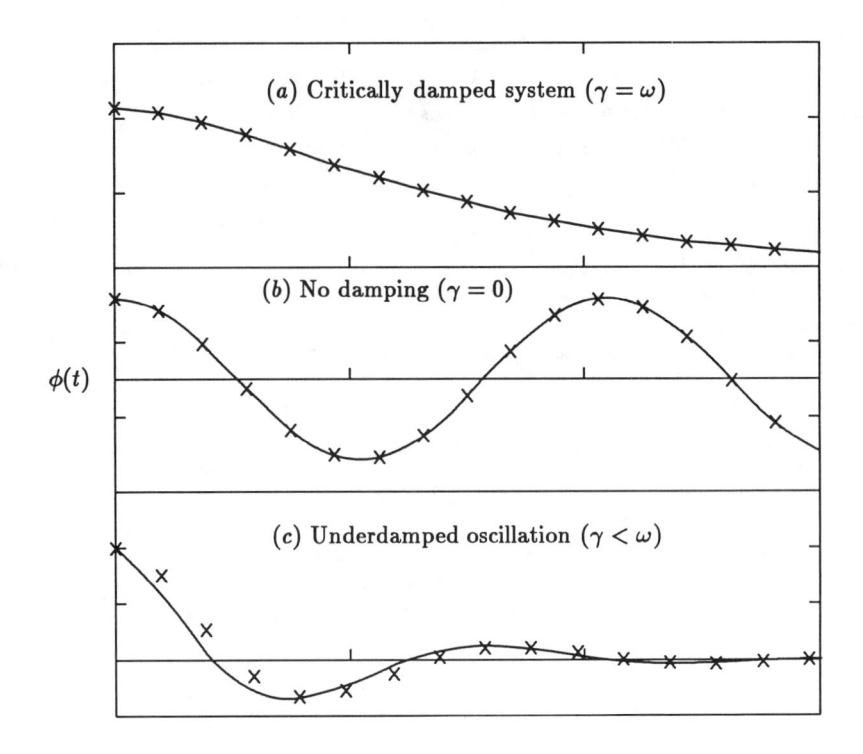

Fig. 8-3 Solutions of the second-order differential equation for a damped harmonic oscillator. Case (a) is for critical damping with $\gamma = \omega_0$. The results without the damping term, $\gamma = 0$, are shown in (b) and those for underdamping, $\gamma^2 < \omega_0^2$, are shown in (c). In each case, the solid curve comes from an analytical solution and the crosses indicate the results obtained by numerical solution using a fourth-order Runge-Kutta method. The agreements between the two sets of results are good except for (c), arising from the fact that the equation for the underdamped case is a "stiff" one.

At $t = 0$, we shall assume that the oscillator is displaced by an amount ϕ_0 and released with some velocity $(d\phi/dt)_{t=0}$. This gives us the initial conditions

$$y_1(t = 0) = \phi_0 \qquad y_2(t = 0) = \left.\frac{d\phi}{dt}\right|_{t=0}$$

For the numerical solution, we shall construct a mesh of points dividing the time interval into a number of steps. For simplicity, we shall use a constant step size. The distance between two adjacent mesh points depends on the numerical accuracy required as well as the total number of steps we wish to take and the time span in which we are interested.

It is well known that there are three possible types of solutions for (8-39), depending on the values of γ and ω_0. In the case of heavy damping with $(\gamma^2 > \omega_0^2)$, the system does not oscillate. The amplitude decays with two different time constants, $(\gamma - \zeta)$ and $(\gamma + \zeta)$, where

$$\zeta = \sqrt{\gamma^2 - \omega_0^2}$$

is a real and positive quantity. The solution of the differential equation takes on the form

$$\phi(t) = \frac{\phi_0 + \xi}{2} e^{-(\gamma-\zeta)t} + \frac{\phi_0 - \xi}{2} e^{-(\gamma+\zeta)t}$$

with the factor ξ given by

$$\xi = \frac{1}{\zeta}\left\{\left.\frac{d\phi}{dt}\right|_{t=0} + \gamma\phi_0\right\}$$

We shall not be concerned with this case, as there is nothing of interest to us here in terms of numerical solutions.

The decay is most rapid when $\gamma = \omega_0$. This is known as the case of critical damping and the results are shown in Fig. 8-3(a). The analytical solution is given by

$$\phi(t) = e^{-\gamma t}\left\{\phi_0 + \left(\left.\frac{d\phi}{dt}\right|_{t=0} + \gamma\phi_0\right)t\right\} \tag{8-40}$$

We see that the numerical solution, represented by the crosses in the figure, is in good agreement with that of (8-40), shown as a solid curve.

If we let $b = 0$, there is no damping and we return to the case of simple harmonic oscillation. The analytical solution for the initial condition of $d\phi/dt = 0$ at $t = 0$ has the form

$$\phi(t) = \phi_0 \cos \omega_0 t$$

This is indicated by the solid curve in Fig. 8-3(b). Again, the numerical solution of (8-37), shown as crosses, is in good agreement with the well-known analytic results.

The case of underdamping ($\gamma^2 < \omega_0^2$) is shown in Fig. 8-3(c). The analytical solution in this case is given by

$$\phi(t) = e^{-\gamma t}\{\phi_0 \cos \omega t + \xi \sin \omega t\}$$

where

$$\omega = \sqrt{\omega_0^2 - \gamma^2} \qquad \xi = \frac{1}{\omega}\left\{\left.\frac{d\phi}{dt}\right|_{t=0} + \gamma\phi_0\right\}$$

Here, we see that the accuracy of the numerical solution is poor, as can be seen by the departures from the analytical values. This is an example of a "stiff" differential equation, arising from the fact that there are two time constants in the solution, ω and γ, and their values are rather different. As a result, the numerical solution becomes unstable, as if it does not know which of the two time scales to follow. We shall return to the stiff equation question §8-11.

Variable step size. The efficiency of the Runge-Kutta method outlined above may be improved further if we change the step size in each region according to the behavior of the functions $f(y, t)$. Similar to methods of numerical integration discussed in Chapter 2, it is possible to take much larger steps if $f(y, t)$ is rather smooth in a region. On the other hand, if the function varies rapidly in some intervals, smaller step sizes are necessary to achieve the same accuracy. In a constant step size approach, the size of h is usually selected in such a way that no region has an error larger than some given amount. To achieve this goal, h must be chosen as the smallest value required by accuracy considerations throughout the entire interval. This can be rather wasteful and, for this reason, a variable step size method is preferred in many cases. The basic principle behind variable step size methods is to take as large a step as possible to achieve the desired accuracy in a given region. Obviously, this is profitable only if it is easy to find the optimum step size for a given region.

There is also an additional advantage in taking a variable step size approach. In any realistic application, it is useful to have an idea of the errors involved during the calculations. To achieve this aim, we must make some estimate of the truncation error (and, hence, the optimum step size) from the results obtained. One way to do this is to start with a reasonable trial value for h and calculate $y(t)$ in two different ways, once using two steps of h each and once using one step of $2h$. The difference between the results in these two approaches provides us with an indication of whether we are in a smooth region where large step sizes can be tolerated or whether we are in a region where only small step sizes are acceptable. If the truncation error associated with the trial value h is acceptable, it can also give us a way to arrive at the value for the size of the next step. On the other hand, if the truncation error is too large, the trial step size are reduced and the calculations are repeated using a smaller value.

We shall not be concerned here with the question of how to arrive at a first estimate of h in a calculation, as any reasonable value is adequate for the purpose. Let us assume that we have just completed step k that ends at $t = t_k$. For the next step, we shall adopt as trial step size the h_t that was used as the step size for the previous step. Let $Y_{2h}(t_k + 2h_t)$ represent the value of $y(t + 2h_t)$ calculated using (8-35) with a step size of $2h_t$. That is,

$$Y_{2h}(t_k + 2h_t) = y(t_k) + \tfrac{1}{3}h_t(p_{2h} + 2q_{2h} + 2r_{2h} + s_{2h}) \qquad (8\text{-}41)$$

where

$$p_{2h} = f(y(t_k), t_k) \qquad\qquad q_{2h} = f(y(t_k) + h_t p_{2h}, t_k + h_t)$$
$$r_{2h} = f(y(t_k) + h_t q_{2h}, t_k + h_t) \qquad s_{2h} = f(y(t_k) + 2h_t r_{2h}, t_k + 2h_t)$$

The value of $y(t)$ at $t = (t_k + 2h_t)$ can also be obtained in the following way. First, we take a step of h_t. This gives us the result

$$z(t_k + h_t) = y(t_k) + \tfrac{1}{6}h_t(p_{h1} + 2q_{h1} + 2r_{h1} + s_{h1}) \qquad (8\text{-}42)$$

with

$$p_{h1} = f(y(t_k), t_k) = p_{2h} \qquad q_{h1} = f(y(t_k) + \tfrac{1}{2}h_t p_{h1}, t_k + \tfrac{1}{2}h_t)$$

$$r_{h1} = f(y(t_k) + \tfrac{1}{2}h_t q_{h1}, t_k + \tfrac{1}{2}h_t) \qquad s_{h1} = f(y(t_k) + h_t r_{h1}, t_k + h_t)$$

Note that p_{h1} is the same as p_{2h} of (8-41) and does not have to be evaluated again here. Starting from $z(t_k + h_t)$, we take another step of size h_t,

$$Y_{1h}(t_k + 2h_t) = z(t_k + h_t) + \tfrac{1}{6}h_t(p_{h2} + 2q_{h2} + 2r_{h2} + s_{h2}) \qquad (8\text{-}43)$$

this time with

$$p_{h2} = f(z(t_k + h_t), t_k + h_t)$$

$$q_{h2} = f(z(t_k + h_t) + \tfrac{1}{2}h_t p_{h2}, t_k + \tfrac{3}{2}h_t)$$

$$r_{h2} = f(z(t_k + h_t) + \tfrac{1}{2}h_t q_{h2}, t_k + \tfrac{3}{2}h_t)$$

$$s_{h2} = f(z(t_k + h_t) + h_t r_{h2}, t_k + 2h_t)$$

Since both $Y_{2h}(t_k + 2h_t)$ and $Y_{1h}(t_k + 2h_t)$ are obtained using a fourth-order Runge-Kutta method, we expect the difference between them is on the order h_t^5:

$$\Delta(t_k, h_t) = Y_{1h}(t_k + 2h_t) - Y_{2h}(t_k + 2h_t) \sim O(h_t^5)$$

That is, if we carry out a Taylor series expansion of $y(t)$ at $t = t_k$, we find that $\Delta(t_k, h_t)$ are proportional to $h_t^5 y^{(5)}(t_k)/(5!)$, where $y^5(t_k)$ are the fifth-order derivatives of $y(t)$ at $t = t_k$.

The value of $\Delta(t_k, h_t)$ is useful for two related purposes. The first is that we can compare $\Delta(t_k, h_t)$ with $\mathcal{E}_{\max}$, the maximum truncation error we can tolerate. If $|\Delta(t_k, h_t)| \leq \mathcal{E}_{\max}$, the step size of $2h_t$ is acceptable and the value of $Y_{1h}(t_k + 2h_t)$ may be taken as the value of $y(t_{k+1})$ at $t_{k+1} = (t_k + 2h_t)$. A second use of the value of $\Delta(t_k, h_t)$ is to estimate the optimum trial step size h_0 for the next subinterval $[t_{k+1}, t_{k+2}]$. Ideally, the value of h_0 should be such that

$$\Delta(t_{k+1}, h_0) = \mathcal{E}_{\max}$$

Since we have only the value of $\Delta(t_k, h_t)$ for the interval $[t_k, t_{k+1}]$ available to us, we can only find the optimum step size h_0 for the preset step. If the solutions do not contain rapid fluctuations, we expect that the change in the size h_0 is small from one step to another. As a result, it is possible to use the value of h_0 for the present step as the estimate for the optimum step size for the next step.

We saw in the previous paragraph that $\Delta(t_k, h_t) \sim h_t^5$. As a result, we have the approximate relation

$$\left| \frac{\mathcal{E}_{\max}}{\Delta(t_k, h_t)} \right| \approx \left(\frac{h_0}{h_t} \right)^5$$

> **Box 8-2**
>
> **Subroutine** RK_ODE(NPTS,T_BEG,T_END,T_ARY, BFY,ND_EQN,ND_PTS)
>
> **Runge-Kutta propagation of a system of n-coupled initial-value, first-order ODEs with step size control**
>
> Argument list:
>   NPTS: Number of steps taken.
>   T_BEG: Beginning of the interval.
>   T_END: End of the interval.
>   T_ARY: One-dimensional array for t_k.
>   BFY: Two-dimensional array for y.
>   ND_EQN: Number of differential equations.
>   ND_PTS: Maximum number of points.
>
> Subprograms used:
>   RK_STP: Fourth-order Runge-Kutta calculation for step sizes h and $2h$ (Box 8-1).
>
> Initialization:
>   (a) Define the maximum tolerable truncation error $\mathcal{E}_{\max}$.
>   (b) Set up arrays for the results of two steps of h each and one step of $2h$.
>   (c) Make an estimate for the initial step size h_t.
>   (d) Impose the initial conditions for y.
>
> 1. Check if the step size is below numerical accuracy.
> 2. Calculate the results of one step of $2h$ and two steps of h each using RK_STEP:
>   (a) Calculate $Y_{2h}(t_k + 2h_t)$ of (8-41):
>   (i) From the values of t_k and y_k, evaluate p_{2h}, q_{2h}, r_{2h}, and s_{2h}.
>   (ii) Evaluate $Y_{2h}(t_k + 2h_t)$ from p_{2h}, q_{2h}, r_{2h}, and s_{2h}.
>   (b) Calculate $Y_{1h}(t_k + 2h_t)$ with (8-42) and (8-43):
>   (i) From the value of $t = t_k$ and y_k, evaluate q_{h1}, r_{h1}, and s_{h1}.
>   (ii) Evaluate $z(t_k + h_t)$ with (8-42) using p_{2h}, q_{h1}, r_{h1}, and s_{h1}.
>   (iii) From $t = (t_k + h_t)$ and $z(t_k + h_t)$, calculate p_{h2}, q_{h2}, r_{h2}, and s_{h2}.
>   (iv) Evaluate $Y_{1h}(t_k + 2h_t)$ from p_{h2}, $q_{h2}(i)$, $r_{h2}(i)$, and s_{h2}.
> 3. Find the maximum truncation error R of (8-46):
>   (a) If the maximum truncation error is less than $\mathcal{E}_{\max}$, accept the step.
>   (i) Check if the maximum number of points is exceeded.
>   (ii) Store the results in the arrays for t and y.
>   (iii) Find the next optimum step size using (8-45).
>   (b) If the maximum error is too large, decrease h and try again.
>   (i) Take care that h_t does not become insignificantly small.
>   (ii) Stop if too many iterations in step size control.
> 4. Return t_k in T_ARY and y_k in BFY.

From this, we obtain the relation

$$h_0 \approx h_t \left| \frac{\mathcal{E}_{\max}}{\Delta(t_k, h_t)} \right|^{1/5} \tag{8-44}$$

This is only an approximate result; however, it is adequate to serve as a guide for estimating the optimum step size for the next subinterval.

If, on the other hand, the trial step size h_t produces a truncation error that is larger than $\mathcal{E}_{\max}$, we must start all over again from (8-41) with a smaller value for h_t for the present step. It is also possible to make use of (8-44) for finding the optimum value for h_t. Let us represent the new value as h_0. Since $\Delta(t_k, h_t)$ is larger than $\mathcal{E}_{\max}$, we need a value of h_0 that is smaller than h_t from which we started. Once this is available, (8-41) to (8-43) may be used to calculate Y_{2h}, Y_{2h}, and Δ with the value of h_0 as the step size. This process is repeated until the condition $\Delta(t_k, h_t) \leq \mathcal{E}_{\max}$ is satisfied. As a practical matter, one must guard against the situation that, as we reduce the step size, the value of h_0 becomes smaller than the least significant figure of t (due to the limited word length of the computer memory) and is essentially zero as far as the numerical calculation is concerned. If this happens, we have a situation equivalent to $h_t = 0$ and the calculation becomes meaningless.

Because the power on the right side of (8-44) is small, the value of h_0 obtained is not very sensitive to those of $\Delta(t_k, h_t)$ and $\mathcal{E}_{\max}$. This is a good thing, as the relation is only an approximate one, coming from estimates of the truncation errors. It is recommended in Press et al., (*Numerical Recipes*, Cambridge University Press, Cambridge, 1986) to use a slightly modified form

$$h_0 = R_f h_t \left| \frac{\mathcal{E}_{\max}}{\Delta(t_k, h_t)} \right|^\eta \tag{8-45}$$

where R_f is a factor on the order of 0.9 and the power is $\eta = 1/4$ if $\Delta(t_k, h_t)$ is greater than $\mathcal{E}_{\max}$ and $\eta = 1/5$ otherwise. This is implemented in the algorithm outlined in Box 8-2.

The estimate of h_0 is slightly more complicated if we have a system of n-coupled equations. Since there are n different dependent variables, $y_1(t), y_2(t), \ldots, y_n(t)$, the tolerance for error may be different for each of them. As a result, we have a set of n values for $\mathcal{E}_{\max}$ and a different value of $\Delta(t_k, h_t)$ for each of the n-coupled first-order ODEs. To determine the optimum step size for the independent variable t, therefore, requires a choice among the n different estimates for the next optimum step size. There are no unique answers to this question. One possible choice is to measure $\Delta(t_k, h_t)$ by the average of $Y_{1h}(t_k, 2h)$ and $Y_{2h}(t_k, 2h)$. In this case, we may define an error in a step as

$$R = \max \left\{ \left| 2 \frac{Y_{1h}(t_k + 2h_t) - Y_{2h}(t_k + 2h_t)}{Y_{1h}(t_k + 2h_t) + Y_{2h}(t_k + 2h_t)} \right| \right\} \tag{8-46}$$

where the numerator is nothing but $\Delta(t_k, h_t)$. Since R defined in this way is "dimensionless," we can specify the tolerance for error $\mathcal{E}_{\max}$ in the same way, such as a value of 10^{-4} in single precision calculations.

§8-3 Predictor-corrector methods

Historically, predictor-corrector methods have been used widely to solve initial value problems. For this reason they occupy a very important place in the literature on differential equations, and a large repertory based on these methods exists in standard subroutine libraries. However, as practical methods, they are not as simple and convenient to use as the one-step Runge-Kutta methods described in the previous section. The accuracies one can achieve with predictor-corrector methods are better in general. However, for the same accuracies, extrapolation methods described in the next section are more efficient. For these reasons, the importance of predictor-corrector methods is decreasing among modern numerical methods for solving differential equations. Nevertheless, a short account of the general principles behind the methods is useful here so that we can be aware of their place in the literature.

For the purpose of discussion, we shall assume that a constant step size h is used to solve a set of n first-order differential equations in the interval $[t_b, t_d]$. Again we shall take the standard form to be

$$\frac{d\boldsymbol{y}}{dt} = \boldsymbol{f}(\boldsymbol{y}, t) \tag{8-47}$$

Where there is no possibility of confusion, we shall simplify the notation by adopting the shorthand

$$\boldsymbol{y}_k \equiv \boldsymbol{y}(t_k) \qquad \boldsymbol{f}_k \equiv \boldsymbol{f}(\boldsymbol{y}(t_k), t_k)$$

If the values of $\boldsymbol{y}_{k-1}, \boldsymbol{y}_{k-2}, \ldots$, are known for a number of points, we can express $\boldsymbol{y}(t)$ using a polynomial expression of the form

$$\boldsymbol{y}^p(t) = \sum_{i=1}^{u} c_i \boldsymbol{y}_{k-i} \tag{8-48}$$

where u is the order of the polynomial and the superscript p is used to distinguish the results from those of the corrector step to be described later. The values of the coefficients c_i may be found from the known values at points $t = t_{k-1}, t_{k-2}, \ldots, t_{k-u}$. We shall come back to this point later. Once the coefficients are known, the expression may be used to find the value of $\boldsymbol{y}(t_k)$ by extrapolating to $t = t_k$. This is known as the *predictor* step for reasons we shall soon see. An important point to realize here is that, so far, we have not yet made any use of (8-47) at $t = t_k$. As a result, $\boldsymbol{y}^p(t_k)$, the predicted value of $\boldsymbol{y}(t_k)$ given by (8-48), may not satisfy (8-47).

We can correct the results of the predictor step in the following way. Since $\boldsymbol{y}^p(t_k)$ represents a good estimate of the values of $\boldsymbol{y}(t_k)$, we can use it to calculate

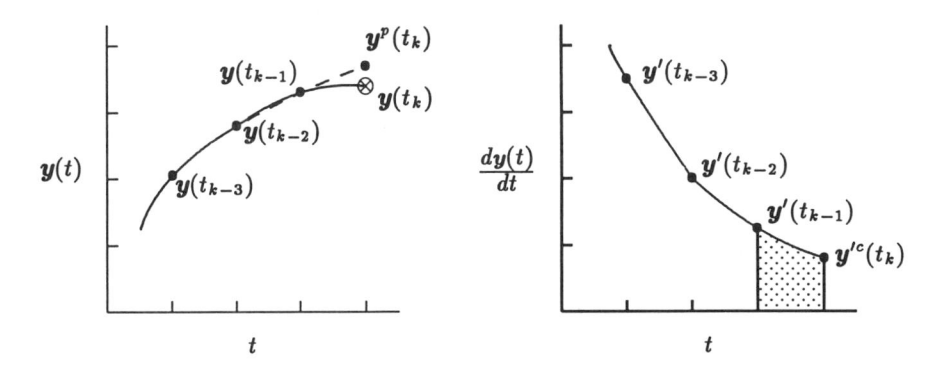

Fig. 8-4 Predictor-corrector method of solving a system of ODEs. The value of $y(t_k)$ is first predicted from a polynomial fit to the known values at $t < t_k$ (dashed curve). The value obtained, $y^p(t_k)$, is then used to find dy/dt at $t = t_k$. The result is, in turn, taken in conjunction with the values of dy/dt at $t < t_k$ to find a polynomial expression for dy/dt to produce a corrected value, $y^c(t_k)$. The accuracy of the contribution from the dotted area may be improved by applying the correction step over and over again.

an approximate value of $f(y,t)$ at $t = t_k$. This, in turn, gives us the value of dy/dt at $t = t_k$:

$$\left.\frac{dy}{dt}\right|_{t=t_k} = f\left(y^p(t_k), t_k\right) \tag{8-49}$$

We can put this new result together with the values of dy/dt at earlier times to form a polynomial expression for the derivatives of $y(t)$ in the form

$$\frac{dy^c(t)}{dt} = \sum_{i=0}^{v-1} d_i \left.\frac{dy}{dt}\right|_{t=t_{k-i}} \tag{8-50}$$

Note also that the summation on the right side now starts from $i = 0$, since we have a value of dy/dt at $t = t_k$, obtained from (8-49), that can be put into use.

The expression for dy/dt may be employed to find an improved or *corrected* value of $y(t_k)$ through integration:

$$y^c(t_k) = y(t_{k-1}) + \int_{t_{k-1}}^{t_k} \frac{dy^c}{dt}\, dt \tag{8-51}$$

In fact, we can apply the corrector step more than once by feeding the results of the above expression back into the right side of (8-49). In this way, we obtain a new set of values for $dy(t)/dt$ in (8-50) and, thus, a new set of values for $y^c(t_k)$ from (8-51). Each time we iterate the calculations given by (8-49) to (8-51), we obtained a better approximation of the value of $y(t)$ at $t = t_k$. This is illustrated schematically in Fig. 8-4.

Polynomial expression for the predictor step. A variant of the Newton's method described in §3-6 may be applied to construct a formula to extrapolate the value of $y(t)$ at $t = t_k$ from those at $t = t_{k-1}, t_{k-2}, \ldots, t_{k-u}$. Instead of the forward differences $\Delta y(t) = y(t + h) - y(t)$ of (2-50), we shall use the backward differences of (2-51):

$$\nabla y(t) = y(t) - y(t - h)$$

For the convenience of discussion, we shall also make use of the recursion relation

$$\nabla^{(r+1)} y(t) \equiv \nabla^r y(t) - \nabla^r y(t - h)$$

to define the symbol for the higher-order backward differences. Similar to (3-96), we have

$$y(t) = y_m + \frac{t - t_m}{h} \nabla y(t_m) + \frac{(t - t_m)(t - t_{m-1})}{2! h^2} \nabla^2 y(t_m) + \cdots$$
$$+ \frac{(t - t_m)(t - t_{m-1}) \cdots (t - t_{m-j+1})}{j! h^j} \nabla^j y(t_m)$$
$$+ \frac{(t - t_m)(t - t_{m-1}) \cdots (t - t_{m-j})}{(j+1)!} y^{(j+1)}(\xi)$$

where $y^{(j+1)}(\xi)$ is the $(j+1)$th derivative of $y(t)$ evaluated at $t = \xi$ in the interval. Similarly, we can have a polynomial expression for $f(t)$:

$$f(y, t) = f_m + \frac{t - t_m}{h} \nabla f_m + \frac{(t - t_m)(t - t_{m-1})}{2! h^2} \nabla^2 f_m$$
$$+ \cdots + \frac{(t - t_m)(t - t_{m-1}) \cdots (t - t_{m-j+1})}{j! h^j} \nabla^j f_m$$
$$+ \frac{(t - t_m)(t - t_{m-1}) \cdots (t - t_{m-j})}{(j+1)!} f^{(j+1)}(\xi) \qquad (8\text{-}52)$$

where $\nabla^j f_m \equiv \nabla^j f(y(t_m), t_m)$ and $f^{(j+1)}(\xi)$ is the $(j+1)$th derivative of $f(t)$ evaluated at $t = \xi$.

We can now integrate both sides of (8-47) from $t = t_{k-1}$ to t_k and obtain

$$y_k = y_{k-1} + \int_{t_{k-1}}^{t_k} f(y, t) \, dt$$

It is possible to make use of the polynomial expression (8-52) for the integrand on the right side. However, at this stage of the calculation, we have the values of $y(t)$ only at mesh points up to $t = t_{k-1}$. As a result, the best we can do is to take $m = (k - 1)$ in (8-52). An order u polynomial approximation for $y(t_k)$ then takes on the form

$$y^p(t_k) = y_{k-1} + h \sum_{i=1}^{u} c_{u,i} f_{k-i} \qquad (8\text{-}53)$$

This is the predicated value of $y(t)$ at $t = t_k$ we had earlier in (8-48) and this way of obtaining the value of $y(t_k)$ is known as the Adams-Bashforth method. It is a multistep method, since the values of $y(t)$ at several different times, t_{k-1}, $t_{k-1}, \ldots, t_{k-p}$, enter as the input.

Let us see how the values of the coefficients $c_{u,i}$ may be obtained. For this purpose, we can start from a Taylor series expansion of y_k in terms of $y(t)$ and its derivatives, all evaluated at $t = t_{k-1}$:

$$y_k = y_{k-1} + h\frac{dy}{dt}\bigg|_{t=t_{k-1}} + \frac{h^2}{2!}\frac{d^2y}{dt^2}\bigg|_{t=t_{k-1}} + \frac{h^3}{3!}\frac{d^3y}{dt^3}\bigg|_{t=t_{k-1}} + \cdots$$

If we are satisfied with an expression that is accurate up to order h, we have a first-order expression for $y^p(t_k)$ with $u = 1$ in (8-53). Using the approximation

$$\frac{dy}{dt}\bigg|_{t=t_{k-1}} \approx f_{k-1}$$

we obtain the result

$$y^p(t_k) = y_{k-1} + hf_{k-1}$$

That gives us the value of $c_{1,1} = 1$.

The coefficients for $u = 2$ may be obtained in a similar way. The general form of a second-order expression for $y^p(t_k)$ is

$$y^p(t_k) = ay_{k-1} + hb_1 f_{k-1} + hb_2 f_{k-2} \qquad (8\text{-}54)$$

where the coefficients a, b_1, and b_2 may be determined by eliminating terms up to order h^2 in the truncation error. For this purpose, we can expand the three quantities on the right side, y_{k-1}, f_{k-1}, and f_{k-2}, in terms of $y(t)$ and its derivatives, all evaluated at $t = t_k$. This may be done using the following Taylor series expansions:

$$y_{k-1} = y_k - hy'_k + \frac{h^2}{2!}y''_k + O(h^3)$$

$$f_{k-1} = f_k - hf'_k + O(h^2) = y'_k - hy''_k + O(h^2)$$

$$f_{k-2} = f_k - 2hf'_k + O(h^2) = y'_k - 2hy''_k + O(h^2)$$

where the primes indicate derivatives with respect to time. When these results are substituted into (8-54), we obtain the relation

$$y^p(t_k) = a\{y_k - hy'_k + \frac{h^2}{2!}y''_k + O(h^3)\}$$
$$+ hb_1\{y'_k - hy''_k + O(h^2)\}$$
$$+ hb_2\{y'_k - 2hy''_k + O(h^2)\}$$

On equating the coefficients of h^0, h^1, and h^2 on both sides of the equation, we obtain a set of three algebraic equations for the three unknown quantities a, b_1, and b_2:

$$y_k = a y_k$$
$$0 = (-a + b_1 + b_2) y'_k$$
$$0 = (\tfrac{1}{2}a - b_1 - 2b_2) y'_k$$

The solution is $a = 1$, $b_1 = 3/2$, and $b_2 = -1/2$ or, in terms of the coefficients defined in (8-53),

$$c_{2,1} = \tfrac{3}{2} \qquad c_{2,2} = -\tfrac{1}{2}$$

For higher orders, the coefficients can be found by a more general method given in Gear (*Numerical Initial Value Problems in Ordinary Differential Equations*, Prentice Hall, Englewood Cliffs, New Jersey, 1971). A list of the values of $c_{u,i}$ for $u \leq 6$ is given in Table 8-1.

Table 8-1
Coefficients for the polynomial expansion of $y^p(t_k)$ in the predictor step.

u	$i=1$	2	3	4	5	6
$c_{1,i}$	1					
$2c_{2,i}$	3	-1				
$12c_{3,i}$	23	-16	5			
$24c_{4,i}$	55	-59	37	-9		
$720c_{5,i}$	1901	-2774	2616	-1274	251	
$1440c_{6,i}$	4277	-7923	9982	-7298	2877	-475

Once we have an estimate of the value of y_k given by $y^p(t_k)$ above, we can make use of it to evaluate f_k using the polynomial expression

$$y^c(t_k) = y(t_{k-1}) + h \sum_{i=0}^{v} d_{v,i} f_{k-i}, \qquad (8\text{-}55)$$

This is the corrected result for y_k of (8-51). The method is an implicit one, as an estimate of y_k is required to calculate the value of y_k itself. As mentioned earlier, we can apply such a correction iteratively, each time using a better approximation to y_k in calculating the values of f_k needed in the first term of the summation on the right side. Note also that the major difference from the predictor step of (8-53) is that the right side now involves an estimated value of f_k. As a result, the approximation is one order higher than the value of v for the step, and the number of coefficients for a given order is one less than the predictor step. The values of the coefficients $d_{v,i}$ for $v = 1$ to 5 (orders 2 to 6) are given in Table 8-2 and a derivation for them can also be found in Gear.

Table 8-2
Coefficients for the polynomial expansion of $y^c(t_k)$ in the corrector step.

v	$i=0$	1	2	3	4
$d_{1,i}$	1				
$2d_{2,i}$	1	1			
$12d_{3,i}$	5	8	-1		
$24d_{4,i}$	9	19	-5	1	
$720d_{5,i}$	251	646	-264	106	-19

Fourth-order predictor-corrector example. It is useful here to see an example of using the predictor-corrector approach to solve a system of n first-order ODEs. To be specific, we shall use fourth-order approximations in both the predictor and the corrector steps.

In one-step approaches, such as the Runge-Kutta methods, the solutions are *self-starting*. That is, we can propagate the solutions forward using only the initial conditions that are given to us by the problem. This is not true for the multistep method we are using here. For a fourth-order predictor-corrector approach, we need to have the values of both $y(t)$ and $f(t)$ at four earlier points before we can apply the predictor step of (8-53). For simplicity, we shall use the fourth-order Runge-Kutta of method (8-35) to provide us with the necessary starting values.

Using the numerical values of the coefficients given in Table 8-1, the fourth-order predictor step of (8-53) has the explicit form

$$y^p(t_k) = y_{k-1} + \frac{h}{24}\left\{55f_{k-1} - 59f_{k-2} + 37f_{k-3} - 9f_{k-4}\right\} \tag{8-56}$$

Similarly, by using the coefficients in Table 8-2, we can write the corrector step of (8-55) for the fourth order in the form

$$y^c(t_k) = y_{k-1} + \frac{h}{12}\left\{5f_k + 8f_{k-1} - f_{k-2}\right\} \tag{8-57}$$

To improve the accuracy, we shall carry out the corrector step a second time using the value of $y^c(t_k)$ obtained in the first corrector step to evaluate the value of $f(y, t)$ at $t = t_k$. The algorithm is outlined in Box 8-3.

As an application, we shall treat the case of a harmonic oscillator,

$$\frac{d^2\phi}{dt^2} + \omega^2\phi(t) = 0$$

with initial conditions

$$\phi(t=0) = 1 \qquad \left.\frac{d\phi}{dt}\right|_{t=0} = 0$$

> **Box 8-3** Subroutine PC(N,H,T,Y,F,ND_PTS,ND_EQN,ERR_MAX)
> **Fourth-order predictor-corrector propagation**
> For a set of n first-order, initial value ODEs
>
> **Argument list:**
>   N: Number of steps already calculated.
>   H: Step size h.
>   T: Value of the independent variable t.
>   Y: Two-dimensional array for the solution y_k.
>   F: Two-dimensional array for the value of f_k.
>   ND_PTS: Total number of steps.
>   ND_EQN: Number of first-order ODEs.
>   ERR_MAX: Maximum error in the solution.
>
> **Subprograms used:**
>   FVAL: Calculate the value of $f(y,t)$ for a given value of $y(t)$ and t.
>
> **Initialization:**
> (a) In the calling program:
>   (i) Choose a step size H.
>   (ii) Specify the initial conditions at $t = t_b$.
>   (iii) Use a fourth-order Runge-Kutta method to propagate the solution forward three steps. Store the value of f_k for $k = 1$ to 4.
> (b) Use the coefficients for the predictor step using Table 8-1 and the corrector step given in Table 8-2.
>
> 1. Predictor step:
>   (a) Find $y^p(t+h)$ with (8-56).
>   (b) Store the results in a separate array.
> 2. First corrector step:
>   (a) Calculate $f(y^p(t+h), t+h)$ with F_VAL using the value of $y^p(t+h)$ obtained.
>   (b) With this and the values of f for earlier steps, calculate $y^c(t+h)$ using (8-57).
>   (c) Store the value.
> 3. Second corrector step:
>   (a) Call FVAL to calculate $f(y^p(t+h), t+h)$ using $y^c(t+h)$ obtained.
>   (b) With this and f of earlier steps, calculate $y^c(t+h)$ using (8-57).
>   (c) Estimate the error from the difference between the two corrector steps and record the maximum value.
> 4. Repeat steps 1 to 3 until the end of the interval.
> 5. Return solution $y(t)$ and the maximum error.

Table 8-3 lists the values of $y^p(t_i)$, $y^c(t_i)$, and those of $y(t_i)$ given by (8-57). For a comparison of the final result, the values for the exact solution $\cos(\omega t)$ are given in the last column. A check of the values of $y^c(t_k)$ against those of $y(t_k)$ provides us with a measure of the realistic errors in the calculation.

    One drawback of predictor-corrector methods is that it is not easy to adapt the algorithm to variable step sizes. As we saw in the previous section for Runge-Kutta methods, this is useful on many occasions. If better accuracies are needed

Table 8-3 Results of the predictor and first and second corrector steps for a harmonic oscillation example.

t	Predictor	First corrector	Second corrector	Exact
0.5	0.88584630	0.87927280	0.87755670	0.87758260
0.6	0.83760830	0.82696510	0.82528490	0.82533560
0.7	0.78100480	0.76638030	0.76475290	0.76484220
0.8	0.71657550	0.69812430	0.69656590	0.69670670
0.9	0.64497590	0.62287970	0.62140580	0.62160990
1.0	0.56691920	0.54139890	0.54002430	0.54030220
1.1	0.48318620	0.45449700	0.45323540	0.45359600
1.2	0.39461430	0.36304310	0.36190730	0.36235760
1.3	0.30208960	0.26795210	0.26695330	0.26749860
1.4	0.20653760	0.17017530	0.16932360	0.16996690
1.5	0.10891430	0.07069089	0.06999480	0.07073697

in a predictor-corrector calculation, one must either reduce the step size h or go to higher-order equations for both the predictor and corrector steps.

§8-4 Solution of initial value problems by extrapolation

In the Romberg method of integration described in §3-5, we saw an example of reducing the truncation errors due to finite step size through the use of extrapolation techniques. Here we shall apply a similar approach to solve a system of differential equations. The main advantage over the Runge-Kutta methods of §8-2 is the higher degree of accuracy that can be achieved with a relatively small amount of additional calculations. On the other hand, since extrapolation is safe only with smooth functions, the method is applicable only in regions where there are no singularities and the functions $f(y,t)$ do not have rapid variations.

Our ultimate goal here is the same as that of the previous two sections: to find the solution of a system of n-coupled first-order ODEs of the form

$$\frac{dy}{dt} = f(y,t)$$

given a set of initial values $y(t)$ at time $t=t_0$. The basic steps in the calculation remain essentially the same as those in Runge-Kutta methods: the solution at time t is used to obtain the solution at a later time $(t+H)$. However, instead of small time steps, as used earlier, the step size H we shall be taking here may be large. To maintain good accuracy, we shall divide H into η smaller steps, each of size $h = H/\eta$. The accuracy of the results therefore depends ultimately on the size

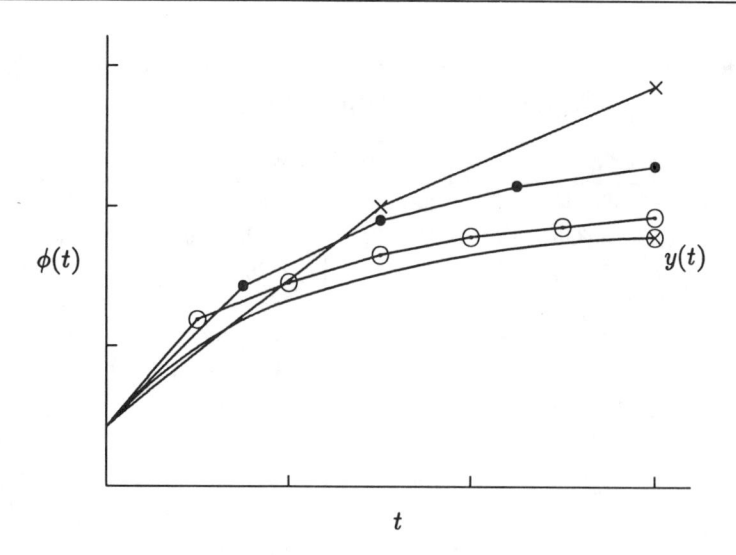

Fig. 8-5 Schematic illustration of the different approximations to solve an ODE by extrapolating to zero step size. The exact solution is represented by the smooth curve, labeled $y(t)$. If the differential equation is solved by dividing the interval into two, the values obtained are represented by the crosses (×). When the same equation is solved by dividing the interval into four, a more accurate set of results, shown as bullets (•), is obtained. Finer subdivisions, represented schematically by the ⊙ symbols, give even better results. By extrapolating to an infinite number of intervals, it is expected that the calculated results will coincide with the exact values.

of h or the value of η used. Clearly, the results improve with larger η, as illustrated schematically in Fig. 8-5. The advantage provided by the extrapolation technique is that we can find the value in the limit of $h = 0$ without actually having to carry out the calculations to $\eta \to \infty$. This is the principle of *Richardson's deferred approach to the limit* we saw earlier in §3-5.

The strategy of the extrapolation method is quite simple. For each interval H, we start with an h obtained with some small value for η, like 2, and gradually decrease h by increasing η. When there are several values of $y(t+H)$ available for different step size h, an extrapolation is carried out to find the value in the limit of $h = 0$. In principle, we can calculate the value of $y(t+H)$ for a given h using any of the methods described earlier. As a practical matter, it is more efficient to use one designed for the purpose.

Modified midpoint method. Consider a function $z(t,h)$ defined in the following way on a discrete set of points, $t_k = t_0 + kh$ for $k = 0, 1, \ldots, \eta$. Starting from $t = t_0$, the values of $z(t,h)$ for the first two points are given by

$$z(t_0; h) = y(t_0)$$
$$z(t_0 + h; h) = y(t_0) + h f(y(t_0), t_0)$$

Once we have these two to start with, the values of $z(t, h)$ at the other points are defined by the relation

$$z(t + h; h) = z(t - h; h) + 2hf(z(t; h), t) \tag{8-58}$$

Because of truncation errors, $z(t; h)$ are not likely to be the same as $y(t)$ for any finite h. The difference between them may be expressed in terms of a series consisting of only even powers of h,

$$z(t; h) = y(t) + \sum_{k=1}^{\infty} h^{2k} \{u_k(t) + (-1)^{(t-t_0)/h} v_k(t)\} \tag{8-59}$$

where the functions $u(t)$ and $v(t)$ are independent of h. As we shall soon see, the forms of these two functions are not important in actual calculations. The $(-1)^{(t-t_0)/h} v_k(t)$ term in the sum is, however, troublesome, as it oscillates in sign. As shown in Stoer and Bulirsch (*Introduction to Numerical Analysis* [English translation by R. Bartels, W. Gautschi, and C. Witzgall], Springer-Verlag, New York, 1980), it can be eliminated in the leading order h^2 by taking the linear combination

$$S(t; h) = \tfrac{1}{2}\{z(t; h) + z(t - h; h) + hf(z(t; h), t)\}$$

The result is a well-behaved series at least up to order h^2:

$$S(t; h) = y(t) + h^2\left\{u_1(t) + \frac{1}{4}\frac{d^2 y}{dt^2}\right\} + \sum_{k=2}^{\infty} h^{2k}\{u_k(t) + (-1)^{(t-t_0)/h} v_k(t)\}$$

In the limit $h \to 0$, we have $S(t; h) = y(t)$. If several values of $S(t + H, h)$ are available for different (finite) values of h, we are in a position to extrapolate its value at $h = 0$ and, hence, the value of $y(t + H)$.

To implement the algorithm, we start by dividing the interval H into η equal steps. To make clear the dependence of the step size on η, we shall add a subscript to h in the form

$$h_\eta = \frac{H}{\eta}$$

For a given h_η, the value of $S(t + H; h_\eta)$ is given by

$$S(T; h_\eta) = \tfrac{1}{2}\{z(T; h_\eta) + z(T - h_\eta; h_\eta) + hf(z(T; h_\eta), T)\} \tag{8-60}$$

where we have used the symbol $T = (t + H)$ to simplify the notation. To obtain the values of $z(T - h_\eta; h_\eta)$ and $z(T; h_\eta)$, we follow (8-58) and use the value of $y(t)$ already in our possession to define

$$z(t, h_\eta) = y(t) \qquad z(t + h_\eta; h_\eta) = z(t; h_\eta) + h_\eta f(z(t, h_\eta), t)$$

Box 8-4

Subroutine MID_PTM(Y_IN,Y_OUT,T,H_BIG,NSTPS,NEQS,FVAL)

Modified midpoint method for a system of n first-order ODEs

Argument list:
  Y_IN: Input value of $y(t)$.
  Y_OUT: Output value of $y(t+H)$.
  T: Value of t.
  H_BIG: Value of H.
  NSTPS: Number of steps η.
  NEQS: Number of equations.
  FVAL: External subprogram to calculate $f(y, t)$.

1. Define the step size h_η to be H/η.
2. Define ZM1 as $z(t; h_\eta)$ and ZM as $z(t + h_\eta; h_\eta)$.
3. Propagate using (8-61) starting from $m = 1$ until $t = $ (T+H_BIG):
    (a) Calculate $y(t+(m+1)h_\eta)$ using (8-61) and store the results temporarily as TEMP.
    (b) Store the value of ZM in ZM1.
    (c) Move the value of TEMP into ZM.
    (d) Increase m by 1 until $m = \eta$.
4. Return the value of $\frac{1}{2}\{ZM + ZM1 + h_\eta f(ZM, t + H)\}$ as $y(t + H)$.

All the later values of $z(t + mh_\eta; h_\eta)$ for $m > 1$ are produced by the recursion relation

$$z(t+(m+1)h_\eta; h_\eta) = z(t+(m-1)h_\eta; h_\eta) + 2h_\eta f\bigl(z(t+mh_\eta; h_\eta), t+mh_\eta\bigr) \tag{8-61}$$

obtained using (8-58). The necessary input to the right side of (8-60) is obtained from the calculated results for $m = \eta$ and $m = (\eta - 1)$.

It is basically a midpoint method, since the solutions to the differential equations for most of the steps from t to $(t + H)$ are propagated from $t = (m-1)h$ to $t = (m+1)h$ using the value of $f(y, t)$ at a point half way between the beginning and the end of the interval for the step. However, the first and last steps are different. For this reason, it is known as the *modified midpoint* method. A summary of the algorithm is given in Box 8-4. In principle, the method can give a complete solution for a set of ODEs. However, it is not an efficient method by itself and is used here only as the inner core for the extrapolation method.

For the purpose of extrapolation, we need a sequence of values, $S(T; h_1)$, $S(T; h_2), \ldots$, calculated with different values of h_k and η_k, for $k = 1, 2, \ldots$. The obvious choice of $\eta_k = 2^k$ for the series is not practical here, as the size of h_k diminishes too quickly to be useful as a sequence for extrapolations. The alternative is the sequence $\eta_k = \{2, 4, 6, 8, 12, 16, \ldots, \}$, which may be written as

$$\eta_k = 2\eta_{k-2} \qquad \text{for} \qquad k > 3$$

The three starting values for this sequence are $\eta_1 = 2$, $\eta_2 = 4$, and $\eta_3 = 6$. For reasonable values of H, it is seldom necessary to go to very large k values.

Once the values of $y(t + H)$ are available for a number of different step sizes $h_1, h_2, \ldots$, we can extrapolate the corresponding value for $h \to 0$. For the convenience of discussion, let us represent the value of $y(t + H)$ calculated with step size $h_k = H/\eta_k$ as $F(x_k)$. Because of (8-59), the independent variable x_k for $F(x_k)$, as far as the extrapolation is concerned, is h_k^2. In other words, we have the relation

$$\lim_{h \to 0} F(h^2) = y(t + H) \tag{8-62}$$

Similar to extrapolations discussed in §3-5, we can regard $F(x)$ as a function of x, with values known to us only at $x = x_1, x_2, \ldots$, corresponding to $h^2 = h_1^2, h_2^2, \ldots$. The polynomial extrapolation method of §3-5 may be used if we express $F(x)$ as a polynomial in x. This is left as an exercise.

Rational polynomial extrapolation. For better accuracy, we shall express $F(x)$ as a ratio of two polynomials and apply the technique of rational polynomial interpolation to find the value of $y(t+H)$ in the limit of $h = 0$. The extrapolation may be carried out using essentially the same algorithm as outlined in Box 3-2. As we saw in Chapter 3, the basic calculations involved in extrapolation are very similar to those for interpolation and, as a result, it is possible to adopt Neville's algorithm for rational function interpolation for our needs here.

We should not forget the fact that we are dealing with a set of n dependent variables here, $y(t + H) = \{y_1(t + H), y_2(t + H), \ldots, y_n(t + H)\}$. As far as extrapolation is concerned, a calculation must be carried out for each of these n quantities. There is no difficulty in doing this, as the n different dependent variables represented by $y(t+H)$ may be treated here as if they were unrelated to each other, and the extrapolation may be carried out as if they were n independent quantities. For each of them, we shall assume that there are ξ different values $F(x)$, calculated with successive smaller step sizes h_k. Similar to (3-32) and (3-33), we can define two sets of auxiliary quantities Θ_{mi} and Δ_{mi}, where the subscript $m = 0, 1, 2, \ldots, (\xi - 1)$ indicates the order of extrapolation, and the subscript $i = 1, 2, \ldots, (\xi - m)$ distinguishes the different elements in each order.

Because we are interested only in extrapolating to $x \equiv h^2 \to 0$, the recursion relations are somewhat simpler than those given in (3-32). They may be put in the form

$$\Theta_{(m+1)k} = \frac{(\Delta_{m(k+1)} - \Theta_{mk})\Delta_{m(k+1)}}{\frac{x_k}{x_{m+k+1}}\Theta_{mk} - \Delta_{m(k+1)}}$$

$$\Delta_{(m+1)k} = \frac{\frac{x_k}{x_{m+k+1}}\Theta_{mk}(\Delta_{m(k+1)} - \Theta_{mk})}{\frac{x_k}{x_{m+k+1}}\Theta_{mk} - \Delta_{m(k+1)}} \tag{8-63}$$

The starting values of Θ and Δ are given by $F(x_k)$

$$\Theta_{0k} = \Delta_{0k} = F(x_k)$$

the values of $y(t+H)$, more precisely, those of $S(T, h_k)$ in (8-60), calculated for $h_k = H/\eta_k$. Let us represent the best estimate of the extrapolated value for order m as $F_m(0)$. For $m = 0$, the value is given by the one calculated with the smallest step size. As a result, we have

$$F_0(0) = F(x_\xi)$$

where $x_\xi = h_\xi^2$ and h_ξ is the smallest step size used to calculate $y(t+H)$. Since we are carrying out an extrapolation, the correction for order m is given by $\Theta_{m(\xi-m)}$, the last element of Θ for that order. As a result, the best estimate of the value for $h^2 \equiv x = 0$ for order m is given by

$$F_m(0) = F_{m-1}(0) + \Theta_{m(\xi-m)}$$

Similar to what we did in §3-2, an estimate of the error of the extrapolation process is provided by the correction term used for the highest order. In the present case, it is $\Theta_{\xi-1,1}$.

The algorithm outlined in Box 8-5 concentrates mainly on getting the value of $y(t+H)$ starting with known values of $y(t)$. To start with, the modified midpoint method is used to obtain $y(t+H)$ for the first two values of η_k. Once these are available, an attempt is made to extrapolate to zero step size. If the estimated error in the extrapolation is larger than the tolerance, one more set of the values of $y(t+H)$ is calculated for the next smaller size for h_η and the extrapolation is carried out again. This is repeated until the errors in the extrapolated results are sufficient small. The values are returned as the solution for $y(t+H)$ and the calculation proceeds to the next subinterval of size H.

The number of extrapolation steps actually taken at a given t gives us an estimate of the size of H to be used for the next subinterval. If the desired accuracy is reached with a small number of η_k, it is possible to make the next H larger than the present one. On the other hand, if we cannot reach the required numerical accuracy within a reasonable number of subdivisions into h_η, it is wise to use a smaller value for H for the next step. From the point of view of computational efficiency, we may want to carry out the calculation with a minimum number of steps in terms of H. This may not always be desirable. On many occasions, we may wish to have a feeling for how the solution behaves as a function of the independent variable. In this case a minimum size must be imposed on H.

Another practical consideration is that there is no point in taking the extrapolation process to too high an order. For this reason, the algorithm puts an upper limit ξ for the order of extrapolation. That is, if there are more $(\xi+1)$ different sets of values available, the extrapolation makes use only of the last $(\xi+1)$.

Box 8-5
Subroutine EXT_STP(T_B,T_D,T_ARY,BFY,EPS,N_BSTP,ND_PTS,ND_EQN)
Solution of first-order ODEs by extrapolating to zero step size

Argument list:
  T_B: Beginning of the interval for t.
  T_D: End of the interval for t.
  T_ARY: One-dimensional array for t.
  BFY: Two-dimensional array for y.
  EPS: Maximum error allowed (ϵ).
  N_BSTP: Number of big steps H taken.
  ND_PTS: Maximum number of big steps dimensioned for in the calling program.
  ND_EQN: Number of first-order ODEs in the system.

Subprogram used:
  MID_PTM: Modified midpoint of Box 8-4.
  EXT_ZRO: Extrapolate to zero step size using Neville's algorithm (Box 3-2).

Initialization:
  (a) Store the initial conditions as the values of the first points of the solution and set up the error tolerance in the calling program.
  (b) Let the maximum order of the extrapolation be 11, the maximum number of tries be N_{try} and extrapolations be N_{ext}, and the number of big steps be 10.
  (c) Use $\{2, 4, 6, 8, 12, 16, \ldots\}$ as the sequence for the values of η_k.

1. Set the size of big step H.
2. Use MID_PTM to calculate $y(t + H)$:
   (a) Divide H into η_k steps of size h_η each.
   (b) Store h_η^2 and the value of $y(t + H)$ obtained for this h_η.
3. Call EXT_ZRO to extrapolate the values of $y(t + H)$ for $h^2 = 0$.
4. Find the maximum uncertainty $\mathcal{E}_{max}$ among the ND_EQN extrapolated values.
5. If $\mathcal{E}_{max} \leq \epsilon$, accept the extrapolated result:
   (a) Store the extrapolated values as $y(t + H)$.
   (b) If t is less than T_D:
       (i) Adjust the size of H for the next step if necessary.
       (ii) Go back to step 2 for the next time interval.
   (c) If $t \geq$ T_D, return the values of t and y.
6. If $\mathcal{E}_{max}$ is greater than ϵ, carry out further extrapolations:
   (a) If no convergence after N_{ext} extrapolations, H is too big. Reduce H and try again.
   (b) If no convergence after N_{try} reductions, abort the calculation with an error message.

Since we are interested only in the value $y(t + H)$ for $h \to 0$, in principle, we can make the extrapolation calculation even more efficient than that used in §3-5. However, for the sake of simplicity and ease in understanding the code, only a minimum amount of changes is made. One must also be cautious about the rational extrapolation process itself. It is possible that the procedure may fail on certain occasions. In such cases, one must go back to the simpler polynomial extrapolation given in §3-5.

§8-5 Solution of boundary value problems by shooting methods

We saw earlier in §8-1 that, in general, it is more complicated to find the numerical solution of a boundary value problem than that of an initial value problem. The difference comes primarily from the fact that we do not have enough information at either end of the interval to propagate the solution. The philosophy of shooting methods is to make estimates of the missing information so as to be able to start the propagation. The numerical calculations are now reduced to those similar to an initial value problem and any of the methods described in the previous three sections may be used.

The projectile problem posed in Problems 8-1 and 8-2 gives a good illustration of the differences between initial value and boundary value problems. Consider an object with mass μ, sent from the surface of the earth with some initial velocity v_0 and at an angle of elevation θ_0 above the horizon. In the absence of any friction caused, for example, by air resistance, the place where the object lands on the surface of the earth depends both on v_0 and θ_0. The path followed by the object may be put into the form of the variation of y, the height above the surface of the earth, as a function of x, the horizontal distance from the starting point. If we are interested in the trajectory of the object for a given initial velocity and angle of elevation, we have an initial value problem. On the other hand, if we wish to find the angle of elevation for a given initial velocity such that the projectile lands at a specific point along the x-axis, we have a boundary value problem.

The starting point of shooting methods is to make an estimate of the missing information for us to start the propagation of the solution. In terms of the projectile problem, this is equivalent to making some guess at the angle of elevation for the projectile of a given initial velocity v_0 to land at a specific point x_d. It is likely that the estimate is not accurate enough and we must either increase or decrease θ_0 somewhat for the projectile to land at x_d. The analogous situation in our boundary value problem is that the estimate is not sufficiently accurate to provide us with a solution to the differential equation that satisfies the boundary condition at the other end of the interval. The main role of shooting methods is to provide ways to improve our estimates so that we can converge quickly to the correct solution.

Quantum mechanical scattering from a square-well potential. Before going into the question of numerical solutions to boundary value problems using a shooting method, it is helpful to have a nontrivial physical problem in mind. This is useful, in part, to provide us with some guidance to the type of solution we are seeking and the accuracy we want to achieve. A good example is potential scattering in quantum mechanics. The interest here is to find the wave function of a particle that interacts with a potential $V(r)$. The Schrödinger equation has the form

$$-\frac{\hbar^2}{2\mu}\nabla^2\psi(\boldsymbol{r}) + V(r)\psi(\boldsymbol{r}) = E\psi(\boldsymbol{r}) \qquad (8\text{-}64)$$

where μ is the reduced mass of the scattering particle. For a scattering solution, the energy E is positive.

For simplicity, we shall assume that the potential is a central one, meaning that it is a function of the radial coordinate r only. As a result, angular momentum is a good quantum number. The wave function may be expanded in terms of components, each with a definite angular momentum ℓ:

$$\psi(\mathbf{r}) = \sum_{\ell=0}^{\infty} \frac{1}{r} u_\ell(r) Y_{\ell,m}(\theta, \phi) \qquad (8\text{-}65)$$

where $Y_{\ell,m}(\theta, \phi)$ are the spherical harmonics. As we saw in §4-2, the angular dependence of the wave function is given by $Y_{\ell,m}(\theta, \phi)$. Because of this simplification, the only part of the Schrödinger equation (8-64) that must be solved for the wave function of the scattering problem is the modified radial wave function $u_\ell(r)$ for different partial waves, each having a definite angular momentum ℓ.

Since the spherical harmonics are eigenfunctions of the angular momentum operator with eigenvalue $\ell(\ell+1)\hbar^2$, as shown in §4-2, the Schrödinger equation of (8-64) reduces to a differential equation for $u_\ell(r)$:

$$-\frac{\hbar^2}{2\mu}\left\{\frac{d^2 u_\ell}{dr^2} - \frac{\ell(\ell+1)}{r^2} u_\ell(r)\right\} + V(r)\, u_\ell(r) = E\, u_\ell(r) \qquad (8\text{-}66)$$

As an illustration, we shall restrict ourselves to the angular momentum $\ell = 0$ case. The equation for the modified radial component of the wave function reduces to a second-order ODE:

$$-\frac{\hbar^2}{2\mu}\frac{d^2 u_0}{dr^2} + V(r)\, u_0(r) = E\, u_0(r) \qquad (8\text{-}67)$$

For the $\ell > 0$ case, we can replace $V(r)$ by an effective potential given by

$$\tilde{V}(r) = V(r) + \frac{\hbar^2}{2\mu}\frac{\ell(\ell+1)}{r^2}$$

In this way, a radial equation of the form of (8-67) may be used to find $u_\ell(r)$ for $\ell > 0$ as well.

For our illustration, we shall assume that $V(r)$ is a square-well potential of height V_0,

$$V(r) = \begin{cases} V_0 & \text{for } r \leq r_g \\ = 0 & \text{for } r > r_g \end{cases} \qquad (8\text{-}68)$$

where r_g is the range of the potential. If V_0 is positive, the potential is repulsive, as shown in Fig. 8-6(a), and, if V_0 is negative, the potential is attractive, as shown of Fig. 8-6(b). Such a simple form of the potential is useful here, as analytical solutions are available for us with which to make comparisons. However, the real interest of numerical solutions is for equations where exact solutions are impossible, such as for the case of a Woods-Saxon potential shown in Fig. 8-6(c):

$$V(r) = \frac{V_0}{1 + \exp\{\frac{r-r_g}{a}\}} \quad (8\text{-}69)$$

Here, the exponential rise in the radial dependence, measured by the diffuseness parameter a, is a more realistic representation of, for example, the interaction between a nucleon and a nucleus than the sharp cutoff offered by a square-well potential.

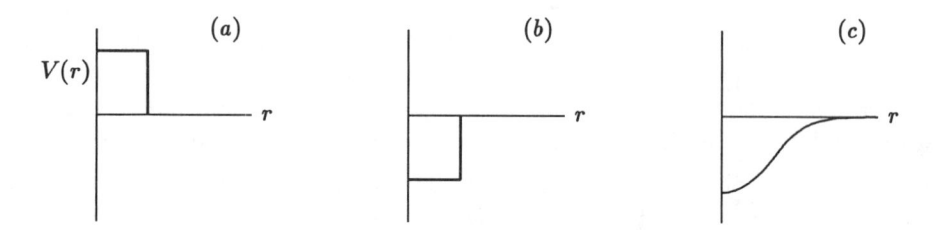

Fig. 8-6 Potential well for scattering: (a) a repulsive square well, (b) an attractive square well, and (c) a Woods-Saxon well that approximates the shape of, for example, the potential between a nucleon and a nucleus.

Since we are dealing with a second-order differential equation, two boundary conditions are required. The first comes from the fact that the wave function $\psi(r)$ must be finite at the origin. As a result, $u_0(r)$ vanishes at the origin:

$$u_0(r=0) = 0 \quad (8\text{-}70)$$

The second boundary condition comes from the radial wave function at large r. Outside the potential well, $V(r) = 0$, the particle is a free one and the modified radial wave function is given by the equation

$$-\frac{\hbar^2}{2\mu}\frac{d^2 u_0(r)}{dr^2} = E u_0 \qquad \text{for} \qquad r \gg r_g$$

This can be put into a form similar to (8-2):

$$\frac{d^2 u_0(r)}{dr^2} + k^2 u_0 = 0$$

where k is real and has the value

$$k = \frac{1}{\hbar}\sqrt{2\mu E} \quad (8\text{-}71)$$

The solution of this equation is familiar, and we shall write it in the form

$$u_0(r) = A\sin(kr + \delta_0) \qquad \text{for} \qquad r \gg r_g \quad (8\text{-}72)$$

Here A is the amplitude, related to the normalization of the wave function, and is of no interest to our problem here. All the information of physical importance is contained in the phase factor δ_0, generally known as the *phase shift*. A schematic illustration of the role of δ_0 is given in Fig. 8-7.

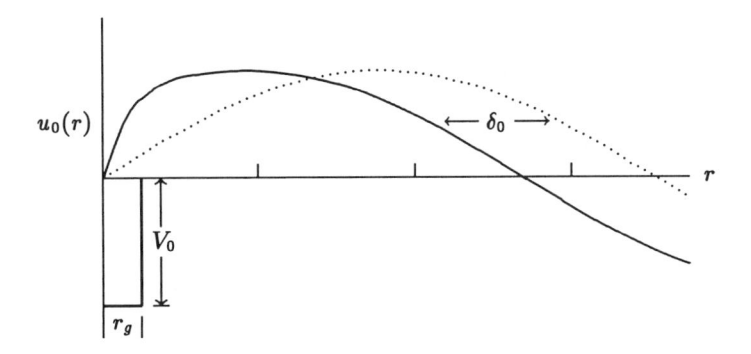

Fig. 8-7 Phase shift δ_0 in the wave function of a particle as a result of scattering by an attractive square-well potential. In the absence of a potential, the wave function is schematically represented by the dotted curve. The shift in the phase of the scattered wave function, represented by the solid curve, depends on the incident energy as well as the potential well that causes the scattering.

The solution of the scattering problem for a square-well potential is well known in quantum mechanics. Let us restrict ourselves to an attractive potential. The phase shift is given by the expression

$$\delta_0 = m\pi - kr_g + \tan^{-1}\left(\frac{k}{\kappa}\tan \kappa r_g\right) \qquad (8\text{-}73)$$

where m is an integer and

$$\kappa = \frac{1}{\hbar}\sqrt{2\mu(E - V_0)} \qquad (8\text{-}74)$$

Since $E > 0$ in a scattering problem and $V_0 < 0$ for an attractive potential, κ is real. At low energies, $m = 0$, and we shall confine ourselves to such cases.

The main interest of our numerical solution for (8-67) is to obtain the value of δ_0. Before attempting to solve this problem, it is instructive to find first the solution of the region inside the square well. For this purpose, we shall make use of the value of δ_0 given by (8-73). Later, we shall return to the problem of finding δ_0 by a numerical solution of the scattering equation when we try to solve the differential equation in the full range of r.

The inside region. To solve (8-67) for the region inside the potential well, $r = [0, r_g]$, we need to impose a boundary condition at $r = r_g$. For purposes of illustration, it may be obtained from $u_0(r)$ given by (8-72)

$$u_0(r_g) = \sin(kr_g + \delta_0) \qquad (8\text{-}75)$$

where, for the convenience of discussion, we have taken the amplitude A to be unity. Alternatively, we could have used the requirement that the logarithmic derivative of the modified radial wave function at $r = r_g$,

$$L(r_g) \equiv \frac{1}{u_0(r_g)} \frac{du_0}{dr}\bigg|_{r=r_g} \qquad (8\text{-}76)$$

be continuous across the boundary between the inside and outside regions of the potential well. We shall follow such an approach later.

For our second-order ODE in the interval $r = [0, r_g]$, the two boundary conditions are then

$$u_0(0) = 0 \qquad u_0(r_g) = \sin(kr_g + \delta_0) \qquad (8\text{-}77)$$

where δ_0 is given by (8-73) with $m = 0$. The differential equation (8-67) may be simplified to the form

$$\frac{d^2 u_0}{dr^2} + \frac{\hbar^2}{2\mu}\{E - V(r)\} u_0(r) = 0 \qquad (8\text{-}78)$$

The numerical solution is now transformed into one for a simple two-point boundary value problem with one boundary condition specified at each end of the interval. For a square-well potential, $V(r) = V_0$ in the whole range of interest to us at the moment, and the factor in front of $u_0(r)$ in (8-78) may be replaced by a constant κ^2, with κ given by (8-74). In the more general case, such as the Woods-Saxon potential of (8-69), it is a function of r.

In using a shooting method to find the solution, the second-order ODE of (8-78) is solved as an initial value problem. Before we can do this, we need a second "initial" condition at $r = 0$, in addition to the one supplied by (8-70). This may be provided for us in the form of the value of du_0/dr at $r = 0$. Since this is not given to us by the problem, we must make an estimate for its value. As a first try, let

$$\frac{du_0}{dr}\bigg|_{r=0} = \beta_1 \qquad (8\text{-}79)$$

where β_1 is some reasonable value we can think of based on physical grounds. The differential equation can now be solved using any one of the methods for initial value problems, such as the extrapolation method of the previous section.

The solution we obtain for $u_0(r)$ depends on the value of β_1 we have chosen. Let us use the symbol $u_0(\beta_1; r)$ to represent this particular solution. When we propagate the solution to $r = r_g$, it is unlikely that the value of $u_0(\beta_1; r)$ at

$r = r_g$ is equal to that given by the boundary condition (8-77). Let us indicate the difference between them as $R(\beta_1)$:

$$R(\beta_1) = u_0(\beta_1; r_g) - u_0(r_g) \tag{8-80}$$

Our aim now is to find a value of $\beta = du_0/dr$ at $r = 0$ such that $R(\beta)$ vanishes.

For this purpose, we shall repeat the calculation with a different estimate β_2 that is also a reasonable choice. On solving the differential equation again as an initial value problem, we obtain

$$R(\beta_2) = u_0(\beta_2; r_g) - u_0(r_g)$$

In general, the functional dependence of R on β is complicated and nonlinear. However, if both our estimates are sufficiently close to the true solution, we can approximate $R(\beta_i)$ to be linear in β. This is possible at least in a small region where $R(\beta) \sim 0$. In this case,

$$\beta = \beta_2 - R(\beta_2)\frac{\beta_2 - \beta_1}{R(\beta_2) - R(\beta_1)} \tag{8-81}$$

satisfies the requirement

$$R(\beta) = 0$$

If this is true, we have solved our boundary value problem using techniques designed for initial value problems.

In general, our linear extrapolation for β used above in deriving (8-81) is not sufficiently accurate. As a remedy, we can iterate the process by defining a new value of β in the following way:

$$\beta_{i+1} = \beta_i - \Delta\beta_i \tag{8-82}$$

where the change in the value of β in iteration i is given by

$$\Delta\beta_i = R(\beta_i)\frac{\beta_i - \beta_{i-1}}{R(\beta_i) - R(\beta_{i-1})} \tag{8-83}$$

In general, if we start from a value of β that is close to the solution, we have the guarantee that

$$|R(\beta_{i+1})| < |R(\beta_i)|$$

As a result, repeated applications of (8-82) will eventually lead to a value of $R(\beta)$ that is less than the error that can be tolerated in the solution. The algorithm is outlined in Box 8-6.

For ODE with order higher than two, the more common situation is one of having n_1 boundary conditions given at one end of the interval and n_2 boundary conditions given at the other end, with (n_1+n_2) equal to the number of equivalent first-order ODEs that must be solved. In this case, we must make n_2 estimates of the values at the start of the interval before we can apply the methods of initial

> **Box 8-6   Program SHT_SQR**
> **Shooting method to obtain the scattering solution**
> For a one-dimensional square-well potential
>
> Subprograms used:
>   EXT_STP: Solution of a system of n first-order ODEs by extrapolation (Box 8-5).
>   EXT_ZRO: Extrapolation to zero step size, using Neville's algorithm (Box 3-2).
>   MID_PTM: Modified midpoint method for a set of n first-order ODEs (Box 8-4).
>   FVAL: Value of $f(y,t)$ for a square well.
> Initialization:
>   (a) Map the interval to $x = [0, 1]$.
>   (b) Let the two initial estimates be $\beta_1 = 1$ and $\beta_2 = 2$.
>   (c) Reduce the second-order ODE to a system of two first-order ones by defining $y_1(x) = \phi(x)$ and $y_2(x) = dy_1/dx$.
> 1. Input energy E and well depth V_0.
> 2. Calculate the constant factors k of (8-71) and κ of (8-74).
> 3. Evaluate the phase shift δ_0 of (8-73) and define the boundary condition at $x = 1$.
> 4. Solve the ODE by extrapolation using EXT_STP:
>   (a) Set up the boundary value at $x = 0$.
>   (b) Make a first estimate β_1 as the value of $y_2(x_b)$.
>   (c) Use EXT_STP to carry out the extrapolation to zero step size.
>   (d) Calculate the difference $R(\beta_1)$ of (8-80).
> 5. Solve the ODE again with a second estimate β_2.
> 6. Improve the estimate:
>   (a) Obtain a new value of β using (8-82).
>   (b) Check for convergence by comparing $|\Delta\beta/\beta|$ with the tolerance for error:
>   (i) If larger, solve the ODE again with the new value of β.
>   (ii) Otherwise, output the solution.

value problems to obtain a solution. A more sophisticated form to improve upon the estimates by iterative calculations can be found in Press and others (1986).

For our illustrative example of the scattering wave function inside a square-well potential, the procedure given by (8-82) is more than adequate. Since $u_0(r) = 0$ at $r = 0$, our estimate β for the value of du_0/dr at $r = 0$ is equivalent to specifying the normalization of the wave function. At the other end of the interval, only the logarithmic derivative $L(r_g)$ of (8-76) is of physical significance. Since $L(r_g)$ is independent of the normalization, the solution, except for the normalization, is independent of the value of β used. The reason for the simplicity comes from the fact that we have made use of the value of the phase shift δ_0 given by the analytical solution of (8-73). In more realistic situations, several iterations are usually needed, as we shall see later.

Solution for the outside region. To find the value of the phase shift δ_0 itself by numerical solution, we need to solve the differential equation also for the region

$r = [r_g, \infty]$ that is outside the potential well. The boundary condition for large r is given by the requirement that the wave function must vanish at infinity,

$$\lim_{r \to \infty} \psi(\mathbf{r}) = 0$$

From (8-65), we see that

$$\psi(\mathbf{r}) \propto \frac{1}{r} u_0(r)$$

As a result, $u_0(r)$ must be finite at large r.

Since $V(r) = 0$ in this region, $u_0(r)$ has the form given by (8-72). Furthermore, the normalization A is immaterial, as we have seen for the inside region. This is especially true if, at the boundaries, we deal with the logarithmic derivatives defined by (8-76). As a result, the boundary condition at large r becomes an arbitrary constant. Let us use r_d as the location where the boundary condition for large r is given. For the convenience of calculation, we shall take

$$u_0(r_d) = 1 \qquad (8\text{-}84)$$

It is not always easy to determine what is a large r in a numerical calculation and we must seek guidance from the physics of the problem. For our present case, we can take r to be effectively infinite when we are far away from the potential well. As a practical matter, we shall take here $r_d = 10 r_g$ as the point where r is essentially "infinite" as far as our physical interest in the scattering problem is concerned. In principle, we can take an even larger value; the price one pays is the loss of efficiency in the solution since we have to cover a larger range of the value of r.

For the boundary condition at other end of the interval, $r = r_g$, we shall make use of the value of the logarithmic derivative $L(r_g)$ defined by (8-76). For the time being, we shall assume that the value of $L(r_g)$ is given by the solution of the inside region, $r \leq r_g$. From the form of $u_0(r)$ outside the potential well given by (8-72), we see that

$$L(r_g) = \frac{1}{u_0(r_g)} \frac{du_0}{dr} \bigg|_{r=r_g} = k \cot(k r_g + \delta_0)$$

The expression tells us also that, if the value of $L(r_g)$ is given by our numerical solution, we can deduce from it the value of the phase shift δ_0. We shall carry out this part of the calculations later when we match the solution from the inside region to that of the outside region.

To make use of the shooting method for the outside region, we start from $r = r_d$ and work backward toward $r = r_g$. For this purpose, we need to have, in addition to the condition given by (8-84), the value of a second "boundary" condition at $r = r_d$. Similar to (8-79) for the inside region, we need to make an estimate of a second "initial" condition at $r = r_d$ to start our numerical solution. For this purpose, we shall take

$$\left.\frac{du_0}{dr}\right|_{r=r_d} = \gamma$$

and vary γ such that the logarithmic derivative at $r = r_g$ matches the one from the inside region.

We can use essentially the same procedure as that for the inside region to solve for $u_0(r)$ for $r = [r_g, r_d]$. Since the initial value problem must now be solved backward with the dependent variable decreasing rather than increasing, some care is needed in the actual coding; however, there are no basic changes in the algorithm. For a given value of γ, the solution is propagated from $r = r_d$ to $r = r_g$. At $r = r_g$, the result is compared with $L(r_g)$ from the inside region and a new estimate of γ is obtained to improve the match. It is a relatively simple matter to replace u_0 by $L(r_g)$ in arriving at a similar expression as (8-83). This is left as an exercise.

Bidirectional shooting. We have now the method to solve the radial equation (8-67) for both the inside and the outside region together. By joining the two solutions, we obtain the scattering wave function for the entire region of $r = [0, r_d]$. For the combined inside and outside regions, the two boundary conditions are

$$u_0(r = 0) = 0 \qquad u_0(r_d) = 1$$

In solving the problem earlier as two separate regions, $r = [0, r_g]$ and $r = [r_g, r_d]$, we artificially introduced a boundary condition at the matching radius $r = r_g$ so that each solution has a value to "shoot" for. We shall continue with this basic approach here. As before, the "boundary condition" at the matching radius for one region is provided by the solution from the other region. To be the wave function for the whole region of $r = [0, r_d]$, the two separate pieces must join onto each other smoothly at the matching radius. This condition may be satisfied by requiring that the logarithmic derivatives from the two regions be equal to each other at $r = r_g$. A schematic illustration of this point is given in Fig. 8-8.

In principle, we could have used the shooting method to cover the entire range $[0, r_d]$. That is, we start the solution at one end, for example $r = 0$, with some estimate of the necessary second "initial" condition and propagate the solution forward all the way. At the other end, the solution is compared with the boundary condition given at $r = r_d$ and the difference is used to adjust the estimate, in the same way as we did above for each of the two subintervals. However, there are several advantages to dividing the region $[0, r_d]$ into two parts and using the bidirectional shooting method. In the first place, the whole interval may be too large for us to propagate the solution all the way and still expect good numerical accuracy. In the second place, the small and large r subintervals may have quite different properties. For example, in our scattering example, the

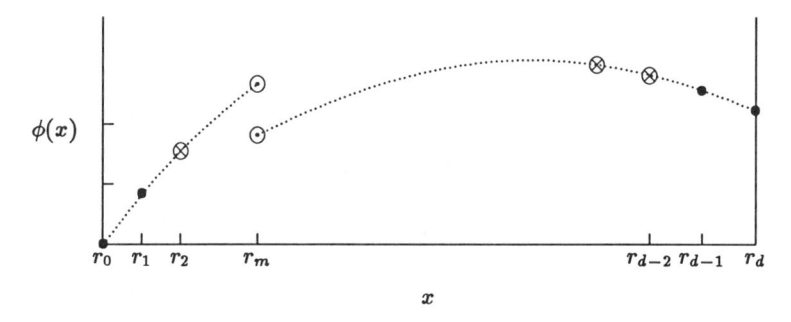

Fig. 8-8 Illustration of the solution of a boundary value problem using the bidirectional shooting method. The differential equation is treated as an initial value problem from both boundaries, at $r = r_0$ and $r = r_d$, by starting with estimates of the missing "initial" conditions. The estimates are improved, step by step, by comparing the two solutions at the matching radius r_m.

step size can be taken to be much larger in the outside region, where there is no interaction, and the division provides us with a natural way to do so. Finally, it may happen that, in certain problems, the truncation errors may accumulate and, as a result, it is not possible to propagate the solution all the way from one end of the interval to the other. In this case, it is helpful to start from both ends so that no single solution is propagated beyond where truncation errors become a serious problem.

The basic principle behind the method of bidirectional shooting is quite straightforward. In each subinterval, the differential equation is solved as an initial value problem. In addition to the boundary condition provided by the physical problem, we introduce a second one based on estimates. Instead of the interval $[0, r_d]$ for our scattering problem, we can think more generally of an interval $r = [r_b, r_d]$ and divide it into two parts $[r_b, r_m]$ and $[r_m, r_d]$, with r_m as the matching radius. Let us use the symbol β to represent the supplemental condition we must have in the subinterval $[r_b, r_m]$ and the symbol γ for the corresponding quantity in the other subinterval $[r_m, r_d]$. At $r = r_m$, we have the logarithmic derivative $L^\ell(\beta)$ from the trial solution in the subinterval $[r_b, r_m]$ on the "left." Similarly, from the trial solution in the subinterval $[r_m, r_d]$ on the "right," we have the logarithmic derivative $L^r(\gamma)$ at the same point. If

$$L^\ell(\beta) = L^r(\gamma) \tag{8-85}$$

we have a solution for the whole region. The only question remaining now is to find a way to improve upon the estimated values of both β and γ so that the two logarithmic derivatives match at $r = r_m$.

Matching the logarithmic derivatives. The actual implementation of the matching condition at $r = r_m$ requires some care. We shall use a simple algorithm here that relies on the fact that, for most physical problems, we have the freedom of choosing the matching point at a place where the solution is well behaved. Furthermore, we can also depend on the physics to provide us with good initial estimates of the values of β and γ. Let us begin with two different estimates, β_1 and β_2, of the values of du_0/dr at $r = r_b$ and, similarly, two different estimates, γ_1 and γ_2, at $r = r_d$. It is unlikely that we have chanced upon a set of choices that satisfy (8-85). As a result, we need a recipe to find successively better estimates for the values of β and γ.

The simple method described in Box 8-7 is constructed in such a way that, for each of the two subintervals $[r_b, r_m]$ and $[r_m, r_d]$, we can use the method of Box 8-6 to arrive at the next set of values for β and γ. For this purpose, each region requires a value at $r = r_m$ for the solutions to "shoot" for. As far as each subinterval is concerned, this is the boundary condition at the other end that the solution must attempt to meet. There are several ways to provide such an artificial boundary condition. A simple approach is to take, as a start, the average of the logarithmic derivatives calculated in the subinterval $[r_b, r_m]$ with estimates β_1 and β_2. This gives us the value

$$B_m = \tfrac{1}{2}\{L^\ell(\beta_1) + L^\ell(\beta_2)\} \tag{8-86}$$

as the "boundary condition" for the solution from the other subinterval, $r = [r_m, r_d]$, to shoot for. Similarly, from the values of $L^r(\gamma_1)$ and $L^r(\gamma_2)$ we can obtain the value $\Delta\gamma$ from the differences

$$R(\gamma_j) = L^r(\gamma_j) - B_m \tag{8-87}$$

in an analogous manner as in (8-83). This gives us a new estimate γ_{j+1} for the value of du_0/dr at $r = r_d$.

We can now use the value of $L^r(\gamma_{j+1})$, obtained from solving the equation in the subinterval $[r_m, r_d]$, as the value B_m for the solution of the $[r_b, r_m]$ region to shoot for. Using the logarithmic derivatives $L^\ell(\beta_i)$ for $i = 1$ and 2 we already have, a new value of β_{i+1} may be obtained. This is used to obtain a new value $L^\ell(\beta_{i+1})$, which can serve as the value of B_m for the other subinterval. The process of iterating between the two subintervals $[r_b, r_m]$ and $[r_d, r_m]$ continues until we achieve convergence as defined by (8-85).

For our square-well example, we have seen that the solution in the subinterval $[r_b, r_m]$ is very stable. Consequently, one can arrive at the "final" value of the logarithmic derivative L^ℓ without much effort. The main part of the work involved is to find a value of γ such that $L^r(\gamma) = L^\ell$. The calculation depends very much on the question of how well we can estimate the values of γ_1 and γ_2 to start with. If the values are too far off from those required for convergence, the method of finding the next γ using (8-82) may fail and the calculation diverges. Fortunately,

---

Box 8-7 Subroutine MATCH

Solution of second-order ODEs by bidirectional shooting

Example of scattering from a potential well

Subprograms used:
  EXT_STP: Solution of n first-order ODEs by extrapolation (Box 8-5).
  EXT_ZRO: Extrapolation to zero step size, using Neville's algorithm (Box 3-2).
  MID_PTM: Modified midpoint method for a system of n first-order ODEs.
    FOUT: Value of $f(y,t)$ for the outside region.
    FINS: Value of $f(y,t)$ for the inside region.

Initialization:
  (a) Divide the region into two, $[r_b, r_m]$ and $[r_m, r_d]$, by selecting a matching radius r_m.
  (b) Set the boundary conditions as $y_0 = 0$ and $y_d = 1$.
  (c) Make estimates for the values of β_1, β_2, γ_1, and γ_2.

1. Two initial solutions for the inside region $r = [r_b, r_m]$ using EXT_STP:
  (a) Obtain a solution using y_0 and β_1 as the initial conditions.
  (b) Obtain another solution with y_0 and β_2 as the initial conditions.
  (c) Calculate the logarithmic derivatives for both cases at r_m.
  (d) Calculate the value of B_m of (8-86).

2. Two solutions for the outside region $r = [r_m, r_d]$ using EXT_STP:
  (a) Obtain a solution using y_d and γ_1 as the initial conditions.
  (b) Obtain another solution using y_0 and γ_2 as the initial conditions.
  (c) Calculate the logarithmic derivatives for both cases at r_m.
  (d) Find the difference $R(\gamma_i)$ of (8-87).
  (e) Use a modified form of (8-83) to calculate $\Delta\gamma$ so as to improve the match.

3. If $\Delta\gamma$ is larger than the tolerance for error, obtain a solution for the outside region with the new γ and calculate the logarithmic derivative $L^r(\gamma)$ again.

4. Improve the estimate for the region $[r_b, r_m]$ using $B_m = L^r(\gamma)$ as the boundary condition at $r = r_m$.
  (a) Find $\Delta\beta$ required to match B_m.
  (b) If $\Delta\beta$ is larger than the tolerance for error, obtain a solution with the new β value.
  (c) Calculate the logarithmic derivative $L^\ell(\beta)$ and use this value as the boundary condition at $r = r_m$ for the region $[r_m, r_d]$.

5. Check if $L^\ell(\beta_i)$ and $L^r(\gamma_j)$ are sufficiently close to each other.
  (a) If so, convergence is achieved.
    (i) Normalize the solutions in two subintervals by setting $y^\ell(r_m) = y^r(r_m)$.
    (ii) Return the results as the solution for the complete interval $[r_b, r_d]$.
  (b) If not, repeat steps 3 and 4 with the new values of B_m until convergence is achieved or the maximum number of iterations is exceeded.

---

we can usually rely on the physics to guide us to good estimates. For example, if we wish to use the Woods-Saxon well of (8-68) for our scattering problem, we can make use of the known analytical solution of an equivalent square well as a guide. As can be seen from Fig. 8-9, the two solutions can be made similar to each other in the region of small r.

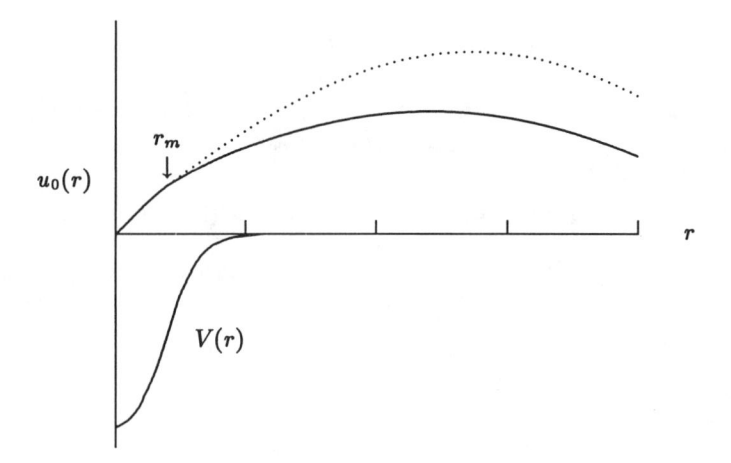

Fig. 8-9 Radial wave function for a Woods-Saxon potential (solid curve) compared with that of a square-well potential (dotted curve). The logarithmic derivatives of both functions are roughly the same at the matching radius r_m and the shapes of the wave functions are essentially indistinguishable for $r < r_m$. The differences outside are exaggerated since the two wave functions are normalized differently so that they more or less overlap each other in the inside region.

The idea of subdividing a region into two or more parts is, in general, a useful one for solving differential equations using the shooting methods. There is no reason for us to stop at two subdivisions, except that the physical problem we used as an example does not really require it. A description of such "multiple" shooting methods can be found in Sewell (*The Numerical Solution of Ordinary and Partial Differential Equations*, Academic Press, San Diego, California, 1988).

§8-6 Relaxation methods

In the previous section, a way was found to solve boundary value problems using the methods for initial value problems. An alternative is to treat the finite difference equation generated from the differential equation as a set of algebraic equations and use methods similar to those discussed in Chapter 5 to find the roots. To take advantage of this approach of solving an ordinary differential equation, we need again to construct a fixed mesh of $(N+1)$ points for the independent variable x. To facilitate the discussion, we shall take these points to be x_0, x_1, ..., x_N, in the interval $[x_0, x_N]$ of interest to us. The values of the dependent variable, $\phi(x_0)$, $\phi(x_1)$, ..., $\phi(x_N)$, at each of these points are treated as the unknowns of a system of algebraic equations. There are several ways to find the roots of this set of equations, and we shall concentrate on relaxation methods. Among other advantages, these methods can be extended in a straightforward manner to handle partial differential equations.

To be specific, we shall use, as the prototype for our discussion, the second-order differential equation

$$\frac{d^2\phi}{dx^2} + g(x)\phi(x) = 0 \tag{8-88}$$

The function $g(x)$ can be either a constant or a function of x. The former corresponds to the case of (8-15), for which the solution is related to a sinusoidal wave form, and the latter is represented by (8-67) for the modified radial wave function of scattering from a Woods-Saxon potential. If $g(x)$ is a function of $\phi(x)$ or its derivatives, or both, it is a nonlinear differential equation, and we shall delay any discussion of such cases till §8-10.

To convert the second-order derivative in (8-88) to finite differences, we can make use of the central difference formula of (2-52). For the case of a constant step size, $h = (x_{k+1} - x_k)$, the result is

$$\left.\frac{d^2\phi}{dx^2}\right|_{x=x_k} \longrightarrow \frac{1}{h^2}\{\phi(x_{k+1}) - 2\phi(x_k) + \phi(x_{k-1})\} \tag{8-89}$$

For unequal step sizes, the result is slightly more complicated:

$$\left.\frac{d^2\phi}{dx^2}\right|_{x=x_k} \longrightarrow \frac{2}{h_{k+1}h_k(h_{k+1}+h_k)}\{h_k\phi(x_{k+1}) - (h_{k+1}+h_k)\phi(x_k) + h_{k+1}\phi(x_{k-1})\} \tag{8-90}$$

where $h_k = (x_k - x_{k-1})$ and we have approximated $(x_{k+1/2} - x_{k-1/2})$ with $\frac{1}{2}(h_{k+1}+h_k)$. In both cases, the $(N+1)$ mesh points are labeled from x_0 to x_N, as illustrated in Fig. 8-10. Since there are two boundary conditions for a second-order ODE, the number of unknowns in this set of equations is $(N-1)$.

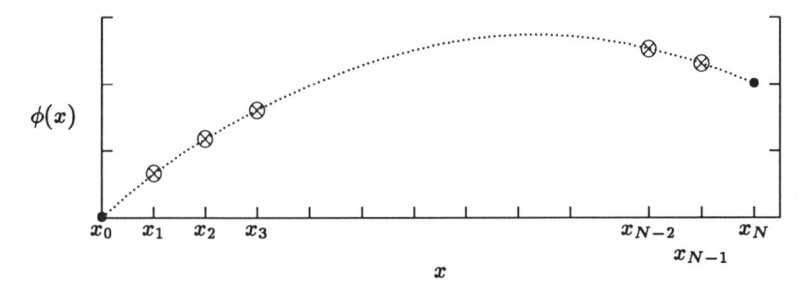

Fig. 8-10 Changing a differential equation into a set of finite difference algebraic equations. The values of the dependent variable $\phi(x)$ are calculated only at a discrete set of points $x_0, x_2, \ldots, x_N$.

To simplify the discussion, we shall write most of the formulas in this section for the case of a constant step size. However, there is no difficulty in adapting the results to the case of variable step sizes. In terms of a set of FDEs, (8-88) takes on the form

$$\phi(x_{k+1}) - 2\phi(x_k) + \phi(x_{k-1}) + h^2 g(x_k)\phi(x_k) = 0 \tag{8-91}$$

Altogether there are $(N-1)$ such equations, as k can take on values from 1 to $(N-1)$. The total number of different $\phi(x_k)$ is $(N+1)$ as $k = 0, 1, \ldots, N$. Instead of using the two boundary conditions to reduce the number of unknown quantities to $(N-1)$, it is more convenient for us to treat them as two additional equations. For example, they may be written in the form

$$\phi(x_0) = \alpha_0 \qquad\qquad \phi(x_N) = \alpha_N \tag{8-92}$$

As a result, we have a system of $(N+1)$ equations for the same number of unknown quantities. There is, therefore, no problem in principle in obtaining all the roots $\phi(x_0), \phi(x_1), \ldots, \phi(x_N)$. However, N can be a large number here, and techniques such as those discussed in Chapter 5 cannot be used without some modification.

To devise a method to solve the FDE, it is useful to write the $(N-1)$ equations of (8-91) together with the two boundary conditions of (8-92) in matrix notation:

$$\phi(x_0) = \alpha_0$$
$$-\phi(x_{k-1}) + \{2 - h^2 g(x_k)\}\phi(x_k) - \phi(x_{k+1}) = 0 \qquad \text{for} \quad k = 1, 2, \ldots, (N-1)$$
$$\phi(x_N) = \alpha_N$$

In terms of matrices, this may be written as

$$AY = b \tag{8-93}$$

Here A is a square matrix of dimension $(N+1)$,

$$A = \begin{pmatrix} 1 & 0 & 0 & 0 & \cdots & 0 \\ -1 & \eta_1 & -1 & 0 & \cdots & 0 \\ 0 & -1 & \eta_2 & -1 & \cdots & 0 \\ \vdots & \vdots & \vdots & \vdots & \ddots & \vdots \\ 0 & \cdots & \cdots & -1 & \eta_{N-1} & -1 \\ 0 & \cdots & \cdots & \cdots & 0 & 1 \end{pmatrix}$$

with

$$\eta_k \equiv \{2 - h^2 g(x_k)\}$$

Both Y and b are column matrices with $(N+1)$ elements,

$$Y = \begin{pmatrix} \phi(x_0) \\ \phi(x_1) \\ \phi(x_2) \\ \vdots \\ \phi(x_{N-1}) \\ \phi(x_N) \end{pmatrix} \qquad b = \begin{pmatrix} \alpha_0 \\ 0 \\ 0 \\ \vdots \\ 0 \\ \alpha_N \end{pmatrix}$$

The main feature to note here is that, although N may be a large number, the fraction of nonvanishing matrix elements in each row of A is small, no more than three in the particular example we have here. The matrix A is, therefore, sparse. In general, the structure of A may be more complicated than our prototype here, but the fact that the matrix is sparse remains. Relaxation methods, and others based on them, take advantage of this feature of the matrix equation to solve boundary value problems.

Iterative method for solving a system of linear algebraic equations. An efficient way to solve a system containing a large number of linear algebraic equations is to use an iterative approach. It is convenient to illustrate first the basic principles involved using a system of only four equations. In terms of numerical solutions to differential equations, this corresponds to the unrealistic situation of having only four points on the mesh. This is not a problem, as it is easy to extend the idea to the case of an arbitrary number of equations, as we shall see below.

The general form of a system of four equations of the form of (8-93) may be written as

$$A_{1,1}Y_1 + A_{1,2}Y_2 + A_{1,3}Y_3 + A_{1,4}Y_4 = b_1$$
$$A_{2,1}Y_1 + A_{2,2}Y_2 + A_{2,3}Y_3 + A_{2,4}Y_4 = b_2$$
$$A_{3,1}Y_1 + A_{3,2}Y_2 + A_{3,3}Y_3 + A_{3,4}Y_4 = b_3$$
$$A_{4,1}Y_1 + A_{4,2}Y_2 + A_{4,3}Y_3 + A_{4,4}Y_4 = b_4 \qquad (8\text{-}94)$$

It is always possible to arrange the order of the equations in such a way that the diagonal elements of A are nonvanishing. We shall therefore proceed with such an assumption in the following discussions. This gives us the possibility of rewriting (8-94) in the form

$$Y_1 = \frac{1}{A_{1,1}}\{b_1 - A_{1,2}Y_2 - A_{1,3}Y_3 - A_{1,4}Y_4\}$$
$$Y_2 = \frac{1}{A_{2,2}}\{b_2 - A_{2,1}Y_1 - A_{2,3}Y_3 - A_{2,4}Y_4\}$$
$$Y_3 = \frac{1}{A_{3,3}}\{b_3 - A_{3,1}Y_1 - A_{3,2}Y_2 - A_{3,4}Y_4\}$$
$$Y_4 = \frac{1}{A_{4,4}}\{b_4 - A_{4,1}Y_1 - A_{4,2}Y_2 - A_{4,3}Y_3\} \qquad (8\text{-}95)$$

This is the starting point for the various methods of solving (8-93) by iterative procedures.

All the methods of solution in this category begin with a set of initial guesses $Y^{(0)}$. In physical problems, this set may come from the analytical solution of a similar situation. For example, if we are interested in the problem of the previous section of obtaining the modified radial wave function in quantum mechanics for a particle scattering off a Woods-Saxon potential, we can use the solution for an equivalent square-well potential as $Y^{(0)}$. Unless we are dealing with an ill-conditioned case, the final solution is, in principle, independent of the initial estimate. On the other hand, a good starting point is essential if we want to minimize the number of iterations required to reach a given accuracy. The iterative methods do not concern themselves with the question of the initial set of "trial" solutions. Their main mission is to give us a way by which we can obtain an improved solution $Y^{(i+1)}$, given an approximate one $Y^{(i)}$. Starting from the trial solution $Y^{(0)}$, we can obtain a solution $Y^{(1)}$ that is a step closer to the final solution. In turn, $Y^{(1)}$ may be used to obtain $Y^{(2)}$, which is a further improvement, and so on. Convergence is achieved when the difference between $Y^{(i+1)}$ and $Y^{(i)}$ is less than some small, positive number ϵ, determined by the requirement of the problem and the accuracy that can be achieved with a particular type of computer.

There are several ways to characterize the differences between two successive iterations and, hence, the manner in which we want the iterative calculation to converge. For example, we can require that the average absolute values of the differences in each term be less than some small, positive number ϵ. Alternatively, we can require the root-mean-square (rms) difference be less than ϵ. As a third possibility, we may demand that the difference at any point in the interval be less than ϵ. In the last case, the maximum difference among all the points becomes our criterion for convergence. The choice between these three methods, and other variants of them, should be determined by the physical requirement of the problem, and we shall not elaborate on the point here. For our example in this section we shall adopt the first one: the average of the absolute difference between $Y^{(i+1)}(x_j)$ and $Y^{(i)}(x_j)$ for all the points x_j should be less than ϵ.

The simplest type of iterative formula we can construct based on (8-95) is

$$Y_1^{(i+1)} = \frac{1}{A_{1,1}}\left\{b_1 - A_{1,2}Y_2^{(i)} - A_{1,3}Y_3^{(i)} - A_{1,4}Y_4^{(i)}\right\}$$

$$Y_2^{(i+1)} = \frac{1}{A_{2,2}}\left\{b_2 - A_{2,1}Y_1^{(i)} - A_{2,3}Y_3^{(i)} - A_{2,4}Y_4^{(i)}\right\}$$

$$Y_3^{(i+1)} = \frac{1}{A_{3,3}}\left\{b_3 - A_{3,1}Y_1^{(i)} - A_{3,2}Y_2^{(i)} - A_{3,4}Y_4^{(i)}\right\}$$

$$Y_4^{(i+1)} = \frac{1}{A_{4,4}}\left\{b_4 - A_{4,1}Y_1^{(i)} - A_{4,2}Y_2^{(i)} - A_{4,3}Y_3^{(i)}\right\} \qquad (8\text{-}96)$$

That is, all the input for calculating the next set of approximate results $\{Y_j^{(i+1)}\}$ comes from those of the previous iteration $\{Y_j^{(i)}\}$. This is known as the Jacobi method. For the more realistic situation of N algebraic equations, where N may be some number much larger than 4, the same set of equations may be written as

$$Y_j^{(i+1)} = \frac{1}{A_{j,j}}\left\{b_j - \sum_{k=1}^{j-1} A_{j,k} Y_k^{(i)} - \sum_{k=j+1}^{N} A_{j,k} Y_k^{(i)}\right\} \qquad (8\text{-}97)$$

A proof of the convergence of this method can be found in Dahlquist and Björck (*Numerical Methods*, Prentice Hall, Englewood Cliffs, New Jersey, 1974).

We can make a slight improvement on the Jacobi method in the following way. Once $Y_1^{(i+1)}$ is found when we finish the calculations for the first one of the four equations of (8-96), the result may be used in solving the other three equations. Similarly, $Y_2^{(i+1)}$ may be used for solving the last two equations, and so on. The resulting equations for our $N = 4$ example are

$$Y_1^{(i+1)} = \frac{1}{A_{1,1}}\left\{b_1 - A_{1,2} Y_2^{(i)} - A_{1,3} Y_3^{(i)} - A_{1,4} Y_4^{(i)}\right\}$$

$$Y_2^{(i+1)} = \frac{1}{A_{2,2}}\left\{b_2 - A_{2,1} Y_1^{(i+1)} - A_{2,3} Y_3^{(i)} - A_{2,4} Y_4^{(i)}\right\}$$

$$Y_3^{(i+1)} = \frac{1}{A_{3,3}}\left\{b_3 - A_{3,1} Y_1^{(i+1)} - A_{3,2} Y_2^{(i+1)} - A_{3,4} Y_4^{(i)}\right\}$$

$$Y_4^{(i+1)} = \frac{1}{A_{4,4}}\left\{b_4 - A_{4,1} Y_1^{(i+1)} - A_{4,2} Y_2^{(i+1)} - A_{4,3} Y_3^{(i+1)}\right\}$$

This is known as the Gauss-Seidel method. The form for an arbitrary N may be written as

$$Y_j^{(i+1)} = \frac{1}{A_{j,j}}\left\{b_j - \sum_{k=1}^{j-1} A_{j,k} Y_k^{(i+1)} - \sum_{k=j+1}^{N} A_{j,k} Y_k^{(i)}\right\} \qquad (8\text{-}98)$$

Since the values of $Y_j^{(i+1)}$ are closer to the final values of Y_j, we expect a better convergence property than the Jacobi method as a result.

A more popular way to solve FDEs obtained from a set of differential equations is to use one of the relaxation methods. This may be derived by adding $Y_j^{(i)}$ to the right side of (8-98) outside the braces and subtracting the same amount from inside. The results for our $N=4$ example take on the form

$$Y_1^{(i+1)} = Y_1^{(i)} + \frac{1}{A_{1,1}}\left\{b_1 - A_{1,1}Y_1^{(i)} - A_{1,2}Y_2^{(i)} - A_{1,3}Y_3^{(i)} - A_{1,4}Y_4^{(i)}\right\}$$

$$Y_2^{(i+1)} = Y_2^{(i)} + \frac{1}{A_{2,2}}\left\{b_2 - A_{2,1}Y_1^{(i+1)} - A_{2,2}Y_2^{(i)} - A_{2,3}Y_3^{(i)} - A_{2,4}Y_4^{(i)}\right\}$$

$$Y_3^{(i+1)} = Y_3^{(i)} + \frac{1}{A_{3,3}}\left\{b_3 - A_{3,1}Y_1^{(i+1)} - A_{3,2}Y_2^{(i+1)} - A_{3,3}Y_3^{(i)} - A_{3,4}Y_4^{(i)}\right\}$$

$$Y_4^{(i+1)} = Y_4^{(i)} + \frac{1}{A_{4,4}}\left\{b_4 - A_{4,1}Y_1^{(i+1)} - A_{4,2}Y_2^{(i+1)} - A_{4,3}Y_3^{(i+1)} - A_{4,4}Y_4^{(i)}\right\}$$

For an arbitrary N, the corresponding equations may be written as

$$Y_j^{(i+1)} = Y_j^{(i)} + \frac{1}{A_{j,j}}\left\{b_j - \sum_{k=1}^{j-1} A_{j,k}Y_k^{(i+1)} - \sum_{k=j}^{N} A_{j,k}Y_k^{(i)}\right\} \qquad (8\text{-}99)$$

We may regard the second of the two terms on the right side as the correction to $Y_j^{(i)}$ for us to improve our solution for Y_j. As a result, it may by possible to modify the rate of convergence by adding a *relaxation* parameter ω as the weighting factor to the quantities inside the braces:

$$Y_j^{(i+1)} = Y_j^{(i)} + \frac{\omega}{A_{j,j}}\left\{b_j - \sum_{k=1}^{j-1} A_{j,k}Y_k^{(i+1)} - \sum_{k=j}^{N} A_{j,k}Y_k^{(i)}\right\} \qquad (8\text{-}100)$$

For $\omega < 1$, it is known as the underrelaxation method. For $\omega = 1$, we return to the Gauss-Seidel method. The main interest here is in $\omega > 1$, the overrelaxation method. Since it often happens that the successive corrections to Y_j are of the same sign, it is quite likely that overrelaxation, with $1 < \omega < 2$, provides a faster way toward convergence than the Gauss-Seidel method. For this reason, successive overrelaxation (SOR) is a method of choice for boundary value problems and the factor ω is also called the *acceleration* parameter. The algorithm, as applied to an example of scattering from a Woods-Saxon potential, is given in Box 8-8.

Further Developments. To develop a method for solving the algebraic equations, let us decompose the square matrix $\boldsymbol{A}$ in (8-93) into three parts: a diagonal matrix $\boldsymbol{D}$, consisting of the diagonal elements of $\boldsymbol{A}$, a lower diagonal matrix $\boldsymbol{L}$, made of the negative of all the elements of $\boldsymbol{A}$ below the diagonal, and an upper diagonal matrix $\boldsymbol{U}$, made of the negative of all the elements of $\boldsymbol{A}$ above the diagonal. In terms of the elements $A_{i,j}$ of the matrix $\boldsymbol{A}$, we have

$$\boldsymbol{D} = \begin{pmatrix} A_{1,1} & 0 & 0 & \cdots & 0 \\ 0 & A_{2,2} & 0 & \cdots & 0 \\ \vdots & \vdots & \ddots & \cdots & \cdots \\ 0 & \cdots & \cdots & A_{N-1,N-1} & 0 \\ 0 & \cdots & \cdots & 0 & A_{N,N} \end{pmatrix}$$

$$\boldsymbol{L} = -\begin{pmatrix} 0 & \cdots & \cdots & \cdots & 0 \\ A_{2,1} & 0 & \cdots & \cdots & 0 \\ A_{3,1} & A_{3,2} & 0 & \cdots & 0 \\ \vdots & \vdots & \ddots & \vdots & \vdots \\ A_{N,1} & A_{N,2} & \cdots & A_{N,N-1} & 0 \end{pmatrix}$$

$$\boldsymbol{U} = -\begin{pmatrix} 0 & A_{1,2} & A_{1,3} & \cdots & A_{1,N} \\ 0 & 0 & A_{2,3} & \cdots & A_{2,N} \\ \vdots & \vdots & \ddots & \vdots & \vdots \\ 0 & \cdots & \cdots & \cdots & A_{N-1,N} \\ 0 & \cdots & \cdots & \cdots & 0 \end{pmatrix}$$

Since, by definition,
$$\boldsymbol{A} = \boldsymbol{D} - \boldsymbol{L} - \boldsymbol{U}$$

we can rewrite (8-93) in the form

$$\boldsymbol{DY} = (\boldsymbol{L} + \boldsymbol{U})\boldsymbol{Y} + \boldsymbol{b} \tag{8-101}$$

In terms of $\boldsymbol{D}$, $\boldsymbol{L}$, and $\boldsymbol{U}$, the Jacobi method of (8-97) may now be expressed as

$$\boldsymbol{DY}^{(i+1)} = (\boldsymbol{L} + \boldsymbol{U})\boldsymbol{Y}^{(i)} + \boldsymbol{b}$$

Since $\boldsymbol{D}$ is diagonal, it can be easily inverted. Using the inverse of $\boldsymbol{D}$, we have

$$\boldsymbol{Y}^{(i+1)} = \boldsymbol{D}^{-1}(\boldsymbol{L} + \boldsymbol{U})\boldsymbol{Y}^{(i)} + \boldsymbol{D}^{-1}\boldsymbol{b} \tag{8-102}$$

The product of matrices,
$$\boldsymbol{J} \equiv \boldsymbol{D}^{-1}(\boldsymbol{L} + \boldsymbol{U}) \tag{8-103}$$

is known as the point Jacobi iteration matrix, as it is related to the solution of a continuous function $\phi(x)$ by its values at discrete points. Note that all the elements of $\boldsymbol{b}$ are known and are not changed in value from one iteration to another. Furthermore, for linear equations, the matrix elements of $\boldsymbol{A}$ are not dependent on those of $\boldsymbol{Y}$, and the matrices $\boldsymbol{D}$, $\boldsymbol{L}$, and $\boldsymbol{U}$, are also not changed from one iteration to another. However, for FDEs produced from nonlinear differential equations in general, this is not true.

In terms of D, L, and U, the Gauss-Seidel method of (8-98) may be written as

$$DY^{(i+1)} = LY^{(i+1)} + UY^{(i)} + b \qquad (8\text{-}104)$$

The analogous form to (8-102) is

$$Y^{(i+1)} = (D-L)^{-1}UY^{(i)} + (D-L)^{-1}b \qquad (8\text{-}105)$$

where the product

$$G \equiv (D-L)^{-1}U$$

is known as the point Gauss-Seidel iteration matrix.

The matrix form of the overrelaxation method may be constructed from (8-104) by multiplying both sides by D^{-1}. To the right side, we can add and subtract the term $Y^{(i)}$, the same as we did to obtain (8-99). After including the relaxation parameter ω, we have the equivalent relation to (8-100):

$$Y^{(i+1)} = Y^{(i)} + \omega D^{-1}\{LY^{(i+1)} + UY^{(i)} - DY^{(i)} + b\} \qquad (8\text{-}106)$$

To construct the point iteration matrix here, we can move all the terms involving $Y^{(i+1)}$ to the left side and obtain the result

$$(1 - \omega D^{-1}L)Y^{(i+1)} = \{(1-\omega)1 + \omega D^{-1}U\}Y^{(i)} + \omega D^{-1}b$$

Analogous to (8-102), we have the relation

$$Y^{(i+1)} = R(\omega)Y^{(i)} + \omega(1 - \omega D^{-1}L)^{-1}D^{-1}b \qquad (8\text{-}107)$$

where the SOR iteration matrix

$$R(\omega) \equiv (1 - \omega D^{-1}L)^{-1}\{(1-\omega)1 + \omega D^{-1}U\}$$

is a function of the relaxation parameter ω.

All three iterative equations, (8-102), (8-105), and (8-107), may be put into the general form

$$Y^{(i+1)} = PY^{(i)} + q \qquad (8\text{-}108)$$

where P is the iteration matrix for each case. As we saw earlier, the matrix q is obtained from b and does not change from one iteration to another. For our needs later, we shall put this iterative equation into a form similar to an eigenvalue equation:

$$Y = PY + q$$

This is equivalent to reversing the steps leading from (8-94) to (8-95). The relation is essentially the same as that given by (8-93).

Convergence. For a discussion on the convergence criteria, we can work with the difference

$$e^{(i)} \equiv Y - Y^{(i)} \tag{8-109}$$

where Y is the exact solution and $Y^{(i)}$ is the result of iteration i. In terms of $e^{(i)}$, we may write (8-108) in the form

$$e^{(i+1)} = P e^{(i)} \tag{8-110}$$

as q is unchanged by the iterative calculations. This relation may be interpreted in the following way. Starting with an initial estimate of the solution $Y^{(0)}$ and the associated difference $e^{(0)}$, we can generate $Y^{(1)}$ and, hence, $e^{(1)}$ using (8-110). When this process is repeated k times, we obtain the result

$$e^{(k)} = P e^{(k-1)} = P^2 e^{(k-2)} = \cdots = P^k e^{(0)}$$

We can demonstrate that the process is convergent if we can show that

$$\lim_{k \to \infty} e^{(k)} = 0 \tag{8-111}$$

Since $Y^{(0)}$, and hence $e^{(0)}$, is arbitrary the only way that this limiting relation can be satisfied is that

$$\lim_{k \to \infty} P^k = 0$$

This is, then, the convergence criterion we are seeking for the iterative methods to be used for solving boundary value problems in this section.

It is easier to restate the same criterion in terms of the eigenvalues of the matrix P. Let λ_j be an eigenvalue of P and v_j be the corresponding eigenvector. That is,

$$P v_j = \lambda_j v_j \tag{8-112}$$

This is true for $j = 0, 1, 2, \ldots, N$. Since $v_0, v_1, v_2, \ldots, v_N$ form a complete set of states in the $(N+1)$-dimensional space, we can express the difference between the exact solution and our initial estimate in terms of them. The result is

$$e^{(0)} = \sum_{\ell=0}^{N} c_\ell^{(0)} v_\ell$$

where the expansion coefficients $c_\ell^{(0)}$ are constants whose actual values are not of immediate concern to us at the moment. Using this expansion, together with the relation of (8-112), we have

$$e^{(1)} = P e^{(0)} = \sum_{\ell=0}^{N} c_\ell^{(0)} P v_\ell = \sum_{\ell=0}^{N} c_\ell^{(0)} \lambda_\ell v_\ell$$

By repeating the step k times, we find that

$$e^{(k)} = \sum_{\ell=0}^{N} c_\ell^{(0)} (\lambda_\ell)^k v_\ell$$

The only way for this equation to satisfy the requirement of (8-111) is that

$$|\lambda_\ell| < 1$$

for all ℓ. An alternative statement to the same effect is to define a quantity for the matrix P called *spectral radius*, given by the maximum absolute value of the eigenvalues:

$$\rho(P) = \max_{\ell=0,1,\ldots,N} |\lambda_\ell| \qquad (8\text{-}113)$$

The condition of convergence for the iterative methods of this section may be stated now as the requirement for the spectral radius of the corresponding point iteration matrix to be less than unity.

Optimum relaxation parameter. The condition of convergence, however, does not give us any indication of the rate for a solution to reach convergence. For SOR, it is known that the optimum value of the relaxation parameter is

$$\omega_x = \frac{2}{1 + \sqrt{1 - \rho^2(J)}} \qquad (8\text{-}114)$$

where $\rho(J)$ is the spectral radius of the point Jacobi iteration matrix J of (8-103). A discussion on how to arrive at this result is given in Smith (*Numerical Solution of Partial Differential Equations: Finite Difference Methods*, Oxford University Press, Oxford, 1985).

For our prototype equation (8-88), the matrix A is tridiagonal as can be seen from (8-93). The diagonal elements of D are 1, η_1, η_2, ..., η_{N-1}, and 1. Since it is a diagonal matrix by definition, the inverse D^{-1} is also diagonal and consists of matrix elements 1, $1/\eta_1$, $1/\eta_2$, ..., $1/\eta_{N-1}$, and 1. The structures of matrices L and U are even simpler, with the only nonvanishing elements equal to -1. For L, they are located just below the diagonal, and for U, just above the diagonal. The Jacobi matrix J is tridiagonal with the diagonal elements equal to zero. However, the matrix is not symmetric and has the form

$$J = - \begin{pmatrix} 0 & 0 & \cdots & & & & \\ \frac{1}{\eta_1} & 0 & \frac{1}{\eta_1} & 0 & \cdots & & \\ 0 & \frac{1}{\eta_2} & 0 & \frac{1}{\eta_2} & \cdots & & \\ \vdots & \vdots & \vdots & \ddots & \vdots & \vdots & \vdots \\ & & & \cdots & \frac{1}{\eta_{N-1}} & 0 & \frac{1}{\eta_{N-1}} \\ & & & & \cdots & 0 & 0 \end{pmatrix}$$

> **Box 8-8 Program SOR_WS**
> **Successive overrelaxation method for boundary value problems**
> Example of scattering from a Woods-Saxon potential
>
> Subprograms used:
>   MESH: Construct a mesh of constant step size.
>   INI_ARY: Initialize the array η of (8-93) and construct a trial solution $Y^{(0)}(x_i)$.
>   RELAX: Solve the FDE for a Woods-Saxon potential using SOR.
>
> Initialization:
>   Use MESH to construct a fixed mesh of points $x_0, x_1, x_2, \ldots, x_N$.
>
> 1. Input energy E of the scattering particle and parameters V_0, r_g, and a for the Woods-Saxon potential of (8-69).
> 2. Use INI_ARY to:
>   (a) Calculate the diagonal elements η_k of A in (8-93).
>   (b) Construct a trial solution $Y^{(0)}(x_i)$.
> 3. Input the relaxation parameter ω of (8-100).
> 4. Use RELAX to carry out the relaxation calculations:
>   (a) Set up the tolerance for error and the maximum number of iterations.
>   (b) Update the solution Y according (8-100) and calculate the estimated error.
>   (c) Check the error,
>   (i) If larger than the tolerance and the total number of iterations is less than the maximum allowed, iterate the solution once more.
>   (ii) Otherwise, return the solution.

A slight modification of the bisection technique described in §3-6 and §5-5 may be used to find the eigenvalues of this matrix. If the step size is constant for the entire interval, the tridiagonal matrix may be mapped to the form

$$T = \begin{pmatrix} a & b & 0 & \cdots & & 0 \\ c & a & b & 0 & \cdots & 0 \\ 0 & c & a & b & \cdots & 0 \\ \vdots & \ddots & \ddots & \ddots & \ddots & \vdots \\ 0 & \cdots & 0 & c & a & b \\ 0 & \cdots & & 0 & c & a \end{pmatrix}$$

The eigenvalues, $\lambda_0, \lambda_1, \ldots, \lambda_N$, of such a matrix are given by

$$\lambda_k = a + 2\sqrt{bc}\, \cos \frac{k\pi}{N+2} \qquad (8\text{-}115)$$

where $(N+1)$ is the dimension of the matrix. For our matrix J, we have $a = 0$ and $b = c \approx \frac{1}{2}$. If there are 40 equal-size steps in the interval, the largest eigenvalue is

$\lambda_1 = \cos \pi/41 = 0.997$. This gives a value of $\omega_x \approx 1.9$ for the optimum relaxation parameter. A more precise calculation using the exact values of b and c gives the value 1.87. The proof of (8-115) is also given in Smith. A plot of the actual number of iterations taken in the calculation outlined in Box 8-8 is given in Fig. 8-11.

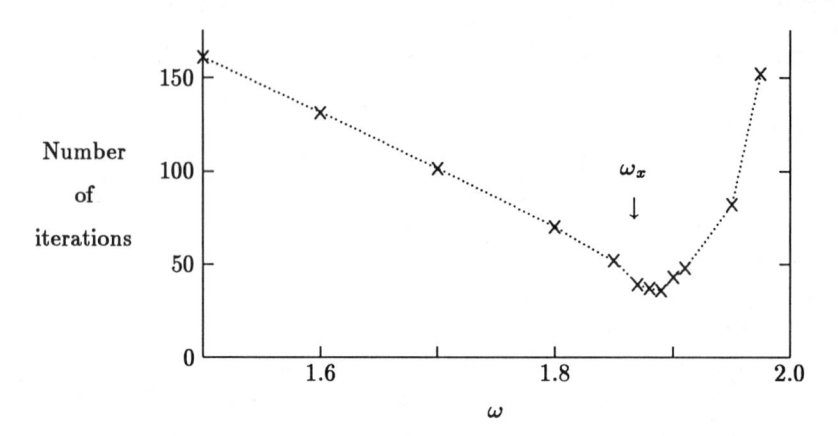

Fig. 8-11 Number of iterations as a function of the relaxation parameter ω required to reach an accuracy of $\epsilon = 9 \times 10^{-5}$ for the differential equation of (8-67). The potential is the Woods-Saxon well of (8-69). An evenly spaced grid of 40 points in the interval $[0, 3]$ is used. The location of the optimum value of $\omega_x = 1.87$, given by (8-114), is indicated by the arrow. Calculation using a variable ω, given by (8-116), requires about the same number of iterations as the minimum value of 36 reached at $\omega = 1.89$. The dotted curve is drawn to guide the eye.

In general, the matrix $\boldsymbol{J}$ is sparse but may be more complicated in form than the tridiagonal one in our example here. Nevertheless, it is still quite possible to find the maximum absolute value of the eigenvalues relatively easily. Furthermore, since we are only using the spectral radius as a way to locate the optimum value of ω in the range $1 \leq \omega < 2$, we do not need highly accurate results.

In practice, the value of ω_x given above provides the fastest convergence only toward the end of a calculation using SOR. At the beginning of the iterative process, a smaller value of $\omega \approx 1$ will actually be more efficient. A method, given in Press and coauthors, of varying the value of ω for each iteration in the following manner,

$$\omega^{(i)} = \begin{cases} \dfrac{1}{1 - \frac{1}{2}\rho^2(\boldsymbol{J})} & \text{for } i = 1 \\ \dfrac{1}{1 - \dfrac{\omega^{(i-1)}}{4}\rho^2(\boldsymbol{J})} & \text{for } i > 1 \end{cases} \tag{8-116}$$

is also found to be a very efficient way of minimizing the number of iterations required.

§8-7 Boundary value problems in partial differential equations

The relaxation methods developed in the last section may also be used in cases where there is more than one independent variable. The equations we wish to solve are now partial differential equations (PDEs). As our prototype in this section, we shall use the Poisson equation

$$\nabla^2 V(\mathbf{r}) = -\rho(\mathbf{r})$$

Such an equation arises, for example, in electrostatics when we wish to find the potential $V(\mathbf{r})$ due to a charge distribution $\rho(\mathbf{r})$. For the convenience of discussion, we have absorbed into the definition of $\rho(\mathbf{r})$ a factor of 4π in cgs units or a factor ϵ_0^{-1} in SI units, with ϵ_0 as the permittivity of free space. In terms of the classifications of PDEs, it is an elliptic equation, as we saw in §8-1.

For simplicity we shall restrict ourselves to two spatial dimensions. In a Cartesian coordinate system, the PDE takes on the form

$$\frac{\partial^2 V}{dx^2} + \frac{\partial^2 V}{dy^2} = -\rho(x,y) \tag{8-117}$$

An equation of this form is obtained if the charge distribution is independent of the z-direction as, for example, in the case of an infinite straight line of charge that extends from $-\infty$ to $+\infty$ along the z-axis. To simplify the boundary conditions, we shall place the charge distribution in the middle of a hollow tube made of conducting material, also infinite in length along the z-direction. For convenience, we shall consider the cross-sectional area of the tube to be a square with a length of 2ℓ on each side. If we put the origin of our two-dimensional space at the center of the square, the boundary condition for this problem may be specified as

$$V(x = \pm\ell, y) = V(x, y = \pm\ell) = V_0$$

where V_0 is a constant, and may be taken to be zero by having the conducting tube grounded. This is known as the Dirichlet type of boundary condition as the values of $V(x,y)$ are given on the boundary. Alternatively, as we have seen earlier in the discussions on ODEs, we can specify the boundary condition in terms of the derivatives at the boundary. Other forms are also possible, including linear combinations of $V(x,y)$ and its derivatives.

Using the same procedure as (8-89), we can replace both second-order derivatives in (8-117) by their equivalent finite differences. For this purpose, we must first adopt a two-dimensional mesh of points, such as the one shown schematically in Fig. 8-12. Our discussion is much simpler if we take an evenly spaced grid of size Δx along the x-direction and Δy along the y-direction. The finite difference form of the Poisson equation of (8-117) is then

$$\frac{1}{\Delta x^2}\{V(x_{i+1}, y_j) - 2V(x_i, y_j) + V(x_{i-1}, y_j)\}$$
$$+ \frac{1}{\Delta y^2}\{V(x_i, y_{j+1}) - 2V(x_i, y_j) + V(x_i, y_{j-1})\} = -\rho(x_i, y_j) \tag{8-118}$$

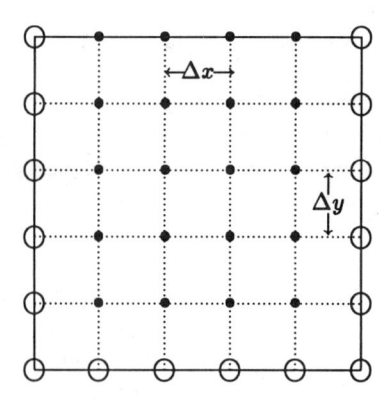

Fig. 8-12 A two-dimensional mesh of points to transform a PDE of two independent variables into a set of FDEs. The size of Δx and Δy may be taken to be different in different regions. The circles represent points where the values of the dependent variable are given by the boundary conditions and dots are points where the values must be obtained by solving the FDEs.

If we follow the procedure of (8-90) instead, we can also obtain a set of FDEs with variable step sizes. The forms of the equations in this case are more complicated, but the methods to solve them essentially remain the same.

If Δy is taken to be equal to Δx, a very simple form of the FDE in (8-118) emerges

$$V(x_{i+1}, y_j) + V(x_{i-1}, y_j) + V(x_i, y_{j+1}) + V(x_i, y_{j-1}) - 4V(x_i, y_j) = -h^2 \rho(x_i, y_j)$$
(8-119)

where $h = \Delta x = \Delta y$. This is the two-dimensional analog of (8-91). The difference between them may be seen by comparing the two diagrams in Fig. 8-13. In the case of a single independent variable, each of the equations represented by (8-91) involves the values of the dependent variable at three adjacent points on the mesh, $\phi(x_{k-1})$, $\phi(x_k)$, and $\phi(x_{k+1})$. We shall call such a group of points a "computational molecule" or "stencil," as shown schematically in Fig. 8-13(a). For the two-dimensional Poisson equation, each of the set of algebraic equations represented by (8-119) involves five points, (i, j), $(i-1, j)$, $(i+1, j)$, $(i, j-1)$, and $(i, j+1)$, centered around the point at (i, j), as shown in Fig. 8-13(b). This is called a five-point stencil. We shall see later that different forms of the PDE, as well as different methods to solve PDEs, require different stencils.

It is fairly straightforward to rewrite (8-119) as an recursive equation in the same way as we did in the previous section. The analog of (8-97) using the Jacobi method is

$$V^{(k+1)}(x_i, y_j) = \frac{1}{4}\Big\{V^{(k)}(x_{i+1}, y_j) + V^{(k)}(x_{i-1}, y_j) + V^{(k)}(x_i, y_{j+1})$$
$$+ V^{(k)}(x_i, y_{j-1}) + h^2 \rho(x_i, y_j)\Big\} \quad (8\text{-}120)$$

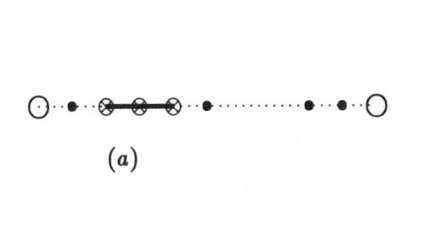

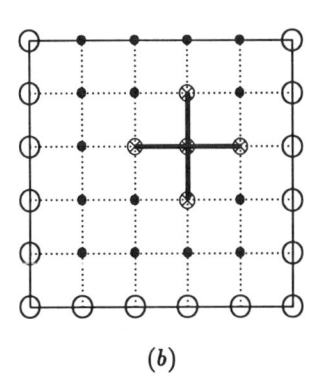

Fig. 8-13 The computational molecule for solving boundary value problems. The ODE of (8-91) is shown in (a), where each FDE expresses the relation between the values of the dependent variable at three adjacent points in the one-dimensional mesh. For the two-dimensional PDE of (8-119), the relation is between five points arranged in the way shown in (b). Circles indicate points where the values are given by boundary conditions.

The relation implies that the value of $V(x,y)$ at the point $(x,y) = (x_i, y_j)$ in the next iteration is given by the average of its four nearest neighbors in the present iteration, modified by the value of $\rho(x,y)$ at the (x_i, y_i).

To find the recursion equations for Gauss-Seidel and SOR (successive overrelaxation) methods, we have a choice of several slightly different approaches, depending on the order in which we wish to "update" the elements in the calculations. The basic philosophy is to make use of the updated values, that is, those of the present iteration, for two of the five elements involved in our five-point stencil to calculate the value of $V(x,y)$ at a particular mesh point for the present iteration. This is identical in spirit to (8-100) where, for the three-point stencil of our second-order ODE, the value of one of the three elements involved is of the present iteration and the other two belong to the previous iteration. For our purpose here, we shall take the simplest approach by working on all the elements of one row at a time, from left to right in the order of increasing index i, and then proceeding to the next row (increasing j). The resulting equation has the form

$$V^{(k+1)}(x_i, y_j) = V^{(k)}(x_i, y_j) - \frac{\omega}{4}\Big\{ h^2 \rho(x_i, y_j) + V^{(k+1)}(x_{i-1}, y_j) \\ + V^{(k+1)}(x_i, y_{j-1}) - 4V^{(k)}(x_i, y_j) \\ + V^{(k)}(x_{i+1}, y_j) + V^{(k)}(x_i, y_{j+1}) \Big\} \quad (8\text{-}121)$$

We recover the Gauss-Seidel method from this equation by putting the relaxation parameter ω to unity.

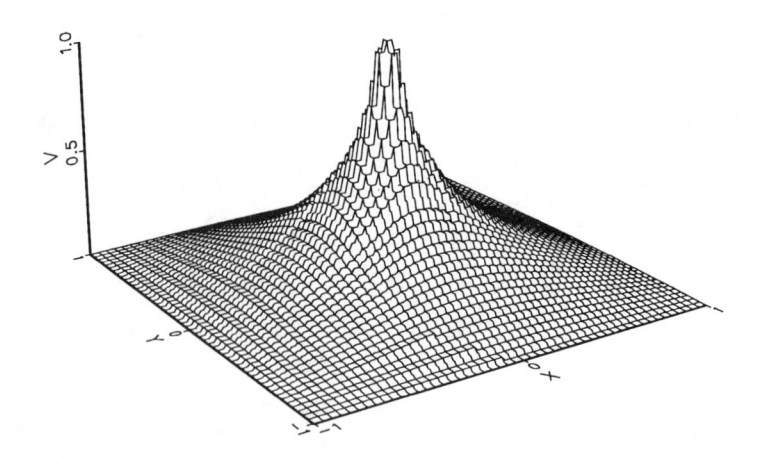

Fig. 8-14 The potential distribution $V(x,y)$ in two dimensions for a point charge in the center of a square, subject to the boundary condition $V = 0$ at the edges. The solution is obtained by solving the PDE of (8-117) using SOR with 31 mesh points on each side. The value of the relaxation parameter ω is 1.816, obtained using (8-114) with $\rho(J)$ given by (8-122). A total of 52 iterations is required to reach an average difference of 10^{-5} between two successive iterations. The number is reduced to 46 by using a variable value of ω according to (8-116).

An alternate procedure may be found by noting that the "corrections" to $V(x_i, y_j)$ for even i or j come, respectively, from the neighboring points with odd i or j, and vice versa. This gives us the possibility of separating each iteration into two parts, one for the odd values of the indexes and the other for the even values. There are certain advantages in taking such an approach but we shall not be going into this subject here.

The only thing remaining before embarking on the actual numerical calculations is the trial solution used as the starting point for the iterative calculations. The Poisson equation is sufficiently stable so that we do not need to be overly concerned with the starting values $V^{(0)}(x,y)$ for the solution. In fact, the convenient choice of

$$V^{(0)}(x_i, y_j) = 0$$

for all i and j, is usually a good one. The resulting potential surface for a point charge of magnitude q located at $x = y = 0$,

$$\rho(x,y) = \begin{cases} q & \text{for } x = y = 0 \\ 0 & \text{otherwise} \end{cases}$$

is shown in Fig. 8-14.

Box 8-9 Program SOR_POS
Solution of partial differential equation using SOR
Example of a two-dimensional Poisson equation

Subprograms used:
  INI_POS: Initializations for solving the Poisson equation using SOR.
  SOR_ITR: SOR iterations for two-dimensional PDE.
Initialization:
  (a) Choose a mesh of 31 points on each side and an error tolerance of $\epsilon = 10^{-5}$.
  (b) Use INI_POS to:
      (i) Set the initial trial solution as zero everywhere.
      (ii) Define the charge density to be zero everywhere except the middle.
1. Input the relaxation parameter ω.
2. Use SOR_ITR to carry out the following calculations:
  (a) Use (8-121) to calculate the difference of $V(x_i, y_j)$ between two iterations.
  (b) Improve the solution by including the difference.
  (c) Store the largest absolute value of the difference as an estimate of the error.
  (d) If the error is larger than ϵ, iterate the solution again.
      Stop if the maximum number of allowed iterations is exceeded.
3. Return $V(x, y)$ as the solution.

For SOR, we also need to determine the optimum value of the relaxation parameter ω_x so as to "accelerate" the convergence as much as possible. The equivalent matrix to A of (8-93) has, in general, five nonvanishing elements in each row. It is still a sparse matrix, but its actual form depends on the order in which we wish to carry out the iterative calculation. It is possible to use a matrix diagonalization to find out the spectral radius of the point Jacobi iteration matrix and thence the value of ω_x using (8-114). Alternatively, we can adopt a variable value for ω that changes from one iteration to the next as in (8-116). For the Poisson equation on a rectangular mesh of sides with lengths $(ph \times qh)$, the spectral radius for a set of Dirichlet boundary conditions is shown in Smith (1985) to be

$$\rho(J) = \frac{1}{2}\left(\cos\frac{\pi}{p} + \cos\frac{\pi}{q}\right) \tag{8-122}$$

For other cases, it may be possible to approximate the problem to a known one and use the spectral radius of its Jacobi matrix. Such topics belong to more advanced treatments on PDEs and we shall not be concerned with them here. The steps used in carrying out the calculations for our example for this section are summarized in Box 8-9.

§8-8 Parabolic partial differential equations

A good example of a parabolic partial differential equation is the diffusion equation

$$\frac{\partial \phi}{\partial t} = \nabla \cdot D\nabla \phi(\boldsymbol{r}, t) \tag{8-123}$$

The differential equation arises in transport phenomena, for example, in describing the concentration $\phi(\boldsymbol{r}, t)$ of a certain type of gas molecule as a function of location and time. If the system is not in equilibrium, the concentration at a given point may vary with time. From the continuity equation,

$$\nabla \cdot \boldsymbol{j}(\boldsymbol{r}, t) + \frac{\partial \phi}{\partial t} = 0 \tag{8-124}$$

we find that the changes are related to the divergence of the current density $\boldsymbol{j}(\boldsymbol{r}, t)$. On the other hand, the current density itself is proportional to the gradient of the concentration:

$$\boldsymbol{j}(\boldsymbol{r}, t) = -D\nabla \phi(\boldsymbol{r}, t) \tag{8-125}$$

On substituting (8-125) into (8-124), we obtain the diffusion equation (8-123). In general, the diffusion coefficient D is a function of the coordinates $\boldsymbol{r}$ and time t. However, for a uniform medium, it may be taken as a constant.

In addition to equations related to the diffusion equation, such as heat conduction and fluid flow through porous medium, the time-dependent Schrödinger equation

$$i\hbar \frac{\partial \Psi}{\partial t} = -\frac{\hbar^2}{2\mu} \nabla^2 \Psi(\boldsymbol{r}, t) + V(\boldsymbol{r}) \Psi(\boldsymbol{r}, t) \tag{8-126}$$

is also an example of parabolic differential equations. From a physics point of view, parabolic (as well as hyperbolic) equations describe the time development of a system under a given set of boundary conditions. Since it involves both initial values and boundary conditions, it is an initial value boundary problem.

For a discussion of the numerical methods to solve such equations, it is more instructive to use (8-123) and simplify it to an equation involving only one of the three spatial coordinates. Furthermore, if D is a constant, we have a parabolic equation in one of its simplest forms:

$$\frac{\partial \phi(x, t)}{\partial t} = D \frac{\partial^2 \phi(x, t)}{\partial x^2} \tag{8-127}$$

Again, we shall transform the differential equation into a finite difference equation by setting up a fixed two-dimensional mesh of points, $j = 0, 1, 2, \ldots, N_x$ for x, and $k = 0, 1, 2, \ldots, N_t$ for t.

Although it makes very little difference to a computer program that the two coordinates in our mesh are referring to two different physical quantities, space and time, the problems we face are quite different for the two. In the present

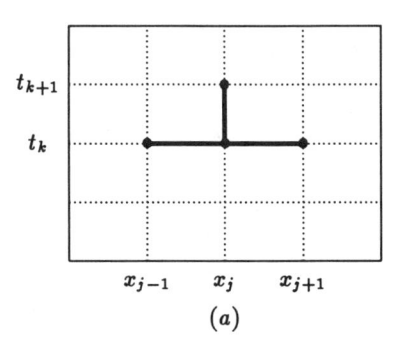

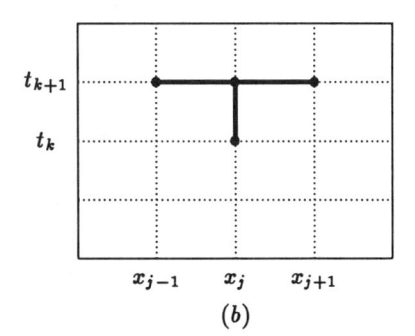

Fig. 8-15 Implicit and explicit methods to find the solution of a PDE that is first order in t and second order in x. In (a), the explicit scheme calculates the values at t_{k+1} using only those already known at t_k. In (b), the implicit scheme involves values of the solution at t_{k+1} that are yet to be found.

case, this results mainly from the fact that the derivatives with respect to the two independent variables are of different order. In addition, the time dependence in the problem requires an input set of initial conditions and the spatial dependence requires a set of boundary conditions. For example, the initial conditions may be given in terms of the values $\phi(t_0, x_j)$ at all the spatial locations $x_0, x_1, \ldots, x_{N_x}$ at time $t = t_0$. For the dependence on x, we note that, since the derivative is second order, two boundary conditions are required for each t_k to determine the problem uniquely. These may be given in terms of the values of $\phi(t, x)$ at both $x = x_0$ and x_{N_x}, one pair for each of the discrete set of values $t_0, t_1, \ldots, t_{N_t}$.

Explicit and implicit methods. Following the same approach as we did in arriving at (8-89), we shall make use of a central difference to convert the second-order derivative in (8-127) to the finite difference:

$$\left.\frac{\partial^2 \phi}{\partial x^2}\right|_{x=x_j} \longrightarrow \frac{1}{(\Delta x)^2}\{\phi(x_{j+1}, t) - 2\phi(x_j, t) + \phi(x_{j-1}, t)\} \tag{8-128}$$

Similarly, we can use a forward difference to change the time derivative to

$$\left.\frac{\partial \phi}{\partial t}\right|_{t=t_k} \longrightarrow \frac{1}{\Delta t}\{\phi(x, t_{k+1}) - \phi(x, t_k)\} \tag{8-129}$$

To put the two relations together to form the equivalent FDE for (8-127), we need to evaluate the right side of (8-128) for a given t and the corresponding quantities of (8-129) for a given x. There is no difficulty associated with x. Since, for (8-128), we have used a central difference centered around x_j, we have good reason to evaluate all the quantities on the right side of (8-129) at $x = x_j$.

There are, however, two different ways we can associate the value of t in (8-128) with those on the mesh. The most obvious is to evaluate all the quantities at $t = t_k$. The advantage of this scheme is that the FDE for (8-127) takes on the form

$$\frac{\phi(x_j, t_{k+1}) - \phi(x_j, t_k)}{\Delta t} = D \frac{\phi(x_{j+1}, t_k) - 2\phi(x_j, t_k) + \phi(x_{j-1}, t_k)}{(\Delta x)^2} \quad (8\text{-}130)$$

As a result, it is possible to calculate $\phi(x_j, t_{k+1})$ directly from the three input quantities $\phi(x_{j+1}, t_k)$, $\phi(x_j, t_k)$, and $\phi(x_{j-1}, t_k)$, as shown schematically in Fig. 8-15(a). This is an explicit method, as all the input quantities belong to an earlier time step and are available when they are needed. By starting from a set of initial conditions at $t = t_0$, all $\phi(x_j, t_k)$ for $k = 1$ may be calculated. Once this is done, we can proceed to $k = 2$, and so on. The disadvantage of this scheme is that the stability of the FDE is restricted to

$$\frac{2D\Delta t}{(\Delta x)^2} \leq 1$$

The derivation of this condition is given in standard references on numerical solutions of PDEs. Here, we shall only try to provide a feeling for how such a restriction arises.

It is quite easy to see the reason why one cannot take big steps in t in an explicit method. For the example above, the numerical solution at a point (x_j, t_k) depends only on those for the three points x_{j-1}, x_j, and x_{j+1} at the previous time step $t = t_{k-1}$. The values at these three points are, in turn, obtained from those at the five points x_{j-2}, x_{j-1}, x_j, x_{j+1}, and x_{j+2} at time $t = t_{k-2}$. When we continue to trace back in time in this way, we find that all the earlier points on which the value at (x_j, t_k) depends fall within the pyramid shown in Fig. 8-16. If the step size in time is too large compared with the step size in the spatial coordinate, the pyramid is narrow at the base and points at the end of the time interval depend only on a subset of the points at $t = 0$. From a physics point of view, it is difficult to think of many situations in which the solution at any time is only a function of a part of the initial conditions. It is far more likely that every point in a proper solution at the end of the time interval involves all the points at $t = 0$. In this way we can see that the numerical solution obtained with the large step size may not correspond to the true solution.

An alternative to the method described above is to evaluate all the quantities on the right side of (8-128) at $t = t_{k+1}$. The resulting FDE for the differential equation of (8-127) takes on the form

$$\frac{\phi(x_j, t_{k+1}) - \phi(x_j, t_k)}{\Delta t} = D \frac{\phi(x_{j+1}, t_{k+1}) - 2\phi(x_j, t_{k+1}) + \phi(x_{j-1}, t_{k+1})}{(\Delta x)^2}$$

$$(8\text{-}131)$$

The advantage of this approach is that the FDE is unconditionally stable. Again, we shall not give the proof here. The method is, however, an *implicit* one. As

$\phi(x_j, t_k)$

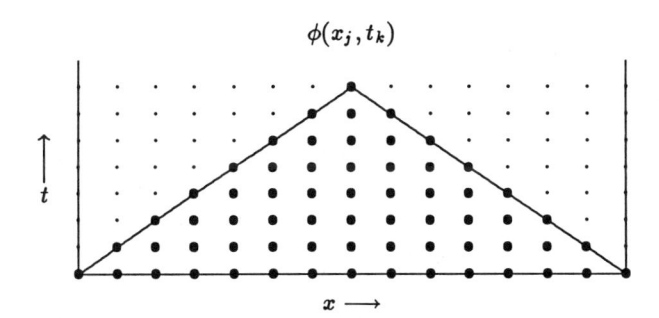

Fig. 8-16 Schematic illustration of the domain of dependence for explicit methods. The solution at a point x_j and time $t = t_k$ depends only on those of three points, x_{j-1}, x_j, and x_{j+1}, at $t = t_{k-1}$. These three points are, in turn, generated from those at x_{j-2}, x_{j-1}, x_j, x_{j+1}, x_{j+2} at $t = t_{k-2}$, and so on. As a result, only the points within the triangle can influence the value at (x_j, t_k) and they form the domain of dependence for $\phi(x_j, t_k)$.

shown schematically in Fig. 8-15(b), the stencil for this method involves the four quantities $\phi(x_j, t_k)$, $\phi(x_{j+1}, t_{k+1})$, $\phi(x_j, t_{k+1})$, and $\phi(x_{j-1}, t_{k+1})$. As a result, the value of any one of the three quantities at $t = t_{k+1}$ cannot be calculated unless the other two are known.

Let us assume that we are going to carry out the calculations in the following order. For each row, we solve for $\phi(x_j, t_k)$ starting from the left and proceed according to increasing value of the index j. Once the values of a complete row of points are obtained, we move on to the next time interval represented by the points in the next row according to the order of increasing value of the index k. In this scheme, we have only two of the four quantities in the computational module, $\phi(x_j, t_k)$ and $\phi(x_{j-1}, t_{k+1})$, available to us at any stage of the calculation; the other two, $\phi(x_j, t_{k+1})$ and $\phi(x_{j+1}, t_{k+1})$, have not yet been calculated. However, the total number of algebraic equations we must solve and the total number of unknown quantities in the implicit scheme are unchanged from those in the explicit scheme. There is, therefore, no problem in principle in obtaining the solutions. The only difference is that the FDE must now be solved as a set of algebraic equations.

In terms of matrices, the set of equations we encounter in the implicit scheme may be put into the form

$$Ay = b \qquad (8\text{-}132)$$

the same as we did earlier in (8-93). The matrix A is again very sparse, tridiagonal in our example here. As a result, it is relatively easy to obtain a solution. We shall defer a discussion of the numerical method to solve (8-132) till later in this section.

Crank-Nicholson method. It is also possible to use a linear combination of explicit and implicit methods by taking an average, weighted by some factor if so desired, of the right sides of (8-130) and (8-131). The result is known as the *Crank-Nicholson method*. For example, by taking equal weights, we obtain

$$\frac{\phi(x_j, t_{k+1}) - \phi(x_j, t_k)}{\Delta t} = \frac{D}{2} \left\{ \frac{\phi(x_{j+1}, t_k) - 2\phi(x_j, t_k) + \phi(x_{j-1}, t_k)}{(\Delta x)^2} \right. $$
$$\left. + \frac{\phi(x_{j+1}, t_{k+1}) - 2\phi(x_j, t_{k+1}) + \phi(x_{j-1}, t_{k+1})}{(\Delta x)^2} \right\}$$
(8-133)

This is the method we shall adopt for all the numerical calculations to be carried out in the rest of this section.

There is also another good reason to use the Crank-Nicholson method for parabolic equations. This can be seen by using the time-dependent Schrödinger equation of (8-126) as an example. It is convenient to transform the differential equation into a dimensionless form first. That is, we shall express energy, length, and time in our equation, respectively, in units of MeV (million electron volts), $\hbar/\{2\mu \times \text{MeV}\}^{1/2}$, and $\hbar/\text{MeV}$. Thus, one unit of time is approximately 6.6×10^{-22} s, and one unit of energy is 1 MeV (1.6×10^{-13} J). The value of one unit of length depends on the mass μ of the particle with which we wish to deal. For a nucleon ($\mu c^2 \approx 10^3$ MeV), it is 4.5×10^{-15} m, and for an electron ($\mu c^2 \approx 0.5$ MeV), it is 2×10^{-13} m.

The resulting Schrödinger equation remains the same if we put $\hbar = 1$ and $2\mu = 1$ in (8-126), a practice used often in the literature to simplify the form of such equations. The basic points we wish to illustrate here may be carried out by restricting ourselves to one spatial dimension. In this limit, our PDE takes on the form

$$i\frac{\partial \Psi}{\partial t} = \hat{H}\Psi(x, t) \tag{8-134}$$

where the Hamiltonian operator is given by

$$\hat{H} = -\frac{\partial^2}{\partial x^2} + V(x)$$

As shown earlier in (7-75), a formal solution to (8-134) may be written as

$$\Psi(x, t) = e^{-i(t-t_0)\hat{H}} \Psi(x, t_0) \tag{8-135}$$

In other words, we can regard $\exp\{-i(t-t_0)\hat{H}\}$ as the operator that "propagates" the wave function $\Psi(x, t_0)$ at time t_0 to time t.

The meaning of having an operator $\hat{H}$ in the argument of the exponential function may be understood in terms of the infinite series expansion

$$e^{-i(t-t_0)\hat{H}} = 1 + \{-i(t-t_0)\hat{H}\} + \frac{1}{2!}\{-i(t-t_0)\hat{H}\}^2 + \frac{1}{3!}\{-i(t-t_0)\hat{H}\}^3 + \cdots$$

$$= \sum_{\ell=0}^{\infty} \frac{1}{\ell!}\{-i(t-t_0)\hat{H}\}^\ell \qquad (8\text{-}136)$$

If we divide the time interval $[t_0, t]$ into N_t equal parts, it is easy to verify that the formal solution of (8-135), with the exponential function expressed in the series form, is identical to our FDE in the limit $N_t \to \infty$.

For an increment in time of Δt from t to $t + \Delta t$, (8-135) takes on the form

$$\Psi(x, t + \Delta t) = e^{-i(\Delta t)\hat{H}} \Psi(x, t) \qquad (8\text{-}137)$$

On expressing all the quantities in terms of their values on the mesh points, we have

$$\Psi(x_j, t_{k+1}) = e^{-i(\Delta t)\hat{H}} \Psi(x_j, t_k)$$

$$= \left[1 + \{-i(\Delta t)\hat{H}\} + \frac{1}{2!}\{-i(\Delta t)\hat{H}\}^2 + \cdots\right] \Psi(x_j, t_k) \qquad (8\text{-}138)$$

for $j = 0, 1, 2, \ldots, N_x$. In terms of this expansion, the explicit method corresponds to truncating the series after the linear term in $\hat{H}$:

$$\Psi(x_j, t_{k+1}) \approx \left[1 + \{-i(\Delta t)\hat{H}\}\right] \Psi(x_j, t_k)$$

$$= \Psi(x_j, t_k) - i\Delta t \left\{ \frac{\Psi(x_{j+1}, t_k) - 2\Psi(x_j, t_k) + \Psi(x_{j-1}, t_k)}{(\Delta x)^2} - V(x_j)\Psi(x_j, t_k) \right\}$$

The implicit method corresponds to the same approximation except that, instead of (8-137), we start with the propagation relation

$$\Psi(x_j, t_k) = e^{+i(\Delta t)\hat{H}} \Psi(x_j, t_{k+1})$$

For a Hermitian $\hat{H}$, the operator $\exp\{\pm(\Delta t)\hat{H}\}$ is unitary. However, any approximations of the operator by truncating the series are not unitary in general. This causes several difficulties, the most serious being that, if the wave function $\Psi(x, t)$ is normalized at $t = t_k$, the normalization may not be preserved when it is propagated in time. This is true with either explicit or implicit methods. Unitarity must therefore be imposed as an additional condition in the calculation, for example, by normalizing the wave function after each time step. In terms of the diffusion problem, the same difficulties appear in the form that the flux is not conserved.

We shall see that the Crank-Nicholson approach is an approximation that preserves unitarity. For this purpose, we shall replace the exponential function by the Cayley form, as suggested in Goldberg, Schey, and Schwartz (*Am. J. Phys.* **35** [1967] 177):

$$e^{-i(\Delta t)\hat{H}} \approx \frac{1 - \frac{1}{2}i(\Delta t)\hat{H}}{1 + \frac{1}{2}i(\Delta t)\hat{H}} = \{1 - \tfrac{1}{2}i(\Delta t)\hat{H}\}\{1 - \tfrac{1}{2}i(\Delta t)\hat{H} + [\tfrac{1}{2}i(\Delta t)\hat{H}]^2 + \cdots\} \quad (8\text{-}139)$$

Comparing this result with the series expansion of the exponential function given by (8-136), we see that the Cayley form is identical to the exact result up to the second power in the argument.

Since our FDE corresponds to retaining only terms up to the first power, there is no additional loss in the accuracy by starting with the Cayley form instead of the series expansion. In place of (8-137), the propagation equation becomes

$$\Psi(x_j, t_{k+1}) = \frac{1 - \tfrac{1}{2}i(\Delta t)\hat{H}}{1 + \tfrac{1}{2}i(\Delta t)\hat{H}} \Psi(x_j, t_k)$$

When both sides of this relation are multiplied by the denominator on the right side, we obtain

$$\{1 + \tfrac{1}{2}i(\Delta t)\hat{H}\}\Psi(x_j, t_{k+1}) = \{1 - \tfrac{1}{2}i(\Delta t)\hat{H}\}\Psi(x_j, t_k) \quad (8\text{-}140)$$

On rearranging the terms, we obtain a form that is exactly the same as that given by the Crank-Nicholson method. Since the Cayley form of the operator is unitary, we have no difficulty in preserving the normalization of the wave function in this case.

Solving the FDE by Gaussian elimination. Let us address the problem of finding the roots of the FDE for both implicit and Crank-Nicholson methods at the same time. Returning to the diffusion equation in one spatial dimension, the FDE of (8-131) for the implicit method may be written in the form

$$\phi(x_{j-1}, t_{k+1}) - (2 + \eta)\phi(x_j, t_{k+1}) + \phi(x_{j+1}, t_{k+1}) = -\eta\phi(x_j, t_k) \quad (8\text{-}141)$$

where

$$\eta = \frac{(\Delta x)^2}{D \Delta t}$$

For the Crank-Nicholson method of (8-133), we have

$$\phi(x_{j-1}, t_{k+1}) - (2 + \eta')\phi(x_j, t_{k+1}) + \phi(x_{j+1}, t_{k+1})$$
$$= -\phi(x_{j-1}, t_k) + (2 - \eta')\phi(x_j, t_k) - \phi(x_{j+1}, t_k) \quad (8\text{-}142)$$

with

$$\eta' = \frac{2(\Delta x)^2}{D \Delta t}$$

The only differences between the two forms of FDE are on the right side. Since all the quantities pertaining to $t = t_k$ are known by the time we come to solve the

FDE at $t = t_{k+1}$, the differences are minor. As a result, it is possible to address both types of FDE together for the most part.

Since the boundary conditions are given at $x = x_0$ and $x = x_{N_x}$, the FDEs for $j = 1$ and $j = (N_x - 1)$ are slightly different in form from those away from the boundaries. Consider first (8-141). Since $\phi(x,t)$ is defined only in the interval $[x_0, x_{N_x}]$, the equation has no meaning for $j < 1$. For $j = 1$, we have

$$\phi(x_0, t_{k+1}) - (2+\eta)\phi(x_1, t_{k+1}) + \phi(x_2, t_{k+1}) = -\eta\phi(x_1, t_k)$$

However, $\phi(x_0, t_{k+1})$ is given by the boundary condition. On rearranging the equation by putting all the terms involving unknown quantities on the left side, we have

$$-(2+\eta)\phi(x_1, t_{k+1}) + \phi(x_2, t_{k+1}) = -\eta\phi(x_1, t_k) - \phi(x_0, t_{k+1})$$

Similarly, for $j = (N_x - 1)$, we have

$$\phi(x_{N_x-2}, t_{k+1}) - (2+\eta)\phi(x_{N_x-1}, t_{k+1}) = -\eta\phi(x_{N_x-1}, t_k) - \phi(x_{N_x}, t_{k+1}) \quad (8\text{-}143)$$

The same differences in form exist also for the Crank-Nicholson method. Again, only the right sides are different from what we have above for the implicit method and we shall not give them here. Note that, since the values of $\phi(x, t_{k+1})$ are given by the boundary conditions at $x = x_0$ and $x = x_{N_x}$, there are a total of $(N_x - 1)$ unknown quantities $\phi(x_1, t_{k+1})$, $\phi(x_2, t_{k+1}), \ldots, \phi(x_{N_x-1}, t_{k+1})$ for each t. With an equal number of equations given by either (8-141) or (8-142), there is no basic problem in obtaining the solution.

Both (8-141) and (8-142) may represented by the following set of equations:

$$\begin{aligned} b_1 y_1 + c_1 y_2 &= d_1 \\ a_j y_{j-1} + b_j y_j + c_j y_{j+1} &= d_j \quad \text{for} \quad j = 2, 3, \ldots, (N-1) \\ a_N y_{N-1} + b_N y_N &= d_N \end{aligned} \quad (8\text{-}144)$$

where $N \equiv (N_x - 1)$. The values of a_j, b_j, c_j, and d_j are slightly different depending on whether we are using the implicit method or the Crank-Nicholson method. In each case, they can be obtained by comparing with (8-141) or (8-142), as the case may be.

In matrix notation, these equations may be written as

$$\boldsymbol{Ay = d}$$

The square matrix $\boldsymbol{A}$, with dimension $(N \times N)$, is tridiagonal for our examples. That is,

$$\boldsymbol{A} = \begin{pmatrix} b_1 & c_1 & 0 & \cdots & \cdots & 0 \\ a_2 & b_2 & c_2 & 0 & \cdots & 0 \\ 0 & a_3 & b_3 & c_3 & \cdots & 0 \\ \vdots & \vdots & \vdots & \ddots & \vdots & \vdots \\ 0 & \cdots & 0 & a_{N-1} & b_{N-1} & c_{N-1} \\ 0 & \cdots & 0 & 0 & a_N & b_N \end{pmatrix} \quad (8\text{-}145)$$

Box 8-10   Program DIFFUS

Crank-Nicholson solution of parabolic PDE

Diffusion equation as example of initial value boundary problem

Subprograms used:
  INIT: Initialize the wave function and other arrays.
  CALC_D: Calculate the matrix elements of d from y and η.
  PROP: Propagate the parabolic FDE by one time step (cf. Box 3-10).
  OUTPUT: Output the solution for a given time i_t.

Initialization:
  (a) Set the ranges $x = [0, x_d]$ and $t = [0, t_d]$, and the number of steps in each.
  (b) Construct a two-dimensional mesh with step sizes Δx and Δt.

1. Input the diffusion coefficient D.
2. Call INIT to:
   (a) Zero the array $\{y_i\}$ as the initial condition.
   (b) Define the arrays $\{a_i\}$, $\{b_i\}$, and $\{c_i\}$ of (8-144) using (8-143).
   (c) Specify the boundary conditions, one end at temperature $100°$ C the other at $0°$ C.
3. Propagate the solution forward by one time step:
   (a) Use CALC_D to calculate the values of d_j of (8-144) from $y(x, t)$.
   (b) Use PROP to:
      (i) Forward eliminate of the subdiagonal elements according to (3-127).
      (ii) Find y_j at $(t + \Delta t)$ by back substitution using (3-129).
4. Increase time from t to $(t + \Delta t)$ and output the solution $y(t)$ at regular intervals.

In general, the form may be more complicated but the matrix remains sparse. For most of the problems involving second-order partial differential equations, the tridiagonal form is the common one.

The form of (8-144) is identical to that of (3-124), where we were solving a set of algebraic equations in calculations associated with the topic of cubic spline. As a result, we can use the same Gaussian elimination method to find the solution. Technically, we can improve the numerical accuracy by including pivoting, as we did in §5-1. This is left as an exercise.

As an illustration, we shall solve (8-127) for the temperature distribution along a one-dimensional conducting rod. The algorithm used is summarized in Box 8-10. As the boundary conditions, we shall maintain one end of the rod at temperature $100°$ C and the other end at $0°$ C. Initially, the entire rod is at temperature $0°$ C. The parabolic differential equation (8-127) is solved using the Crank-Nicholson method and the results are shown in Fig. 8-17 for time $t = 1, 10, 50$, and infinity. Both the spatial coordinate and time are measured in arbitrary units, and, in these units, the diffusion coefficient is taken to be $D = 0.2$.

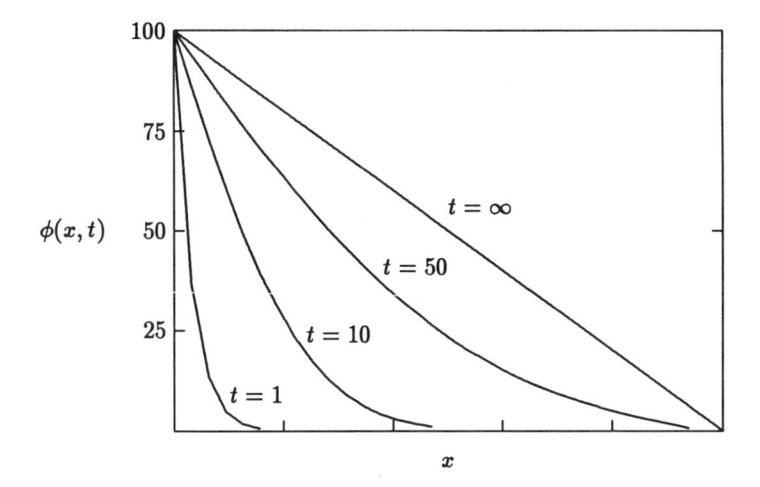

Fig. 8-17 Temperature distribution along a one-dimensional conducting rod as a function of time from $t = 0$ and distance from one end. One end of the rod is kept at 100° C and the other end at 0° C. The solutions are obtained by solving (8-127) for the diffusion coefficient $D = 0.2$.

Scattering of a wave packet. As a second example, we shall apply the methods developed here to solve the problem of scattering from a square-well potential. Instead of calculating the phase shift, as we did earlier in §8-5, we are now interested in the time development of the wave packet as it moves toward the potential from some large distance away. When the packet gets close to the potential, it is influenced by the interaction and the wave function is changed as a result. In one spatial dimension, this is the problem discussed in Goldberg, Schey and Schwartz, and the results appear in several textbooks of quantum mechanics as an illustration for the behavior of a wave packet in scattering.

The wave function of a particle traveling along the x-direction is governed by the Schrödinger equation (8-134). For the scattering problem, we shall assume that the particle is initially at $x = -x_c$, far away from where the potential $V(x)$ is nonzero. The square-well potential is centered at $x = 0$ and has the form

$$V(x) = \begin{cases} V_0 & \text{for } |x| \leq \tfrac{1}{2} r_g \\ 0 & \text{otherwise} \end{cases}$$

For $|x_c| \gg r_g$, the particle is a free one. At this stage, the wave function, as well as the velocity of the particle, is completely given by the initial conditions we wish to impose on the problem. The boundary conditions, on the other hand, come from the requirement of quantum mechanics that the wave function must vanish at $x = \pm\infty$.

There are several possibilities in selecting the initial form of the wave function. The recommendation of Goldberg, Schey, and Schwartz is to use a Gaussian

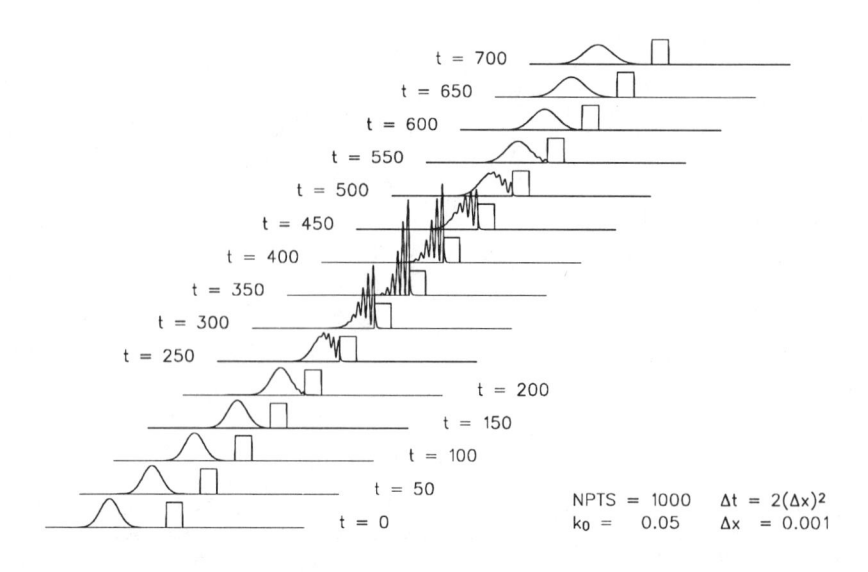

Fig. 8-18 Scattering of a quantum mechanical particle from a one-dimensional repulsive ($V_0 > 0$) square-well barrier centered at $x = L/2$. The interval $x = [0, L]$ is divided into 1000 equal parts ($\Delta x = L/1000$) and the time step is taken to be $\Delta t = 2(\Delta x)^2$. The particle starts as a Gaussian wave packet centered at $x_c = L/4$ with $k_0 = 50\pi$. Since the barrier is high ($V_0 = 2k_0^2$ and $r_g = 0.064L$) in this example, the wave packet is reflected back because of the interaction.

wave packet:

$$\Psi(x, t=0) = e^{ik_0 x - (x-x_c)^2/2\sigma_0^2} \tag{8-146}$$

where x_c is the location of the center and σ_0 is the width of the wave packet at $t = 0$. The wave number k_0 is related to the kinetic energy of the wave packet and, hence, to the velocity it travels along the (positive) x-direction. The physics of the problem is completely determined if we are given the values of V_0 and r_g for the potential well and k_0, x_c, and σ_0 for the initial wave packet.

To solve the resulting differential equation by the Crank-Nicholson method, we need to construct a fixed mesh of points so as to convert the PDE into a set of FDEs. The method follows closely that used for the diffusion problem above and we shall not go into the details here. For a repulsive well ($V_0 > 0$), the wave functions obtained from the numerical solution at different times are shown in Fig. 8-18. To see the rapid fluctuations in the wave function when the center of the packet is near the potential well, the step size in x must be taken to be fairly small. For this reason, the amount of computation is nontrivial but can still be carried out on a fast personal computer. Extension of the scattering problem to two spatial dimensions is given in Galbraith, Ching, and Abraham, (*Am. J. Phys.* **52** [1984] 60).

§8-9 Hyperbolic partial differential equations

As a class of initial value boundary problems, the hyperbolic partial differential equations are different in nature from the parabolic equations discussed in the previous section. Examples of physical problems in this class are transport equations and wave equations. Hyperbolic equations can often be ill-behaved and must be treated with caution. We shall encounter examples of such behavior later in this section.

Let us use $\phi(\mathbf{r},t)$ to represent the density of a certain type of particle at time t and spatial point $\mathbf{r}$. If the distribution is not in equilibrium, a net flow of the particles can take place between different regions. The changes in $\phi(\mathbf{r},t)$ in this case are described by the Boltzmann transport equation,

$$\frac{\partial \phi}{\partial t} = -\nabla \cdot \{\mathbf{v}\phi(\mathbf{r},t)\} + f(\mathbf{r},t) \tag{8-147}$$

where $\mathbf{v}$ is the velocity of the flow and $f(\mathbf{r},t)$ is the source term.

Wave equations, on the other hand, describe phenomena such as the propagation of electromagnetic waves in space. In one spatial dimension, they take on the form

$$\frac{\partial^2 \phi}{\partial t^2} - \frac{1}{v^2}\frac{\partial^2 \phi}{\partial x^2} = 0 \tag{8-148}$$

Equations of this type occur also, for example, in describing the small amplitude vibrations of a stretched string, as we saw earlier in (8-8). Here, $\phi(x,t)$ is the displacement of the string at location x and time t, and v is the propagation velocity of the vibration along the string. If we increase the number of spatial dependent variables to two, the corresponding equation governs, for example, the vibrations of a drum. To simplify our discussions, we shall consider only one spatial coordinate.

Transport equation in one spatial dimension. To appreciate some of the difficulties faced by hyperbolic equations, let us consider the transport equation of (8-147) in the limit of one spatial dimension and constant velocity. The resulting equation may be written as

$$a\frac{\partial \phi}{\partial t} + b\frac{\partial \phi}{\partial x} = c \tag{8-149}$$

By defining

$$p = \frac{\partial \phi}{\partial t} \qquad\qquad q = \frac{\partial \phi}{\partial x}$$

the equation takes on the form

$$ap + bq = c \tag{8-150}$$

Along a curve C in the x-t plane, small changes in $\phi(x,t)$ may be expressed in terms of p and q:

$$d\phi = \frac{\partial \phi}{\partial t}dt + \frac{\partial \phi}{\partial x}dx = p\,dt + q\,dx \qquad (8\text{-}151)$$

where dt/dx is the tangent to C. We can eliminate p between (8-150) and (8-151) and obtain

$$a\,d\phi - c\,dt = q(a\,dx - b\,dt)$$

If we choose the curve C such that

$$a\,dx - b\,dt = 0 \qquad (8\text{-}152)$$

we have $\phi(x,t)$ as the solution of the differential equation

$$a\,d\phi - c\,dt = 0 \qquad (8\text{-}153)$$

along C. The curve is known as a characteristic curve. Physically, it may be interpreted as the trajectory followed by a particle in the convectional flow given by the transport equation.

The method of characteristics. Before applying the usual method of finite differences, it is instructive here to see how a hyperbolic differential equation can also be solved by the method of characteristics. For this purpose, we shall regard (8-152) and (8-153) as a pair of first-order ODEs,

$$\frac{dx}{dt} = \frac{b}{a} \qquad \frac{d\phi}{dt} = \frac{c}{a}$$

that replaces the original first-order PDE of (8-149). The equation for the characteristic curve is given by the first of these two equations. In the simple case of $b/a = v$ being a constant, the characteristic curves may be obtained by integration

$$x = \alpha + \int_{t_0}^{t} \frac{b}{a}\,dt = x_0 + v(t - t_0)$$

where α is the constant of integration and may be taken to be x_0, the location of the characteristic curve at $t = t_0$.

If $c = 0$, corresponding to the case without a source term in the transport equation (8-147), the solution $\phi(x,t)$ is a constant along a characteristic curve. That is, if the initial value of $\phi(x,t_0)$ at a point $x = x_0$ is given by

$$\phi(x_0, t_0) = G(x_0)$$

all the subsequent values of $\phi(x,t)$ along a characteristic curve C that intersects the point (x_0, t_0) in the x-t plane are given by the equation

$$\phi_C(x,t) = G(x_0) = G(x - v(t - t_0))$$

Since the original PDE (8-149) is a first-order differential equation, we must have an initial value for every spatial point x at the starting time. In general, the initial condition for the entire interval of x may be given by the values of $\phi(x,t)$ along a curve in the x-t plane. Let us label this curve as Γ. From every point x along Γ we can carry out an integration to find all the values of $\phi(x,t)$ along the characteristic curve that originates from the point x on Γ. The picture is very similar to the streamlines in a wind tunnel made visible by marking the flow with colored dye.

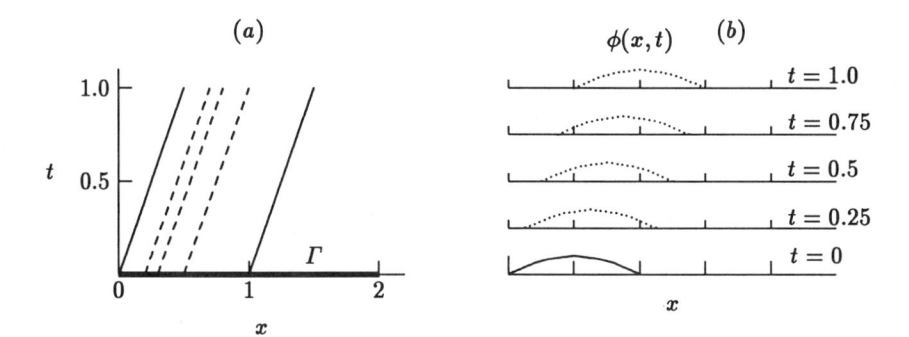

Fig. 8-19 Example of a simple transport equation with constant velocity. The initial values are specified along Γ. Two examples of characteristic curves are shown as solid lines in (a). The solution along any one of the dotted curves may be obtained by integrating from a point on Γ. Examples for the solution of (8-154) for four different times are shown schematically in (b).

A simple example may be useful here to illustrate the point. Consider the following first-order PDE:

$$\frac{\partial \phi}{\partial t} + 2\frac{\partial \phi}{\partial x} = 0 \tag{8-154}$$

It has the same form as (8-149) with $a = 1$, $b = 2$, and $c = 0$. We shall be interested only in the intervals $x = [0,2]$ and $t = [0,1]$. The characteristic curves are given by

$$x = 2t + \alpha$$

They are straight lines in the x-t plane with a constant slope of $v = 2$, and different curves are distinguished by different values of α. Two such curves, $x = 2t$ and $x = 2t + 1$, are shown in Fig. 8-19(a). The initial condition may be specified as

$$\phi(x, t=0) = \begin{cases} A\sin(\pi x) & \text{for } 0 < x < 1 \\ 0 & \text{for } 1 \leq x \leq 2 \end{cases}$$

In this case, the "curve" Γ on which the initial conditions are specified is a straight line extending from $x = 0$ to $x = 2$ along the x-axis, as shown in the figure. Since

(8-154) is a linear PDE, the shape of $\phi(x,t)$ as a function of x is unchanged with time except that a point of constant $\phi(x,t)$ value is shifted in x by an amount vt in time t. In other words, ϕ is a function of a single independent variable $\xi = (x - vt)$. The forms of $\phi(x,t)$ as a function of x are shown in Fig. 8-19(b) for $t = 0$, 0.25, 0.5, 0.75, and 1.0.

Numerical solution. More generally, the three coefficients, a, b, and c, in (8-149) may be functions of x, t, as well as $\phi(x,t)$. In this case, the characteristic curves are no longer simple straight lines, as we have for (8-154), and it may not be possible to carry out the calculations analytically. Numerically, the solution may be obtained in the following way. Starting from any point (x_r, t_r) along Γ, the solution to (8-149) along a characteristic curve C may be found by successive approximation using (8-152) and (8-153). First, we choose a point t_s near t_r; that is, $\Delta t \equiv (t_s - t_r)$ is a small quantity, as shown schematically in Fig. 8-20. Let us find the value of x_s such that (x_s, t_s) is a point along a characteristic curve C originating from a point (x_r, t_r) along Γ. In principle, this can be done using (8-152). However, since both a and b may also be functions of $\phi(x,t)$, whose values are known only along Γ at this moment, the calculation must be approximate, using the values of a and b at the point (x_r, t_r) to approximate those needed at (x_s, t_s). As a result, the value of x_s obtained in this way can only serve as a first estimate. In a similar way, we can use (8-153) to obtain the value of $\phi(x_s, t_s)$ to the same accuracy. Once these first estimates are available, we can put them back into (8-152) and (8-153) to obtain better approximations for x_s and $\phi(x_s, t_s)$. The process is repeated until the solutions converge. The final values of (x_s, t_s) and $\phi(x_s, t_s)$ obtained may now be used as the starting point to find the corresponding values for another point nearby. By taking a number of such small steps, it is possible to trace the complete characteristic curve C that starts at (x_r, t_r), and the values of $\phi(x,t)$ along the curve. Once this is done, we can move to another point along the initial condition curve Γ and find the solution for another characteristic curve.

It is obvious that, to cover the complete x-t plane, we must have an appropriate initial value curve Γ that contains enough information to generate all the characteristic curves of interest. The importance of this requirement may be seen by the following counterexample. If Γ happens to coincide with one of the characteristic curves, it will not be possible to find a unique solution anywhere other than those that are given. This point is quite obvious if we use the method of characteristics to find the solution. However, the difficulty is one particular to hyperbolic PDEs, not the method of solution. Unless a proper set of initial values is available, it may not be possible to find the solution to the differential equation everywhere in the domain of interest.

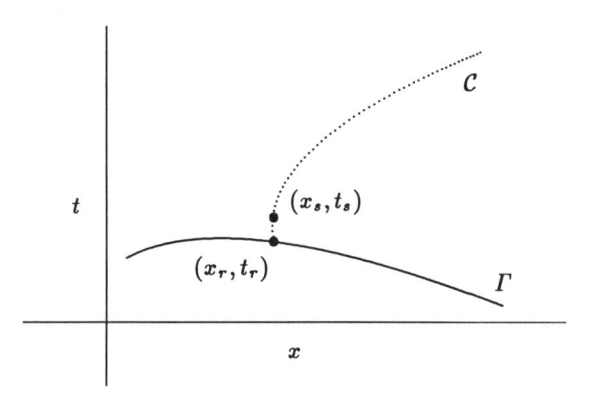

Fig. 8-20 Schematic representation of the method of obtaining the solution for $\phi(x,t)$ along a characteristic curve C for an arbitrary one-dimensional transport equation. The initial conditions are specified by giving the values of $\phi(x,t)$ along the curve Γ. By starting from a point (x_r, t_r) on Γ, the solution $\phi(x,t)$ along a characteristic curve, and the curve itself, may be found by taking small steps, such as the one from (x_r, t_r) to (x_s, t_s).

Two-dimensional PDEs. The method of characteristics can also be extended to second-order PDEs. In general, such equations appear in the form

$$a\frac{\partial^2 \phi}{\partial t^2} + b\frac{\partial^2 \phi}{\partial t \partial x} + c\frac{\partial^2 \phi}{\partial x^2} + e = 0 \tag{8-155}$$

where the coefficients a, b, c, and e may be functions of $\partial \phi/\partial t$, $\partial \phi/\partial x$, ϕ, x, and t. As we shall see in the next section, they may not contain any of the second-order derivatives of $\phi(x,t)$ if we wish to restrict ourselves to quasilinear equations.

Similar to what we did for (8-150) in the case of a first-order equation, the form of the PDE may be simplified to

$$ar + bs + cu + e = 0 \tag{8-156}$$

by defining

$$r = \frac{\partial^2 \phi}{\partial t^2} \qquad s = \frac{\partial^2 \phi}{\partial t \partial x} \qquad u = \frac{\partial^2 \phi}{\partial x^2} \tag{8-157}$$

Along a characteristic curve, r, s, and t satisfy the following two relations:

$$dp = \frac{\partial p}{\partial t}dt + \frac{\partial p}{\partial x}dx = r\,dt + s\,dx$$

$$dq = \frac{\partial q}{\partial t}dt + \frac{\partial q}{\partial x}dx = s\,dt + u\,dx \tag{8-158}$$

where

$$p = \frac{\partial \phi}{\partial t} \qquad q = \frac{\partial \phi}{\partial x}$$

From (8-156), (8-158), and the definitions given by (8-157), we can eliminate r and u from these equations. This gives us the result

$$s\left\{a\left(\frac{dx}{dt}\right)^2 - b\left(\frac{dx}{dt}\right) + c\right\} - \left\{a\frac{dp}{dt}\frac{dx}{dt} + c\frac{dq}{dt} + e\frac{dx}{dt}\right\} = 0 \quad (8\text{-}159)$$

The characteristic curves are then given by the solutions of the equation

$$a\left(\frac{dx}{dt}\right)^2 - b\left(\frac{dx}{dt}\right) + c = 0 \quad (8\text{-}160)$$

From this, we see that (8-159) is independent of s along these curves.

There are two possible roots for the quadratic equation of (8-160). Being a hyperbolic equation, $(b^2 - 4ac) > 0$, the two roots are different from each other. Consequently, at each point in the x-t plane, there are two characteristic curves. For the convenience of discussion, we shall call one of these two curves the α-characteristic curve and the other one the β-characteristic curve. Along either one of these two curves, (8-159) reduces to the form

$$a\frac{dp}{dt}\frac{dx}{dt} + c\frac{dq}{dt} + e\frac{dx}{dt} = 0 \quad (8\text{-}161)$$

It is possible to use (8-160) and (8-161) to find the solution $\phi(x,t)$ for (8-155) in the same way as we did in using (8-152) and (8-153) to solve the first-order PDE represented by (8-147).

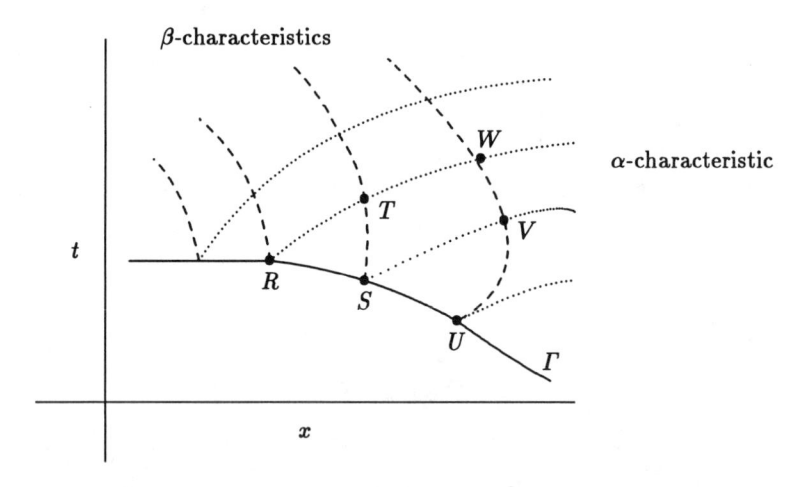

Fig. 8-21 Characteristic curves for a second-order hyperbolic PDE. To solve the PDE using the method of characteristics, we start with three adjacent points, R, S, and U, along Γ. Two of them are used to find the location of T, where the characteristic curves from the pair meet, as well as the value of $\phi(x,t)$ at the intersection. Another pair is used to find the corresponding quantities at V. From T and V, the location and the solution at W are obtained. The solution at other points may be found in the same way by starting with different points along Γ until the whole region of interest is covered.

Since we are dealing with second-order PDEs, we need two conditions specified at each point along the initial curve Γ. We can take these as the values of p and q. For two points R and S that are very near each other on Γ, the α-characteristic curve starting from R and the β-characteristic curve starting from S may intersect each other at some point T in the x-t plane, as shown schematically in Fig. 8-21. The location of T may be found by successive approximation, in a similar way to what was done earlier for first-order PDE. Once this is done, the same process can be repeated for another pair of points S and U along Γ and this gives us V. From the two points T and V we can obtain a point W, and so on. In this way, the value of $\phi(x,t)$ may be mapped out, point by point, in the x-t space of interest to us.

Finite difference method for transport equations. It is also possible to propagate the solution of, for example, a transport equation, forward in time using the method of finite difference. For the convenience of discussion, we shall divide the x-t space into an evenly spaced rectangular mesh of points with Δt as the step size in time and Δx as the step size along x. Since the derivative with respect to x is first order, we need one boundary condition for each t value on the mesh. Let us assume that this is given to us in terms of the values of $\phi(x,t)$ at $x = x_0$ and the domain of interest is in $x \geq x_0$. In this case, the most logical difference scheme is to take a forward difference in t and a backward difference in x. The resulting FDE for (8-149) may be written in the form

$$\frac{\phi(x_i, t_{k+1}) - \phi(x_i, t_k)}{\Delta t} + v \frac{\phi(x_i, t_k) - \phi(x_{i-1}, t_k)}{\Delta x} = f(x_i, t_k)$$

where $v = b/a$ and $f = c/a$. The three-point stencil used in such an approach is shown in Fig. 8-22(a).

This is an explicit method, as the unknown quantity at $t = t_{k+1}$ is given directly in terms of those at $t = t_k$. We can see this more directly by rewriting (8-148) in the form

$$\phi(x_i, t_{k+1}) = (1 - \eta)\phi(x_i, t_k) + \eta \phi(x_{i-1}, t_k) + (\Delta t) f(x_i, t_k)$$

where

$$\eta = \frac{v(\Delta t)}{\Delta x} \tag{8-162}$$

For the solution to be stable, it is necessary for the maximum value of η to be less than or equal to unity. This is quite easy to understand if we examine the domain of dependence, in the same way we did in the previous section. Here, if Δt is larger compared with Δx, the solution at a point (x_i, t_k) is calculated from those at a small number of points at earlier times. On the other hand, the physical information propagates at the velocity v. For $\eta > 1$, the numerical calculation is moving forward at a rate faster than v. As a result, only a smaller number of earlier points are used than the number required by the transport phenomenon itself. It

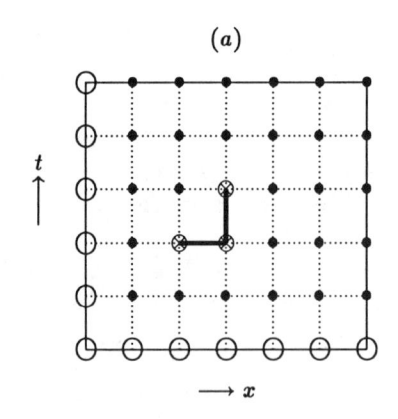

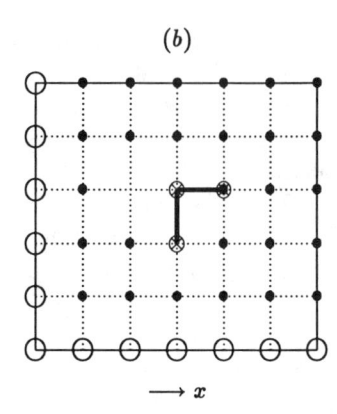

Fig. 8-22 Finite difference methods for solving transport equations of one spatial dimension. The stencil (a) is an explicit method, which is stable if $v\Delta t \leq \Delta x$. The implicit method, given in (b), is always stable but more complicated to implement in general. The initial and boundary conditions are assumed to be supplied at the points indicated by circles.

is obvious that, in this case, the solution cannot be physically meaningful. This shows up in our calculation in the form of an unstable solution. A mathematical derivation of this conclusion may be found in standard references on PDEs.

We can also design an implicit method to solve (8-149). If the initial and boundary conditions are supplied at those mesh points indicated by circles in Fig. 8-22, it is possible to use forward differences for both time and space to obtain the finite difference equation

$$\frac{\phi(x_i, t_{k+1}) - \phi(x_i, t_k)}{\Delta t} + v \frac{\phi(x_{i+1}, t_{k+1}) - \phi(x_i, t_{k+1})}{\Delta x} = f(x_i, t_k)$$

In terms of the η defined in (8-162), this may be put into the form

$$\eta \phi(x_{i+1}, t_{k+1}) + (1 - \eta)\phi(x_i, t_{k+1}) = \phi(x_i, t_k) + (\Delta t) f(x_i, t_k)$$

The stencil for this case is shown in Fig. 8-22(b). Since the calculation, in general, involves solving a set of simultaneous equations, it is not as easy to implement as the explicit methods. However, there are no fundamental difficulties in carrying out the work, as we saw in previous sections.

Wave equations. For wave equations, both the time and spatial derivatives are second order. For each set of points along the x-axis, two initial values must be specified. We shall assume that these are given to us in terms of the values of $\phi(x, t)$ at $t = t_0$ and $t = (t_0 + \Delta t)$, as indicated schematically in Fig. 8-23. For each set of points along the t-axis, two boundary values must be provided by the physics of the problem. We shall assume that they are given to us in terms of the values of $\phi(x,t)$ at $x = x_b$ and $x = x_d$, the two ends of the interval for x. Using central differences for both derivatives in (8-148), we obtain a FDE of the form

$$\frac{\phi(x_i, t_{k+1}) - 2\phi(x_i, t_k) + \phi(x_i, t_{k-1})}{(\Delta t)^2}$$
$$- \frac{1}{v^2} \frac{\phi(x_{i+1}, t_k) - 2\phi(x_i, t_k) + \phi(x_{i-1}, t_k)}{(\Delta x)^2} = 0$$

The five-point stencil is shown in Fig. 8-23(a). Since this is an explicit scheme, the FDE can be put into the form of the solution for the only unknown quantity in the equation

$$\phi(x_i, t_{k+1}) = -\phi(x_i, t_{k-1}) + \eta \phi(x_{i+1}, t_k) + 2(1-\eta)\phi(x_i, t_k) + \eta \phi(x_{i-1}, t_k)$$

(8-163)

where $\eta = (\Delta t)^2/(v \Delta x)^2$. The condition of stability, $\eta \leq 1$, may be obtained from arguments based on the domain of dependence, as we did earlier for transport equations.

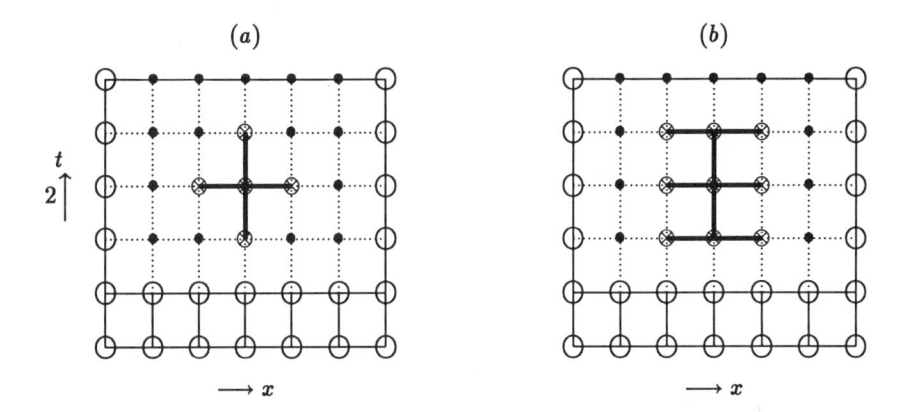

Fig. 8-23 Finite difference methods for solving wave equations. The stencil in (a) pertains to an explicit method with stability condition $v(\Delta t) \leq \Delta x$. The implicit method given in (b) involves nine points of the mesh and may be used where explicit methods fail. Circles indicate values supplied by initial and boundary conditions.

An implicit FDE equation for the wave equation may be derived using the Crank-Nicholson scheme:

$$\frac{1}{(\Delta t)^2} \delta_t^2 \phi(x_i, t_k) - \frac{1}{4v^2(\Delta x)^2} \{\delta_x^2 \phi(x_i, t_{k+1}) + 2\delta_x^2 \phi(x_i, t_k) + \delta_x^2 \phi(x_i, t_{k-1})\} = 0$$

(8-164)

where, to simplify the notation, we have used

$$\delta_t^2 \phi(x_i, t_k) \equiv \phi(x_i, t_{k+1}) - 2\phi(x_i, t_k) + \phi(x_i, t_{k-1})$$

$$\delta_x^2 \phi(x_i, t_k) \equiv \phi(x_{i+1}, t_k) - 2\phi(x_i, t_k) + \phi(x_{i-1}, t_k) \tag{8-165}$$

There are different ways to "average" the three finite differences in space at times $t = t_{k-1}$, t_k, and t_{k+1} on the right side of (8-164), and we have chosen the one that puts twice the weight on the middle point compared with each of the two on the side. The relation given by (8-164) may be regarded as an expression that gives the three quantities $\phi(x_{i+1}, t_{k+1})$, $\phi(x_i, t_{k+1})$, and $\phi(x_{i-1}, t_{k+1})$ at $t = t_{k+1}$ in terms of the six quantities at $t = t_k$ and t_{k-1}. In matrix form, the equations are very similar to the parabolic example of the previous section, and the same method of reducing the tridiagonal matrix on the left side of the equation may be used.

Box 8-11   Program WAVE

Explicit method to solve the wave equation in (x, t)

An example of second-order hyperbolic PDE

Initialization:
 (a) Set up a mesh of $(N_x + 1)$ points for x and $(N_t + 1)$ for t.
 (b) Set up the range $[x_b, x_d]$ and boundary conditions.
 (c) Select a propagation velocity v.
 (d) Obtain the step size Δx.
1. Input η of (8-162).
2. Calculate Δt, the step size in time t.
3. Specify the initial conditions.
4. Propagate forward in time using (8-163).
5. Output $\phi(x_i, t_k)$ once every 50 time steps.

Vibrations of a stretched string.   As an example, we shall solve the wave equation of (8-148) for a stretched string. To fix the boundary conditions, we shall anchor both ends of the string to the walls of the laboratory. In other words,

$$\phi(x_b, t) = \phi(x_d, t) = 0$$

where x_b and x_d are the locations of the two ends of the string and our domain of interest lies in the interval $x = [x_b, x_d]$. Without losing any generality, we can let $x_b = -1$ and $x_d = 1$. Initially, at $t = 0$, the string is at rest,

$$\left.\frac{d\phi}{dt}\right|_{t=0} = 0$$

To set the string into vibration, we introduce a sinusoidal pulse near the end at $x = x_d$ in the form

$$\phi(x, t = 0) = \begin{cases} A\sin(2\pi x) & \text{for } 0.5 \leq x \leq 1 \\ 0 & \text{otherwise} \end{cases}$$

The subsequent behavior of the string is described by (8-148) and may be calculated using the FDE of (8-163). The algorithm is outlined in Box 8-11.

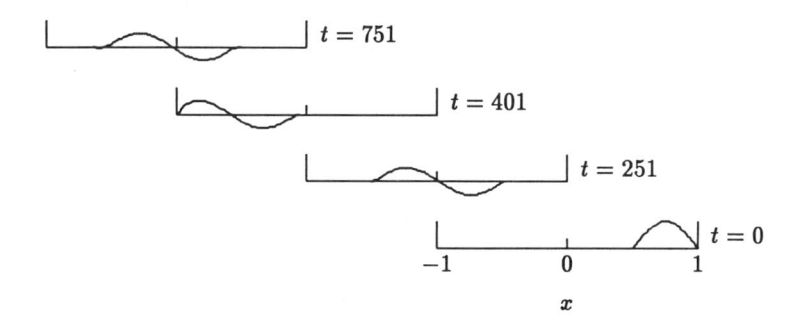

Fig. 8-24 Propagation of a wave with $\eta = v(\Delta t)/(\Delta x) = 0.25$ along a stretched string fixed at both ends. The initial shape of the string is a sine wave in the region $x = [0.5, 1]$, as shown by the diagram labeled $t = 0$. At $t = 251$ in arbitrary units, the wave is more or less at the center of the string. It reaches the end away from its starting point at $t = 501$. The wave is then reflected back at the $x = -1$ end and travels in the opposite direction, as shown by its location at $t = 751$. At $t = 1000$ (not shown), the wave essentially returns to its position at $t = 0$. The oscillation travels between the two ends of the string in this way, as there is no damping term to dissipate the energy.

Since no damping is assumed in the problem, the energy represented by the initial pulse is conserved. For the particular one we have selected, it travels back and forth along the string between the two fixed ends with constant amplitude, as shown schematically in Fig. 8-24. In our representation of the string by a finite number of points in the FDE, only a finite number of normal modes can be present in the problem. If a sharp pulse is introduced into the system, instead of the smooth sine wave used here, the pulse shape will be modified, as it requires more normal modes to represent a sharp pulse than our mesh can provide. The selection of a finite number of points along the x-direction for our FDE is equivalent to replacing the string by N_x number of discrete "masses" coupled together by tension in the connecting (massless) string. In the usual way of deriving the wave equation, an approach that is almost opposite to the present one is used. For example, in standard texts for wave phenomena, it is common to obtain (8-148) by starting with N-coupled oscillators. The wave equation is obtained on invoking the limit of $N \to \infty$. Here, we take the opposite approach and solve the wave equation numerically by replacing the continuous string with a finite number of mesh points.

§8-10 Nonlinear differential equations

So far, we have been dealing mainly with linear differential equations. In terms of the standard form of a second-order PDE given in (8-9),

$$p\frac{\partial^2 \phi}{\partial x^2} + q\frac{\partial^2 \phi}{\partial x \partial y} + r\frac{\partial^2 \phi}{\partial y^2} + s\frac{\partial \phi}{\partial x} + t\frac{\partial \phi}{\partial y} + u\phi + v = 0$$

we find that the coefficients p, q, r, s, t, u, and v are functions of the independent variables x and y only. If any of them involve $\phi(x, y)$ or its derivatives, the equation becomes *nonlinear*. There are many possible forms of nonlinear differential equations and, in general, they are more difficult to solve. In physics, we are more likely to encounter a special class of nonlinear differential equations, called *quasilinear equations*. Here, the highest-order derivatives are linear in the equation. For example, in the second-order PDE shown above, the coefficients p, q, r, s, t, u, and v do not involve any of the three second-order derivatives, $\partial^2\phi/\partial x^2$, $\partial^2\phi/\partial x \partial y$, and $\partial^2\phi/\partial y^2$, but they may be functions of $\partial\phi/\partial x$, $\partial\phi/\partial x$, and $\phi(x, y)$, as well as x and y.

On the other hand, if only the higher powers of the independent variables appear in the coefficients, the differential equation remains linear. For example, the one-dimensional Schrödinger equation for a particle of mass μ in an anharmonic potential well, as we saw earlier in (5-54),

$$-\frac{\hbar^2}{2\mu}\frac{d^2\phi}{dx^2} + \left\{\frac{1}{2}\mu\omega^2 x^2 + \epsilon\hbar\omega\left(\frac{\mu\omega}{\hbar}\right)^2 x^4\right\}\phi(x) = 0$$

is a linear ODE, as only x appears in higher powers than unity.

As an example, consider a pendulum with angular displacement $\phi(t)$, as shown in Fig. 8-25. The oscillation is governed by the second-order ODE

$$\frac{d^2\phi}{dt^2} + \omega_0^2 \sin\phi(t) = 0 \tag{8-166}$$

where $\omega_0 = \sqrt{g/\ell}$, with g being the acceleration due to gravity and ℓ the length of the pendulum. It is a quasilinear equation, since $\phi(t)$ enters the equations as the argument of a transcendental function. If the displacement is sufficiently small, we can approximate $\sin\phi(t)$ by its first term in the series expansion

$$\sin\phi = \phi - \frac{1}{3!}\phi^3 + \cdots$$

In this limit, we obtain the result

$$\frac{d^2\phi}{dt^2} + \omega_0^2 \phi(t) = 0$$

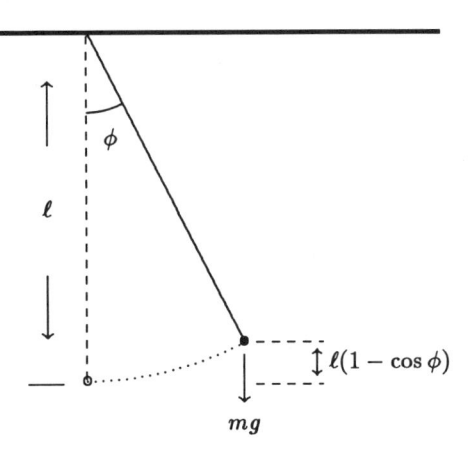

Fig. 8-25 Pendulum with a finite angular displacement ϕ. The potential energy is $mg\ell(1-\cos\phi)$ and the force acting on the pendulum is $mg\ell\sin\phi$, which may be approximated as $mg\ell\phi$ if ϕ is small.

the usual linear form of the ODE for simple harmonic motion. As we shall soon see, the appearance of terms that are nonlinear in the function $\phi(t)$ and its derivatives complicates the solution to the corresponding FDE.

Compared with nonlinear differential equations, quasilinear differential equations are, in general, easier to solve. This can be seen from a finite difference approximation to (8-166). Using a central difference scheme, we have the result

$$\phi(t_{k+1}) - 2\phi(t_k) + \phi(t_{k-1}) + (\Delta t)^2 \omega_0^2 \sin\phi(t_k) = 0 \qquad (8\text{-}167)$$

Because of the $(\Delta t)^2$ factor, the influence of the nonlinear term on the FDE is relatively weak. Consequently, reasonable approximations may be used. However, we still have to face some practical problems.

Quasilinear initial value problem. For linear differential equations, the FDEs are also linear in the unknown quantities. Because of this, we were able, in the previous sections, to make use of efficient techniques to obtain, for example, the roots of the linear algebraic equations. The presence of the nonlinear term, as for example the $\sin\phi(t)$ term in (8-167), changes the situation. Even when the term is small, we cannot simply adapt the methods designed for linear systems to solve the equation without some additional considerations. There are two possible approaches to take. The first is to calculate the contributions of $\sin\phi(t_k)$ using the best estimate of the value of $\phi(t_k)$ we have for the point. In this way, we remove the nonlinear term from the list of unknown quantities for the algebraic equation. The rest of the equation is linear and can therefore be solved by one of the methods we discussed earlier for a system of linear algebraic equations.

The second approach is to use an explicit scheme for the FDEs. If, for the moment, we ignore questions concerning stability and convergence, we have no difficulty in propagating the solution from a set of initial conditions in the presence of the nonlinear term. Consider (8-167) as an example. Since it is an initial value problem, we may regard it as an equation to find $\phi(t_{k+1})$ from the values of $\phi(t_k)$ and $\phi(t_{k-1})$. The equation is second order and we have two initial values, which may be taken to be those for $\phi(t_0)$ and $\phi(t_1)$. With these two values as the input, we can obtain $\phi(t_2)$ using (8-167) by putting $k = 1$. With $\phi(t_1)$ and $\phi(t_2)$, we can solve for $\phi(t_3)$ by putting $k = 2$, and so on, as we did earlier in (8-14). In other words, we ignore the fact that the equation is nonlinear and solve the FDEs using one of the standard methods for initial value problems. The stability of these methods, such as the Runge-Kutta or extrapolation methods discussed earlier, is difficult to establish for nonlinear and quasilinear ODEs in general. On the other hand, for the example of (8-166), it is easy to see by actual calculation that either one of the two methods is able to yield fairly stable solutions in regions of interest.

Let us solve (8-166) for a pendulum with finite initial amplitude as an example. The period τ is known from analytical calculations to be given by the integral

$$\tau = \frac{2}{\omega_0} \int_{-\pi/2}^{+\pi/2} \frac{d\beta}{\sqrt{1 - \sin^2 \frac{\phi_0}{2} \sin^2 \beta}}$$

$$= \frac{2\pi}{\omega_0} \left\{ 1 + \frac{1}{4} \sin^2 \frac{\phi_0}{2} + \frac{9}{64} \sin^4 \frac{\phi_0}{2} + \frac{25}{256} \sin^6 \frac{\phi_0}{2} + \cdots \right\} \quad (8\text{-}168)$$

where ϕ_0 is the initial amplitude. We shall make use of this result as a check for the accuracy of our numerical solution for (8-166).

Table 8-4 Period of a pendulum with finite initial amplitude.

| Initial amplitude || Period ||
Radians	Degrees	Numerical	Series
0.1	5.7	6.287	6.287
0.25	14.3	6.308	6.308
0.5	28.6	6.383	6.382
1.0	57.3	6.700	6.698
1.5	85.9	7.300	7.265

Using either of the methods described above, we can find the roots for the FDE given in (8-167). The solution appears in the form of a table of the values of $\phi(t_k)$ for a range of the discrete values of t_k. To make use of the results of (8-168) as a check, we must find the period of $\phi(t)$ from such a table of numerical values. One way to do this is to use an inverse interpolation to find the zeros of the oscillatory function. Once this is done, we can define the period τ as twice the difference between two consecutive zeros in the results. The Bessel inverse interpolation technique of §3-6 can be easily adapted to this task and the results are shown in Table 8-4 for various initial amplitudes. The values listed in the last column, labeled "series," are calculated using (8-168) up to the $\sin^6(\phi_0/2)$ term given explicitly. The small discrepancies for large values of ϕ_0 between the numerical and series results are likely to be coming from the need for additional terms in the series expansion. The comparison provides us some confidence that the numerical solution to (8-166) has the required accuracy.

Quasilinear boundary value problem. As an example of quasilinear boundary value problems, we shall consider the "kink" solution of the differential equation

$$\frac{d^2\phi}{dx^2} + \phi(x) - \phi^3(x) = 0 \qquad (8\text{-}169)$$

The equation arises in field theory when the potential takes on the form

$$V(\phi) = \frac{\lambda}{4}\left\{\phi^2(x) - \frac{\mu^2}{\lambda}\right\}^2$$

instead of the usual $\phi^2(x)$ dependence given by, for example, a quadratic approximation shown schematically in Fig. 4-1. It represents one of the simplest extensions we can make on the basic $\phi^2(x)$ potential and, as a result, is widely used in a variety of investigations.

To derive (8-169), we start with the Lagrangian density for a field ϕ that is a function of x and t only:

$$\mathcal{L}(x,t) = \frac{1}{2}\left(\frac{\partial\phi}{\partial t}\right)^2 - \frac{1}{2}\left(\frac{\partial\phi}{\partial x}\right)^2 - V(\phi)$$

From this, we obtain, using the Lagrange equation of motion, the differential equation

$$\frac{\partial^2\phi}{\partial t^2} - \frac{\partial^2\phi}{\partial x^2} = \mu^2\phi(x,t) - \lambda\phi^3(x,t)$$

The constants μ^2 and λ may be absorbed into x and $\phi(x)$ by the replacements

$$x \longrightarrow \frac{x}{\mu} \qquad\qquad \phi \longrightarrow \frac{\mu}{\sqrt{\lambda}}\phi$$

On ignoring the time dependence, we obtain (8-169). Among other interests, the equation is one of the few whose solitary wave, or soliton, solutions are known analytically. As a differential equation, it is quasilinear, since the nonlinear term is in $\phi(x)$ and the second-order derivative appears linearly.

We shall be concerned only with a particular solution of (8-169), generally known as the static "kink" solution. There are several different possible forms of the solution depending on the boundary conditions we wish to impose on the system. The one shown in Fig. 8-26 is obtained with the conditions

$$\phi(-\infty) = -1 \qquad \phi(+\infty) = +1$$

Alternatively, we can have an "antikink" solution by imposing, instead, the conditions $\phi(\pm\infty) = \mp 1$. The kink solution is also known from analytical calculations and has the form

$$\phi(x) = \tanh\left(\frac{x-x_0}{\sqrt{2}}\right) \tag{8-170}$$

Here x_0 is the point where the value of $\phi(x)$ changes sign. For our numerical solution, the value of x_0 depends on the initial trial solution used to solve the problem, as the differential equation and the boundary conditions do not contain any explicit dependence on x_0.

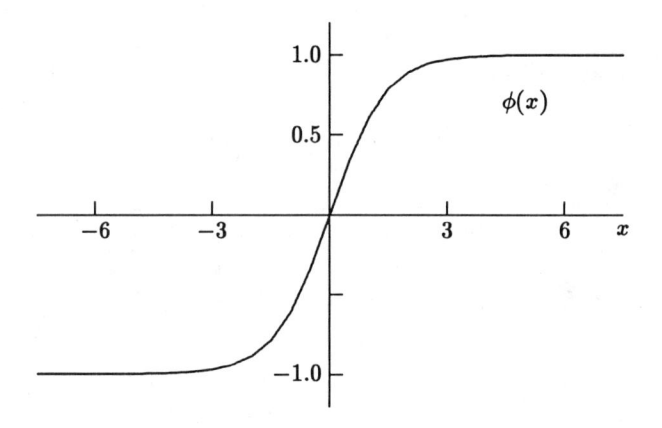

Fig. 8-26 The kink solution as an example of a soliton. It is a solution of the second-order quasilinear ODE given by (8-169). Analytically, it has the form $\phi(x) = \tanh(x/\sqrt{2})$, one of the few cases where such a solution is known.

Using a mesh of $(N+1)$ points, $x_0, x_1, x_2, \ldots, x_N$, equally spaced along the x-axis, the corresponding FDE for (8-169) takes on the form

$$\frac{1}{(\Delta x)^2}\left\{\phi(x_{i+1}) - 2\phi(x_i) + \phi(x_{i-1})\right\} + \phi(x_i) - \phi^3(x_i) = 0 \tag{8-171}$$

where Δx is the step size. As a boundary value problem, we shall assume that the values of $\phi(x_0)$ and $\phi(x_N)$ are given to us by the physics. The rest of the values of $\phi(x_i)$ must be obtained by solving the set of $(N-1)$ equations given by (8-171). However, since the equations are no longer linear in the unknown quantities $\phi(x_i)$, we cannot apply directly the methods described earlier for solving boundary value problems in linear differential equations.

There are two ways to overcome the difficulties caused by the nonlinear terms in the FDE. The first is to *linearize* the problem by making a Taylor series expansion of the nonlinear terms around some estimated values of the solution at each mesh point and retain only first terms of the series. For example, the $\phi^3(x_i)$ term in (8-171) may be expanded by a series of the form

$$\phi^3(x_i) \approx \tilde{\phi}^3(x_i) + 3\tilde{\phi}^2(x_i)\{\phi(x_i) - \tilde{\phi}(x_i)\} + \cdots \qquad (8\text{-}172)$$

where $\tilde{\phi}(x_i)$ is an estimated value of $\phi(x_i)$. By truncating the series after the second term, we have a linear approximation for $\phi^3(x_i)$. Initially, the estimate may be obtained from an approximate solution of the original differential equation. Once a first approximation is available, we can improve the results by making use of the approximate solution. This is carried out iteratively until the solution converges, in the same way as we have done on many earlier occasions. However, we shall not pursue this particular method of solution here.

Newton's method to solve the FDE. A second approach for solving the FDE of (8-171) is to use Newton's method. As in (8-172), this method also involves linearizing the problem by approximating the nonlinear term with the first two terms of a Taylor series. To illustrate the method, consider first the question of finding the roots of an algebraic equation of the form

$$f(s) = 0$$

where $f(s)$ is a nonlinear function of s. Let us assume the root is located at $s = s^{(r)}$. Near the root, we can make a series expansion of the function $f(s)$ in a similar manner to what was done in (8-172). If $s^{(m)}$ is the mth estimate of the root, we have the relation

$$f(s) \approx f(s^{(m)}) + \left.\frac{df}{ds}\right|_{s=s^{(m)}} \{s - s^{(m)}\} \qquad (8\text{-}173)$$

An improved estimate may be obtained by realizing that, if $s^{(m+1)}$ is the root, then

$$f(s^{(m+1)}) = 0 \approx f(s^{(m)}) + \left.\frac{df}{ds}\right|_{s=s^{(m)}} \{s^{(m+1)} - s^{(m)}\} \qquad (8\text{-}174)$$

From this, we obtain the next estimate for the value of the root:

$$s^{(m+1)} = s^{(m)} - \frac{f(s^{(m)})}{\left.\frac{df}{ds}\right|_{s=s^{(m)}}}$$

Since the relation given by (8-174) is only approximate, we do not expect $s^{(m+1)}$ to be the same as the solution $s^{(r)}$. However, it should represent a better estimate for $s^{(r)}$ than $s^{(m)}$. The process may be iterated until the value of $f(s^{(m)})$ is sufficiently close to zero for $s^{(m)}$ to be acceptable as the root. Again, we need a starting value $s^{(0)}$, which can be obtained from, for example, an approximate analytical solution to the original differential equation. In general, the convergence of the method is quadratic near the minimum. As a result, the approach is efficient if a good starting value is available.

To use the method to solve FDE arising from nonlinear differential equations, we need to generalize it for a set of n functions $f_1(s), f_2(s), \ldots, f_n(s)$ that depend on n variables $s = \{s_1, s_2, \ldots, s_n\}$. In the place of (8-173), we now have the relation

$$f(s) \approx f(s^{(m)}) + \mathcal{J}(s^{(m)})\{s - s^{(m)}\} \qquad (8\text{-}175)$$

where $\mathcal{J}(s^{(m)})$ is the Jacobian matrix

$$\mathcal{J}(s) \equiv \begin{pmatrix} \frac{\partial f_1}{\partial s_1} & \frac{\partial f_1}{\partial s_2} & \cdots & \frac{\partial f_1}{\partial s_n} \\ \frac{\partial f_2}{\partial s_1} & \frac{\partial f_2}{\partial s_2} & \cdots & \frac{\partial f_2}{\partial s_n} \\ \vdots & \vdots & \ddots & \vdots \\ \frac{\partial f_n}{\partial s_1} & \frac{\partial f_n}{\partial s_2} & \cdots & \frac{\partial f_n}{\partial s_n} \end{pmatrix} \qquad (8\text{-}176)$$

Starting from $s^{(m)}$ for the mth-order approximation, we can find the value for the next order from the relation

$$s^{(m+1)} = s^{(m)} - \Delta^{(m)} \qquad (8\text{-}177)$$

where $\Delta^{(m)}$ is the solution to the set of equations

$$\mathcal{J}(s^{(m)})\Delta^{(m)} = f(s^{(m)}) \qquad (8\text{-}178)$$

In general, the Jacobian matrices arising from FDE are sparse and the solution to (8-177) may be found with relative ease.

For our FDE of (8-171), there are $(N+1)$ points on the mesh. Since two boundary conditions are supplied by the problem, we are left with a set of $(N-1)$ unknown quantities, $\phi(x_1), \phi(x_2), \ldots, \phi(x_{N-1})$. It is possible to treat (8-171) as a set of $(N-1)$ equations, and the $(N-1)$ unknown $\phi(x_i)$ for $i = 1, 2, \ldots, (N-1)$ are the roots of this set of simultaneous equations. A typical equation in the set has the form

$$f_i(\phi) = \frac{1}{(\Delta x)^2}\{\phi(x_{i+1}) - 2\phi(x_i) + \phi(x_{i-1})\} + \phi(x_i) - \phi^3(x_i) \qquad (8\text{-}179)$$

We can use the approximation of (8-175) to locate the roots using the recursive approach given by (8-177) and (8-178). The only nonzero matrix elements of the Jacobian matrix are

$$\mathcal{J}_{i,i-1} = \frac{\partial f_i}{\partial \phi(x_{i-1})} = \frac{1}{(\Delta x)^2}$$

$$\mathcal{J}_{i,i} = \frac{\partial f_i}{\partial \phi(x_i)} = \frac{-2}{(\Delta x)^2} + 1 - 3\phi^2(x_i)$$

$$\mathcal{J}_{i,i+1} = \frac{\partial f_i}{\partial \phi(x_{i+1})} = \frac{1}{(\Delta x)^2} \qquad (8\text{-}180)$$

Chapter 8]   Ordinary and Partial Differential Equations   583

Box 8-12   Program STATIC_KINK

Solution of the kink equation by Newton's method

Subprograms used:
   INIT: Construct an equally spaced mesh and give an initial estimate of the solution.
   SLV_TRI: Gaussian elimination for a tridiagonal matrix (cf. Box 3-10).

Initialization:
   Select the error tolerance ϵ, the number of mesh points N, and the maximum number of iterations allowed.

1. Input the range and normalization factor.
2. Use INIT to:
   (a) Construct an equally spaced mesh of $(N+1)$ points.
   (b) Give an initial estimate of the solution as a step function at $x = 0$.
3. Iterate the solution:
   (a) Calculate $f_i(\phi)$ of (8-179) and the elements of the Jacobi matrix $\mathcal{T}_{i,i}$ of (8-180).
   (b) Use SLV_TRI to solve the tridiagonal Jacobi matrix equation (8-178).
   (c) Update the solution:
       (i) Improve the solution with (8-177) using the value of Δ obtained.
       (ii) Estimate the average error $\mathcal{E}$ by averaging over $|\Delta|$.
4. Iterate the solution again if $\mathcal{E}$ is larger than the tolerance of error ϵ.
5. Output the solution if $\mathcal{E}$ is sufficiently small, or terminate the calculation if the maximum number of iterations is exceeded.

---

for $i = 1, 2, \ldots, (N-1)$. Since $\phi(x_0)$ and $\phi(x_N)$ are given by the boundary conditions, we have
$$\mathcal{J}_{1,0} = \mathcal{J}_{N-1,N} = 0$$
The Jacobian matrix is therefore a tridiagonal matrix of dimension $(N-1)$ and has the same structure as (8-145). As a result, we can find the solution to (8-178) using the same Gaussian elimination technique, as was done in §3-7 and §8-8.

The procedure to solve the nonlinear FDE of (8-171) may be summarized in the following way. The starting point is an initial estimate of the values of $\phi(x_i)$ for $i = 1, 2, \ldots, (N-1)$. Using these values, we construct the Jacobian matrix of (8-180), solve (8-178) by Gaussian elimination for Δ, and obtain a better set of estimates using (8-177). These three steps are repeated, if necessary, using the improved estimates until the differences in two successive iterations are smaller than the error that can be tolerated. The procedure is outlined in Box 8-12.

For the calculated results shown in Fig. 8-26, the following step function is used as the initial estimate:
$$\phi^{(0)}(x_i) = \begin{cases} -1 & \text{for } x_i < 0 \\ +1 & \text{otherwise} \end{cases}$$
This simple form ensures that the point for $\phi(x) = 0$ is located at $x = 0$. By shifting the point where $\phi^{(0)}(x)$ changes from -1 to $+1$ in such a trial solution,

it is possible to produce kink solutions corresponding to different values of x_0 in (8-170).

§8-11 Stiff problems

In §8-2, where we were solving an initial value problem for a damped harmonic oscillator, it was found that the numerical accuracy for the underdamped case is much worse than that produced by the same computer program for the corresponding overdamped and critically damped oscillations. The reason for the difference in the numerical accuracies is that, in the two latter cases, the amplitude of the "oscillation" decays exponentially, essentially with a single time constant γ. In contrast, the amplitude in the underdamped system contains two components, one decaying exponentially with time constant γ and the other oscillating with a different time constant that is related to the angular frequency ω. In solving the initial value problem, the role of each of the two time constants is determined by the initial conditions — any errors introduced in the numerical calculations can easily distort the relative importance of the two components in the solution, resulting in large numerical errors. This is known as the problem of *stiffness*, and differential equations displaying such a symptom are called stiff equations.

A good illustration of the stiffness problem is provided by the second-order equation

$$\frac{d^2\phi}{dt^2} - L^2\phi(t) = 0 \qquad (8\text{-}181)$$

where L is some large, positive quantity to be specified later. Such a form of the ODE is encountered in a variety of physical problems. The general solution is

$$\phi(t) = Ae^{-Lt} + Be^{+Lt} \qquad (8\text{-}182)$$

where A and B are two constants of integration that must be determined from the initial or boundary conditions. For our illustration, we shall take L to be a constant, for example,

$$L^2 = 100$$

and the initial conditions are

$$\phi(t=0) = 1 \qquad \left.\frac{d\phi}{dt}\right|_{t=0} = -L \qquad (8\text{-}183)$$

The solution of the ODE under these conditions is well known:

$$\phi(t) = e^{-10t} \qquad (8\text{-}184)$$

The $\exp(+Lt)$ term does not appear here because of the initial conditions. For purposes of illustration, let us solve the same equation numerically using the method of Euler given earlier in §8-1. It will be clear from the discussion that the

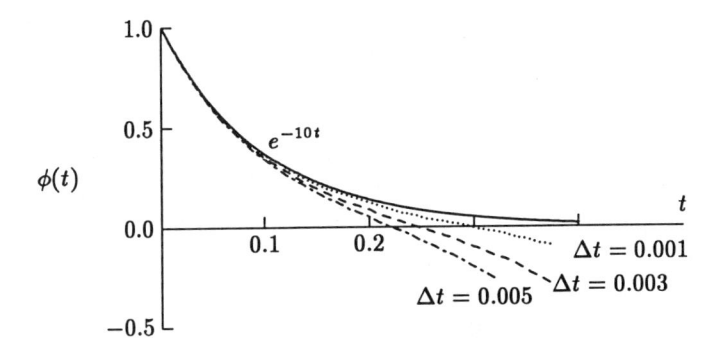

Fig. 8-27 Solution of the stiff equation (8-181) using Euler's method. The analytical solution e^{-10t} is shown by the solid curve. The numerical results with step size of $\Delta t = 0.001$, 0.003, and 0.005 are shown, respectively, as the dotted, dashed, and dash-dot curves.

difficulties associated with stiffness are present in other explicit methods as well and are not limited to the particular way we solve the problem.

In the same way as we did in going from (8-10) to (8-11), we can construct an evenly spaced set of points, $t_0, t_1, t_2, \ldots, t_N$, and convert (8-181) into a FDE of the form

$$\phi(t_{k+1}) = \{2 + (\Delta t)^2 L^2\}\phi(t_k) - \phi(t_{k-1})$$

The initial conditions of (8-183) may be introduced into the calculation in the form

$$\phi(t_1) = 1.0 \qquad \frac{\phi(t_2) - \phi(t_1)}{\Delta t} = -10 \qquad (8\text{-}185)$$

where we have used a forward difference scheme to approximate the first derivative of $\phi(t)$ at $t = 0$. The calculated results using different step sizes Δt are shown in Fig. 8-27.

It is quite clear from examining the figure that the numerical results depart quickly from the exact value of $\exp(-Lt)$, shown as the solid curve, even for a fairly small step size Δt. Furthermore, the larger the value of Δt, the earlier in time for the departure to become noticeable. The discrepancies from the exact results may be viewed in the following way. The general solution to the differential equation has two different time scales, $\exp(-Lt)$ and $\exp(+Lt)$, as we saw in (8-182). It is through the initial conditions that the $\exp(+Lt)$ term is suppressed in (8-184). However, in the numerical solution, truncation and other errors introduce a small admixture of the unwanted term. Let us denote the amplitude of the $\exp(+Lt)$ term by ϵ. In the exact solution, $\epsilon = 0$. Our numerical solution, on the other hand, takes on the form

$$\phi(t) = e^{-Lt} + \epsilon e^{+Lt}$$

The size of ϵ depends, in part, on the value of Δt used. Regardless of how we may reduce the numerical errors, the unwanted $\exp(+Lt)$ term will eventually dominate the solution if $\epsilon \neq 0$, as the function $\exp(+Lt)$ rises rapidly with time. This is the main cause for the departure of the numerical solution from the exact value. One may argue that, in our illustration, a large part of the error comes from the forward difference approximation used in imposing the initial condition in (8-185). This is not entirely correct. We can test this point by replacing (8-185) with $\phi(t_1) = 1$ and $\phi(t_2) = \exp(-10t_2)$ as the initial conditions. The difficulty with the numerical solution persists, though with somewhat reduced magnitude.

Solutions of stiffness problem. There are ways to overcome the stiffness problem in some cases and we shall briefly mention two. Let us return to the damped harmonic oscillator example given earlier in (8-39). We recall that the second-order ODE has the form

$$\frac{d^2\phi}{dt^2} + 2\gamma \frac{d\phi}{dt} + \omega_0^2 \phi(t) = 0 \tag{8-186}$$

For the underdamped case, where the problem of stiffness occurs, we can transform the equation into one involving only a single time scale by defining a function $y(t)$ through the relation

$$\phi(t) = e^{-\gamma t} y(t) \tag{8-187}$$

On substituting this form of $\phi(t)$ into (8-186), we obtain an equation for $y(t)$,

$$\frac{d^2 y}{dt^2} + (\omega_0^2 - \gamma^2) y(t) = 0$$

This equation has only one time scale, given by $\sqrt{\omega_0^2 - \gamma^2}$, and we should have no difficulty in solving it numerically using any of the techniques discussed earlier for initial value problems.

Transformations similar to that of (8-187) can often be an effective way to eliminate the problem of stiffness. On the other hand, to apply the transformation, we must have beforehand a knowledge of one of the time scales involved. This is not always possible. As an alternative for a large variety of problems, we can apply the Riccati transformation

$$\frac{d\phi}{dt} = y(t)\phi(t) \tag{8-188}$$

which turns the differential equation from one for $\phi(t)$ into one for $y(t)$. On differentiating both sides of (8-188) with respect to t, we obtain

$$\frac{d^2\phi}{dt^2} = \frac{dy}{dt}\phi(t) + y(t)\frac{d\phi}{dt} = \frac{dy}{dt}\phi(t) + y^2(t)\phi(t)$$

Using this transformation, (8-181) is reduced to a first-order, but quasilinear, equation in $y(t)$:

$$\frac{dy}{dt} + y^2(t) = L^2$$

This may be solved using one of the methods discussed in the previous section. Through the Riccati transformation, the damped harmonic oscillator equation of (8-186) may be changed into the form

$$\frac{dy}{dt} + y^2(t) + 2\gamma y(t) = -\omega_0^2$$

Again, there is no difficulty in solving this quasilinear equation. However, for more complicated systems, the transformation can sometimes lead to a system of equations for which it is difficult to find a solution.

Another way to handle the problem of stiffness is to adopt an implicit method to solve the FDE. All the numerical solutions we have discussed so far in this section are explicit methods, as $\phi(t_{k+1})$ is given explicitly in terms of $\phi(t)$ at earlier times. To see why this method can fail, it is simpler to think instead in terms of a first-order ODE of the form

$$\frac{d\phi}{dt} = \eta\phi(t) \qquad (8\text{-}189)$$

We shall first take η to be a constant, but we shall return later to the more general case of a quasilinear system where η may a function of $\phi(t)$ and t.

Using a forward difference,

$$d\phi/dt \to \frac{\phi(t_{k+1}) - \phi(t_k)}{\Delta t}$$

we can put (8-189) into a FDE of the form

$$\frac{\phi(t_{k+1}) - \phi(t_k)}{\Delta t} = \eta\phi(t_k)$$

This may be put in terms of an algebraic equation,

$$\phi(t_{k+1}) = \{1 + (\Delta t)\eta\}\phi(t_k)$$

If $(\Delta t)\eta < 2$, the value of $\phi(t)$ increases in magnitude and oscillates in sign from one point on the mesh to the next. This is clearly unacceptable as a solution. On the other hand, if we use a backward difference approximation,

$$d\phi/dt \to \frac{\phi(t_k) - \phi(t_{k-1})}{\Delta t}$$

we have the result

$$\phi(t_k) = \phi(t_{k-1}) + (\Delta t)\eta\phi(t_k) \qquad (8\text{-}190)$$

This is an implicit equation for $\phi(t_k)$. For η a constant, we obtain a solution in the form

$$\phi(t_k) = \frac{\phi(t_{k-1})}{1 + \eta\Delta t}$$

which is stable for any step size Δt. In general, implicit methods are better suited for stiff problems.

If η is a function involving $\phi(t)$, we must apply techniques such as Newton's method discussed in the previous section to solve the quasilinear equation corresponding to (8-190). However, for higher-order equations, it may be easier to find a suitable implicit method to solve the problem. There does not seem to be any prospect at the moment for devising a method to handle the stiffness problem in general and, as a result, it remains one of the challenges of numerical methods.

§8-12 Finite element methods

In addition to finite difference methods, differential equations can also be solved using finite element methods (FEM), as mentioned in §8-1. We shall not try to define what a FEM is until later, after we have seen an example. In general, FEMs are more complicated to program, but they have several important advantages. For example, a large number of problems call for variable step sizes because of properties inherent in the problem. For many of the finite difference methods, it is not easy to incorporate this feature into the solution, as we have seen in the earlier sections of this chapter. On the other hand, it is relatively easy to tailor FEM to the physical problem in hand. Partly for this reason, a large part of the technical details of FEM often depend on the problem one wishes to solve. The subject of FEM is vast and we can only give an introduction here. For more details, see, for example, Burnett (*Finite Element Analysis*, Addison-Wesley, Reading, Massachusetts, 1988).

Let us start our discussion with an example. Consider the simple first-order differential equation

$$\frac{d\phi}{dt} + \lambda\phi(t) = 0 \qquad (8\text{-}191)$$

that occurs frequently in physics. For instance, $\phi(t)$ may represent the number of radioactive nuclei in a sample at time t. In this case, the equation describes the decay of the sample with decay constant λ. For the convenience of discussion, we shall restrict our interest in the time interval $[0, 1]$ and take

$$\lambda = 1$$

We can treat this differential equation as an initial value problem, with the condition that

$$\phi(t=0) = 1 \qquad (8\text{-}192)$$

The analytical solution for this case is a familiar one:

$$\phi(t) = e^{-t}$$
$$= 1 - t + \frac{1}{2!}t^2 - \frac{1}{3!}t^3 + \cdots \qquad (8\text{-}193)$$

As we shall see soon, it is helpful, in our introduction to FEMs, to think in terms of a series form for the solution.

Trial solution. Let us solve (8-191) using a power series expansion consisting of n terms as the trial solution

$$\tilde{\phi}_n(t) = a_0 + a_1 t + a_2 t^2 + \cdots + a_n t^n \qquad (8\text{-}194)$$

The tilde over $\phi(t)$ is to emphasize that this is a only a trial solution, and the subscript n, known as the degrees of freedom of the approximation, indicates the order of the power series used. To satisfy the initial condition of (8-192), it is necessary that

$$a_0 = 1$$

The other coefficients $a_1, a_2, \ldots, a_n$ must be found from the differential equation.

For the convenience of discussion, let us take $n = 2$. In this case, our trial solution assumes the simple form

$$\tilde{\phi}_2(t) = 1 + a_1 t + a_2 t^2 \qquad (8\text{-}195)$$

Obviously, the result is going to be quite different from the analytic solution given in (8-193). However, our interest is only in the restricted domain of $t = [0, 1]$ and, as we shall soon see, it is possible to find a set of values for a_1 and a_2 such that $\tilde{\phi}_2(t)$ represents a very good approximation of the function $\exp(-t)$ in the domain. In other words, we shall treat the coefficients $a_1, a_2, \ldots, a_n$ in (8-194) as parameters, and they are adjusted in such a way that $\tilde{\phi}_n(t)$ is as close to the exact solution as possible within the domain of interest.

Let us substitute the trial solution (8-195) into the differential equation (8-191). Since it is not the true solution, it is unlikely that $d\tilde{\phi}_2/dt + \tilde{\phi}_2(t)$ vanishes everywhere in $[0, 1]$. We shall define a quantity $R(t; a_1, a_2)$ to measure the difference from zero:

$$R(t; a_1, a_2) = \frac{d\tilde{\phi}_2}{dt} + \tilde{\phi}_2(t) = 1 + (1+t)a_1 + (2t+t^2)a_2 \qquad (8\text{-}196)$$

This is known as the *residual*. It is obvious that it is a function of t as well as the parameters a_1 and a_2. More generally, if there are n parameters instead of 2 in (8-195), the residual may be defined as

$$R(t; \boldsymbol{a}) \equiv \frac{d\tilde{\phi}_n}{dt} + \tilde{\phi}_n(t)$$

where we have used a bold-faced letter to denote a set of n parameters, $\boldsymbol{a} \equiv \{a_1, a_2, \ldots, a_n\}$.

The collocation method. One possible way to determine the values of the parameters is to require the residual to vanish at n points $t_1, t_2, \ldots, t_n$, within the domain of interest,

$$R(t_i; \boldsymbol{a}) = 0 \qquad \text{for} \quad i = 1, 2, \ldots, n \qquad (8\text{-}197)$$

For our $n = 2$ approximation, we can take any two points t_1 and t_2 within $[0, 1]$, for example, $t_1 = 1/3$ and $t_2 = 2/3$. As a result, the interval is divided into three parts. However, it is not essential for the subintervals to be equal in size for the method to work.

At $t = t_1$, the residual is obtained from (8-196) by putting $t = 1/3$. This gives us the relation

$$R(\tfrac{1}{3}; \boldsymbol{a}) = 1 + \frac{4}{3}a_1 + \frac{7}{9}a_2$$

Similarly, at $t = t_2$, we have

$$R(\tfrac{2}{3}; \boldsymbol{a}) = 1 + \frac{5}{3}a_1 + \frac{16}{9}a_2$$

By imposing the requirement that both $R(\tfrac{1}{3}; \boldsymbol{a})$ and $R(\tfrac{2}{3}; \boldsymbol{a})$ must vanish at the same time, we obtain a set of two simultaneous equations for a_1 and a_2:

$$12a_1 + 7a_2 = -9$$
$$15a_1 + 16a_2 = -9 \qquad (8\text{-}198)$$

This gives us

$$a_1 = -\frac{27}{29} \qquad a_2 = \frac{9}{29}$$

The $n = 2$ approximate solution to the differential equation obtained by the collocation method for the interval $[0, 1]$ is then

$$\phi_2^{\text{col}}(t) = 1 - \frac{27}{29}t + \frac{9}{29}t^2 \qquad (8\text{-}199)$$

Although this result appears on the surface to be quite different from that for the exact solution given by (8-193), we see in Fig. 8-28 that, within the domain of interest, the differences are, in fact, quite small.

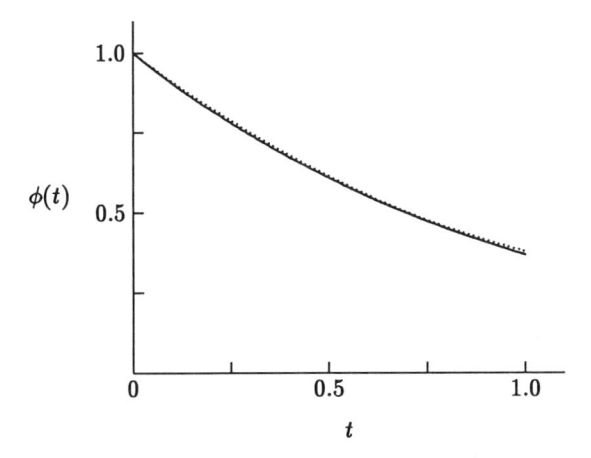

Fig. 8-28 Comparison of the $n = 2$ approximation solution to (8-191), obtained using the collocation method (dotted curve) and the exact result (solid curve).

The subdomain method. Instead of insisting that the residual vanishes at two points in the domain, we can also obtain the necessary equations to solve for the parameters a_1 and a_2 by setting the average value of $R(t, \boldsymbol{a})$ in each of two different parts of the domain, Δt_1 and Δt_2, to vanish. That is, we demand that

$$\frac{1}{\Delta t_i} \int_{\Delta t_i} R(t; \boldsymbol{a})\, dt = 0 \qquad (8\text{-}200)$$

where the integration is carried out within the subinterval, or subdomain, Δt_i. This is known as the *subdomain* method. For the general case of n parameters, we need n such subdomains or *elements*, $\Delta t_1, \Delta t_2, \ldots, \Delta t_n$. It is not essential that all the elements be of equal size and they may even overlap each other. We shall see later that there are several ways to determine the parameters $\boldsymbol{a}$ by dividing the domain of interest into a number of finite elements; hence the name *finite element methods*.

For our two-parameter approximation, we shall, for simplicity, take two equal subdomains, $\Delta t_1 = [0, \frac{1}{2}]$ and $\Delta t_2 = [\frac{1}{2}, 1]$. The two average residuals are

$$\frac{1}{\Delta t_1} \int_0^{1/2} R(t; \boldsymbol{a})\, dt = 2\left\{\frac{1}{2} + \frac{5}{8}a_1 + \frac{7}{24}a_2\right\}$$

$$\frac{1}{\Delta t_2} \int_{1/2}^1 R(t; \boldsymbol{a})\, dt = 2\left\{\frac{1}{2} + \frac{7}{8}a_1 + \frac{25}{24}a_2\right\}$$

The equations for a_1 and a_2, analogous to (8-198), are obtained by requiring both average residuals to vanish:

$$5a_1 + \frac{7}{3}a_2 = -4$$

$$7a_1 + \frac{25}{3}a_2 = -4$$

This gives us

$$a_1 = -\frac{18}{19} \qquad a_2 = \frac{6}{19}$$

The $n = 2$ approximation solution to (8-191) using the subdomain method is then

$$\phi_2^{\text{sub}}(t) = 1 - \frac{18}{19}t + \frac{6}{19}t^2 \tag{8-201}$$

The results are compared with the exact values in Fig. 8-29. It is seen that the two results are almost indistinguishable from each other within the domain of interest. Outside the interval $t = [0, 1]$, the $n = 2$ approximation can be quite different from the exact solution. However, this is not of any concern to us.

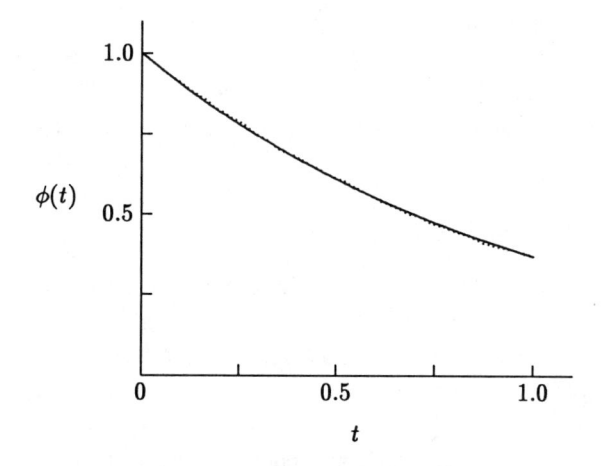

Fig. 8-29 Comparison of the $n = 2$ approximation solution to (8-191) obtained using the subdomain method (dotted curve) and the exact result (solid curve).

The least-squares method. Instead of requiring the average value of the residual within each subdomain to vanish, we can also use least-squares techniques to find the optimum values of the parameters, $a_1, a_2, \ldots, a_n$, that produce the smallest average residual $R^2(t; \boldsymbol{a})$ in the entire domain. Using the idea of maximum likelihood discussed in §6-3, the optimum values are obtained by solving the following set of n equations:

$$\frac{\partial}{\partial a_i} \int_{t_b}^{t_d} R^2(t; \boldsymbol{a})\, dt = 2 \int_{t_b}^{t_d} R(t; \boldsymbol{a}) \frac{\partial R}{\partial a_i}\, dt$$

where $i = 1, 2, \ldots, n$. For our $n = 2$ approximation, the values of a_1 and a_2 produced are different from those found using the subdomain method, but the calculated results for $\phi(t)$ are very similar. Derivation of the necessary formulas and the details of the calculations are left as an exercise.

Table 8-5

Weighting functions used in different finite element methods of (8-202).

Method		Weighting function
Collocation	$R(t_i; \boldsymbol{a}) = 0$	$W_i(t) = \delta(t_i)$
Subdomain	$\frac{1}{\Delta t_i} \int_{\Delta t_i} R(t; \boldsymbol{a}) \, dt = 0$	$W_i(t) = \begin{cases} 1 & t \text{ within } \Delta t_i \\ 0 & \text{otherwise} \end{cases}$
Least-squares	$\int_{t_b}^{t_d} R(t; \boldsymbol{a}) \frac{\partial R}{\partial a_i} \, dt = 0$	$\frac{\partial}{\partial a_i} R(t; \boldsymbol{a})$
Galerkin	$\int_{t_b}^{t_d} R(t; \boldsymbol{a}) \psi_i(t) \, dt = 0$	$\psi_i(t)$

The Galerkin method. The three methods discussed above in making use of the residual $R(t; \boldsymbol{a})$ to find the values of the parameters $\boldsymbol{a}$ are very similar to each other. They belong to a class of more general methods called weighted residual methods. In terms of the integral

$$\int_{t_b}^{t_d} R(t; \boldsymbol{a}) W_i(t) \, dt = 0 \tag{8-202}$$

for $i = 1, 2, \ldots, n$, the only difference between the three methods is in the weighting function $W_i(t)$. In the collocation method the weighting function is a delta function $\delta(t_i)$. Since, for a delta function,

$$\int_{-\infty}^{\infty} f(x) \delta(x - x_0) \, dx = f(x_0)$$

the integral of (8-202) can be carried out explicitly and the result is identical to (8-197). The subdomain method, on the other hand, uses a step function as the weight, and the least-squares method uses the partial derivative $\partial R/\partial a_i$. These differences are summarized in Table 8-5.

Other forms of the weighting function can also be used. In fact, most finite element calculations these days follow the Galerkin method, which uses elements of the trial solution themselves as $W_i(t)$. To see this, let us reformulate the trial solution in a more general way. Instead of the power series used in (8-194), we can expand the trial solution $\tilde{\phi}_n(t)$ in terms of $(n+1)$ linearly independent functions

$\psi_0(t)$, $\psi_1(t)$, $\psi_2(t),\ldots,\psi_n(t)$. We may regard these functions as a set of basis functions with which we can express our trial solution:

$$\tilde{\phi}_n(t) = \psi_0(t) + \sum_{i=1}^{n} a_i \psi_i(t) \tag{8-203}$$

It is convenient to treat the first term $\psi_0(t)$ on the right side slightly differently from the others and consider it as a function with part of or all the initial and boundary conditions of the problem incorporated into it. For this reason, it enters $\tilde{\phi}_n(t)$ without having a parameter associated with it. This is equivalent to putting into the $\psi_0(t)$ term everything that can be fixed by the initial and boundary conditions. For our $n = 2$ trial function in our example of (8-191), one possible choice in this approach is to take

$$\psi_0(t) = 1 \qquad \psi_1(t) = t \qquad \psi_2(t) = t^2 \tag{8-204}$$

A different set of basis functions may actually be better here, but this is not the place to pursue the topic. Our main interest is to see the way that the basis functions $\{\psi_i(t)\}$ may be used as the weighting functions to find the values for the parameters a_i.

The condition of (8-202) now takes on the form

$$\int_{t_b}^{t_d} R(t; \mathbf{a})\psi_i(t)\, dt = 0 \tag{8-205}$$

for $i = 1, 2,\ldots, n$. From these n equations, we obtain a set of values for the n parameters, in the same way as we did in (8-197), (8-200), and (8-202) for, respectively, the collocation, subdomain, and least-squares methods.

We shall carry out a simple demonstration to show that the Galerkin method of (8-205) also gives good results for our example of (8-191). For an $n = 2$ approximation using the basis functions provided by (8-204), the two Galerkin equations are

$$\int_0^1 \{1 + a_1(1+t) + a_2(2t+t^2)\}t\, dt = \frac{1}{2} + \frac{5}{6}a_1 + \frac{11}{12}a_2 = 0$$

$$\int_0^1 \{1 + a_1(1+t) + a_2(2t+t^2)\}t^2\, dt = \frac{1}{3} + \frac{7}{12}a_1 + \frac{7}{10}a_2 = 0$$

The values of a_1 and a_2 satisfying this set of algebraic equation are $-32/35$ and $2/7$, respectively. The resulting solution to the differential equation is then

$$\phi_2^G(t) = 1 - \frac{32}{35}t + \frac{2}{7}t^2 \tag{8-206}$$

Within the domain of interest, the results are very close to those of the exact solution of (8-194), as can be seen from Fig. 8-30.

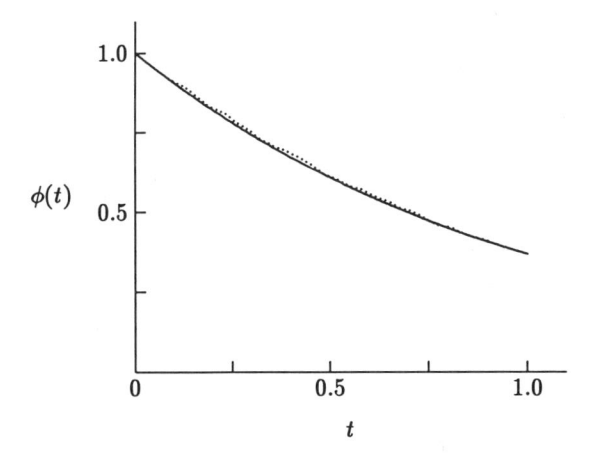

Fig. 8-30 Comparison of the $n = 2$ approximation solution for (8-191), obtained using the Galerkin method (dotted curve) and the exact values (solid curve). The results are essentially the same as for the subdomain method shown in Fig. 8-29.

Relation with variational approaches. It is also possible to adopt a variational approach to FEMs. We shall not try to carry out a derivation for the method. It is, however, of interest to give a short account here, in part, to see how historically FEMs enter into solving physical problems. Many differential relations, such as equations of motion in mechanics, may be obtained by applying the variational principle. For example, the energy E of a particle, of mass μ, at the end of a massless spring, with spring constant k, is a sum of the kinetic energy T and potential energy V:

$$E = T + V$$

As functions of time, T and V may be expressed as

$$T = \frac{\mu}{2}\left(\frac{d\phi}{dt}\right)^2 \qquad V = \frac{k}{2}\phi^2(t)$$

where $\phi(t)$ is the displacement of the particle from its equilibrium position at time t. The total energy is then the sum of these two terms:

$$E = \frac{\mu}{2}\left(\frac{d\phi}{dt}\right)^2 + \frac{k}{2}\phi^2(t) \qquad (8\text{-}207)$$

It is a function of $\phi(t)$, as well as the velocity:

$$\dot{\phi}(t) \equiv \frac{d\phi}{dt}$$

Since the energy is a conserved quantity, any variation of it vanishes

$$\delta E(\phi, \dot{\phi}) = \frac{\partial E}{\partial \phi}\frac{d\phi}{dt} + \frac{\partial E}{\partial \dot{\phi}}\frac{d\dot{\phi}}{dt} = 0$$

The partial derivatives of E with respect to ϕ and $\dot{\phi}$ may be obtained using (8-207). This gives us the familiar equation of motion for a harmonic oscillator:

$$\mu \frac{\partial^2 \phi}{\partial t^2} + k\phi(t) = 0$$

An alternate derivation of the same equation of motion may be obtained through a variation of the action

$$S = \int_{t_b}^{t_d} \mathcal{L}\, dt \qquad (8\text{-}208)$$

where $\mathcal{L} = (T-V)$ is the Lagrangian. More generally, we can also incorporate into the integral S some or all of the boundary conditions. However, this particular point is not essential for our purpose here.

In our example, the action is a function of $\phi(t)$ and $\dot{\phi}(t)$. In the more general case, it may be written as a function of t together with other generalized coordinates. If we adopt a set of basis functions, $\psi_1(t), \psi_2(t), \ldots, \psi_n(t)$, we can construct a trial function $\tilde{\phi}_n(t)$, as we did earlier in (8-203). The functions $\phi(t)$ and $\dot{\phi}(t)$ on the right side of (8-208) are now replaced by the trial function and its derivative. In this way, the integral S becomes a *functional* of the basis functions. In the same spirit as we did earlier to produce the equation of motion by applying the variational principle to the total energy, we can find an optimum set of parameters $\boldsymbol{a}$ by varying the action S:

$$\delta S = \frac{\partial S}{\partial a_1} \delta a_1 + \frac{\partial S}{\partial a_2} \delta a_2 + \cdots + \frac{\partial S}{\partial a_n} \delta a_n$$

Since variations on different a_i are independent of each other, the only way for δS to vanish for a set of arbitrary variations is that

$$\frac{\partial S}{\partial a_i} = 0 \qquad \text{for} \qquad i = 1, 2, \ldots, n$$

This is known as the Ritz method. A demonstration that the Ritz method for a specific case actually leads to the same result as the Galerkin method can be found in Burnett (1988).

Many applications of FEMs are based on the Ritz method and are often derived from physical rather than mathematical considerations. In spite of this difference, the Galerkin method gives very similar results as the Ritz method for the same trial solution. For this reason, many of the recent developments in FEMs center around the Galerkin method.

Convergence criteria and error estimates. Since we are using a specific functional form $\tilde{\phi}_n(t)$ to provide the best approximation for the solution of a differential equation, we have no way of knowing directly whether the approximation is a good one. What we have done so far is to compare with the exact solution. Except for the purpose of illustration, this is not a practical thing to do. In fact, our interests in the method are usually for cases where the exact solutions are not available.

One way to provide us with a feeling for the errors in the approximate solution, that is, the difference from the (unknown) true solution, is to carry out the calculations with different numbers of degrees of freedom and compare the results. We have used this approach several times earlier, including the extrapolation method of §8-4. Let $\phi_n(t)$ be the approximate solution for n degrees of freedom. In terms of the basis function $\{\psi_i(t)\}$, it may be written as

$$\phi_n(t) = \psi_0(t) + \sum_{i=1}^{n} a_i \psi_i(t) \qquad (8\text{-}209)$$

If we repeat the calculation with one more degree of freedom, we obtain a slightly different approximation:

$$\phi_{n+1}(t) = \psi_0(t) + \sum_{i=1}^{n+1} a'_i \psi_i(t)$$

If the difference between $\phi_{n+1}(t)$ and $\phi_n(t)$ becomes smaller as n increases, we can be reasonably sure that the approximations are converging to the true solution $\phi(t)$. For this purpose, we can adopt as our convergence condition the criterion of pointwise convergence

$$\lim_{n \to \infty} |\phi_{n+1}(t) - \phi_n(t)| = 0 \qquad (8\text{-}210)$$

That is, we shall assume that the solution converges if the average of the absolute differences calculated at a large number of points in the interval of interest is less than some small, positive number ϵ. The choice of ϵ depends on the error that can be tolerated, the machine accuracy, and the complexity involved in increasing the number of degrees of freedom.

It is also possible that the above criterion has nothing to do with the real convergence condition we want; that is, the difference from the true solution $\phi(t)$ is small. This condition may be stated as

$$|\phi(t) - \phi_n(t)| \leq \epsilon \qquad (8\text{-}211)$$

For example, if a wrong set of basis functions is used, it is quite possible that the approximate solution will converge to some function other than $\phi(t)$. In this case, convergence according to (8-210) does not imply that (8-211) is true. It is possible to formulate a set of conditions that will lead to true convergence, but we shall not go into the proofs here. For our purpose, we shall adopt the attitude that the physics behind the problem will usually guide us to a correct set of basis functions and, consequently, we can take (8-210) as the criterion for convergence.

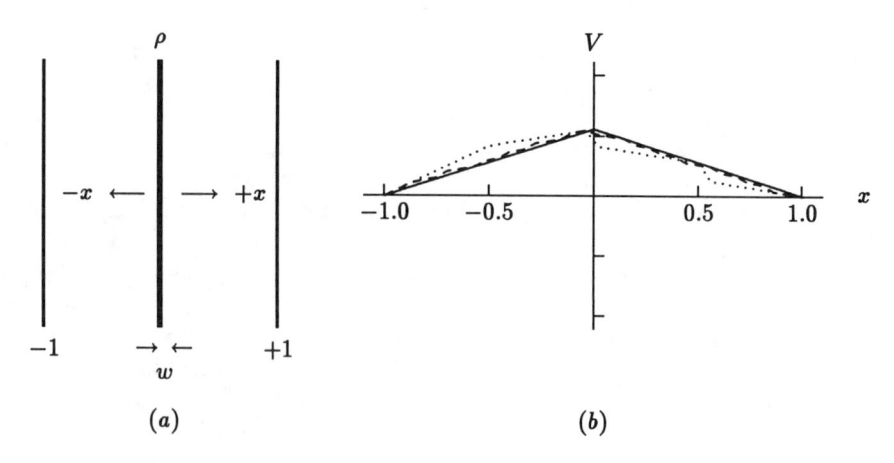

Fig. 8-31 One-dimensional electric potential distribution obtained by solving the Poisson equation (8-212). The charge distribution, with width w much smaller than the height, is in the center of a grounded, conducting cylinder with a rectangular-shaped cross section as shown in (a). The potential is essentially independent of y and z, and its variations as a function of x are shown in (b). The dotted curve is obtained with the Galerkin method using $n = 3$ degrees of freedom, the dashed line is obtained with $n = 13$, and the solid triangular-shaped curve is the exact result.

Example of one-dimensional Poisson equation. As an application, let us solve the Poisson equation in one dimension using the Galerkin method,

$$\frac{d^2V}{dx^2} = -\rho_0 \, \delta(x) \tag{8-212}$$

The physical problem, shown schematically in Fig. 8-31, corresponds to a flat beam of charged particles moving along the z-axis. The cross-sectional area of the beam is rectangular, with the height much greater than the width w. The linear charge density is ρ_0. Again, we have absorbed into the definition of ρ_0 a factor of 4π in cgs units or ϵ_0^{-1} in SI units. The boundary conditions are obtained by assuming that the beam is at the center of a grounded beam pipe made of conducting material. The cross section of the pipe is also rectangular with width of L, which we shall take to be 2 in arbitrary units, and height $H \gg L \gg w$. This gives us the boundary conditions

$$V(x) = 0 \qquad \text{for} \qquad x = \pm 1 \tag{8-213}$$

For our numerical calculations, we shall take $\rho_0 = 1$ in arbitrary units.

As the basis functions, we shall use a set of triangular functions defined in the following way:

$$\psi_i(x) = \begin{cases} \frac{1}{\Delta x}(x - x_{i-1}) & \text{for } x_{i-1} \leq x \leq x_i \\ \frac{1}{\Delta x}(x_{i+1} - x) & \text{for } x_i \leq x \leq x_{i+1} \\ 0 & \text{elsewhere} \end{cases} \quad (8\text{-}214)$$

for $i = 1, 2, \ldots, n$. To simplify the discussion, we shall take

$$\Delta x = x_{i+1} - x_i$$

to be a constant, and the interval $x = [-1, +1]$ is divided into $(n+1)$ equal segments, each of length Δx. Each basis function is, therefore, an isosceles triangle, displaced from the previous one by a distance equal to one-half of the length of the base line. The first-order derivatives of the basis functions are then

$$\frac{d\psi_i}{dx} = \begin{cases} \frac{1}{\Delta x} & \text{for } x_{i-1} \leq x \leq x_i \\ -\frac{1}{\Delta x} & \text{for } x_i \leq x \leq x_{i+1} \\ 0 & \text{elsewhere} \end{cases}$$

A schematic diagram, showing the behavior of $\psi_i(x)$ and $d\psi_i/dx$, is given in Fig. 8-32.

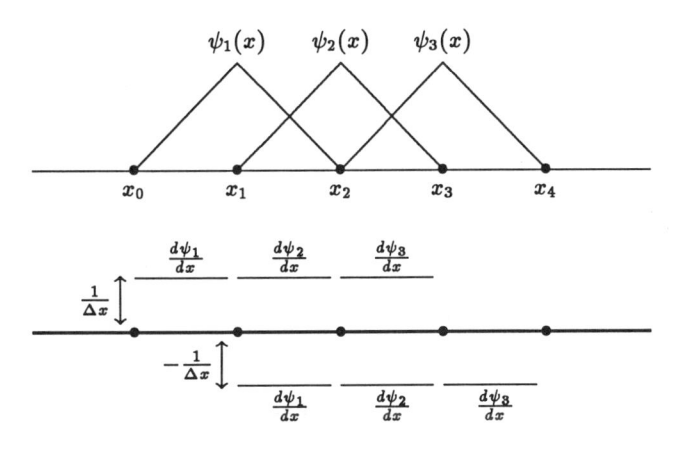

Fig. 8-32 Basis functions $\psi_i(x)$ and their derivatives used to solve the one-dimensional Poisson equation (8-212) for $n = 3$ degrees of freedom.

If we let $x_0 = -1$ and $x_{n+1} = +1$, that is, x_0 and x_{n+1} mark the two ends of our domain of interest, all the basis functions vanish on both boundaries. In this way, every ψ_i satisfies the boundary conditions for the problem. Our trial solution with n degrees of freedom takes on the form

$$V(x) \approx \tilde{\phi}_n(x) = \sum_{i=1}^{n} a_i \psi_i(x) \quad (8\text{-}215)$$

On substituting this expression into (8-212), we obtain the residual

$$R(x; a) = \sum_{i=1}^{n} a_i \frac{d^2\psi_i}{dx^2} + \rho_0 \delta(x)$$

The Galerkin equation becomes

$$\sum_{i=1}^{n} a_i \int_{x_0}^{x_{n+1}} \frac{d^2\psi_i}{dx^2} \psi_j(x)\, dx + \rho_0 \int_{x_0}^{x_{n+1}} \delta(x)\psi_j(x)\, dx = 0 \qquad (8\text{-}216)$$

The first term on the left side may be simplified using integration by parts:

$$\int_{x_0}^{x_{n+1}} \frac{d^2\psi_i}{dx^2} \psi_j(x)\, dx = \left.\frac{d\psi_i}{dx}\psi_j(x)\right|_{x=x_0}^{x=x_{n+1}} - \int_{x_0}^{x_{n+1}} \frac{d\psi_i}{dx}\frac{d\psi_j}{dx}\, dx$$

$$= - \int_{x_0}^{x_{n+1}} \frac{d\psi_i}{dx}\frac{d\psi_j}{dx}\, dx \qquad (8\text{-}217)$$

where we have invoked the fact that all the basis functions $\psi_j(x)$ vanish at the boundaries.

On inserting the final result of (8-217) into (8-216), we obtain a matrix equation for the parameters a in the form

$$Ga = f \qquad (8\text{-}218)$$

where the n-dimensional square matrix G is made of elements

$$G_{i,j} = \int_{x_0}^{x_{N+1}} \frac{d\psi_i}{dx}\frac{d\psi_j}{dx}\, dx = \begin{cases} \frac{2}{\Delta x} & \text{for } i = j \\ -\frac{1}{\Delta x} & \text{for } |i - j| = 1 \\ 0 & \text{otherwise} \end{cases} \qquad (8\text{-}219)$$

Similarly, the n-dimensional column matrix f consists of elements

$$f_j = \rho_0 \int_{x_0}^{x_{n+1}} \delta(x)\psi_j(x)\, dx \qquad (8\text{-}220)$$

Explicitly, the set of equations for the parameters a in (8-218) has the form

$$\frac{1}{\Delta x}\begin{pmatrix} 2 & -1 & 0 & 0 & \cdots & 0 \\ -1 & 2 & -1 & 0 & \cdots & 0 \\ 0 & -1 & 2 & -1 & \cdots & 0 \\ \vdots & \vdots & \vdots & \ddots & \vdots & \vdots \\ 0 & \cdots & 0 & -1 & 2 & -1 \\ 0 & \cdots & 0 & 0 & -1 & 2 \end{pmatrix} \begin{pmatrix} a_1 \\ a_2 \\ a_3 \\ \vdots \\ a_{n-1} \\ a_n \end{pmatrix} = \begin{pmatrix} f_1 \\ f_2 \\ f_3 \\ \vdots \\ f_{n-1} \\ f_n \end{pmatrix}$$

Again, we are dealing with a tridiagonal matrix, and the Gaussian elimination techniques discussed earlier in §3-7 and §8-8 may be used to find the values of the parameters $a \equiv \{a_1, a_2, \ldots, a_n\}$.

A little care is needed in treating the delta function on the right side of (8-220). Analytically, there is no ambiguity. However, we are dealing with a slightly different situation in numerical calculations. Let us assume that n is odd and

$$n = 2m - 1$$

with $m = 1, 2, \ldots$. The middle point of the domain of interest to us is at $x = 0$, where the charge is located. In our mesh, this point is labeled as x_m. For such a mesh, the only basis function that is nonzero at $x = 0$ is

$$\psi_m(x) = \begin{cases} \frac{1}{\Delta x}(\Delta x + x) & \text{for } -\Delta x \leq x \leq 0 \\ \frac{1}{\Delta x}(\Delta x - x) & \text{for } 0 \leq x \leq \Delta x \\ 0 & \text{elsewhere} \end{cases}$$

Since the value of $f_m(x)$ is 1 at $x = 0$, we have the result

$$f_i = \begin{cases} \rho_0 & \text{for } i = m \\ 0 & \text{otherwise} \end{cases} \tag{8-221}$$

This implies that we have lost the information of having the charge confined only in a very narrow region at the center. Physically, this is equivalent to solving a problem with a charge density distribution that is uniform and nonvanishing in the element m but zero everywhere else. This is not the same as the statement of the problem we gave ourselves at the start. There are two ways to get around this difficulty. The first is take a step size equal to w, the width in the spread of the beam. This is not a very practical approach, since w is small and, as a result, a large number of elements is needed if we insist on an evenly spaced grid. The second is to adopt a variable "step size," in the same way as done for some finite difference methods. This is possible in FEMs but we shall not try this method here.

A problem related to the size of the interval Δx may also arise when we try to get an estimate of the error by increasing the number of degrees of freedom n. When n is changed to $(n + 1)$, the value of the central element of $\boldsymbol{f}$ remains unchanged if we use (8-221). However, the extent of the charge distribution is reduced by such a move. As a result, we have a different charge distribution when n is modified. This is not acceptable, as it ruins our convergence test. The way around this difficulty is to realize that the delta function charge distribution is only a mathematical convenience. However, it turns into a difficulty here. Physically, the charge is distributed in a narrow region w, but w is not really zero as implied by the delta function. In other words, the charge distribution has the form

$$\rho(x) = \begin{cases} \frac{\rho_0}{w} & \text{for } |x| < \frac{w}{2} \\ 0 & \text{elsewhere} \end{cases}$$

In our finite element analysis, the smallest unit is Δx and it may be assumed to be larger than w. In this case, the appropriate charge distribution in our FEM is

$$\rho(x) = \begin{cases} \frac{\rho_0}{\Delta x} & \text{for } |x| < \frac{1}{2}\Delta x \\ 0 & \text{elsewhere} \end{cases}$$

Using this charge distribution for the right side of (8-214), the values of the matrix elements of $\boldsymbol{f}$ in (8-221) become

$$f_i = \begin{cases} \frac{\rho_0}{\Delta x} & \text{for } i = m \\ 0 & \text{otherwise} \end{cases}$$

This is a better choice for the matrix elements of f_i than that of (8-221). We can now make use of G_{ij} of (8-219) and carry out the rest of the calculations to solve the matrix equation (8-218). The details of the algorithm are outlined in Box 8-13.

Box 8-13   Program FEM_POS

Finite element solution of a one-dimensional Poisson equation

Using the Galerkin method and triangular basis functions

Subprograms used:
  INI_FEM: Initialize the one-dimensional Poisson equation with charge ρ_0 in the center.
  SLV_TRI: Gaussian elimination for a tridiagonal matrix (cf. Box 3-10).
  ERR_EST: Calculate $\phi(x)$ and estimate the error according to (8-222).

Initialization:
  Define charge ρ_0, maximum number of iterations, and convergence criterion ϵ.

1. Input the number of parameters to be used as a start.
2. Use INI_FEM to calculate:
   (a) Superdiagonal, diagonal, and subdiagonal elements of (8-219).
   (b) Matrix elements f_i of (8-221).
3. Use SLV_TRI to solve (8-218).
4. Use ERR_EST to estimate the error:
   (a) First pass:
       (i) Construct an array x_i at where the values of $\phi(x)$ are evaluated.
       (ii) Zero the array for $\phi(x)$.
   (b) Calculate $\phi(x_i)$ from the parameters and the triangular basis functions.
   (c) Calculate the average error $\mathcal{E}$ using (8-222).
   (d) Store the value of $\phi(x_i)$ for the error calculation in the next iteration.
   (e) Return the average error $\mathcal{E}$.
5. Compare $\mathcal{E}$ with the convergence criterion ϵ:
   (a) If $\mathcal{E} > \epsilon$, increase the number of parameters by 2 and iterate the solution again.
   (b) Terminate the calculation if either convergence is achieved or the maximum number of iterations is exceeded.

The results of the calculation are shown in Fig. 8-31(b). The plot is for

$$V(x) = \sum_{i=1}^{n} a_i \psi_i(x)$$

using the values of a obtained by solving (8-218). The analytic solution of this problem is also known:

$$V(x) = \rho_0 (1 - |x|) \qquad \text{for} \qquad |x| \leq 1$$

This is shown as a solid curve in the figure. Because of the triangular shapes of the basis functions, the $n = 3$ results, shown as the dotted curve in the figure, do not have a smooth slope. By the time n is increased to 13, shown as the dashed curve, the differences from the exact result become quite small. At $n = 35$, the average value of the error, defined as

$$\mathcal{E} = \frac{1}{2} \int_{-1}^{+1} |\phi_{n+2}(t) - \phi_n(t)| dt \approx \frac{1}{n-2} \sum_{k=1}^{n-2} |\phi_{n+2}(t_k) - \phi_n(t_k)| \qquad (8\text{-}222)$$

is less than 5×10^{-3}.

The main reason for choosing the triangular basis functions for our example here is that the actual equations to be solved are identical to those in the case of a finite difference method. In fact, if we only want the values of $\phi(x)$ at $x = x_1$, $x_2, \ldots, x_n$, we have the result that

$$\phi_n(x_i) = a_i$$

since, at these points, the values of $\psi_i(x_i) = 1$ and all the other basis functions vanish. In this case, the parameters $a_1, a_2, \ldots, a_n$ have the same roles as $\phi(x_1)$, $\phi(x_2), \ldots, \phi(x_n)$, the values of the solution at, respectively, $x_1, x_2, \ldots, x_N$, in finite difference methods. This is not surprising, since the equations to solve for both sets of quantities are identical for the particular set of basis functions used here. In general, we could have chosen a more suitable basis functions for the problem we have here. However, for purposes of illustration, there is hardly any point to do so.

For one-dimensional problems, there is not much advantage in taking a finite element approach over that of finite differences. However, in two and higher dimensions, the superiority of finite element methods becomes clear. In particular, it is attractive to be able to select the mesh and basis functions in accordance with the requirements of the physical problem in hand. In addition to the ability of keeping a closer contact with the physics, such approaches can also be very useful in reducing the computational complexity of the problem. A review of some of the FEMs used in physical problems can be found in a collection of review articles appeared as a special issue in volume 6 of the journal *Computer Physics Report* (1987).

Problems

8-1 A projectile starts from the surface of the earth at $x = x_0$ with an initial velocity v_0 and at an angle of elevation θ_0. If we ignore air resistance, the only force acting on the projectile is gravity along the y-direction. Set up the differential equation describing the trajectory of the projectile assuming that the x-axis is in the horizontal plan. Solve the equation as an initial value problem using v_0 and θ_0 as the initial values. Let x_d be the point where the projectile lands on the surface of the earth again. For a given v_0, plot the value of $d = x_d - x_0$ as a function of θ_0 and show that the maximum occurs at $\theta_0 = 45°$.

8-2 The projectile motion of Problem 8-1 can be turned into a boundary value problem if we wish the projectile to land at a specific point x_s on the surface of the earth. The solution may be obtained using the same computer program as Problem 8-1 by mimicking one of the shooting methods. Try this approach by varying the angle of elevation θ_0 until the $y = 0$ point occurs at x_t for $|x_s - x_t|$ less than some small value ϵ.

8-3 Rewrite (8-83) in terms of logarithmic derivatives at $r = r_g$.

8-4 Find the scattering solution of (8-67) for the $\ell = 0$ modified radial wave function inside a square-well potential using the method of successive over-relaxation (SOR). Explore the dependence of convergence on the relaxation parameter ω and the number of intervals. The shape of the potential well is given by (8-68). Use $u_0(0) = 0$ and $u_0(r_g) = 1$ as the boundary conditions and $u_0(r) = r/r_g$ as the initial trial solution.

8-5 For the solution of the two-dimensional Poisson equation of §8-7, the potential at the location of a point charge is singular, given roughly by $V(\eta) \propto q/\eta$ where η is the distance to the point charge. What happens to this singularity in the solution to the finite difference equation?

8-6 Write a computer program to calculate the solution for the wave equation of (8-148) for a stretched string with the end at $x = 0$ fixed and the other end at $x = 1$ moving with a sinusoidal time dependence. The initial condition may be taken to be such that the displacement and velocity of the string are zero everywhere.

8-7 Besides the equation for the kink solution, the static sine-Gordan equation

$$\frac{d^2\phi}{dx^2} - \sin\phi(x) = 0$$

is also a nonlinear ordinary differential equation with known analytical solitary wave solution

$$\phi(x) = 4\tan^{-1}\{\exp(x - x_0)\}$$

Solve the differential equation numerically using Newton's method and compare the results with the analytical form given above.

8-8 Use a least-squares method to derive the equations satisfied by a_1 and a_2 for an $n = 2$ approximation to the solution of (8-191). Calculate the values of $\tilde{\phi}_2(t)$ in the interval $t = [0,1]$ and compare the results with the exact solution given by the exponential function of (8-193).

8-9 Solve the differential equation of (8-191) with three degrees of freedom using the Galerkin method. Calculate the average absolute values of the differences from the $n = 2$ solution of (8-206) for a number of points in the interval $t = [0,1]$. Compare the result with the same average difference from the exact solution of (8-193).

8-10 In cylindrical coordinates, the Laplacian operator has the form
$$\nabla^2 \phi(\mathbf{r}) = \frac{1}{r}\frac{\partial}{\partial r}\left(r\frac{\partial \phi}{\partial r}\right) + \frac{1}{r}\frac{\partial^2 \phi}{\partial \theta^2} + \frac{\partial^2 \phi}{\partial z^2}$$
For an infinitely long line of charge at the center of an equally long, hollow cylinder made of conducting material, the electric field is independent of both θ and z. Use the Galerkin method to solve for the radial dependence of the potential $\phi(r)$ inside the cylinder, assuming that it has a radius of $r_g = 1$ m and is grounded. The linear charge density is $\lambda = 1$ and is confined to a radius $a \ll r_g$.

8-11 What are the major differences between an eigenvalue problem, such as that given by (4-90) for a harmonic oscillator potential, and a two-point boundary value problem, such as that represented by (8-66)? Devise a numerical method to solve the eigenvalue problem of (4-90) using one of the methods for boundary value problems.

Chapter 9
Graphical Presentation of Results

In numerical calculations, the results often appear as a table of numbers and, for a complicated system, such a table can be an extensive one. A good way to gain an overall feeling for a large set of numbers is to view them in the form of a plot. The possibility and ease of making graphical presentations on modern workstations and personal computers are quite remarkable and should perhaps be ranked as important a development in computational physics as the progress made in carrying out the calculations themselves.

For most applications, the conversion of numerical results to a graph on the screen or a sheet of output page is carried out using a set of instructions, or subroutines, made specially for such purposes. However, because of the fast development in the output devices, a diversity of hardware is used. Partly for this reason, very little standardization is found at the moment among the different software packages. As a result, portability of computer programs involving graphics from one type of machine to another can sometimes be a rather difficult task. Until the emerging standards take effect, it is not possible to give a meaningful treatment of the various ways graphics are done. For this reason also, we shall restrict our discussion here to the general background of computer graphics and to provide some of the basic knowledge in using the available plotting packages. The algorithms presented in this section are for illustrative purposes only. In general, they are not necessarily the best methods for the task and some of them are not even expected to work for all occasions.

§9-1 Basic ingredients of computer graphics

Consider the simplest form of line graphs commonly encountered in physical problems. If we wish to show the variation of a dependent variable y as a function of an independent variable x in the form
$$y = f(x)$$
we can plot the value of y for a range of x as a graph. For this purpose, we can imagine the computer screen, or the output page, as a sheet of graph paper, with the horizontal direction representing x values and the vertical direction y values.

Text-mode plotting. The simplest way to display a set of numerical results in the form of a graph is to use the *text* or *character* mode of the output device. The basic idea is the same as making a plot on a sheet of graph paper. At the start, we have a blank screen, the equivalent of a clean sheet of graph paper. To make a plot, we put dots, or some other symbols, at the appropriate locations on the screen, the same way as we put marks on a sheet of graph paper to display a graph. Since we are already able to send characters to the output device using, for example, the PRINT statement in FORTRAN, there is no difficulty in displaying the plot on the screen. The only thing we need to do is to convert our numerical results to the equivalent character output. This is a convenient way to display graphical output, as no special graphics capability is required of the computer. Later, we shall see that to improve the resolution, better methods must be used.

In the character mode, most computer screens can display 25 lines with 80 characters in each line. This means that our graph paper has a maximum of 80 divisions for the horizontal axis and 25 divisions for the vertical axis, without any possibility of using a fraction of a division. The situation is slightly better with a printer. Most line printers can print a page consisting of 60 lines with 80 characters in each line. Even this is still a very crude sheet of graph paper. However, there is no need for us to be overly concerned with this particular limitation at this moment. Later, we shall see that it is possible to do much better in the graphics mode. In spite of the low resolution, the convenience of text-mode plotting makes it a very useful tool on many occasions.

To make the discussion easier, let us assume that we wish to display the relation

$$y = \sin(x) \tag{9-1}$$

for x in the range $[0, \pi]$. For simplicity, we shall limit ourselves to the computer screen as the output device. To obtain as much detail as possible, we shall make use of all the 80 horizontal divisions to represent values in the range $[0, \pi]$. This means that each division on our x-axis represents a value

$$\Delta x = \frac{1}{80-1}(\pi - 0)$$

The reason that 79, rather than 80, appears in the denominator comes from the fact that our first division represents $x = 0$ and our 80th division $x = \pi$. The possible values of y for this range of x are in the interval $[0, 1]$. If we make use of all 25 vertical divisions to represent this range of y values, each division along the y-axis represents a value

$$\Delta y = \frac{1}{25-1}(1 - 0)$$

The rest of the plotting may proceed in a way not very different from what we normally do using a sheet of ordinary graph paper.

The first thing is to decide on the location of the origin and the directions of the x- and y-axes. For our example, the most natural place to put the origin of the coordinate system is the lower-left corner of the screen, with the value of x increasing horizontally to the right and the value of y increasing vertically upward. The bottom line on the screen then corresponds to the x-axis and the leftmost column, the y-axis. Each of the (80×25) positions in the plot area may be labeled by two indexes (m, n), with m as the x-coordinate measured in number of characters, starting from the y-axis as $m = 0$, and n as the y-coordinate measured in the number of lines, starting with the x-axis as $n = 0$. Thus, the range of m is $[0, 79]$ and that of n is $[0, 24]$. Each of the (80×25) locations on our graph paper is now associated with a particular pair of (x, y) values. It is convenient for our discussion to assume that the information associated with these (80×25) positions is stored in the form of a two-dimensional array of characters. At this stage, our graph paper is blank except for our coordinate axes. This may be represented by putting blanks, or the null character, in all locations in the two-dimensional array.

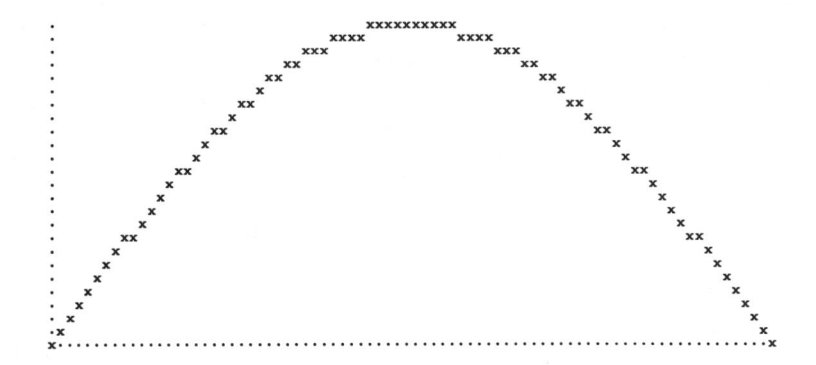

Fig. 9-1 A plot of the function $y = \sin(x)$ for $x = [0, \pi]$ using only the character output facility. The resolution is 80 characters horizontally and 25 lines vertically. The x- and y-axes are marked, respectively, by a horizontal line and a vertical column of dots.

Starting from $x = 0$, we calculate the value of y for each value of x according to (9-1). For $x = 0$, our function gives $y = 0$. We can put a mark, for example, the character X, in the location $(m, n) = (0, 0)$ of the two-dimensional array to represent this result and leave the rest of the leftmost column blank [that is, $(m, n) = (0, j)$ for $j = 1$ to 24]. The second division on the x-axis corresponds to $x = \Delta x \approx 0.04$. The value of y for this value of x is $\sin(\Delta x) \approx 0.04$. Since $y/\Delta y \approx 1$, the location $(m, n) = (1, 1)$ is now changed to the character X. Next we go to $x = 2\Delta x$ and calculate the location of y corresponding to this value of x, and so on. The general result is that, for $x = k\Delta x$, we replace the location (k, ℓ) with the character X. The value of ℓ is obtained from the relation

$$\ell = \frac{\sin(k\Delta x)}{\Delta y}$$

The calculation finishes when we reach the end of the interval in which we are interested at $x = \pi$. At this stage we can list the array with a simple formatted PRINT statement to output the 25 lines of 80 characters in each line. The result is shown in Fig. 9-1.

Aspect ratio and resolution. Our simple example above demonstrates the ease in providing a rough sketch of the variation of y as a function of x without having to invoke any special graphics features of the computer. The resulting plot, however, suffers from several important limitations. The first is that the vertical height and horizontal width of each unit in our plot are not the same. In other words, we are using a sheet of graph paper that is divided into little rectangles rather than squares. The ratio of the width to the height of the smallest element of a plotting device is known as the *aspect ratio*. On a computer screen, the width of each character is usually smaller than its height. Consider a typical 14-inch monitor, one whose diagonal length of the screen is around 14 inches (35 cm). The actual display area is about 25 cm (10 in.) horizontally and 18 cm (7 in.) vertically. In displaying 25 lines of 80 characters each, the width and height of the space occupied by each character are, respectively, 0.31 cm (1/8 in.) and 0.72 cm (9/32 in.). The aspect ratio is therefore 0.43.

For most plots, departure of the aspect ratio from unity is not a problem, as the horizontal and vertical axes usually refer to different quantities and, hence, have different units any way. However, in some cases the geometric shape of the plot is of importance, for example, when we wish to draw a semicircle using the relation

$$y = \sqrt{r^2 - x^2} \qquad \text{for} \qquad |x| \leq |r| \tag{9-2}$$

The simple algorithm outlined above to display the sine curve will give half an ellipse. To display a circle on the screen, we need to use, instead, the formula

$$y = Y_{asp} \sqrt{\frac{r^2}{X_{asp}^2 + Y_{asp}^2} - \frac{x^2}{X_{asp}^2}}$$

where X_{asp}/Y_{asp} is the aspect ratio of the output device. Note that different output devices are likely to have different aspect ratios. For example, a printer set at 10 characters an inch and 6 lines to the inch will have an aspect ratio of 3/5. Unless we change to the proper aspect ratio in each case, a program that prints a circle on one type of output device may not produce a circle on anther. We see here an example of the difficulties encountered in writing plotting packages for use on a variety of output devices.

The second problem in a character plot is the limited resolution. The problem is especially serious in the vertical direction. The reason is quite clear. Since we have only a total of 25 divisions at our disposal, we can only resolve one part in 25 for the range of y values in which we are interested. The resolution may be improved by going to the graphics mode. For example, on a personal computer equipped with a good graphics device (for example, a screen and video board), it is relatively easy to achieve a resolution of one part in 1024 horizontally and one part in 768 vertically. On a normal laser printer, the resolution is 1/300 in. (300 dots per inch or 120 dots per cm) in both x and y directions. Printers capable of 500 dots per in. (200 dots per cm) and screens having 4096 dots per line or better are also quite common. It is clear that with such high resolutions we have much greater latitude in visualizing a computer output, and the impact of such a possibility on computational physics is yet to be fully appreciated.

The graphics mode. If we examine the way a character is displayed on a computer screen or a page of the printed output, we will likely find that it is made of small dots. The composite nature of each character is most obvious on electronic display boards, such as those used in many airports to announce flight schedules. Each letter on the board is shown as a rectangular grid of dots or lights. As an example, a (7×7)-dot representation of the digits 0 to 9 is given in Fig. 9-2. If square dots are used instead of the rectangular-shaped ones in the diagram, a (7×9)-dot representation is preferred. This is necessary so as to produce the rectangular-shaped characters with which we are familiar.

Fig. 9-2 A (7×7) rectangular-dot representation of the digits 0 to 9.

To make the display more pleasing to the eye, many more dots are needed to make a smooth trace of each character and the dots are packed closer to each other so that they form a continuous band. The actual representation used on a computer screen depends on the monitor used, as well as the way the monitor is controlled. Since a large variety of monitors and video boards are in use these days, it is not possible to give any reasonable overall description of the way characters, and more importantly, graphics are displayed on a computer screen. For our purpose, we shall treat the computer screen as made of a number of dots, or *pixels*. Each pixel is a basic unit of our display that cannot be subdivided into smaller ones. Let the number of available pixels in a horizontal line be M and the number of lines on a screen be N. The total number of pixels on a screen is then the product $(M \times N)$. The actual values of M and N depend on the hardware. For example, on a good video graphics adaptor for personal computers, one can get $M = 1024$ and $N = 768$. For some of the better workstations, values such as $M = N = 4096$ or better are possible. Just as in character plots, we shall regard the screen as a sheet of graph paper, consisting of M horizontal divisions and N

vertical divisions. If instead of to the screen we wish to output the graph to a printer or a plotter, the appropriate values of M and N for the hard copy device must be used instead.

To display the $(M \times N)$ pixels on a screen, there must be somewhere in the computer a space to store these $(M \times N)$ pieces of information. This is the function of the video memory. A little calculation will show that this is not a trivial part of the computer. If $M = 1024$ and $N = 768$, we need 768K bits (recall that 1K = 1024) of memory just to record whether each pixel is on or off. This is the minimum amount of video memory we need and it can only give us a monochrome display. To show color, we need also information concerning what color each pixel should be. For a modest value of 16 different colors, 4 bits ($2^4 = 16$) of storage is required for each pixel, or a minimum of 384K bytes for the video memory. This is not a small number. In fact, it is bigger than the memory of some of the largest computers available in the early 1970s. For many workstations designed for graphics applications, the amount of video memory may even exceed that for the rest of the operations of the computer.

In the graphics mode, the programmer has the possibility of addressing, meaning storing values into, each location of the video memory. The process is quite simple, but it requires some background in the computer architecture for a full understanding. To avoid any discussion concerning hardware, we shall assume that we are provided a subroutine called SET_PXL that has three arguments i, j, and k. The sole function of this subroutine is to change the pixel at location (i, j) to color k. Obviously, the first two arguments are restricted, respectively, to values $0 \leq i < M$ and $0 \leq j < N$. The third one, k, is a number that ranges from 0 to the maximum number of available colors, for example, 16. On personal computers, this subprogram is similar to the function **putpixel** or **_setpixel** of some of the C compiler, the subroutine SETPIX of some FORTRAN compilers, and the function PSET in the BASIC language. As long as there is any graphics capability on the computer, it is not difficult to write such an elementary plotting function, but we shall not be going into this question here. What we shall do now is to make use of this basic plotting tool to demonstrate the principles behind displaying graphics on a computer screen. Although it is possible to make a large number of relatively complicated plots with a few subroutines built upon SET_PXL, we are a long way away from having a complete plotting package. For any serious graphics programming, it is usually more convenient to use, for example, one of the commercially available subroutine packages.

Example of a rainbow. Let us illustrate the basic elements of plotting by simulating a rainbow on the screen (see also Olson and others, *The Physics Teacher* April 1990, page 226). The physics involved here is that, when a light ray enters a raindrop, it suffers refractions and reflections. Since the index of refraction

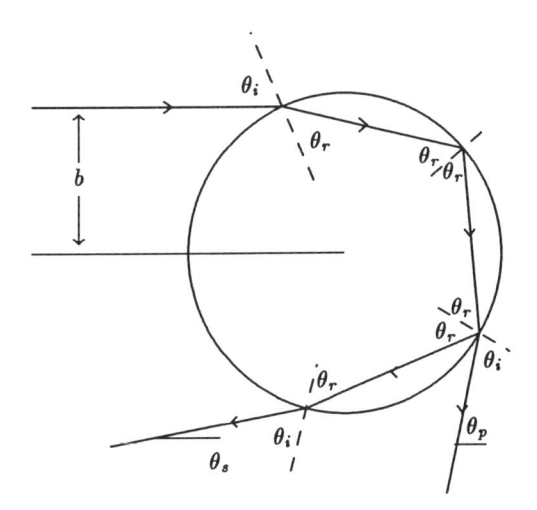

Fig. 9-3 Schematic diagram of the path of a light ray refracted by a rain drop. The ray incidents at the raindrop with an impact parameter b. If it suffers only one internal reflection, it emerges as a primary ray at an angle θ_p with respect to the incident direction. If there is one more internal reflection, it emerges as a secondary ray at angle θ_s.

n depends on the color, or wavelength, light of different color emerges from a raindrop at different angles. The net effect is that rainbows are formed.

For simplicity, we shall use only three colors, red, green, and blue, with indexes of refraction 1.33, 1.34, and 1.35, respectively. To simulate the fact that sunlight contains a mixture of many colors, we shall take that the color of each light ray is a random sample of one of these three colors. That is, for each light ray, a random number c, with uniform distribution in the interval $[0,3]$ (three times a uniform random number in $[0,1]$), is generated and a color is assigned to it according to

$$\text{color} = \begin{cases} \text{blue} & \text{for } 0 \leq c < 1 \\ \text{green} & \text{for } 1 \leq c < 2 \\ \text{red} & \text{for } 2 \leq c < 3 \end{cases} \quad (9\text{-}3)$$

The angle of incidence, θ_i, for the light ray depends only on the impact parameter b, as shown in Fig. 9-3. If we put the origin of our coordinate system at the center of a raindrop and take the z-axis to be parallel to the direction of incidence of the light ray, we have

$$b = \sqrt{x^2 + y^2} \quad (9\text{-}4)$$

where (x,y) are the coordinates of the incident ray when it is still outside the raindrop.

The shape of the raindrop may be assumed to be spherical with a radius equal to unity in some arbitrary units. To simulate all the possible angles of incidence, x and y are taken to be two independent random numbers, each with a uniform distribution in the interval $[0, 1]$. The relation between the angle of incidence θ_i and the impact parameter b is given by

$$\theta_i = \sin^{-1} b \qquad (9\text{-}5)$$

as can be seen from Fig. 9-3. Using the index of refraction n, we obtain the angle of refraction as

$$\theta_r = \sin^{-1} \frac{b}{n} \qquad (9\text{-}6)$$

It is obvious from the figure that, if the light ray is totally reflected only once inside the raindrop, it emerges at an angle

$$\Theta_p = (\theta_i - \theta_r) + (\pi - 2\theta_r) + (\theta_i - \theta_r) = \pi - 4\theta_r + 2\theta_i$$

with respect to the incident ray. Relative to the z-axis, the value of the same angle is

$$\theta_p = \pi - \Theta_p = 4\theta_r - 2\theta_i$$

This is the source of primary rainbow. Similarly, if there is a second total internal reflection, the corresponding angle of emergence relative to the z-axis is

$$\theta_s = 6\theta_r - 2\theta_i - \pi$$

This is the source of secondary rainbow.

The intensities I_p and I_s of, respectively, the primary and secondary rays are given by standard transmission and reflection factors for electromagnetic waves across the boundary of two media. The results are

$$I_p = \tfrac{1}{2}\{s(1-s)^2 + p(1-p)^2\}$$
$$I_s = \tfrac{1}{2}\{s^2(1-s)^2 + p^2(1-p)^2\} \qquad (9\text{-}7)$$

where the two factors p and s are given by

$$p = \frac{\sin^2(\theta_i - \theta_r)}{\sin^2(\theta_i + \theta_r)} \qquad s = \frac{\tan^2(\theta_i - \theta_r)}{\tan^2(\theta_i + \theta_r)}$$

In principle, these intensities should be translated into the intensities of the display for the emerging rays. This is not easy to do except on very elaborate graphics workstations. For the purpose of our simulation, it is really unnecessary. Instead, we can turn these intensity factors into probabilities for the rays to be observed. To achieve this, we can generate a random number ℓ that is uniformly distributed in $[0, 1]$. The ray emerging from the raindrop is displayed only if the intensity is greater than $(f \times \ell)$. The factor f is related to the amount of light coming out of

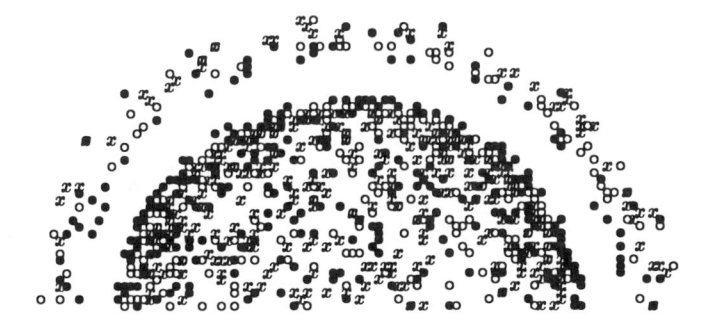

Chapter 9] Graphical Presentation of Results 615

Fig. 9-4 A black and white rendition of a simulated rainbow. The dots (•) represent red color pixels, circles (o) green color, and crosses (x) blue color. Only 1000 points are shown here. Larger numbers may be used on a monitor. Areas where all three colors overlap should be considered as white.

the raindrop and is taken to be 0.04 for the primary ray and 0.02 for the secondary ray.

To complete the simulation, we should assign a relative probability for the incident light to emerge as a primary ray or a secondary ray. Since there is some loss of intensity at each reflection and refraction, the sum of the two probabilities should be less than unity. However, we shall ignore this point here. The entire simulation consists of repeating the process for many light rays, each with a random color, a random pair of (x, y) coordinates, a random probability to be absorbed after the first internal reflection, and another random probability to be absorbed after the second internal reflection.

Our primary interest here is in displaying the color output. We shall assume that the incident light is horizontal, a fairly realistic condition for rainbows to be observed. Each pixel of the computer screen may be taken as a small raindrop. For the light refracted from a particular drop to be seen, the ray must emerge from the drop at the correct range of angles above the horizon and the correct range of angles to the left or right of the observer. This depends on the color and the x- and y-coordinates of the ray incident on the rain drop. To speed up the simulation, we shall not assign the position of the raindrop until we find out its exit angles, θ_p for primary rays and θ_s for secondary rays. Once these pieces of information are known, the raindrop is placed at the proper location on the screen and colored accordingly. If θ_p or θ_s is too large, the raindrop will be too high in the sky to make its color visible to the observer. We can therefore set an upper limit of $\pi/3 = 60°$ as the cutoff angle. The location of a raindrop in terms of pixels with incident coordinates (x, y) and $\theta = \theta_p$ or θ_s is then

$$x_p = H \times \theta \times \frac{3}{\pi} \times \frac{x}{b} + X_0$$

$$y_p = V \times \theta \times \frac{3}{\pi} \times \frac{|y|}{b} + Y_0 \qquad (9\text{-}8)$$

Box 9-1 Program RAINBOW

Simulation of a rainbow on the screen of three colors

Using the basic plot command SET_PXL

Subprograms used:
  PLT_DOT: Put a dot at (x, y) to a particular color.
  SET_PXL: Compiler-dependent subroutine to move the present position to a particular pixel.
  RSUB: Uniform random number generator in $[0, 1]$.

Initialization:
  (a) Input a random number seed and the maximum number of dots to be plotted.
  (b) Use SET_PXL to initialize the graphics screen.

1. Repeat the following steps for each dot:
2. The incoming light ray:
   (a) Determine the impact parameter b using (9-4) by generating two uniform random numbers in $[0, 1]$ as the values of x and y.
   (b) Generate a uniform random number in $[0, 3]$ and determine the color using (9-3).
3. Calculate the following quantities for the light ray:
   (a) The index of refraction for the color.
   (b) The angle of incidence θ_i and angle of refraction θ_r using (9-5) and (9-6).
   (c) The intensities I_p and I_s for the primary and secondary rays according to (9-7).
4. Primary ray:
   (a) Generate a uniform random number x in $[0, 1]$. If $I_p > fx$, with $f = 0.04$, display the primary ray using PLT_DOT.
      (i) Convert x, y, θ_p into the location of a pixel using (9-8).
      (ii) Use SET_PXL to set the pixel to the color of the incident ray.
5. Repeat step 4 for the secondary ray using $f = 0.02$.
6. Stop the simulation if the maximum number of dots is reached.

---

where H and V are, respectively, the horizontal and vertical scale factors, and X_0 and Y_0 are factors that shift the display to the central part of the screen. The values of H and V depend on those of M and N of the display. They are adjusted so that the picture fills most of the screen.

The calculations involved in carrying out such a simulation are outlined in Box 9-1. We assume that, at the start, the screen is cleared and set to black color. To display the result of a drop, all we need to do is to use the subroutine SET_PXL to go to the pixel at (x_p, y_p) and turn it to the appropriate color, red, green, or blue. The calculations required for each raindrop are quite simple and can be done in a very small amount of computer time. After a few thousand drops, the screen looks very much like a rainbow. In black and white, the picture is shown in Fig. 9-4. The simulation may also be carried out with more colors and with different indexes of refraction. Some of these extensions of the basic ideas are given in Olson and coauthors.

§9-2 Functions with one independent variable

On a graphics output device, we can, in principle, display any picture we wish up to the resolution given by the density of the pixels. Let us consider a simple plot by assuming a monochrome screen. That is, the screen can only display two contrasting colors which we shall assume to be black and white. The status of each pixel is therefore controlled by a single bit in the video memory. If the value of the bit is zero or "off," the color of the pixel is black, and if the value of the bit is one or "on," the color is white. At the start, we shall assume that the entire screen is blank, meaning that all the pixels are black and the entire video memory is stored with zeros. A picture may be put on such a screen by turning the appropriate bits to the on position.

Constructing a map of the pixels. As an example, let us try to draw the same sine curve

$$y = \sin(x) \qquad \text{for} \qquad x = [0, \pi] \qquad (9\text{-}9)$$

as we did in the previous section. Instead of using characters, we shall make use of the graphics mode here.

Again, we shall assign the lowest horizontal line of pixels on the screen as the x-axis. If there are M pixels in a line, each corresponds to an increment in the value of x of the amount

$$\Delta x = \frac{\pi - 0}{M - 1}$$

In other words, because of the discrete nature of the pixels, we are dividing the x-axis into a mesh of M points, each corresponding to a step size of Δx. Similarly, if N is the number of vertical lines and we wish to make use of all the available lines to display the range of y values between 0 and 1, the step size in y is

$$\Delta y = \frac{1 - 0}{N - 1}$$

To display the sine cure of (9-9), we shall start with $x = 0$ and calculate the corresponding value of y for all the values of x in $[0, \pi]$. The values (x, y) are converted to the location of pixels by the formulas

$$h = (M - 1) \times \frac{x}{\pi} \qquad\qquad v = (N - 1) \times \sin x$$

where h is the horizontal position of the pixel and v is its vertical position. As we did in the previous section, the origin of our coordinate system is at the bottom-left corner of the screen. Once we locate the pixel that represents the function for $x = 0$, we can turn it on. Next we move to $x = \Delta x$ and locate the corresponding pixel and turn it on. The steps are repeated for each point on the horizontal grid at $x = 2\Delta x$, $3\Delta x$, ..., until we come to the end of the interval at $x = \pi$. The result is shown in Fig. 9-5. It is basically the same curve as that shown in Fig. 9-1, except we have greatly improved the resolution.

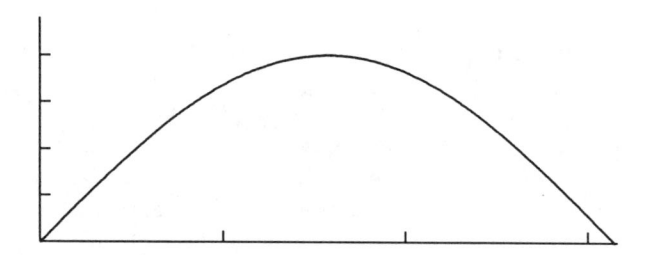

Fig. 9-5 A plot of the function $y = \sin(x)$ for $x = [0, \pi]$ by addressing each pixel.

Plotting straight lines. The ability to address each pixel provides us with the basic element in making a plot, as we have seen in the sine function example above. However, the process is very inefficient, as the function $y = f(x)$ must be evaluated once for each of the M pixels along the x direction. This is not always feasible. Often, the function is given to us as a table of pairs of (x, y) values, such as the results from the numerical solution of a differential equation. In this case, it is not economical for us to go back and find the solution with a very fine mesh just for the purpose of displaying the results. A more serious difficulty arises when we want to plot a set of measured quantities for which we have no way of carrying out measurements for any of the intermediate points. One essential feature of a plotting package is, therefore, the ability to interpolate between the supplied values.

A simple way to interpolate between two points is to use a straight line. Under the appropriate circumstances, for which we shall give examples later, it is quite adequate to use this simplest form of interpolation for displaying a set of results graphically. Unless we are using an analog display device, all curves are, in essence, made of collections of small straight-line segments. This is especially true for a computer screen or a simple dot-matrix printer (meaning one without the capability of partially overlapping one dot with another), as all images must be constructed using dots on a fixed grid.

Our problem is, however, a slightly different one. What we have is a table of (x, y) values that are not necessarily close together in value. When this table is converted to pixel positions, only a finite number of them can be determined directly from the values supplied. A simple way of joining two such pixels separated by more than a few "blank" ones is to use a straight line. Let the locations of two such points be (x_i, y_i) and (x_{i+1}, y_{i+1}), with $x_{i+1} \neq x_i$. Along a straight line from (x_i, y_i) to (x_{i+1}, y_{i+1}), the value of y for a point x in the interval $[x_i, x_{i+1}]$ is given by

$$y = y_i + \frac{x - x_i}{x_{i+1} - x_i}(y_{i+1} - y_i) \tag{9-10}$$

For $x_{i+1} = x_i$, the straight line joining (x_i, y_i) and (x_{i+1}, y_{i+1}) is parallel to the y-axis and no interpolation is necessary. All that has to be done in this case is to

---

Box 9-2 ST_LINE(IX_D,IY_D,KOL)
Straight line interpolation between two points

Argument list:
IX_D,IY_D: New (x, y) position in terms of pixels.
KOL: color index of the line. No line is drawn if KOL< 0.
1. If KOL $\geq$ 0, carry out the following steps starting from the present pixel position (IX_B,IY_B)
   (a) If IX_B$\neq$IX_D:
       (i) For every integer value x between IX_B and IX_D, use (9-10) to find the value of y along a straight line from (IX_B, IY_D) to (IX_D,IY_D).
       (ii) Use SET_PXL to put the pixel at (x, y) to color KOL.
   (b) If IX_B=IX_D, put all the pixels at (x, y) with x =IX_B and y between IY_B and IY_D to color KOL.
2. Set the present pixel position as (IX_D,IY_D).

---

turn on all the pixels (x_i, y) for all y between y_i and y_{i+1}. Similarly, for $y_i = y_{i+1}$, the straight-line segment is parallel to the x-axis, as can be seen from (9-10).

We can write a subroutine, which we shall name as ST_LINE, to carry out the process of interpolating between two points by a straight line and put all the pixels along the line to a specific color. To conform as much as possible to a subroutine existing in several commonly available plotting packages, we shall construct this subroutine in such a way that it draws a straight line from the present location on the screen to a location (x_d, y_d). In addition, we shall allow the possibility of specifying the color of the line by a third argument k. As we did earlier in SET_PXL, the integer k is a color index, representing one of the available colors of our output device. The present position is given by the end point of the previous call to ST_LINE. To allow for a change of the location without actually drawing a line, we shall use the fact that only the positive value of k corresponds to true colors. As a result, we can design the subroutine in such a way that, if $k \geq 0$, a line of the appropriate color will be drawn. On the other hand, if $k < 0$, no line will be drawn to the new location (x_d, y_d). In this way, ST_LINE may be used to move to any point on the screen and make the point as the "present location," for example, for the starting point of a new line. Our subroutine has, therefore, three integer arguments x_{i+1}, y_{i+1}, and k. The actual steps are summarized in Box 9-2.

With ST_LINE, it is a fairly simple matter to draw the x- and y-axis on the screen. As far as the computations are concerned, the x-axis is simply a horizontal straight line between two points and the y-axis is a corresponding vertical line. To put in the tick marks, we need to add at regular intervals short vertical lines for the x-axis and short horizontal lines for the y-axis. This is outlined in Box 9-3. We shall not be going into the question of labeling the axis, however, as it will require techniques that are best left to standard packages.

> **Box 9-3a Subroutine** X_AXIS(IX_B,IX_D,IY,N_TICKS,KOL)
> **Draw x-axis from** IX_B **to** IX_D **with** N_TICKS **ticks and color** KOL
>
> Argument list:
>   IX_B: Starting location of the horizontal axis.
>   IX_D: End location of the horizontal axis.
>   IY: Location of the axis in the vertical direction.
>   N_TICKS: Number of tick marks.
>   KOL: Color index of the axis.
> Subprogram used:
>   ST_LINE: Draw a straight line between two points (Box 9-2).
> 1. Draw the axis:
>   (a) Use ST_LINE to move the present pixel position to (IX_B,IY).
>   (b) Use ST_LINE to draw a straight line from (IX_B,IY) to (IX_D,IY).
> 2. Add in the tick marks:
>   (a) Calculate the distance between two tick marks.
>   (b) Use ST_LINE to bring the present pixel position to the location of the first tick mark.
>   (c) Use ST_LINE to draw a straight line 7 pixels long in the vertical direction.
>   (d) Repeat steps (b) and (c) for the remaining tick marks.
>
> ---
>
> **Box 9-3b Subroutine** Y_AXIS(IX,IY_B,IY_D,N_TICKS,KOL)
> **Draw y-axis from** IY_B **to** IY_D **with** N_TICKS **ticks and color** KOL
>
> Argument list:
>   IX: Location of the axis in the horizontal direction.
>   IY_B: Starting location of the vertical axis.
>   IY_D: End location of the vertical axis.
>   N_TICKS: Number of tick marks.
>   KOL: Color index of the axis.
> Subprogram used:
>   ST_LINE: Draw a straight line between two points (Box 9-2).
> 1. Draw the axis:
>   (a) Use ST_LINE to move the present pixel position to (IX,IY_B).
>   (b) Use ST_LINE to draw a straight line from (IX,IY_B) to (IX,IY_D).
> 2. Add in the tick marks:
>   (a) Calculate the distance between two tick marks.
>   (b) Use ST_LINE to bring the present pixel position to the location of the first tick mark.
>   (c) Use ST_LINE to draw a straight line 7 pixels long in the horizontal direction.
>   (d) Repeat steps (b) and (c) for the remaining tick marks.

Smoothing. Instead of joining two input points by a straight line, we can use a polynomial interpolation. This can be done by replacing (9-10) with an expression of the form

$$y = a_0 + a_1 x + a_2 x^2 + a_3 x^3 + \cdots \tag{9-11}$$

where the coefficients $a_0, a_1, \ldots$, may be found by one of the techniques discussed

> **Box 9-4 Program CUBIC_LN**
>
> **Plot a curve with smoothing – using sine curve as an example**
>
> Subprograms used:
> SMTH_PLT: Use CUBIC_SPLINE to generate a smooth plot.
> CUBIC_SPLINE: Calculate the second-order derivatives required for cubic spline smoothing (Box 3-10).
>   ST_LINE: Draw a straight line between two points (Box 9-2).
>   X_AXIS: Draw the x-axis (Box 9-3a).
>   Y_AXIS: Draw the y-axis (Box 9-3b).
>   SET_PXL: Compiler-dependent subroutine to move the present position to a definite pixel.
>
> Initialization:
>   (a) Set the number of input points to M and ranges $x = [0, \pi]$ and $y = [0, 1]$.
>   (b) Generate the input arrays.
> 1. Input N, the number of straight-line segments used in smoothing.
> 2. Use SMTH_PLT to plot a smooth curve with cubic spline from the M input points.
>   (a) Set the second-order derivatives f'' at both ends to be zero.
>   (b) Use CUBIC_SPLINE to generate the remaining derivatives.
>   (c) Initialize the plot:
>   (i) Use SET_PXL to turn on the graphics mode.
>   (ii) Use X_AXIS to draw the x-axis.
>   (iii) Use Y_AXIS to draw the y-axis.
>   (d) Plot the smooth curve:
>   (i) Set the first output point to be the same as the first one in the input array.
>   (ii) Obtain the step size Δx in the input array.
>   (iii) Obtain the step size δx in the interpolated array.
>   (iv) Increase the x value by δx and find out the input interval it belongs to.
>   (v) Calculate the value of y with (3-120).
>   (vi) Scale the values of x and y to pixel positions and use ST_LINE to draw a line to the pixel.
>   (vii) Repeat the above steps until the end of the curve.
> 3. Terminate the plot.

in Chapter 3. Since we are only interested in the visual effect here, there is no point in going to very high orders in the polynomial. The most common procedures are quadratic interpolation, where we stop at the x^2 term in (9-11), and cubic interpolation, where we include, in addition, the x^3 term. We shall adopt the latter, since the method of cubic spline has already been discussed in §3-7.

For our purpose here, we may regard interpolation as the process by which we find pairs of values, $(\tilde{x}_1, \tilde{y}_1), (\tilde{x}_2, \tilde{y}_2), \ldots$, between two consecutive inputs points (x_i, y_i) and (x_{i+1}, y_{i+1}), such that the set of points $\{(\tilde{x}_k, \tilde{y}_k)\}$ produced by the interpolation has the same behavior as the input set $\{(x_i, y_i)\}$. Our requirement is slightly different from that in §3-7, where we were more interested in the actual numerical values of $\{(\tilde{x}_k, \tilde{y}_k)\}$. Here, all we wish to achieve is to join the points

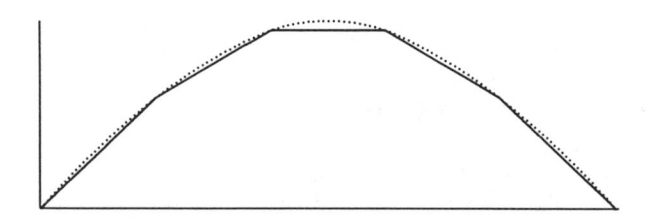

Fig. 9-6 A plot of the function $y = \sin(x)$ for $x = [0, \pi]$ with six entries. The solid curve is made by joining the calculated values with straight lines and the dotted curve is made using a cubic interpolation.

(x_i, y_i), $(\tilde{x}_1, \tilde{y}_1)$, $(\tilde{x}_2, \tilde{y}_2)$, ..., (x_{i+1}, y_{i+1}), by short straight-line segments, such that the complete curve appears as a smooth one. As we can see from Fig. 9-6, it is quite adequate to represent a sine wave in the interval $x = [0, \pi]$ with six calculated values and obtain the rest of the points to form a smooth curve using cubic interpolation. The steps involved in making the plot are outlined in Box 9-4.

§9-3 Functions with two independent variables

The subject of displaying functions with two or more independent variables is an interesting one in itself. Let us concentrate on the simplest case of two independent variables, x and y. A function of these two variables, for example,

$$z = f(x, y) \tag{9-12}$$

is a three-dimensional object, since z changes if we vary either x or y. A good way to visualize the behavior of z is to think in terms of the variations in the height of a piece of land. In this case, z represents the altitude of the surface and it changes as we move in any horizontal direction. For simplicity, let us imagine the terrain is a semisphere of radius r sitting on top of a horizontal plane. The surface of the semisphere is given by the equation

$$z = \begin{cases} \sqrt{r^2 - x^2 - y^2} & \text{for } x^2 + y^2 \leq r^2 \\ 0 & \text{elsewhere} \end{cases} \tag{9-13}$$

where r is the radius of the sphere. A three-dimensional view of this object is given in Fig. 9-7. One difficulty in showing a picture of such an object is that our output devices, graph papers, computer screens, and hard-copy outputs are two-dimensional. As a result, some "transformations," or projections, of x, y, and z to a two-dimensional surface must be carried out.

One aim of a three-dimensional plot is to give a photographlike image of the object from a particular view point. To simulate a photographic picture of the object, it is necessary to add shading and other renditions to the computer image. It is perhaps one of the biggest challenges of the computer industry to produce

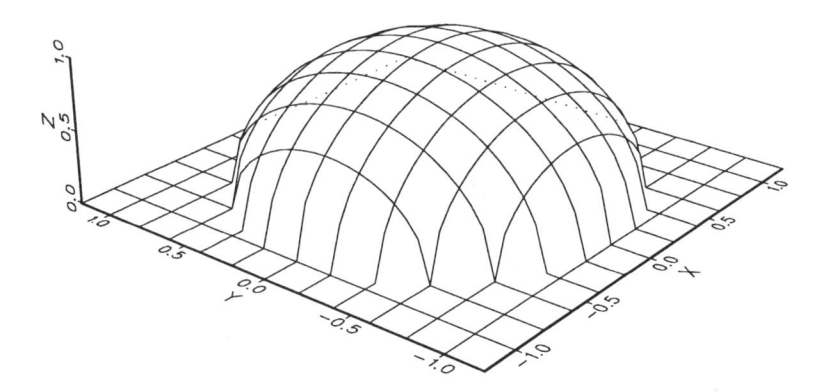

Fig. 9-7 Representation of a three-dimensional object. The surface of the upper half of a sphere is shown in the form of a wire frame.

graphics workstations that can give lifelike pictures of objects with a minimum amount of human effort and computer time. Such pictures are useful, for example, in animation and in computer-aided designs. On the other hand, in general, it is not easy to obtain quantitative information from such an image. Perhaps for this reason fancy computer graphics have not been as popular in computational physics as, for example, in engineering. We shall return later to consider some of the technical points of making such three-dimensional plots.

Contour plots. Instead of a photograph, it is also possible to represent a three-dimensional object on the computer screen or a sheet of paper by making a contour plot. Here, our graph paper becomes the xy-plane. Points with the same value of z in (9-12) are joined together to form a *contour* line. An easy way to visualize these contour lines is to think again in terms of the height of land in an area. If we move along a given contour line, there is no change in the gravitational potential energy.

Let us consider the problem of using a contour plot to show the topology of an area consisting of a few islands scattered in a calm sea. As the reference point for altitude, we can take the present sea level as zero in a scale based on some arbitrary units. The shorelines then form the contour lines for these islands at height zero. If the sea level rises a hundred units above its present level, the shore lines of our islands will change, reflecting the shapes of the islands at the height of 100 units. These new shorelines then form our contour lines for 100 units. The process may be repeated for each additional 100 units. Each time we get a new set of shorelines and a new group of contours lines, until all the landmasses are submerged under the water. A map made of all these contour lines is therefore

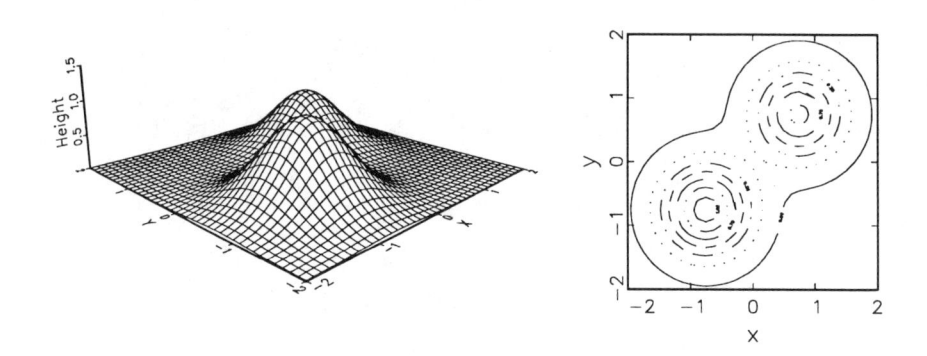

Fig. 9-8 A simulation of two islands in a perfectly calm sea. The height z as a function of x and y is given here by the relation $z = a_1 \exp\{(x - c_x)^2 + (y - c_y)^2\} + a_2 \exp\{(x - d_x)^2 + (y - d_y)^2\}$. A three-dimensional view of the islands is given on the left and a contour map of the same islands is given on the right.

a useful way to indicate quantitatively the variation of the height of the land in this area. This is illustrated by the plots in Fig. 9-8.

Mathematically, a contour line for z equal to some constant α may be regarded as, a curve in the xy-plane given by the relation

$$f(x,y) = \alpha \tag{9-14}$$

In general, the functional form of $f(x,y)$ is not known. The main calculations involved in constructing a contour plot are then to use interpolation techniques to find the values of x and y that satisfy (9-14). The input to such calculations are the values of z for a number of points at (x_1, y_1), (x_2, y_2),.... In terms of our example of islands, it is equivalent to the situation of constructing a contour map from a knowledge of the heights at a finite number of locations on the islands, (x_1, y_1), (x_2, y_2),.... For simplicity, we shall consider the intervals to be equally spaced at distance Δx along the x-direction and Δy along the y-direction. This is often referred to as regular grid data and an example is shown in Fig. 9-9. A good contour plot program can usually carry out the task for a set of randomly selected points, or scattered data, as well. This is essential, as it is not always possible to obtain input data at regular intervals. For our discussions, we shall restrict ourselves to the simpler case of regular grid data.

There are several ways to plot a contour map from a set of evenly spaced input. The choice among them depends to a large extent on the accuracy required. As an illustration, we shall adopt a four-point approximation. In other words, the local value of the function $f(x,y)$ is assumed to be given by an equation of the form

$$f(x,y) \approx a_1 + a_2 x + a_3 y + a_4 xy \tag{9-15}$$

Chapter 9]   Graphical Presentation of Results   625

where a_1, a_2, a_3, and a_4 are coefficients to be determined later. Since it is a very simple formula, we cannot expect it to represent the function for all values of x and y. However, within a rectangular area bounded by four adjacent points on the mesh, (x_{i-1}, y_{j-1}), (x_{i-1}, y_j), (x_i, y_{j-1}), and (x_i, y_j), as shown in Fig. 9-9, a reasonable approximation can be obtained if the values of the four coefficients are determined from those of z at the four corners.

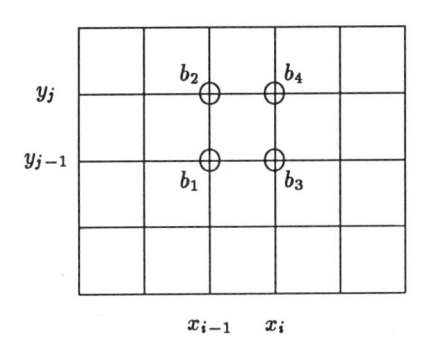

Fig. 9-9 A set of regular grid data for a contour plot. The values of the function $z = f(x,y)$ are known at each of the points, for example, b_1, b_2, b_3, and b_4, at respectively $(x, y) = (x_{i-1}, y_{j-1})$, (x_{i-1}, y_j), (x_i, y_{j-1}), (x_i, y_j).

Consider the square cell in Fig. 9-9 that is bounded by the four points at (x_{i-1}, y_{j-1}), (x_{i-1}, y_j), (x_i, y_{j-1}), and (x_i, y_j). Let the values supplied to us at these four points be

$$b_1 = f(x_{i-1}, y_{j-1}) \qquad b_2 = f(x_{i-1}, y_j) \qquad b_3 = f(x_i, y_{j-1}) \qquad b_4 = f(x_i, y_j)$$

The coefficients a_1, a_2, a_3, and a_4 may be obtained by requiring that (9-15) produce the correct values b_1, b_2, b_3, and b_4 at these four corners. That is,

$$a_1 + a_2 x_{i-1} + a_3 y_{j-1} + a_4 x_{i-1} y_{j-1} = b_1$$
$$a_1 + a_2 x_{i-1} + a_3 y_j + a_4 x_{i-1} y_j = b_2$$
$$a_1 + a_2 x_i + a_3 y_{j-1} + a_4 x_i y_{j-1} = b_3$$
$$a_1 + a_2 x_i + a_3 y_j + a_4 x_i y_j = b_4 \qquad (9\text{-}16)$$

The solution to this set of four equations is

$$a_1 = \frac{(b_1 y_j - b_2 y_{j-1}) x_i - (b_3 y_j - b_4 y_{j-1}) x_{i-1}}{(x_{i-1} - x_i)(y_{j-1} - y_j)}$$

$$a_2 = \frac{(b_2 - b_4) y_{j-1} - (b_1 - b_3) y_j}{(x_{i-1} - x_i)(y_{j-1} - y_j)}$$

$$a_3 = \frac{(b_3 - b_4) x_{i-1} - (b_1 - b_2) x_i}{(x_{i-1} - x_i)(y_{j-1} - y_j)}$$

$$a_4 = \frac{b_1 - b_2 - b_3 + b_4}{(x_{i-1} - x_i)(y_{j-1} - y_j)} \tag{9-17}$$

Using these values, the equation for a given contour $z = \alpha$ has the form

$$a_1 + a_2 x + a_3 y + a_4 xy = \alpha$$

This is equivalent to the relation

$$y = \frac{(\alpha - a_1) - a_2 x}{a_3 + a_4 x} \quad \text{or} \quad x = \frac{(\alpha - a_1) - a_3 y}{a_2 + a_4 y} \tag{9-18}$$

The problem of the contour plot for $z = \alpha$ is now reduced to be the same as one of making a two-dimensional plot of y as a function of x, or vice versa.

If there are no degeneracies within a cell, meaning that the contour line $z = \alpha$ passes only once through it, we can use either one of the two relations given by (9-18) to draw the contour line in the cell. The only other consideration is which of the two relations to use, that is, whether we choose x or y as the independent variable. One way to make the selection is to use a criterion based on the average slope of z along each of the two directions. If

$$\overline{\frac{\Delta z}{\Delta y}} = \frac{b_2 + b_4 - b_1 - b_3}{y_j - y_{j-1}}$$

is larger than

$$\overline{\frac{\Delta z}{\Delta x}} = \frac{b_3 + b_4 - b_1 - b_2}{x_i - x_{i-1}}$$

we shall vary x in small steps, and for each x, the value of y is calculated using (9-18). Conversely, if $\overline{\Delta z/\Delta y}$ is smaller than $\overline{\Delta z/\Delta x}$, the value of y is varied in small steps and the value of x is calculated for each y.

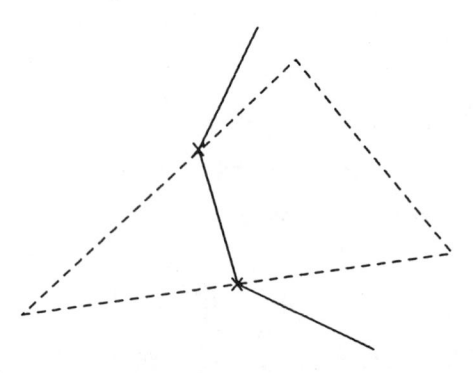

Fig. 9-10 A triangular element for contour plots. For a sufficiently small triangle, a contour $z = \alpha$ entering the triangle at one side must exit through one of the other two sides. The contour inside the triangle may be approximated by a straight line joining the entry and exit points.

In a general-purpose contour plotting package, it is more efficient to divide the area to be plotted into a number of small triangles, rather than squares as we have done in the example above. For each triangle, if the contour $z = \alpha$ passes through one edge, it must leave through another one, as shown in Fig. 9-10. The advantage over a square is that we have one less side to work with. If each of the triangles is sufficiently small, we can simply use a straight line to join the two points where the contour enters and leaves, without having to be concerned with the problem of degeneracy. The visual effect of a contour plot can also be enhanced by giving different colors to different zones, that is, the regions between two adjacent contour lines. Again, we shall leave these more advanced techniques to standard plotting packages.

Three-dimensional plots. A popular way for representing three-dimensional objects on a screen or a sheet of paper is to simulate a photograph, as we saw in Fig. 9-7. For simplicity, we shall consider only a monochrome display device. In this limit, the only way to construct an image is to use different shades of gray. In fact, if we examine a black and white photograph carefully, we will find that it is made of a regular grid of small black dots of different sizes. This is especially noticeable in pictures with coarse resolution, such as those printed in a newspaper. Since these dots appear in regular intervals, the only degree of freedom in simulating different shades of gray is to vary the size of the individual dots. Although this is a very simple and clever way of constructing a black and white image, it is not easy to make use of the technique on a computer, as the size of each pixel is fixed. To simplify the discussion, we shall also ignore the possibility for a pixel to display different shades of gray.

For our purpose here, it is useful to regard the process that constructs a photographic image from an object as made of a two-dimensional grid of light rays, each originating from a small area of the object and ending on the photograph. Although we shall not make use of the possibility available on some computers to vary the intensity of each pixel, the density of pixels is usually much higher than on a typical page of a newspaper. As a result, we can make use of the variations in the density of the dots to simulate different gray scales. A simple illustration of this technique is given in Fig. 9-11.

The rendition of digital results into a visual image is perhaps one of the more important applications of computer graphics. In addition to the ability of varying the size of the image, it is also possible to obtain the view from different directions by rotating the image. The importance of such capability is quite obvious in computer-aided design and related fields. This perhaps explains the reason for the explosion of graphics workstations in the market. However, the use of the more sophisticated computer graphics for three-dimensional images in computational physics is much more subdued. This is due largely to the fact that the usual quantities we wish to represent are often not three-dimensional objects we can feel and touch as, for example, in engineering. In fact, we deal more

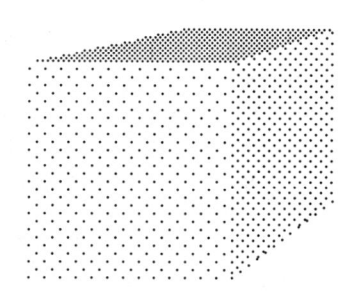

Fig. 9-11 The use of shading and perspective to represent a three-dimensional object. Instead of dots of different sizes at regular intervals, dots of the same size at different densities are used to indicate the differences in lighting from different surfaces of a cube.

commonly with phase spaces and Hilbert spaces for quantities such as partition functions and wave functions, rather than the Euclidean space for real objects. Consequently, any assignments of color and shade are usually false and are used only to help us assimilate the large amount of information. Such a rendition of the calculated results is important for visual effect but is not as crucial or as helpful as in engineering. For this reason, it is usually quite adequate to show only the outlines, or a wire-frame representation, of the object. In this case, the object is replaced by a three-dimensional frame constructed of wires joining the important points on the object. Only the image of the wire frame is projected onto the screen. Such a representation of a semisphere, given in equation form by (9-13), was shown earlier in Fig. 9-7.

The use of a wire frame to represent a solid object means that only points where two or more wires meet are important. In projecting the object onto a two-dimensional surface, it is necessary only to carry out the calculations for these points, as the wires are nothing more than lines joining together these points. The construction of a two-dimensional image for a three-dimensional object is, in general, a tedious process. The reduction to a relatively small number of points is therefore useful in making the process much faster.

Perspective and hidden line removal. The apparent size of an object depends on the angle it subtends at the viewing point. For this reason, two parallel lines of the same length will seem to be different if they are at different distances from the viewer. To lend realism, the same apparent difference in length as a function of the distance from the viewer must also be present in our computer image. This may be illustrated by the diagram given in Fig. 9-12. For simplicity, the object consists of only two points A and B at different distances from the viewer. The image we wish to construct is represented by the projection of these two points on a plane called the *image plane*, indicated as I in the diagram. It is convenient to take the center of the image plane as the origin of our coordinate system and the

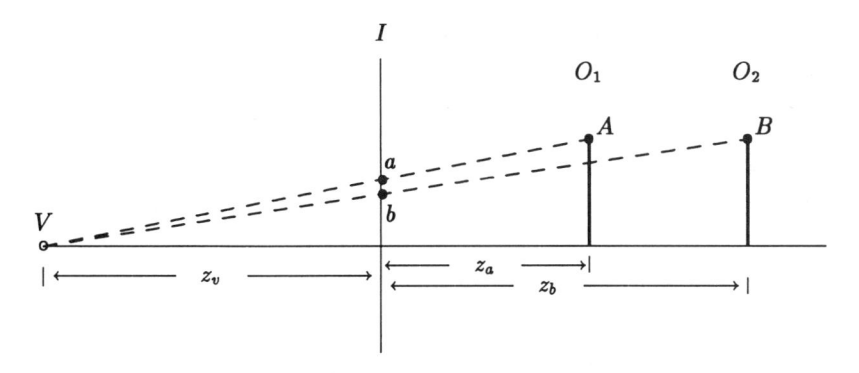

Fig. 9-12 Perspective transformations of two objects O_1 and O_2, viewed from the point marked V. The point A in O_1 is projected to the point a on the image plane located at I, between V and O_1, and the point B in O_2 is projected to the point b. Since O_2 is farther away than O_1, b is lower than a, even though B and A are at the height.

line from the origin to the viewer, located at the point V in the diagram, as the z-axis. In this way, the image plane is the same as the xy-plane of our coordinate system and is perpendicular to the z-axis.

Let the distance between the viewer and the image plane be z_v and the horizontal distances from the image plane to A and B be, respectively, z_a and z_b. A light ray originating from A and reaching the viewer at V passes through the image plane at the point a. Similarly, a light ray arriving at the same viewer from B passes through the image plane at the point b. If the vertical heights from A and B to the z-axis are H_a and H_b, respectively, the distances of h_a and h_b from a and b to the origin are given by

$$\frac{h_a}{z_v} = \frac{H_a}{z_v + z_a} \qquad \frac{h_b}{z_v} = \frac{H_b}{z_v + z_b}$$

If $H_a = H_b$, the ratio of h_a to h_b is $(z_v + z_a)/(z_v + z_b)$, the ratio of the distances from viewer to A and B. This way of making the projection is called the central projection. Since the calculations must be carried out for every visible point of the object, in practice, one may have to resort to even simpler methods so as to reduce the amount of calculations involved.

In wire-frame representations, a three-dimensional object is usually taken to be opaque. This is also useful in giving a feeling of distance in a convenient way. For an opaque object, points away from the viewer are not visible and, by displaying only the visible lines and surfaces, the clarity and impact of the image are enhanced. Hiding the invisible part of an object in an image is known as the hidden-line, or more generally, the hidden-surface problem. The diagrams in Fig. 9-13 provide a simple illustration of this point. A cube is defined by giving the locations of its eight corners, and a wire frame representation of the cube is made by joining each corner with its three neighbors by straight lines. There are a total of 12 lines in this way. If the cube is considered to be transparent and viewed directly above one of its corners, the wire frame image takes on the appearance shown in Fig. 9-13(a). This hardly transmits a sense of being the image of a solid object. However, if we remove three interior lines, representing the three from the corner directly away from the viewer, the image becomes either that shown in part (b) or in part (c), depending on whether the cube is viewed from the top or the bottom.

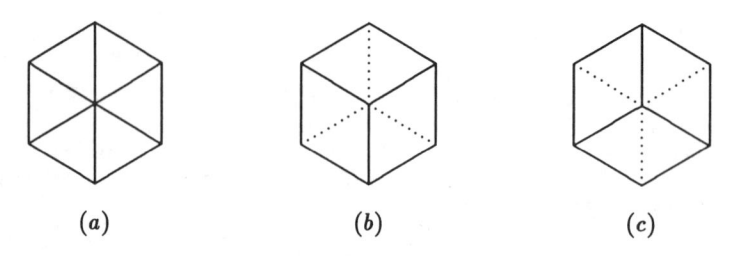

Fig. 9-13 Demonstration of the importance of hidden-line removal. If the hidden lines are shown, the image of a cube is not very different from an eight-sided polygonal, as we can see in (a). However, when three of the internal lines, shown in (b) as dotted lines, are removed, the image appears as a cube viewed from an angle above. Similarly, when the other three internal lines are removed instead, the object appears in (c) as a cube viewed from an angle below. For simplicity, no perspective transformation has been carried out in constructing any of these images.

There are several ways to remove hidden lines and hidden surfaces. All of them are sufficiently complicated that no single algorithm dominates the field. The various methods may be divided into two categories, depending on whether the removal is done at the object or in the image plane.

The principle behind eliminating the invisible parts at the object plane is very simple and may be described in terms of ray tracing. If the ray of light coming to the viewer from a particular point of the object is obscured by another part, the point is invisible and should not appear as a part of the image. This can happen, for example, if there is another point that is closer to the view occupying the same path. To find out whether there are such intervening parts of the object, the rays from different parts must be sorted according to their distance to the viewer, with the proper perspective dictated by the location of the viewpoint. Such a calculation, however, can be extremely time consuming. The problem is not simplified very much by using a wire-frame representation. This comes from the necessity that one must consider not only all the points but also all the lines joining any two points. For example, even when both points are visible, the line joining them can still be partially blocked by other parts of the object. Similarly, one must also determine how much of a line going from a visible point to an invisible point is actually visible. This means that each line must be divided into several small segments and each segment examined by itself to determine whether it is visible.

The principle behind hidden-surface removal at the image plane is also quite simple. Consider the case of an image plane made of a number of pixels. In constructing the image, only the visible part of the object can be recorded in the video memory that controls these pixels. A simple but practical way to achieve this is to use the following method. Let us divide the object into thin layers parallel to the image plane. The process of constructing the image for this object starts from the layer that is farthest away. Once this is done, the second layer of the object is projected onto the same image plane. The crucial point here is that all the pixels in the image of the first layer that are covered by the second layer are reset according to the requirement of the latter. In this way, any part of the first layer that overlaps the second layer is replaced by the image of second layer and thus becomes invisible to the viewer. In other words, the part of the surface of the first layer that is hidden by the second layer is removed in the process of constructing the image of the second layer. Once the image of the second layer is completed, we can proceed to the third layer, and so on. This process of erasing the overlapping areas in the image plane as we build a new one continues layer by layer until we come to the last layer, the one that is closest to image plane. This is illustrated schematically in Fig. 9-14 with a cube. Again the amount of computation involved is large, as each layer of the object must be made sufficiently thin in order to produce a good image. Some saving can be achieved by considering small regions of each layer as a unit, rather than individual pixels. The size of each region depends on the coherence of the object, that is, how nearby points are related to each other. However, we shall not get into such detailed discussions here.

Fig. 9-14 Hidden-surface removal at the image plane. The object is divided into thin layers and projection to the image plane starts with the layer that is farthest away. The next layer is then projected on top of the image of the previous layer. Any part of the image of the old layer covered by that of the new layer is erased and replaced by the new layer.

Because of the large amount of computation involved in projecting a three-dimensional object onto a two-dimensional image plane, many approximations are used in practice. In most cases, it is not possible to detect any visible defects in the image constructed by these methods. In particular, such approximations are quite adequate for most purposes in computational physics. In many problems of interest, fancy rendition of the object is seldom called for, and most of the standard plotting packages are able to produce images that are satisfactory for our needs. On the other hand, there are many physical problems involving more than two independent variables. For example, to show the results of the solutions for the collision of two particles, three spatial coordinates and one time coordinate are required. To visualize the time development of the scattering process, movies are often made of the calculated results, with each frame of the movie showing the results for a given time step. The rendition of such complicated results often takes an amount of computer time comparable to that required for solving the scattering equations themselves. For the more complicated situations involving an even larger number of independent variables, the projection problem becomes very difficult and special methods for making graphical presentation of such results are usually required.

Problems

9-1 Carry out a plot for the semicircle of (9-2) on a screen using the character mode and without considering the aspect ratio of a character on the computer screen. Plot again after allowing for the fact that the aspect ratio is not unity. What happens if the graph is now printed out on a sheet of paper?

9-2 Use the equivalent of the SET_PXL subprogram described in §9-1 on your computer to plot a straight line. From this, develop a subroutine to plot a horizontal or a vertical axis including tick marks.

9-3 Use the equivalent of the SET_PXL function described in §9-1 on your computer to draw numbers 0 to 9 on the screen.

9-4 Draw the sine curve of Fig. 9-6 with more straight-line segments than the six used in the figure. What is roughly the number needed to give the feeling of a smooth curve? Try the same thing on a normal distribution curve $y = (2\pi)^{-1/2} \exp\{-x^2/2\}$ in the interval $x = [-3, +3]$. Why are many more straight-line segments needed here than in the sine curve?

9-5 In land surveying, often only the altitudes at a regular grid of points are measured. Use the function for a semisphere given in (9-12) to generate such a set of data. Use the method outlined in §9-3 to make a contour map for the semisphere landscape. The result should be a set of concentric circles.

Chapter 10
Computer Algebra

The possibility of using machines to carry out complicated algebraic manipulations, commonly referred to as computer algebra, was recognized from the early days of computers. In the 1960s, the Lisp language was introduced, in part, to process symbolic data. The development of Reduce (Rand Corporation), one of the major symbolic manipulation programs today, was started in 1963. By now, there are over half a dozen different packages for algebraic calculations, such as Derive (Soft Warehouse), Macsyma (Symbolics Inc.), Maple (Symbolic Computation Group, University of Waterloo), Mathematica (Wolfram Research, Inc), and MuMath (Soft Warehouse). A concise comparison of some of these packages was published by Foster and Bau (*Science* **243** (1989) 679) and a very informative article, "Computer Algebra," written by a group of authors involved with the Macsyma project, can be found in the December 1981 issue of *Scientific American*. Other software packages, such as Eureka (Borland) and MathCad (MathSoft), can also perform a variety of algebraic calculations.

In this chapter, we shall give a short introduction to the subject of symbolic manipulation (SM) on a computer, providing some idea of its potentials as well as limitations. The algorithms for carrying out various types of algebraic tasks are as specialized as those for numerical calculations. A comprehensive description of them is more suited to a volume dedicated specially to the task and we shall not attempt it here. Our basic philosophy will be the same as that of the previous chapter, to illustrate the basic principles involved. In general, the potentials of SM have not been fully utilized in physics and one of the aims here is to encourage readers to try out one of the available SM packages.

Elementary algebra operations. In numerical calculations, the expression

$$C = A + B$$

means that the value of C is the sum of A and B. If either A or B is not given a numerical value, the expression becomes meaningless. In fact, a good FORTRAN compiler will usually warn the programmer if any quantities on the right side of an equation are "undefined," meaning not associated with a numerical value. In algebraic computations, A and B are symbols whose meanings can be anything the user wishes to assign. Our interest here is to obtain algebraic results by manipulating these symbols according to some well-defined set of rules. Very often, these rules are quite simple to write down, but the work involved in applying them can be tedious. The use of a computer to carry out the operations is therefore very attractive from the point of view of time as well as accuracy (meaning not making any mistakes here).

A simple illustration is provided by the problem of proving the identity

$$\sum_{k=1}^{n} k = \frac{n(n+1)}{2} \qquad (10\text{-}1)$$

One way to carry out the proof is to use induction. It is obvious that the relation holds for $n = 1$. The next step in the argument is to assume that the relation is true up to $n = (m - 1)$. That is, we take it for granted that

$$\sum_{k=1}^{m-1} k = \frac{(m-1)m}{2}$$

For the convenience of discussion, we shall use the symbols $S_L(m-1)$ and $S_R(m-1)$ for, respectively, the left and right sides of the equation. That is,

$$S_L(m-1) \equiv \sum_{k=1}^{m-1} k \qquad S_R(m-1) \equiv \frac{(m-1)m}{2} \qquad (10\text{-}2)$$

Now let us test the relation (10-1) for $n = m$, using the fact that $S_L(m-1) = S_R(m-1)$ from our assumption. The expressions for $S_L(m-1)$ and $S_L(m)$ are not known for an arbitrary m. However, their difference is well defined and has the value

$$S_L(m) - S_L(m-1) = m$$

The proof of the identity (10-1) is complete once we show that the difference between $S_R(m)$ and $S_R(m-1)$ is also m.

The calculations involved may be carried out using a computer. Let us define S_R by the algebraic relation

$$S_R = N * (N+1)/2 \qquad (10\text{-}3)$$

where N is an algebraic symbol. If we let $N = (m-1)$, then S_R takes on the value specified by (10-2) with N replaced by $(m-1)$. This gives us the value of $S_R(m-1)$. If we reassign the value of N to be m, the value of S_R becomes that of $S_R(m)$. To prove the identity of (10-1), we need to check if it is true that

$$S_R(N=m) - S_R(N=m-1) - m = 0$$

If so, the identity of (10-1) holds for all $n \geq 1$, as the value of m is arbitrary here (as long as $m > 0$).

It is also possible to demonstrate the identity (10-1) using numerical calculations. For example, we can calculate the sum $S_L(m)$ numerically to some large value of n, such as 10, and we find that the result is 55. For $n = 10$, the right side of (10-1) is $S_R(5) = 10 * 11/2 = 55$. This verifies that the identity (10-1) holds for $n = 10$. We can repeat the calculation for other values of n for which we have numerical accuracy. The net result is that we can be fairly well convinced of the correctness of the relation (10-1). On the other hand, from the point of view of logic, all we have done by numerical calculations is to show that there is no contradiction to (10-1) within certain limits. It will be difficult to convince any mathematician that the demonstration constitutes a proof of the identity.

The example clearly shows the power of algebraic calculations, a point that hardly needs to be emphasized. On the other hand, it is perhaps too simple an example to demonstrate that the calculations can be complicated enough for a computer to be useful. As an exercise, the interested reader can try to prove the following identity by induction:

$$\sum_{k=1}^{n} k^7 = n^2(n+1)^2(3n^4 + 6n^3 - n^2 - 4n + 2)/24 \qquad (10\text{-}4)$$

The tedium involved should convince anyone that the symbolic manipulation capability of computers is an important asset.

Algebraic programming. The example of proof by induction given above serves also to illustrate the basic differences between numerical and algebraic programming. If we take the simplistic view that numerical calculations on a computer consist primarily of additions, subtractions, multiplications, and divisions, then algebraic manipulations are made of pattern recognitions and logical operations. For example, in (10-3), the computer is instructed to make the association of S_R with the expression $N(N+1)/2$, with N being an algebraic symbol. If we let $N = m$, then S_R takes on the form $m(m+1)/2$, and if $N = (m-1)$, S_R takes on the form $(m-1)m/2$. Again, m is another algebraic symbol. Let us store the results $m(m+1)/2$ and $(m-1)m/2$ as $S_{R,m}$ and $S_{R,(m-1)}$, respectively. If we next let $m = k$, we expect that $S_{R,m}$ and $S_{R,(m-1)}$ are changed into $k(k+1)/2$ and $(k-1)k/2$, without having to assign any specific meaning or numerical value to k. It is clear that high-level languages, such as FORTRAN, designed to carry out numerical calculations in a convenient way, are not suitable for the kind of symbolic manipulation work illustrated here. As a result, most of the software packages for computer algebra are written either in the Lisp language or in languages that are based on C, with features developed specifically for algebraic calculations.

We shall not make any attempt here to describe the Lisp language or any other SM languages. The interested reader should consult references that specialize in these topics. Part of the reason for taking such an approach is that all the SM packages on the market these days make a good attempt to shelter the reader from the basic operations upon which they are built. If a single language is to be picked as the popular language for SM, one should perhaps take the Lisp language. This is analogous to the situation that FORTRAN is the language most widely, but by no means exclusively, used in numerical calculations. However, the FORTRAN language is fairly well standardized since most compilers are successful in meeting their claims to be following the latest ANSI (American National Standards Institute) specifications. This is not true for the Lisp language. In fact, there is a family of Lisp dialects, such as MacLisp, Interlisp, and Franz Lisp, just to mention a few. Even Common Lisp and Standard Lisp are not necessarily what their names may imply. Although it is relatively simple to switch from one dialect to another, it can be confusing to the beginner. Until the Common Lisp, or some other form of the Lisp language, becomes the standard, it is not easy to give a short summary of the programming language that is beneficial to the average reader and we shall not attempt it here.

Some of the more comprehensive SM packages, such as Macsyma, Mathematica, and Reduce, also serve as programming languages. That is, one can develop procedures, meaning self-contained sets of instructions, to carry out special types of calculations. This is similar to the way in which numerical programmers develop subroutines for such specific tasks as fitting curves and diagonalizing matrices. Because of the diversity of algebraic calculations for which a computer can be helpful, different SM packages tend to put their emphases on different aspects of the topic. For this reason, no single package can claim to be superior in every respect to the others. In addition to the basic set of commands provided by each package, there is also a large collection of procedures supplied by the users. Some of these are published in journals such as *Computer Physics Communications* (North Holland). Just as in numerical work, several extensive public-domain libraries are available through computer networks, such as those given in §1-4. Similar to numerical calculations, these user-supplied libraries have made SM much more convenient to use.

Because of the diversity of commands used in different SM packages, it is not very meaningful for us to follow any one them in our discussion. For our purpose, we shall attempt to make the discussion as independent of any of them as possible. Readers who have not been exposed to computer algebra before are, however, urged to try out any one of the SM packages mentioned at the beginning of the chapter so as to appreciate some of things they can do.

Evaluation of a determinant. As an illustration of computer algebra, we shall evaluate a determinant of order n:

$$D = \det \begin{vmatrix} A_{1,1} & A_{1,2} & \cdots & A_{1,n} \\ A_{2,1} & A_{2,2} & \cdots & A_{2,n} \\ \vdots & \vdots & \ddots & \vdots \\ A_{n,1} & A_{n,2} & \cdots & A_{n,n} \end{vmatrix} \qquad (10\text{-}5)$$

The elements $A_{i,j}$ are assumed to be stored in a two-dimensional array. If all the elements $A_{i,j}$ are given numerical values, we can use one of the methods described in §5-1 to find the numerical value of D. This is not our aim here. Our interest is in carrying out the calculation algebraically. For this purpose, $A_{i,j}$ are taken to be algebraic symbols. For example, for $n = 2$, we have

$$D = \det \begin{vmatrix} a & b \\ c & d \end{vmatrix}$$

That is

$$A_{1,1} = a \qquad A_{1,2} = b \qquad A_{2,1} = c \qquad A_{2,2} = d$$

where a, b, c, and d are algebraic symbols without being assigned any numerical values. The value of D in this case is

$$D = ad - bc \tag{10-6}$$

For larger values of n, the same way of carrying out the calculation can be quite tedious. For our illustration, we shall develop a method whereby the work may be done on a computer. Since we wish only to demonstrate the principle of algebraic programming, we shall not complicate the problem unnecessarily by introducing a new programming language at the same time. For this reason, we shall give the instructions to the computer in terms of the FORTRAN language, even though it may not be the most convenient thing to do.

As we have seen in §5-1, there are several ways to evaluate a determinant. One method is to expand D as a linear combination of the products of the elements of one row (or one column) with their corresponding minors, as we did in (5-7). The minor $M_{i,j}$ of a determinant $\mathbf{A}$ of order n was defined there as a determinant of order $(n-1)$, constructed out of the original determinant except that we remove all the elements in column i and row j of $\mathbf{A}$. For example, in the case of $n = 3$, we have

$$D = \det \begin{vmatrix} A_{1,1} & A_{1,2} & A_{1,3} \\ A_{2,1} & A_{2,2} & A_{2,3} \\ A_{3,1} & A_{3,2} & A_{3,3} \end{vmatrix}$$

$$= A_{1,1} \det \begin{vmatrix} A_{2,2} & A_{2,3} \\ A_{3,2} & A_{3,3} \end{vmatrix} - A_{1,2} \det \begin{vmatrix} A_{2,1} & A_{2,3} \\ A_{3,1} & A_{3,3} \end{vmatrix} + A_{1,3} \det \begin{vmatrix} A_{2,1} & A_{2,2} \\ A_{3,1} & A_{3,2} \end{vmatrix}$$

$$= A_{1,1}\left(A_{2,2}A_{3,3} - A_{2,3}A_{3,2}\right)$$

$$- A_{1,2}\left(A_{2,1}A_{3,3} - A_{2,3}A_{3,1}\right)$$

$$+ A_{1,3}\left(A_{2,1}A_{3,2} - A_{2,2}A_{3,1}\right) \tag{10-7}$$

where we have evaluated all the order 2 determinants in the intermediate stage using the result of (10-6). For $n = 4$, the minors are determinants of order 3, and (10-7) may be used to evaluate each of them first. In this way, the calculation becomes a recursive one, with each step reducing the determinant to one order lower. This process ends when the order is 2, where we can apply (10-6) directly.

Since it is not easy to carry out recursive calculations in the FORTRAN language, we shall, instead of using the method of minors, expand the determinant directly as a linear combination of products of n elements, as we did in (5-4):

$$D = \sum_{i,j,k,\ldots} \epsilon_{i,j,k,\ldots} A_{1,i} A_{2,j} A_{3,k} \cdots \qquad (10\text{-}8)$$

where $\epsilon_{i,j,k,\ldots}$ is the Levi-Civita symbol:

$$\epsilon_{i,j,k,\ldots} = \begin{cases} +1 & \text{for even permutations of the } n \text{ subscripts } (1,2,\ldots,n) \\ -1 & \text{for odd permutations of the } n \text{ subscripts } (1,2,\ldots,n) \\ 0 & \text{if any two or more of the subscripts are equal} \end{cases} \qquad (10\text{-}9)$$

Again, the elements $A_{i,j}$ are symbolic data here. In terms of the FORTRAN language, we can think of them as a two-dimensional array of characters. For example, in our $n = 2$ example given in (10-6), $A_{1,1}$ is a, $A_{1,2}$ is b, and so on. However, the subscripts are numerical information. As a result, the FORTRAN language may be used in a convenient way to manipulate these indexes.

The summation in (10-8) means that all the possible values of the n indexes i, j, k, ..., must be taken. There is, therefore, a total number of n^n possible terms in the sum. However, because of the Levi-Civita symbol, only terms with $i \neq j \neq k \neq \cdots$ can contribute to D. In other words, among the n^n terms, only $n!$ are different from zero. It is therefore more efficient for us to find out first which are the nonvanishing terms before carrying out any calculations with them.

Levi-Civita symbol. We shall begin first with the simpler problem of evaluating the Levi-Civita symbol. As shown in (10-9), there are three possible values of $\epsilon_{i,j,k,\ldots}$: 0, -1, or $+1$, depending on the order of the indexes. The n indexes $(i, j, k, \ldots)$ are numerical data, each an integer in the range $[1, n]$. An efficient way to find the value of $\epsilon_{i,j,k,\ldots}$ is to permute these n quantities so that they appear in an ascending order. If we find any two of them to be equal to each other in the process of permutation, a zero is returned. Otherwise, the value of $\epsilon_{i,j,k,\ldots}$ is either $+1$ or -1, depending on whether the number of permutations required is even or odd in bringing the set of n integers $(i, j, k, \ldots)$ to the order $(1, 2, 3, \ldots, n)$.

Consider again the example of $n = 3$. There are $3! = 6$ nonvanishing terms in the sum on the right side of (10-8) in this case. That is,

$$D = \epsilon_{123} A_{1,1} A_{2,2} A_{3,3} + \epsilon_{132} A_{1,1} A_{2,3} A_{3,2}$$
$$+ \epsilon_{213} A_{1,2} A_{2,1} A_{3,3} + \epsilon_{231} A_{1,2} A_{2,3} A_{3,1}$$
$$+ \epsilon_{312} A_{1,3} A_{2,1} A_{3,2} + \epsilon_{321} A_{1,3} A_{2,2} A_{3,1}$$

The values of ϵ_{132} and ϵ_{213} are -1, as it takes one permutation to bring the indexes in each case to the order (123). Similarly, $\epsilon_{321} = -1$, as it take three permutations of neighboring indexes to bring them to order 123. The values of the other Levi-Civita symbols are $+1$, since the numbers of permutations required are even.

Box 10-1   Function LEVI_C(LARY,N)
Levi-Civita symbol for n indexes

Argument list:
   LARY: One-dimensional array of indexes.
   N: Number of indexes.
1. Initialize $C = +1$.
2. Search for the smallest element among $\ell_i, \ell_{i+1}, \ldots, \ell_n$, starting from $i = 1$:
   (a) Begin by assuming ℓ_i to be the minimum.
   (b) Compare elements ℓ_{i+1} to ℓ_n with the minimum found so far.
      (i) Return zero if two indexes are equal to each other.
      (ii) Replace the minimum if a smaller index is found.
   (c) If the minimum is not ℓ_i,
      (i) Exchange ℓ_i with the minimum.
      (ii) Change the sign of C.
3. Return C as the value of $\epsilon_{\ell_1,\ell_2,\ldots,\ell_n}$.

Instead of actually permuting any two neighboring indexes, we can replace the process by interchanging an arbitrary pair of indexes in such a way that the smaller one of the pair ends up in front of the larger one. In interchanging any two indexes ℓ and m, without disturbing the order of the r indexes in between them, the number of permutations required is always odd, independent of the value of r (see Problem 10-4). Using this result, we can design a way to find the value of a Levi-Civita symbol by a computer program. We begin by examining all n indexes and finding out which is the smallest. If the smallest is not the first one, it is interchanged with the first one. Otherwise, no interchange takes place. The process is then carried out for the remaining $(n-1)$ indexes, and so on, until only the last one is left. If the total number of interchanges made is even, a $+1$ is returned. If the number is odd, a -1 is returned. On the other hand, if, at any stage of the permutation, any two indexes are found to be equal to each other, the search stops and a zero is returned. The algorithm is summarized in Box 10-1.

Generate an ordered list of n quantities.   Once we solve the problem of finding the value of the Levi-Civita symbol, the main part of the work remaining in evaluating a determinant by (10-8) is to find the set of n indexes $(i,j,k,\ldots)$ for the product $A_{1,i} A_{2,j} \cdots A_{n,\ell}$ in each of the $n!$ contributing terms. The problem is equivalent to finding all the $n!$ possible permutations of the n quantities

$(1, 2, \ldots, n)$. For $n = 2$, this is trivial and the result is $(1, 2)$ and $(2, 1)$. For $n > 2$, it is convenient to use a method by which the list of permutations is generated in an ascending (or descending) order. That is, we shall regard $(i, j, k, \ldots)$ as a single integer of n digits (or double digits if $n > 9$, triple digits if $n > 99$, and so on). In an ordered list of $n!$ members, the place a specific permutation occupies depends on the value of this composite integer. The particular order we wish to adopt for the $n!$ permutations is such that all the entries in the corresponding list of $n!$ composite integers are in ascending order. Again, for our example of $n = 3$, the list of six members in ascending order is 123, 132, 213, 231, 312, and 321.

For any n, the smallest number in the list is $123\cdots(n-1)n$. The next smallest number of n digits generated by a permutation of the n integers is $123\cdots n(n-1)$. To find the next smallest number in the same list, we must move n to the third last place and the result is $123\cdots n(n-2)(n-1)$. It is quite obvious now how we can proceed further and generate the complete ordered list one by one until we come to the largest number $n(n-1)\cdots 321$. The work can be tedious even for a modest size of $n = 5$. On the other hand, it is easy to design an algorithm for a computer to take over the calculations.

This can be done in the following way. At some stage of the generation, let us assume that we have just obtained the member $(\ell_1 \ell_2 \ldots \ell_{n-1} \ell_n)$. Note that $\ell_i \neq \ell_j$ for $i \neq j$ since they are produced from permutations of $(12 \ldots n)$. Furthermore, we shall assume that $\ell_n > \ell_{n-1}$. To produce a new member, we shall first interchange the last pair, ℓ_{n-1} and ℓ_n. The result is $(\ell_1, \ell_2, \ldots, \ell_n, \ell_{n-1})$. This is a new member and, in terms of the order we have defined, it comes after $(\ell_1, \ell_2, \ldots, \ell_{n-1}, \ell_n)$, as $\ell_n > \ell_{n-1}$.

Next, we shall generate another member in the list that is later in order than these two. To do this, let us examine the value of ℓ_{n-2}. Let

$$L = \ell_t + 1$$

with $t = (n-2)$ at this stage. Our aim is to see if L can be the index for position t, without having to change any of the indexes $\ell_1, \ell_2, \ldots, \ell_{t-1}$ in front of ℓ_t. If $L > n$, it cannot be a possible value for an index, and we shall, therefore, reject it. Instead, we shall go back one more place and try $t = (n-3)$. This process of searching for the next index by decreasing the value of t continues until we find a value of L that is less or equal to n. If this condition cannot be met (subject to the proviso given in the next paragraph) when we come to $t = 1$, it means that we have exhausted all the permutations and the calculation stops. In our $n = 3$ example above, this happens when the first digit is 3, and we know that the member 321 is the last one in the list of ascending order.

If for some value of t for $1 \leq t \leq (n-2)$ we have that $L \leq n$, we have the possibility of a new member. To establish this, we shall examine next whether this L is equal to one of the indexes $\ell_1, \ell_2, \ldots, \ell_{t-1}$. If so, we must also reject it, as this cannot come from a permutation of n different indexes. A new value of

L may be obtained from the old one by increasing L by 1 once more. Again, we must check whether this new value is greater than n or equal to one of the indexes earlier than t. If an L value is found that satisfies both of these two conditions, we have the beginning of a new member of the list. To complete the construction, we must assign the values $\ell_{t+1}, \ell_{t+2}, \ldots, \ell_n$ with numbers in the set $(1, 2, \ldots, n)$ that are not used already in $\ell_1, \ell_2, \ldots, \ell_t$. Furthermore, we must arrange them in ascending order, such that $\ell_{t+1} < \ell_{t+2} < \cdots < \ell_n$. In this way, the new member produced is the smallest one that can be generated for a set with the first t "digits" unchanged. This completes the algorithm for generating all the possible members of a list $n!$ members made of permutations of $1, 2, \ldots, n$ in ascending order. A summary is given in Box 10-2.

To complete the calculation of D using (10-8) for a given value of n, we start with the two-dimensional array $A_{i,j}$ for $i = 1, 2, \ldots, n$ and $j = 1, 2, \ldots, n$, filled with the correct algebraic symbols. Next, we generate each of the $n!$ sets of n indexes using the method given in Box 10-2. For each set of indexes, $(i, j, k, \ldots)$, the value of the Levi-Civita symbol (± 1 here) is calculated using the method outlined in Box 10-1. The complete output consists of $n!$ terms, each with a sign given by the Levi-Civita symbol and an algebraic value given by the product of n quantities $A_{1,i}, A_{2,j}, A_{3,k}, \ldots$.

Symbolic manipulation packages. The example given above to evaluate a determinant is a trivial but useful application of SM. The ability to handle determinants forms the basis of solving a system of linear equations, as we saw in §5-1. However, this is only a small part of the algebraic operations for which we have need in physics. We expect the coverage of an SM package to span a much broader range of basic algebraic calculations, such as finding the roots of an equation, manipulation of vectors and matrices, working with polynomials, rational functions, trigonometry functions, and commutation relations. Most of the available packages also include the possibility of carrying out some numerical calculations as well as differentiation of algebraic expressions.

One major difference between the different packages is in their capability to perform indefinite integrals. One should not be surprised at this. Most of the methods for carrying out integrations work only for a very limited range of integrands. The use of computers cannot make any fundamental difference here. Computers are very useful in carrying out a set of well-defined instructions. However, for many indefinite integrals, we do not have a sure way to arrive at the final result. For such cases, the use of a computer may still be helpful at times, but it cannot perform any of the work for which we have no idea of how to proceed ourselves. As a result, indefinite integrals often become a source of frustration in SM, resulting in many criticisms that are not well justified.

The usefulness of SM packages has been greatly enhanced by some of the nonalgebraic functions they have been programmed to carry out. The most important feature is perhaps the possibility offered by some of the more extensive

---

Box 10-2   Subroutine PM_ARY(LARY,N,KEY,IARY)

Generate the next permutation of n quantities in ascending order

Argument list:
    LARY: One-dimensional array of indexes.
    N: Number of indexes.
    KEY: $= 0$, first call,
          > 0, a new permutation,
          < 0, no more permutation.
    IARY: One-dimensional array of input indexes.

Initialization:
    Set up a logical variable XCHG to signal if the last two indexes are in ascending order.

First call to the subroutine:
  (a) Set indexes $(\ell_1, \ell_2, \ldots, \ell_n)$ to ascending order $(1, 2, \ldots, n)$.
  (b) Set XCHG to .FALSE..
  (c) Return the indexes.

All subsequent calls:
1. If XCHG is .FALSE.,
  (a) Interchange ℓ_n with ℓ_{n-1} in LARY.
  (b) Set XCHG to .TRUE. and $t = (n - 1)$.
  (c) Return LARY.
2. If XCHG is .TRUE., generate a new permutation by increasing the value in a previous cell:
  (a) Move the cell position t back by 1.
  (b) If $t = 0$, all possible permutation is exhausted. Return KEY$= -1$ to signal the end.
  (c) If not, define $L = \ell_t + 1$.
  (d) Go back to step (a) if $L > n$.
  (e) Otherwise, check if L is used in an earlier cell.
  (f) If so, go back to (c) to increase the value of L again.
  (g) Otherwise, define $\ell_t = L$ and fill the rest of the cell in ascending order using indexes not found in the earlier part of the list.
  (h) Set XCHG to .FALSE..
3. Return the indexes.

---

packages to act also as programming languages. As we saw earlier, this feature allows the possibility of writing a sequence of the commands of the package as a semiindependent unit to carry out a specific task. This is very similar to a subprogram in numerical calculations and can be called upon whenever required by a calculation. This seems to be the design philosophy of Reduce in which the basic ingredients of SM form the core. Special needs of many common applications come as separate entities, which the user can input when needed. For example, even the elementary trigonometry sum rule $\sin^2 x + \cos^2 x = 1$ is not necessarily in the core but can be put in easily. Macsyma, on the other hand, has a large number of built-in functions. This makes the system much easier to use for many ordinary calculations. The price one pays for the convenience is that it requires large computers on which to run.

An important feature of modern SM packages is their input and output facilities. All the packages are designed for interactive use. For the ease of input, good editing commands are essential. This usually requires some integration with the hardware on which it is run and therefore is still a problem in some cases. There are several different forms of output that are desirable in an SM package. An important one is graphics. As emphasized in the previous chapter, a good way to have a feeling for the results of many calculations is to see a plot. In this respect, Mathematica is able to make better use of the capability of modern computers, perhaps because it is developed explicitly for the hardware that is available now. On the other hand, simple packages, such as Derive, are also doing a very good job in this respect.

In many incidences, the output of an algebraic calculation forms the starting point of a numerical calculation. For this purpose, the possibility of output in terms of statements in FORTRAN and other high-level languages is very useful. For example, a solution of a system of four linear equations is given in (9-16) to find the approximate value of a function of two independent variables. Although it is a very simple algebraic calculation, an SM package was used to carry out the work and to produce the statements that form the core of a subroutine to calculate the approximate numerical value of the function $z = f(x,y)$. One advantage in taking such an approach is that the chance of error is greatly reduced.

In general, SM on computers has not been fully exploited in solving physical problems. Part of the reason is that the more sophisticated packages require a large computer on which to run. However, with progress in affordable workstations, this is changing very fast. Another reason for the limited use is, in part, due to misunderstandings on what SM package can and cannot do. Just as with numerical calculations, algebraic computing is very useful but can only solve certain types of problems. It is only through intelligent use that computers become useful, if not crucial, in helping us with many of our calculations, both numerical and algebraic.

Finally, FORTRAN and other high-level languages have evolved into a very user friendly environment. This is especially true when they are combined with many of the program development tools associated with them. Such developments have only begun to take place in SM packages, and much more remains to be done. Besides manuals and instructions that accompany each package, the interested user may also find the following books helpful: Davenport, Siret, and Tournier (*Computer Algebra, Systems and Algorithms for Algebraic Computation*, Academic Press, London, 1988), Rayna, (*REDUCE, Software for Algebraic Computation*, Springer-Verlag, New York, 1987), Wolfram (*Mathematica, A System for Doing Mathematics by Computer*, Addison-Wesley, Reading, Massachusetts, 1988), and Wooff and Hodgkinson (*muMATH: A Microcomputer Algebra System*, Academic Press, London, 1987).

Problems

10-1 Use a computer algebra software package to prove the identity of (10-4) by induction.

10-2 Calculate a determinant of order $n = 4$ using minors according to (5-7). This is the method used in (10-7).

10-3 Repeat the calculation of the previous exercise using (10-8) instead.

10-4 Show that, in interchanging any two indexes ℓ and m with r indexes in between them, the number of permutations required is always odd.

10-5 Design an algorithm to carry out integrations over the independent variable x with the integrand containing the factor $\sqrt{a^2 - x^2}$. What are the limitations of the integrals that can be carried out?

10-6 Generate by hand an ordered list of all the permutations of $n = 5$ quantities.

Appendix A

Synopsis of

the FORTRAN Language

A summary of the FORTRAN language is given here primarily for the purpose of helping readers who are familiar with other high-level languages to understand some of the programming examples given in the book. Only the commonly used statements are described here. Furthermore, we shall limit ourselves, for the most part, to those parts of the language that are actually used in the examples. The major omissions are the details of the input-output statements, including format, and elements that are not in common use, such as direct access. Certain features of FORTRAN are left over from FORTRAN II, an older version of the language, and are not used very much nowadays. These include IF statements that transfer the computation to different places in the program depending on whether the argument is less than, equal to, or greater than zero. We shall skip these aspects as well. For a complete description of the language, one should consult references such as Meissner and Organick (*FORTRAN 77*, Addison-Wesley, Reading, Massachusetts, 1980). A description of a later standard of the language can be found in Metcalf and Reid (*FORTRAN 90 Explained*, Oxford University Press, Oxford, 1990).

Partly because of the influence from other high-level languages, FORTRAN is undergoing a number of changes. Some of these are now incorporated as a part of the new standardization of the language, FORTRAN 90. Features, such as DO (*statements*) ENDDO constructions, and DO WHILE statements, that have been, for some time, part of certain compilers as additions, are now part of the core. However, it will take some time until all the compilers are converted to the new standard. For this reason, we have restricted ourselves to FORTRAN 77 for all the examples in this volume, in spite of strong temptations otherwise.

The FORTRAN Language. Each FORTRAN statement consists of an optional label in columns 1 to 5. Instructions can begin in column 7 or after a TAB. FORTRAN statements are not case sensitive, meaning that upper- and lowercase letters are treated as the same. Normally, a statement should not extend beyond column 72. Long statements can occupy more than one line and continuation lines are indicated by putting a number or certain other characters in column 6. The statement label is usually omitted, unless required by the flow of the program logic. Comments are inserted in the program by lines with the character C in column 1. Since extra blanks in a statement are ignored, they may be used to make the program easier to read by indenting a block of instructions that form a natural group in the logical flow of the calculation.

For our brief description of the most commonly used elements of the language, we shall use the convention that a set of angle brackets $\langle \cdots \rangle$ implies a group of statements of a program. Brackets $[\cdots]$ indicate options that may be included if needed. Words in italics are to be substituted with names of variables or subprograms.

Basic mathematical operations

Symbol	Name	Example	Meaning
=	equal	$a = b$	a is equal to b
+	addition	$a + b$	sum of a and b
−	subtraction	$a - b$	b subtracted from a
*	multiplication	$a * b$	product of a and b
/	division	a/b	a divided by b
**	power	$a**b$	a to the power b

Logical operations

Symbol	Name	Example	Meaning
.EQ.	equal	a.EQ.b	a is identical to b
.NE.	not equal	a.NE.b	a is not the same as b
.GT.	greater than	a.GT.b	$a > b$
.GE.	greater or equal	a.GE.b	$a \geq b$
.LT.	less than	a.LT.b	$a < b$
.LE.	less or equal	a.LE.b	$a \leq b$
.AND.	logical and	a.AND.b	both a and b are true
.OR.	logical or	a.OR.b	either a or b is true
.NOT.	logical not	.NOT.a	a is not true

Program declarations

PROGRAM *pname*   beginning of a program

Declares the name of the program to be *pname*. May be omitted.

SUBROUTINE *sname*[(a[, b[, c, ...]]])   beginning of a subroutine

Declares an independent subprogram unit identified by the name *sname*. Execution is transferred to the subprogram through a CALL statement. Communication with the rest of the program takes place through arguments a, b, c, ..., and common areas. Each subprogram must contain one or more RETURN statements, and the last statement must be END.

[*type*] FUNCTION *fname*[(a[, b[, c, ...]]])   beginning of a function

Same as SUBROUTINE, except that a single piece of result may be returned to the calling program as the value of the function. Example, r =FUNC$(a, b, c, ...)$ puts the value of the function FUNC calculated with arguments a, b, c, ... into the location for variable r. The type of variable, integer, real, complex, or character, returned by the function may be declared using *type*. If omitted, standard FORTRAN conventions for naming variables apply to *fname*.

ENTRY *sname*[(a[, b[, c, ...]]])   entry point

Alternate entry point for a function or subroutine. Certain restrictions apply to the arguments an entry point can have.

BLOCK DATA [*bname*]   data initialization subprogram

A subprogram used to initialize the values of variables stored in common areas. The subprogram cannot contain any executable statements.

EXTERNAL $[f_1[, f_2[, f_3, ...]]]$ external subprograms

Indicates to the compiler that f_1, f_2, f_3, ... are functions that must be supplied externally. It can also be used to indicate that certain functions in the intrinsic library are to be replaced by those supplied by the users. Required also if the function name is to be used as an argument in an subprogram.

INTRINSIC $[f_1[, f_2[, f_3, ...]]]$ intrinsic functions

Indicates to the compiler that intrinsic functions f_1, f_2, f_3, ... are used as arguments in subprograms.

RETURN   return to the calling program

In general, every subroutine and function must contain one or more RETURN statements to transfer the calculation back to the calling program.

STOP [*message*]   stop the execution

Stops the execution and issues an optional message.

PAUSE [*message*]   suspend the execution

The execution is suspended until the return key on the console is pressed. Useful for examining intermediate results.

END   end of a program section

This is not an executable statement. It signals to the compiler that this is the end of a program or a subprogram.

Data declarations

PARAMETER $(a = v_1[, b = v_2[, c = v_3, \ldots]])$ assign values to variables

Assigns value v_1 to variable a, value v_2 to b, value v_3 to c, and so on. The values cannot be changed during execution.

DATA $a[, b[, c, \ldots]]/v_1[, v_2[, v_3, \ldots]]/$ assign values to data

Assigns value v_1 to variable a, value v_2 to b, value v_3 to c, and so on. The values may be changed during execution.

DIMENSION $a(n_1)[, b(n_2)[, c(n_3), \ldots$ size of arrays

Declares the size of array a to be n_1 locations, array b to be n_2 locations, array c to be n_3 locations, and so on, in a program or a subprogram. Arrays may be more than one dimensional.

COMMON $[cname]/a[, b[, c, \ldots]]/$ Declare common area

Designates variables a, b, c, ... to reside in a storage area named *cname*. Common areas of the same name in different subprograms share the same storage locations. If the name is omitted, it is assumed to be in a special common area called blank common.

EQUIVALENCE $(a, b, \ldots)[, (c, d, \ldots)]$ assign same location

Variables a, b, ..., share the same storage location and variables c, d, ... share another common storage location. Allows memory locations to be referred to by different names.

Type declarations

INTEGER $i_1[, i_2[, i_3, \ldots]]$ integer data

Declares i_1, i_2, i_3, ... as integers. Unless declared otherwise, FORTRAN assumes any variable with name beginning with letters I, J, K, L, M, and N as integers and the others as floating numbers.

REAL $a[, b[, c \ldots]]$ floating number data

Declares variables a, b, c, ... to be floating numbers. Each can be either single precision (REAL*4) or double precision (REAL*8). Unless declared otherwise, FORTRAN assumes that variables with names beginning with letters A to H and O to Z are single-precision variables.

COMPLEX $c_1, c_2, \ldots$ complex number data

Declares variables $c_1, c_2, \ldots$ to be complex numbers. Complex numbers can be either single precision (COMPLEX*8), with each variable occupying eight bytes (four bytes for the real part and four bytes for the imaginary part), or double precision (COMPLEX*16), with both real and imaginary parts of each variable occupying eight bytes.

CHARACTER*n $a[, b[, c, \ldots]]$ character data

Declares variables a, b, c, ... to be character variables each of length n bytes. Character variables are used for storing ASCII data.

LOGICAL $a[, b[, c, \cdots]]$ logical data

Declares variable a, b, c, ... to be logical variables, storing either .TRUE. or .FALSE. as their values.

IMPLICIT $dtype$ $(range)$ implicit variable type

Declares that all variables having names beginning with letters in the $range$ to be of type $dtype$, where $dtype$ is one of the data types, INTEGER, REAL, COMPLEX, CHARACTER, and LOGICAL.

Looping and branching

GO TO ℓ branch to label ℓ

Transfers the execution of the program from here to the statement labeled ℓ.

GO TO $(\ell_1, \ell_2, \ldots, \ell_n)$ k computed go to

Branches to the statement labeled ℓ_1 if $k = 1$, to ℓ_2 if $k = 2$, and so on. The value of k may be assigned during execution.

GO TO ℓ $(\ell_1, \ell_2, \ldots, \ell_n)$ assigned go to

Branches to the statement labeled ℓ. The value of ℓ is specified by an earlier ASSIGN statement and it must be one of the values in the list $\ell_1, \ell_2, \ldots, \ell_n$.

ASSIGN ℓ TO m assign value

Gives the value of the statement label ℓ to integer m.

DO ℓ $k = i_1, i_2, i_3$ ⟨statements⟩ ℓ ⟨statement⟩   do loop

Carries out the following group of statements until (and including) the one labeled ℓ. The execution starts with index $k = i_1$ and k is increased by i_3 each time until (and including) $k = i_2$. The increment i_3 can be either a positive or negative integer. If $i_3 = 1$ it may be omitted.

CONTINUE   continue statement

A statement that performs no execution. It usually has a statement label associated with it and is inserted to clarify the flow of the program, for example, to mark the end of a DO loop.

IF ($\mathcal{L}$) ⟨statement⟩   logical if

The ⟨statement⟩ is executed if the logical expression $\mathcal{L}$ is true.

IF ($\mathcal{L}$) THEN ⟨statement group 1⟩ [ELSE ⟨statement group 2⟩] ENDIF
                                                                                 conditional branch

If the logical expression $\mathcal{L}$ is true, execute the statements in group 1 between THEN and ELSE (or ENDIF if ELSE absent). Otherwise, execute statements in group 2 between ELSE and ENDIF. The ELSE may be omitted if there are no statements in group 2, meaning nothing to be done if a is false.

IF ($\mathcal{L}_1$) THEN ⟨statement group 1⟩ [ELSEIF ($\mathcal{L}_2$) THEN ⟨statement group 2⟩]
        [ELSE ⟨statement group 3⟩] ENDIF
                                                                                      Nested branch

If the logical expression $\mathcal{L}_1$ is true, execute statements in group 1 between THEN and ELSE (or ELSEIF, or ENDIF). Otherwise, if logical express $\mathcal{L}_2$ is true, execute statements in group 2 following ELSEIF ($\mathcal{L}_2$) THEN until the next ELSEIF, ELSE, or ENDIF. Execute the statements in group 3 between ELSE and ENDIF if none of the logical expressions above is true. More than one ELSEIF ($\mathcal{L}_i$) ⟨···⟩ block may be included. The THEN ⟨···⟩ block may be omitted if there are no statements in group 3.

CALL $sname[(a\ [, b\ [, c, \ldots]])]$ subroutine call

Transfer the execution to the subroutine named $sname$. Upon returning from the subroutine, the execution continues with the statement following the subroutine CALL.

$v = fname([a [, b [, c, \cdots]]])$ function call

Transfers the execution to function named $fname$. The value returned by the function is stored in v and the execution continues with the statement following the function call.

Input and output statements

READ *, $[a [, b [, c, \ldots]]]$ free format read

Reads alphanumeric information from the standard input unit without having to provide a format. Each variable is transferred from the input unit to the storage area according to the type (integer, real, complex, character, or logical) specified for each of the variables. If no variable is specified, one line of the input is read but no information is transferred, equivalent to skipping a line from the standard input unit.

READ ℓ, $[a [, b [, c, \ldots]]]$ formatted read

Reads alphanumeric information from the standard input unit according to the format specified in the statement labeled ℓ.

READ (k, ℓ) $[a [, b [, c, \ldots]]]$ formatted read

Reads alphanumeric information from input unit k according to the format given by the statement labeled ℓ.

READ (k) $[a [, b [, c, \ldots]]]$ unformatted read

Reads the values of a, b, c, and so on, that are stored in binary format from unit k. No format is needed here since the input is assumed to be in the same form as in the computer memory.

PRINT *, $[a [, b [, c, \ldots]]]$ free format print

Outputs to the standard output unit. The values of the variables a, b, c, ... are to be converted from the binary format in the computer memory to the default output format according to the data types specified for the variables.

PRINT ℓ, $[a [, b [, c, \ldots]]]$ formatted print

Outputs to the standard output unit. The conversion from binary values of the variables a, b, c, ..., in the computer memory is to be done according to the format given by the statement labeled ℓ.

WRITE *, [a [, b [, c, ...]]]　　　　　　　　　free format write
　　Outputs to the standard output unit in the format that is set as the default for the type of variable to be printed.

WRITE ℓ, [a [, b [, c, ...]]]　　　　　　　　　formatted write
　　Outputs to the standard output unit using format given by the statement labeled ℓ.

WRITE (k, ℓ) [a [, b [, c, ...]]]　　　　　　　　formatted write
　　Outputs to unit k using format given by the statement labeled ℓ.

WRITE (k) [a [, b [, c, ...]]]　　　　　　　　　unformatted write
　　Outputs binary information to unit k.

Other input-output commands

OPEN (UNIT=k[, ···])　　　　　　　　　　open an input-output unit
　　Connects a file to unit k in the program for input or output. The standard input and output units are usually opened by default at the start of the execution. All other I/O units must be explicitly opened before reading or writing.

CLOSE (UNIT=k[, ···])　　　　　　　　　close an input-output unit
　　Disconnects the file associated with unit k from the program.

FORMAT ($\langle specifications \rangle$)　　　　　　　　　format specification
　　Specifies the format to translate between binary information in the computer memory and alphanumeric information of the input or output unit.

ENDFILE (k)　　　　　　　　　　　　　　end file
　　Puts an end-of-file mark on output unit k.

BACKSPACE (k)　　　　　　　　　　　　go back one record
　　Goes back one record on unit k.

REWIND (k)　　　　　　　　　　　　　go back to the beginning
　　Goes back to the beginning of unit k.

Intrinsic library functions (see Table 1-1).

Appendix B
List of Programs

The FORTRAN programs used as examples in the various chapters were developed using the F77L Compiler (Lahey Computer Systems, Inc.) under MS DOS (Microsoft, Inc.). In particular, the graphics subroutines are compiler dependent. Except for graphics, the programs have been tested with a NDP FORTRAN-386 compiler (Microway, Inc.) and other FORTRAN-77 compilers under several UNIX operating systems. For system where local variables in subroutines are not saved by default, the appropriate switch should be used during compilation.

	Box	
Chapter 1		
PRIME	1-1	Generator for the first K prime numbers
PRM_DRV	1-2	Prime number decomposition of an integer
BINOMIAL	1-3	Binomial coefficient using prime number decomposition
ADD_PRM	1-4	Add two rational fractions using prime number decomposition
Chapter 2		
TRAPZ_INT	2-1	Trapezoidal rule integration (used in ROMBERG)
SIMPSON	2-2	Simpson's rule for numerical integration
GL_QUAD	2-3	Gauss-Legendre quadrature for $\int_{-a}^{+a} e^x dx$
ONE_D	2-4	Normal probability function by Monte Carlo integration
THREE_D	2-5	Example of three-dimensional Monte Carlo integration
Chapter 3		
NEV_DRV	3-1	Neville's algorithm for interpolation
RAT_INT	3-2	Rational function interpolation using Neville's algorithm
PROB_Q	3-3	Probability integral using continued fraction of $\gamma(a,x)$
BIT_REV	3-4	Generate a list of 2^n items in bit-reversed order
FFT	3-5	Fast Fourier transform
RICHARD	3-6	Integration by extrapolation using Richardson's approach
ROMBERG	3-7	Romberg integration uses NEVILLE interpolation
BES_INTP	3-9	Inverse interpolation using Bessel's formula
SPLINE	3-10	Cubic spline with Gaussian elimination
Chapter 4		
HERMITE	4-1	Driver for coefficients of Hermite polynomials
LEG_YLM	4-2	Driver for coefficients of Legendre polynomials, and
	4-3	Coefficients of associated Legendre polynomials, and
	4-4	Coefficients of spherical harmonics
SPH_BES	4-5	Coefficients of spherical Bessel function
VSPH_BES	4-6	Value of spherical Bessel function (included in SPH_BES)
PROP_BES	4-7	Values of spherical Bessel function by propagation
LAG_ASC	4-8	Coefficients of associated Laguerre polynomials
GAMMA8	4-9	Gamma function using an eight-term approximation
CG_COEF	4-11	Clebsch-Gordan coefficients

List of programs

Chapter 5
DETERM	5-1	Value of a determinant by Gauss-Jordan elimination
LIN_EQN	5-2	Solve a system of linear equation by Gauss-Jordan elimination
MAT_INV	5-3	Matrix inversion by Gauss-Jordan elimination
LIN_LU	5-4	Solution of linear equation using LU-decomposition and
	5-5	Forward and back substitution FBLU_SBS
JCB_DRV	5-6	Jacobi diagonalization of a real, symmetric matrix
TRI_DIAG	5-7	Householder tridiagonalization of a real, symmetric matrix
BSCT_DRV	5-8	Eigenvalues of a tridiagonal matrix by bisection
TRI_QL	5-9	Eigenvalues and eigenvectors of a tridiagonal matrix (cf. TRI_DIAG)
JCB_NSYM	5-10	Eigenvalues and eigenvectors of a real nonsymmetric matrix

Chapter 6
LLSQ	6-1	Linear least-squares fit to a straight line
PROB_ATV	6-2	Probability integral using incomplete beta function
MGRS_DRV	6-3	Multiple regression analysis
NLN_DRV	6-4	Nonlinear least-squares fit with Marquardt method

Chapter 7
FREQNCY	7-3	Frequency test for random numbers produced by
	7-1	RSHFL Improving the linear congruence method by shuffling, and
	7-2	RSUB Random number generator using subtraction method
RUN_UP	7-4	Run-up test of a sequence of random numbers
PER_CNT	7-5	Percolation on a square lattice, using
	7-6	SWEEP, a sweep through the lattice using Metropolis algorithm
ISING	7-7	Two-dimensional Ising model
PATH_INTEGRAL	7-8	Path integral calculation of harmonic oscillator wave function
MANDL_PLOT	7-9	Mandelbrot plot
DLA	7-10	Simulation of diffusion-limited aggregation

Chapter 8
RG_KUTTA	8-1	Driver for RNG_KTT4, fourth-order Runge-Kutta method
RG_KT_VH	8-2	Driver for Runge-Kutta method with step size control
PRE_COR	8-3	Driver for fourth-order predictor-corrector method
ODE_EXT	8-5	Driver for EXT_STP, solution of ODEs by extrapolation
	8-4	MID_PTM, modified midpoint method for first-order ODEs
SHT_SQR	8-6	Shooting method to obtain the scattering solution
SHT_MCH	8-7	Driver for MATCH, match the solution from two subintervals
SOR_WS	8-8	Successive overrelaxation for boundary value problems
SOR_POS	8-9	Solution of partial differential equation using SOR
DIFFUS	8-10	Crank-Nicholson solution of parabolic PDE
PDE_WV	8-11	Driver for explicit method to solve the wave equation
KINK	8-12	Driver for solving the nonlinear static kink equation
FEM_POS	8-13	Finite element solution of a 1-dimensional Poisson's equation

Chapter 9
RAINBOW	9-1	Simulation of a rainbow of three colors
CUBIC_LN	9-4	Plot a curve with smoothing, using
	9-2	ST_LINE - Straight line interpolation to (x_d, y_d)
	9-3	X_AXIS - draw x-axis from x_b to x_d with N_TICKS ticks
		Y_AXIS - draw y-axis from y_b to y_d with N_TICKS ticks

Chapter 10
DET_ALG		Evaluate a determinant algebraically, using
	10-1	LEVI_C - Levi-Civita symbol for n indexes
	10-2	PM_ARY - generate the next permutation of n quantities

References

Books on Mathematical Physics, Mathematics, and Statistics

Arfken, G. *Mathematical Methods for Physicists*. Academic Press, New York, 1970.

Courant, R., and Hilbert, D. *Methods of Mathematical Physics*. Interscience Publishers, New York, 1962.

Cramer, H. *Mathematical Methods of Statistics*. Princeton University Press, Princeton, N.J., 1946.

Kendall, M., and Stuart, A. *The Advanced Theory of Statistics*, vol. 1. Macmillan, New York, 1977.

Lyons, L. *Statistics for Nuclear and Particle Physicists*. Cambridge University Press, Cambridge, 1986.

Mathews, J., and Walker, R.L. *Mathematical Methods of Physics*. Benjamin, New York, 1965.

Morse, P.M., and Feshbach, H. *Methods of Theoretical Physics*. McGraw-Hill, New York, 1953.

Riordan, J. *Combinatorial Identities*. Wiley, New York, 1968.

von Mises, R. *Mathematical Theory of Probability and Statistics*. Academic Press, New York, 1964.

Books on Numerical Methods, Algebraic Computing, and Computational Physics

Bevigton, P.R. *Data Reduction and Error Analysis for the Physical Sciences*. McGraw-Hill, New York, 1969.

Burnett, D.S. *Finite Element Analysis, From Concepts to Applications*. Addison-Wesley, Reading, Mass., 1988.

Cullum, J.K., and Willoughby, R.A. *Lanczos Algorithms for Large Symmetric Eigenvalue Computations*, vols. I and II, Birkhäuser, Boston, 1985.

Dahl, O.-J., Dijkstra, E.W., and Hoare, C.A.R. *Structured Programming*. Academic Press, London, 1972.

Dahlquist, G., and Björck, Å. *Numerical Methods*, English translation by N. Anderson. Prentice Hall, N.J., 1974.

Davenport, J.H., Siret, Y., and Tournier, E. *Computer Algebra, Systems and Algorithms for Algebraic Computation*. Academic Press, London, 1988.

Dewey, B.R. *Computer Graphics for Engineers*. Harper & Row, New York, 1988.

Dongarra, J.J. *LINPACK User's Guide*. SIAM, Philadelphia, 1979.

Fox, L., and Mayers, D.F. *Computing Methods for Scientists and Engineers*. Clarendon Press, Oxford, 1968.

Fröberg, C.-E. *Numerical Mathematics*. Benjamin/Cummings, Menlo Park, Calif., 1985.

Gear, C. William. *Numerical Initial Value Problems in Ordinary Differential Equations.* Prentice Hall, Englewood Cliffs, N.J., 1971.

Hildebrand, F.B. *Introduction to Numerical Analysis.* McGraw-Hill, New York, 1956.

Knuth, D.E. *The Art of Computer Programming,* vol. II, 2nd ed., Addison-Wesley, Menlo Park, Calif., 1981.

Koonin, S.E. *Computational Physics.* Addison-Wesley, Redwood City, Calif., 1986.

MacKeown, P.K., and Newman, D.J. *Computational Techniques in Physics.* Adam Hilger, Bristol, 1987.

Meissner, L.P., and Organick, E.I. *Fortran 77, Featuring Structured Programming.* Addison-Wesley, Reading, Mass., 1980.

Metcalf, M., and Reid, J. *Fortran 90 Explained.* Oxford University Press, Oxford, 1990.

Phillips, J. *The NAG Library: A Beginner's Guide.* Clarendon Press, Oxford, 1986.

Press, W.H., Flannery, B.P., Teukolsky, S.A., and Vetterling, W.T. *Numerical Recipes, The Art of Scientific Computing.* Cambridge University Press, Cambridge, 1986.

Rayna, G. *REDUCE, Software for Algebraic Computation.* Springer-Verlag, New York, 1987.

Sewell, G. *The Numerical Solution of Ordinary and Partial Differential Equations.* Academic Press, San Diego, Calif., 1988.

Smith, B.T., Boyle, J.M., Garbow, B.S., Ikebe, Y., Klema, V.C., and Moler, C.B. *Matrix Eigensystem Routines – EISPACK Guide,* Lecture Notes in Computer Science, vol. 6, 2nd ed., Springer-Verlag, Berlin, 1974. *Matrix eigensystem routines: EISPACK guide extension,* Lecture Notes in Computer Science, vol. 51, Springer-Verlag, Berlin, 1977.

Smith, G.D. *Numerical Solution of Partial Differential Equations: Finite Difference Methods,* 3rd ed., Clarendon Press, Oxford, 1985.

Stoer, J., and Bulirsch, R. *Introduction to Numerical Analysis* (English translation by R. Bartels, W. Gautschi, and C. Witzgall). Springer-Verlag, New York, 1980.

Wilkinson, J.H. *The Algebraic Eigenvalue Problem.* Oxford University Press, Oxford, 1965.

Wilkinson, J.H., and Reinsch, C. *Linear Algebra,* Handbook for Automatic Computation, vol. II, edited by F.L. Bauer and others, Springer-Verlag, Berlin, 1971.

Wolfram, S. *Mathematica, A System for Doing Mathematics by Computer.* Addison-Wesley, Reading, Mass., 1988.

Wooff, C., and Hodgkinson, D. *muMATH: A Microcomputer Algebra System.* Academic Press, London, 1987.

Handbooks and Tables

Abramowitz, M., and Segun, I.A., editors. *Handbook of Mathematical Functions.* Dover, New York, 1965.

Background information and standard physics textbooks

Anderson, E.E. *Modern Physics and Quantum Mechanics.* W.B. Saunders Co., Philadelphia, 1971.

Brink, D.M., and Satchler, G.R. *Angular Momentum*, 2nd ed., Clarendon Press, Oxford, 1968.

Cohen-Tannoudji, C., Diu, B., and Laloë, F. *Quantum Mechanics*, English translation by S.R. Hemley, N. Ostrowsky, and D. Ostrowsky. Wiley, New York, 1977.

Domb, C., and Lebowitz, J.L., editors, *Phase Transitions and Critical Phenomena.* Academic Press, London, 1972-.

Fetter, A.L., and Walecka, J.D. *Quantum Theory of Many-Particle Systems.* McGraw-Hill, New York, 1971.

Feynman, R.P., and Hibbs, A.R. *Quantum Mechanics and Path Integrals.* McGraw-Hill, New York, 1965.

Huang, K. *Statistical Mechanics*, 2nd ed., Wiley, New York, 1987.

Lee, T.D. *Particle Physics and Introduction to Field Theory.* Harwood, Chur, Switzerland, 1981.

Ma, S.K. *Modern Theory of Critical Phenomena.* Benjamin, Reading, Mass. 1976.

Merzbacher, E. *Quantum Mechanics*, 2nd ed., Wiley, New York, 1970.

Pfeuty, P., and Toulouse, G. *Introduction to the Renormalization Group and to Critical Phenomena*, English translation by G. Barton. Wiley, London, 1977.

Porter, C.E. *Statistical Theory of Spectra: Fluctuations.* Academic Press, New York, 1965.

Ring, P., and Schuck, P. *The Nuclear Many-Body Problem.* Springer-Verlag, New York, 1980.

Ryder, L.H. *Quantum Field Theory.* Cambridge University Press, Cambridge, 1985.

Schiff, L.I. *Quantum Mechanics*, 3rd ed., McGraw-Hill, New York, 1968.

Stauffer, D. *Introduction to Percolation Theory.* Taylor and Francis, London, 1985.

Wong, S.S.M. *Introductory Nuclear Physics.* Prentice Hall, Englewood Cliffs, N.J., 1990.

Journal articles and conference proceedings

Borwein, J.M., and Borwein, P.B. *Scientific American* **260** (February 1988) 112, "Ramamujan and Pi."

Bowdler, H., Martin, R.S., Reinsch, C., and Wilkinson, J.H., in Wilkinson, J.H., and Reinsch, C. *Linear Algebra*, Handbook for Automatic Computation, vol. II, edited by F.L. Bauer and others, Springer-Verlag, Berlin, 1971, "The QR and QL Algorithms for Symmetric Matrices."

Bracewell, R.N. *Scientific American* **260** (June 1989) 86, "The Fourier Transform."

Brody, T.A., Flores, J., French, J.B., Mello, P.A., Pandey, A., and Wong, S.S.M. *Rev. Mod. Phys.* **53** (1982) 385, "Random Matrix Physics, Spectrum and Strength Fluctuations."

Cannon, T.M., and Hunt, B.R. *Scientific American* **245** (October 1981) 214, "Image Processing by Computer."

Chaitin, G.J. *Scientific American* **232** (May 1975) 47. "Randomness and Mathematical Proof."

Chiu, T.W., and Guu, T.S. *Comp. Phys. Comm.* **47** (1987) 129, "A Shift-Register Sequence Random Number Generator Implemented on the Microcomputers with 8088/8086 and 8087."

Dewdney, A.K. *Scientific American* **260** (February 1989) 108; **260** (June, 1989) 125; and references therein,"Computer Recreations."

Diaconis, P., and Efron, B. *Scientific American* **248** (May 1983) 116, "Computer-Intensive Methods in Statistics."

Dyson, F.J., and Mehta, M.L. *J. Math. Phys.* **4** (1963) 701, "Statistical Theory of the Energy Levels of Complex Systems IV."

Eberlein, P.J., and Boothroyd, J., in Wilkinson, J.H., and Reinsch, C. *Linear Algebra*, Handbook for Automatic Computation, vol. II, edited by F.L. Bauer and others, Springer-Verlag, Berlin, 1971, "Solution to Eigenproblem by a Norm Reducing Jacobi Type of Method."

Feldman, J.A. *Scientific American* **241** (December 1979) 94, "Programming Languages."

Foster, K.R., and Bau, H.H. *Science* **243** (1989) 679, "Symbolic Manipulation Programs for the Personal Computer."

Galbraith, I., Ching, Y.S., and Abraham, E. *Am. J. Phys.* **52** (1984) 60, "Two-Dimensional Time-Dependent Quantum-Mechanical Scattering Event."

Goldberg, A., Schey, H.M., and Schwartz, J.L. *Am. J. Phys.* **35** (1967) 177, "Computer Generated Motion Pictures of One-Dimensional Quantum-Mechanical Transmission and Reflection Phenomena."

Hoshen, J., and Kopelman, R. *Phys. Rev.* **B14** (1976) 3438, "Percolation and Cluster Distribution: I. Cluster Multiple Labeling Technique and Critical Concentration Algorithm."

Knuth, D.E. *Scientific American* **236** (April 1977) 63, "Algorithms."

Lawande, S.V., Jensen, C.A., and Sahlin, H.L. *Comp. Phys.* **3** (1969) 416, "Monte Carlo Integration of the Feynman Propagator in Imaginary Time"; **4** (1969) 451, "Monte Carlo Evaluation of Feynman Path Integrals in Imaginary Time and Spherical Polar Coordinate."

Lubimov, V.A., Novikov, E.G., Nozik, V.Z., Tretyakov, E.F., and Kosik, V.S. *Phys. Lett.* **94B** (1980) 266, "An Estimate of the ν_e Mass from the β-Spectrum of Tritium in the Valine Molecule"; and in *Neutrino Mass and Related Topics:* Proc. of XVI INS Intl. Symp., Tokyo, 1988, editors S. Kato and T. Ohshima, World Scientific, Singapore, 1988.

Mannheim, P.D. *Am. J. Phys.* **51** (1983) 328, "The Physics behind Path Integrals in Quantum Mechanics."

Marsaglia, G., Narasimhan, B., and Zaman, A. *Comp. Phys. Comm.* **60** (1990) 345, "A Random Number Generator for PC's."

Metropolis, N., Rosenbluth, A.W., Rosenbluth, M.N., and Teller, A.H. *J. Chem. Phys.* **21** (1953) 1087, "Equation of State Calculations by Fast Computing Machines."

Olson, D., Brozovich, C., Carr, J., Hatton, H., Miles, G., Jr., and Zwicke, G. *The Physics Teacher* (April 1990) 226, "Monte Carlo Computer Simulation of a Rainbow."

Pavelle, R., Rothstein, M., and Fitch, J. *Scientific American* **245** (December 1981) 136, "Computer Algebra."

Pomerance, C. *Scientific American* **247** (December 1982) 136, "The Search for Prime Numbers."

Press, W.H., and Teukolsky, S.A. *Comput. Phys.* **5** (1991) 68, "Numerical Calculation of Derivatives."

Racah, G. *Phys. Rev.* **62** (1942) 438, "Theory of Complex Spectra II."

Renfrew, C. *Scintific American*, **261** (Oct. 1989) 106, "The Origins of Indo-European Languages."

Schulman, L.S., and Seiden, P.E. *Science* **233** (1986) 425, "Percolation and Galaxies."

Schulman, L.S., and Seiden, P.E. *J. Stat. Phys.* **27** (1982) 83, "Percolation Analysis of Stochastic Models of Galactic Evolution."

Sebe, T., and Nachamkin, J. *Ann. Phys.* (N.Y.) **51** (1969) 100, "Variational Buildup of Nuclear Shell Model Bases."

Steffen, R.M. *Adv. Phys.* **4** (1955) 293, "Extranuclear Effects on Angular Correlations of Nuclear Radiations."

Wigner, E.P. *Ann. Math.* **62** (1955) 548, "Characteristic Vectors of Bordered Matrices with Infinite Dimensions"; *SIAM* **9** (1978) 1, "Random Matrices in Physics."

Wilson, K.G. *Phys. Rev.* **D10** (1974) 2445, "Confinement of Quarks."

Computer Physics Report **6** (1987) (Review of finite element methods).

Index

A

abscissas and weight factors, 43-47
acceleration
   due to gravity, 487
   parameter, 542
action, 459-60, 462, 596
active space, 288
Adams-Bashforth method, 513
additive generator, 389
aggregates, 479
algebraic
   calculations, 6
   equations, 495, 536
   operations, 19, 633-35
   programming, 635-36
ALGOL language, 6
algorithm, 12, 18-21, 26
alternating current circuit, 502
angle of
   incidence, 614
   refraction, 614
angular correlation, 360, 361
angular momentum, 204, 525
   eigenfunction, 158
   eigenvalue, 158, 167
   raising and lowering, 167
anharmonic
   oscillator, 146, 238, 293-96, 313, 316
   potential well, 576
   term, 293
annihilation operators, 299, 449
APL language, 7
arc tangent, 92
aspect ratio, 610
assembly language, 2
associated
   Laguerre polynomial, 186-94
   Legendre equation, 164
   Legendre polynomial, 164-66, 168

average spacing, 451
averaging operator, 129
Avogadro number, 421, 434
axis, 619
azimuthal angle, 155, 157, 164, 169

B

back substitution, 140, 217, 222, 228, 230, 236
backward difference, 63, 64, 510, 568, 585
   operator, 63
band matrices, 273, 310, 313
BASIC language, 7
basis
   function, 594
   states, 237, 283, 445
Bernoulli numbers, 36
Bessel
   differential equation, 171, 178
   formula, central difference, 128, 130
   formula for interpolation, 129
   function, 171-79
   inverse interpolation, 579
beta-decay, 416
bidirectional shooting, 532-36
binary
   number, 8
   representation, 2, 8, 109
binomial
   coefficient, 13, 15, 27, 162, 164, 326
   distribution, 324-26, 328, 407
   expansion, 77
   series, 162
   theorem, 326

bisection, 124-25, 266-69, 275, 292, 296
bit, 8
bit-reversed order, 109-12
Bohr radius, 188
Boltzmann
   factor, 436
   transport equation, 565
Boolean operations, 409
boundary condition
   Dirichlet, 549, 553
   periodic, 440
boundary value problem, 491, 494-95, 524-48, 580
   partial differential equation, 549-53
bra-ket notation, 237, 456-57, 466
Brownian motion, 407-12
byte, 8

C

C language, 2, 6
canonical ensemble, 403, 436, 438
Cartesian system, 456
Cayley form, 560
central
   difference, 62-64, 131, 132, 492, 498, 500, 537, 555, 575, 577
   difference operator, 63, 129
   limit theorem, 327, 330, 342, 400
   moments, 380
   potential, 179, 525
   projection, 629
centroid, 321
chaos, 1, 455
character
   mode, 608
   plot, 611
characteristic
   curve, 566
   equation, 240
   method of, 566-69
charge distribution, 56
chi-square (χ^2), 343
   -distribution, 350-51, 365
   probability integral, 94, 96, 330, 351
   reduced, 349
   -statistic, 353
   test, 348-53, 362-64

see also least-squares
Cholesky decomposition, 315
classical
   action, 460-61, 463, 464
   mechanics, 297, 315, 459
Clebsch-Gordan coefficients, 13, 203-08
closure property, 290
cluster, 426
cobalt, 434
coefficients of
   associated Laguerre polynomial, 193-94
   associated Legendre polynomial, 169
   Hermite polynomial, 153
   interpolation polynomial, 134-36
   Laguerre polynomial, 192-93
   Legendre polynomial, 168-69
   spherical Bessel function, 175-77
   spherical harmonics, 170
cofactor of a determinant, 215, 316
collective
   degrees of freedom, 297
   excitations, 296
   phenomena, 421
collocation method, 590, 593, 594
column matrix, 254
comments in program, 4
Common Lisp, 636
commutator, 299
complementary error function, 195
compile, 6
compiler, 2, 22
complete elliptic integral, 437
complex
   matrices, 301, 310-12
   number, 27, 472, 473, 475
   plane, 103
   rotation, 306
computational molecule, 550
computer, 1
   memory, 8
   simulations, 426
   word, 9, 208

computer algebra, 633-43
computer languages, 6-7
    see also programming language
condensed matter, 421
Condon and Shortley convention, 205
confidence
    coefficient, 414
    interval, 413-21
    level, 413-21
    limit, 415
    region, 413, 414, 415
configurations, 436
conservation laws, 444
consistency, 496
continued fraction, 88-96, 354
    arc tangent, 92-93
    incomplete gamma function, 94, 354
    recursion relation, 93-94
continuity equation, 554
contour
    lines, 623
    plot, 623
convectional flow, 566
convergence, 288, 496
    criteria, 540, 545, 597
coordinate representation, 100, 456, 457
corrector step, 511
correlated ground state, 297-99
correlation, 394, 439
    angular, 360, 361
    coefficient, 347, 364-65, 453
    length, 424
    linear, 393
    test, 394-95
Coulomb potential, 179, 182-83
coupled first-order ODE, 489
covariance, 335, 346
    matrix, 359, 373, 375
Crank-Nicholson method, 558-60, 573
creation operators, 299, 449
critical
    damping, 505
    exponent, 424-26
    phenomenon, 422-34
    point, 424, 443
    temperature, 425, 437
cubic
    interpolation, 621
    spline, 135-42, 621
Curie temperature, 434
curvature matrix, 359, 372, 373, 375, 379

D

damped harmonic oscillator, 502-05, 584, 586, 587
damping term, 488
data simulation, 412-20
debugging, 3
decay
    constant, 324
    width, 444
decision matrix, 438
decomposition
    LU-, 272
    LR-, 228, 230
    prime number, 16
    QL-, 273
    QR-, 272
de-excitation operator, 298, 299, 301
defective matrix, 309
definite integral, 21, 29, 30
degeneracy, 270, 275, 453
degrees of freedom, 349, 589
delta function, 98-99, 601
Δ_3-statistic, 454-55
density of eigenvalues, 454
derivative, 62
    logarithmic, 588, 533
Derive, 633
detailed balance, 405
determinant, 135, 214-23, 636-38
diagonal matrix, 240
diagonalization, 240-316, 446
    Givens method, 251-53
    Householder method, 255-64
    Jacobi method, 243-50, 306-09
    Lanczos method, 283-96
    tridiagonal matrix, 265-83

difference
 downward, 75
 equation, 494
 operator, 62
 upward, 75
differential equation, 458, 487
 Bessel, 171, 178
 boundary value problem, 491, 494-95, 524-48, 580
 elliptic, 490, 549
 hyperbolic, 490, 565
 initial value boundary problem, 491, 554
 initial value problem, 491, 499-23
 Laguerre, 180, 183
 Legendre, 157-58
 linear, 490
 nonlinear, 490, 496, 544, 576-84
 ordinary, 488-89
 parabolic, 490, 554-64
 partial, 489-90, 536, 549, 554-76
 quasilinear, 569, 576, 577, 586, 588
differentiation, 29, 62-66
 errors in, 66
diffuseness parameter, 526
diffusion, 407, 557
 coefficient, 490, 554
 equation, 490, 554-55, 562
 -limited aggregation (DLA), 479-83, 485
dimension, 476-78
 fractional, 476, 478, 479
Dirac
 bra-ket notation, 237, 456-57, 466
 delta function, 98-99, 601
 notation, 239
direct current circuit, 211
Dirichlet boundary condition, 549, 553
discrete Fourier transform, 100
distribution
 binomial, 324-26, 328, 407
 charge, 56
 χ^2, 350-51, 365
 eigenvalue, 445
 exponential, 453
 F-, 366
 Gaussian (see also normal), 323, 327, 445

level-density, 450, 455
Lorentzian, 323, 328, 378
normal, 317, 323, 327-28, 333, 380, 392, 399, 402, 411, 446, 448, 484
parent, 324
Poisson, 326-27, 333, 391, 397
probability, 324, 415
random numbers, 50, 397, 399-400
semicircular, 401, 402, 445, 447
spacing, 450-53
Student's t-, 367
Yukawa, 56
divided differences, 125-27
divisible, 11
DLA (diffusion-limited aggregation), 479-83, 485
domain of dependence, 557, 573
double factorial, 201
duplication formula, 178
Dyson's F-test, 484

E

effective potential, 525
efficiency, 21, 496
eigenfunction, 147, 237, 444
eigenvalue, 147, 183, 237, 239, 283, 444
 angular momentum operator, 158
 distribution, 445
 problem, 237-40, 310, 314-16
eigenvector, 239, 283
 left and right, 303-05
 representation, 240
electromagnetic wave, 565
electrostatics, 159, 549
elliptic
 equation, 490, 549
 integral, 437
end-point energy, 416
energy
 harmonic oscillator, 149
 -level density, 447, 448, 455
 spectrum, 447

Index

ensemble, 445, 446
  average, 440-44, 446
  canonical, 403, 436, 438
equation of motion, 314
equilateral triangle, 477
ergodic, 436
error (*see also* uncertainty)
  function, 73, 119, 122, 195, 202, 329
  integral, 195-96
  systematic, 320
  truncation, 10, 287, 496, 499, 506
Euclidean
  dimension, 478
  norm, 307
  space, 628
Euler-Maclaurin formula, 35, 61, 117, 119, 120, 200
Euler's
  constant, 198
  method, 491-94, 499, 584
Eureka, 633
excess, 323
excitation operator, 298, 299, 301
explicit method, 497, 499, 555-57, 559, 571, 572-73, 577, 587
exponent, 10
exponential
  decay law, 337
  distribution, 453
  function, 47, 149, 559
  integral, 195, 202
extrapolation, 37, 42, 69, 70, 115-22, 517-23, 528

F

F
  -distribution, 366
  Dyson's test, 484
  -test, 356-57, 366
factorial, 13, 20 178, 196, 207
  double, 201
  function, 196 (*see also* gamma function)
  logarithm of, 17-18, 208
fast Fourier transformation (FFT), 100, 104-14
FDE (finite difference equation), 493, 496, 499, 536, 604
Fermi
  energy, 297
  function, 416
  -gas level-density formula, 447
  gas model, 447
ferromagnetic material, 421
ferromagnetism, 421, 434
Feynman path integral, 460
FFT (fast Fourier transform), 100, 104-14
finite difference, 29, 37, 62, 125, 492, 499, 549, 571
  equation, 493, 496, 499, 536, 604
  method, 66, 571, 603
  operator, 63
finite element, 499, 591
  collocation method, 590, 593, 594
  Galerkin method, 594, 596, 598, 605
  least-squares method, 592-93
  Ritz method 596
  subdomain method, 591-92, 593
  variational approach, 595-96
  weighted residual, 593
finite size, 434
  effect, 428, 432-34, 440, 443, 482
first-order
  derivative, 66
  ODE, 488
  phase transition, 422
Fisher's F-test, 356-57
five-point stencil, 550
floating number, 8-10
  representation, 17
flow chart, 20, 25
fluctuations, 450, 454, 455
fluid flow, 554
forced harmonic oscillator, 489
FORTRAN language, 2, 6
forward
  difference, 62, 63, 126, 127, 129, 496, 555, 571, 585, 586, 587
  operator, 63
  substitution, 229, 230, 236
four-byte integer, 9

Fourier
  coefficients, 97, 99, 101
  fast transform, 100, 104-14
  inverse transform, 99, 104
  series, 96
  transform, 96-114, 457
fractals, 471-79
  dimension, 478-79
  growth, 479
fractional dimension, 476, 477, 479
Franz Lisp, 636
free particle, 456, 457-60
frequency test, 390-93, 483
friction, 488
full width at half-maximum, 323, 328
function, 3

G

Galerkin
  equation, 600
  method, 594, 596, 598, 605
gamma function, 178, 195, 196-202
  incomplete, 88, 94-96, 202-03, 329
  polynomial approximation, 199-200
Gauss
  -Jordan elimination, 218, 222, 225, 228, 233, 316
  -Legendre integration, 42, 44, 46, 48, 55, 60
  -Legendre quadrature, 42-48, 67
  -Seidel iteration matrix, 544
  -Seidel method, 541, 542, 551
Gaussian
  distribution (see also normal distribution), 323, 327, 445
  elimination, 140-42, 217, 220, 560-62, 583
  orthogonal ensemble (GOE), 445-47
  quadrature, 42-48
  wave packet, 563
generalized
  coordinates, 314, 596
  eigenvalue problem, 310, 314-16
generating function
  associated Laguerre polynomial, 186-87
  Bernoulli number, 36
  Bessel function, 178

Hermite polynomial, 151-52
  Laguerre polynomial, 184-85
  Legendre polynomial, 159-61, 163
geometric optics, 461
Givens method, 251-53
global property, 450
GOE (Gaussian orthogonal ensemble), 445-47
gradient method, 373-74
Gram
  -Charlier expansion, 146
  -Schmidt orthogonalization, 284, 309
graphical presentations, 607
graphics, 607-32
  mode, 611-12, 617
  output device, 617
  packages, 7
grid, 37, 54, 467
ground state
  correlated, 297-99
  energy, 288
  wave function, 464

H

half-life, 324, 337
half-width, 323
Hamiltonian, 146, 205, 380, 445, 455
  matrix, 239
  operator, 147, 239, 241, 283, 444
Hankel function, 171
harmonic oscillation, 505
  simple, 493
harmonic oscillator, 146, 237, 238, 313, 328, 464-71, 488, 491-94, 502-05, 595
  damped, 502-05, 584, 586, 587
  forced, 489
  isotropic, 180-81, 190-92
  length parameter, 180
  potential, 146, 314
  wave function, 147-49, 150, 190-92, 293

harmonics, 99
heat conduction, 554
heavy damping, 505
Heisenberg
　model, 435
　uncertainty principle, 320, 321, 328
helium, 415
Helmholtz equation, 155, 171, 180
　radial dependence, 171
Hermite
　integration method, 60
　polynomial, 60, 67, 145-54, 238, 294
Hermitian, 241, 286, 311, 445
　matrices, 309
Hessenberg form, 305
hexadecimal number, 8, 9
hidden
　-line, 628-31
　-surface, 628-31
high-level computer languages, 2-3
Hilbert space, 287, 628
hollow pipe, 494
Householder, 253
　transformation, 305
　tridiagonalization, 255-64
Huygens's principle, 461
hydrogenlike atom, 182-83, 187-90
hyperbolic partial differential equation, 490, 562-73
hypothesis testing, 412-20

I

ideal gas, 297
image
　plane, 628, 631
　processing, 100
impact parameter, 613
implicit
　equation, 587
　method, 497, 514, 555, 556, 557, 559, 560, 572, 587
improper integral, 49, 53, 58, 58
IMSL, 179
incomplete
　beta function, 354-55

gamma function, 88, 94-96, 202-03, 329
indefinite integral, 27, 30
index of refraction, 613
induction, 634
inertial tensor, 314
infinite
　integrals, 59
　series expansion, 24, 559
infinity, 59
initial value
　boundary problem, 491, 554
　curve, 568
　problem, 491, 499-523
inner product, 254
instrumental uncertainty, 341
integer, 8-9
　calculation, 10-17, 26
　mode, 11
　random, 384-86
integrable singularity, 58
integral
　definite, 21, 29, 30
　exponential function, 195, 202
　indefinite, 27, 30
　Riemann, 30
integrand, 29
integration, 29-61, 644
　Gaussian quadrature, 42-48
　improper, 58-62
　interval, 49
　Laguerre, 60
　Monte Carlo, 49-53, 55-58, 60, 330, 390
　multidimensional, 54-58
　rectangular rule, 32-35, 49, 54, 463
　Romberg, 37, 42, 120-22, 517
　Simpson's rule, 36-42, 49, 54, 67
　trapezoidal rule, 35-37, 49, 55, 61, 67, 121

interaction, 434
Interlisp, 636
intermediate states, 457
internal
   energy, 421, 436, 437, 442
   representation, 8
interpolation, 69-87, 122, 618
   Bessel formula, 129
   inverse, 122-35, 579
   Newton's, 125-27
   polynomial, 69-76, 87, 125-27, 135, 620
   rational function, 77-87, 116, 123, 521, 523
   trigonometric, 100
interpreter, 6
intrinsic
   functions, 5, 22
   library, 5
   limitations, 65
inverse
   Fourier transform, 99, 104
   interpolation, 122-35, 579
   matrix, 102, 224
inverted difference, 89
iron, 425, 434
Ising model, 434-44, 463, 467, 483
   square lattice, 438-44
isotropic harmonic oscillator, 180-81, 190-92
iterative
   approach, 39, 540 (*see also* recursion relation)
   method, differential equations, 540
iteration matrix, 544
   Gauss-Seidel, 544
   Jacobi, 544
   SOR, 544, 553

J

Jacobi
   diagonalization, 243-50, 306-09
   iteration matrix, 544
   matrix, 553
   method, 243-50, 306-09, 540, 544, 550
Jacobian, 398, 399

matrix, 582
Julia set, 472-73, 485

K

k-body interaction, 448
kinetic energy, 237
Kronecker delta, 97, 237, 239, 375
Kurie plot, 416
kurtosis, 323

L

Lagrange
   equation of motion, 579
   method, 71
   polynomial interpolation, 70
Lagrangian, 314, 315, 459, 464, 579, 596
Laguerre
   differential equation, 180, 183
   integration method, 60
   polynomial, 60, 67, 183-94
Lanczos method, 283-96
language (*see* programming language)
Laplace expansion theorem, 215
laser printer, 611
lattice gauge calculations, 471
least-squares
   finite element, 592-93
   gradient method, 373-74
   linear, 342-48, 358-68
   linearization approximation, 373
   Marquardt method, 373-77
   multiple regression, 358-68
   nonlinear, 368-80
   parabolic approximation, 371-73
   straight line, 342-48
left eigenvectors, 303-05
Legendre
   associated polynomial, 164-66, 168
   equation, 157-58
   polynomial, 42, 158-64, 360-61

Leibnitz formula, 166
length parameter, 180
level
  density, 447, 451
  -density distribution, 450, 455
  -density parameter, 447
  repulsion, 450, 453
Levi-Civita symbol, 214, 638-39
L'Hospital rule, 59
library
  IMSL, 179
  intrinsic, 5
  NAG, 179
  private, 5
  program, 4-6
  subroutine, 5, 7, 22-23, 487
lifetime, 337
likelihood function, 338-41
line
  graphs, 607
  printers, 608
linear
  congruence generator, 386, 387, 388, 394
  correlation, 393
  differential equations, 490
  equations, 212
  least-squares, 342-48, 358-68
  regression, 344
linearization
  approximation, 373
  method, 581
linkage editor, 6
linker, 6
Lisp language, 2, 6, 633, 635
local, 71
  property, 450
  value, 370
logarithmic derivative, 528, 533
logarithms, 17
  factorials, 17-18, 208
logical operations, 635
long-range order, 421
Lorentzian distribution, 323, 328, 378
  mean and variance, 329
lower triangular matrix, 228, 306
LR-decomposition, 272

LU-decomposition, 228, 230

M

machine
  language, 2
  word, 384
MacLisp, 636
Macsyma, 633
magnetic
  field, 421, 425, 436
  susceptibility, 421, 425, 437, 443
magnetism, 421
magnetization, 421, 425, 436, 442
Mandelbrot set, 472-76
mantissa, 10
many-body
  interactions, 450
  state, 297
  systems, 296, 421
Maple, 633
Marquardt method, 373-80
matching radius, 532, 533
MathCad, 633
Mathematica, 633
matrix, 211
  band, 273, 310, 313
  column, 254
  complex, 301, 310-12
  covariance, 359, 373, 375
  curvature, 359, 372, 373, 375, 379
  decision, 438
  defective, 309
  diagonal, 240
  diagonalization, 240-316, 446
  elements, 19
  Hermitian, 309
  Hessenberg form, 305
  inversion, 224-28, 359
  method, 237, 444
  multiplication, 213
  normal, 309
  notation, 212, 495

matrix (cont'd)
  positive definite, 315
  random, 317, 444-55
  real, nonsymmetric, 303, 306, 309
  real, symmetric, 240, 296, 450
  rectangular, 254
  skew-symmetric, 311
  sparse, 310, 313, 539, 553, 557
  square, 224
  transformation, 241, 261, 304
  triangular, 272, 305
  tridiagonal, 253, 258, 275, 285, 574, 583
  types of, 214
  unit, 214, 226
  upper diagonal, 139
  upper triangular, 221, 228, 272, 305
maximum likelihood, 338, 592
  method of, 337-41
mean, 321, 380
  free path, 407
  life, 337
mean and variance of distributions
  binomial, 326, 380
  χ^2, 351, 380
  Lorentzian, 328
  normal, 327
  Poisson, 326
median, 322
medium, 380
mesh, 37, 494
method of
  Adams-Bashforth, 513
  bisection, 124-25, 266-69
  characteristics, 566-69
  Crank-Nicholson, 558-60, 573
  explicit, 497, 499, 555-57, 559, 571, 572-73, 577, 587
  implicit, 497, 514, 555, 556, 557, 559, 560, 572, 587
  least-squares, 342-80
  logarithm, 17
  maximum likelihood, 337-41
  middle-square, 385
  midpoint, 499, 502, 518-21
  multistep, 497, 513, 515
  one-step, 497

Metropolis algorithm, 403-06, 438, 467, 470
middle-square method, 385
midpoint method, 499, 502, 518-21
minor of a determinant, 215, 316, 637
modified
  midpoint method, 518-21
  radial wave function, 525
modulus, 386
molecular
  diffusion, 407
  dynamics, 412
  spin, 434
momentum
  representation, 100, 456
  space, 456
monitor, 610, 611
monochrome display device, 627
Monte Carlo, 55, 61, 67, 68, 463, 479
  calculation, 3, 57, 383-483
  data, 419
  integration, 49-53, 55-58, 60, 330, 390
  simulation, 446
most probable value, 322, 380
multidimensional integrals, 49, 54-58
multiparticle, multihole excitation, 300
multiple
  correlation coefficient, 364-66
  regression, 358-68
multistep method, 497, 513, 515
musical notes, 99

N

NAG library, 179
natural line shape, 328
nearest-neighbor spacing, 450
  distribution, 450-53
Neumann function, 171-79
neutrino, 415
  mass, 415-20
Neville's algorithm, 71-74, 521
Newtonian mechanics, 487

Newton's
  interpolation, 125-27
  method for nonlinear FDE, 581-84, 588
nickel, 434
nonintegrable singularity, 58
nonlinear
  differential equation, 490, 496, 544, 576-84
  least-squares method, 368-80
  ordinary differential equation, 490, 496, 544, 576-84
nonnormal matrices, 309
nonsymmetric matrix, 305
norm, 254, 257
normal
  coordinates, 297
  distribution, 317, 323, 327-28, 333, 380, 392, 399, 402, 411, 446, 448, 485
  matrix, 309
  modes, 575
  probability function, 51-53, 67, 132, 195, 329-30
  probability integral, 51, 132, 202, 329
nucleon-nucleus potential, 526
numerical
  computations, 6
  differentiation, 29, 62, 64
  experiments, 427
  integration, 29, 37, 38, 59, 116
  quadrature, 31
  simulations, 448
  solutions, 19

O

object code, 2, 6
octal representation, 8, 9
one-body
  force, 447
  interactions, 448
one-dimensional harmonic oscillator, 145
one-particle, one-hole ($1p1h$), 297
one-step
  method, 499
  process, 497

operator, 63, 147
  annihilation, 449
  averaging, 129
  Boolean, 409
  creation, 449
  de-excitation, 298, 299, 301
  difference, 62
  excitation, 298, 299, 301
  Hamiltonian, 147, 239, 241, 283, 444
  Hermitian, 241
order parameter, 421, 424
ordinary differential equation (ODE), 488-89
  nonlinear, 604
orthogonal
  polynomials, 42
  property, Legendre polynomials, 42, 45
  transformation, 445
orthogonality relation
  associated Laguerre polynomials, 186
  Clebsch-Gordan, 206
  eigenvectors, 305
  Hermite polynomials, 150
  Laguerre polynomials, 185
  Legendre polynomials, 161
  spherical harmonics, 167
  $3j$-symbols, 206
outer product 254
overrelaxation method, 542, 544, 548, 604

P

Padé approximation, 77-80
parabolic
  approximation, 371-73
  partial differential equation, 554-64
parent
  distribution, 324
  function, 353
parity, 157

partial
  denominator, 88, 91, 355
  derivative, 489
  differential equation (PDE), 489-90, 536, 549, 554-76
  numerators, 88, 355
  waves, 525
partition function, 436, 463, 465
Pascal, 7
path integral, 461-63
  method, 455-71
pattern recognition, 635
Pauli
  exclusion principle, 296, 449
  spin matrix, 435
PDE (partial differential equation), 489-90, 536, 549, 554-76
pendulum, 577-79
percolation, 421, 426-34
  phase transition, 421, 424
  threshold, 427
percolitis, 422-24
period, random number, 384
periodic
  boundary condition, 440
  function, 96, 157
permutation, 640-41
perspective, 629
  transformation, 629
perturbation, 450
  methods, 293, 426, 463, 471
phase
  polynomial, 97
  shift, 527, 530, 563
  transition, 421, 427, 433, 442
photographic image, 623
physical observables, 19
π, 24-26, 67
pivot, 218, 227, 233
pivoting, 222, 233-34
pixel, 473, 611, 617
plane wave, 456, 457
plasma oscillations, 297
plotting (*see also* graphics), 607-32
  contour, 623-27
  straight line, 618-19
  text-mode, 608-09

three-dimensional, 622, 627-32
wire-frame representation, 628
pointwise convergence, 597
Poisson
  distribution, 326-27, 333, 391, 397
  equation, 490, 549, 552, 553, 598-603
polar angle, 155, 157, 164
polynomial, 79
  approximation, 70, 116, 136, 179, 195, 199
  associated Laguerre, 186-94
  associated Legendre, 164-66, 168
  extrapolation, 116, 521, 523
  Hermite, 60, 67, 145-54, 238, 294
  interpolation, 69-76, 87, 125-27, 135, 620
  Laguerre, 60, 67, 183-94
  Legendre, 42, 158-64, 360-61
portability, 22, 23, 389
positive definite matrix, 315
potential, 145
  effective, 525
  energy, 237
  harmonic oscillator, 146, 314
  scattering, 524
  square well, 525, 540, 563
  Woods-Saxon, 526, 536, 537, 540, 542
power series expansion, 148
  exponential function, 149
powers of prime numbers, 13
predictor
  -corrector methods, 510-17
  step, 510
pressure, 421
primary rainbow, 614
prime number, 10-16
  decomposition, 13-15, 168
  list, 11-12
  representations, 208
primitive polynomial method, 409
principal quantum number, 188
private library, 5

probability
  distribution, 321-24, 415
  integral, χ^2-distribution, 94, 96, 330, 351
  machine, 484
probable error, 323
procedures, 636
program library, 4-6
programming, 1-7
programming language, 6-7, 636, 642
  assembly, 2
  ALGOL, 6
  APL, 7
  BASIC, 7
  C, 2, 6
  FORTRAN, 2
  high-level, 2
  Lisp, 2, 6, 633, 635, 636
  machine, 2
  Pascal, 7
projectile problem, 524, 604
projection, 456, 622
propagation, 494, 524, 571
  of error, 334, 381
  of solution, 494, 558
propagator, 458, 464
pseudo-random number, 384

Q

QL
  -algorithm, 272, 292
  -decomposition, 273
QR
  -algorithm, 272
  -decomposition, 272
quadratic
  equation, 27, 570
  interpolation, 621
quantum
  chaos, 455
  chromodynamics, 463, 471
  field theory, 463
  mechanical problem, 19
  mechanical system, 445, 450, 455
  mechanics, 100, 157, 455, 524
quarks, 463, 471
quasilinear

boundary value problem, 579-81
initial value problem, 577-79
differential equation, 569, 576, 576-84, 586, 588

R

radial
  equation, 182
  wave function, 183, 187, 525
radioactive
  decay, 324, 337, 588
  nuclei, 337
radioactivity, 377, 416
rainbow, 613-16
random
  bit generator, 408-10
  digits, 383
  integer, 384-86
  matrix, 317, 444-55
  phase approximation, 296, 301-03
  sample, 49
  walk, 407-12, 484
random number, 3, 56, 383, 387, 445, 613
  additive generator, 389
  correlations, 384
  frequency test, 390-93, 483
  generator, 3, 383-406
  given distribution, 397-98
  linear congruence method, 386, 387, 388, 394
  middle-square method, 385
  normal distribution, 399-400, 446
  period, 384
  rejection method, 400-03
  run test, 394-97
  serial correlation test, 394-95
  subtractive generator, 389
  uniform distribution, 50, 384-90, 397, 613
randomness, 383
  test, 390

rational
  function, 77
  function interpolation, 77-87, 116
    123, 521, 523
  polynomial approximation, 87, 330
  polynomial extrapolation, 116, 521-23
Rayleigh formulas, spherical Bessel
  function, 174
real
  nonsymmetric matrices, 296, 303,
    306, 309
  number, 11
  symmetric matrix, 240, 296, 450
reciprocal differences, 90, 91
rectangular
  matrix, 254
  rule of integration, 32-35, 49, 54, 463
recursion relation
  associated Laguerre polynomial, 193
  Bessel function, 178
  characteristic equation, 265
  continued fraction, 88, 94
  divided difference, 126
  factorial, 20
  FFT, 107-09
  gamma function, 198
  Hermite polynomial, 152
  inverse interpolation, 131
  integration, 118
  Laguerre polynomial, 187
  Legendre polynomial, 160
  logarithm of factorial, 18
  Mandelbrot set, 472
  Monte Carlo integration, 50
  Neville's interpolation, 72, 75
  rational interpolation, 86
  reciprocal difference, 92
  Richardson integration, 118
  spherical Bessel function, 175
Reduce, 23, 633, 636
reduced
  chi-square, 349
  instruction set computer (RISC), 2
  mass, 237
refraction, 613
regular grid data, 624
rejection method, 400-03

relative prime, 386
relaxation
  methods, 536, 539, 541, 549
  parameter, 542, 547-48, 553
  time, 424
residual, 589, 591
restoring force, 488
Riccati transformation, 586
Richardson's deferred approach, 116-20,
  518
Riemann integral, 30
right eigenvectors, 303-05
rigid spectrum, 454
RISC (reduced instruction set computer), 2
Ritz method, 596
Rodrigues formula, 164
Romberg integration, 37, 42, 120-22,
  517
rotation, 243, 306
rotational invariance, 449
row vector, 254
run
  -down test, 394
  test, 394-97
  -up statistic, 396-97
  -up test, 394, 396
Runge-Kutta method, 499-509
  fourth-order, 502
  variable step size, 506-09

S

sampling techniques, 49
scalar, 254
  product, 254
scaling relation, 477, 478
scattered data, 624
scattering, quantum mechanical, 178,
  524-36, 563-64
Schrödinger
  equation, 147, 237, 293, 460, 558-60
  representation, 460
screened charge distribution, 56

second quantization, 449
secondary rainbow, 614
second-order
  derivative, 37, 63, 66, 68
  differential equation, 147, 155, 489, 525, 537
  phase transition, 422, 424, 443
seed, 385, 386
self
  -similarity, 478-79
  -starting solution, 515
semicircular distribution, 401, 402, 445, 447
separation of variables, 156
serial correlation test, 393-94, 397
shear, 306
shooting methods, 495, 524-36, 604
  matching radius, 532, 533
short-range interaction, 421
shuffling, 388, 397
Sierpinski gasket, 477-78
sign bit, 9
significant figures, 26
similarity transformation, 240, 242, 256, 306
simple harmonic oscillations, 493
Simpson's rule, 37-42, 42, 49, 54, 67
sine-Gordan equation, 604
single-particle energies, 448
singular value decomposition, 359
singularity, 59
  nonintegrable, 59
skew
  -Hermitian, 309
  -symmetric matrix, 311
skewness, 323
smooth function, 136
smoothing, 620-22
solitary wave, 579, 604
soliton, 579
SOR (successive overrelaxation), 542, 547-48, 604
source code, 2, 6, 23
space
  active, 288
  Hilbert, 287, 628
spacing distribution, 450-53

sparse matrix, 310, 313, 539, 553, 557
special functions, 145-208
specific heat, 421, 425, 437, 443
spectral radius, 546, 547, 553
spherical
  Bessel function, 171-79
  coordinate system, 456
  Hankel functions, 171
  harmonics, 166-70, 525
  Neumann function, 171-79
spin orientation, 434
spline, 135
spontaneous
  magnetic moment, 434
  magnetization, 437, 442
spring constant, 488
spurious states, 309
square
  lattice, 427, 428-34
  matrix, 214, 224
  root, 27
  well potential, 525, 540, 563
stability, 233, 496
stable solution, 496
standard
  deviation, 323, 335
  Lisp, 636
  subroutine library, 23
state vector, 456
statistical
  analyses, 319
  behaviors, 444
  description of data, 319
  field theory, 463
  measure, 348, 390, 413
  mechanics, 403, 421, 445, 463, 490
  test, 348-57
statistics, 348, 421
stencil, 550
step size, 32, 492
stiff differential equation, 505, 581-85
stiffness problem, 584

Stirling's formula, 200
stretched string, 565, 604
strong interactions, 471
structured program, 3
Student's
   t-distribution, 367
   t-statistic, 353
   t-test, 353-54
Sturm
   -Liouville problems, 315
   sequence, 265-69, 317
subdiagonal, 253
subdomain method, 591-92, 594
subinterval, 32
subordinate matrix norm, 269
subprograms, 3
subroutine, 3
   library, 5, 7, 22-23, 487
subtractive generator, 389, 394
successive overrelaxation (SOR), 542, 547-48, 604
   acceleration parameter, 542
   iteration matrix, 544, 553
superdiagonal, 253
   matrix elements, 285
symbolic
   data, 633
   manipulation (SM), 6, 633
symmetric matrices, 272, 316
symmetry, 444, 453
   relations, angular momentum, 205
system
   clock, random integer, 385
   linear equations, 211, 219-36, 239, 359
   subroutines, 4
systematic errors, 320

T

Tamm-Dancoff approximation (TDA), 299
tangent, 62
Taylor series, 34, 371
   expansion, 64, 151, 334, 496, 513
TDA, 299
temperature, 407, 421

text-mode plotting, 608-09
thermalization, 406, 440, 470
thermodynamics, 436
three-dimensional
   integral, 55
   objects, 627
   plots, 623, 627-32
$3j$-symbol, 205
three-point stencil, 550
time
   constants, 505
   -dependent Schrödinger equation, 554, 558-60
   development, 563
   evolution, 455
   reversal, 445
   reversal invariance, 444
transformation
   of basis, 254
   of probability, 398
   matrix, 241, 261, 304
   similarity, 240, 242, 256, 306
   unitary, 242, 272
transition
   probability amplitude, 458
   rate, 444
transport
   equation, 565, 571
   phenomena, 554
transpose, 254
trapezoidal rule, 35-37, 49, 54, 61, 66, 121
trial
   path, 469
   solution, 589
triangle, 273
triangular
   basis functions, 603
   form, 305
   functions, 598
   lattice, 427
   matrix, 272, 305

tridiagonal, 139
  basis, 287, 289
  form, 256, 259, 283
  matrix, 253, 258, 275, 285, 574, 583
tridiagonalization, 251-64, 446
trigonometric
  functions, 24, 96-97, 243, 247
  interpolation, 100
tritium, 415
  beta decay, 417
true value, 320
truncation errors, 10, 287, 496-97, 499, 506
two-body
  forces, 447
  interactions, 448
  matrix element, 449
  potential, 449
  random ensemble (TBRE), 447-49
two-byte integer, 9
two-dimensional
  integral, 53
  rotation, 243
two-point boundary value problem, 528

U

uncertainty (*see also* error), 319, 324, 330-36
  in the parameter, 362, 379
  propagation, 334-36
uncorrelated, 335
under
  -damping, 505
  -relaxation method, 542
underflow, 10
uniform
  deviates, 384
  random number, 50, 384-90, 397, 613
unit matrix, 214, 226
unitary, 559
  matrix, 242, 272, 275, 309
  transformation, 242, 272
universality, 426, 445
unstable
  particle, 337
  solution, 496, 572

upper triangular matrix, 221, 228, 272, 305

V

value of π, 24, 68
Vandermonde determinant, 135
variable step size, 506
variance (*see also* mean and variance), 322
variational
  approach, 595
  principle, 288, 595
vector
  addition coefficient, 204, 205
  notation, 498
  row, 254
vibrating string, 489
video
  graphics adaptor, 612
  memory, 612
viewpoint, 622, 630
von Koch snowflake curve, 484

W

wave
  equations, 565, 572-75
  form, 99
  function, 19, 147, 150, 525
  packet, 563
  partial, 525
Weierstrass convergence theorem, 70
weighted residual methods, 593
width, 329
Wigner
  surmise, 451, 453
  $3j$-symbol, 205
wire-frame representation, 628
Woods-Saxon potential, 526, 536, 540, 542
word length, 384, 509

Y

Yukawa charge distribution, 56

Dear Customer:

If you are unable to use the 3½" diskette provided with this book, kindly return it with the order form below to the following address to receive a 5¼" diskette:

>Professional and Technical Reference Division
>Prentice Hall
>440 Sylvan Avenue, Route 9W
>Englewood Cliffs, NJ 07632
>Attn: Maureen Diana

Replacement diskettes will be provided on an exchange basis only. You should receive your replacement diskette within four weeks of returning the reply card with the original diskette bound into this book.

ORDER FORM

Complete the order form below and send to the address above:

Ship to:

Name _____

Address _____

_____

_____

_____

Phone (___) _____